T0201317

ENVIRONMENTAL ASSESSMENT ON ENERGY AND SUSTAINABILITY BY DATA ENVELOPMENT ANALYSIS

Operations Research and Management Science (ORMS) is a broad, interdisciplinary branch of applied mathematics concerned with improving the quality of decisions and processes and is a major component of the global modern movement towards the use of advanced analytics in industry and scientific research. The Wiley Series in Operations Research and Management Science features a broad collection of books that meet the varied needs of researchers, practitioners, policy makers, and students who use or need to improve their use of analytics. Reflecting the wide range of current research within the ORMS community, the Series encompasses application, methodology, and theory and provides coverage of both classical and cutting edge ORMS concepts and developments. Written by recognized international experts in the field, this collection is appropriate for students as well as professionals from private and public sectors including industry, government, and nonprofit organizations who are interested in ORMS at a technical level.

Related Titles

Environmental Assessment on Energy and Sustainability by Data Envelopment Analysis Toshiyuki Sueyoshi and Mika Goto

Big Data and Differential Privacy: Analysis Strategies for Railway Track Engineering Nii O. Attoh-Okine June 2017

Advances in DEA Theory and Applications: With Extensions to Forecasting Models Kaoru Tone (Editor) June 2017

Sustainable Operations and Supply Chain Management Valeria Belvedere, Alberto Grando January 2017

Healthcare Analytics: From Data to Knowledge to Healthcare Improvement Hui Yang (Editor), Eva K. Lee (Editor) July 2016

Decision Science for Housing and Community Development: Localized and Evidence-Based Responses to Distressed Housing and Blighted Communities Michael P. Johnson, Jeffrey M. Keisler, Senay Solak, David A. Turcotte, Armagan Bayram, Rachel Bogardus Drew October 2015

Cost Estimation: Methods and Tools Gregory K. Mislick, Daniel A. Nussbaum April 2015

Discrete-Event Simulation and System Dynamics for Management Decision Making Sally Brailsford (Editor), Leonid Churilov (Editor), Brian Dangerfield (Editor) April 2014

Elements of Random Walk and Diffusion Processes Oliver C. Ibe September 2013

Game Theory: An Introduction, Set, 2nd Edition E. N. Barron April 2013

Game Theory: An Introduction, 2nd Edition N. Barron April 2013

ENVIRONMENTAL ASSESSMENT ON ENERGY AND SUSTAINABILITY BY DATA ENVELOPMENT ANALYSIS

TOSHIYUKI SUEYOSHI
New Mexico Institute of Mining and Technology
New Mexico, USA

MIKA GOTO
Tokyo Institute of Technology
Tokyo, Japan

This edition first published 2018

Registered Offices
John Wiley & Sons, Inc., 111 River Street, Hoboken, NJ 07030, USA
John Wiley & Sons Ltd, The Atrium, Southern Gate, Chichester, West Sussex, PO19 8SQ, UK

Editorial Office
9600 Garsington Road, Oxford, OX4 2DQ, UK

For details of our global editorial offices, customer services, and more information about Wiley products visit us at www.wiley.com.

Wiley also publishes its books in a variety of electronic formats and by print-on-demand. Some content that appears in standard print versions of this book may not be available in other formats.

Library of Congress Cataloging-in-Publication data applied for

Hardback ISBN: 9781118979341

Cover Design: Wiley
Cover Images: (Top Image) © Fiona McAllister Photography/Gettyimages; (Bottom Image) © D3Damon/Getty Images

Set in 10/12pt Times by SPi Global, Pondicherry, India
Printed and bound in Malaysia by Vivar Printing Sdn Bhd

10 9 8 7 6 5 4 3 2 1

CONTENTS

PREFACE

Global warming and climate change are now a very serious issue around the world. The climate change problem, due to global warming, implies an increase in average global temperature regarding air, sea and land. Natural events and economic activities, including industrial developments and business activities, contribute to the increase in average global temperature. Such a climate change is primarily caused by an increase in greenhouse gases such as carbon dioxide. In addition, we are now facing various environmental difficulties, such as how to handle nuclear and industrial wastes, all of which are byproducts of our economic and industrial developments.

To combat the environmental issues, this book discusses the importance of both economic success and environmental protection for sustainability enhancement. The underlying philosophy of this book is that we need to develop eco-technology innovation and managerial challenges to support a progress for reducing an amount of greenhouse gas emissions. In challenging toward such a research direction, this book proposes a new use of "data envelopment analysis (DEA)," as a holistic approach, to assess various aspects concerning sustainability development. In the sense, the new methodology proposed in this book is referred to as "DEA environmental assessment."

An important feature of this book is that it focuses upon "energy sectors" because they are closely associated with environmental problems. Therefore, this book is not interested in a conventional use of DEA for performance assessment, rather discussing the new approaches for energy-related sustainability development. In discussing these new approaches for energy and environmental assessment, it is necessary for us to clearly specify important concerns to be discussed in this book. Some of these concerns are summarized as follows:

(a) *History*: Many DEA researchers have long believed that the first DEA publication was the article prepared by Professor W. W. Cooper and coworkers in 1978. Viewing DEA as an extension of goal programing, along with fractional programming and a historical linkage between L1 regression and goal programming, this book considers that DEA has an analytical linkage with L1 regression. In this view, the history of DEA was connected in a roundabout fashion with the development of science in the eighteenth century, as manifested in the work of Laplace and Gauss, because they attempted to develop algorithms for the L1 regression.

(b) *Methodological Bias*: DEA is not a perfect methodology for performance assessment. Many different models have been proposed since the initial publication. DEA researchers and users need to understand the existence of methodological bias in their applications. Said simply, different methodologies produce different results. Therefore, it is necessary for us to examine several different DEA models to examine the methodical validity to prepare business and policy implications.

(c) *Measures*: It is usually believed among researchers and users that DEA is a methodology for efficiency assessment. Acknowledging the importance of DEA-based efficiency assessment, this book is different from the conventional belief because DEA can provide us with not only the efficiency assessment but also other different measures such as scale measures (e.g., returns to scale and damages to scale), substitution measures (e.g., marginal rate of transformation and rate of substitution) and other various managerial measures (e.g., future prediction). Thus, it is not sufficient to examine only the level of efficiency regarding various organizations.

(d) *Undesirable Outputs*: Conventional DEA incorporates multiple components of the input vector and the desirable output vector. The previous approach had only two production factors. Meanwhile, DEA environmental assessment additionally incorporates multiple components of the undesirable output vector, thus having three production factors.

(e) *Disposability Concepts*: The proposed environmental assessment utilizes two disposability concepts. One of the two concepts is "natural disposability" in which operational performance is measured as the first priority and environmental performance is measured as the second priority. The other disposability concept is "managerial disposability" which has an opposite priority on operational and environmental performance measures. Here, the concept of disposability indicates the elimination of inefficiency sources.

(f) *Congestion*: This book discusses a possible occurrence of congestion that is classified into two categories: (f-1) undesirable congestion under natural disposability and (f-2) desirable congestion under managerial disposability. The proposed DEA approach incorporates a possible occurrence of undesirable congestion and that of desirable congestion into the environmental assessment. We discuss how to measure an occurrence of desirable congestion, or eco-technology innovation, in a comparison with that of undesirable congestion. The identification of undesirable congestion is important, for example, in avoiding a cost increase due to a shortage of transmission or a limit of transportation capacity in a whole production system. However, the identification of desirable congestion is more important than that of undesirable congestion because we are interested in reducing an amount of various pollutions, so developing a sustainable society.

(g) *Input Direction*: The proposed environmental assessment incorporates an analytical capability to increase or decrease the components of an input vector. The input increase implies an "economic or corporate growth" under managerial disposability, while the input decrease implies these "stabilities" under natural disposability. The input increase has an upper limit on an efficiency frontier related to undesirable outputs, while the input decrease has a lower limit in an efficiency frontier related to desirable outputs. The analytical feature is very different from a conventional use of DEA that incorporates only the direction of an input decrease along with an increase in some components of the desirable output vector. The direction of an input vector becomes an important component in examining and developing social or corporate sustainability. This book will explore the methodological issue from the perspective of DEA-based sustainability development.

This book consists of two sections. Section I describes a conventional framework of DEA which provides us with a mathematical basis for understanding the proposed research direction toward environmental assessment and sustainability development. Section II, which is the gist of this book, is related to its conceptual and methodological extensions toward environmental assessment in energy and other industrial sectors.

In preparing this book, the authors have reused figures, tables and related descriptions from their original publications. They have obtained copyright permissions concerning the reuses from publishers (e.g., Elsevier, IEEE and John Wiley & Sons) via the Copyright Clearance Center (Danvers, Mass.). The authors realize that their original works no longer belong to them, rather belonging to the publishers after publishing their articles in journals.

The authors acknowledge that this book has been financially supported by Japan Society for the Promotion of Science (JSPS) Grant-in-Aid for Scientific Research (KAKENHI) 26285050 and 16K01236.

At the end of this preface, it is important to note that comments and constructive criticisms should be directed to the first author of this book. After spending four years, he can finally escape from the painful effort of producing this book.

Now, the first author will be able to reply to positive inquiries, not negative ones, about the book. All errors and mistakes related to this book are his responsibility alone.

Finally, it is hoped that this book will make a contribution for developing new DEA models and applications in energy and other industrial sectors. We look forward to seeing many research extensions of the approaches discussed in this book.

Toshiyuki Sueyoshi
New Mexico Institute of Mining and Technology

Mika Goto
Tokyo Institute of Technology

SECTION I

DATA ENVELOPMENT ANALYSIS (DEA)

1

GENERAL DESCRIPTION

1.1 INTRODUCTION

It is important to keep in mind that the purpose and interest of this book are not a conventional use of data envelopment analysis (DEA) for efficiency measurement and performance analysis. Rather, this book will direct our research attention and concerns toward a new use of DEA on environment assessment and sustainability development. This chapter[1] is designed to discuss a new research direction for DEA.

This book consists of two sections (I and II). As an initial step, this chapter starts by reviewing fundamental research concepts for a conventional use of DEA. Such a conventional use will be discussed in all chapters of Section I. Then, Section II will extend it from the perspective of policy and business implications concerning environmental assessment and sustainability development. The methodology used for the newly proposed research is referred to as "DEA environmental assessment." In addition to the environmental assessment, this book focuses upon energy sectors in the world because they are closely associated with various types of industrial pollution. An important environmental issue to be discussed in this book is how to challenge global warming and climate change in the world.

[1] This chapter is partly based upon the article: Ijiri, Y. and Sueyoshi, T. (2010) Accounting essays by Professor William W. Cooper: revisiting in commemoration of his 95th birthday. *ABACUS: A Journal of Accounting, Finance and Business Studies*, **46**, 464–505.

Environmental Assessment on Energy and Sustainability by Data Envelopment Analysis, First Edition. Toshiyuki Sueyoshi and Mika Goto.

Of course, we clearly understand that it is not easy to solve these climate issues by using only the proposed DEA environmental assessment and its applications to energy sectors. Rather, this book will attempt to investigate the global difficulty from the perspectives of business, policy and economics, so that we can assist technology development to avoid serious consequences such as heat waves, droughts, floods and food crisis, as well as other damage to human, social and economic systems. Thus, we will challenge various issues due to the climate change by utilizing the analytical capabilities of DEA environmental assessment, newly proposed in this book. This book will also attempt to change the profit-driven business logic used in a conventional use of DEA in such a manner that it can fit within the global trend for developing a sustainable society.

This chapter is organized in the following manner: Section 1.2 describes the structure of this book. Section 1.3 summarizes contributions on Sections I and II. Section 1.4 specifies abbreviations and nomenclature used in this book. Finally, Section 1.5 summarizes this chapter.

1.2 STRUCTURE

First of all, we need to mention that DEA is not a perfect methodology, rather it is an approximation approach for the performance assessment of many organizations in the public and private sectors. See Chapter 6 for methodical comparisons between conventional DEA models. However, it is true that DEA can provide corporate leaders and policy makers with an empirical guideline to assist their decision makings. Such a guideline is very important in assessing environmental issues and sustainability developments. To attain the research direction discussed here, the two sections of this book (Sections I and II) contain the following chapters:

Section I: Conventional DEA

Chapter 1 (General Description): This chapter provides a general description on the structure of this book.

Chapter 2 (Overview): This chapter conveys the message that DEA can serve as a very useful methodology in terms of not only conventional performance assessment but also practical and academic purposes in guiding organizations in public and private sectors. However, it is true that DEA is not a perfect methodology, rather being an approximation approach for performance assessment. The review in this chapter provides us with an intuitive description, or rationale, concerning why various DEA applications can be used for examining performance assessment.

Chapter 3 (History): This chapter returns to the eighteenth century to describe the origin of L1 regression and its analytical linkage to DEA. It is possible for us to consider that various DEA models are methodologies for multi-objective optimization which have been originated from the development of goal programming (GP). The history of GP started from the development of L1 regression. Thus, this chapter describes an analytical linkage among L1 regression, GP and DEA.

Chapter 4 (Radial Measurement): This chapter discusses two radial models that are used to measure a level of operational efficiency based on the Debreu–Farrell criterion. These models are classified into two categories under variable or constant returns to scale (RTS).

Chapter 5 (Non-radial Measurement): This chapter discusses non-radial models and their variations, as methodological alternatives to the two radial models, based on the Pareto–Koopmans criterion.

Chapter 6 (Desirable Properties): This chapter discusses nine desirable properties for the measurement of operational efficiency. It is better for each DEA model to satisfy such desirable properties from the perspective of production economics and optimization. Seven radial and non-radial models are theoretically compared from the perspective of nine criteria.

Chapter 7 (Strong Complementary Slackness Conditions): It is widely known that DEA has four difficulties in the applications. First, multiple projections occur in DEA. Second, multiple references occur in DEA, as well. Third, DEA cannot handle zero and/or negative values in a data set. Finally, an occurrence of zero may be usually found in dual variables. The occurrence of the fourth problem indicates that production factors, corresponding to dual variables with zero, are not fully utilized in DEA assessment. This chapter discusses the new use of strong complementary slackness conditions (SCSCs) to deal with the first, second and fourth difficulties related to DEA. The third difficulty will be discussed in Chapters 26 and 27, later.

Chapter 8 (Returns to Scale): This chapter discusses the concept and type of RTS in a unified framework of DEA production and cost analyses under the assumption of a unique optimal solution (e.g., a unique projection and a unique reference set). Dual relationships between production-based and cost-based RTS measures are discussed in this chapter.

Chapter 9 (Congestion): A possible occurrence of congestion serves as a very important concept for environmental assessment. Therefore, this chapter reviews the implication of the occurrence within a conventional framework of DEA. The occurrence of congestion indicates a capacity limit on part or all of a whole production facility. This chapter reassesses previous discussions on a possible occurrence of congestion. The concept discussed in this chapter will be later extended into a new development on eco-technology innovation in Chapter 21.

Chapter 10 (Network Computing): To deal with a large data set on various DEA assessments, this chapter highlights the architecture of network computing that is designed to coordinate a simultaneous use of multiple personal computers and other types of computing devices. This chapter provides a

DEA-based computational structure and algorithmic uniqueness. The proposed network computing can fit with modern computer technology, or a super computer, that has multi-processors for parallel computation.

Chapter 11 (DEA-Discriminant Analysis): Discriminant analysis (DA) is a classification method that can predict the group membership of a newly sampled observation. This chapter discusses a new type of non-parametric DA approach to provide a set of weights for a discriminant function, consequently yielding an evaluation score for group memberships. The non-parametric DA is referred to as data envelopment analysis–discriminant analysis (DEA-DA) because it maintains a discriminant capability by incorporating the non-parametric feature of DEA into DA. DEA-DA is very useful in assessing financial performance in the private sector.

Chapter 12 (Literature Study on DEA): This chapter lists previous research efforts on DEA, along with a link to environmental assessment. This chapter also summarizes software sources that can be used for the computation of DEA.

Section II: DEA Environmental Assessment

Chapter 13 (World Energy): This chapter describes a recent world-wide energy trend. Energy is separated into primary and secondary categories. Primary energy is classified into fossil and non-fossil fuels. The fossil fuels include oil, natural gas and coal, while the non-fossil ones include nuclear and renewable energies (e.g., solar, wind, biomass, water and others). In this chapter, electricity is considered as a representative of secondary energy.

Chapter 14 (Environmental Protection): This chapter discusses a historical review of various policy efforts to prevent industrial pollution in the four regions: the European Union, Japan, China and the United States. This review of environmental issues is important in understanding a historical reason concerning why we are now facing different types of industrial pollution (e.g., air, water, soil and others) and serious pollution issues (e.g., the global warming and climate change) along with their industrial development and economic growth in the world.

Chapter 15 (Concepts): This chapter describes conceptual frameworks that serve as an analytical basis for developing DEA environmental assessment. The chapter incorporates desirable and undesirable outputs in addition to inputs in the proposed computational framework. It is necessary for us to attain economic prosperity in all nations to support the development of social welfare for people and to improve their living standards. Simultaneously, we are now facing various pollution risks on the Earth (e.g., heat waves, droughts, floods and food crises, as well as damages to human, social and economic systems). To handle the global difficulties in the world, this chapter introduces natural and managerial disposability concepts and then links them to other economic and strategic concepts used for sustainability development. Our research effort will be explored by using the new conceptual framework concerning DEA.

Chapter 16 (Non-Radial Approach for Unified Efficiency Measures): This chapter extends economic and corporate strategies for sustainability development by presenting mathematical formulations for a non-radial DEA approach. This chapter starts with a description on non-radial models because desirable and undesirable outputs are more easily unified for the proposed environmental assessment than a radial approach. This chapter applies the proposed non-radial approach to evaluate the performance of national and international petroleum firms in the world.

Chapter 17 (Radial Approach for Unified Efficiency Measures): This chapter shifts our methodological description from non-radial to radial measurement for environmental assessment. The natural and managerial disposability concepts discussed in Chapter 16 are incorporated into the proposed radial approach. This chapter applies the proposed radial approach to compare the performance of coal-fired power plants under an independent system operator (ISO) and a regional transmission organization (RTO) with that of the other power plants not belonging to ISO and RTO in the United States.

Chapter 18 (Scale Efficiency): This chapter describes how to measure scale efficiency under natural or managerial disposability. The scale efficiency indicates the level of a managerial capability regarding how each organization can control its operational size in terms of enhancing a level of unified (operational and environmental) efficiency. This chapter discusses how to measure a degree of scale efficiency within both radial and non-radial measurements. This chapter applies the proposed approach to measure the performance of coal-fired power plants in the north east region of the United States.

Chapter 19 (Measurement in Time Horizon): This chapter discusses the use of DEA environmental assessment in a time horizon because most data sets on energy and environment are often structured by time series. In applying the proposed assessment to such time-series data, this chapter needs to classify production factors in a time horizon. A unique feature of the proposed assessment is that it incorporates the Malmquist index and its subcomponents in order to examine the occurrence of a frontier shift among multiple periods. This chapter utilizes the proposed approach to examine the relationship between a fuel mix for electricity generation and CO_2 emission in ten industrial nations.

Chapter 20 (Returns to Scale and Damages to Scale): As an extension of scale efficiency discussed in Chapter 18, this chapter describes how to measure RTS under natural disposability and damages to scale (DTS) under managerial disposability. In this chapter, we extend the concept of RTS, discussed in Chapter 8, by additionally incorporating undesirable outputs into a computational framework of DEA under natural disposability. The concept of DTS extends RTS to an analytical framework for measuring scale-related relationship between inputs and undesirable outputs under managerial disposability.

Business implications from RTS and DTS are discussed for an application to the Japanese chemical industry.

Chapter 21 (Desirable and Undesirable Congestions): This chapter classifies a possible occurrence of congestion into two categories: undesirable congestion (UC) under natural disposability and desirable congestion (DC) under managerial disposability. An occurrence of UC implies a capacity limit on part or all of a whole production facility. The phenomenon of UC can be found in all business sectors, including energy industries. Therefore, the identification of UC is important for energy sectors. Meanwhile, an occurrence of DC implies eco-technology innovation. It is easily imagined that DC is more important than UC in terms of environment protection. This chapter discusses policy implications obtained from such a possible occurrence of UC and DC for coal-fired power plants in the United States.

Chapter 22 (Marginal Rate of Transformation and Rate of Substitution): This chapter discusses how to measure marginal rate of transformation (MRT) and rate of substitution (RSU) among three production factors. This chapter examines the two measures along with the occurrence of DC, or eco-technology innovation, under managerial disposability. This chapter discusses "explorative analysis," along with a data adjustment, as a new multiplier restriction method in measuring a degree of MRT and that of RSU. This chapter compares the performance of countries in European Union and North America based upon the measurement of MRT and RSU.

Chapter 23 (Returns to Damage and Damages to Return): This chapter discusses a possible occurrence of UC under natural disposability and that of DC under managerial disposability. Then, considering the two disposability concepts, this chapter compares between returns to damage (RTD) under UC and damages to return (DTR) under DC. This chapter applies the proposed approach to examine Chinese energy policy and regional planning. In the application, the occurrence of UC can be considered as an economic difficulty that has been recently observed in the world economy. UC has such an important implication in economics, different from the capacity limit on a production facility discussed in a context of production economics. Meanwhile, DC implies the potential of eco-technology capability in each region. Both RTD and DTR are new economic concepts extended from RTS and DTS.

Chapter 24 (Disposability Unification): This chapter discusses how to unify natural and managerial disposability concepts in radial and non-radial measurements. An important assumption, which is incorporated in the unification, is that "undesirable outputs are considered as byproducts of desirable outputs." As a result of such an assumption, the desirable and undesirable outputs have a functional similarity (i.e., convex functions) between them. This chapter unifies the two disposability concepts and then extends them to identify a possible occurrence of DC, or eco-technology innovation, under managerial disposability. The proposed approach is used for examining the performance of independent and integrated petroleum companies in the United States.

Chapter 25 (Common Multipliers): This chapter discusses the combined use of DEA with SCSCs, proposed in Chapter 7, and DEA-DA from Chapter 11 in order to conduct an "efficiency-based rank analysis" of energy firms. The proposed approach is useful in preparing performance assessment where a single organization is efficient and the remaining others have some level of inefficiency. The proposed approach, equipped with SCSCs, first classifies firms into efficient or inefficient group based upon their efficiency scores. Then, the proposed approach utilizes DEA-DA to obtain common multipliers. The second stage measures an adjusted efficiency score for each organization. This chapter applies the proposed approach to examine the performance of Japanese electric power firms.

Chapter 26 (Translation Invariance to Handle Zero and Negative Values): This chapter discusses the property of "translation invariance" in non-radial measurement. The property indicates that an efficiency measure should be not influenced even if inputs, desirable and undesirable outputs are shifted toward a same direction by adding or subtracting a specific real number. The property makes it possible that we can evaluate the performance of organizations, whose production factors contain many zeros and/or negative values in a data set. The proposed approach is used for relatively comparing the energy structures among 33 industrial nations in the world.

Chapter 27 (Handling Zero and/or Negative Values in Radial Measurement): This chapter discusses a new use of radial measurement by incorporating the analytical capability to handle zero and/or negative values in a data set. The approach can handle the data set within a framework of radial and non-radial measurements. The proposed approach provides us with not only new quantitative assessment on unified performance but also information regarding how to invest for eco-technology innovation for abating industrial pollutions. This chapter applies the proposed approach to United States industrial sectors by paying attention to both successful companies with positive net incomes and unsuccessful companies with negative net incomes, so being able to measure all aspects of their performance components in a short-term horizon.

Chapter 28 (Literature Study on DEA Environmental Assessment): This chapter summarizes previous research efforts on DEA applied to energy and environment. This chapter also examines a recent research trend on applications from the 1980s to the 2010s. The applications contain 693 articles in total, most of which are published in well-known international journals, listed in Science Citation Index or Social Science Citation Index.

1.3 CONTRIBUTIONS IN SECTIONS I AND II

First, it is necessary to clearly describe that DEA was first proposed by Professor William Wager Cooper[2] (23 July 1914 to 29 June 2012) and his associates. He was

[2]According to Ijiri and Sueyoshi (2010), Professor William W. Cooper was born in Birmingham, Alabama, in 1914. His father was a bookkeeper and later a distributor for Anheuser–Busch. When William was three years old, the family moved to Chicago where his father owned a chain of gasoline stations that he lost in the Great Depression. Professor Cooper continued in high school only until the end of his sophomore year. With his father in ill health and no family revenue, he had to work at whatever he could find. This included everything from professional boxing to spotting pins in bowling alleys and caddying at golf courses. While hitchhiking to a golf course one day, he met Hall of Fame member in accounting, Eric Kohler, who thereafter became his life-long mentor. This included a loan of funds which enabled him to start a non-degree track at the University of Chicago. He quickly grew to like the academic atmosphere and soon took the college entrance examinations, intending to become a physical chemist because that seemed to offer the best chance of a job. At about this time, Kohler, then a principal with Arthur Andersen, asked him to look over the mathematics used in a patent infringement suit in which Andersen had been retained by the defendant. He found errors in the mathematics used by the plaintiff's engineers and Andersen hired him full-time in the summer and part-time during the school year. This awakened his interest in accounting so that he changed his major at the University of Chicago from chemistry to economics, and Kohler helped him to learn accounting. He graduated Phi Beta Kappa in economics in 1938. Kohler had by then left Anderson and assumed the position of Controller for the Tennessee Valley Authority (TVA). Kohler brought him to the TVA to head up work on "procedural auditing" (what would now be called "performance auditing") and to advise Kohler on the mathematics of cost allocation and other disputed matters in which the TVA was involved. This included helping Kohler to prepare testimony on these and other matters to be investigated by a Joint House–Senate Investigation Committee. Most of the work was completed by mid-1940 so that he left to become a PhD candidate at Columbia University where he had been awarded a doctoral fellowship in the School of Business. After passing his "prelims" in 1942, he again left academia to join the Division of Statistical Standards at the United States Bureau of Budget (now the OMB) where, as part of the United States war effort, he was placed in charge of coordinating all of the Federal Government's accounting and accounting-related statistics program. By late 1944, with the war coming to an end, he left to teach at the University of Chicago. In 1946, he returned to Washington to chair a committee to decide the fate of various war-time programs in financial statistics. He then transferred to Carnegie Institute of Technology (now Carnegie Mellon University, CMU) where he helped found, first, the Graduate School of Industrial Administration and, later, the School of Urban and Public Affairs. There was time out, however, to develop "end-use" audits that Kohler wanted to institute as Comptroller of the Marshal Plan. In 1976, after 30 years at CMU, Professor Cooper went to the Harvard Business School to help reorient their doctoral programs while holding the chair in accounting named for Hall of Fame member Arthur Lowes Dickinson. This task was completed in 1980, when he went to the University of Texas at Austin where he was initially appointed Professor of Management, Accounting and Management Science, and Information Systems, and was the Foster Parker Professor of Finance and Management (Emeritus) and the Nadya Kozmetsky Scott Centennial Fellow in the IC Institute. In 1945, he received an award for the most valuable article on accounting, the first ever awarded by the American Institute of Accountants (now the American Institute of Certified Professional Accountants, AICPA). A fellow of the Econometric Society, he was founding president of the Institute of Management Sciences, and he was also president of the Accounting Researchers International Association. He was the Director of Publications for the American Accounting Association. In 1990, he was named an Outstanding Accounting Educator by the same organization. He was Visiting International Lecturer for the American Accounting Association (AAA), traveling abroad in 1986 to lecture on accounting topics and visit with scholars in Latin America. In 1982, he was co-recipient of the John Von Neumann Theory Medal, jointly awarded by

the father of DEA. Since the original development, his research direction and concerns have been long influencing many proceeding works on DEA until now. Therefore, this chapter briefly reviews his contributions. Then, we specify the proposed research rationale concerning why and how DEA is important for environmental assessment in energy sectors.

According to Ijiri and Sueyoshi (2010), the first published article of Professor Cooper was an economic analysis entitled "*The yardstick of public utility regulation*" that appeared in June 1943 in *The Journal of Political Economy*, vol. **51**, part 3, pp. 258–262. In fact, still earlier in 1938, he published a proceedings paper for the Committee on Capital Gains Taxation of the National Tax Association under which he coauthored with E.L. Kohler. He coauthored an article with E.L. Kohler entitled "*Costs, prices and profits – accounting in the war program*," and published it in July 1945, in *The Accounting Review*, vol. **20**, part 3, pp. 267–308. On 31 August 1945, the American Institute of Accountants (currently the American Institute of CPAs) established a new award and chose his article as the most significant contribution to accounting in the year of its publication.

He had been a catalyst of change on a world-wide basis for more than five decades. His research inspired teaching, he was an editor for many periodicals and was a consultant to private, governmental and public institutions. As a prodigious author, his writings often focused on quantitative and creative approaches to management. With his long-time collaborator, a mathematician called A. Charnes, Professor Cooper was known everywhere as "Mr. Linear Programming," partly because they developed whole new areas such as GP, DEA and chance constrained programming, all of which originate from linear programming.

In reviewing his contributions, this chapter finds two important concerns related to DEA applications to energy and environment. The first concern is that the contribution of Professor Cooper could be summarized in short as "public accounting" and "public economics." His research interests included research on public sectors. Therefore, the first DEA article by Charnes, *et al.* (1978) was related to performance assessment on education units, so belonging to a public sector application on education. The other concern was that Professor Cooper was very proud of his development on GP. The rationale was because GP could solve L1 regression which none had solved from the eighteenth century. See Chapter 3 for a detailed description on the fact. It is possible for us to consider that various DEA models are methodologies for multi-objective optimization which have been originated from the development of GP. The history of GP has started from the development of L1 regression originated from science in the eighteenth century. Figure 1.1 depicts such a historical view on L1 regression, GP and DEA.

the Operations Research Society of America and the Institute of Management Sciences. In 1988, he received the Distinguished Service to Auditing Award from the Auditing Section of the AAA as well as an award for serving as the founding editor of Auditing, *A Journal of Practice and Theory.* He has received three McKinsey Foundation Awards for the most valuable article of the year on a management topic, and he has been a consultant to more than 200 institutions including the Marshall Plan, the United States General Accounting Office and others.

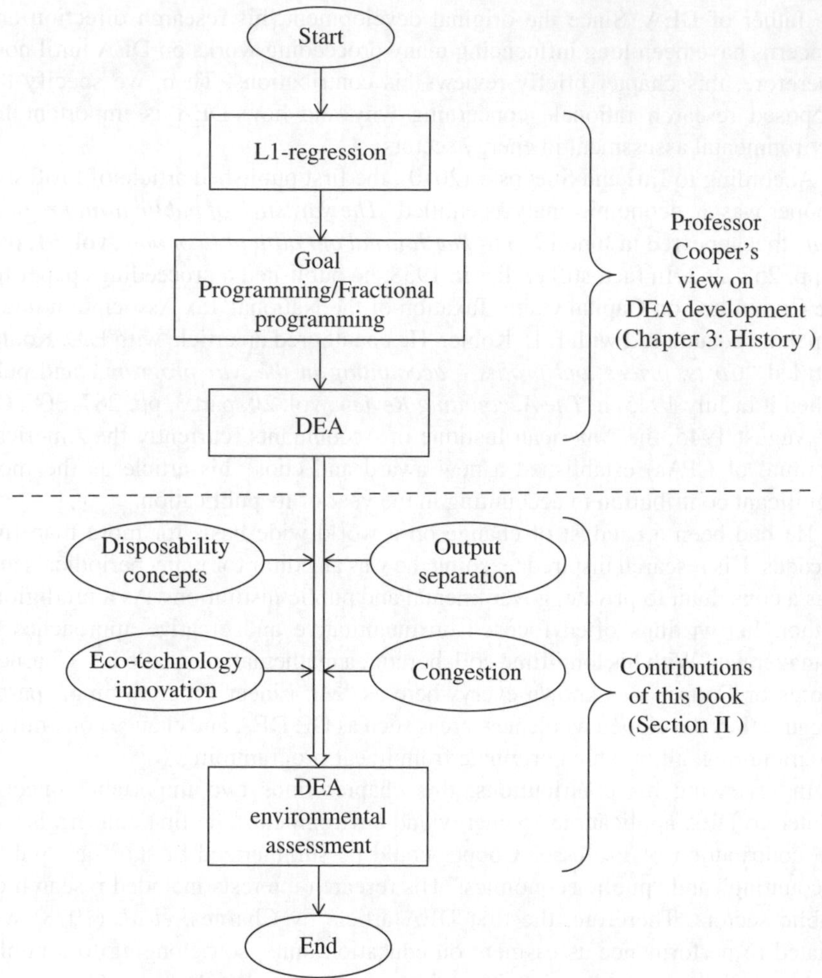

FIGURE 1.1 Developments and contributions.

Figure 1.1 displays an influence from the development of "fractional programming" that has changed the ratio model (i.e., the sum of weighted outputs divided by the sum of weighted inputs), which is non-linear programming, to its linear programming equivalent. As a result of such a structural change, we can now solve the ratio model by linear programming. The original formulation was often referred to "DEA ratio form" because of the analytical reason. In this book, we consider that such a series of DEA developments have been greatly guided by the contribution of Professor Cooper and his associates.

A problem of conventional DEA is that there is a difficulty in applying it to environmental assessment because undesirable outputs are not incorporated in the original formulation and its variations. Therefore, this book needs to shift the structure of conventional DEA to environmental assessment by adding undesirable outputs. The restructuring process needs to consider four important research changes, including (a) output separation into desirable and undesirable categories, (b) concepts on natural or managerial disposability that are linked to different treatments on inputs, (c) a possible occurrence of congestion and (d) a conceptual implication of eco-technology to reduce various pollutions. All of these have not been explored in previous DEA studies. This book will introduce all such new aspects and will explore them through DEA environmental assessment for energy and other industrial sectors.

1.4 ABBREVIATIONS AND NOMENCLATURE

1.4.1 Abbreviations Used in This Book

AAA	American Accounting Association
AEc	allocative and scale efficiency
AEv	allocative efficiency
AFD	acquisition, funding and development
AI	aggregation index
AICPA	American Institute of Certified Professional Accountant
AM	additive model
AR	assurance region
AZ	Altman's Z
BACT	best available control technology
BIT	bituminous coal
BOE	barrel of oil equivalent
BP	British Petroleum
CAA	Clean Air Act
CARB	California's Air Resources Board
CDP	Carbon Disclosure Project
CDTR	constant damages to return
CEc	cost and scale efficiency
CEv	cost efficiency
CEQ	Council on Environmental Quality
CEQA	California Environmental Quality Act
CHP	combined heat and power
CH_4	methane
CMU	Carnegie Mellon University
COI	classification and overlap identification
CO_2	carbon dioxide
CPB	Central Planning Bureau
CPC	Communist Party of China
CPU	central processing unit

CPUC	California Public Utilities Commission
CR	cone ratio
CRTD	constant returns to damage
CRTS	constant returns to scale
CSAPR	Cross State Air Pollution Rule
CSCs	complementary slackness conditions
CSE	cost-based scale efficiency
CWA	Clean Water Act
DA	discriminant analysis
DAM	day ahead market
DC	desirable congestion
DDTR	decreasing damages to return
DEA	data envelopment analysis
DEA-DA	data envelopment analysis–discriminant analysis
DG	directorate general
DgRTD	degree of returns to damage
DgDTR	degree of damages to return
DME	di-methyl ether
DMU	decision making unit
DOC	diesel oxidation catalyst
DRTD	decreasing returns to damage
DRTS	decreasing returns to scale
DTR	damages to return
DTS	damages to scale
EAP	environmental action program
EC	European Community
EF	efficiency frontier
EG	efficiency growth
EGUs	electric generating units
EIA	energy information administration
EIS	environmental impact statement
EP	equilibrium point
EPA	Environmental Protection Agency
ERGM	enhanced Russell graph measure
ES	elasticity of substitution
ESA	Endangered Species Act
EU	European Union
FERC	Federal Energy Regulatory Commission
FIFRA	Federal Insecticide, Fungicide, and Rodenticide Act
FIT	feed-in tariff
FWCA	Fish and Wildlife Coordination Act
GDP	gross domestic product
GGTR	greenhouse gas tailoring rule
GHG	greenhouse gas

GP	goal programming
GRP	gross regional product
GTL	gas to liquids
GW	gigawatt
GWe	gigawatt-electrical
GWh	gigawatt hour
HAPs	hazardous air pollutants
HD	hierarchical decomposition
HEW	health, education and welfare
HO	handling overlap
IC	input-oriented congestion
IDTR	increasing damages to return
IEA	International Energy Agency
IEEA	International Electricity Association
IM	Malmquist index under managerial disposability
IMC	IM with crossover
IN	Malmquist index under natural disposability
INC	IN with crossover
IOCs	international oil companies
IPCC	Intergovernmental Panel on Climate Change
IRTD	increasing returns to damage
IRTS	increasing returns to scale
ISO	independent system operator
IUEM	inter-temporal unified efficiency under managerial disposability
IUEN	inter-temporal unified efficiency under natural disposability
IUIM	inter-temporal unified index under managerial disposability
IUIN	inter-temporal unified index under natural disposability
JCC	Japan crude cocktail
KW	kilowatt
KWh	kilowatt hour
LAV	least absolute value
Lb	pound
LNG	liquefied natural gas
LMPs	locational marginal prices
LOO	leave one out
M&A	mergers and acquisitions
MACT	maximum achievable control technology
MATS	mercury and air toxics standards rule
MCF	magnitude control factor
MCP	market clearing price
MEP	Ministry of Environmental Protection
MIP	mixed integer programming
mmBtu	million British thermal units
MMT	methylcyclopentadienyl manganese tricarbonyl

MPSS	most productive scale size
MRT	marginal rate of transformation
MS	milli-second
MW	megawatt
NAPCA	National Air Pollution Control Administration
N_2O	nitrous oxide
NEPA	National Environmental Policy Act
NESHAP	National Emissions Standards for Hazardous Air Pollutants
NGO	non-governmental organization
NO_2	nitrogen dioxide
NOCs	national oil companies
NR	non-radial
NYSE	New York Stock Exchange
OC	output-oriented congestion
OE	operational efficiency
OEc	operational and scale efficiency
OECD	Organization for Economic Co-operation and Development
OEv	operational efficiency under variable RTS
OLS	ordinary least squares
O&M	operating and maintenance
OR and MS	operations research and management science
ORSA	Operations Research Society of America
PB	period block
PC	personal computer
PM	particulate matter
PoPS	pollution possibility set
Pr&PoPS	production and pollution possibility set
PrPS	production possibility set
PSE	production-based scale efficiency
PV	photovoltaic
PVPS	photovoltaic power systems
Q.E.D.	*quod erat demonstrandum*
R	radial
RAM	range-adjusted measure
RCRA	Resource Conservation and Recovery Act
R&D	research and development
RGGI	Regional Greenhouse Gas Initiative
Rh	right-hand side
RICE NESHAP	Reciprocating Internal Combustion Engines and National Emission Standards for Hazardous Air Pollutants
RM	radial model
RM(c)	radial model under constant RTS
RM(v)	radial model under variable RTS
RP	reserve-production

RS reference set
RSU rate of substitution
RTM real time market
RTD returns to damage
RTO regional transmission organization
RTS returns to scale
SARM slack-adjusted radial measure
SBM slack-based measure
SCSCs strong complementary slackness conditions
SCI Science Citation Index
SDA scale damages
SDWA Safe Drinking Water Act
SE scale efficiency
SEA Single European Act
SEC scale economies
SEM scale efficiency under managerial disposability
SEN scale efficiency under natural disposability
SEPA state environmental protection administration
SFA stochastic frontier analysis
SMR super ministry reform
SO sulfur oxide
SOCP second-order cone programming
SSCI Social Science Citation Index
s.t. subject to
SUB subbituminous coal
TA total average
TCP/IP transmission control protocol / Internet protocol
TD treatment for dominance
TFP total factor productivity
TIMS The Institute of Management Science
TSCA Toxic Substance Control Act
TVA Tennessee Valley Authority
TVEs township and village enterprises
TWh terawatt hour
UC undesirable congestion
UE unified efficiency
UEM unified efficiency under managerial disposability
UEN unified efficiency under natural disposability
UENM unified efficiency under natural and managerial disposability
UK United Kingdom
UN United Nations
URS unrestricted
US United States
U3O8 triuranium octoxide

VOCs volatile organic compounds
VRTS variable returns to scale
W window
WTI West Texas Intermediate

1.4.2 Nomenclature Used in This Book

$\boldsymbol{A}$	a matrix of observed inputs and desirable outputs of all DMUs
$A_j = \left(a_{1j}, a_{2j}, \ldots, a_{mj}\right)^{Tr}$	a column vector for restriction on inputs
$\boldsymbol{B} = \left(B_1, B_2, \ldots, B_n\right)$	a $h \times n$ matrix of observed undesirable outputs of all DMUs
b_{fj}	an observed value of the f-th undesirable output on the j-th DMU
$\bar{b}_{fj} = b_{fj} + \delta_f$	an adjusted value of the f-th undesirable output
b_{fjt}	an observed value of the f-th undesirable output on the j-th DMU in the t th period
$B_j = \left(b_{1j}, b_{2j}, \ldots, b_{hj}\right)^{Tr}$	a column vector of observed h undesirable outputs of the j-th DMU
$B_k = \left(b_{1k}, b_{2j}, \ldots, b_{hk}\right)^{Tr}$	a column vector of observed h undesirable outputs of the k-th DMU
c	a value of cost
c (sub or superscript)	this expresses constant RTS or DTS
cf	a discriminant score for cutting-off between groups
$C_j = \left(c_{1j}, c_{2j}, \ldots, c_{sj}\right)^{Tr}$	a column vector for restriction on desirable outputs
c_z (network)	the number of the z-th client
C_1	a partial group of G_1 which are clearly above an estimated discriminant function in Stage 1
C_2	a partial group of G_2 which are clearly below an estimated discriminant function in Stage 1
$d^b = \left(d_1^b, d_2^b, \ldots, d_h^b\right)^{Tr}$	a column vector of slacks related to h undesirable outputs
d_f^b	a slack related to the f-th undesirable output
$d^g = \left(d_1^g, d_2^g, \ldots, d_s^g\right)^{Tr}$	a column vector of slacks related to s desirable outputs
d_r^g	a slack related to the r-th desirable output
d_r^{g-}	a part of d_r^g under natural disposability ($r = 1,\ldots, s$)
d_r^{g+}	a part of d_r^g under managerial disposability ($r = 1,\ldots, s$)
$\boldsymbol{D}^{\mathbf{g}}$	a positive diagonal desirable output matrix
$\boldsymbol{D}^{\mathbf{x}}$	a positive diagonal input matrix
$d^x = \left(d_1^x, d_2^x, \ldots, d_m^x\right)^{Tr}$	a column vector of m slacks related to inputs
d_i^x	a slack related to the i-th input
d_i^{x-}	a part of d_i^x under natural disposability ($i = 1,\ldots, m$)

d_q^{x-} a part of d_i^x under unified (natural and managerial) disposability ($q = 1,\ldots,m^-$)

d_q^{x+} a part of d_i^x under unified (natural and managerial) disposability ($q = 1,\ldots,m^+$)

d_i^{x+} a part of d_i^x under managerial disposability ($i = 1,\ldots,m$)

D_1 a partial group of G_1 which are above an estimated discriminant function in Stage 2

D_2 a partial group of G_2 which are below an estimated discriminant function in Stage 2

e a row vector whose components are all 1

e_b an undesirable output-based scale elasticity

e_c a cost-based scale elasticity

e_g a desirable output-based scale elasticity

e_1 a unit vector whose first component is 1

E an efficient DMU group in J_n whose members have unity in their efficiency measures and zero in all slacks and unique optimal solution

E' an efficient DMU group in J_n whose members have unity in their efficiency measures and have zero in all slacks and multiple optimal solutions

ECD a group of DMUs which may become efficiency-candidates implying that they may belong to $E \cup E'$

f a subscript of the f-th undesirable output ($f = 1,\ldots,h$)

$G = (G_1, G_2, \ldots, G_n)$ a $s \times n$ matrix of observed desirable outputs of all DMUs

$G_j = (g_{1j}, g_{2j}, \ldots, g_{sj})^{Tr}$ a column vector of observed s desirable outputs of the j-th DMU

$G_k = (g_{1k}, g_{2k}, \ldots, g_{sk})^{Tr}$ a column vector of observed s desirable outputs of the k-th DMU

$\bar{g}_{rj} = g_{rj} + \beta_r$ an adjusted value of the r-th desirable output

$g_{rj}^+ = \max(g_{rj}, \varepsilon_s)$ an adjusted value of the r-th desirable output to express a positive part

$g_{rj}^- = \max(g_{rj}, \varepsilon_s) - g_{rj} + \varepsilon_s$ an adjusted value of of the r-th desirable output to express a negative part

g_{rj} an observed value of the r-th desirable output on the j-th DMU

g_{rjt} an observed value of the r-th desirable output on the j-th DMU in the t-th period

g_{rj}^+ a positive part of g_{rj} after disaggregation

g_{rj}^- a negative part of g_{rj} after disaggregation

G_1 a set of observations in the first group above an estimated discriminant function

G_2 a set of observations in the second group below an estimated discriminant function

G_ς	a set of observations in the ς -th group (for $\varsigma = 2,\ldots,q-1$)
i	a subscript of the i-th input ($i = 1,\ldots,m$)
IE	an inefficient DMU group in J_d whose efficiency measures are less than unity and some intensity variable(s) in E is positive on optimality
IE'	an inefficient DMU group in J_n whose members have efficiency measures that are less than unity and some intensity variable(s) in E is positive on optimality
IF	an inefficient DMU group in J_d whose members have unity in their efficiency measures but they have at least one positive slack on optimality
IF'	an inefficient DMU group in J_n whose members have unity in their efficiency measures but at least one positive slack and multiple optimal solutions
j	a subscript of the j-th DMU ($j = 1,\ldots,n$)
J	a whole DMU set
J_d	a dominated DMU group
J_n	a non-dominated DMU group
J_t	all DMUs in the t-th period
k	a subscript of the specific k-th DMU ($k = 1,\ldots, n$) whose performance is measured by DEA
lb	a lower bound
$\underline{lu}_r$	a lower bound of u_r/u_1
$\underline{lv}_i$	a lower bound of v_i/v_1
M	a prescribed large number
M (superscript)	managerial disposability
N (superscript)	natural disposability
NR (superscript)	non-radial measurement
NY	a group of DMUs which are not yet examined at the current stage
ob_j	the j-th goal (or objective) in GP
p_j	a dual variable for the j-th observation ($\mu_j = np_j - 1$) in regression and GP
q	a subscript of the q-th input ($q = 1,\ldots, m^+$) under managerial disposability
r	a subscript of the r-th desirable output ($r = 1,\ldots, s$)
R (superscript)	radial measurement
R_f^b	a data range related to the f-th undesirable output
R_r^g	a data range related to the r-th desirable output
R_h	a column vector of the right hand side of a formulation
R_i^x	a data range related to the i-th input
RS_k	a reference set of the k-th DMU
Tr	a vector transpose
t	the t-th period

$U=(u_1,u_2,\ldots,u_s)$	a raw vector of s dual variables for desirable outputs
ub	an upper bound
u_r	a dual variable related to the r-th desirable output
$\overline{uu}_r$	an upper bound of u_r/u_1
$\overline{uv}_r$	an upper bound of v_i/v_1
v (subscript)	this expresses variable RTS or DTS
$V=(v_1,v_2,\ldots,v_m)$	a raw vector of m dual variables for inputs
v_i	a dual variable related to the i-th input
v_i^- (natural and managerial disposability)	a dual variable related to the i-th input
v_q^+ (natural and managerial disposability)	a dual variable related to the q-th input
$W=(w_1,w_2,\ldots,w_h)$	a raw vector of h dual variables for undesirable outputs
w_f	a dual variable related to the f-th undesirable output
w_r^g	a weight assigned to the r-th desirable output-based slack
w_i^x	a weight assigned to the i-th input-based slack
$\mathbf{X}=(X_1,X_2,\ldots,X_n)$	a $m\times n$ matrix of observed inputs of all DMUs
x_i	an unknown variable for the i-th input quantity to be used to find cost minimum
x_{ij}	an observed value of the i-th input on the j-th DMU
$\bar{x}_{ij}=x_{ij}+\alpha_i$	an adjusted value of the i-th input
x_{ijt}	an observed value of the i-th input on the j-th DMU in the t th period
$X_j=(1,x_{1j},\ldots,x_{mj})$	the j-th observation on the m-th independent variable in regression
$X_j=(x_{1j},x_{2j},\ldots,x_{mj})^{Tr}$	a column vector of observed m inputs on the j-th DMU
$X_k=(x_{1k},x_{2k},\ldots,x_{mk})^{Tr}$	a column vector of observed m inputs of the k-th DMU
y_j	the j-th observation on a dependent variable in regression
z_{ij}	the j-th observation on the i-th independent factor
z_{ik}	the newly sampled k-th observation on the i-th independent factor
α	θ under a possible occurrence of congestion in DEA
β	τ under a possible occurrence of congestion in DEA
$\beta=(\beta_0,\beta_1,\ldots,\beta_m)^{Tr}$	a column vector of parameters on unknown independent variables in regression
γ_i^+	an unknown binary variable related the i-th positive slack
γ_i^-	an unknown binary variable related the i-th negative slack
δ_j^+ and δ_j^-	positive and negative deviations (errors) of the j-th observation from an estimated regression hyperplane in regression and GP

Δ_x	an increased column vector of m inputs of the k-th DMU
Δ_g	an increased column vector of s desirable outputs of the k-th DMU
ε	a small number (e.g., 0.1 and 1) for MCF that is prescribed by a DEA user
ε_s	a very small number (e.g., 0.00001) to be prescribed by a DEA user
ε_n	a non-Archimedean small number
η	an unknown decision variable to be used for satisfying SCSCs
θ	an efficiency score measured by an input-oriented radial model
θ_i	an efficiency score measured by the i-th input (Russell measure)
$\bar{\vartheta}^*$ and $\underline{\vartheta}^*$	upper and lower bounds on an adjusted intercept for determining the type of RTS
ζ_i	a binary variable regarding the i-th parameter
ζ_r	a binary variable regarding the r-th parameter
κ	a functional form for a supporting hyperplane
$\lambda = (\lambda_1, \lambda_2, \ldots, \lambda_n)^{Tr}$	a column vector of intensity (or structural) variables for all DMUs
λ_i (discrimination)	an unknown weight estimate on the i-th factor
λ_j	an intensity or structural variable related to the j-th DMU
λ_{jt}	an intensity or structural variable related to the j-th DMU in the t th period
μ	an arbitrary positive number
μ_j	a binary variable of the j-th observation to express correct or incorrect classification
$\pi = [\pi_1, \ldots, \pi_m]^{Tr}$	an m-dimensional non-negative column vector
$\Pi = [\pi, \pi, \ldots, \pi]$	a non-negative matrix
ρ	a variable to express an absolute distance between the discriminant score and a discriminant function
ρ_j	an adjusted efficiency of the j-th DMU
σ	a dual variable regarding the constraint (i.e. the sum of all λ_j is unity)
τ	an efficiency score measured by a desirable output-oriented radial model
τ_r	an efficiency score measured by the r-th desirable output (Russell measure)
υ_i^+ and υ_i^-	binary variables regarding the i-th parameter
$\bar{\upsilon}^*$ and $\underline{\upsilon}^*$	upper and lower bounds on an adjusted intercept for determining the type of DTS
ϕ	an empty set
φ_j	an intensity or structural variable to substitute for λ_j
ξ	an inefficiency score measured by DEA environmental assessment

$\psi = \left[\psi_1,\ldots,\psi_s\right]^{Tr}$ a s-dimensional nonnegative column vector
$\Psi = \left[\psi,\psi,\ldots,\psi\right]$ a non-negative matrix
ω a value on [0, 1]
ϖ the minimum value on $\sum_{j=1}^{n}\left(\sum_{i=1}^{m} x_{ij} - \sum_{r=1}^{s} g_{rj}\right)\lambda_j$
ω_j a weight to express the importance of each j-th goal in regression and GP

1.4.3 Mathematical Concerns

This book has three mathematical concerns, all of which need to be discussed in this chapter.

(a) *Small numbers*: This book uses three types of small number (ε). First, ε_n is a non-Archimedean small number that has been long used by DEA researchers. However, none knows what it is in reality. See Chapters 7 and 25 for a description on how to handle the difficulty. Second, ε is used to express a small number (e.g., 0.1, 1.0 or 2.0). Finally, ε_s is used to express a very small number (e.g., 0.0001). The last two types of ε are utilized for formulations on DEA environmental assessment in Section II. The three small numbers are our mathematical conveniences. Chapters in Section I follow a traditional description on very small numbers. The research spirit to incorporate such small numbers is very important in obtaining reliable DEA results in which all dual variables (multipliers) should be positive. Otherwise, production factors in a data set are not fully utilized in DEA as they are in most studies in production economics. Such a treatment is unacceptable, in particular in DEA environmental assessment, which pays attention to not only a degree of efficiency but also other important measures (e.g., DC, UC, RTS, DTS, MRT, RSU, RTD and DTR) under natural and managerial disposability concepts.

(b) *Italic expression*: This book uses an italic expression if a symbol is related to a mathematical formulation. Furthermore, we often use italics to emphasize important concepts or visual expressions related to the mathematical formulation in each chapter.

(c) *Augmented form*: All chapters in Section I use "ordinary" linear programming formulations. A unique feature of these formulations is that their constraints include an inequality/equality sign (e.g., ≥ and ≤) so that slacks are often excluded from these formulations. Meanwhile, chapters in Section II use "augmented" formulations that incorporate slack variables so that their constraints are expressed by only equality constraints. In the augmented formulations, the existence of a slack(s) implies an "inequality" constraint(s) even in their formulations. The rationale regarding why we use the augmented form is that a weight is assigned to each slack in the objective function and each constraint is associated with a dual variable (i.e., a multiplier). Each dual should be positive because it represents a weight assigned to each production factor. Thus, all production factors (inputs, desirable and undesirable outputs) are fully utilized in DEA formulations because they need to be structured by a full dimension of positive dual variables or multipliers. This type of research concern was first discussed in the first DEA article by Charnes *et al.* (1978). See Chapter 4, as well. Note that all chapters in Section II will document the augmented formulations that are different from their original structure.

1.5 SUMMARY

This chapter introduces a proposal for a new use of DEA, as a holistic approach, to assess various aspects of energy and environmental issues. As discussed in this chapter, this book consists of two sections. The first of these sections, or Section I, is related to the conventional use of DEA and its related framework, which serves as a mathematical basis for environmental assessment in energy sectors and in other industry sectors. Section II, which is the gist of this book, is related to DEA-based conceptual and methodological extensions for environmental assessment in energy sectors.

Here, it is important to note that Chapter 12 summarizes the references cited in Section I (Chapters 1–12). The references are structured in the following manner:

> Cooper, W.W. (2005) Origins, uses of, and relations between goal programming and data envelopment analysis. *Journal of Multi-Criteria Decision Analysis*, **15**, 3–11.

Similarly, Chapter 28 summarizes the references cited in Section II (Chapters 13–27) which classifies previous research efforts based upon energy categories discussed in Chapter 13 and sustainability. Thus, to clearly specify the classification group to which each article belongs, all references in Chapter 28 have a number within [ ] added after the article details, as follows:

> Sueyoshi, T., Goto, M. (2016) Undesirable congestion under natural disposability and desirable congestion under managerial disposability in U.S. electric power industry measured by DEA environmental assessment. *Energy Economics*, **55**, 173–188. [520]

Here, the additional number [520], as an illustrative example, indicates the order of each article. The number is used to specify the group classification summarized in Chapter 28. In addition to the reference format summarized in Chapter 28, other published research works, not directly related to DEA environmental assessment, are listed in the footnotes of each chapter in Section II.

2

OVERVIEW

2.1 INTRODUCTION

This chapter is designed to provide managers, policy makers, researchers and individuals interested in DEA with an overview on the proposed approach. DEA is a holistic methodology for evaluating performance in various organizations (decision making units: DMUs) in private and public sectors. It is indeed true that DEA is very useful in both practical and academic concerns within the context of conventional production economics. However, DEA is not a perfect methodology, in particular for environmental assessment, because the conventional framework of DEA does not fit with various pollution issues. For example, the traditional use of DEA does not consider the existence of undesirable outputs (e.g., pollution of water and air). Eco-technology innovation was not sufficiently discussed in previous DEA studies. Research issues originating from various environmental issues will be explored both from modern business strategy and from industrial policy, including energy and environmental concerns, as discussed in the chapters of Section II.

As an initial step of this book, it is necessary for us to pay attention to the strengths and drawbacks of DEA within its conventional frameworks and usages. The review in this chapter provides an intuitive description of various DEA applications for performance assessment in public and private sectors. The description contains several illustrative examples.

Environmental Assessment on Energy and Sustainability by Data Envelopment Analysis, First Edition. Toshiyuki Sueyoshi and Mika Goto.

The remainder of this chapter is organized as follows: Section 2.2 intuitively describes what DEA is as a holistic methodology for performance assessment. Section 2.3 summarizes remarks on a conventional use of DEA and its applications. Section 2.4 describes a reformulation process from an original formulation (i.e., ratio form) to an equivalent linear programming model. Section 2.5 discusses analytical implications of a reference set for each organization to be examined. Section 2.6 documents illustrative examples on how to compute DEA problems by linear programming. Section 2.7 summarizes this chapter.

2.2 WHAT IS DEA?

Charnes *et al.* (1978)[1] published the first article on DEA in the *European Journal of Operational Research* after their presentation at an international conference which was organized at Hawaii in 1977, supported by ORSA/TIMS (where ORSA and TIMS stand for Operations Research Society of America and The Institute of Management Science, respectively). See Chapter 3 for a detailed description on the historical aspect.

To discuss an overview on DEA, this chapter first returns to the description prepared by Charnes *et al.* (1978) regarding how to evaluate DMUs in a public sector. The conventional use of DEA performance assessment considers that there are n DMUs ($j=1,\ldots,n$) to be evaluated[2]. Each DMU utilizes m inputs ($i=1,\ldots,m$)

[1] Professor A. Charnes was an extremely short-tempered individual who did not have any interpersonal skills for communicating with other people. He was always very busy in maintaining his research institute within the University of Texas at Austin (so, financially supporting many Ph.D. students from all over the world), so that he did not have any time to prepare manuscripts. It might also be true that he did not have any language proficiency at the level of journal publication in business and economics. It is true that he was a mathematician in business and economics. Most of his research works were prepared by Professor W.W. Cooper, who was a gentleman in ordinary life. However, Professor Cooper often became a short-tempered old man, as when he discussed research with his students. He was very serious on research. The original idea on DEA came from Professor Cooper and E. Rhodes who was Professor Cooper's Ph.D. student at that time. Then, their idea was forwarded to Professor Charnes who added the non-Archimedean small number (ε_n) in the original DEA model (i.e., ratio form). They first submitted their first manuscript version to *Management Science* but it was rejected by the journal. Their revised version was resubmitted to *European Journal of Operational Research.* That was the first journal publication on DEA in 1978. See Chapter 1 for a description of Professor Cooper. See Chapter 3, as well.

[2] DMUs conventionally stand for hospitals, public agencies and schools in the public sector as well as companies, plants and industries in the private sector, whose efficiency measures are relatively compared and determined by DEA. In this chapter, the treatment of DMUs is separated by a concept of "social sustainability" in the public sector and another concept of "corporate sustainability" in the private sector. The conventional use of DEA originated from "social economics" and "social accounting" for the public sector. See Ijiri and Sueyoshi (2010) for these economic concepts based upon which Professor Cooper developed DEA with his associates at that time for the initial development. In this book, DEA environmental assessment discusses the concept of social and corporate sustainability along with these different economic and strategic perspectives, all of which have been not explored in previous developments of DEA.

to yield s desirable outputs[3] ($r=1,\ldots,s$). The components of input and desirable output vectors are referred to as "production factors" in this book. The original DEA model, often referred to as "a ratio form," has the following fractional structure for efficiency measurement, which is later specified as operational efficiency (OE):

$$\begin{aligned}\text{Efficiency} &= (\text{sum of total weighted desirable outputs})/\\&\quad(\text{sum of total weighted inputs}),\\&=\left(\sum_{r=1}^{s}u_r g_{rj}\right)\Big/\left(\sum_{i=1}^{m}v_i x_{ij}\right)\end{aligned} \tag{2.1}$$

where u_r ($r=1,\ldots,s$) and v_i ($i=1,\ldots,m$) are "weights" assigned, respectively, to g_{rj} (the r-th desirable output) and to x_{ij} (the i-th input of the j-th DMU). Both weights are often referred to as "multipliers" in DEA[4]. They are measured as "dual variables" in DEA formulations. Therefore, this chapter uses these three terms: weights, multipliers and dual variables, depending upon their uses and contexts in DEA performance assessment.

The level of efficiency may be considered as "productivity" in production economics and business. It is important to note that if a multiplier(s), in particular v_i, is zero for some i, then Equation (2.1) may have a high likelihood that it produces a strange result, such as an infinite objective value. Thus, Equation (2.1) requires that all multipliers and components of production factors should be strictly positive[5].

Table 2.1 exhibits an illustrative example to discuss the analytical structure of DEA in the case of a single input (e.g., the number of employees; unit: 100 people) and a single desirable output (e.g., the amount of sales; unit: US$ 1 million).

The table contains eight organizations from {A} to {H} and they are considered as DMUs in the proposed DEA context. The performance (e.g., labor productivity) of these DMUs is summarized at the bottom of the table. The level of efficiency indicates the relationship that {C} = 1 > {G} = 0.8 > {A} = 0.75 > {B} = 0.667 > {E} = 0.625 > {D} = {F} = 0.5, where {C} attains the status of full efficiency and the other DMUs have some levels of inefficiency. For example, DMU

[3] In production economics, typical inputs include the amount of labor, capital and materials. Desirable outputs include the amount of goods, services, revenue and sales. The selection of inputs and desirable outputs depends upon social and corporate sustainability concepts in a DEA environmental assessment. In a conventional use of DEA, the selection depends upon whether its application is concerned with the public or private sector.

[4] The multipliers are measured as dual variables in DEA.

[5] Many researchers previously produced DEA results without considering any occurrence of zero in a data set and multipliers. However, if such results were acceptable, they were just due to their accidental fortunes. This chapter intentionally drops the requirement on positive multipliers by setting them to be unrestrictive in their signs. We also discuss the occurrence of zero and/or negative values in a data set in the computational framework of DEA environmental assessment. See Chapters 26 and 27 on how to handle an occurrence of zero and/or negative values in a data set.

TABLE 2.1 Performance assessment: one input and one desirable output

DMU	A	B	C	D	E	F	G	H
Input	4	3	3	2	8	6	5	5
Desirable Output	3	2	3	1	5	3	4	2
Efficiency	0.75	0.667	1	0.5	0.625	0.5	0.8	0.4

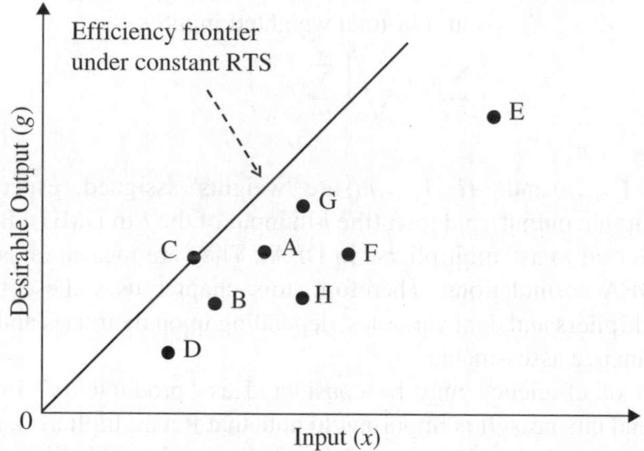

FIGURE 2.1 Efficiency frontier and performance assessment (a) An efficiency frontier under constant RTS (returns to scale) passes from the origin to {C}. The constant RTS implies that a unit increase in an input propositionally increases a desirable output, as depicted in the straight line for the efficiency frontier. See Chapter 8 for a description on RTS. (b) The {C} is efficient and the remaining others have some level of inefficiency under such a production condition.

{A} exhibits 75% in an efficiency measure. The remaining 25% indicates the level of inefficiency[6]. Thus, by using a percentage expression, DEA can provide us with a "quick and easy" assessment.

[6]A production function, defined in economics, assumes an efficient status of producers and does not consider any production inefficiency. However, it is reasonable to consider that there is some level of inefficiency in reality. For example, it is often observed that a company can produce a certain level of desirable outputs, using a lower amount of inputs, compared to another company that produces the same amount of desirable outputs. Professor Cooper has often expressed in his class for graduate students that none can work without any rest and error. He has always insisted that economic theory assumes that all organizations operate on an efficiency frontier without any inefficiency. The assumption regarding the frontier operation is because we can take a derivative of a functional form for optimal allocation among production factors. We clearly understand the importance of such an assumption because it enhances mathematical tractability in order to assess the performance of various economic activities. However, it is true that no economic entity can operate perfectly without any loss. An important feature of DEA is that it can identify a level of inefficiency by relatively comparing the achievements of multiple organizations. Thus, it is possible for us to consider that DEA is an empirical methodology for performance analysis. This book will use the term "disposability" to indicate the elimination of inefficiency.

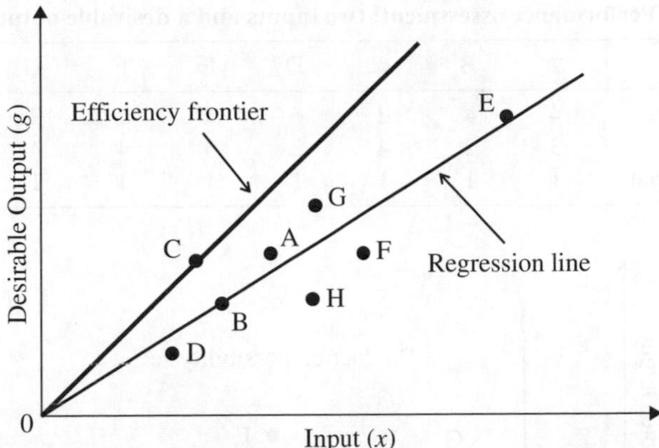

FIGURE 2.2 Efficiency frontier and regression line (a) An efficiency frontier passes from the origin to {C}. This DMU is efficient and the remaining others have some level of inefficiency. (b) The regression line locates on the center of all observations (i.e., DMUs). (c) The regression analysis is used for future prediction and the efficiency frontier is used for performance assessment.

Figure 2.1 visually describes the location of eight DMUs listed in Table 2.1 on the x–g coordinates (input and desirable output). An important feature of DEA is that it can incorporate an analytical capability to identify an efficiency frontier.

The line from the origin to {C} is one such efficiency frontier, based on which DEA evaluates the efficiency level of all the other DMUs. The efficiency status of the other DMUs, locating below the efficiency frontier, is determined by relatively comparing their observed achievements with their projections on the efficiency frontier. The relative comparison between a DMU and the others makes it possible that DEA empirically identifies an efficiency frontier, based on which it measures a distance between the location of its observation and the frontier consisting of some DMUs.

Figure 2.2 depicts a difference between the efficiency frontier and a regression line. The efficiency frontier from the origin to {C} locates on or above all observations {A–H}. As mentioned previously, the frontier provides an evaluation basis for performance analysis. Meanwhile, the regression line, passing through the middle of all observations, is used to predict the estimated value ($\hat{g}$) of a dependent variable on the line, given an unknown (future) value of an independent variable (x).

Table 2.2 exhibits an illustrative example for input-oriented measurement in the case of two inputs (e.g., the number of employees; unit: 100 people and number of offices) and a single desirable output (e.g., the amount of sales; unit: US$ 1 million). Figure 2.3 visually summarizes the computational result measured by DEA. The horizontal axis indicates the first input (x_1) divided by the desirable output (g) and the vertical axis indicates the second input (x_2) divided by the desirable output.

TABLE 2.2 Performance assessment: two inputs and a desirable output

DMU	A	B	C	D	E	F	G	H	I
Input 1	4	4	2	6	7	7	3	8	5
Input 2	3	2	4	2	3	4	4	1	3
Desirable Output	1	1	1	1	1	1	1	1	1

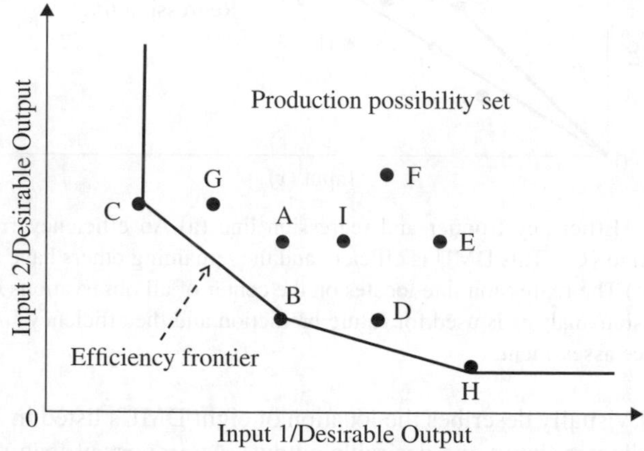

FIGURE 2.3 Efficiency frontier and performance assessment: two inputs and one desirable output (a) An efficiency frontier consists of piece-wise linear contour line segments by connecting {C–B–H}. The other DMUs are inefficient. They need to improve their performance levels so that they can attain the efficiency frontier. (b) The production possibility set locates in the north-east area of the figure where all efficient and inefficient DMUs can exist under their production technology.

In Figure 2.3, when DMUs use a lower amount on inputs, it is considered that they can attain better performance. As visually specified in this figure, three DMUs (i.e., {B}, {C} and {H}) consist of an efficiency frontier and the remaining others exhibit some level of inefficiency. This region, locating at the north-east corner from the efficiency frontier, is often referred to as a "production possibility set." The set indicates a possible combination between two inputs and a desirable output in which DMUs can exist with some level of inefficiency and zero inefficiency (and so, full efficiency).

In Figure 2.4, let us consider that there are three DMUs {A, B, C} to be measured by DEA. To measure the level of efficiency on the inefficient DMU {A}, DEA identifies a reference set of {A} that contains other DMUs {B and C}. To attain the status of efficiency, the inefficient DMU {A} needs to decrease the amount of inputs at the level that it can attain an efficiency frontier on a line segment between the two DMUs {B and C}.

For example, A_1 is a hypothetical achievement which is measured by decreasing an amount of the first input. Since A_1 is a projected point on an efficiency frontier,

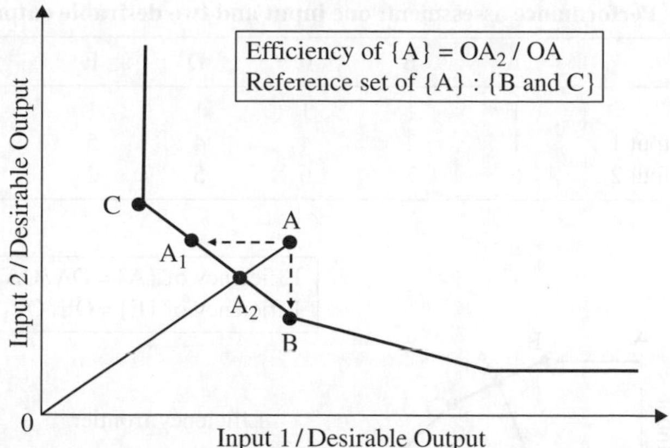

FIGURE 2.4 Efficiency measurement for DMU {A} (a) DMU {A} needs to reduce an amount of the two inputs. For example, {A} can shift to any projected points between A_1 and {B}. The point A_2 is one of such possible projected points in the case where the two inputs are reduced. (b) An efficiency frontier for {A} consists of part of a line segment between the two DMUs {B and C}. (c) The efficiency measure of {A} is determined by the ratio of the two distances (i.e., OA_2/OA) measured on the L2 norm for our visual description. See Chapter 3 about the L2 norm distance measurement. Such a treatment is for our visual convenience. This chapter understands that the reality of DEA efficiency measurement should be discussed by the L1 norm distance.

it is not a member of the reference set of {A}. Another projected point, for example, can be found on {B} by decreasing an amount of the second input. Both are acceptable in terms of attaining an efficiency frontier. The third alternative for projection can be found on A_2 where {A} needs to reduce an amount of both inputs. Any projection points between A_1 and {B} are possible for DEA. DEA selects one projection from many on the line segment B–A_1 in the manner that {A} can project itself toward an efficiency frontier. For example, A_2 may be such a possible projection because it is on a directional vector to improve the efficiency increase. In the case of A_2, the degree of efficiency on {A} is determined by the distance of OA_2 divided by that of OA.

Table 2.3 exhibits the output-oriented efficiency measurement, along with Figure 2.5 which visually summarizes a computational result by DEA. The table has DMUs whose performance measures have three production factors: a single input and two desirable outputs. The horizontal axis indicates the first desirable output divided by the input and the vertical axis indicates the second desirable output divided by the input.

In Figure 2.5, when DMUs produce a larger amount of desirable outputs, they attain better performance. As visually specified in the figure, four DMUs {B–C–F–G} consist of an efficiency frontier and the remaining others exhibit some level of inefficiency. The south-west region from the efficiency frontier indicates

TABLE 2.3 Performance assessment: one input and two desirable outputs

DMU	A	B	C	D	E	F	G
Input	1	1	1	1	1	1	1
Desirable Output 1	1	2	4	4	5	6	7
Desirable Output 2	6	7	6	5	2	4	2

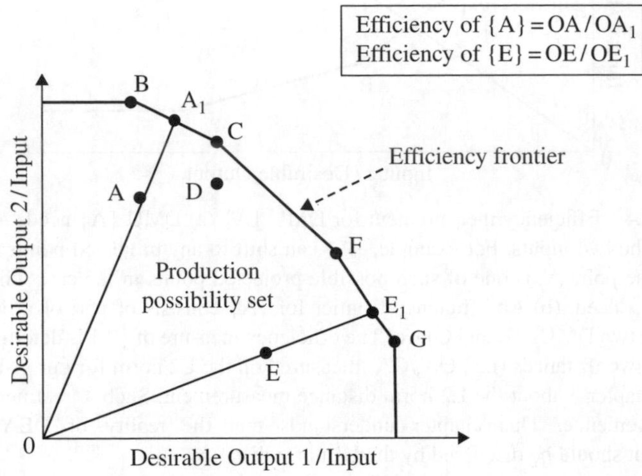

FIGURE 2.5 Efficiency frontier and performance assessment in one input and two desirable outputs (a) DMU {E} needs to increase the amount of the two desirable outputs. For example, the DMU can shift to any projected points between {F} and {G}. The point E_1 is one such possible projected point in the case where the two desirable outputs are increased. (b) An efficiency frontier consists of line segments between the DMUs {B–C–F–G}. (c) The level of efficiency of {E} is determined by the ratio of the two distances (i.e., OE/OE_1) in the L2 norm distance measurement. Such is just for our visual convenience, as mentioned previously. This chapter understands that DEA belongs to the L1 norm measurement. See Chapter 3 on the L1 and L2 distance measurements.

a production possibility set in the case, indicating a possible combination between the two desirable outputs and the input in which inefficient DMUs may exist with some level of inefficiency.

As depicted in Figure 2.5, for example, the reference set of DMU {E} contains efficient DMUs {F and G}, based upon which DEA determines a degree of the efficiency. To attain the status of efficiency, {E} needs to increase the amount of desirable outputs at the level that it can attain an efficiency frontier on a line segment between DMUs {F and G}. Any point between the two DMUs is acceptable for the projection. For example, the point E_1 indicates such a possible projection. The degree of efficiency on {E} is institutively determined by the distance of OE divided by that of OE_1. Such an efficiency measurement can also be applied to DMU {A}.

2.3 REMARKS

Many researchers and practitioners often believe that DEA is a perfect method that they can apply to any real data sets. Unfortunately, the belief is not correct. It is indeed true that DEA is useful as an evaluation methodology whose practicality can be documented in many previous applications. However, the truth is by far different from the reality.

The following concerns may be important to keep in our minds before applying DEA to many performance assessments (Sueyoshi and Goto, 2013a):

(a) *Data independence*: All DMUs independently operate each other so that their input and desirable output vectors have independence among them. Thus, no rank deficit occurs in a data domain for inputs and desirable outputs.

(b) *Zero and negative values in data*: DEA usually treats a data set with the condition that it does not contain any zero and negative values. Such negative data (e.g., negative growth and net income) can be easily found in many financial and economic analyses. As mentioned in Thompson *et al.* (1993), DEA needs an additional treatment to handle the zero in a data set. This book will discuss how to deal with zero and/or negative values in a data set within the computational framework of DEA environmental assessment. Chapters 26 and 27 discuss how to handle an occurrence of such a data set. See Chapter 7, as well. It is often observed that a data set containing zero in the observations does not disturb DEA computation. However, such a case is just a lucky accident.

(c) *Selection of inputs and desirable outputs*: It is not easy to determine which inputs and desirable outputs are used for DEA performance analysis. Professor Cooper has always suggested that the best strategy for us is to ask a decision maker(s) (e.g., a director in a public agency and a president in a firm) about the selection of their production factors. However, it is often found that a DEA user(s) cannot access information from a decision maker(s). In many cases, the selection of production factors depends upon his subjective decision of a user(s). The subjectivity may not reflect the operational reality of DMUs in many cases. Furthermore, the DEA user(s) needs to keep in mind that the less input generally produces the better performance in a conventional DEA. An opposite case is found in the desirable output selection. The selection logic on inputs and desirable outputs works on only a conventional use of DEA, but it does not work on DEA environmental assessment because outputs are classified into two categories (i.e., desirable and undesirable outputs).

(d) *Ratio variables*: It is often recommended that a data set should not contain any ratio variable (e.g. return on assets and return on equity). A rationality of such recommendation is that DEA has originated from the ratio form, or

Equation (2.1). The other rationality is that the ratio variables may often have a common numerator or denominator so that these variables may violate the property of data independency. However, the DEA models, particularly non-radial models, discussed in this book do not originate from the ratio form. Consequently, it may be possible for us to use the ratio variable as long as it maintains the property of data independency among production factors. The difficulty is solved by a methodology proposed by Chapters 11 and 25. The methodology, referred to as "DEA-discriminant analysis (DA)" can handle a data set that is structured by financial data, or often financial ratios, and other types of data which are more useful than production factors in corporate assessment.

(e) *Outlier in data*: If an outlier exists in a data set examined in DEA, it is often necessary to drop it from the data because the outlier may destroy the shape of an efficiency frontier so that DEA does not produce a reliable result. See, for example, Tran *et al*. (2008) that discusses how to identify the outlier. However, such a decision depends upon the researcher and/or user being interested in DEA applications.

(f) *Data adjustment*: When an input(s) or a desirable output(s) is much larger than the remaining other production factors, the large factor often dominates the whole computational process of DEA. In this case, it is necessary for us to adjust the data set by dividing all observations on each production factor by the factor average. Such a data adjustment produces more reliable DEA results than those without adjustment. See, for example, Chapters 22 and 23 for a description on how such a data adjustment can be linked to a new approach, referred to as "explorative analysis." Here, it is important to note that the data adjustment should not produce any zero and/or negative values because DEA cannot handle them. Therefore, normalization [i.e., (observed value – average value)/standard deviation], widely used in statistics and econometrics, does not work because it produces zero and negative values. See Chapters 26 and 27 for a special approach that can handle such an occurrence of zero and/or negative values in a data set. See Chapter 6 for a description on the property of "unit invariance." The property indicates that the unit of inputs and outputs should not influence an efficiency measure. The property originates from a mathematical rationale, but not a computational perspective on a computer code.

(g) *Imprecise number*: It is trivial that DEA has a difficulty in handling an imprecise number. Although we can mathematically handle the number, it is impossible for us to apply DEA to a data set with such an imprecise number on a computer code. Moreover, the fractional number (e.g., 1/3) suffers from a rounding-off error in the DEA computation. Thus, the number is mathematically acceptable but unacceptable by DEA. Some DEA users replace it by 0.333 and the others may use 0.334. See Chapter 7 that lists this type of problem in DEA. To avoid the specification error, it is necessary for us to specify a data range between 0.333 and 0.334 to express 1/3 for running a computer code for DEA. Consequently, DEA may have multiple solutions in handling the imprecise number. See Cooper *et al.* (1999) for a detailed description on how to handle the imprecise data. This chapter clearly understands that their research has intended to deal with a much larger range on data impreciseness (e.g., a data range between 0.2 and 0.5). It is easily envisioned that the DEA capability to handle the imprecise data is important for predicting future performance by DEA applied to energy and environment.

2.4 REFORMULATION FROM FRACTIONAL PROGRAMMING TO LINEAR PROGRAMMING

An analytical rationale regarding why many researchers and practitioners pay attention to DEA is that it can be solved by linear programming. Such computational tractability is an important feature of DEA. To attain the computational feasibility, this chapter needs to describe the reformulation of Equation (2.1) to linear programming equivalence.

The reformulation is important because the original equation is a ratio form and thus is a problem of nonlinear programming. The research difficulty was mainly solved by the effort of Professor Charnes in his collaboration with Professor Cooper. Their collaboration on the reformulation was first published in the research of Charnes and Cooper (1962). Then, they applied the research effort to the first publication on DEA in Charnes *et al.* (1978). Professor Charnes made a contribution in transforming from fractional programming to linear programming equivalence in the initial stage of DEA development.

To explain their reformulation, this chapter returns to Equation (2.1) and reorganizes it by the following ratio form that measures the level of efficiency on the specific k-th DMU ($k = 1,\ldots,n$) to be examined by DEA:

$$
\begin{aligned}
&\text{Maximize } \sum_{r=1}^{s} u_r g_{rk} \Big/ \sum_{i=1}^{m} v_i x_{ik} \\
&s.t. \quad \sum_{r=1}^{s} u_r g_{rj} \Big/ \sum_{i=1}^{m} v_i x_{ij} \le 1 \; (j = 1,\ldots,n), \\
&\qquad u_r \ge 0 \; (r = 1,\ldots,s) \text{ and } v_i \ge 0 \; (i = 1,\ldots,m),
\end{aligned}
\tag{2.2}
$$

where *s.t.* stands for "subject to." The objective function of Model (2.2) indicates that the performance, or operational efficiency, of the specific k-th DMU is expressed by the ratio computed from the sum of total weighted desirable outputs divided by the sum of total weighted inputs. Furthermore, as formulated in the constraints, the performance of all DMUs ($j = 1,\ldots,n$) is expressed by the ratio that is equal or less than unity (so, 100%). An important feature of Model (2.2) is that it can evaluate the performance of all DMUs by a percentage expression, where 100% indicates the status of full efficiency, while 0% indicates the status of full inefficiency.

Here, it is important to note that this chapter uses a subscript "j" to express all DMUs ($j=1,\ldots,n$) and another subscript "k" to indicate a specific DMU to be measured by DEA. Both are used hereafter in this book. Since we need to evaluate the performance of the specific k-th DMUs to be assessed, the subscript ($k=1,\ldots,n$) is used in this book.

As found in the objective function, a unique feature of DEA computation is that it repeats the computational process of Model (2.2) by n times ($k=1,\ldots,n$) by changing the k-DMU. It is easily imagined that DEA is computationally time-consuming and tedious. However, a modern personal computer can easily handle the computational burden related to DEA by recent computational technology. See Chapter 10 on DEA network computing.

To intuitively describe the reformulation from Model (2.2) to the linear programming equivalence, this chapter reorganizes Model (2.2) as follows:

$$\begin{aligned} &\text{Maximize} \sum_{r=1}^{s} u_r g_{rk} \\ &s.t. \quad \sum_{i=1}^{m} v_i x_{ik} = 1 \\ &\quad -\sum_{i=1}^{m} v_i x_{ik} + \sum_{r=1}^{s} u_r g_{rj} \le 0 \left(j=1,\ldots,n\right), \\ &\quad u_r \ge 0 \left(r=1,\ldots,s\right) \text{ and } v_i \ge 0 \left(i=1,\ldots,m\right). \end{aligned} \tag{2.3}$$

Charnes and Cooper (1962) have proved that Model (2.3) is equivalent to Model (2.2). The reformulation can be intuitively acceptable to us. See their detailed mathematical description. Model (2.3) is equivalent to the following linear programming formulation[7]:

$$\begin{aligned} &\text{Maximize} \sum_{r=1}^{s} u_r g_{rk} \\ &s.t. \quad \sum_{i=1}^{m} v_i x_{ik} = 1 \\ &\quad -\sum_{i=1}^{m} v_i x_{ij} + \sum_{r=1}^{s} u_r g_{rj} \le 0 \ \left(j=1,\ldots,n\right), \\ &\quad u_r \ge 0 \left(r=1,\ldots,s\right) \text{ and } v_i \ge 0 \left(i=1,\ldots,m\right). \end{aligned} \tag{2.4}$$

[7] The model formulates that the production activity of the k-th DMU is designed to maximize the amount of desirable outputs, given a certain amount of inputs.

Model (2.4) has the following dual formulation:

$$
\begin{aligned}
&\text{Mimimize } \theta \\
&s.t. \quad -\sum_{i=1}^{m} x_{ij}\lambda_j + \theta x_{ik} \geq 0 \quad (i=1,\ldots,m), \\
&\qquad \sum_{r=1}^{s} g_{rj}\lambda_j \geq g_{rk} \, (r=1,\ldots,s), \\
&\qquad \lambda_j \geq 0 \, (j=1,\ldots,n) \text{ and } \theta\text{: URS.}
\end{aligned}
\tag{2.5}
$$

Here, θ is a dual variable in the objective function and λ_j $(j=1,\ldots,n)$ are the dual variables of the second group of constraints of Model (2.4), respectively. Model (2.5) is usually considered as a linear programming equivalence to the original fractional programming problem in Model (2.2). The remaining chapters in this book consider Model (2.5) as a primal formulation because it has an efficiency score (θ) in the formulation and is thus intuitive to us in many DEA applications.

If the efficiency measure (θ*) becomes unity on the optimality of Model (2.5), then the k-th DMU is evaluated as full efficiency under the additional requirement that all slacks are zero. In other words, $\theta^* = 1$ does not immediately imply the status of full efficiency by Model (2.5). It is necessary for us to confirm whether all slacks are zero or not. The slacks are specified as follows:

$$
d_i^x = \theta x_{ik} - \sum_{j=1}^{n} x_{ij}\lambda_j \, (i=1,\ldots,m) \text{ and } d_r^g = \sum_{j=1}^{n} g_{rj}\lambda_j - g_{rk} \, (r=1,\ldots,s). \tag{2.6}
$$

To examine whether a slack(s) exists on the optimality of Model (2.5), use of the following model is recommended in the DEA community:

$$
\begin{aligned}
&\text{Maximize } \sum_{i=1}^{m} d_i^x + \sum_{r=1}^{s} d_r^g \\
&s.t. \quad \sum_{i=1}^{m} x_{ij}\lambda_j + d_i^x = \theta^* x_{ik} \quad (i=1,\ldots,m), \\
&\qquad \sum_{r=1}^{s} g_{rj}\lambda_j - d_r^g = g_{rk} \quad (r=1,\ldots,s), \\
&\qquad \lambda_j \geq 0 \, (j=1,\ldots,n),\ d_i^x \geq 0 \, (i=1,\ldots,m) \text{ and} \\
&\qquad d_r^g \geq 0 \, (r=1,\ldots,s).
\end{aligned}
\tag{2.7}
$$

Here, θ* is obtained from the optimal objective function of Model (2.5). The slack identification process consists of a two-stage process: Stage 1: solve

Model (2.5); and Stage 2: solve Model (2.7) after incorporating the optimal objective value, or an efficiency score, determined by Model (2.5). See Cooper *et al*. (2006).

It is widely known in the DEA community (e.g., Cooper *et al*., 2006) that the optimal solution of Model (2.5), combined with the optimal solution of Model (2.7), consists of the optimal solution of DEA. The computational process is applicable for the ratio-based measurement, later referred to as "radial models" in Chapter 4. In contrast, the other type of models, later referred to as "non-radial models," does not need such a computational treatment because an efficiency measure is determined only by slacks on optimality. It is often observed that most previous studies have not utilized the proposed two-stage treatment even for radial measurement because they look for computational tractability on their DEA applications.

Finally, this chapter needs to reconfirm that the status of DEA efficiency in the radial measurement is confirmed by two results: (a) the efficiency score θ^* is unity and (b) all slacks are zero on the optimality of DEA models. However, confirmation is not necessary for the non-radial measurement. See Chapter 10 on a network computing on the computation on radial measurement.

2.5 REFERENCE SET

An important feature of DEA is that it determines the level of efficiency concerning the k-th DMU by examining a group of efficient DMUs, often referred to as "a reference set." The set consists of some DMUs on an efficiency frontier based upon which DEA evaluates the performance of the k-th DMU by relative comparison. Returning to Model (2.5), the reference set (RS_k) is identified by

$$RS_k = \{ j | \lambda_j > 0 \text{ for some } j \in J \} \quad \text{where} \quad J = \{\text{all DMUs}\}. \tag{2.8}$$

Assuming the reference set is unique, we can identify how to improve the performance of the k-th DMU. That is, Equation (2.6) can be rewritten on the optimality of Model (2.5) as follows:

$$\begin{aligned} d_i^{x*} &= \theta^* x_{ik} - \sum_{j \in RS_k} x_{ij} \lambda_j^* \left(i = 1,\ldots,m\right) \text{ and } d_r^{g*} \\ &= \sum_{j \in RS_k} g_{rj} \lambda_j^* - g_{rk} \left(r = 1,\ldots,s\right). \end{aligned} \tag{2.9}$$

The composite (ideal) performance $(\bar{x}_{ik}, \bar{g}_{rk})$ of the k-th DMU measured by Model (2.5) is identified on an efficiency frontier as follows:

$$\bar{x}_{ik} \Rightarrow \sum_{j \in RS_k} x_{ij}\lambda_j^* = \theta^* x_{ik} - d_i^{x*} \ (i = 1,\ldots,m) \text{ and}$$
$$\bar{g}_{rk} \Rightarrow \sum_{j \in RS_k} g_{rj}\lambda_j^* = g_{rk} + d_r^{g*} \ (r = 1,\ldots,s). \tag{2.10}$$

Thus, the reference set indicates a basis for relative comparison to determine the efficiency level of the k-th DMU. However, Model (2.5) does not guarantee that the optimal solution is always unique. The identification process by Equations (2.10) is limited on the unique optimal solution of Model (2.5). Such uniqueness on optimality is a problem of DEA. See Chapter 7 for a discussion on how to handle an occurrence of multiple solutions (e.g., multiple reference sets and multiple projections).

2.6 EXAMPLE FOR COMPUTATIONAL DESCRIPTION

Table 2.4 exhibits an example with two inputs and one desirable output. The data set consists of six DMUs (from {A} to {F}) whose desirable outputs are all unity. Figure 2.6 visually describes the data set listed in the table.

Model (2.4), to determine the efficiency of DMU A, is formulated as follows:

$$\begin{aligned} &\text{Maximize}\, u \\ &s.t. \quad 4v_1 + 2v_2 = 1, \\ &\quad -4v_1 - 2v_2 + u \le 0, \\ &\quad -4v_1 - 3v_2 + u \le 0, \\ &\quad -4v_1 - v_2 + u \le 0, \\ &\quad -3v_1 - 2v_2 + u \le 0, \\ &\quad -2v_1 - 4v_2 + u \le 0, \\ &\quad -6v_1 - v_2 + u \le 0, \\ &\quad v_1 \ge 0, v_2 \ge 0 \text{ and } u \ge 0. \end{aligned} \tag{2.11}$$

The optimal solution is $v_1^* = 0.167$, $v_2^* = 0.167$ and $u^* = 0.833$. The efficiency score becomes 0.833.

Model (2.11), corresponding to Model (2.4), has the following dual formulation:

$$\begin{aligned} &\text{Mimimize } \theta \\ &s.t. \quad -4\lambda_A - 4\lambda_B - 4\lambda_C - 3\lambda_D - 2\lambda_E - 6\lambda_F + 4\theta \ge 0, \\ &\quad -2\lambda_A - 3\lambda_B - \lambda_C - 2\lambda_D - 4\lambda_E - \lambda_F + 2\theta \ge 0, \\ &\quad \lambda_A + \lambda_B + \lambda_C + \lambda_D + \lambda_E + \lambda_F \ge 1, \\ &\quad \lambda_j \ge 0\ (j = \text{A},\text{B},\ldots,\text{F}) \text{ and } \theta\text{: URS.} \end{aligned} \tag{2.12}$$

TABLE 2.4 Illustrative example for computation

DMU	A	B	C	D	E	F
Input 1	4	4	4	3	2	6
Input 2	2	3	1	2	4	1
Desirable Output	1	1	1	1	1	1

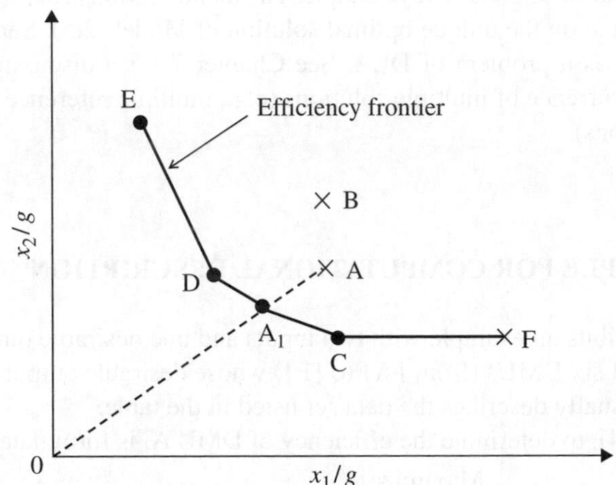

FIGURE 2.6 Efficiency measurement of DMU {A} (a) The example is for the input-oriented measurement under variable RTS. (b) An efficiency frontier consists of {E}-{D}-{C}. They are efficient on the frontier, but the other DMUs exhibit some level of inefficiency. (c) DMU{A} needs to reduce the amount of two inputs to attain the status of efficiency, specified by A_1, while maintaining the same level of an output. (d) DMU{F} is on the frontier so that the efficiency score is unity. However, the DMU contains a slack related to the first input on optimality. Therefore, the DMU is inefficient.

The optimal solution is $\lambda_A^* = 0$, $\lambda_B^* = 0$, $\lambda_C^* = 0.333$, $\lambda_D^* = 0.667$, $\lambda_E^* = 0$, $\lambda_F^* = 0$ and $\theta^* = 0.833$. The solution implies that:

0.833 (input of {A}) = 0.333 (input of {C}) + 0.667 (input of {D}) for the two inputs and desirable output of {A} = 0.333 (desirable output of {C}) + 0.667 (desirable output of {D}).

(a) DMU {A} is inefficient because of two efficient DMUs {C and D} and DMU {B} is inefficient because of two efficient DMUs {D and E} even if all slacks are zero. DMU {F} is unity in its efficiency measure, but being inefficient because of the first input slack is positive. This type of problem often occurs only in radial measures, but not in non-radial measures. However, this chapter cannot immediately claim that non-radial measures outperform radial measures as DEA models because the non-radial models have other types of problems, not found in the radial models (e.g., the non-radial measures are often larger than the radial measures so that the former measures cannot document their discriminant capability in terms of their efficiency scores). See Chapter 5 for a detailed description on the aspect.

(b) The computational result may be unstable because DMUs {C, E and F} have zero in dual variables. They should be all positive to obtain reliable results. Otherwise, it is difficult to guarantee the reliability of DEA computational results. This clearly indicates a drawback concerning a conventional use of DEA. See Chapters 4 and 7 for a description on how to handle the type of difficulty. Unfortunately, this type of difficulty has been long neglected by previous DEA studies because they were only interested in the status of efficiency, not other important measures such as RTS and marginal rate of transformation. See Chapters 20–23 that describe these practical implications from the perspective of environmental assessment and sustainability.

Next, we use the following formulation to determine the magnitude of slacks:

$$
\begin{aligned}
\text{Maximize} \quad & d_1^x + d_2^x + d_1^g \\
s.t. \quad & 4\lambda_A + 4\lambda_B + 4\lambda_C + 3\lambda_D + 2\lambda_E + 6\lambda_F + d_1^x = 4(0.833), \\
& 2\lambda_A + 3\lambda_B + \lambda_C + 2\lambda_D + 4\lambda_E + \lambda_F + d_2^x - 2(0.833), \\
& \lambda_A + \lambda_B + \lambda_C + \lambda_D + \lambda_E + \lambda_F - d_1^g = 1, \\
& \lambda_j \geq 0\,(j = \text{A,B,...,F}), d_1^x \geq 0, d_2^x \geq 0 \text{ and } d_1^g \geq 0.
\end{aligned}
\tag{2.13}
$$

Note that the first two constraints of Model (2.13) originated from Model (2.5). Consequently, the optimal solution pair on DMU {A} contains $v_1^* = 0.167$, $v_2^* = 0.167$ and $u^* = 0.833$ from Model (2.11), $\lambda_A^* = 0$, $\lambda_B^* = 0$, $\lambda_C^* = 0.333$, $\lambda_D^* = 0.667$, $\lambda_E^* = 0$, $\lambda_F^* = 0$ and $\theta^* = 0.833$ from Model (2.12) and $d_1^{x*} = 0$, $d_2^{x*} = 0$ and $d_1^{g*} = 0$ from Model (2.13). The solution implies that DMU {A} is inefficient and the level of efficiency is 0.833. Table 2.5 summarizes the computational results for all six DMUs.

As mentioned previously, a unique feature of DEA is that it is necessary for us to repeat its computation n times for the efficiency evaluation of all DMUs ($k = 1,\ldots,n$). In addition, it is important to note the following concern on DEA computation. The reference set of DMU {A} contains efficient DMUs {C and D} from $\lambda_C^* = 0.333$ and $\lambda_D^* = 0.667$, so implying that the level of efficiency is determined by a virtual A_1, which exists between efficient DMUs {C and D} on the frontier {E–D–C}, as depicted in Figure 2.6. The identification of a reference set gives us a simple but effective computational strategy. That is, if we solve the

TABLE 2.5 Computational summary

Company	DEA Efficiency	Reference Set	v_1^*	v_2^*	u^*	d_1^{x*}	d_2^{x*}	d_1^{g*}
A	0.833	$C(\lambda_C^* = 0.333)$, $D(\lambda_D^* = 0.667)$	0.167	0.167	0.833	0.000	0.000	0.000
B	0.727	$D(\lambda_D^* = 0.909)$, $E(\lambda_E^* = 0.091)$	0.182	0.091	0.727	0.000	0.000	0.000
C	1	$C(\lambda_C^* = 1)$	0.000	1	1	0.000	0.000	0.000
D	1	$D(\lambda_D^* = 1)$	0.250	0.125	1	0.000	0.000	0.000
E	1	$E(\lambda_E^* = 1)$	0.500	0.000	1	0.000	0.000	0.000
F	1	$C(\lambda_C^* = 1)$	0.000	1	1	2.000	0.000	0.000

DEA model on DMU {A} and identify its reference set {C and D}, then it is not necessary for us to solve the efficiency score of the two DMUs because we have already known that their measures are unity, so being efficient. No additional computation is necessary for DEA assessment on DMUs {C and D}. Thus, it is possible for us to skip their performance measurement. The DMU elimination can reduce the computational time and burden regarding Model (2.5) by examining a reference set of an efficient DMU. This feature will be fully utilized in Chapter 10 on DEA network computing.

Next, let us discuss two additional concerns associated with DEA radial models, for example, Model (2.5), discussed in this chapter. One of the two issues is that {F} has $\lambda_A^* = 0$, $\lambda_B^* = 0$, $\lambda_C^* = 1$, $\lambda_D^* = 0$, $\lambda_E^* = 0$, $\lambda_F^* = 0$ and $\theta^* = 1$ on Model (2.5) as well as $d_1^{x*} = 2.000$, $d_2^{x*} = 0$ and $d_1^{g*} = 0$ from Model (2.7) in the two models. This indicates that although $\theta_F^* = 1$, {F} is not efficient because of $d_1^{x*} = 2.000$. The computational result can be easily confirmed in Figure 2.6 because {C} dominates {F}. The DMU needs to reduce the amount of the first input by two units to attain the status of efficiency. Thus, it is important to examine both whether the efficiency score is unity and whether all slacks are zero on optimality, as mentioned previously on the radial measurement. Finally, Table 2.5 summarizes all computational results on six DMUs. The other concern is that DMUs {C}, {E} and {F} (where {C} and {F} are efficient and {F} is inefficient) have zero in these dual variables. For example, the two DMUs ({C} and {F}) have $v_1^* = 0$, indicating that the first input is not fully used in their DEA evaluations. Consequently, the computational result in Table 2.5 becomes unstable because other optimal solutions may exist on dual variables and slacks. There is a possibility that different computer codes may produce different results. In the result summarized in Table 2.5, such multiple solutions may occur on dual variables, but not on their efficiency measures. To avoid this difficulty, it is necessary for us to incorporate additional information, expressed by side constraints, on multipliers (i.e., dual variables) in the dual formulation. Later, Chapter 4 will discuss how to handle the second problem and other types of problem due to an occurrence of zero in dual variables of efficient DMUs[8]. Chapter 7 will discuss the problem from a relationship between primal and dual formulations of linear programming. Chapters 22 and 23 will explore the problem of zero in dual variables from "explorative analysis" that provides restriction on them.

[8]An occurrence of zero in multipliers (i.e., dual variables) is not influential on inefficient DMUs because they do not consist of an efficiency frontier based upon which DEA evaluates the performance of all DMUs. It is necessary for us to pay attention on the multipliers of efficient DMUs. See Chapter 7.

2.7 SUMMARY

This chapter provides an overview on DEA. The DEA approach is indeed a useful methodology in both practical and academic perspectives. Unfortunately, it is true that DEA is not perfect in many applications. It is necessary for us to pay attention to the strengths and drawbacks in various DEA applications when we examine the performance of organizations. As an initial step for directing toward DEA environmental assessment, this chapter started with a description of methodological aspects of DEA in terms of theoretical discussions and simple applications. The description of DEA in this chapter will be extendable to a fundamental use of DEA in Section I. Then, DEA will be further explored for pollution prevention and sustainability development in Section II.

3

HISTORY

3.1 INTRODUCTION

This chapter[1] discusses the mathematical linkage among L1 regression, goal programming (GP) and DEA. As mentioned in Chapter 1, the purpose of this book is to discuss how to prepare a use of DEA environmental assessment, not a conventional use of DEA itself. However, it is true that this book cannot exist without previous research efforts on DEA. In particular, this chapter needs to mention that all the methodological descriptions in this book are due to the efforts of Professor William W. Cooper who was a father of DEA. See Figure 1.2. It is easily envisioned in the figure that the description in this chapter will be further extendable to DEA in Section I and DEA environmental assessment in Section II.

Before starting this chapter, it is necessary for us to describe two concerns. One of the two concerns is that the contribution of Professor Cooper in economics and accounting can be summarized, in short, as the conceptual development

[1] This chapter is partly based upon the two articles: (a) Glover, F. and Sueyoshi, T. (2009), Contributions of Professor William W. Cooper in operations research and management science. *European Journal of Operational Research*, **197**, 1–16 and (b) Sueyoshi, T. and Chang, Y. (1989) Goal programming approach for regression median. *Decision Sciences*, **20**, 4.

Environmental Assessment on Energy and Sustainability by Data Envelopment Analysis, First Edition. Toshiyuki Sueyoshi and Mika Goto.

of "social economics" and "social accounting,"[2] based upon which he developed DEA[3] for performance assessment of organizations in a public sector. This book will extend his two concepts (i.e., social economics and social accounting) to the development of "social sustainability" for a public sector, further connecting to "corporate sustainability" for a private sector. The other concern is that many DEA researchers believe that DEA originated from the ratio form, or Model (2.2). However, we disagree with such a view on the origin of DEA. To correct the conventional view, this chapter discusses a methodological linkage among L1 regression, GP and DEA, returning to statistics from the eighteenth century. The description in this chapter is based upon the article prepared by Cooper (2005). It is easily imagined that his disciples should know the historical concern in Professor Cooper's mind, but unfortunately none has documented his view on DEA. Therefore, this chapter attempts to describe his view on the development of DEA, returning to statistics from the eighteenth century.

The remainder of this chapter is organized as follows: Section 3.2 discusses the origin of L1 regression. Section 3.3 extends our description from L1 regression to GP. The article on L1 regression was the first research effort of GP that identified the formulation, subsequently referred to as GP. It is widely known that GP serves as an important methodological basis for multi-objective optimization. Section 3.4 reviews an analytical property of L1 regression. Section 3.5 describes how L1 regression is mathematically reorganized to L2 regression and GP. Section 3.6 discusses the origin of DEA. Section 3.7 describes the relationship between GP and DEA. Section 3.8 visually describes a historical flow from L1 regression, GP and DEA. Section 3.9 summarizes this chapter.

3.2 ORIGIN OF L1 REGRESSION

To describe the history concerning how GP arose and was extended to DEA, this chapter returns to the statistics of the eighteenth century and then traces forward to consider how GP influences modern statistics and DEA. Cooper (2005) has already described the relationship between GP and DEA. In addition to his description, this chapter covers other important aspects on L1 regression that

[2] The contributions of Professor Cooper in accounting, economics and public policy were discussed by Ijiri and Sueyoshi (2010).

[3] Professor Cooper increased his publications in OR and MS after the first DEA article was published in 1978. The development of DEA gave him an opportunity to increase his publication rate. At that time, Professor Cooper was already recognized as an internationally well-known researcher in OR and MS so that many researchers paid attention to DEA. Furthermore, the practicality of DEA also invited much research interests and activities among many researchers in OR, MS, economics, accounting, management and other business-related areas in different countries. The contribution of DEA can be found in many places. For example, the research of Charnes *et al.* (1978) was selected as the most influencing article in *European Journal of Operational Research* celebrating the 30th anniversary of the Euro.

provide an analytical linkage between GP and DEA not examined in his article. Of course, Professor Cooper clearly understood the additional aspects, discussed here, on L1 regression, GP and DEA. Thus, the description in this chapter is due to his suggestion and discussion when the first author of this book was his Ph.D. student at the University of Texas at Austin.

According to a historical review work on statistics prepared by Harter (1974, 1975), regression analysis was initially investigated as far back as the sixteenth century. In the eighteenth century, Roger J. Boscovich (1711–1787) first established L1 regression analysis as a research methodology. To understand his idea, let us consider a data set that contains an independent variable (x) and a dependent variable (y). Let us consider that the data set has n observations ($j=1,\ldots,n$). Then, we fit a regression model, or $y=\beta_0+\beta_1 x$, to the data set. Both β_0 and β_1 are unknown parameters to be estimated. Boscovich first proposed the minimization of a sum of absolute deviations as a regression criterion that was mathematically expressed by the following equation:

$$\text{Minimize} \sum_{j=1}^{n} \left| y_j - \left(\beta_0 + \beta_1 x_j\right) \right|. \tag{3.1}$$

Using Equation (3.1), Boscovich measured the meridian near Rome in 1757. See Stigler (1973) for a detailed description of Equation (3.1). The criterion was the original form of L1 regression, or often referred to as "least absolute value (LAV) regression" in modern statistics and econometrics.

Influenced by Boscovich's research, Laplace (Pierre Simon Marquis de Laplace: 1749–1827) considered Equation (3.1) to be the best criterion for regression analysis and used it until approximately 1795. At that time, Laplace undertook to serve as a diplomat for Bonaparte Napoleon (1769–1821). However, Laplace thereafter discontinued his examination of the topic because his algorithm for the criterion had a computational difficulty (Eisenhart, 1964).

To describe Laplace's algorithm for solving Equation (3.1), let us assume that $\beta_0=0$ and $x_j>0$ ($j=1, \ldots, n$). Then, it is possible for us to reorganize the data set in such a manner that $y_1/x_1 \geq y_2/x_2 \geq \ldots \geq y_n/x_n$. If there is an integer number (ς) that satisfies the following condition:

$$x_1+x_2+\cdots x_{\varsigma-1} < x_\varsigma + x_{\varsigma+1} + \cdots x_n \quad \text{and} \quad x_1+x_2+\cdots x_\varsigma > x_{\varsigma+1}+x_{\varsigma+2}+\cdots x_n,$$

then the estimate of β_1 is measured by $\hat{\beta}_1 = y_\varsigma/x_\varsigma$[4]. A drawback of Laplace's algorithm is that it can solve only a very small L1 regression problem.

[4]The equation indicates that L1 regression determines parameter estimates on a "median" of an observed data set. Consequently, the L1 estimate has robustness to an outlier. In contrast, L2 regression is very easily influenced by the outlier because each parameter estimate is a "mean." Many researchers such as Boscovich and Laplace in the eighteenth century understood the property of robustness of L1 regression. Therefore, they first investigated the L1 criterion based upon their experiences, so not the L2 criterion, in the initial stage of regression analysis before the eighteenth century.

Consequently, Laplace stopped studying the L1 regression after 1795 because of the computational difficulty.

According to a series of articles prepared by Harter (1974, 1975), the first researcher to overcome the computational difficulty on regression, encountered by Laplace, was Gauss (Carl Friedrich Gauss: 1775–1855). In 1795, when aged only 20, Gauss first proposed the minimization of a sum of squared deviations, which was mathematically expressed by

$$\text{Minimize} \sum_{j=1}^{n} \left[y_j - \left(\beta_0 + \beta_1 x_j \right) \right]^2 . \tag{3.2}$$

The regression criterion is nowadays referred to as "L2 regression" or "ordinary least squares (OLS)" method. The most important feature of Equation (3.2) is that it is continuous and differentiable. The feature of L2 criterion is different from L1 criterion because the latter is not differentiable and Equation (3.1) needs to be solved as a nonlinear programming problem. Using the L2 criterion, we can easily obtain parameter estimates of a regression line by minimizing Equation (3.2) even if the criterion is applied to a data set with a large sample size. At that time, Gauss believed that the regression criterion was a trivial contribution in science. So, he did not publish the least squares method until 1821. According to Eisenhart (1964), the first researcher who published the method of least squares was Legendre (Adrien Marie Legendre: 1752–1833), who introduced both the regression criterion and the well-known "least squares normal equations" in 1805.

After the discovery of the least squares method, Gauss studied the probability distribution of errors from 1797 to 1798 and found the "normal distribution." Gauss proved that least squares estimates became "maximum likelihood estimates" if errors follow the normal distribution. Maximum likelihood estimation was already studied by Daniel Bernoulli (1700–1782) at that time. Unfortunately, Gauss did not publish his findings until 1809. After Gauss published his research results in 1809, Laplace published immediately the "central limit theorem" (Stigler, 1973). Note that the L1 regression produces maximum likelihood estimates under the double exponential distribution, or Laplace distribution. See, for example, the study of Norton (1984).

As a result of the discovery of OLS, normal distribution, normal equations to solve OLS and the central limit theorem, the L2 regression (e.g. least squares method and various extensions) became a main stream of regression analysis in modern statistics and econometrics. The L1 regression had almost disappeared from the history of statistics because of its computational difficulty due to the lack of differentiability as found for the computation of OLS. Most of scientists "religiously" believed that the method of least squares under a normal distribution was the best regression criterion. Even today, none makes any question on the

TABLE 3.1 An illustrative example

Line speed (feet per minute)	Defective parts (number)
10	15
20	25
30	40
40	50
50	75
60	80

methodological validity of the least squares method. Almost all statistical textbooks at the undergraduate and graduate levels discuss only the least squares method for regression analysis, just neglecting the existence of alternative approaches such as L1 regression[5].

The following illustrative example describes differences between L1 regression and L2 regression (i.e., OLS). A supervisor of a manufacturing process believes that an assembly-line speed (in feet per minute) affects the number of defective parts found during on-line inspection. To test his belief, management has a uniform batch of parts inspected visually at a variety of line speeds. The data in Table 3.1 are collected for examination.

First, let us consider an estimated regression equation ($y = \beta_0 + \beta_1 x$) that relates the line speed (x) to the number (y) of defective parts. Using the two estimated equations measured by L1 regression and L2 regression (i.e., OLS), we forecast the number of defective parts found for a line speed of 90 feet per minute. Here, it is not necessary to describe how to apply the L2 regression because it can be easily solved using any statistical code, including Excel. Rather, this chapter documents the following formulation using Equation (3.1):

[5]There is another well-known criterion, referred to as the Chebychev regression, or "L_∞ regression." See Appa and Smith (1973), Arthanari and Dodge (1981) and Dielman and Pfaffenberger (1984) for their detailed descriptions on the L_∞ criterion on regression analysis. This chapter is not designed for mathematical programming approaches for regression analysis, and therefore we do not describe the type of regression in detail, rather we focus on a description on the relationship among L1 regression, GP and DEA. We clearly understand that much research effort has been expended on the methodology in the history of statistics. For example, Wagner (1959) showed a linear programming formulation for L1 regression. A survey study can be found in Narula and Wellington (1982). Armstrong and Kung (1980) and Barrodale and Roberts (1978) developed computational algorithms for L1 regression. The most important contribution on the L1 regression was found in Bassett and Koenker (1978, 1982) and Koenker and Bessett (1978). They expressed an asymptotic property of L1 regression so that we could conduct a statistical test on parameter estimates of L1 regression.

$$\begin{aligned}
&\text{Minimize} \sum_{j=1}^{6}\left(\delta_j^+ + \delta_j^-\right)\\
&s.t.\quad \beta_0 + 10\beta_1 + \delta_1^+ - \delta_1^- = 15,\\
&\qquad\ \ \beta_0 + 20\beta_1 + \delta_2^+ - \delta_2^- = 25,\\
&\qquad\ \ \beta_0 + 30\beta_1 + \delta_3^+ - \delta_3^- = 40,\\
&\qquad\ \ \beta_0 + 40\beta_1 + \delta_4^+ - \delta_4^- = 50,\\
&\qquad\ \ \beta_0 + 50\beta_1 + \delta_5^+ - \delta_5^- = 75,\\
&\qquad\ \ \beta_0 + 60\beta_1 + \delta_6^+ - \delta_6^- = 80,\\
&\qquad\ \ \beta_0 : \text{URS}, \beta_1 : \text{URS},\\
&\qquad\ \ \delta_j^+ \geq 0\,(j = 1,\ldots,6) \text{ and}\\
&\qquad\ \ \delta_j^- \geq 0\,(j = 1,\ldots,6),
\end{aligned}$$

where the above formulation for L1 regression produces an estimated regression line by $y = 0.000 + 1.333x$, where 0.000 is a very small number close to zero. Meanwhile, L2 regression, or OLS, produces $y = -1 + 1.386x$. The estimated number of defected parts at the line speed of 90 feet per minute becomes 119.97 under L1 regression and 123.74 under L2 regression. Thus, the two regression analyses have slightly different predictions on the independent variable.

3.3 ORIGIN OF GOAL PROGRAMMING

The computational difficulty related to L1 regression was first overcome by Charnes, Cooper and Ferguson (1955) who transformed the L1 regression into an equivalent linear programming formulation. The computer algorithm for linear programming, referred to as the "simplex method", was developed in the beginning of 1950s. Along with the development of a computer code using the simplex method at that time, their research first proposed the mathematical formulation for L1 regression.

To describe their reformulation, this chapter generalizes Equation (3.1) in the following vector expression:

$$\text{Minimize} \sum_{j=1}^{n}\left|y_j - X_j\beta\right|, \tag{3.3}$$

where the dependent variable (y_j) is to be fitted by a reference to a row vector of m independent variables $X_j = (1, x_{1j}, \ldots, x_{mj})$ for all observations ($j = 1,\ldots,n$). The regression model is expressed by $X_j\,\beta$ where $\beta = (\beta_0, \beta_1, \ldots, \beta_m)^{Tr}$ is a column vector of $m+1$ unknown parameters that are to be estimated by Equation (3.3). The superscript (Tr) is a vector transpose.

Following the research effort of Charnes and Cooper (1977), positive and negative parts of each error are introduced to solve Equation (3.3) in the following manner:

$$\delta_j^+ = 1/2\left\{\left|y_j - X_j\beta\right| + \left(y_j - X_j\beta\right)\right\} \quad \text{and}$$
$$\delta_j^- = 1/2\left\{\left|y_j - X_j\beta\right| - \left(y_j - X_j\beta\right)\right\}\left(j = 1,\ldots,n\right). \tag{3.4}$$

Here, the two equations indicate positive and negative parts of the j-th error, respectively. Based upon Equations (3.4), the two deviations of the j-th error become as follows:

$$\delta_j^+ + \delta_j^- = \left|y_j - X_j\beta\right| \quad \text{and} \quad \delta_j^+ - \delta_j^- = y_j - X_j\beta\left(j = 1,\ldots,n\right). \tag{3.5}$$

Using Equations (3.4) and (3.5), the L1 regression, expressed by Equation (3.3), is reformulated by the following original GP model:

$$\begin{aligned} &\text{Minimize} \sum_{j=1}^{n}\left(\delta_j^+ + \delta_j^-\right) \\ &s.t. \quad X_j\beta + \delta_j^+ - \delta_j^- = y_j\left(j = 1,\ldots,n\right), \\ &\qquad \beta:\text{URS}, \ \delta_j^+ \geq 0\left(j = 1,\ldots,n\right) \text{ and} \\ &\qquad \delta_j^- \geq 0\left(j = 1,\ldots,n\right). \end{aligned} \tag{3.6}$$

An important feature of Model (3.6)[6] is that it can incorporate prior information on parameter estimates into the proposed formulation as additional side constraints. Such an estimation capability cannot be found in conventional regression methods in statistics. For example, Charnes *et al.* (1955) incorporated such side constraints in order to incorporate hierarchy-based salary consensus (e.g., the salary of the president is not lower than that of any secretary). See also Sueyoshi and Sekitani (1998a, b) for L1 regression applied to time series analysis that contains serial correlation among different periods.

Model (3.6) is a special case of the GP model[7], which incorporates weights (indicating the importance among goals) in the objective function. A weighted GP model can be expressed by

$$\begin{aligned} &\text{Minimize} \sum_{j=1}^{n}\left(\omega_j^+\delta_j^+ + \omega_j^-\delta_j^-\right) \\ &s.t. \quad X_j\beta + \delta_j^+ - \delta_j^- = ob_j\left(j = 1,\ldots,n\right), \\ &\qquad \beta:\text{URS}, \ \delta_j^+ \geq 0\left(j = 1,\ldots,n\right) \text{ and} \\ &\qquad \delta_j^- \geq 0\left(j = 1,\ldots,n\right), \end{aligned} \tag{3.7}$$

[6]The formulation in Model (3.6) was initially referred to as an "inequally constrained regression" in Charnes *et al.* (1955) and was subsequently changed to GP in Charnes and Cooper (1961).

[7]Charnes *et al.* (1963) made GP a very well-known approach as a managerial methodology for accounting, finance and other business research. The first research article in 1963 was based upon the Ph.D. thesis of Professor Yuji Ijiri. Dr. Ijiri was a major contributor in the history of accounting.

where the j-th observation on the dependent variable (y_j) is replaced by the j-th goal or objective (ob_j). Each deviation has a weight (ω) to express the importance of each goal in Model (3.7). The vector β implies not only unknown parameters of a regression model but also other types of unknown decision variables.

The GP terminology was first introduced in Charnes and Cooper (1961) and soon the GP model became widely used as one of the principal methods in the general realm of multi-objective optimization. See Charnes *et al.* (1963) and Charnes and Cooper (1975). An important application of the L1 regression can be found in Charnes *et al.* (1988) that was an extension of their previous works (e.g., Charnes *et al.*, 1986). Their research in 1988 discussed the existence of a methodological bias in empirical studies. The methodological bias implies that "different methods often produce different results." It is necessary for us to examine different methods to make any suggestion for large decisional issues. The research concern is very important in particular when we make a policy suggestion from empirical evidence for guiding large policy issues such as global warming and climate change. A few previous studies have mentioned an existence of the methodological bias in energy and environmental studies.

Hereafter, let us discuss how to apply Model (3.6) for GP by using an illustrative example. The example is summarized as follows: There are many leaking tanks in gas stations which often reach underground water sources and produce serious water pollution. An environmental agency needs to replace old leaking tanks by new ones. As an initial step, the agency attempts to determine its location in the area covered by existing four gas stations to be fixed. Using horizontal and vertical coordinates, the gas stations containing leaking tanks are located at (1,7), (5,9), (6,25) and (10,20).

To solve the problem by GP, let β be $(hc, vc)^{Tr}$ where hc and vc stand for unknown decision variables that indicate the location of agent's office in horizontal and vertical coordinates, respectively. The GP model is formulated as follows:

$$\text{Minimize} \sum_{j=1}^{8} \left(\delta_j^+ + \delta_j^- \right)$$

$$\begin{aligned}
s.t. \quad & hc + \delta_1^+ - \delta_1^- = 1, \\
& vc + \delta_2^+ - \delta_2^- = 7, \\
& hc + \delta_3^+ - \delta_3^- = 5, \\
& vc + \delta_4^+ - \delta_4^- = 9, \\
& hc + \delta_5^+ - \delta_5^- = 6, \\
& vc + \delta_6^+ - \delta_6^- = 25, \\
& hc + \delta_7^+ - \delta_7^- = 10, \\
& vc + \delta_8^+ - \delta_8^- = 20, \\
& hc \geq 0, vc \geq 0, \\
& \delta_j^+ \geq 0 \left(j = 1,\ldots,8 \right) \text{ and} \\
& \delta_j^- \geq 0 \left(j = 1,\ldots,8 \right),
\end{aligned}$$

where all weights (ω_j) are set to be unity because all goals are equally important in the location problem.

The optimal solution of the above GP problem is $hc^* = 6$ and $vc^* = 20$. In other words, the best location of the agent is (6, 20) on the horizontal and vertical coordinates. This chapter needs to note two concerns on the optimal solution. One of the two concerns is that both hc and vc are non-negative, as formulated above, because the GP formulation is for a locational problem, not the regression analysis as discussed previously. The other concern is that the optimal solution may suffer from an occurrence of multiple solutions. This type of difficulty often occurs because the problem is formulated and solved by linear programming.

3.4 ANALYTICAL PROPERTIES OF L1 REGRESSION

Returning to Model (3.6) for L1 regression, this chapter discusses the dual formulation and the implication of dual variables. First, Model (3.6) is reorganized by the following GP model for L1 regression that is more specific than the original formulation:

$$\begin{aligned}
&\text{Minimize } \sum_{j=1}^{n}\left(\delta_j^+ + \delta_j^-\right) \\
&s.t. \quad \beta_0 + \sum_{i=1}^{m}\beta_i x_{ij} + \delta_j^+ - \delta_j^- = y_j\left(j=1,\ldots,n\right), \\
&\qquad \beta_i : \text{URS}\left(i=0,1,\ldots,m\right), \delta_j^+ \geq 0\left(j=1,\ldots,n\right) \text{ and} \\
&\qquad \delta_j^- \geq 0\left(j=1,\ldots,n\right).
\end{aligned} \tag{3.8}$$

Here, it is important to reconfirm that β_0 is an unknown parameter to express the constant of a regression model and β_i $(i=1,\ldots,m)$ are unknown parameters to be estimated. They indicate the slope of the regression model $(\beta_0 + \sum_{i=1}^{m}\beta_i x_i)$. The subscript ($j$) stands for the j-th observation.

The dual formulation of Model (3.8) becomes

$$\begin{aligned}
&\text{Maximize} \sum_{j=1}^{n} y_j \mu_j \\
&s.t. \quad \sum_{j=1}^{n}\mu_j = 0, \\
&\qquad \sum_{j=1}^{n} x_{ij}\mu_j = 0 \ \left(i=1,\ldots,m\right), \\
&\qquad 1 \geq \mu_j \geq -1 \ \left(j=1,\ldots,n\right) \text{ and} \\
&\qquad \mu_j : \text{URS} \quad \left(j=1,\ldots,n\right),
\end{aligned} \tag{3.9}$$

where μ_j stands for the dual variable for the j-th observation. The complementary slackness conditions, that is, $\delta_j^+(1-\mu_j)=0$ and $\delta_j^-(\mu_j+1)=0$ for all j, of linear programming indicates the following relationship between Models (3.8) and (3.9):

if $\mu_j^*=1$, then $\delta_j^{+*}>0$ (the j-th observation locates above the regression hyperplane),

if $1>\mu_j^*>-1$, then $\delta_j^{+*}=\delta_j^{-*}=0$ (the j-th observation locates on the regression hyperplane) and

if $\mu_j^*=-1$, then $\delta_j^{-*}>0$ (the j-th observation locates below the regression hyperplane).

Thus, the optimal dual variable (μ_j^*) related to the j-th observation describes the locational relationship between an estimated regression hyperplane and the observation.

Following the research of Sueyoshi and Chang (1989), this chapter may obtain an interesting transformation that changes the j-th dual variable by $\mu_j = np_j - 1$. Then, Model (3.9) becomes

$$\begin{aligned}
&\text{Maximize} \sum_{j=1}^{n} y_j(np_j-1)\\
&s.t. \quad \sum_{j=1}^{n}\left(np_j-1\right)=0,\\
&\qquad \sum_{j=1}^{n} x_{ij}\left(np_j-1\right)=0 \ \left(i=1,\dots,m\right),\\
&\qquad 1\geq\left(np_j-1\right)\geq-1 \quad \left(j=1,\dots,n\right) \text{ and}\\
&\qquad p_j\geq 0 \qquad \left(j=1,\dots,n\right).
\end{aligned} \tag{3.10}$$

Model (3.10) is equivalent to the following formulation:

$$\begin{aligned}
&\text{Maximize} \sum_{j=1}^{n} y_j p_j\\
&s.t. \quad \sum_{j=1}^{n} p_j = 1,\\
&\qquad \sum_{j=1}^{n} x_{ij}p_j = \frac{\sum_{j=1}^{n} x_{ij}}{n} \ \left(i=1,\dots,m\right) \text{ and}\\
&\qquad 2/n \geq p_j \geq 0 \quad \left(j=1,\dots,n\right).
\end{aligned} \tag{3.11}$$

An important feature of Model (3.11) is that the dual variable (p_j) for the j-th observation may be considered as a "probability" that the observation locates above an estimated regression hyperplane. In other words,

(a) if $p_j = 2/n$, then the observation locates above the regression hyperplane,
(b) if $2/n > p_j > 0$, then it locates on the hyperplane and
(c) if $p_j = 0$, then it locates below the regression hyperplane.

3.5 FROM L1 REGRESSION TO L2 REGRESSION AND FRONTIER ANALYSIS

3.5.1 L2 Regression

This chapter can reorganize Model (3.2), as an extension of Model (3.8), by slightly changing the objective function as follows:

$$\begin{aligned}
&\text{Minimize } \sum_{j=1}^{n}\left(\delta_j^{+2}+\delta_j^{-2}\right) \\
&s.t. \quad \beta_0+\sum_{i=1}^{m}\beta_i x_{ij}+\delta_j^{+}-\delta_j^{-}=y_j\ (j=1,\ldots,n), \qquad (3.12)\\
&\beta_i : \text{URS}\ (i=0,1,\ldots,m), \delta_j^{+}\geq 0\ (j=1,\ldots,n) \text{ and}\\
&\delta_j^{-}\geq 0\ (j=1,\ldots,n).
\end{aligned}$$

The difference between Models (3.8) and (3.12) is that the objective function of the latter model minimizes the sum of squared errors, measured by slacks, so that it corresponds to the OLS. One problem is that Model (3.12) belongs to nonlinear programming so that it may have limited computational practicality. The contribution of Gauss in the eighteenth century was that he solved Model (3.12) by a matrix operation so that his approach could deal with a large data set for regression analysis, as mentioned previously.

3.5.2 L1-Based Frontier Analyses

Another important extension of Model (3.8) is that it is possible for us to use it for frontier analysis, which is closely related to DEA, along with minor modifications. There are two alternatives for frontier analysis. One of the two alternatives is for production analysis that is formulated as follows:

$$\begin{aligned}
&\text{Minimize } \sum_{j=1}^{n}\delta_j^{+} \\
&s.t. \quad \beta_0+\sum_{i=1}^{m}\beta_i x_{ij}+\delta_j^{+}=y_j\ (j=1,\ldots,n), \qquad (3.13)\\
&\beta_i : \text{URS}\ (i=0,1,\ldots,m) \text{ and}\\
&\delta_j^{+}\geq 0\ (j=1,\ldots,n).
\end{aligned}$$

Model (3.13) produces an estimated (lower) regression hyperplane that locates on or below all the observations. One-sided slacks (δ_j^+) are considered in the objective function of Model (3.13). This type of frontier analysis is useful for estimating a cost function.

The other alternative is formulated as follows:

$$
\begin{aligned}
\text{Minimize} \ & \sum_{j=1}^{n} \delta_j^- \\
s.t. \quad & \beta_0 + \sum_{i=1}^{m} \beta_i x_{ij} - \delta_j^- = y_j \left(j = 1,\ldots,n \right), \\
& \beta_i : \text{URS} \left(i = 0,1,\ldots,m \right) \text{ and} \\
& \delta_j^- \geq 0 \left(j = 1,\ldots,n \right).
\end{aligned} \tag{3.14}
$$

Model (3.14) indicates an opposite case to that of Model (3.13). One-sided slacks (δ_j^-) are considered in the objective function of Model (3.14). An estimated (upper) regression hyperplane locates on or above all the observations. This type of frontier analysis is useful for estimating a production function.

Methodological strengths and drawbacks of Models (3.13) and (3.14), compared with DEA, are summarized as the following four differences: First, Models (3.13) and (3.14) can be used for efficiency assessment that compares observations by two estimated frontiers, respectively. Functionally, the proposed frontier analyses work like DEA in their efficiency assessments. Second, these models can directly handle zero and/or negative values in a data set. A problem is that the frontier analyses measure the performance of observations by only a single desirable output. In contrast, DEA can handle multiple desirable outputs, but it cannot directly handle zero and/or negative values. See Chapters 26 and 27 for a description on how to handle zero and/or negative values by DEA and its environmental assessment. Third, another difference between DEA and the frontier analyses is that DEA can avoid specifying a functional form. In contrast, the frontier analysis requires the functional specification. Finally, both L1-based frontier analyses and DEA can evaluate the performance of observations (i.e., DMUs). The L1-based frontier analyses look for parameter estimates of two frontier hyperplanes, while DEA measures weights on production factors (i.e., inputs and desirable outputs). That is a structural difference between L1-based frontier analyses and DEA.

Table 3.2 lists an illustrative example for L1-based frontier analyses. The data set contains ten observations (DMUs), each of which has an independent variable (x) and a dependent variable (y). Figure 3.1 compares three regression lines which are estimated by L1 regression and two L1-based frontier analyses, respectively. The L1 regression is measured by Model (3.8). The two frontier analyses are measured by Models (3.13) and (3.14), respectively. Model (3.8) estimates L1

TABLE 3.2 An illustrative example for L1 frontier analyses

Observation	x	y
1	10	15
2	3	11
3	5	10
4	2	6
5	4	9
6	4	4
7	12	15
8	7	13
9	6	8
10	8	12

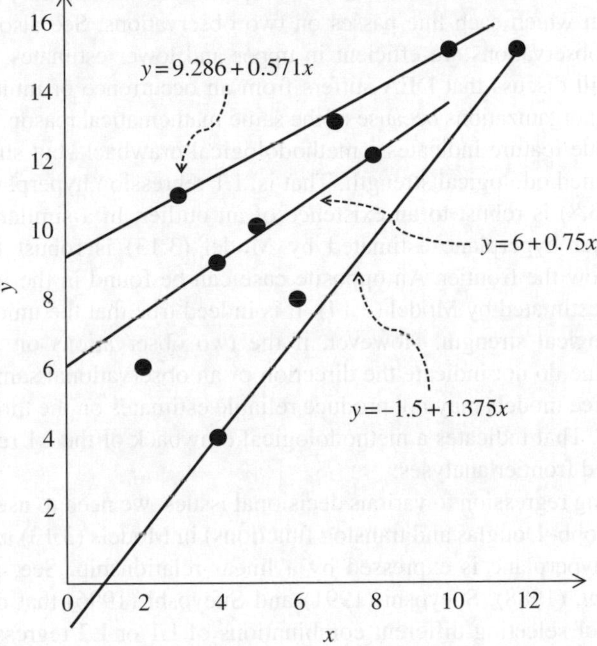

FIGURE 3.1 Three regression lines: L1 regression and two frontier analyses (a) Each dot (•) indicates an observation. (b) The L1 regression, locating in the middle of observations, is measured by Model (3.8). The L1-based lower frontier line is measured by Model (3.13) and the L1-based upper frontier line is measured by Model (3.14).

regression line as $y=6+0.75x$. Models (3.13) and (3.14) estimate the two L1-based frontier lines as $y=-1.5+1.375x$ (as a lower bound) and $y=9.286+0.571x$ (as an upper bound), respectively.

Table 3.3 summarizes upper and lower estimates determined by the two L1-based frontier analyses. The upper estimates are listed in the fourth row of Table 3.3 and the corresponding efficiency scores are measured by $y_j/\hat{y}_j$, where y_j is the observed dependent variable of the j-th DMU and $\hat{y}_j$ is its upper estimate of the j-th observation. The lower estimates and efficiency scores are listed in the bottom of Table 3.3. The efficiency scores are measured by $\hat{y}_j/y_j$ where $\hat{y}_j$ is the lower estimate of the j-th observation.

It is necessary to add two concerns on the L1 regression and two L1-based frontier analyses. One of the two concerns is that, as found in Figure 3.1 and Table 3.3, these methodologies need multiple constraints to identify an optimal solution by linear programming formulations. See Models (3.12) to (3.14) in which each observation corresponds to each constraint in these formulations. The optimality is found on a "vertex" that consists of multiple (e.g., at least two) constraints. Each vertex is a corner point in a feasible region of linear programming. Thus, the vertex on optimality needs multiple constraints, so multiple observations. See, for example, Figure 3.1 in which each line passes on two observations. See also Table 3.3 in which two observations are efficient in upper and lower estimates. Later in this book, we will discuss that DEA suffers from an occurrence of multiple efficient (often many) organizations because of the same mathematical reason.

The unique feature indicates a methodological drawback, but simultaneously implying a methodological strength. That is, L1 regression hyperplane estimated by Model (3.8) is robust to an existence of an outlier. In a similar manner, the upper frontier hyperplane estimated by Model (3.13) is robust to an outlier locating below the frontier. An opposite case can be found in the lower frontier hyperplane estimated by Model (3.14). It is indeed true that the unique feature is a methodological strength. However, if the two observations on an estimated regression line do not indicate the direction of an observational sample, then the proposed three models may not produce reliable estimates on the three regression hyperplanes. That indicates a methodological drawback of the L1 regression and two L1-based frontier analyses.

In applying regression to various decisional issues, we need to use a functional form (e.g, Cobb–Douglas and translog functions) in Models (3.13) and (3.14) if a repression hyperplane is expressed by a linear relationship. See, for example, Charnes *et al.* (1988), Sueyoshi (1991) and Sueyoshi (1996) that discussed the importance of selecting different combinations of L1 or L2 regression method,

TABLE 3.3 Efficiency measures by L1-based Frontier analyses

DMU	1	2	3	4	5	6	7	8	9	10
x	10	3	5	2	4	4	12	7	6	8
y	15	11	10	6	9	4	15	13	8	12
Upper Estimate	15.00	11.00	12.14	10.43	11.57	11.57	16.14	13.28	12.71	13.85
Efficiency	1.00	1.00	0.82	0.58	0.78	0.35	0.93	0.98	0.63	0.87
Lower Estimate	12.25	2.63	5.38	1.25	4.00	4.00	15.00	8.13	6.75	9.50
Efficiency	0.82	0.24	0.54	0.21	0.44	1.00	1.00	0.63	0.84	0.79

regression analysis or frontier analysis, along with a selection of various functional forms, for guiding large public policy issues (e.g, the divestiture of a telecommunication industry in industrial nations such as the United States and Japan). Their concern is still useful and important in most current empirical studies. That it, it is necessary for researchers to examine different methodology combinations (i.e., regression vs frontier, L1 vs L2 and various functional forms) in order to avoid a methodological bias on empirical studies in energy and environment. It is indeed true that the L2 regression is very important and widely used by many studies, but it is only one of many regression analyses. It is also necessary for us to understand that there is no perfect methodology. Hence, we constantly make an effort to improve the quality of our methodology, looking for methodological betterment, until we can guide various policy and strategic issues by providing reliable empirical evidence.

3.6 ORIGIN OF DEA

The GP served as a methodological basis for the development of DEA with a linkage via "fractional programming" (Charnes and Cooper, 1962, 1973). The reformulation from a fractional model[8] to a linear programming equivalence was first proposed by Charnes *et al.* (1978, 1981) and the reformulation was widely used in the OR and MS literature. The reformulation was also used to develop the first DEA model, often referred to as a "ratio form," because of the relationship to fractional programming as discussed in Chapter 2. See also Chapter 1 for a description on DEA. See also Figure 3.2.

The ratio form (for input-oriented measurement) has the following formulation to determine the level of efficiency[9] on the k-th organization:

$$
\begin{aligned}
&\text{Minimize } \theta - \varepsilon_n \left(\sum_{i=1}^{m} d_i^x + \sum_{r=1}^{s} d_r^g \right) \\
&\text{s.t.} \quad -\sum_{j=1}^{n} x_{ij}\lambda_j - d_i^x \qquad + \theta x_{ik} = 0 \quad (i = 1,\ldots,m), \\
&\qquad\quad \sum_{j=1}^{n} g_{rj}\lambda_j \qquad - d_r^g \qquad = g_{rk} \quad (r = 1,\ldots,s), \\
&\qquad\quad \theta : \text{URS}, \ \lambda_j \geq 0 \ (j = 1,\ldots,n), \\
&\qquad\quad d_i^x \geq 0 \ (i = 1,\ldots,m) \text{ and } d_r^g \geq 0 \ (r = 1,\ldots,s).
\end{aligned}
\tag{3.15}
$$

[8] See Chapter 2. See also Schaible (1996) that has provided the large literature on fractional programming evolved after Charnes and Cooper (1962).

[9] The term "technical efficiency" has been widely used in the DEA community. This chapter uses "operational efficiency" instead of technical efficiency because this book discusses technology innovation on environment. To clearly distinguish between the conventional term on efficiency and technology innovation, we use the term "operational efficiency," hereafter, in the whole book context.

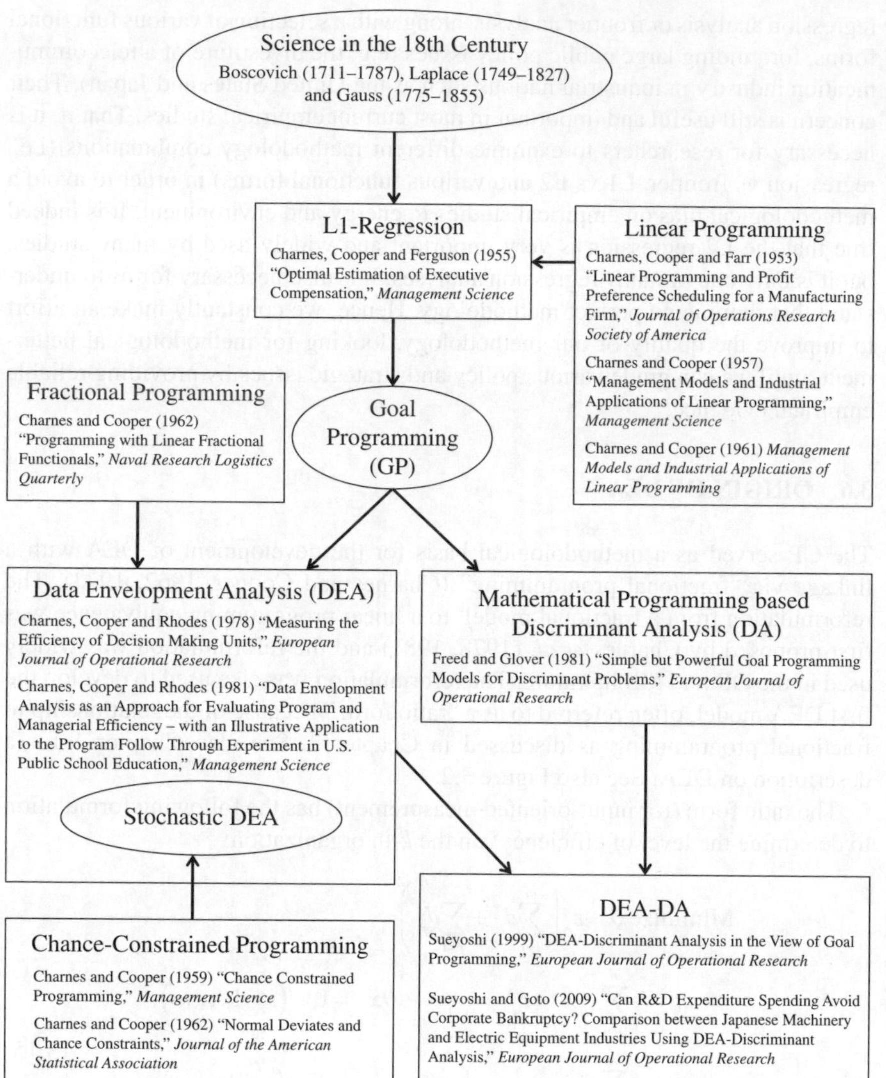

FIGURE 3.2 Progress from L1 regression to DEA (a) Source: Glover and Sueyoshi (2009).

Model (3.15) evaluates the performance of the *k*-th organization by relatively comparing it with those of *n* organizations ($j=1,\ldots,n$) in relation to each other. Hereafter, each organization, referred to as a decision making unit (DMU) in DEA, uses *m* inputs ($i=1,\ldots,m$) to produce s desirable outputs ($r=1,\ldots,s$). The x_{ij} is an observed value related to the *i*-th input of the *j*-th DMU and the g_{rj} is an observed value related to its *r*-th desirable output. Slacks (i.e., deviations) related to inputs and desirable outputs are d_i^x and d_r^g, respectively. The symbol (ε_n) stands for a non-Archimedean small number. The scalar (λ_j), often referred to as a "structural" or "intensity" variable, is used to make a linkage among DMUs in a data domain of inputs and desirable outputs. An efficiency score is measured by an unknown variable (θ) that is unrestricted in Model (3.15). The status of operational efficiency is confirmed by Model (3.15) when both $\theta^*=1$ and all slacks are zero on optimality, as discussed in Chapter 2. An important feature of Model (3.15), to be noted here, is that it is input-oriented. It is possible for us to change Model (3.15) into a desirable output-oriented formulation.

At the end of this section, we need to mention that Chapter 2 does not incorporate the non-Archimedean small number (ε_n) because of our descriptive convenience. However, this chapter incorporates the very small number in the DEA formulation. It is true that if we are interested in measuring only the level of efficiency, then the incorporation may not produce a major influence on the magnitude of efficiency. Furthermore, the non-Archimedean small number is just our mathematical convenience in the context of production economics. In reality, none knows what it is.

Clearly acknowledging such concerns on ε_n, this chapter needs to mention the two other analytical concerns. One of the two concerns is that if we drop ε_n in the objective function, then dual variables (multipliers) often become zero in these magnitudes. This is problematic because the result clearly implies that corresponding inputs or desirable outputs are not utilized in DEA assessment. The other concern is that if we are interested in not only an efficiency level but also other types of measurement (e.g., marginal rate of transformation and rate of substitution), then dual variables (multipliers) should be strictly positive in these signs. Otherwise, the computational results obtained from DEA are often useless, so not conveying any practical implications to users. Thus, the incorporation of such a very small number is important for DEA as first proposed in the original ratio model (Charnes *et al.*, 1978, 1981) at this stage. See Chapters 23, 26 and 27 for detailed discussions on various DEA problems due to an occurrence of zero in dual variables.

3.7 RELATIONSHIPS BETWEEN GP AND DEA

DEA researchers have long paid attention to the additive model, first proposed by Charnes *et al.* (1985), because it has the analytical structure based upon Pareto optimality. See Chapter 5 for a detailed description on the additive model. Although Chapter 2 does not clearly mention it, Model (2.7) belongs to the additive model that is designed to identify the total amount of slack(s) on optimality. See also the research of Cooper *et al.* (2006) that discussed the relationship between the ratio form and the additive model. An important feature of the additive model is that it aggregates input-oriented and desirable output-oriented measures to produce a single measure for operational efficiency (*OE*). Acknowledging that Cooper (2005) has already discussed the relationship between GP and the additive model, this chapter does not follow his description, rather it discusses another path to describe the differences between them.

First of all, it is important for us to confirm that the efficiency of the *k*-th DMU is measured by the following additive model:

$$
\begin{aligned}
&\text{Maximize} \sum_{i=1}^{m} d_i^x + \sum_{r=1}^{s} d_r^g \\
&\text{s.t.} \quad \sum_{j=1}^{n} x_{ij}\lambda_j + d_i^x = x_{ik} \ (i=1,\dots,m), \\
&\qquad \sum_{j=1}^{n} g_{rj}\lambda_j - d_r^g = g_{rk} \ (r=1,\dots,s), \\
&\qquad \sum_{j=1}^{n} \lambda_j = 1, \\
&\qquad \lambda_j \ge 0 \ (j=1,\dots,n), d_i^x \ge 0 \ (i=1,\dots,m) \text{ and} \\
&\qquad d_r^g \ge 0 \ (r=1,\dots,s).
\end{aligned}
\tag{3.16}
$$

Since Model (3.16) does not have any efficiency score (θ), as found in Model (3.15), it measures a level of *OE* by the total amount of slacks. The status of efficiency is measured as follows:

(a) Full efficiency ↔ all slacks are zero and

(b) Inefficiency ↔ at least one slack is non-zero.

See Chapters 5 and 6 that provide detailed descriptions on the additive model.

To discuss an analytical linkage between the additive model and the GP, this chapter returns to Models (3.13) and (3.14) as special cases of GP. This chapter considers the following four steps to identify such an analytical linkage between them:

Step 1 (objective change from minimization to maximization): The frontier analysis models, or Models (3.13) and (3.14), are reorganized as follows:

$$
\begin{aligned}
&\text{Maximize} \sum_{j=1}^{n} \delta_j^- \\
&\text{s.t.} \quad \beta_0 + \sum_{i=1}^{m} \beta_i x_{ij} - \delta_j^- = y_j \ (j=1,\dots,n), \\
&\qquad \beta_i : \text{URS} \ (i=0,1,\dots,m) \text{ and} \\
&\qquad \delta_j^- \ge 0 \ (j=1,\dots,n),
\end{aligned}
\tag{3.17}
$$

and

$$\text{Maximize } \sum_{j=1}^{n} \delta_j^+$$

$$\begin{aligned} s.t. \quad & \beta_0 + \sum_{i=1}^{m} \beta_i x_{ij} + \delta_j^+ = y_j \left(j=1,\ldots,n \right), \\ & \beta_i : \text{URS} \left(i=0,1,\ldots,m \right) \text{ and} \\ & \delta_j^+ \geq 0 \left(j=1,\ldots,n \right), \end{aligned} \tag{3.18}$$

where we change the objective function from minimization to maximization in Models (3.13) and (3.14). Accordingly, the sign of slacks (δ_j^-) is changed from negative to positive in Model (3.17). In contrast, the sign of slacks (δ_j^+) is changed from positive to negative in Model (3.18). Thus, the combination between Models (3.13) and (3.14) is equivalent to that of Models (3.17) and (3.18). However, the relationship between lower and upper bounds in the former models becomes opposite in the latter models.

Step 2 (from parameters to weights): The purpose of the two frontier analyses models is to determine the upper and lower frontier hyperplanes, both of which connect these parameter estimates, because they belong to regression analysis. In contrast, the purpose of the additive model is to identify an efficiency frontier that consists of inputs and desirable outputs regarding n DMUs (so, being observations) and to determine the level of their efficiency scores by the total sum of slacks. Therefore, we set $\beta_0 = 0$ (so, no constant) in Models (3.17) and (3.18) and then replace the unknown parameters (β_i for $i=1,...,m$) in Models (3.17) and (3.18) by other unknown intensity variables (λ_j for $j=1,\ldots,n$). In this case, it is necessary to exchange $j=1,...,n$ and $r=1,\ldots,s$ in Model (3.17) as well as $j=1,\ldots,n$ and $i=1,...,m$ in Model (3.18).

Step 3 (from dependent variables to inputs and desirable outputs on the k-th DMU): We incorporate two different groups into Models (3.17) and (3.18) to obtain GP formulations. One of the two groups (for desirable outputs) has s goals (g_{rk}: $r=1,\ldots,s$) of the k-th DMU to replace y_j ($j=1,\ldots,n$) in Model (3.17). The other group (for inputs) has m goals (x_{ik}: $i=1,\ldots,m$) of the k-th DMU to replace y_j ($j=1,\ldots,n$) in Model (3.18). Furthermore, an input-oriented frontier is less than or equal x_{ik}. In contrast, a desirable output-oriented frontier is more than or equal g_{rk}. By paying attention to these analytical features, this chapter reorganizes Models (3.17) and (3.18) as follows:

$$\text{Maximize } \sum_{r=1}^{s} d_r^g$$

$$\begin{aligned} s.t. \quad & \sum_{j=1}^{n} g_{rj} \lambda_j - d_r^g = g_{rk} \left(r=1,\ldots,s \right), \\ & \lambda_j \geq 0 \left(j=1,\ldots,n \right) \text{ and } d_r^g \geq 0 \left(r=1,\ldots,s \right). \end{aligned} \tag{3.19}$$

and

$$\begin{aligned}
&\text{Maximize } \sum_{i=1}^{m} d_i^x \\
&s.t. \quad \sum_{j=1}^{n} x_{ij}\lambda_j + d_i^x = x_{ik} \; (i=1,\ldots,m), \\
&\qquad \lambda_j \ge 0 \; (j=1,\ldots,n) \text{ and } d_i^x \ge 0 \; (i=1,\ldots,m).
\end{aligned} \tag{3.20}$$

Here, the observation (j) in Model (3.17) is replaced by the goal (r) so that Equation (3.19) has $\delta_j^- = d_r^g$ on the left hand side. In a similar manner, the observation (j) in Model (3.18) is replaced by the goal (m) so that Model (3.20) has $\delta_j^+ = d_i^x$ on the left hand side.

Step 4 (normalization): We combine Models (3.19) and (3.20) and then incorporates the condition for "normalization" on unknown intensity variables (i.e., $\sum_{j=1}^{n} \lambda_j = 1$) in the combined formulation. The normalization makes it possible that parameters estimates of regression are changed to weights among production factors as formulated in DEA. As a result of such a combination, we can obtain the formulation of Model (3.16).

3.8 HISTORICAL PROGRESS FROM L1 REGRESSION TO DEA

Figure 3.2 visually describes the historical development flow from L1 regression to DEA. The figure reorganizes the part of Professor Cooper's view on DEA development, as depicted in the upper part of Figure 1.1. This chapter specifies it more clearly in such a manner that it can fit within the context of this book.

3.9 SUMMARY

The first publication of DEA was generally regarded as the article prepared by Charnes *et al.* (1978) although Professor Cooper and his associates presented DEA at the TIMS Hawaii conference in 1977. Viewing DEA as an extension of GP, along with fractional programming and in consideration of the historical linkage between L1 regression and GP, this book considers that DEA has an analytical linkage with L1 regression. In this concern, the history of DEA was connected in a roundabout fashion with developments of science in the eighteenth century, as manifested in the work of Laplace and Gauss, because they attempted to develop algorithms for the L1 regression. That was the view on the DEA development of Professor Cooper.

Following Professor Cooper's view on GP and DEA, which has been never documented in his previous publications, this chapter has reviewed these historical linkages. For example, DEA has many methodological benefits as a non-parametric approach[10] that does not assume any functional form on specification between inputs and desirable outputs. Furthermore, DEA can be solved by linear programming. The computational tractability attracts many users in assessing performance of organizations in private and public sectors. However, it may be true that DEA has often misguided many previous applications. To avoid the misuse found in many previous research efforts, it is necessary for us to pay attention to a careful examination on unique features of DEA performance analysis. As an extension of a conventional use of DEA, the environmental assessment, which is the main research concern of this book, we will discuss methodological benefits and drawbacks of DEA from the perspective of environmental assessment in Section II.

[10]According to a well-known statistician such as Huber (1981, p.6) "A procedure is called non-parametric if it is supposed to be used for a broad, non-parametric set of underlying distributions. For instance, the sample mean and the sample median are the non-parametric estimates of the population mean and the sample mean, respectively. A test is called distribution-free if the probability of falsely rejecting the null hypothesis is the same for all possible underlying continuous distributions." This chapter considers that DEA is non-parametric because it measures weights (λ^*) of a piece-wise linear production function, not parameter estimates of the production function, in the case of variable returns to scale (RTS). Thus, the definition of "non-parametric" in this chapter is different from the definition of conventional statistics.

4

RADIAL MEASUREMENT

4.1 INTRODUCTION

This chapter[1] discusses radial models for efficiency measurement. Efficiency is referred to as "technical efficiency" in conventional DEA and is "operational efficiency" in this chapter and the whole book. The concept of technical efficiency was first axiomatically proposed by Debreu (1951, 1959). Following the work of Debreu (1951)[2], Farrell (1957)[3] used a well-known "activity analysis approach" with

[1] This chapter is partly based upon the article: Sueyoshi, T. (1997) Measuring scale efficiencies and returns to scale of Nippon Telegraph and Telephone in production and cost analyses. *Management Science*, **43**, 779–796.

[2] Gérard Debreu (4 July 1921 to 31 December 2004) was a French economist and mathematician, who came for United States citizenship. Best known as a professor of economics at the University of California, Berkeley, where he began working in 1962, Gerard Debreu's contributions were in general equilibrium theory – a highly abstract theory about whether and how each market reached equilibrium. In a famous paper, coauthored with Kenneth Arrow and published in 1954, Debreu proved that, under fairly unrestrictive assumptions, prices may exist for bringing markets into equilibrium. In his 1959 book, *"The Theory of Value,"* Debreu introduced more general equilibrium theory, using complex analytic tools from mathematics – set theory and topology – to prove his theorems. In 1983, Debreu received the Nobel Prize for incorporating new analytical methods into economic theory and for his rigorous reformulation of the theory of general equilibrium.

[3] Michael James Farrell (1926–1975) was an applied economist at the University of Cambridge, United Kingdom. Since his contribution has been long supported by many production economists, this chapter

Environmental Assessment on Energy and Sustainability by Data Envelopment Analysis, First Edition. Toshiyuki Sueyoshi and Mika Goto.

matrix inversions. Based upon the advice of Alan Hoffman (an operational researcher), who was one of the commentators on Farrell (1957), Farrell and Fieldhouse (1962) proposed a linear programming formulation for measuring a level of so-called "technical efficiency," or operational efficiency (*OE*) in this book. Unfortunately, almost no attention had been given to their works on efficiency measurement for a long time. The efficiency measurement was extensively investigated after Charnes *et al.* (1978) proposed DEA, as discussed in the preceding chapters. It is easily imagined that methodologies for environmental assessment discussed

needs to review his contributions from the perspective of DEA. Our review discussion is based upon three articles (Farrell, 1954, 1957; Farrell and Fieldhouse, 1962). The first article (Farrell, 1954: *An application of "activity analysis" to the theory on the firm*) was prepared when he visited Yale University where he could meet T.C. Koopmans and J. Tobin. On page 292 of the article, he discussed activity analysis, proposed by Koopmans, which could explore the corporate behavior of a firm by an application of "linear programming." In his article, the production relationship between production factors could be expressed by a static model in multiple periods. As a result, linear programming was applicable to the assessment of corporate behavior. The second article (Farrell, 1957: *The measurement of productive efficiency*) was innovative and closely related to the classical DEA development by providing the methodology with a conceptual basis. The article discussed an efficient production function, inspired by the activity analysis of linear programming (Farrell, 1957, p. 11) and started discussing an efficiency measure, referred to as "technical efficiency," which was first discussed in Debreu's "coefficient of resource utilization" (Debreu, 1951). In addition to the concept of technical efficiency, according to his article (Farrell, 1957, p. 255), "an efficient production function was expressed by a theoretical function specified by engineers." However, such an engineering-based empirical function was complicated and practically impossible in measuring the theoretical efficiency function from the perspective of production economics. This chapter pays attention to the fact that Farrell (1957) has used the term "technical efficiency" because of his awareness on the engineering perspective, following Debreu (1951). Here, we may have simple questions such as "what is *engineering*" and "what is *type of technology*" in his economics context. It is very clear to us that the production technology in the middle of the twentieth century is by far different from the current century. Fully acknowledging his contribution in production economics, this chapter does not use the term "technical efficiency" to avoid our confusion with "technology innovation" on industrial pollution that is the gist of this book. The second article (Farrell, 1957, p. 255, 260) also discussed "price efficiency" and "overall efficiency" under increasing and diminishing RTS. These economic concepts have long provided production economists with a conceptual basis on DEA. No wonder why many studies have discussed his contribution as a staring study of DEA even if he did not mention anything on DEA. Finally, the third article (Farrell and Fieldhouse, 1962: *Estimating efficient production function under increasing returns to scale*) extended Farrell's (1957) study by discussing a linear programming structure that was solved by the simplex method of linear programming (Farrell, 1957, pp. 265–266). Their study documented two interesting concerns from our perspective. One of the two interesting concerns was that they knew an occurrence of degeneracy, or multiple solutions, if a data set contained zero in linear dependency among production factors. The other concern was that they discussed the importance of a "dual formulation," not discussed by production economists even nowadays. As discussed by Glover and Sueyoshi (2009), it is easily imagined that their appendix on the method of computation (see Farrell, 1957, pp. 264–267) was guided by Alan Hoffman, as a reviewer of their manuscript, who was an operations researcher. Consequently, their description on computation is still useful in modern DEA algorithmic development. See Chapter 10. It may be true that many DEA researchers have been long discussing the concept of technical efficiency, due to Farrell's engineering concern, but not paying serious attention to its dual formulation, as discussed by their works. As documented in their study, the collaboration between production economics and OR and MS is essential in extending new research dimensions on DEA and its applications on environmental assessment.

in Section II of this book can be considered as extensions from a conventional use of DEA. See Chapter 1. The DEA environmental assessment, discussed in chapters of Section II, is concerned with energy and sustainability as a research focus, along with conceptual and structural developments from DEA.

Generally speaking, the previous DEA research proposed two groups of models. One of the two groups was referred to as "radial measures," while the other was "non-radial measures." This chapter discusses radial measures and Chapter 5 discusses non-radial measures. The two radial models were first proposed by Charnes *et al.* (1978) and Banker *et al.* (1984), respectively, depending upon the type of returns to scale (RTS): constant RTS or variable RTS. See Chapter 8 for an analytical framework on RTS. The radial measure proposed by Charnes *et al.* (1978) was formulated under constant RTS and the other model proposed by Banker *et al.* (1984) was formulated under variable RTS. Both radial models belong to the "Debreu–Farrell measures." This chapter clearly understands that DEA has other variations on radial models, but they do not have practicality at the level of the two types of radial models.

It is important to note that the two radial models have a unique feature in measuring a degree of *OE*. That is, they measure an efficiency score in the objective function of their linear programming formulations. In contrast, non-radial measures (e.g., an additive model: Charnes *et al.*, 1985) determine a level of *OE* by examining only a total amount of slacks on optimality, because each non-radial measure does not have any efficiency score in the objective function. The latter type of measures belongs to the "Pareto[4]–Koopmans[5] measure" (Koopmans, 1951; Russell, 1985, 1990). See Chapter 5 on Pareto optimality.

[4] Vilfredo Federico Damaso Pareto (15 July 1848 to 19 August 1923) was an Italian engineer, sociologist, economist, political scientist and philosopher. Pareto is best known for two concepts that are named after him. The first and most familiar is the concept of "Pareto optimality." See Chapter 10. A Pareto-optimal allocation of resources is achieved when it is not possible to make anyone better off without making someone else worse off. The second is Pareto's law of income distribution. This law, which Pareto derived from British data on income, showed a linear relationship between each income level and the number of people who received more than that income. Pareto found similar results for Prussia, Saxony, Paris and some Italian cities. Although Pareto thought that his law should be "provisionally accepted as universal," he realized that exceptions were possible.

[5] Tjalling Charles Koopmans (August 28, 1910 to February 26, 1985) was a Dutch American mathematician and economist. He began his university education at the Utrecht University at the age of 17, specializing in mathematics. Three years later, in 1930, he switched to theoretical physics. In 1933, he met Jan Tinbergen, the 1969 Bank of Sweden prize winner, and moved to Amsterdam to study mathematical economics under him. Koopmans shared the 1975 Nobel Prize with Leonid Kantorovich "for their contributions to the theory of optimum allocation of resources." In 1938, he succeeded Jan Tinbergen at the League of Nations in Geneva and then left in 1940 when Hitler invaded the Netherlands. In the United States, Koopmans became a statistician with the Combined Shipping Adjustment Board in Washington where he tried to solve the practical problem of how to reorganize shipping to minimize transportation costs. The problem was complex because the variables included thousands of merchant ships, millions of tons of cargo and hundreds of ports. He solved it. The technique he developed to do so was called "activity analysis" and is now called linear programming. His first write-up of the analysis was in a 1942 memorandum. His techniques were very similar to those used by Kantorovich, whose work was discovered only much later.

The remainder of this chapter[6] is organized as follows: Section 4.2 describes input-oriented radial models. Section 4.3 discusses desirable output-oriented radial models. Section 4.4 compares these radial models. Section 4.5 describes multiplier restriction and cross-reference approaches. Section 4.6 describes cost models that are conceptually linked to the input-oriented radial models. The section also summarizes various efficiency measures, along with three illustrative examples, within the proposed radial measurement. Section 4.7 summarizes this chapter.

4.2 RADIAL MODELS: INPUT-ORIENTED

Radial models (RMs) are very popular and widely accepted by many DEA users because these models have an efficiency score in these formulations, so that the performance of each DMU to be examined is expressed by a percentage. Furthermore, they can be specified by a visual description in a L2 norm expression, which appeals to us more intuitively than a L1 norm. Such a use of L2 norm on projection from the status of inefficiency to an efficiency frontier is our visual convenience. As mentioned in Section 4.1, the radial formulations are classified into two categories under variable RTS and constant RTS, respectively. This chapter starts from an input-oriented RM(v) where (v) stands for variable RTS.

4.2.1 Input-Oriented RM(v) under Variable RTS

By duplicating Model (3.15) and adding an additional constraint of $\sum_{j=1}^{n} \lambda_j = 1$ into the formulation, the radial model[7], or input-oriented RM(v), has the following mathematical structure to measure an efficiency score (θ) of the specific k-th DMU:

[6] Since Chapter 5 discusses the non-radial measures in detail, this chapter does not discuss them further, rather focusing upon a detailed description on the radial measures. We clearly acknowledge that there are many other types of model variations in DEA. Most of them are originated from either radial or non-radial models, all of which are examined in Chapters 4 and 5. Unfortunately, this book cannot cover all of them. However, it is believed that our description in this book is sufficient enough to extend them to the proposed DEA environmental assessment.

[7] This type of radial model is often referred to as "Banker–Charnes–Cooper; BCC)" in conventional DEA. This chapter refers to the model as RM(v). Moreover, "Charnes–Cooper–Rhodes; CCR)" is used to indicate RM(c) in many computer codes. This book clearly understands the contribution of Professors A. Chaners and W.W. Cooper. However, DEA approaches should be extended into more different research dimensions, in particular in environmental assessment that has many unique aspects which cannot be found in a conventional use of DEA. Acknowledging their contributions, this book intentionally use the terms RM(v) and RM(c), instead of BCC and CCR.

$$
\begin{aligned}
&\text{Minimize } \theta - \varepsilon_n \left(\sum_{i=1}^{m} d_i^x + \sum_{r=1}^{s} d_r^g \right) \\
&s.t. \quad -\sum_{j=1}^{n} x_{ij}\lambda_j - d_i^x \qquad + \theta x_{ik} = 0 \quad (i = 1,\ldots,m), \\
&\qquad \sum_{j=1}^{n} g_{rj}\lambda_j \qquad - d_r^g \qquad = g_{rk} \; (r = 1,\ldots,s), \\
&\qquad \sum_{j=1}^{n} \lambda_j \qquad\qquad = 1, \\
&\qquad \theta\text{: URS}, \; \lambda_j \geq 0 \, (j = 1,\ldots,n), \\
&\qquad d_i^x \geq 0 \, (i = 1,\ldots,m) \text{ and } d_r^g \geq 0 \, (r = 1,\ldots,s),
\end{aligned}
\tag{4.1}
$$

where the side constraint $\left(\sum_{j=1}^{n} \lambda_j = 1\right)$ is additionally incorporated into Model (4.1). Thus, the surface of a production possibility set for Model (4.1) is shaped by a convex polyhedral cone under variable RTS. It is possible for us to drop the side constraint from Model (4.1), as formulated in Model (3.15). In such a case, a production possibility set is characterized by constant RTS. Thus, Model (4.1) becomes input-oriented RM(c) where (c) indicates the status of constant RTS.

The term $-\varepsilon_n \left(\sum_{i=1}^{m} d_i^x + \sum_{r=1}^{s} d_r^g \right)$ becomes zero or negative on the optimality of Model (4.1). The objective value consists of an efficiency score[8] and a total sum of slacks, where the latter is used to adjust the level of efficiency, or *OE*. Here, it is important to note that, if the slack adjustment term represents a level of inefficiency, then the sign before the term should be negative in the objective function in order to maintain the efficiency range between zero and unity. In contrast, Model (4.9), along with Equation (4.10), indicates a desirable output-oriented model for RM(v) that is used for measuring a level of inefficiency in the objective function. In the case, the sign becomes positive. See Chapter 17, as well.

To describe the methodological strengths and drawbacks of Model (4.1), this chapter prepares the dual formulation of Model (4.1) that is mathematically specified as follows:

$$
\begin{aligned}
&\text{Maximize} \quad \sum_{r=1}^{s} u_r g_{rk} + \sigma \\
&s.t. \quad -\sum_{i=1}^{m} v_i x_{ij} + \sum_{r=1}^{s} u_r g_{rj} + \sigma \leq 0 \; (j = 1,\ldots,n), \\
&\qquad \sum_{i=1}^{m} v_i x_{ik} = 1, \\
&\qquad v_i \geq \varepsilon_n \, (i = 1,\ldots,m), \; u_r \geq \varepsilon_n \, (r = 1,\ldots,s) \text{ and} \\
&\qquad \sigma\text{: URS}.
\end{aligned}
\tag{4.2}
$$

[8] If it represents an inefficiency score, then the sign before the sum of slacks becomes positive in the objective function.

Here, $V = (v_1,\ldots,v_m)$ and $U = (u_1,\ldots,u_s)$ are two row vectors of dual variables related to the first and second sets of constraints in Model (4.1). An unrestricted dual variable (σ) is derived from the third constraint of Model (4.1). In transforming from Model (4.1) to Model (4.2), this chapter implicitly treats that $-v_i \leq -\varepsilon_n$ and $-u_r \leq -\varepsilon_n$ are changed to $v_i \geq \varepsilon_n$ for all i and $u_r \geq \varepsilon_n$ for all r in Model (4.2).

It is important to note that the objective function of Model (4.1) equals that of Model (4.2) on optimality. Thus, the level of *OE* is determined by

$$\theta^* - \varepsilon_n \left(\sum_{i=1}^{m} d_i^{x*} + \sum_{r=1}^{s} d_r^{g*} \right) = \sum_{r=1}^{s} u_r^* g_{rk} + \sigma^* \tag{4.3}$$

on the optimality of Models (4.1) and (4.2).

A methodological benefit of Model (4.1) is that all dual variables ($v_i \geq \varepsilon_n$ for all i and $u_r \geq \varepsilon_n$ for all r) are strictly positive in their signs, so being able to function as multipliers on production factors. As a result, their information is fully utilized in performance evaluation although some dual variables may become very small as multipliers in such a manner that $v_i \geq \varepsilon_n$ for some i and $u_r \geq \varepsilon_n$ for some r. This analytical feature is very important in measuring a type of RTS, a marginal rate of transformation (MRT), a rate of substitution (RSU) between production factors and other scale related measures, all of which will be discussed in Chapters 8, 20 and 22.

4.2.2 Underlying Concept

This book conveys a message that DEA can be used for the measurement of not only an efficiency score but also other important measures (e.g., RTS and other scale measures). Thus, DEA can be considered as a holistic approach for performance assessment. Consequently, it is necessary for us to satisfy various conditions on DEA solutions. One of such conditions, for example, is that all dual variables should be strictly positive. Of course, the requirement may drop in many decisional cases, depending upon the type of these applications. See Chapters 9, 21, 22 and 23 in which some dual variables are unrestricted in their signs.

A drawback of Model (4.1) is that ε_n is a prescribed small number, often referred to as a non-Archimedean small number. The specification[9] of ε_n incorporates a subjective decision by a DEA user(s). In reality, none knows what the small number is. As a result, many users often drop the non-Archimedean small number in their efficiency assessments. Model (4.1) without the very small number becomes

[9] The non-Archimedean small number was proposed by Professor Charnes, not Professor Cooper. This was a significant contribution, indeed. Unfortunately, none have been paying attention to the importance of this finding for a long time.

$$\begin{aligned} &\text{Minimize} \quad \theta \\ &s.t. \quad -\sum_{j=1}^{n} x_{ij}\lambda_j - d_i^x \quad + \theta x_{ik} = 0 \quad (i=1,\ldots,m), \\ &\qquad \sum_{j=1}^{n} g_{rj}\lambda_j \quad - d_r^g \quad = g_{rk} \; (r=1,\ldots,s), \\ &\qquad \sum_{j=1}^{n} \lambda_j \quad = 1, \\ &\qquad \theta\text{: URS, } \lambda_j \geq 0 \; (j=1,\ldots,n), \\ &\qquad d_i^x \geq 0 \; (i=1,\ldots,m) \text{ and } d_r^g \geq 0 \; (r=1,\ldots,s). \end{aligned} \tag{4.4}$$

As mentioned previously, when we measure the level of *OE* concerning the *k*-th DMU, Models (4.1) and (4.4) may not produce any major difference between them because ε_n is the very small number. The elimination is seemingly acceptable. However, a serious problem may occur in the dual model whose multipliers, obtained from dual variables, may be zero. Such a result is unacceptable in this book on environmental assessment because it is impossible to measure RTS and the other type of scale measures (e.g., rate of substitution). See, for example, Chapters 22 and 23.

The dual formulation of Model (4.4) has the following structure:

$$\begin{aligned} &\text{Maximize} \quad \sum_{r=1}^{s} u_r g_{rk} + \sigma \\ &s.t. \quad -\sum_{i=1}^{m} v_i x_{ij} + \sum_{r=1}^{s} u_r g_{rj} + \sigma \leq 0 \; (j=1,\ldots,n), \\ &\qquad \sum_{i=1}^{m} v_i x_{ik} \quad = 1, \\ &\qquad v_i \geq 0 \; (i=1,\ldots,m), u_r \geq 0 \; (r=1,\ldots,s) \text{ and} \\ &\qquad \sigma\text{: URS.} \end{aligned} \tag{4.5}$$

Here, all mathematical descriptions in Model (4.2) are applicable to Model (4.5).

The objective value of Model (4.4) equals that of Model (4.5) on optimality. Thus, the level of *OE* is determined by

$$\theta^* = \sum_{r=1}^{s} u_r^* g_{rk} + \sigma^* \tag{4.6}$$

on the optimality of Models (4.4) and (4.5).

In addition to the above description, Models (4.4) and (4.5) have the other two important features. One of the two features is that Model (4.5) has $\sum_{i=1}^{m} v_i x_{ik} = 1$ because θ is unrestricted (URS) in Model (4.4). In other words, the upper bound of θ^* is unity in Model (4.4), indicating the status of full efficiency, because the first constraints indicate $\sum_{r=1}^{s} u_r g_{rk} + \sigma \leq \sum_{i=1}^{m} v_i x_{ik} = 1$ in Model (4.5) for the *k*-th

DMU. The left hand side of the equation is the objective value of Model (4.5). Hence, both Models (4.4) and (4.5) indicate $\theta^* = \sum_{r=1}^{s} u_r^* g_{rk} + \sigma^* \leq \sum_{i=1}^{m} v_i^* x_{ik} = 1$ on the optimality of the k-th DMU. There is no lower bound on θ^* because the efficiency score is expected to be unrestricted in Model (4.4). However, a conventional use of the efficiency score, applied to a cross-sectional data set, becomes non-negative[10] because Model (4.5) is structured by a maximization problem within an efficiency frontier.

The other concern is that $v_i \geq 0$ $(i = 1,\ldots,m)$ and $u_r \geq 0$ $(r = 1,\ldots,s)$ in Model (4.5) may have a possible occurrence of $v_i = 0$ for some i and/or $u_r = 0$ for some r. This clearly indicates that some production factors are not utilized in the efficiency assessment, so being problematic as mentioned previously. Therefore, the original DEA study (i.e., Charnes *et al.*, 1978) incorporates $\varepsilon_n \left(\sum_{i=1}^{m} d_i^x + \sum_{r=1}^{s} d_r^g \right)$ into the objective function of Model (4.1). As a result of such incorporation, Model (4.2) has $v_i \geq \varepsilon_n$ $(i = 1,\ldots,m)$ and $u_r \geq \varepsilon_n$ $(r = 1,\ldots,s)$ so that all dual variables, or multipliers, are strictly positive in their signs. However, as mentioned previously, a problem of the non-Archimedean small number (ε_n) is that none knows what it is in reality. Consequently, a subjective decision is included in the selection process of ε_n. Chapter 7 discusses how to handle the difficulty by incorporating strong complementary slackness conditions (SCSCs) into Models (4.4) and (4.5). The approach can overcome the difficulty related to ε_n so that it can avoid our selection subjectivity.

4.2.3 Input-Oriented RM(c) under Constant RTS

Models (4.1) and (4.2) are formulated under variable RTS. If Model (4.1) drops $\sum_{j=1}^{n} \lambda_j = 1$, then the dual variable (σ) becomes zero in Model (4.2). In the case, the primal and dual models imply the status of constant RTS on their production possibility sets. The input-oriented RM(c) under constant RTS has the following primal and dual formulations:

$$
\begin{aligned}
\text{Minimize} \quad & \theta - \varepsilon_n \left(\sum_{i=1}^{m} d_i^x + \sum_{r=1}^{s} d_r^g \right) \\
\text{s.t.} \quad & -\sum_{j=1}^{n} x_{ij}\lambda_j - d_i^x \qquad + \theta x_{ik} = 0 \quad (i = 1,\ldots,m), \\
& \sum_{j=1}^{n} g_{rj}\lambda_j \qquad - d_r^g \qquad = g_{rk} \quad (r = 1,\ldots,s), \\
& \theta: \text{URS}, \ \lambda_j \geq 0 \ (j = 1,\ldots,n), \\
& d_i^x \geq 0 \ (i = 1,\ldots,m) \text{ and } d_r^g \geq 0 \ (r = 1,\ldots,s),
\end{aligned}
\tag{4.7}
$$

[10] The unrestricted θ in its sign is very important in measuring a dynamic change of the efficiency score in a time horizon. Technology innovation is identified by a negative value on θ. The aspect will be discussed in Section II (DEA environmental assessment).

and:

$$
\begin{aligned}
&\text{Maximize} && \sum_{r=1}^{s} u_r g_{rk} \\
&s.t. && -\sum_{i=1}^{m} v_i x_{ij} + \sum_{r=1}^{s} u_r g_{rj} \le 0 \quad (j=1,\ldots,n), \\
& && \sum_{i=1}^{m} v_i x_{ik} = 1, \\
& && v_i \ge \varepsilon_n \,(i=1,\ldots,m) \text{ and } u_r \ge \varepsilon_n \,(r=1,\ldots,s).
\end{aligned}
\tag{4.8}
$$

Here, all mathematical descriptions on dual variables in Model (4.2) are applicable to Model (4.8).

As mentioned previously, it is possible to drop $\varepsilon_n\left(\sum_{i=1}^{m} d_i^x + \sum_{r=1}^{s} d_r^g\right)$ in the objective function of Model (4.7) as found in many research efforts on production economics. In the case, it is necessary for us to examine whether all dual variables are positive or not on the optimality of Model (4.8). The objective value of Model (4.7) equals that of Model (4.8) on optimality.

4.3 RADIAL MODELS: DESIRABLE OUTPUT-ORIENTED

4.3.1 Desirable Output-oriented RM(v) under Variable RTS

As a methodological alternative of the input-oriented RM(v) and RM(c), previous DEA studies have paid attention to the following desirable output-oriented model, or RM(v), to estimate an efficiency score (θ) regarding the k-th DMU:

$$
\begin{aligned}
&\text{Maximize} && \tau + \varepsilon_n\left(\sum_{i=1}^{m} d_i^x + \sum_{r=1}^{s} d_r^g\right) \\
&s.t. && \sum_{j=1}^{n} x_{ij}\lambda_j + d_i^x = x_{ik} \quad (i=1,\ldots,m), \\
& && -\sum_{j=1}^{n} g_{rj}\lambda_j + d_r^g + \tau g_{rk} = 0 \quad (r=1,\ldots,s), \\
& && \sum_{j=1}^{n} \lambda_j = 1, \\
& && \lambda_j \ge 0 \,(j=1,\ldots,n),\ d_i^x \ge 0 \,(i=1,\ldots,m), \\
& && d_r^g \ge 0 \,(r=1,\ldots,s) \text{ and } \tau: \text{URS}.
\end{aligned}
\tag{4.9}
$$

The level of *OE* (θ^*) is measured by an unrestricted measure (τ^*) and slacks on the optimality of Model (4.9) as follows:

$$\theta^* = 1\Big/\left[\tau^* + \varepsilon_n\left(\sum_{i=1}^{m} d_i^{x*} + \sum_{r=1}^{s} d_r^{g*}\right)\right]. \tag{4.10}$$

There are two important concerns associated with Model (4.9). One of these concerns is that the objective value of Model (4.9) becomes more than unity ($\tau^* \geq 1$) on optimality. Therefore, the *OE* measure of desirable output-oriented RM(v) is determined by Equation (4.10). The other concern is that the objective function of Model (4.9) incorporates a positive sign before the term regarding the total sum of slacks like $+\varepsilon_n\left(\sum_{i=1}^{m} d_i^{x*} + \sum_{r=1}^{s} d_r^{g*}\right)$, not $-\varepsilon_n\left(\sum_{i=1}^{m} d_i^{x*} + \sum_{r=1}^{s} d_r^{g*}\right)$ as found in Model (4.7), because τ^* does not directly indicate the level of *OE*. Rather, the level of *OE* (θ^*) is measured by the reciprocal of the inefficiency score via Equation (4.10).

Model (4.9) has the following dual formulation:

$$\begin{aligned}
&\text{Minimize} && \sum_{i=1}^{m} v_i x_{ik} + \sigma \\
&s.t. && \sum_{i=1}^{m} v_i x_{ij} - \sum_{r=1}^{s} u_r g_{rj} + \sigma \geq 0 \quad (j = 1,\ldots,n), \\
& && \sum_{r=1}^{s} u_r g_{rk} = 1, \\
& && v_i \geq \varepsilon_n \left(i = 1,\ldots,m\right), u_r \geq \varepsilon_n \left(r = 1,\ldots,s\right) \text{ and} \\
& && \sigma\text{: URS.}
\end{aligned} \tag{4.11}$$

Here, all mathematical descriptions on dual variables in Model (4.2) are applicable to Model (4.11) by changing inputs to desirable outputs.

As mentioned previously, it is possible to drop the non-Archimedean small number (ε_n) from Model (4.9) and Equation (4.10). In such a case, the desirable output-oriented RM(v) becomes

$$\begin{aligned}
&\text{Maximize} && \tau \\
&s.t. && \sum_{j=1}^{n} x_{ij}\lambda_j + d_i^x = x_{ik} \left(i = 1,\ldots,m\right), \\
& && -\sum_{j=1}^{n} g_{rj}\lambda_j + d_r^g + \tau g_{rk} = 0 \quad \left(r = 1,\ldots,s\right), \\
& && \sum_{j=1}^{n} \lambda_j = 1, \\
& && \lambda_j \geq 0 \left(j = 1,\ldots,n\right), d_i^x \geq 0 \left(i = 1,\ldots,m\right), \\
& && d_r^g \geq 0 \left(r = 1,\ldots,s\right) \text{ and } \tau\text{: URS.}
\end{aligned} \tag{4.12}$$

The level of *OE* (θ^*) is measured by an unrestricted measure (τ^*) by

$$\theta^* = 1/\tau^*. \tag{4.13}$$

Model (4.12) has the following dual formulation:

$$\begin{aligned}
&\text{Minimize} && \sum_{i=1}^{m} v_i x_{ik} + \sigma \\
&s.t. && \sum_{i=1}^{m} v_i x_{ij} - \sum_{r=1}^{s} u_r g_{rj} + \sigma \geq 0 \ \ (j = 1,\ldots,n), \\
& && \sum_{r=1}^{s} u_r g_{rk} = 1, \\
& && v_i \geq 0 \ (i = 1,\ldots,m), u_r \geq 0 \ (r = 1,\ldots,s) \text{ and} \\
& && \sigma\text{: URS.}
\end{aligned} \tag{4.14}$$

Here, all the descriptions for the input-oriented RM(v) in Model (4.5) are applicable to Model (4.14).

4.3.2 Desirable Output-oriented RM(c) under Constant RTS

It is possible to reorganize Model (4.9) under constant RTS that measures the level of *OE* (θ^*) as follows:

$$\begin{aligned}
&\text{Maximize} && \tau + \varepsilon_n \left(\sum_{i=1}^{m} d_i^x + \sum_{r=1}^{s} d_r^g \right) \\
&s.t. && \sum_{j=1}^{n} x_{ij}\lambda_j + d_i^x = x_{ik} \ (i = 1,\ldots,m), \\
& && -\sum_{j=1}^{n} g_{rj}\lambda_j + d_r^g + \tau g_{rk} = 0 \ \ (r = 1,\ldots,s), \\
& && \lambda_j \geq 0 \ (j = 1,\ldots,n), d_i^x \geq 0 \ (i = 1,\ldots,m), \\
& && d_r^g \geq 0 \ (r = 1,\ldots,s) \text{ and } \tau\text{: URS.}
\end{aligned} \tag{4.15}$$

Model (4.15) drops $\sum_{j=1}^{n} \lambda_j = 1$ from Model (4.9). The *OE* (θ^*) is measured by an unrestricted measure (τ^*) via $\theta^* = 1 \Big/ \left[\tau^* + \varepsilon_n \left(\sum_{i=1}^{m} d_i^{x*} + \sum_{r=1}^{s} d_r^{g*} \right) \right]$. The elimination of the side constraint makes it possible to measure the desirable output-oriented RM(c) under constant RTS.

The dual formulation of Model (4.15) has the following formulation:

$$
\begin{aligned}
\text{Minimize} \quad & \sum_{i=1}^{m} v_i x_{ik} \\
s.t. \quad & \sum_{i=1}^{m} v_i x_{ij} - \sum_{r=1}^{s} u_r g_{rj} \geq 0 \quad (j = 1,\ldots,n), \\
& \sum_{r=1}^{s} u_r g_{rk} = 1, \\
& v_i \geq \varepsilon_n \,(i = 1,\ldots,m) \text{ and } u_r \geq \varepsilon_n \,(r = 1,\ldots,s).
\end{aligned}
\tag{4.16}
$$

Here, $V = (v_1, \ldots, v_m)$ and $U = (u_1, \ldots, u_s)$ are two row vectors of dual variables related to the first and second sets of constraints in Model (4.15).

As an extension of Models (4.15) and (4.16), this chapter reorganizes them as follows:

$$
\begin{aligned}
\text{Maximize} \quad & \tau \\
s.t. \quad & \sum_{j=1}^{n} x_{ij} \lambda_j + d_i^x = x_{ik} \,(i = 1,\ldots,m), \\
& -\sum_{j=1}^{n} g_{rj} \lambda_j + d_r^g + \tau g_{rk} = 0 \quad (r = 1,\ldots,s), \\
& \lambda_j \geq 0 \,(j = 1,\ldots,n),\, d_i^x \geq 0 \,(i = 1,\ldots,m), \\
& d_r^g \geq 0 \,(r = 1,\ldots,s) \text{ and } \tau\text{: URS.}
\end{aligned}
\tag{4.17}
$$

The efficiency score (θ^*) is measured by an unrestricted measure (τ^*) via $\theta^* = 1/\tau^*$.

The dual formulation of Model (4.17) has the following formulation:

$$
\begin{aligned}
\text{Minimize} \quad & \sum_{i=1}^{m} v_i x_{ik} \\
s.t. \quad & \sum_{i=1}^{m} v_i x_{ij} - \sum_{r=1}^{s} u_r g_{rj} \geq 0 \quad (j = 1,\ldots,n), \\
& \sum_{r=1}^{s} u_r g_{rk} = 1, \\
& v_i \geq 0 \,(i = 1,\ldots,m) \text{ and } u_r \geq 0 \,(r = 1,\ldots,s).
\end{aligned}
\tag{4.18}
$$

See a description of Model (4.16) regarding $V = (v_1,\ldots,v_m)$ and $U = (u_1,\ldots,u_s)$.

4.4 COMPARISON BETWEEN RADIAL MODELS

4.4.1 Comparison between Input-Oriented and Desirable Output-Oriented Radial Models

There are several differences between the four radial models (RMs), two of which are summarized as follows. One of the two differences is that an efficiency score of the input-oriented RM(v) is different from that of the desirable output-oriented RM(v) under variable RTS. In contrast, the efficiency measure of input-oriented RM(c) equals that of desirable output-oriented RM(c) under constant RTS. See Proposition 4.1 about the mathematical proof. The other difference is shown in Figure 4.1 which depicts a visual difference between efficiency frontiers under constant RTS and variable RTS in the *x*–*g* domain. A straight line from the origin indicates an efficiency frontier under constant RTS, while a piece-wise linear contour line, or {A-B-C-D}, indicates the efficiency frontier under variable RTS.

As depicted in Figure 4.1, the two production possibility sets imply two areas for covering a possible combination between inputs and desirable outputs. They are axiomatically specified as follows:

$$P_v(X) = \left\{ (G, X) : G \le \sum_{j=1}^{n} G_j \lambda_j,\ X \ge \sum_{j=1}^{n} X_j \lambda_j,\ \sum_{j=1}^{n} \lambda_j = 1,\ \lambda_j \ge 0 \right\} \text{ and}$$
$$P_c(X) = \left\{ (G, X) : G \le \sum_{j=1}^{n} G_j \lambda_j,\ X \ge \sum_{j=1}^{n} X_j \lambda_j,\ \lambda_j \ge 0 \right\}. \tag{4.19}$$

The first production possibility set is formulated under variable RTS, or "VRTS," while the second one is formulated by constant RTS, or "CRTS" in

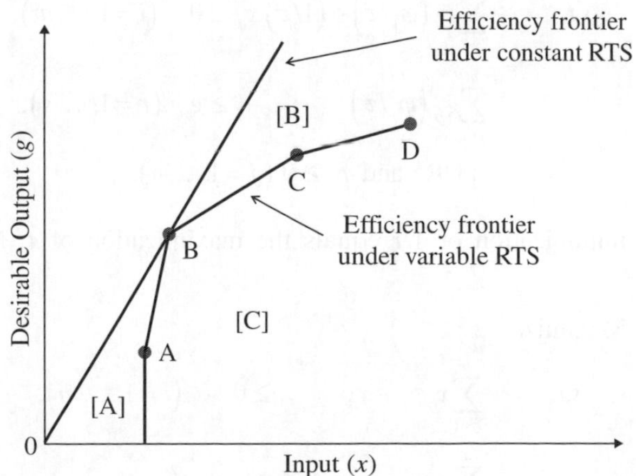

FIGURE 4.1 Efficiency frontiers under constant and variable RTS

Equation (4.19). The difference between the two production possibility sets can be found in only the incorporation of $\sum_{j=1}^{n} \lambda_j = 1$. In Figure 4.1, the first production possibility set under variable RTS consists of the area [C] and the second one under constant RTS consists of the three areas expressed by [A]∪[B]∪[C] where the symbol ∪ indicates a union set.

In examining RM(c), it is necessary for this chapter to prove the following proposition:

Proposition 4.1: The optimal objective value of the input-oriented RM(c) equals that of the desirable output-oriented RM(c) under constant RTS.

[Proof] Model (4.7), or the input-oriented RM(c), is equivalent to the following model without slacks:

$$\begin{aligned} &\text{Minimize} \quad \theta \\ &s.t. \quad -\sum_{j=1}^{n} x_{ij}\lambda_j + \theta x_{ik} \geq 0 \quad (i=1,\ldots,m), \\ &\qquad \sum_{j=1}^{n} g_{rj}\lambda_j \geq g_{rk} \quad (r=1,\ldots,s), \\ &\qquad \theta: \text{URS and } \lambda_j \geq 0 \; (j=1,\ldots,n), \end{aligned} \tag{4.20}$$

where Model (4.20) drops $\varepsilon_n \left(\sum_{i=1}^{m} d_i^x + \sum_{r=1}^{s} d_r^g \right)$ from Model (4.7) for our descriptive convenience. Let $\theta = 1/\tau$ and $\lambda_j = \varphi_j / \tau$ in Model (4.20), then the model becomes

$$\begin{aligned} &\text{Minimize} \quad 1/\tau \\ &s.t. \quad -\sum_{j=1}^{n} x_{ij} \left(\varphi_j / \tau \right) + \left(1/\tau \right) x_{ik} \geq 0 \quad (i=1,\ldots,m), \\ &\qquad \sum_{j=1}^{n} g_{rj} \left(\varphi_j / \tau \right) \geq g_{rk} \quad (r=1,\ldots,s), \\ &\qquad \tau: \text{URS and } \varphi_j \geq 0 \; (j=1,\ldots,n). \end{aligned} \tag{4.21}$$

Since the minimization of $1/\tau$ equals the maximization of τ, Model (4.21) becomes

$$\begin{aligned} &\text{Maximize} \quad \tau \\ &s.t. \quad -\sum_{j=1}^{n} x_{ij}\varphi_j + x_{ik} \geq 0 \quad (i=1,\ldots,m), \\ &\qquad \sum_{j=1}^{n} g_{rj}\varphi_j \geq \tau g_{rk} \quad (r=1,\ldots,s), \\ &\qquad \tau: \text{URS and } \varphi_j \geq 0 \; (j=1,\ldots,n). \end{aligned} \tag{4.22}$$

It is easily identified by comparing between Models (4.20) and (4.22) that the output-oriented Model (4.22) equals the input-oriented Model (4.20) in terms of their optimal efficiency measures. Thus, the efficiency score of the input-oriented RM(c) equals that of the desirable output-oriented RM(c) under constant RTS.

Q.E.D.

4.4.2 Hybrid Radial Model: Modification

As an extension of the radial models, we can control the sum of structural (intensity) variables by adding the lower (*lb*) and upper (*ub*) bounds, or $lb \le \sum_{j=1}^{n} \lambda_j \le ub$ in the proposed RM formulations. Such models are often referred to as "hybrid models" in the DEA community.

To discuss a hybrid model, this chapter considers it by adding the condition on the lower and upper bounds in Model (4.1). The hybrid model for input-oriented measurement becomes

$$
\begin{aligned}
\text{Minimize} \quad & \theta - \varepsilon_n \left(\sum_{i=1}^{m} d_i^x + \sum_{r=1}^{s} d_r^g \right) \\
\text{s.t.} \quad & -\sum_{j=1}^{n} x_{ij}\lambda_j - d_i^x \qquad + \theta x_{ik} = 0 \quad (i = 1,\ldots,m), \\
& \sum_{j=1}^{n} g_{rj}\lambda_j \qquad - d_r^g \qquad = g_{rk} \quad (r = 1,\ldots,s), \\
& lb \le \sum_{j=1}^{n} \lambda_j \le ub, \\
& \theta\text{: URS}, \lambda_j \ge 0 \ (j = 1,\ldots,n), \\
& d_i^x \ge 0 \ (i = 1,\ldots,m) \text{ and } d_r^g \ge 0 \ (r = 1,\ldots,s),
\end{aligned}
\tag{4.23}
$$

where the previous side constraint $\left(\sum_{j=1}^{n} \lambda_j = 1\right)$ for variable RTS in Model (4.1) is replaced by the new one $\left(lb \le \sum_{j=1}^{n} \lambda_j \le ub\right)$ in Model (4.23).

The dual formulation of Model (4.23) becomes as follows:

$$
\begin{aligned}
\text{Maximize} \quad & \sum_{r=1}^{s} u_r g_{rk} + lb\sigma_1 - ub\sigma_2 \\
\text{s.t.} \quad & -\sum_{i=1}^{m} v_i x_{ij} + \sum_{r=1}^{s} u_r g_{rj} + \sigma_1 - \sigma_2 \le 0 \quad (j = 1,\ldots,n), \\
& \sum_{i=1}^{m} v_i x_{ik} \qquad = 1, \\
& v_i \ge \varepsilon_n \ (i = 1,\ldots,m), u_r \ge \varepsilon_n \ (r = 1,\ldots,s), \\
& \sigma_1 \ge 0 \text{ and } \sigma_2 \ge 0.
\end{aligned}
\tag{4.24}
$$

Using Model (4.23), the input-oriented radial measurement is classified into the following three categories, for example: (a) $lb = 0$ and $ub = \infty \leftrightarrow$ RM(c), (b) $lb = 1$ and $ub = 1 \leftrightarrow$ RM(v), and (c) $lb = 0$ and $ub = 1 \leftrightarrow$ RM under increasing RTS, respectively. This type of classification can be applied to the desirable output-oriented radial model, as well. See Tone (1993) for the first discussion on the hybrid expression on RM. See also Sueyoshi (1997) which provided a detailed description on the hybrid expression and its applications.

4.5 MULTIPLIER RESTRICTION AND CROSS-REFERENCE APPROACHES

An empirical difficulty of the proposed radial models is that they often suffer from an occurrence of many efficient DMUs. For example, among 100 DMUs, 90 DMUs are efficient and the remaining 10 DMUs are inefficient. Such an occurrence is mathematically acceptable but managerially unacceptable, because it is difficult for us to determine the rank of efficient DMUs. It is important to note that the type of problem often occurs on all types of DEA models.

4.5.1 Multiplier Restriction Methods

***Assurance Region (AR) Analysis*:** To reduce the number of efficient DMUs by restricting the range of multipliers (i.e., dual variables), Thompson *et al.* (1986), Dyson and Thanassoulis (1988) and others proposed the use of AR analysis, which was discussed at the initial stage of DEA development. AR analysis incorporates the following multiplier restrictions into the dual formulations of the proposed RMs:

$$\underline{lv}_i \le v_i / v_1 \le \overline{uv}_i \left(i = 2,3,\ldots,m\right) \quad \text{and} \quad \underline{lu}_r \le u_r / u_1 \le \overline{uu}_r \left(r = 2,3,\ldots,s\right), \qquad (4.25)$$

where the ratio of the i-th input multiplier to the first one is expressed by its lower and upper bounds. The ratio combinations are applicable to the r-th desirable output to the first one, as specified in Equation (4.25). It is trivial that multipliers are usually reqiured to be positive in many cases, whose magnitudes are expected to be much larger than ε_n.

Equation (4.25) is further extendable by the following input and desirable output weighted measure on the specific k-th DMU:

$$\begin{aligned} &\underline{lv}_i \le x_{ik} v_i \Big/ \sum_{i=1}^{m} x_{ik} v_i \le \overline{uv}_i \left(i = 1,2,\ldots,m\right) \quad \text{and} \\ &\underline{lu}_r \le y_{rk} u_r \Big/ \sum_{r=1}^{s} y_{rk} u_r \le \overline{uu}_r \left(r = 1,2,\ldots,s\right), \end{aligned} \qquad (4.26)$$

where the ratio is determined by the i-th weighted input divided by the sum of the total weighted inputs. In a similar manner, the ratio can be applied to the r-th desirable output as formulated in Equation (4.26). Each equation indicates the importantce of the i-th input and the r-th desirable output among these production factors.

Figure 4.2 visually describes an implication of multiplier restriction in the data domain of two inputs ($x1$ and $x2$). All DMUs from {D1} to {D5}, consisting of an efficiency frontier, have the same magnitude of a desirable output so that the figure drops it. Let us consider A1 and A2 as the two vectors to restrict input multipliers. They are used as the upper and lower bounds for the multiplier restriction, expressed by $\tilde{V}$ in Figure 4.2.

As a result of such multiplier restriction, the number of efficient DMUs decreases from five to one, that is, {D3} because only the DMU can exist only within the multiplier range ($\tilde{V}$) for inputs. This chapter uses $\tilde{V}$ because input multipliers are applied for restriction on v_i ($i = 1,\ldots, m$).

The proposed AR analysis has two difficulties in the practical uses. One of the two difficulities is that the approach depends upon prior information (e.g., expreience and previous emprical evidence) for multiplier restriction. It is often difficult for us to assess such prior information. The other difficulty is that the approach may produce an infeasible solution on optimality as a result of strict restriction. It is necessary for us to carefully think how to restrict multipliers on a use of the AR approach. This type of difficulty will be solved in DEA environmental assessment. See Chapters 22 and 23 which describe a new type of multiplier restriction, referred to as "explorative analysis," which does not use any prior information along with a data adjustment. The AR analisis does not incoporate the data adjustment.

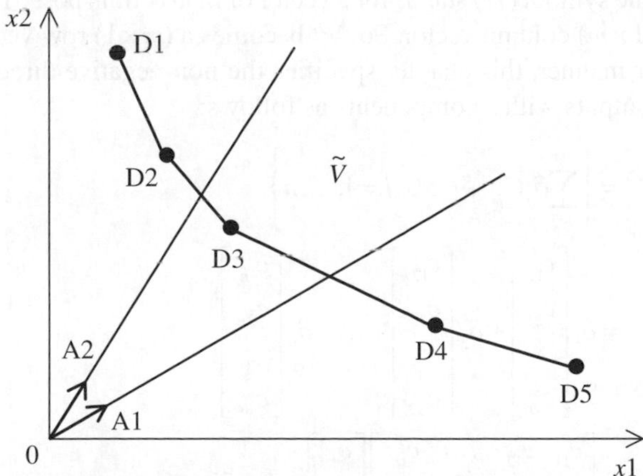

FIGURE 4.2 Multiplier restriction (a) The figure is prepared for our visual convenience. The number of efficient DMUs decreases from five {D1-D2-D3-D4-D5} on the efficiency frontier to one, {D3}. It is necessary for us to understand that strong multiplier restriction often produces an infeasible solution on optimality. The two directional vectors (A1 and A2) are used to restrict the direction of input multipliers. (b) It is assumed that all DMUs produce the same amount of a desirable output.

4.5.2 Cone Ratio Method

As mentioned previously, DEA produces many efficient DMUs. Charnes *et al.* (1989) and Brockett *et al.* (1997) proposed another type of approach for multiplier restriction, referred to as the "cone ratio (CR) method," to reduce the number of efficient DMUs. To describe the methodology, this chapter starts from the following non-negative input vector with m components, where the vector and matrix notations are for our visual convenience:

$$\begin{aligned}\tilde{V} &= \left\{ \sum_{j=1}^{n} \eta_j A_j : \eta_j \geq 0, j = 1,\ldots,n \right\} \\ &= \eta_1 \begin{bmatrix} a_{11} \\ a_{21} \\ \vdots \\ a_{m1} \end{bmatrix} + \eta_2 \begin{bmatrix} a_{12} \\ a_{22} \\ \vdots \\ a_{m2} \end{bmatrix} + \ldots + \eta_n \begin{bmatrix} a_{1n} \\ a_{2n} \\ \vdots \\ a_{mn} \end{bmatrix} \\ &= \begin{bmatrix} a_{11} & a_{12} & \cdots & a_{1n} \\ a_{21} & a_{22} & \cdots & a_{2n} \\ \vdots & \vdots & \vdots & \vdots \\ a_{m1} & a_{m2} & \cdots & a_{mn} \end{bmatrix} \begin{bmatrix} \eta_1 \\ \eta_2 \\ \vdots \\ \eta_n \end{bmatrix} \\ &= A\eta \end{aligned} \tag{4.27}$$

where $\mathbf{A}_j = (a_{1j}, a_{2j}, \ldots, a_{mj})^{Tr}$ is used for a column vector for input restriction ($j = 1,\ldots,n$). The symbol (Tr) stands for a vector or matrix transpose. The restriction vector $\tilde{V}$ is a $(1 \times m)$ column vector. So, $\tilde{V}^{Tr}$ becomes a $(m \times 1)$ row vector.

In a similar manner, this chapter specifies the non-negative directional vector for desirable outputs with s components as follows:

$$\begin{aligned}\tilde{U} &= \left\{ \sum_{j=1}^{n} \delta_j C_j : \delta_j \geq 0, j = 1,\ldots,n \right\} \\ &= \delta_1 \begin{bmatrix} c_{11} \\ c_{21} \\ \vdots \\ c_{s1} \end{bmatrix} + \delta_2 \begin{bmatrix} c_{12} \\ c_{22} \\ \vdots \\ c_{s2} \end{bmatrix} + \ldots + \delta_n \begin{bmatrix} c_{1n} \\ c_{2n} \\ \vdots \\ c_{sn} \end{bmatrix} \\ &= \begin{bmatrix} c_{11} & c_{12} & \cdots & c_{1n} \\ c_{21} & c_{22} & \cdots & c_{2n} \\ \vdots & \vdots & \vdots & \vdots \\ c_{s1} & c_{s2} & \cdots & c_{sn} \end{bmatrix} \begin{bmatrix} \delta_1 \\ \delta_2 \\ \vdots \\ \delta_n \end{bmatrix} \\ &= C\delta \end{aligned} \tag{4.28}$$

where $C_j = (c_{1j}, c_{2j},\ldots,c_{sj})^{Tr}$ is used for a column vector for restriction on desirable outputs (j = 1,…, n). The restriction vector $\tilde{U}$ is a $(1\times s)$ column vector. So, $\tilde{U}^{Tr}$ becomes a $(s\times 1)$ row vector.

Using Equations (4.27) and (4.28), the input-oriented RM(c), or Model (4.8), becomes the following vector notation under the proposed CR method:

$$
\begin{aligned}
&\text{Maximize} \quad UG_k \\
&\quad s.t \qquad -VX_j + UG_j \le 0 \quad (j=1,2,\ldots,n), \\
&\qquad\qquad\; VX_k \qquad\quad = 1, \\
&\qquad\qquad\; V^{Tr} \in \tilde{V} \text{ and } U^{Tr} \in \tilde{U},
\end{aligned}
\tag{4.29}
$$

where $U = (u_1, u_2,\ldots,u_s)$ is a raw vector of s dual variables for desirable outputs, $V = (v_1, v_2,\ldots,v_m)$ is a raw vector of m dual variables for inputs, $X_j = (x_{1j}, x_{2j},\ldots, x_{mj})^{Tr}$ is a column vector of m observed inputs of the j-th DMU, $X_k = (x_{1k}, x_{2k},\ldots, x_{mk})^{Tr}$ is a column vector of m observed inputs of the specific k-th DMU and $G_j = (g_{1j}, g_{2j},\ldots,g_{sj})^{Tr}$ is a column vector of s observed desirable outputs of the j-th DMU.

Model (4.29) can be reorganized by incorporating Equations (4.27) and (4.28). The resulting formulation becomes

$$
\begin{aligned}
&\text{Maximize} \quad \delta^{Tr}\left(\boldsymbol{C}^{Tr}G_k\right) \\
&\quad s.t. \qquad -\eta^{Tr}\left(\boldsymbol{A}^{Tr}X_j\right) + \delta^{Tr}\left(\boldsymbol{C}^{Tr}G_j\right) \le 0 \;(j=1,2,\ldots,n), \\
&\qquad\qquad\;\; \eta^{Tr}\left(\boldsymbol{A}^{Tr}X_k\right) \qquad\qquad\qquad = 1, \\
&\qquad\qquad\;\; \eta \ge 0 \text{ and } \delta \ge 0.
\end{aligned}
\tag{4.30}
$$

Here, $V^{Tr} \in \tilde{V}$ and $U^{Tr} \in \tilde{U}$ in Model (4.29) are replaced by $\tilde{V} = A\eta$ and $\tilde{U} = C\delta$ in Model (4.30).

The dual formulation of Model (4.30) becomes as follows:

$$
\begin{aligned}
&\text{Minimize} \quad \theta \\
&\quad s.t. \qquad -\sum_{j=1}^{n}\left(\boldsymbol{A}^{Tr}X_j\right)\lambda_j + \theta\left(\boldsymbol{A}^{Tr}X_k\right) \ge 0, \\
&\qquad\qquad\;\; \sum_{j=1}^{n}\left(\boldsymbol{C}^{Tr}G_j\right)\lambda_j - \boldsymbol{C}^{Tr}G_k \qquad \ge 0, \\
&\qquad\qquad\;\; \lambda_j \ge 0\;(j=1,2,\ldots,n) \text{ and } \theta\text{: URS.}
\end{aligned}
\tag{4.31}
$$

Like the AR analysis, the proposed CR method has also a difficulty in both accessing prior information and being often infeasible on optimality. Another difficulty of the CR method is that the AR analysis is more intutive and easily used than the CR method. Therefore, Chapters 22 and 23 extend the AR analysis, not the CR method, to prepare "explorative analysis." A benefit of CR is that it requires less computation time than AR because CR has less constraints than AR in the formulation. See Chapter 10 on DEA network computing which takes advantage of the mathematical structure with less constraints in the formulation.

4.5.3 Cross-reference Method

Besides the multilier restriction approaches, a cross-reference method is widely used by many DEA researchers to classify efficient DMUs into two categories. One of the two categories includes efficient ones that locate on a vertex of an efficiency frontier. The other category includes the other efficient ones that locate on the efficiency frontier, but not locating on the vertex. See the DMU classification discussed in Chapter 9 (for algorithm development for network computing). Also, see Charnes *et al.* (1986, 1991) for the original DMU classification, parts of which are used in this chapter. A gist of the cross-reference method is the classification of DMUs on an efficiency frontier.

In this chapter, the three groups of DMUs are characterized by an optimal solution of radial measurement regrading the k-th DMU, for example, by Model (4.20) with slacks. The assumption on constant RTS is necessary so that the proposed cross-reference method, using RM(c), can produce a feasible solution on optimality. In contrast, RM(v) may produce an infeasible solution in the case of variable RTS.

The three groups of DMU classifications considered in this chapter are mathematically specfied in the following manner:

$$E = \left\{ k \in J \mid \theta_k^* = 1, \lambda_k^* = 1, \lambda_j^* = 0 \left(\text{all } j \neq k \in J \right), \text{all zero slacks and unique solution} \right\},$$

$$E' = \left\{ k \in J \mid \theta_k^* = 1, \lambda_k^* \leq 1, \lambda_j^* > 0 \left(\text{some } j \in J \right), \text{all zero slacks and multiple solutions} \right\} \text{ and}$$

$$IF = \left\{ k \in J \mid \theta_k^* = 1 \text{ and at least one positive slack} \right\}. \tag{4.32}$$

Here, all DMUs are expressed by a whole set (J).

The first two groups of efficient DMUs locate on an efficiency frontier. The efficient group of DMUs (E) locates on the vertexes of an efficiency frontier. The efficient DMUs in the other group (E') locate between such efficient DMUs that belong to E. The DMUs in E' may suffer from an occurrence of multiple solutions on a combination(s) of a reference set. The last group (IF) consists of inefficient DMUs, but they exist on the efficiency frontier. They are not efficient because of at least one positive slack.

As mentioned previously, DEA may produce many efficient DMUs in $E \cup E'$. Therefore, it is necessary for us to classify them further if we want to prepare their ranking assessment. That is the purpose of the cross-reference method. See Chapter 25 on the ranking analysis.

To describe the proposed approach, let us select DMU {a}, belonging to $E \cup E'$ and return to input-based RM(c), or Model (4.20), to obtain the computational feasibility. The cross-reference efficiency ($\theta^*_{a,b}$) of the b-th DMU in J is measured by the following radial model (e.g., Doyle and Green, 1994; Green *et al.*, 1996):

$$\begin{aligned}
&\text{Minimize} \quad \theta \\
&s.t. \quad -\sum_{j\in J-a} x_{ij}\lambda_j - d_i^x \quad + \theta x_{ib} = 0 \quad (i = 1,\ldots,m), \\
&\qquad \sum_{j\in J-a} g_{rj}\lambda_j \qquad - d_r^g \qquad = g_{rb} \, (r = 1,\ldots,s), \\
&\qquad \theta\text{: URS}, \lambda_j \ge 0 \, (j \in J - a), \\
&\qquad d_i^x \ge 0 \, (i = 1,\ldots,m) \text{ and } d_r^g \ge 0 \, (r = 1,\ldots,s).
\end{aligned} \tag{4.33}$$

An important feature of Model (4.33) is that if the efficient DMU {a} belongs to E, then the elimination changes the shape of an efficiency frontier. As a result, the degree of $\theta^*_{a,b}$ may be changed from the original efficiency measure. In contrast, if it belongs to E', then the elimination does not make any change on the frontier shape because it does not locate on the vertex of an efficiency frontier. In the case, the degree of $\theta^*_{a,b}$ does not produce any efficiency change. Thus, the cross-reference method can provide us with a new measure on efficient DMUs, based upon which we can classify them. The most important benefit of the proposed approach is that it is possible for us to measure a degree of cross-reference of the a-th DMU in $E \cup E'$ by examining the degree of $D^*_{a,a} = \theta^*_{a,a} - \theta^*_a$. In this case, $\theta^*_{a,a}$ may be more than unity so that $D^*_{a,a}$ becomes positive. Thus, all efficient DMUs in group E on the vertexes of an efficiency frontier are reevaluated by the degree of $D^*_{a,a}$. Thus, the cross-reference method can provide us with an additional measure on efficient DMUs, based upon which we can classify them for ranking analysis.

Finally, it is important to note three concerns on a use of the cross-reference method. First, the approach is feasible under constant RTS, but it may produce an infeasible solution under variable RTS. Second, there is no theoretical rationale on the validity of the cross-reference method although it intuitively appeals to DEA users. Finally, the assessment on efficient DMUs is based upon local evaluation by considering a limited number of efficient DMUs in a reference set. It does not cover the group-wide performance assessment of all DMUs by using a single set of common multipliers (weights) on efficient DMUs. See, for example, Chapter 25 that discusses how to determine common multipliers for the group-wide assessment. Thus, the cross-reference approach has a high level of computational tractability on DMU ranking analysis, but lacking any theoretical support within the framework of RM(c).

4.6 COST ANALYSIS

4.6.1 Cost Efficiency Measures

A methodological benefit of the radial measurement is that it can conceptually connect these models to cost efficiency analysis that minimizes the amount of total production cost (Sueyoshi, 1999b). Assuming that we can access an input price vector, or $P_j = (p_{1j}, p_{2j}, \ldots, p_{mj})$, of all DMUs ($j = 1, \ldots, n$), the cost efficiency of the k-th DMU is measured by the following model:

$$\begin{aligned}
\text{Minimize} \quad & \sum_{i=1}^{m} p_{ik} x_i \\
s.t. \quad & -\sum_{j=1}^{n} x_{ij}\lambda_j + x_i \geq 0 \quad (i = 1,\ldots,m), \\
& \sum_{j=1}^{n} g_{rj}\lambda_j \qquad \geq g_{rk} \; (r = 1,\ldots,s), \\
& \sum_{j=1}^{n} \lambda_j \qquad = 1, \\
& \lambda_j \geq 0 \; (j = 1,\ldots,n) \text{ and } x_i \geq 0 \; (i = 1,\ldots,m).
\end{aligned} \tag{4.34}$$

Here, $X = (x_1, x_2, \ldots, x_m)^{Tr}$ is an unknown input quantity vector, not an observed vector as previously discussed for radial models, or RMs. It is important to note that Model (4.34) does not incorporate slacks into the formulation because it looks for an input vector that attains cost minimum, not an efficiency score. Moreover, dual variables become positive, as discussed below, because an input-related dual vector corresponds to an input price vector. Therefore, this chapter needs to drop a slack vector in the objective function of Model (4.34).

The cost efficiency of the k-th DMU is determined by the following manner (Farrell, 1957):

$$\mu^* = c_k^* \big/ c_k = \sum_{i=1}^{m} p_{ik} x_i^* \Big/ \sum_{i=1}^{m} p_{ik} x_{ik}. \tag{4.35}$$

The cost efficiency implies the mimimized cost divided by an observed cost regrading the k-th DMU.

The dual formulation of Model (4.34) becomes:

$$\begin{aligned}
&\text{Maximize} \quad \sum_{r=1}^{s} u_r g_{rk} + \sigma \\
&\text{s.t.} \quad -\sum_{i=1}^{m} v_i x_{ij} + \sum_{r=1}^{s} u_r g_{rj} + \sigma \le 0 \quad (j = 1,\dots,n), \\
&\qquad\quad v_i \le p_{ik} \, (i = 1,\dots,m), \\
&\qquad\quad v_i \ge 0 \, (i = 1,\dots,m), u_r \ge 0 \, (r = 1,\dots,s) \text{ and} \\
&\qquad\quad \sigma\text{: URS.}
\end{aligned} \tag{4.36}$$

On optimality, both primal and dual models have a cost minimum expressed by $\sum_{i=1}^{m} p_{ik} x_i^* = \sum_{r=1}^{s} u_r^* g_{rk} + \sigma^*$. Moreover, $x_i^* (p_{ik} - v_i^*) = 0 \, (i = 1,\dots,m)$ always exsist between the two models as part of complementary slackness conditions (CSCs) of linear programming. If $x_i^* > 0$ for all i, then Model (4.36) becomes the following condensed formulation:

$$\begin{aligned}
&\text{Maximize} \quad \sum_{r=1}^{s} u_r g_{rk} + \sigma \\
&\text{s.t.} \quad \sum_{r=1}^{s} u_r g_{rj} + \sigma \le \sum_{i=1}^{m} p_{ik} x_{ij} \quad (j = 1,\dots,n), \\
&\qquad\quad u_r \ge 0 \, (r = 1,\dots,s) \text{ and } \sigma\text{: URS,}
\end{aligned} \tag{4.37}$$

because $p_{ik} = v_i^*$ is maintained on the optimality of Model (4.37). Equation (4.35) indicates that the cost minimum of the k-th DMU is less than or equal all the cost estimates measured by its observed input price vector.

4.6.2 Type of Efficiency Measures in Production and Cost Analyses

Previous discussions used the term "efficiency." Hereafter, the concept of efficiency is separated into more specific classifications related to the proposed input-oriented radial measurement. Some of these efficiency meaures were first conceptually specified by Farrell (1957) in the case of a single desirable output with multiple inputs under constant and diminishing RTS. Then, Sueyoshi (1999b) elaborated them by extending these production factors under constant RTS and variable RTS.

The efficiency measures used in this chapter are specified as follows:

(a) Operational efficiency (OE_v) under variable RTS: the degree of opetational efficincy that is measured by the optimal objective value of RM(v), for example, Model (4.1),

(b) Operational and Scale Efficiency (OE_c) under constant RTS: the degree of operational efficiency that is measured by the optimal objective value of RM(c), for example, Model (4.7),

(c) Production-based scale efficiency (PSE): $PSE = OE_c / OE_v$.

(d) Cost efficiency (CE_v) under variable RTS: $CE_v = \mu^*$ from Equation (4.35) via Model (4.34),

(e) Cost and Scale Efficiency (CE_c) under constant RTS: $CE_c = \mu^*$ from Equation (4.35) via Model (4.34) without $\sum_{j=1}^{n} \lambda_j = 1$.

(f) Cost-based scale efficiency (CSE): $CSE = CE_c / CE_v$.

(g) Allocative efficiency (AE_v) under variable RTS: $AE_v = CE_v / OE_v$.

(h) Allocative and scale efficiency (AE_c) under constant RTS: $AE_c = CE_c / OE_c$.

The desirable output-oriented measures have another type of efficiency classifications. This chapter discribes only the input-oriented efficiency ones because they can directly link to cost-based efficiency concepts.

Figure 4.3 visually describes various efficiency measures, all of which are defined by the eight classifications: (a–h) as discussed above. The figure has two efficiency frontiers, depending upon the assumption on RTS (i.e., constant or variable). The efficiency frontier, consisting of a piece-wise linear contour line, or DMUs {F-G-B-H-I-J}, is shaped under variable RTS, while the other curve (I–I′) is shaped under constant RTS. For our descriptive convenience in Figure 4.3, this chapter starts with a visual explanation on OE_v.

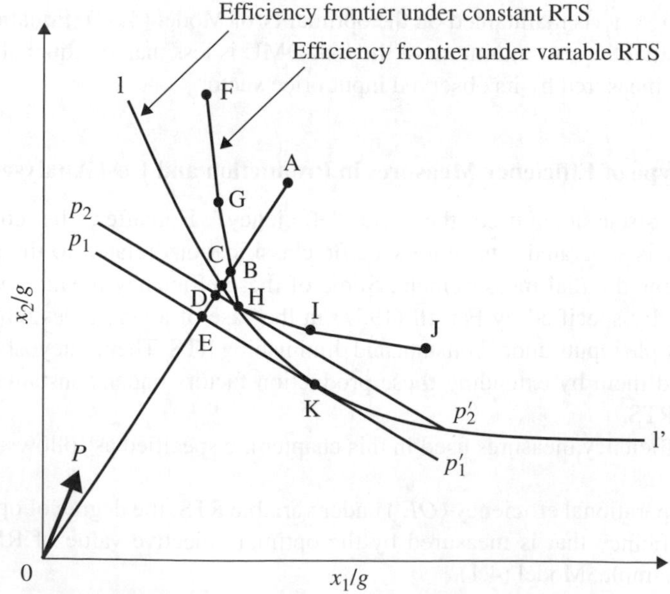

FIGURE 4.3 Efficiency measures (a) Source: Sueyoshi, (1997).

The degree of OE_V is visually specified by the distance OB/OA under variable RTS, where OB stands for a distance between the origin O and DMU {B}. In a simular manner, OA indicates the distance between the origin O and DMU {A}. This chapter drops the term DMU if a specified point in the figure is used to indicate a projected point, so not indicating a DMU. Shifting from variable to constant RTS, OE_c is measured by OC/OA. The difference between the two measures is expressed by $PSE = OE_c/OE_v$ = (OC/OA)/(OB/OA) = OC/OB. For example, DMU {A} needs to reduce the amount of an input vector from {A} to {B} under variable RTS. The level of such input reduction, indicating an amount of inefficiency, is expressed by the distance of (OA–OB)/OA. By controling the operation size, the DMU {A} can attain the input decrease until it attains a projected point, C. The inefficiency due to an inappropriate operation size is expressed by (OB–OC)/OA. Thus, OE_C is separated into two components in the manner of $OE_c = (OE_v)(PSE)$.

The production-based description can be applied to the cost-based one such as CE_v, CE_c and CSE. Given an input price vector (P), H indicates the cost-minimum for DMU {A} in the case of variable RTS. Here, lines p_1–p_1' and p_2–p_2' indicate a level of cost under a fixed ratio of input prices with p_1/p_2. D and H locate on the line (p_2 and p_2'), with both having the same value of cost-minimum. Therefore, CE_v = OD/OA. If the line is closer to the origin, the level of cost is lower. Meanwhile, under constant RTS, K indicates such a cost-minimum. The total production cost becomes same on K and E because both points locate on the line (p_1 and p_1'). Therefore, CE_c = OE/OA. The degree of CSE is determined by the ratio between CE_v and CE_c. Thus, this chapter has $CSE = CE_c/CE_v$ = (OE/OA)/(OD/OA) = OE/OD. The decomposition related to the cost minimization process indicates that DMU {A} needs to attain H by an effort for cost mimimization and further attain K by controlling its operational size.

In addtion to these efficiency concepts, AE_v and AE_c have unique features that cannot be found in the other efficiency measures. For example, the degree of AE_v is determined by $AE_v = CE_v/OE_v$ and AE_V = (OD/OA)/(OB/OA) = OD/OB in Figure 4.3. Given an input price vector (P), the efficiency measure indicates how each DMU controls the ratio of input allocation by itself for cost-mimimization under variable RTS. The degree of AE_C has the same implication under constant RTS.

4.6.3 Illustrative Example

Table 4.1 lists a data set to discuss three efficiency measures related to two radial models. The table contains eight DMUs {A–H}, each of which uses one input to produce one desiable output. Table 4.2 summarizes computational results on OE_v,

TABLE 4.1 An illustrative example

DMU	A	B	C	D	E	F	G	H
Input	2	3	7	10	5	7	8	10
Desirable Output	1	4	7	8	3	6	2	4

TABLE 4.2 Production-based efficiency measures

DMU	A	B	C	D	E	F	G	H
OEv	1	1	1	1	0.533	0.810	0.292	0.300
OEc	0.375	1	0.750	0.600	0.450	0.643	0.188	0.300
PSE	0.375	1	0.750	0.600	0.844	0.794	0.643	1

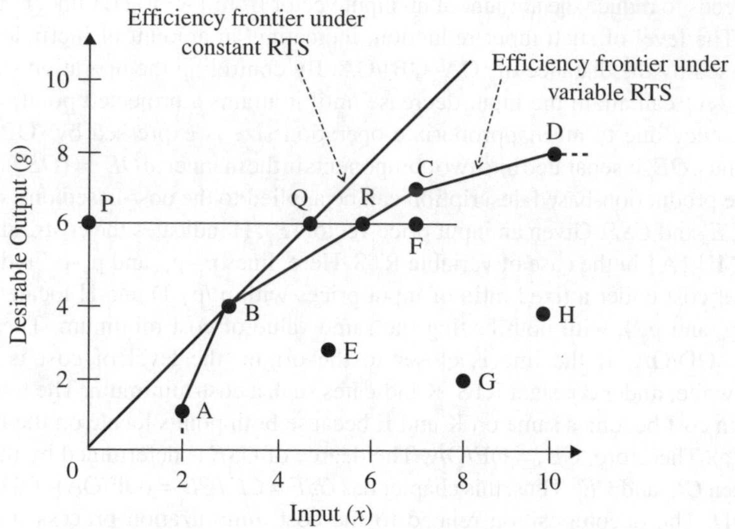

FIGURE 4.4 Input-oriented projections and two efficiency frontiers

OE_c and PSE. Figure 4.4 visually describes OE_v measures under variable RTS and OE_c measures under constant RTS, both are associated with the two efficiency frontiers.

The formulation to measure OE_v of DMU {A}, for example, is formulated as follows:

$$\begin{aligned}
&\text{Minimize} \quad \theta \\
&s.t. \quad -2\lambda_A - 3\lambda_B - 7\lambda_C - 10\lambda_D - 5\lambda_E - 7\lambda_F - 8\lambda_G - 10\lambda_H + 2\theta \geq 0 \\
&\qquad \lambda_A + 4\lambda_B + 7\lambda_C + 8\lambda_D + 3\lambda_E + 6\lambda_F + 2\lambda_G + 4\lambda_H \geq 1 \\
&\qquad \lambda_A + \lambda_B + \lambda_C + \lambda_D + \lambda_E + \lambda_F + \lambda_G + \lambda_H = 1 \\
&\qquad \lambda_j \geq 0 \ (j = A, B, \ldots, H) \text{ and } \theta : \text{URS}.
\end{aligned}$$

It is necessary to repeat the computation process on DMU {A} to measure the other DMUs {B–H}.

As listed in Table 4.2, along with Figure 4.4, only DMU {B} attains the status of efficiency in the three measures. The four DMUs {A, B, C and D} attain the status of efficiency on OE_v under variable RTS, but only DMU {B} become efficient under constant RTS. The other DMUs {E, F, G and H} have exhibited some level of inefficiency on OE_v. To describe the implication of *PSE*, measured by OE_c/OE_v, let us pay attention to the inefficient DMU {F}. Since the computational results of Table 4.2 are determined by the input-oriented measurement, the ideal performance of DMU {F} can be found in a projected point R on the efficiency frontier under variable RTS and another projected point Q under constant RTS.

The degree of OE_v on DMU {F} is measured by PR/PF = 0.810. The degree of OE_c is measured by PQ/PF = 0.643. Based upon the two efficiency measures, the degree of *PSE* on the DMU is determined by OE_c/OE_v = (PQ/PF)/(PR/PF) = PQ/PR = 0.643/0.810 = 0.794. The description of DMU {F} in terms of the three efficiency measures is applicable regarding the other DMUs.

Table 4.3 lists a data set to discuss eight efficiency measures related to production and cost-based efficiency measures. First, let us describe the formulation to measure the level of CE_v on DMU {F}, for example, which is structured as follows:

$$\begin{aligned}
\text{Minimize} \quad & 3x_1 + 4x_2 \\
\textit{s.t.} \quad & -2\lambda_A - 3\lambda_B - 5\lambda_C - 8\lambda_D - 3\lambda_E - 4\lambda_F - 7\lambda_G - 9\lambda_H + x_1 \geq 0, \\
& -8\lambda_A - 4\lambda_B - 2\lambda_C - \lambda_D - 9\lambda_E - 6\lambda_F - 2\lambda_G - 8\lambda_H + x_2 \geq 0, \\
& 10\lambda_A + 9\lambda_B + 10\lambda_C + 9\lambda_D + \lambda_E + 2\lambda_F + \lambda_G + 2\lambda_H \geq 2, \\
& \lambda_A + \lambda_B + \lambda_C + \lambda_D + \lambda_E + \lambda_F + \lambda_G + \lambda_H = 1, \\
& \lambda_j \geq 0 \left(j = A, B, \ldots, H \right) \text{ and } x_i \geq 0 \left(i = 1, 2 \right).
\end{aligned}$$

Table 4.4 summarizes eight efficiency measures related to production and cost analyses. As summarized in the table, DMU {C} is efficient in terms of the eight

TABLE 4.3 An illustrative example

DMU	A	B	C	D	E	F	G	H
Input 1	2	3	5	8	3	4	7	9
Input 2	8	4	2	1	9	6	2	8
Desirable Output	10	9	10	9	1	2	1	2
Price of Input 1	3	4	2	1	2	3	3	4
Price of Input 2	3	1	3	2	2	4	1	2

TABLE 4.4 Production and cost-based efficiency measures

DMU	A	B	C	D	E	F	G	H
OEv	1	1	1	1	0.762	0.727	0.846	0.412
OEc	1	1	1	1	0.078	0.160	0.087	0.088
PSE	1	1	1	1	0.103	0.220	0.103	0.214
CEv	0.700	1	1	0.900	0.583	0.639	0.565	0.385
CEc	0.700	0.900	1	0.810	0.058	0.128	0.061	0.085
CSE	1	0.900	1	0.900	0.100	0.200	0.108	0.222
AEv	0.700	1	1	0.900	0.766	0.878	0.668	0.934
AEc	0.700	0.900	1	0.810	0.744	0.799	0.697	0.971

measures. The DMUs {A, B, D} have mixed efficiency measures in the eight measures and the remaining DMUs {E, F, G, H} are inefficient in these measures.

4.7 SUMMARY

This chapter discussed radial models that were widely used for the measurement of OE_v and other seven efficiency measures (so, becoming a total of eight measures), all of which were obtained from the production and cost-based measurements. The production-based measurement is usually classified into radial measures under variable and constant RTS. This chapter extended production-based radial measures into cost-based measurement. As a result, this chapter provided an analytical linkage among these production and cost-based models from their efficiency measures.

This chapter also described multiplier restriction methods (i.e., AR and CR) that specified the range of dual variables (multipliers). As a result, the proposed methods were expected to produce positive dual variables in the manner that they could reduce the number of efficient DMUs. It is widely known that DEA may produce many efficient DMUs. It is mathematically acceptable but managerially unacceptable because we cannot rank efficient DMUs. However, the multiplier restriction approaches (e.g., AR and CR) discussed in this chapter do not have any mathematical justification that they can always have positive dual variables. In the worst case, they produce an infeasible solution because of too strict restriction. To avoid such a difficulty, Chapter 7 proposes the incorporation of strong complementary slackness conditions (SCSCs) into DEA to obtain positive dual variables on efficient DMUs. It is not necessary to apply SCSCs to inefficient DMUs because they do not influence the degree of efficiency on all DMUs. Note that the use of SCSCs is not perfect in terms of ranking all DMUs. See Chapter 25 that extends a mathematical description on SCSCs for the ranking analysis.

In addition to the multiplier restriction methods, this chapter discussed the cross-reference approach, which was useful in classifying efficient DMUs. A problem of the approach was that it reevaluated an efficient DMU based upon local comparison on part of DMUs on an efficient frontier, so not having any common weights or group-wide evaluation including all DMUs (at least all efficient DMUs). Furthermore, there was no mathematical justification for the cross-reference approach. Such were the drawbacks of the cross-reference approach. See Chapter 25 that discusses how to obtain common weights for group-wide evaluation.

At the end of this chapter, it is necessary for us to describe that there is no perfect DEA approach. Any approach has its methodological strength and weakness. Consequently, it is necessary for us to utilize DEA carefully by understanding such unique features. This concern will be extended to Chapter 6 that discusses a methodological bias issue, implying "different approaches produce different results," on several DEA (radial and non-radial) models.

5

NON-RADIAL MEASUREMENT

5.1 INTRODUCTION

This chapter[1] discusses non-radial models based upon the Pareto–Koopmans measurement to determine the level of operational efficiency (*OE*). The DEA assessment is generally separated into two groups of models. One of the two groups is referred to as "radial measures," or input and desirable output-oriented RM(v) and RM(c), as discussed in Chapter 4. The other group of models includes "non-radial measures," most of which are discussed in this chapter. The radial models, belonging to "Debreu–Farrell measures," can determine the level of *OE* regarding DMUs by examining an efficiency score, along with slacks, in the objective function of their linear programming formulations[2]. In contrast, the

[1] This chapter is partly based upon the three articles: (a) Bardhan, I., Bowlin, W.F., Cooper, W.W. and Sueyoshi, T. (1996a) Models and measures for efficiency dominance in DEA. Part I: Additive models and MED measures, *Journal of the Operations Research Society of Japan*, 39, 322–332. (b) Bardhan, I., Bowlin, W.F., Cooper, W.W. and Sueyoshi, T. (1996b) Models and measures for efficiency dominance in DEA. Part II: Free disposal Hulls and Russell measures, *Journal of the Operations Research Society of Japan*, 39, 333–344 and (c) Sueyoshi, T. and Sekitani, K. (2007b) Computational strategy for Russell measure in DEA: second-order cone programming. *European Journal of Operational Research*, **180**, 459–471.

[2] As discussed in Chapter 4, RM(v) and RM(c) radially project the performance of an inefficient DMU onto an efficiency frontier by L2 norm distance measurement. Meanwhile, non-radial models depend upon the projection by slacks, so being under L1 distance measurement.

Environmental Assessment on Energy and Sustainability by Data Envelopment Analysis,
First Edition. Toshiyuki Sueyoshi and Mika Goto.

non-radial measures (e.g., an additive model: Charnes *et al.*, 1985), discussed in this chapter determine the *OE* level by examining the total amount of slacks, because they do not have any efficiency score in these objective functions. This type of measure belongs to the "Pareto–Koopmans measure" (Koopmans, 1951; Russell, 1985).

In addition to these *OE* measurement criteria, it is important to note that the proposed non-radial measures can avoid specifying the non-Archimedean small number (ε_n). The unique feature is important because none knows what it is in reality. See Chapter 7 for a discussion on another approach to avoid specifying such a small number even in the radial measures.

This chapter is concerned with descriptions on five non-radial models in terms of their *OE* assessments. Section 5.2 compares analytical differences between the Debreu–Farrell type of radial models and the Pareto–Koopmans type of non-radial models from the perspective of *OE* as well as characterization and classifications of DMUs. This type of comparison has been not sufficiently explored in previous studies on radial and non-radial comparison. The proposed comparison provides us with unique features of non-radial models and their implications in DEA. Section 5.3 discusses the "Russell measure" which provides us with a theoretical linkage between radial and non-radial measurements. An important feature of the Russell measure is that it is formulated by non-linear programming because it can unify input-oriented and desirable output-oriented measures. As a consequence of the non-linear programming structure, the Russell measure has a computational difficulty although it has indeed several desirable properties as an *OE* measure, all of which may not be found in radial models. Section 5.4 discusses "an additive model" as the second non-radial model, proposed by Charnes *et al.* (1985). Charnes *et al.* (1982, 1983) linked the additive model to "a multiplicative model," as the third non-radial measure, by changing a data set by a natural logarithm. This chapter considers that the two non-radial models (additive and multiplicative) are same because a difference can be found in only the data manipulation. Section 5.5 discusses "a range-adjusted measure (RAM)" as the fourth non-radial model. Cooper *et al.* (1999) proposed the RAM as an extension of the additive model (Cooper *et al.*, 2000, 2001). Section 5.6 investigates slack-adjusted radial measure (SARM) as an extension of RAM, serving as the fifth alternative. Although the DEA community did not pay serious attention to SARM, the model will serve as a methodological basis for DEA environmental assessment in Section II. Tone (2001) proposed another type of non-radial model, referred to as slack-based measure (SBM) as the sixth non-radial model. SBM is discussed in Section 5.7. Section 5.8 prepares methodological comparison among radial and non-radial models for the conventional use of DEA. Section 5.9 summarizes this chapter.

At the end of this section, we clearly acknowledge that there are many other types of model variations related to non-radial measurement. Most of them are originated from the radial and non-radial models examined in Chapter 4 and this chapter. Unfortunately, this chapter cannot cover all of them because the purpose of this book is the DEA environmental assessment of energy and sustainability. The approach is conceptually and analytically different from a conventional use of DEA as discussed in the chapters of Section II. However, our description in this chapter provides us with an analytical basis on how to extend DEA to the proposed energy and environmental assessment.

5.2 CHARACTERIZATION AND CLASSIFICATION ON DMUs

This chapter characterizes analytical differences between the Debreu–Farrell (for radial) and Pareto–Koopmans (for non-radial) types of efficiency measures from DEA solutions and then classifies all DMUs based upon the characterization. A description on such distinctions is also discussed from economic and analytical linkages with the concept of Pareto optimality because it provides DEA with a theoretical basis for relative efficiency measurements and algorithm developments for non-radial measurement. See Chapter 10, which fully utilizes the characterization and classification of DMUs for algorithmic development on network computing.

In this chapter, we interpret the concept of Pareto optimality as follows: "*Pareto Optimality*: There is no other DMU which outperforms a specific DMU in terms of all components of production factors without changing the structure of an efficiency frontier."

According to today's global economy (https://www.britannica.com/topic/Pareto-optimality), the original definition of Pareto optimality is different from the above. This chapter interprets the optimality concept from a perspective of DEA performance assessment. An important contribution of Pareto optimality in this chapter is that it provides DEA with an underlying definition of relative efficiency. Koopmans (1951) used the concept to connect between inputs and desirable outputs and apply it for measuring productivity and other efficiencies. As a consequence, the efficiency measures based upon Pareto optimality are referred to as "Pareto–Koopmans" measures (or non-radial measures) in this chapter.

To prepare an analytical linkage between the concept of Pareto optimality and the status of *OE*, this chapter starts discussing the DMU characterization and classification first proposed by Charnes *et al.* (1986, 1991). As mentioned above, the classification has been extended into the development of special computer algorithms which can solve various radial and non-radial DEA models. See Chapter 10.

An important condition derived from the Pareto optimality is a concept of "dominance" that is often utilized for DEA algorithmic developments. The concept is specified as follows:

Dominance: The DMU classification (e.g., Sueyoshi, 1990; Sueyoshi and Chang, 1989) partitions a whole set (J) of all DMUs into two subsets: $J = J_d \cup J_n$.

One subset (J_d) is a dominated set and the other (J_n) is a non-dominated set. To specify the two groups, a concept of dominance is mathematically specified by:

$$\left[x_{1j},\ldots,x_{mj},-g_{1j},\ldots,-g_{sj}\right]^{Tr} \geq \left[x_{1j'},\ldots,x_{mj'},-g_{1j'},\ldots,-g_{sj'}\right]^{Tr}. \quad (5.1)$$

Here, "≥" implies that at least one component of the two vectors of production factors has the relationship ">." The production factors are expressed by input and desirable output vectors as specified by the two parts of Equation (5.1), respectively. Using Equation (5.1), the two subsets become as follows:

$$J_d = \left\{ j \in J \,\middle|\, \text{there is a DMU } j' \text{ that satisfies Equation } (5.1) \right\} \text{ and } J_n = J - J_d. \quad (5.2)$$

The concept of dominance is closely related to the Pareto optimality because DMUs, which are on the Pareto optimality, comprise the members of J_n. The opposite is not true. As a result, it is important for DEA to eliminate J_d from an efficiency frontier at the early stage of DEA computation. An algorithm for DEA needs to pay attention to only DMUs in J_n before the computational process of the whole DMUs.

In the case of Pareto–Koopmans (non-radial) measures (e.g., an additive model and RAM), the two subsets are further separated in the following manner (e.g., Sueyoshi and Chang, 1989):

$$J_n = E \cup E' \cup IE' \text{ and } J_d = IE, \quad (5.3)$$

where the four subsets are defined by the following specifications:

$$E = \left\{ k \in J_n \middle| \lambda_k^* = 1, \lambda_j^* = 0 \left(\forall j \neq k \in J_n \right) \text{and all slacks are zero} \right\},$$

$$E' = \left\{ k \in J_n \middle| \lambda_k^* < 1, \lambda_j^* > 0 \left(\exists j \in E \right) \text{and all slacks are zero and multiple solutions} \right\},$$

$$IE' = \left\{ k \in J_n \middle| \lambda_k^* = 0, \lambda_j^* > 0 \left(\exists j \in E \right) \text{and at least one slack is positive} \right\} \text{ and}$$

$$IE = \left\{ k \in J_d \middle| \lambda_k^* = 0, \lambda_j^* > 0 \left(\exists j \in E \right) \text{and at least one slack is positive} \right\}.$$

The symbols "∀" and "∃" means "all" and "some," respectively. In the above classification, efficient DMUs ($E \cup E'$) attain the status of Pareto optimality, but inefficient DMUs ($IE' \in J_n$: not clearly inefficient and $IE \in J_d$: clearly inefficient) do not belong to the optimality status.

Meanwhile, the Debreu–Farrell (radial) measures such as RM(v) and RM(c) may separate the whole data set into the following six subsets (e.g., Sueyoshi, 1992):

$$J_n = E \cup E' \cup IE' \cup IF' \quad \text{and} \quad J_d = IE \cup IF, \quad (5.4)$$

all of which may be redefined by the following six subsets:

$$E = \left\{ k \in J_n \mid \theta_k^* = 1, \lambda_k^* = 1, \lambda_j^* = 0 \left(\forall j \neq k \in J_n \right), \text{ all zero slacks and unique solution} \right\},$$

$$E' = \left\{k \in J_n \middle| \theta_k^* = 1, \lambda_k^* \le 1, \lambda_j^* > 0\left(\exists j \in J_n\right), \text{all zero slacks and multiple solutions}\right\},$$

$$IF' = \left\{k \in J_n \mid \theta_k^* = 1, \text{at least one positive slack and multiple solutions}\right\},$$

$$IE' = \left\{k \in J_n \mid \theta_k^* < 1, \lambda_k^* = 0 \;\&\; \lambda_j^* > 0\left(\exists j \in \mathrm{E}\right)\right\},$$

$$IF = \left\{k \in J_d \mid \theta_k^* = 1 \text{and at least one positive slack}\right\} \text{and}$$

$$IE = \left\{k \in J_d \mid \theta_k^* < 1, \lambda_k^* = 0 \;\&\; \lambda_j^* > 0\left(\exists j \in \mathrm{E}\right)\right\}.$$

The proposed two groups of DMU classifications, that is, Equations (5.3) and (5.4), indicate five important concerns to be discussed here. First, multiple solutions occur on DMUs belonging to E'. Second, the Debreu–Farrell (radial) measures have IF and IF', while the Pareto–Koopmans (non-radial) measures do not have such two subsets because the Pareto optimality excludes them from an efficiency frontier. Such a difference indicates that the computational process for Pareto–Koopmans measures is more tractable and faster than that of Debreu–Farrell measures because the former group consists of only four DMU classifications. Third, the concept of Pareto optimality makes it possible that we measure the level of OE by examining a total amount of slacks on optimality. Fourth, both the Debreu–Farrell (radial) and Pareto–Koopmans (non-radial) measures have inefficient DMU group (IE') whose status is not clearly identified by the concept of dominance (non-dominated but not Pareto optimal). Finally, it is easily imagined that we can effectively solve the whole DMU set by identifying efficient DMUs and eliminating inefficient DMUs at the early stage of DEA computation. The algorithm development, incorporating the Pareto optimality and the concept of dominance, will be discussed in Chapter 10.

5.3 RUSSELL MEASURE

Maintaining the unique features of both radial and non-radial measures, the Russell measure determines a level of OE by aggregating the efficiency scores related to all production factors (Russell, 1985):

The Russell measure on the k-th DMU is measured by the following formulation:

$$\begin{aligned}
&\text{Minimize} \quad \frac{1}{m+s}\left(\sum_{i=1}^{m}\theta_i + \sum_{r=1}^{s}\frac{1}{\tau_r}\right) \\
&s.t. \quad -\sum_{j=1}^{n}x_{ij}\lambda_j + \theta_i x_{ik} \qquad \ge 0\left(i = 1,\ldots,m\right), \\
&\qquad \sum_{j=1}^{n}g_{rj}\lambda_j \qquad - \tau_r g_{rk} \ge 0\left(r = 1,\ldots,s\right), \\
&\qquad \lambda_j \ge 0\left(j = 1,\ldots,n\right), 1 \ge \theta_i \ge 0\left(i = 1,\ldots,m\right) \text{ and} \\
&\qquad \tau_r \ge 1\left(r = 1,\ldots,s\right),
\end{aligned} \tag{5.5}$$

where unknown variables (θ_i and τ_r) indicate the level of each *OE* measure related to the i-th input ($i=1,\ldots,m$) and the r-th desirable output ($r=1,\ldots,s$), respectively. It is possible to incorporate slacks in Model (5.5). This chapter follows the original description prepared by Russell (1985), so not incorporating slacks in Model (5.5).

An important feature of Model (5.5) is that it is formulated by non-linear programming, as found in the objective function of Model (5.5). Sueyoshi and Sekitani (2007b) have proposed a use of "second-order cone programming (SOCP)," which is a new optimization approach originated from an interior point method of linear and non-linear programming, both to compute directly the Russell measure and to obtain the dual formulation. The SOCP approach proposed by their study can provide an exact solution on the Russell measure. Since we do not use the Russell measure for DEA environmental assessment in Section II, this chapter does not describe it further, except noting that interesting readers may refer to the research effort in the following: Sueyoshi and Sekitani (2007b), Computational strategy for Russell measure in DEA: Second-order cone programming. *European Journal of Operational Research*, **180**, 459–471.

It can be easily imagined that an extension of Model (5.5) to DEA environmental assessment is a straightforward matter with a minor modification on SOCP.

To solve Model (5.5) by an approximation (short-cut) approach, Bardhan *et al.* (1996a, b) reformulated it as a linear programming problem. The Russell measure approximation may be expressed by the following input-oriented model on the k-th DMU:

$$\begin{aligned}
&\text{Mimimize} && \sum_{i=1}^{m}\theta_i \\
&s.t. && -\sum_{j=1}^{n}x_{ij}\lambda_j+\theta_i x_{ik}\ge 0 && (i=1,\ldots,m), \\
& && \sum_{j=1}^{n}g_{rj}\lambda_j \ge g_{rk} && (r=1,\ldots,s), \\
& && \lambda_j\ge 0 && (j=1,\ldots,n) \text{ and} \\
& && 1\ge\theta_i\ge 0 && (i=1,\ldots,m).
\end{aligned} \tag{5.6}$$

The original Russell measure is formulated under constant RTS because it does not contain the side constraint $\left(\sum_{j=1}^{n}\lambda_j=1\right)$. It is possible for us to incorporate the side constraint into Models (5.5) and (5.6), so reformulating it under variable RTS.

The Russell measure aggregates all efficiency scores (θ_i) related to all inputs ($i=1,\ldots,m$). Hence, Models (5.5) and (5.6) do not belong to the conventional radial measurement, rather expressing the magnitude of each individual efficiency score in so-called "Pareto improvement" on a specific production factor. In the implication, this chapter considers that the Russell measure exists between the radial measure (indicating the improvement of all production factors) and non-radial measure (indicating the improvement on each production factor). On the optimality of Model (5.6), where we can obtain θ_i^* ($i=1,\ldots,m$) and λ_j^* ($j=1,\ldots,n$), the input-based Russell measure ($Russell_{in}$) on the k-th DMU is determined by

$$Russell_{in}^* = \frac{1}{m}\sum_{i=1}^{m}\theta_i^*, \tag{5.7}$$

where θ_i^* is obtained from the optimality of Model (5.6). Equation (5.7) measures an average of the aggregated input-oriented efficiency scores.

Shifting from the input-oriented measurement, an approximation approach for the Russell measure becomes the following desirable output-oriented model on the k-th DMU:

$$\begin{aligned}
&\text{Maximize} && \sum_{r=1}^{s}\tau_r \\
&s.t. && \sum_{j=1}^{n}x_{ij}\lambda_j \le x_{ik} && (i=1,\ldots,m),\\
& && -\sum_{j=1}^{n}g_{rj}\lambda_j + \tau_r g_{rk} \le 0 && (r-1,\ldots,s),\\
& && \lambda_j \ge 0 && (j=1,\ldots,n) \text{ and}\\
& && \tau_r \ge 1 && (r=1,\ldots,s).
\end{aligned} \tag{5.8}$$

Model (5.8) is formulated by maximizing the total sum of efficiency scores (τ_r) on all desirable outputs. On the optimality of Model (5.8), where we can obtain τ_r^*($r=1,\ldots,s$) and λ_j^* ($j=1,\ldots,n$), the desirable output-based Russell measure ($Russell_{out}$) concerning the k-th DMU is determined by

$$Russell_{out}^* = \frac{1}{s}\sum_{r=1}^{s}\frac{1}{\tau_r^*}. \tag{5.9}$$

An important feature of the proposed Russell measure is that it aggregates the desirable output-oriented measures and Equation (5.9) measures an average of these efficiency scores.

According to Bardhan *et al.* (1996a,b), the combined formulation between inputs and desirable outputs becomes as follows:

$$
\begin{aligned}
\text{Mimimize} \quad & \sum_{i=1}^{m}\theta_i - \sum_{r=1}^{s}\tau_r \\
s.t. \quad & -\sum_{j=1}^{n}x_{ij}\lambda_j + \theta_i x_{ik} \geq 0 \ (i=1,...,m), \\
& \sum_{j=1}^{n}g_{rj}\lambda_j \quad -\tau_r g_{rk} \geq 0 \ (r=1,...,s), \\
& \lambda_j \geq 0 \quad (j=1,...,n), \\
& 1 \geq \theta_i \geq 0 \quad (i=1,...,m) \text{ and} \\
& \tau_r \geq 1 \quad (r=1,...,s).
\end{aligned}
\tag{5.10}
$$

Model (5.10) has positive on the first term and negative on the second term of the objective function because the inputs are minimized while the desirable outputs are maximized in the Russell measurement. The negative sign in minimization implies maximization in the objective function. Thus, the part $\left(\text{Mimimize} - \sum_{r=1}^{s}\tau_r\right)$ in the objective function of Model (5.10) substitutes for the part $\left(\text{Minimize}\sum_{r=1}^{s}\frac{1}{\tau_r}\right)$ of Model (5.5).

After obtaining θ_i^* $(i=1,\ldots,m)$ and τ_r^* $(r=1,\ldots,s)$ on the optimality of Model (5.10), this chapter can determine the Russell measure ($Russell_{both}$) for inputs and desirable outputs as follows:

$$
Russell_{both}^* = \frac{1}{m+s}\left(\sum_{i=1}^{m}\theta_i^* + \sum_{r=1}^{s}\frac{1}{\tau_r^*}\right). \tag{5.11}
$$

On the optimality of Equation (5.11), the *k*-th DMU satisfies the following conditions on an efficiency frontier:

$$
\sum_{j=1}^{n}x_{ij}\lambda_j^* = \theta_i^* x_{ik}\left(i=1,\ldots,m\right) \quad \text{and} \quad \sum_{j=1}^{n}g_{rj}\lambda_j^* = \tau_r^* g_{rk}\left(r=1,\ldots,s\right). \tag{5.12}
$$

Equations (5.12) implies that all the slacks are zero. The Russell measure determines a level of improvement on each production factor and obtains an average regarding those of all production factors. Thus, the Russell measure contains such a non-radial property, so that we can consider as an intermediate model between radial and non-radial measurements.

5.4 ADDITIVE MODEL

DEA researchers have long paid attention to the additive model (AM) as an alternative to the two radial models, because the non-radial model aggregates input-oriented and desirable output-oriented measures to produce a single non-radial measure for *OE*. The AM is mathematically specified as follows:

$$\begin{aligned}
&\text{Maximize} \sum_{i=1}^{m} d_i^x + \sum_{r=1}^{s} d_r^g \\
&s.t \quad \sum_{j=1}^{n} x_{ij}\lambda_j + d_i^x \quad = x_{ik} \; (i=1,\ldots,m), \\
&\quad\quad \sum_{j=1}^{n} g_{rj}\lambda_j \quad - d_r^g = g_{rk} \; (r=1,\ldots,s), \\
&\quad\quad \sum_{j=1}^{n} \lambda_j \quad\quad\quad = 1, \\
&\quad\quad \lambda_j \geq 0 \; (j=1,\ldots,n), \\
&\quad\quad d_i^x \geq 0 \left(i=1,\ldots,m\right) \text{ and } d_r^g \geq 0 \left(r=1,\ldots,s\right).
\end{aligned} \tag{5.13}$$

The objective value indicates the level of inefficiency measured by an amount of total slacks related to all inputs and desirable outputs.

The level of *OE*, or AM efficiency, regarding the *k*-th DMU is determined by

$$AM^* = 1 - \left(\sum_{i=1}^{m} d_i^{x*} + \sum_{r=1}^{s} d_r^{g*} \right), \tag{5.14}$$

where the slack values d_i^{x*} and d_r^{g*} are determined on the optimality of Model (5.13).

In addition to Model (5.13), it is possible for us to change the objective function by adding weights (w_i^x and w_r^g) to slacks in the objective function of Model (5.13) in order to control the magnitude of inefficiency. The weights are also used to express the importance among inputs and desirable outputs. In the case, the objective function of Model (5.13) becomes as follows:

$$\sum_{i=1}^{m} w_i^x d_i^x + \sum_{r=1}^{s} w_r^g d_r^g. \tag{5.15}$$

A practical difficulty of the AM with the objective function, expressed by Equation (5.15), is that none knows how to determine such weights in the objective function. Cooper *et al.* (1999) proposed measuring the *OE* of the *k*-th DMU after changing $w_i^x = \left[(m+s)x_{ik}\right]^{-1}$ and $w_r^g = \left[(m+s)g_{rk}\right]^{-1}$ in Equation (5.15). This type of adjustment is later referred to as a range-adjusted measure (RAM).

The adjustment method is one of many alternatives for weight selection in Equation (5.15). The level of *OE* of the *k*-th DMU is determined by

$$AM^* = 1 - \left(\sum_{i=1}^{m} w_i^x d_i^{x*} + \sum_{r=1}^{s} w_r^g d_r^{g*} \right) \tag{5.16}$$

where the slack values d_i^{x*} and d_r^{g*} are obtained from the optimality of Model (5.13) which is adjusted by Equation (5.15) with $w_i^x = \left[(m+s)x_{ik}\right]^{-1}$ and $w_r^g = \left[(m+s)g_{rk}\right]^{-1}$. Each slack is divided by the corresponding observed input or desirable output and the total number of inputs and desirable outputs. Consequently, the second part of the right hand side of Equation (5.16) indicates a level of total inefficiency. The AM efficiency is determined by subtracting the total inefficiency from unity.

The efficiency status of AM is classified into the following two cases: (a) AM efficiency ↔ all slacks are zero and (b) AM inefficiency ↔ at least one slack is positive on optimality.

Multiplicative Model: If a data set on inputs and desirable outputs is expressed by natural logarithm, the additive model is referred to as "a multiplicative model" that corresponds to a generalization of Cobb-Douglas type of production functions (Cooper *et al.*, 1982, 1983; Sueyoshi and Chang, 1989). Mathematically, the model changes a data set (x_{ij}, g_{rj}) to $(\hat{x}_{ij}, \hat{g}_{rj})$, where $\hat{x}_{ij} = \ln(x_{ij})$ and $\hat{g}_{rj} = \ln(g_{rj})$ for all *i* and all *r* of the *j*-th DMU ($j = 1,\ldots,n$). Thus, the inputs and desirable outputs have the Cobb–Douglas type of relationships such as $x_{ik} = \prod_{j=1}^{n} \hat{x}_{ij}^{\lambda j}\left(e^{d_i^x}\right)$ for $i=1,\ldots,m$ and $g_{rk} = \prod_{j=1}^{n} \hat{g}_{rj}^{\lambda j}\left(e^{-d_r^g}\right)$ for $r=1,\ldots,s$. Model (5.13) for the multiplicative model becomes as follows:

$$\begin{aligned}
&\text{Maximize} && \sum_{i=1}^{m} d_i^x + \sum_{r=1}^{s} d_r^g \\
&s.t && \sum_{j=1}^{n} \hat{x}_{ij}\lambda_j + d_i^x = \hat{x}_{ik} \;\; (i=1,\ldots,m), \\
& && \sum_{j=1}^{n} \hat{g}_{rj}\lambda_j - d_r^g = \hat{g}_{rk} \;\; (r=1,\ldots,s), \\
& && \sum_{j=1}^{n} \lambda_j = 1, \\
& && \lambda_j \geq 0 \; (j=1,\ldots,n), \\
& && d_i^x \geq 0 \; (i=1,\ldots,m) \text{ and } d_r^g \geq 0 \; (r=1,\ldots,s).
\end{aligned} \tag{5.17}$$

Here, it is important to note three concerns on Model (5.17). First, the multiplicative model is the same as the additive model in (5.13) except the data transformation by natural logarithm. Therefore, this chapter considers that there is no difference between the additive model and the multiplicative model. Second, the multiplicative model requires that all inputs and desirable outputs are strictly positive because of the data logarithmic transformation. The property on data is required on not only the multiplicative model but also all DEA models. See Chapters 26 and 27 for a description on how to handle zero and negative values in a data set. Finally, Cooper *et al.* (2006, p. 114) documented the multiplicative model under constant RTS by dropping $\sum_{j=1}^{n} \lambda_j = 1$ from Model (5.17). The model adds the side constraint under variable RTS.

5.5 RANGE-ADJUSTED MEASURE

As discussed in Section 5.4, Cooper *et al.* (1999) proposed the following RAM model that is an extended structure by specifying the weights for slacks regrading all production factors:

$$\begin{aligned}
\text{Maximize} \quad & \sum_{i=1}^{m} R_i^x d_i^x + \sum_{r=1}^{s} R_r^g d_r^g \\
s.t. \quad & \sum_{j=1}^{n} x_{ij}\lambda_j + d_i^x = x_{ik} \; (i=1,\ldots,m), \\
& \sum_{j=1}^{n} g_{rj}\lambda_j - d_r^g = g_{rk} \; (r=1,\ldots,s), \\
& \sum_{j=1}^{n} \lambda_j = 1, \\
& \lambda_j \ge 0 \; (j=1,\ldots,n), \\
& d_i^x \ge 0 \; (i=1,\ldots,m) \text{ and} \\
& d_r^g \ge 0 \; (r=1,\ldots,s),
\end{aligned} \tag{5.18}$$

where the ranges for inputs and desirable outputs are specified as follows:

(a) $R_i^x = (m+s)^{-1}\left(\max\left\{x_{ij} \middle| j=1,\ldots,n\right\} - \min\left\{x_{ij} \middle| j=1,\ldots,n\right\}\right)^{-1}$, indicating a data range related to the i-th input ($i=1,\ldots,m$) and

(b) $R_r^g = (m+s)^{-1}\left(\max\left\{g_{rj} \middle| j=1,\ldots,n\right\} - \min\left\{g_{rj} \middle| j=1,\ldots,n\right\}\right)^{-1}$, indicating a data range related to the r-th desirable output ($r=1,\ldots,s$).

To determine the level of operational efficiency, this chapter obtains slack values (d_i^{x*} and d_r^{g*}) from the optimal solution of Model (5.18). Then, the level of *OE*, or RAM efficiency, is measured by

$$RAM^* = 1 - \left(\sum_{i=1}^{m} R_i^x d_i^{x*} + \sum_{r=1}^{s} R_r^g d_r^{g*}\right). \tag{5.19}$$

The sum of range-weighted slacks indicates the level of inefficiency. The efficiency is measured by subtracting the level of inefficiency from unity. Consequently, if all slacks are zero on optimality, then the k-th DMU is fully efficient. In contrast, if some slack(s) is positive, then the DMU contains some level of inefficiency.

The dual formulation of Model (5.18) is specified as follows:

$$\begin{aligned} \text{Minimize} \quad & \sum_{i=1}^{m} v_i x_{ik} - \sum_{r=1}^{s} u_r g_{rk} + \sigma \\ s.t. \quad & \sum_{i=1}^{m} v_i x_{ij} - \sum_{r=1}^{s} u_r g_{rj} + \sigma \ \geq \ 0 \quad (j = 1,\ldots,n), \\ & v_i \geq R_i^x \qquad\qquad (i = 1,\ldots,m), \\ & u_r \geq R_r^g \qquad\qquad (r = 1,\ldots,s) \text{ and} \\ & \sigma : \text{URS}, \end{aligned} \tag{5.20}$$

where v_i $(i = 1,\ldots, m)$ and u_r $(r = 1,\ldots, s)$ are all dual variables related to the first and second groups of constraints in Model (5.18). The dual variable (σ) is obtained from the third constraint of Model (5.18).

The optimality of Models (5.18) and (5.20) on the k-th DMU maintains the following relationship between their objective values:

$$\sum_{i=1}^{m} R_i^x d_i^{x*} + \sum_{r=1}^{s} R_r^g d_r^{g*} = \sum_{i=1}^{m} v_i^* x_{ik} - \sum_{r=1}^{s} u_r^* g_{rk} + \sigma^*. \tag{5.21}$$

An important feature of RAM is that all dual variables, except σ, are positive in their signs. Thus, all production factors are fully incorporated into the efficiency measurement of RAM. See Chapter 16 that extends Models (5.18) and (5.20) to DEA environmental assessment.

5.6 SLACK-ADJUSTED RADIAL MEASURE

Although RAM provides an important measure for *OE*, it has a practical difficulty in determining the level of efficiency. For example, Aida *et al.* (1998) reported that most inefficient DMUs belonged to a very small inefficiency range. Consequently, a variance of efficiency distribution was small, so being close to unity in many RAM applications, which was mathematically acceptable but unacceptable from a practical perspective of performance assessment. To overcome the practical difficulty, Aida *et al.* (1998) proposed a use of slack-adjusted radial measure (SARM) as follows:

$$\begin{aligned}
&\text{Minimize} \quad \theta - \left(\sum_{i=1}^{m} R_i^x d_i^x + \sum_{r=1}^{s} R_r^g d_r^g \right) \\
&s.t. \quad -\sum_{j=1}^{n} x_{ij}\lambda_j - d_i^x \quad + \theta x_{ik} = 0 \quad (i = 1,\dots,m), \\
&\qquad \sum_{j=1}^{n} g_{rj}\lambda_j \quad - d_r^g \quad = g_{rk} \; (r = 1,\dots,s), \\
&\qquad \sum_{j=1}^{n} \lambda_j \quad = 1, \\
&\qquad \theta : \text{URS}, \lambda_j \geq 0 \; (j = 1,\dots,n), \\
&\qquad d_i^x \geq 0 \; (i = 1,\dots,m) \text{ and } d_r^g \geq 0 \; (r = 1,\dots,s).
\end{aligned} \tag{5.22}$$

From the optimal solution of Model (5.22), this chapter measures the *OE* measure, or SARM efficiency, of the *k*-th DMU as follows:

$$SARM^* = \theta^* - \left(\sum_{i=1}^{m} R_i^x d_i^{x*} + \sum_{r=1}^{s} R_r^g d_r^{g*} \right) \tag{5.23}$$

The SARM measure indicates the level of input-oriented *OE*. The second term of Equation (5.23) can be considered as a slack-based adjustment which is due to the level of inefficiency measured by slacks. If the efficiency measure is unity, then the DMU is fully efficient. Since Model (5.22) is a minimization problem, the second term has a negative sign in the objective function of Model (5.22).

Cooper *et al.* (2001) replaced the objective of Model (5.22) by

$$\theta - \varepsilon_n \left(\sum_{i=1}^{m} R_i^x d_i^x + \sum_{r=1}^{s} R_r^g d_r^g \right), \tag{5.24}$$

where ε_n is a non-Archimedean small number that is smaller than any positive number. Section II of this book will use ε_s for a different implication, so being different from the non-Archimedean small number as proposed by Cooper *et al.* (2001). The small number is used to control the degree of slacks in the determination of *OE*. See Chapter 17 on using the small number.

The dual formulation of Model (5.22) becomes as follows:

$$\begin{aligned}
&\text{Maximize} \quad \sum_{r=1}^{s} u_r g_{rk} + \sigma \\
&s.t. \quad -\sum_{i=1}^{m} v_i x_{ij} + \sum_{r=1}^{s} u_r g_{rj} + \sigma \leq 0 \qquad (j = 1,\dots,n), \\
&\qquad \sum_{i=1}^{m} v_i x_{ik} \quad = 1, \\
&\qquad v_i \geq R_i^x \; (i = 1,\dots,m), u_r \geq R_r^g \; (r = 1,\dots,s) \text{ and} \\
&\qquad \sigma : \text{URS}.
\end{aligned} \tag{5.25}$$

Here, v_i $(i=1,\ldots,m)$ and u_r $(r=1,\ldots,s)$ are all dual variables related to the first and second groups of constraints in Model (5.22). The dual variable (σ) is obtained from the third constraint of Model (5.22).

An important feature of Model (5.25) is that all dual variables are strictly positive so that information on all production factors is fully utilized in the SARM performance assessment. The analytical feature is very different from a conventional use of DEA where many dual variables may become zero, so being not appropriate for performance assessment. The SARM does not have such a methodological problem. See Chapter 17 in which DEA environmental assessment will fully utilize the methodological strength of SARM.

The following relationship always satisfies on the optimality of Models (5.22) and (5.25):

$$\theta^* - \left(\sum_{i=1}^{m} R_i^x d_i^{x*} + \sum_{r=1}^{s} R_r^g d_r^{g*}\right) = \sum_{r=1}^{s} u_r^* g_{rk} + \sigma^*. \tag{5.26}$$

Moreover, the first and second groups of constraints of Model (5.25) indicate

$$\sum_{r=1}^{s} u_r g_{rk} + \sigma \leq \sum_{i=1}^{m} v_i x_{ik} = 1 \tag{5.27}$$

on the k-th DMU. Thus, the SARM measure on the k-th DMU is always less than or equal unity. Thus, the SARM satisfies the property of efficiency requirement. See Chapter 6 on this property.

5.7 SLACK-BASED MEASURE

Using an additive model and RAM, it is possible for us to measure the sum of weighted slacks in the objective function of these non-radial models, so indicating the level of total inefficiency. The efficiency level is measured by subtracting it from unity. In a similar vein, Tone (2001) proposed the slack-based measure (SBM) by paying attention to the amount of slacks, which has the following structure:

$$\left(1 - \frac{1}{m}\sum_{i=1}^{m} \frac{d_i^x}{x_{ik}}\right) \Bigg/ \left(1 + \frac{1}{s}\sum_{r=1}^{s} \frac{d_r^g}{g_{rk}}\right). \tag{5.28}$$

In Equation (5.28), the numerator indicates a level of input-oriented efficiency that is measured by subtracting the sum of weighted input slacks from unity. The denominator indicates a level of desirable output-oriented efficiency that is measured by adding the sum of weighted desirable output-oriented slacks to unity. If all slacks are zero, then Equation (5.28) indicates the status of efficiency.

A unique feature of Equation (5.28) is that the level of input-based efficiency is divided by that of desirable output-based efficiency. Acknowledging the effort on paying attention to the development of slack-based efficiency measurement, this chapter needs to mention that the ratio form of the input-based slacks to the desirable output-based slacks may have a limited practical implication for performance assessment at a level of the other radial and non-radial models, in particular the two radial models.

Considering Equation (5.28), Tone (2001) formulated the following SBM model that has a non-linear programming structure:

$$\begin{aligned}
&\text{Minimize} && \left(1-\frac{1}{m}\sum_{i=1}^{m}\frac{d_i^x}{x_{ik}}\right)\Bigg/\left(1+\frac{1}{s}\sum_{r=1}^{s}\frac{d_r^g}{g_{rk}}\right) \\
&s.t. && \sum_{j=1}^{n}x_{ij}\lambda_j+d_i^x = x_{ik}\ (i=1,\ldots,m), \\
& && \sum_{j=1}^{n}g_{rj}\lambda_j - d_r^g = g_{rk}\ (r=1,\ldots,s), \\
& && \lambda_j \ge 0\,(j=1,\ldots,n), \\
& && d_i^x \ge 0\,(i=1,\ldots,m)\text{ and} \\
& && d_r^g \ge 0\,(r=1,\ldots,s).
\end{aligned} \tag{5.29}$$

Model (5.29) is formulated by non-linear programming so that we need to change it from the ratio form to the corresponding linear programming formulation. For the reformulation, Tone (2001) introduced a scalar (ϕ) to obtain the following restructured formulation:

$$\begin{aligned}
&\text{Minimize} && \phi-\frac{1}{m}\sum_{i=1}^{m}\frac{d_i^x}{x_{ik}} \\
&s.t. && \phi+\frac{1}{s}\sum_{r=1}^{s}\frac{d_r^g}{g_{rk}} = 1, \\
& && \sum_{j=1}^{n}x_{ij}\lambda_j+d_i^x = x_{ik}\ (i=1,\ldots,m), \\
& && \sum_{j=1}^{n}g_{rj}\lambda_j - d_r^g = g_{rk}\ (r=1,\ldots,s), \\
& && \lambda_j \ge 0\,(j=1,\ldots,n), \\
& && d_i^x \ge 0\,(i=1,\ldots,m)\text{ and} \\
& && d_r^g \ge 0\,(r=1,\ldots,s).
\end{aligned} \tag{5.30}$$

The reformulation is due to a structural change from fractional programming to linear programming. See Chapter 2 on the structural change.

Let $\zeta_i^x = \phi d_i^x$, $\zeta_r^g = \phi d_r^g$ and $\eta_j = \phi\lambda_j$, then Model (5.30) becomes the following linear programming equivalent formulation:

$$
\begin{aligned}
&\text{Minimize } \phi - \frac{1}{m}\sum_{i=1}^{m}\frac{\zeta_i^x}{x_{ik}} \\
&s.t. \quad \phi + \frac{1}{s}\sum_{r=1}^{s}\frac{\zeta_r^g}{g_{rk}} = 1, \\
&\sum_{j=1}^{n} x_{ij}\eta_j + \zeta_i^x = \phi x_{ik} \ (i = 1,\ldots,m), \\
&\sum_{j=1}^{n} g_{rj}\eta_j - \zeta_r^g = \phi g_{rk} \ (r = 1,\ldots,s), \\
&\eta_j \geq 0 \, (j = 1,\ldots,n), \\
&\zeta_i^x \geq 0 \, (i = 1,\ldots,m), \\
&\zeta_r^g \geq 0 \, (r = 1,\ldots,s) \text{ and} \\
&\phi \geq 0.
\end{aligned}
\tag{5.31}
$$

The optimal solution of Model (5.31) determines the degree of SBM as follows:

$$SBM^* = \phi^* - \frac{1}{m}\sum_{i=1}^{m}\frac{\zeta_i^{x*}}{x_{ik}}.$$

The other components of the optimal solution are determined by $d_i^{x*} = \zeta_i^{x*}/\phi^*$, $d_r^{g*} = \zeta_r^{g*}/\phi^*$ and $\lambda_j^* = \eta_j^*/\phi^*$. If SBM^* is unity, then it indicates the status of *OE* because $d_i^{x*} = 0$ for all *i* and $d_r^{g*} = 0$ for all *r* in Model (5.29).

Enhanced Russell Graph Measure: To approximate the Russell measure, the research of Paster *et al.* (1999) proposed an "enhanced Russell graph measure (ERGM)" that is formulated by the following non-linear programming:

$$
\begin{aligned}
&\text{Mimimize } \frac{1}{m}\sum_{i=1}^{m}\theta_i \Big/ \frac{1}{s}\sum_{r=1}^{s}\tau_r \\
&s.t. \quad -\sum_{j=1}^{n} x_{ij}\lambda_j + \theta_i x_{ik} \geq 0 \ (i = 1,\ldots,m), \\
&\sum_{j=1}^{n} g_{rj}\lambda_j - \tau_r g_{rk} \geq 0 \ (r = 1,\ldots,s), \\
&\lambda_j \geq 0 \, (j = 1,\ldots,n), \theta_i \geq 0 \, (i = 1,\ldots,m) \text{ and} \\
&\tau_r \geq 1 \, (r = 1,\ldots,s).
\end{aligned}
\tag{5.32}
$$

The efficiency measure obtained by ERGM attains full efficiency if the objective value is unity on optimality. The result is identified when there is no positive slack. See Cooper *et al.* (2006) for a detailed discussion on ERGM. The research effort (Cooper *et al.*, 2006, p. 103, Theorem 4.11) has proved that ERGM is equivalent to SBM and vice versa.

Here, it is important to note the four concerns on SBM and ERGM. First, Cooper *et al.* (2006, p. 103) and Cooper *et al.* (2007a, p. 3), have clearly mentioned the equivalence between SBM and ERGM. Cooper *et al.* (2007a, p. 3 and p. 14) has also noted in the footnote that "SBM is a closely related measure of ERGM." Second, as discussed by Sueyoshi and Sekitani (2007b), ERGM (=SBM) may approximate the Russell measure to reduce the computational burden because they are formulated by linear programming. However, the Russell measure and ERGM are not exactly same in terms of the *OE* measurement. See Chapter 6. Third, only SOCP can exactly solve the Russell measure. Finally, a straightforward use of the three measures (Russell measure, SBM and ERGM) has a difficulty in DEA environmental assessment because they must incorporate undesirable outputs. Therefore, this chapter has introduced them within a conventional framework of DEA.

5.8 METHODOLOGICAL COMPARISON: AN ILLUSTRATIVE EXAMPLE

Table 5.1 lists an illustrative example that consists of 20 DMUs. The performance of each DMU is measured by three inputs and two desirable outputs. The three

TABLE 5.1 An illustrative example

		Inputs			Desirable Outputs	
No.	DMU	Capital (10 Billion Yen)	Number of Branches	Number of Employees	Net Profit (Million Yen)	Deposit (10 Billion Yen)
1	A	859	371	15,788	218,938	28,910
2	B	1043	436	14,930	159,932	29,804
3	C	1040	327	13,567	223,340	27,405
4	D	786	374	17,412	218,989	39,653
5	E	605	369	12,148	88,091	20,146
6	F	843	338	13,020	175,483	28,254
7	G	753	353	14,394	176,477	27,388
8	H	465	193	7315	37,611	9998
9	I	723	281	10,750	118,963	18,546
10	J	71	135	2584	12,765	3286
11	K	49	173	3714	20,308	4753
12	L	132	189	4073	17,666	4986
13	M	107	163	4569	29,830	6610
14	N	185	186	5323	51,154	8648
15	O	121	191	3976	10,194	5289
16	P	91	189	4509	42,982	6578
17	Q	27	115	2862	8633	3749
18	R	52	222	3832	7606	4917
19	S	59	177	4261	9733	5585
20	T	51	194	3492	5765	3763

inputs are: amount of capital, number of branch offices and number of employees. The two desirable outputs are: amount of net profit and amount of deposit.

Table 5.2 summarizes efficiency scores measured by seven different radial and non-radial approaches. They include two radial measures such as input-oriented RM(v) and RM(c) as well as five non-radial models such as Russell measure, additive model, RAM, SARM and SBM. Table 5.2 has methodological implications, all of which are summarized by the following five findings: First, the four DMUs {C, D, P, Q} are evaluated as operationally efficient by the seven approaches. The other 16 DMUs contain some level of inefficiency, depending upon the type of DEA approach. Second, there is a different efficiency distribution between the two radial models because they are structured by different assumptions (constant or variable) on RTS. Third, the status of efficiency or

TABLE 5.2 Comparison among radial and non-radial approaches

		Efficiency Score						
No.	DMU	RM(c)	RM(v)	Russell	Additive	RAM	SARM	SBM
1	A	1.000	1.000	1.000	1.000	1.000	0.965	1.000
2	B	0.877	0.897	0.812	0.849	0.804	0.771	0.686
3	C	1.000	1.000	1.000	1.000	1.000	1.000	1.000
4	D	1.000	1.000	1.000	1.000	1.000	1.000	1.000
5	E	0.728	0.779	0.784	0.746	0.819	0.713	0.561
6	F	0.985	1.000	0.928	1.000	1.000	0.957	0.877
7	G	0.908	0.910	0.905	0.941	0.931	0.903	0.821
8	H	0.600	0.829	0.707	0.643	0.886	0.769	0.409
9	I	0.790	0.852	0.811	0.897	0.897	0.829	0.677
10	J	0.707	1.000	0.684	1.000	1.000	1.000	0.469
11	K	0.998	1.000	0.949	1.000	1.000	0.985	0.915
12	L	0.634	0.805	0.748	0.703	0.937	0.796	0.400
13	M	0.871	0.902	0.889	0.923	0.978	0.902	0.765
14	N	0.849	0.961	0.832	0.977	0.987	0.961	0.717
15	O	0.708	0.858	0.735	0.444	0.934	0.834	0.295
16	P	1.000	1.000	1.000	1.000	1.000	1.000	1.000
17	Q	1.000	1.000	1.000	1.000	1.000	1.000	1.000
18	R	0.902	0.940	0.787	0.614	0.925	0.896	0.444
19	S	0.917	1.000	0.855	1.000	1.000	1.000	0.542
20	T	0.743	0.814	0.708	0.344	0.935	0.787	0.298

(a) RM(c) is solved by Model (4.7), RM(v) is solved by Model (4.1), Russell Measure is solved by Model (5.10), Additive model is solved by Model (5.13) with Equation (5.16), RAM is solved by Model (5.18) with Equation (5.19), SARM is solved by Model (5.22) with Equation (5.23) and SBM is solved by Model (5.31). (b) The illustrative example have been solved without any data adjustment (i.e., divided by each factor average). (c) The additive model does not satisfy the property of "efficiency requirement". That is, its *OE* measure should be between zero and one, where "zero" implies full inefficiency and "one" implies full efficiency. See Chapter 6 on a detail description on the property.

inefficiency has similarity between the Russell measure and SBM. This indicates that the two non-radial models have a close relationship, compared with the other DEA models. The efficiency distribution measured by SBM is larger than that of the Russell measure. Thus, SBM is more sensitive to the existence of slacks than the Russell measure. Fourth, there is no difference in the efficiency/inefficiency status between the additive model and RAM. A difference can be found in the magnitude of their efficiency scores because the slack of RAM has weights in its objective function of Model (5.18) with Equation (5.19), but the additive model does not have any weight assignment on slacks in Model (5.13). The level of efficiency is determined by Equation (5.16).

Finally, the comparison between RAM and SARM indicates that RAM produces DMUs whose efficiency levels are close to unity. The inefficiency scores, measured by RAM, belong to a very small range. The SARM is developed for overcoming such a shortcoming of RAM. Thus, SARM produces a wider range in the inefficiency distribution than RAM.

5.9 SUMMARY

As methodological alternatives to the radial measures discussed in Chapter 4, this chapter described non-radial models, based upon the Pareto–Koopmans measurement. The non-radial measures do not incorporate an efficiency score in their formulations, rather they evaluate the *OE* of DMUs by examining the total amount of slacks. The non-radial models discussed in this chapter, including the Russell measure, an additive model, a multiplicative model, RAM, SARM and SBM. These non-radial models have unique features which cannot be found in the radial measures. It is better for us to measure the performance of DMUs by applying several different methodologies to avoid a methodological bias often occurring in many empirical studies. The non-radial measures discussed in this chapter may serve as such methodological alternatives to the radial measures.

6

DESIRABLE PROPERTIES

6.1 INTRODUCTION

This chapter[1] compares radial and non-radial models, discussed in Chapters 4 and 5, using nine desirable properties to measure the level of operational efficiency (*OE*). As discussed in those two chapters, RM(v) and RM(c) belong to the Debreu–Farrell measures (e.g., Farrell, 1957; Farrell and Fieldhouse, 1962) and non-radial models belong to the Pareto–Koopmans measures (e.g., Koopmans, 1951; Russell, 1985) in *OE*-based performance evaluation. As discussed in Chapter 5, Charnes *et al.* (1985) first proposed an additive model as a non-radial model. Then, Charnes *et al.* (1982, 1983) linked the additive model to a multiplicative model by changing a data set by natural logarithm. Cooper *et al.* (1999) also proposed a non-radial DEA model, referred to as "a range-adjusted measure (RAM)," that was an extension of the additive model (Cooper *et al.*, 2000, 2001). Aida *et al.* (1998) extended RAM by reorganizing it as a radial measure, so-called "slack-adjusted radial measure (SARM)." Meanwhile, Tone (2001) proposed another type of non-radial model referred to as the

[1]This chapter is partly based upon the article: Sueyoshi, T and Sekitani, K. (2009) An occurrence of multiple projections in DEA-based measurement of technical efficiency: theoretical comparison among DEA models from desirable properties. *European Journal of Operational Research*, **196**,764–794.

Environmental Assessment on Energy and Sustainability by Data Envelopment Analysis, First Edition. Toshiyuki Sueyoshi and Mika Goto.

"slacks-based measure (SBM)." In a similar manner, Pastor *et al.* (1999) proposed the "enhanced Russell graph measure (ERGM)," as a non-radial measure, that incorporated the analytical feature of the Russell measure into a framework of SBM. As discussed in Chapter 5, this chapter considers SBM = ERGM because both have the same analytical structure. In such a research trend for computational tractability, Bardhan *et al.* (1996b) first formulated an approximation model for the Russell measure by linear programming. The measure provided an analytical linkage between the radial and non-radial models so that it could eliminate a need for a separate treatment of input-oriented and output-oriented efficiency measures in the Russell measure[2].

Acknowledging the contributions of these previous studies as summarized in Chapters 4 and 5, we have been long wondering which model should be used for DEA-based performance analyses. Can we apply any DEA model to any performance evaluation? The inquiry indicates a research necessity to investigate nine criteria, based upon which we can determine the selection issue of an appropriate DEA model. Of course, such desirable criteria, discussed in this chapter, may not cover all necessities for a conventional use of DEA, but they are useful in examining the applicability of each model.

Here, this chapter clearly acknowledges the existence of other previous research efforts (e.g., Cooper *et al.*, 1999) that have investigated such desirable properties and have compared several DEA models. However, these previous comparisons covered part of important radial and non-radial models from a partial perspective of *OE*. Furthermore, these previous studies did not provide mathematical proofs on the desirable criteria in a unified manner, as found in this chapter. Therefore, following the previous contribution of Sueyoshi and Sekitani (2009), this chapter analytically investigates both what criteria (mathematical properties) are important for the *OE* measurement and how to select an appropriate DEA model under different *OE* criteria. All the criteria need to be mathematically proved by a unified analytical synthesis within a conventional framework of DEA, but not environmental assessment, which is a main research objective of this book. Such a new research extension will be found in chapters of Section II.

At the end of this section, it is necessary for us to mention that the nine desirable properties discussed here are mathematically important and analytically useful from the perspective of a conventional use of DEA. However, we clearly understand that each DEA model should have not only intuitive and visual capabilities to explain its mathematical structure but also fundamental rationales on

[2] For example, DEA can handle imprecise (fuzzy) data from the radial and non-radial models. Furthermore, DEA can control RTS by adding the upper and lower bounds on the sum of lambdas in these models. This study does not discuss these DEA models because the theoretical exploration discussed in this study can be applied to them. See, Cooper *et al.* (2007, pp. 4–5) that explained an origin of the Russell measure. See Chapters 22 and 23 on a description on how to use a new type of multiplier restriction method for setting upper and lower bounds to discuss DEA environmental assessment.

why they are useful as a methodology for performance analysis. For example, dual variables obtained from each DEA model should be positive in these signs. Such is a general rule of DEA. However, it has several exceptions (e.g., congestion measurements in Chapters 9, 21, 22 and 23) in extending DEA for performance analysis and environmental assessment. Furthermore, in the other case where some dual variables are zero in these signs, the result indicates that production factors, corresponding to the dual variables, are not fully utilized for DEA assessment. It may be true that such a concern does not produce any major difference if we are interested in only efficiency assessment. However, it can be easily imagined that the practicality of DEA is not limited to such efficiency measurement, rather having more wide applicability such as innovation identification on production technology and eco-technology to reduce an amount of various industrial pollutions. Thus, the performance of each radial or non-radial model should be discussed based on their applicability and practicality, although this chapter does not discuss all of them here, rather focusing on only their analytical structures and features from optimization. Thus, the purpose of this chapter is to provide us with an analytical basis on a conventional framework of DEA-based production analysis.

The remainder of this chapter is organized as follows: Section 6.2 summarizes desirable properties obtained from previous DEA studies. Section 6.3 reviews radial and non-radial models from the perspective of *OE* criteria. Section 6.4 summarizes previous research efforts based upon the criteria on desirable properties. Section 6.5 proposes a standard model for DEA that covers the analytical features of all radial and non-radial models. The standard model is used for mathematical proofs for examining whether each model satisfies or violates each criterion. Section 6.6 examines theoretically seven DEA models. Section 6.7 summarizes this chapter.

6.2 CRITERIA FOR OE

The *OE* measurement has the following desirable properties on radial and non-radial models within a conventional framework of DEA:

(a) *Homogeneity* (Banker *et al.*, 1984; Russell, 1985; Dmitruk and Koshevoy, 1991; Blackorby and Russell, 1999): An output-based *OE* measure should be homogeneous of degree one in desirable outputs, while an input-based *OE* measure should be homogeneous of degree minus one in the inputs. For example, if we double all the inputs, then the input-based *OE* measure should be cut in half.

(b) *Strict Monotonicity* (Banker *et al.*, 1984; Russell, 1985; Dmitruk and Koshevoy, 1991; Blackorby and Russell, 1999; Cooper *et al.*, 1999): An *OE* measure should be non-decreasing in desirable outputs and non-increasing in inputs, along with an efficient desirable output (input) vector.

(c) *Efficiency Requirement* (Russell, 1985; Banker *et al.*, 1984; Cooper *et al.*, 1999): An *OE* measure should be between zero and one, where "zero" implies full inefficiency and "one" implies full efficiency.

(d) *Unique Projection for Efficiency Comparison* (Russell, 1985; Banker *et al.*, 1984): An *OE* measure compares each observed input (or desirable output) vector with an efficient input (desirable output) one. This property[3] was carefully discussed by Russell (1985, p. 110). The uniqueness of vector comparison is so desirable that this chapter adds the uniqueness as a desirable property. For example, a radial *OE* measure minimizes an input amount that can be shrunk along a projection ray, while holding the desirable output quantities constant. Meanwhile, the non-radial *OE* measures do not have such a ray expansion, so that these measures minimize the total amount of slacks for determining a level of *OE*. Although they are different each other concerning how to adjust the observed amounts of production factors to their optimal amounts, both radial and non-radial measures may suffer from a difficulty of multiple projection vectors and slacks for efficiency comparison. See Chapter 7 for a discussion on how to handle this type of problem related to DEA.

(e) *Aggregation* (Pastor *et al.*, 1999; Blackorby and Russell, 1999; Cooper *et al.*, 2007a): A jointly measured *OE* among all inputs and all desirable outputs is attractive for performance analysis. The property indicates that the aggregation of inputs and desirable outputs should not influence the *OE* measurement of all DMUs.

(f) *Unit Invariance* (Cooper *et al.*, 1999; Lovell and Pastor, 1995): The unit of inputs and outputs should not influence an *OE* measure.

(g) *Invariance to Alternate Optima* (Cooper *et al.*, 1999): An occurrence of multiple solutions on DEA should not influence an *OE* measure.

(h) *Translation Invariance* (Ali and Seiford, 1990; Lovell and Pastor, 1995; Pastor, 1996; Cooper *et al.*, 1999; Pastor and Ruiz, 2007): An *OE* measure should be not influenced even if inputs and/or desirable outputs are shifted toward the same direction by adding or subtracting a specific real number. This property makes it possible to handle zero and negative values in a data set. See Chapters 26 and 27.

(i) *Frontier Shift Measurability* (Färe *et al.*, 1994b): It is desirable that an *OE* measure can measure a frontier shift among different time periods (or different locations). For example, DEA can link the radial-based *OE* measures to the Malmquist index. The combination measures a frontier shift among different periods (Bjurek, 1996; Lambert, 1999). Consequently, the *OE* measure can analyze not only cross-sections but also time-series and panel data sets. See Sueyoshi and Aoki (2001) where the Malmquist index

[3] See Chapter 7 for a discussion on how to handle an occurrence of multiple projections.

was combined with DEA window analysis, referred to as "Malmquist window analysis," to avoid a retreat and/or a cross-over among efficiency frontiers in different periods. See Chapter 19 that discusses a use of Malmquist index for environmental assessment in a time horizon.

Here, it is important to add the following two concerns on desirable properties. One of these two concerns is that the property of "unit invariance" will be discussed from an analytical perspective of DEA models in this chapter, but such a mathematical exploration is different from a computational aspect to solve these models. In short, there is a difference between analytical and computational aspects on DEA solutions. The other concern is that the property of "translation invariance" will be fully utilized in Chapter 26 that discusses how to handle an occurrence of many zeros in production factors of energy sectors. This property is very important in applying DEA for energy and environmental protection by non-radial measurement. However, the scope of the property is limited only within the non-radial measurement, not being extendable to the radial measurement. See Chapter 27 that discusses how to handle an occurrence of zero and negative values within the radial measurement for environmental assessment.

6.3 SUPPLEMENTARY DISCUSSION

To initiate our description on desirable *OE* properties, this chapter first needs to describe supplementary comments on additive model (AM) and slack-adjusted radial measure (SARM).

Additive Model: Assuming that x_{ij} and g_{rj} ($j = 1,\ldots,n$) are all strictly positive in these components, many DEA researchers believe that the additive model is unit invariant as well as translation invariant. They also think that the additive model has the property of translation invariance only when it includes the condition that the sum of λ_j is unity. It is true that their widely accepted beliefs are effective if slacks are uniquely projected on an efficiency frontier. If the additive model does not satisfy the property of unique projection, multiple combinations of optimal slacks (d_i^{x*} and d_r^{g*}) occur in the additive model. Consequently, it has multiple *OE* measures so that it has neither unit invariance nor translation invariance in terms scores in the *OE* measurement. As mentioned previously, Ali and Seiford (1990, p. 404, lines 4–8 in the right column) and Cooper *et al.* (2000, p. 92–94) proved that the additive model is translation invariant to the data shift. They have claimed that "the data shift does not alter the efficient frontier and the classification of DMUs as efficient or inefficient (the objective function value) is invariant to translations of data." Their claim is correct. However, they did not discuss anything regarding whether the additive model is translation invariant in terms of the *OE* measure. That is, they did not pay attention to the additive model from a degree of the *OE* measure. In contrast, this chapter is interested in the unit invariance and the translation invariance in terms of the degree of *OE*. See Example 6.13, which proves that the additive model is not translation invariant in terms of *OE*.

SARM: The *OE* measure of SARM has the following assumption that has been not sufficiently examined by Aida *et al.* (1998):

Proposition 6.1: SARM has an optimal solution if and only if $\left(1-\sum_{i=1}^{m} R_i^x x_{ik}\right) \geq 0$, where R_i^x stands for a data range on inputs. If there is an optimal solution (θ^*, λ^*) in the SARM, then we have $\theta^* = \underset{i=1,\cdots,m}{\text{maximize}} \sum_{j=1}^{n} x_{ij}\lambda_j^* / x_{ik}$.

[Proof] See Appendix of this chapter.

Here, this chapter has two important concerns on the proposition. One of the two concerns is that Cooper *et al.* (2001) replaced the objective function for SARM by $\theta - \varepsilon_n\left(\sum_{i=1}^{m} R_i^x d_i^x + \sum_{r=1}^{s} R_r^g d_r^g\right)$ where R_r^g stands for a data range on desirable outputs and R_i^x is for the data range on inputs. In this case, Proposition 6.1 is not necessary. However, since ε_n is a non-Archimedean small number in their study, the model with the proposed objective in their model is almost same as the radial model under variable RTS (with slacks) in terms of the desirable properties. Therefore, this chapter uses the formulation of SARM, as proposed in Chapter 5, in order to examine whether it satisfies the desirable properties on *OE*. The other concern is that the other DEA models to be examined in this chapter have a feasible solution so that they do not need the optimal condition as summarized in Proposition 6.1.

6.4 PREVIOUS STUDIES ON DESIRABLE PROPERTIES

Following Sueyoshi and Sekitani (2009), Table 6.1 summarizes the relationship between seven DEA models and nine desirable properties in terms of the *OE* measurement. In the table, the symbol "○" indicates that each model satisfies a desirable property specified in each cell of the table. The symbol "×" indicates the opposite case.

Table 6.1 also lists previous research on the examination on *OE* properties at the bottom of each cell. If we cannot identify the previous research, there is no specification on previous research source in the cell. In such cases, this chapter depends upon the effort of Sueyoshi and Sekitani (2009). Based upon our best knowledge, Table 6.1 covers most of the previous studies related to desirable *OE* properties. Of course, this chapter knows that there may be other important studies on *OE*-based desirable properties in previous literature. However, to avoid a descriptive duplication, we select a few example studies in each cell of Table 6.1.

Table 6.1 has two theoretical inconsistencies between previous research and this chapter. One of the two inconsistencies is that, as summarized in Table 6.1, RM(v) and RM(c) are unit invariant. Meanwhile, the research of Lovell and Pastor (1995) have discussed that they are unit invariant with respect to the radial measures, but did not pay attention to the existence of slack components. Furthermore,

TABLE 6.1 Relationship between seven models and nine desirable properties

$X>0$ and $Y>0$	RM(v)	Additive	RAM	SARM	SBM	Russell	RM(c)
Homogeneity	○ [BR], [F]	×	×	×	× [PRS]	×	○ [BR], [F]
Strict monotonicity	×	×	× [○ in CPP]	×	○ [PRS]	○ [PRS]	×
Efficiency requirement	×	×	○ [CPP]	×	○ [PRS]	○ [PRS]	×
Unique efficiency comparison	× [OP]	×	× [SS]	×	×	×	× [OP]
Aggregation	× [BR]	×	×	×	×	×	× [BR]
Unit invariance	○ [P], [LP]	×	○ [CPP], [LP]	○ [S]	○ [PRS], [T]	○	○ [LP]
Invariance to alternate optima	○	×	○ [CPP]	○	○	○	○
Translation invariance	× [LP], [○ in AS]	× [○ in AS]	○ [CPP], [LP]	× [S]	×	×	× [LP]
Frontier shift measurability	× [CRS], [B], [L]	× [CN]	×	×	×	×	○ [CRS], [B], [L]

(a) Source: Sueyoshi and Sekitani (2009). This chapter duplicates their contributions on the table. (b) o indicates that a model satisfies a desirable property and × indicates the opposite case. Reference codes: (c) {AS} Ali and Seiford (1990), [B] Bjurek (1996), [BR] Blackorby and Russell (1999), [CN] Charnes and Neralic (1990), [CRS] Charnes, Rousseau and Semple (1996), [CPP] Cooper, Park and Pastor (1999), [F] Fukuyama (2000), [P] Pastor (1996), [L] Lambert (1999), [LP] Lovell and Pastor (1995), [OP] Olesen and Petersen (2003), [PRS] Pastor, Ruiz and Sirvent (1999), [S] Sueyoshi (2001a), [SS] Sueyoshi and Sekitani (2007a) and [T] Tone (2001).

in terms of the translation invariance, their study (Lovell and Pastor, 1995) has discussed that these models have the property of partial invariance (i.e., invariance with respect to inputs or desirable outputs, but not both, depending upon model orientation). A difference between this chapter and their work is that we are interested in whether the *OE* measure is influenced by a unit change. Meanwhile, in their study (Lovell and Pastor, 1995), all the components of an optimal solution are required to be the same under the property of unit invariance. Partial invariance is attained if part of these optimal components is not influenced by the unit change. From their perspective, the claim of this chapter is looking for partial unit invariance in the *OE* measure. The other inconsistency is that, in this chapter, the additive model is neither unit invariant nor translation invariant. However, Lovell and Pastor (1995) have claimed that a normalized weighted additive model satisfies the two properties. Their normalized (weighted) additive model corresponds to RAM in this chapter. As indicated in Table 6.1, RAM is unit invariant and translation invariant. Thus, our claim is consistent with theirs. Chapter 16 extends RAM for environmental assessment because it has the property of translation invariance. See Chapter 26, as well.

In reviewing the previous studies, the following three concerns are important in examining strict monotonicity. First, the strict monotonicity has been often discussed as a desirable *OE* property in the literature on production economics. See, for example, Dmitruk and Koshevoy (1991). Unfortunately, most of the previous discussions assume a single desirable output in the *OE* measurement. In contrast, this chapter examines the property of strict monotonicity in a framework of multiple desirable outputs and inputs. Our discussion on the desirable property does not have such an assumption on a single desirable output. Second, RAM's strict monotonicity examined in this chapter is different from Cooper *et al.* (1999, p. 17). Later, this chapter will explain why RAM violates the property of strict monotonicity. Finally, in discussing the strict monotonicity, previous studies did not provide a unified mathematical definition on the property. For example, the numerator in SBM is decreasing in an input vector and the denominator is increasing in a desirable output vector, so that the ratio is monotonically decreasing in the two production factors. Meanwhile, the Russell measure, as discussed in Färe *et al.* (1994a), is decreasing in the input vector and the desirable output vector. See Bardhan *et al.* (1996b) and Cooper *et al.* (2006). Thus, it is necessary for us to examine the property of strict monotonicity in a unified analytical synthesis with multiple inputs and multiple desirable outputs.

6.5 STANDARD FORMULATION FOR RADIAL AND NON-RADIAL MODELS

To prove the claims summarized in Table 6.1, this chapter considers a standard DEA model that serves as a mathematical basis for our examination. Given unknown matrixes ($\boldsymbol{X}>0$ and $\boldsymbol{G}>0$) and observed column vectors ($X_k>0$ and $G_k>0$) regarding a specific k-th DMU to be examined, this chapter formulates the standard model by the following vector and matrix expressions for an observed data set ($\boldsymbol{A}$) regarding inputs and desirable outputs:

$$\begin{aligned} &\text{Minimize } f\left(X_k, G_k, \boldsymbol{X}, \boldsymbol{G}, \lambda\right) \\ &s.t. \quad \boldsymbol{A}\lambda \le R_h \text{ and } \lambda \ge 0. \end{aligned} \tag{6.1}$$

The important feature of the standard model, or Model (6.1), is that it is structured by a vector of λ which expresses a set of intensity variables on production factors. R_h stands for the right hand side to express a vector related to inputs and desirable outputs on the k-th DMU. f is a function for production activities. To condense our description, this chapter uses the vector and matrix expressions, not the scalar expression found in most of the other chapters.

Let $Q(X_k, G_k, \boldsymbol{X}, \boldsymbol{G}, \boldsymbol{A}, R_h)$ be the optimal objective value that is obtained from Model (6.1). The optimal value indicates a degree of *OE* measure for all DEA models except the additive model and RAM. In the two models, the optimal objective value indicates the level of inefficiency. The *OE* measure is determined by subtracting the degree of inefficiency from unity. The unique feature on RAM will be later incorporated into all non-radial models used for environmental assessment.

The following three models indicate, as examples, how Model (6.1) can express each of DEA models:

RM under Variable RTS (input-oriented): Reformulation of RM(v) produces the following formulation:

$$\begin{aligned} &\underset{\lambda}{\text{Minimize}} \ \underset{i=1,\dots,m}{\text{maximize}} \ \sum_{j=1}^{n} x_{ij}\lambda_j / x_{ik} \\ &s.t. \quad \boldsymbol{G}\lambda \ge G_k, \\ &\qquad e\lambda = 1 \text{ and } \lambda \ge 0. \end{aligned} \tag{6.2}$$

Here, e is a row vector whose components are all 1. θ, an efficiency measure for *OE*, which is usually incorporated in the objective function of the input-oriented RM(v), is replaced by $\theta = \underset{i=1,\dots,m}{\text{maximize}}\left(\sum_{j=1}^{n} x_{ij}\lambda_j / x_{ik}\right) = Q(X_k, G_k, \boldsymbol{X}, \boldsymbol{G}, \boldsymbol{A}, R_h)$ in the RM(v). Consequently, the standard model (6.1) can express Model (6.2) as follows:

$$\begin{aligned} &f\left(X_k, G_k, \boldsymbol{X}, \boldsymbol{G}, \lambda\right) = \underset{i=1,\dots,m}{\text{maximize}} \sum_{j=1}^{n} x_{ij}\lambda_j / x_{ik}, \\ &\boldsymbol{A}^{Tr} = \left[-\boldsymbol{G}^{Tr}, e^{Tr}, -e^{Tr}\right] \text{ and } R_h^{Tr} = \left[-G_k^{Tr}, 1, -1\right]. \end{aligned} \tag{6.3}$$

Model (6.2) omits slacks, associated with a non-Archimedean small number (ε_n), that are originally incorporated in the RM(v). The superscript (Tr) stands for a vector transpose. For our descriptive convenience, this chapter drops slacks from Model (6.2) so that Model (6.1) can express Model (6.2) in a condensed form.

Additive Model: This chapter can reformulate the additive model as follows:

$$\text{Minimize } -\sum_{i=1}^{m}\left(x_{ik}-\sum_{j=1}^{n}x_{ij}\lambda_j\right)-\sum_{r=1}^{s}\left(\sum_{j=1}^{n}g_{rj}\lambda_j-g_{rk}\right) \tag{6.4}$$
$$s.t. \quad \boldsymbol{X}\lambda \le X_k,\ \boldsymbol{G}\lambda \ge G_k,\ e\lambda = 1 \text{ and } \lambda \ge 0.$$

Hence, the standard model (6.1) can express the additive model as follows:

$$f\left(X_k, G_k, X, G, \lambda\right) = -\sum_{i=1}^{m}\left(x_{ik}-\sum_{j=1}^{n}x_{ij}\lambda_j\right)-\sum_{r=1}^{s}\left(\sum_{j=1}^{n}g_{rj}\lambda_j-g_{rk}\right), \tag{6.5}$$
$$\boldsymbol{A}^{Tr} = \left[\boldsymbol{X}^{Tr}, -\boldsymbol{G}^{Tr}, e^{Tr}, -e^{Tr}\right] \text{ and } R_h^{Tr} = \left[X_k^{Tr}, -G_k^{Tr}, 1, -1\right].$$

SBM (= ERGM): The chapter reformulates SBM (= ERGM) as follows:

$$\text{Minimize } (1/m)\sum_{i=1}^{m}\left(\sum_{j=1}^{n}x_{ij}\lambda_j/x_{ik}\right)\Big/(1/s)\sum_{r=1}^{s}\left(\sum_{j=1}^{n}g_{rj}\lambda_j/g_{rk}\right) \tag{6.6}$$
$$s.t. \quad \boldsymbol{X}\lambda \le X_k,\ \boldsymbol{G}\lambda \ge G_k \text{ and } \lambda \ge 0.$$

Hence, SBM becomes the following standard model:

$$f\left(X_k, G_k, X, G, \lambda\right) = \left(1/m\right)\sum_{i=1}^{m}\left(\sum_{j=1}^{n}x_{ij}\lambda_j/x_{ik}\right)\Big/(1/s)\sum_{r=1}^{s}\left(\sum_{j=1}^{n}g_{rj}\lambda_j/g_{rk}\right), \tag{6.7}$$
$$\boldsymbol{A}^{Tr} = \left[\boldsymbol{X}^{Tr}, -\boldsymbol{G}^{Tr}\right] \text{ and } R_h^{Tr} = \left[X_k^{Tr}, -G_k^{Tr}\right].$$

The above reformulation process, discussed for the three models: RM(v), the additive model and SBM, is applicable to other DEA models. See Table 6.2 for a summary on how the standard model (6.1) can express all the DEA models examined in this chapter.

The following three concerns are important in discussing the standard model (6.1) and Table 6.2. First, a methodological comparison on DEA may be dated back to Ahn *et al.* (1988). Their study combined the two RMs with conventional statistical tests and analyses. Then, they characterized these measures by comparing them with statistical tests. Second, a general DEA model was first proposed by Yu *et al.* (1996a, 1996b) and Wei and Yu (1997). Their studies provided a unified framework for four types of RMs under different conditions on RTS. Their general model incorporated a case (called "K-cone" or "predilection cone") where it restricted multipliers, measured by dual variables, by using prior information. As a result of such multiplier restriction, different DEA models had a restricted projection surface on an efficiency frontier. In contrast, the standard

TABLE 6.2 Reformulation as a standard model (expression by λ): Model (6.1)

Model	$f(X_k, G_k, X, G, \lambda)$	A	R_h
RM(v): input oriented	$\underset{i=1,\cdots,m}{\text{maximize}} \frac{\sum_{j=1}^{n} x_{ij}\lambda_j}{x_{ik}}$	$\begin{bmatrix} -G \\ e \\ -e \end{bmatrix}$	$\begin{bmatrix} -G_k \\ 1 \\ -1 \end{bmatrix}$
Additive model	$-\sum_{i=1}^{m}\left(x_{ik} - \sum_{j=1}^{n} x_{ij}\lambda_j\right) - \sum_{r=1}^{s}\left(\sum_{j=1}^{n} g_{rj}\lambda_j - g_{rk}\right)$	$\begin{bmatrix} X \\ -G \\ e \\ -e \end{bmatrix}$	$\begin{bmatrix} X_k \\ -G_k \\ 1 \\ -1 \end{bmatrix}$
RAM	$-\sum_{i=1}^{m} R_i^x\left(x_{ik} - \sum_{j=1}^{n} x_{ij}\lambda_j\right) - \sum_{r=1}^{s} R_r^g\left(\sum_{j=1}^{n} g_{rj}\lambda_j - g_{rk}\right)$	$\begin{bmatrix} X \\ -G \\ e \\ -e \end{bmatrix}$	$\begin{bmatrix} X_k \\ -G_k \\ 1 \\ -1 \end{bmatrix}$
SARM	$\left(1 - \sum_{i=1}^{m} R_i^x x_{ik}\right)\left(\underset{i=1,\ldots,m}{max} \frac{\sum_{j=1}^{n} x_{ij}\lambda_j}{x_{ik}}\right) + \left[\sum_{i=1}^{m} R_i^x \sum_{j=1}^{n} x_{ij}\lambda_j - \sum_{r=1}^{s} R_r^g\left(\sum_{j=1}^{n} g_{rj}\lambda_j - g_{rk}\right)\right]$	$\begin{bmatrix} -G \\ e \\ -e \end{bmatrix}$	$\begin{bmatrix} -G_k \\ 1 \\ -1 \end{bmatrix}$
RM (c): input-oriented	$\underset{i=1,\ldots,m}{\text{maximize}} \frac{\sum_{j=1}^{n} x_{ij}\lambda_j}{x_{ik}}$	$-G$	$-G_k$
SBM and ERGM	$\left(\frac{1}{m}\sum_{i=1}^{m} \frac{\sum_{j=1}^{n} x_{ij}\lambda_j}{x_{ik}}\right) \Big/ \left(\frac{1}{s}\sum_{r=1}^{s} \frac{\sum_{j=1}^{n} g_{rj}\lambda_j}{g_{rk}}\right)$	$\begin{bmatrix} X \\ -G \end{bmatrix}$	$\begin{bmatrix} X_k \\ -G_k \end{bmatrix}$
Russell measure	$\frac{1}{m+s}\left(\sum_{i=1}^{m} \frac{\sum_{j=1}^{n} x_{ij}\lambda_j}{x_{ik}} + \sum_{r=1}^{s} \frac{g_{rk}}{\sum_{j=1}^{n} g_{rj}\lambda_j}\right)$	$\begin{bmatrix} X \\ -G \end{bmatrix}$	$\begin{bmatrix} X_k \\ -G_k \end{bmatrix}$

model (6.1) proposed in this chapter covers not only radial but also non-radial models. In Model (6.1), six DEA models, except the Russell measure, are structurally simplified and unified into a standard linear programming model. This chapter uses the proposed standard model (6.1) only for mathematical proofs on desirable *OE* properties. The Russell measure is formulated by nonlinear programming. Finally, assuming that SARM has an optimal solution, Proposition 6.1 implies that it has the following objective function:

$$\left(1-\sum_{i=1}^{m}R_i^x x_{ik}\right)\left(\underset{i=1,\ldots,m}{\text{maximize}}\frac{\sum_{j=1}^{n}x_{ij}\lambda_j}{x_{ik}}\right)+\left[\sum_{i=1}^{m}R_i^x\sum_{j=1}^{n}x_{ij}\lambda_j-\sum_{r=1}^{s}R_r^g\left(\sum_{j=1}^{n}g_{rj}\lambda_j-g_{rk}\right)\right] \tag{6.8}$$

Thus, Table 2 has Equation (6.8) as the objective function of SARM.

Here, it is important to note that radial approaches discussed for DEA environmental assessment are based upon SARM and non-radial approach for the assessment are based upon RAM, along with some modifications (e.g., incorporation of undesirable outputs). The two models used for conventional DEA will become an analytical basis for environmental assessment in Section II. The SARM originates from RAM. So, it is possible to consider that all the models used in chapters of Section II are originated from RAM.

We have four rationales concerning why we pay attention to RAM. First, it has the property of "translation invariance" so that it can handle an occurrence of zero and/or negative values in a data set without any difficulty. Chapter 26 describes how a non-radial model, originated from RAM, can utilize the property of translation invariance for dealing with an occurrence of zero and/or negative values in environmental assessment. Chapter 27 discusses a radial model, originated from SARM, can handle zero and/or negative values for environmental assessment. Second, radial and non-radial models, originated from RAM and SARM, can always produce positive multipliers (dual variables) in their signs so that all production factors will be fully utilized in environmental assessment. Third, the proposed models can easily incorporate undesirable outputs and then unify between desirable and undesirable outputs in their formulations. Finally, as documented in Chapters 20, 21 and 23, these models have structural flexibility in the manner that they can measure various scale-related measures such as RTS and damages to scale (DTS). For these measures, some multipliers are set to be unrestricted in their signs.

6.6 DESIRABLE PROPERTIES FOR DEA MODELS

6.6.1 Aggregation

As mentioned by Blackorby and Russell (1999, p. 10), the previous definition of aggregation was considerably "general and too abstract." Therefore, this chapter follows the definition on aggregation referred to as "aggregation index (AI)" in their study (Blackorby and Russell, 1999, p. 11). The AI axiom implies that "all DMUs are efficient if and only if an aggregated DMU is efficient." This chapter redefines the AI axiom as follows: Let us consider the aggregation of input and desirable output vectors as $X_{TA} \equiv \sum_{j=1}^{n} X_j$ and $G_{TA} \equiv \sum_{j=1}^{n} G_j$. Then, this chapter can specify input and desirable output matrixes as $\boldsymbol{X}' \equiv \left[X_{TA}, X_1, \ldots, X_n\right]$ and $\boldsymbol{G}' \equiv \left[G_{TA}, G_1, \ldots, G_n\right]$ after incorporating the aggregated input and desirable output vectors into these original matrixes. Here, the subscript "TA" stands for total aggregation. Suppose that $\left(\boldsymbol{A}', R_h'\right)$ are generated by $(\boldsymbol{X}', \boldsymbol{G}')$ where R_h' stands for the right hand side of an aggregated matrix, or $\boldsymbol{A}' = \left[\boldsymbol{X}', \boldsymbol{G}'\right]$. Here, let $\Omega' \equiv \left\{\lambda' \mid \boldsymbol{A}'\lambda' \leq R_h'\right\}$ and $\Omega \equiv \left\{\lambda \mid \boldsymbol{A}\lambda \leq R_h\right\}$, then the AI axiom has the following definition:

$$\begin{aligned} &\underset{\lambda \in \Omega}{\text{Minimize}}\, f\left(X_k, G_k, \boldsymbol{X}, \boldsymbol{G}, \lambda\right) = \gamma \text{ for all } k \text{ if and only if} \\ &\underset{\lambda' \in \Omega'}{\text{minimize}}\, f\left(X_{TA}, G_{TA}, \boldsymbol{X}', \boldsymbol{G}', \lambda'\right) = \gamma. \end{aligned} \tag{6.9}$$

Here, γ indicates a value of the functional form (f) when a DMU is efficient in terms of *OE*. For example, $\gamma = 0$ stands for the status of efficiency in the additive model and RAM. In the other DEA models, $\gamma = 1$ indicates the status of efficiency.

Proposition 6.2: All the DEA models violate the property on aggregation of inputs and desirable outputs.

[Proof] The following two examples confirm Proposition 6.2. Let us consider first the RM(v).

As listed in Example 6.1, two DMUs {A, B} are efficient, while DMU {C} is inefficient by RM(v). DMU {TA} (= DMU{A} + DMU{B} + DMU{C}) is efficient by the RM(v). The AI axiom in Equation (6.9) requests that if DMU {TA} is efficient, then all DMUs should be efficient in each RM(v)'s individual measurement. However, DMU{C} is inefficient. The result of Example 6.1 illustrates that the RM(v) does not satisfy the AI axiom. Since the status of inefficiency measured by the RM(v) is equivalent to that of other DEA models such as the additive model, RAM and SARM, this chapter confirms that these DEA models violate the property of aggregation, as well, by the following example:

Now, let us consider the case of Russell measure. As listed in Example 6.2, all three DMUs {A, B, C} are efficient by the Russell measure. However, DMU {TA} is inefficient by the Russell measure. The AI axiom indicates that if all DMUs are efficient, then the aggregated DMU {TA} (= DMU{A} + DMU{B} + DMU{C}) should be efficient. Thus, the Russell measure violates the property of aggregation. Next, SBM (= ERGM) measures the *OE* on DMU {TA} as 0.909 and

EXAMPLE 6.1

DMU	Input	Desirable output	RM(v)	Efficiency status
{A}	1	1	1	Efficient
{B}	2	3	1	Efficient
{C}	3	2	0.5	Inefficient
{TA}	6	6	1	Efficient

(a) Source: Sueyoshi and Sekitani (2009).

EXAMPLE 6.2

DMU	Input	Desirable output 1	Desirable output 2	Russell measure	Efficiency status
{A}	1	1	2.5	1	Efficient
{B}	1	2	2	1	Efficient
{C}	1	3	0.5	1	Efficient
{TA}	3	6	5	0.944	Inefficient

(a) Source: Sueyoshi and Sekitani (2009).

RM(c) determines the level of *OE* as 0.933. Thus, the result (i.e., the Russell measure violates the property of aggregation) can be also applied to the RM(c) and SBM (= ERGM)[4].

6.6.2 Frontier Shift Measurability

Using sensitivity analysis directly linked to the frontier shift measurability, this chapter examines a case where DEA shifts inputs and desirable outputs regarding the k-th DMU from the current production level to another one. Let us consider the matrix ($\boldsymbol{A}$) that consists of $\boldsymbol{X}$ and $\boldsymbol{G}$. The data matrix is written as $\boldsymbol{A}(\boldsymbol{X}, \boldsymbol{G})$. Similarly, R_h is expressed by $R_h(X_k, G_k)$ as the right-hand side of DEA models.

When $R_h(X_k, G_k)$ is changed to $R_h(X_k - \Delta_x, G_k + \Delta_g)$, then the standard model (6.1) becomes

$$\begin{aligned} &\text{Minimize } f\left(X_k - \Delta_x, G_k + \Delta_g, \boldsymbol{X}, \boldsymbol{G}, \lambda\right) \\ &s.t. \quad \boldsymbol{A}\left(\boldsymbol{X}, \boldsymbol{G}\right) \le R_h\left(\mathrm{X}_k - \Delta_x, G_k + \Delta_g\right) \text{ and} \\ &\qquad \lambda \ge 0. \end{aligned} \tag{6.10}$$

Here, Δ_x indicates the amount of inputs and Δ_g indicates that of desirable outputs, both are changed from these current production levels.

Model (6.10) cannot guarantee that the new production level $(X_k - \Delta_x, G_k + \Delta_g)$ always belongs to an original production possibility set, shaped by the original production structure of $(\boldsymbol{X}, \boldsymbol{G})$. Consequently, Model (6.10) may produce an

[4] Sueyoshi and Sekitani (2007b) and Tone (2001) discussed the analytical relationship between the Russell measure, SBM and ERGM. It is important to mention a difference between Cooper *et al.* (2007a) and this study. Using Example 6.2, this chapter confirms that ERGM violates the property of aggregation within the original formulation of ERGM (= SBM). Cooper *et al.* (2007a), based upon the same recognition on ERGM violation, extended the original formulation into a mixed integer programming formulation. Then, they have discussed that the new ERGM model satisfies the property of aggregation. Thus, the ERGM (=SBM) model used in this chapter is the original model, not the extended ERGM model proposed by Cooper *et al.* (2007a). Thus, such a difference in the two models leads to different results in the aggregation property of ERGM.

infeasible solution. The following proposition describes analytical relationships among DEA models due to the sensitivity analysis:

Proposition 6.3: Let us consider Model (6.10) under $\boldsymbol{X} > 0$, $\boldsymbol{G} > 0$, $X_k - \Delta_x > 0$ and $G_k + \Delta_g > 0$. Then, the following relationships exist among DEA models:

(a) The input-oriented RM(c) has always an optimal solution.
(b) An optimal solution exists in the additive model if and only if an optimal solution exists in RAM.
(c) If an optimal solution exists in the additive model, then an optimal solution exists in the input-oriented RM(v).
(d) An optimal solution exists in SARM if and only if an optimal solution exists in the input-oriented RM(v) and

$$\left(1-\left[\sum_{i=1}^{m} R_i^x\left(x_{ik}-\Delta_{xi}\right)+\sum_{r=1}^{s} R_r^g\left(g_{rk}+\Delta_{gr}\right)\right]\right)\geq 0.$$

(e) The existence of an optimal solution in SBM is equivalent to that of ERGM.
(f) The existence of an optimal solution in SBM is equivalent to that of the Russell measure.

In claim (c) above, Δ_{xi} and Δ_{gr} are the i-th and r-th components of the two vector changes, respectively.

[Proof] Proposition 6.3 can be directly obtained from Table 6.2, Proposition 6.1 and the research effort of Charnes *et al.* (1996). Q.E.D.

The relationships between seven DEA models due to sensitivity analysis are visually summarized in Figure 6.1. See Cooper *et al.* (2005) for a detailed description on DEA-based sensitivity analysis. Figure 6.1 visually indicates that, if input-oriented RM(v) violates the property of frontier shift measurability, then the three models (i.e., an additive model, SARM and RAM) violate the property. Similarly, if SBM violates the property, then Russell measure and SBM (= ERGM) violate the property. As indicated by Cooper *et al.* (1999), only input-oriented RM(c) can satisfy the property. The property of frontier shift measurability is applicable to output-oriented RM(c) because the radial model does not have the side constraint (i.e., the sum of λ_j for $j=1,\ldots,n$) in its formulation.

The following example illustrates why an optimal solution does not exist in sensitivity analysis of input-oriented RM(v) and SBM:

Let us consider the sensitivity analysis of DMU {C} where it reduces the amount of input from 2 to 3/2 and increases the amount of desirable output from 1/10 to 4.

RM(v): If the sensitivity analysis of input-oriented RM(v) is applied to DMU {C}, then the constraints on λ in Model (6.2) become $2\lambda_A + 3\lambda_B + (1/10)\lambda_C \geq 4$, $\lambda_A + \lambda_B + \lambda_C = 1$ and $\lambda_A, \lambda_B, \lambda_C \geq 0$. Since $2\lambda_A + 3\lambda_B + 0.1\lambda_C \geq 4 = 4\lambda_A + 4\lambda_B + 4\lambda_C$,

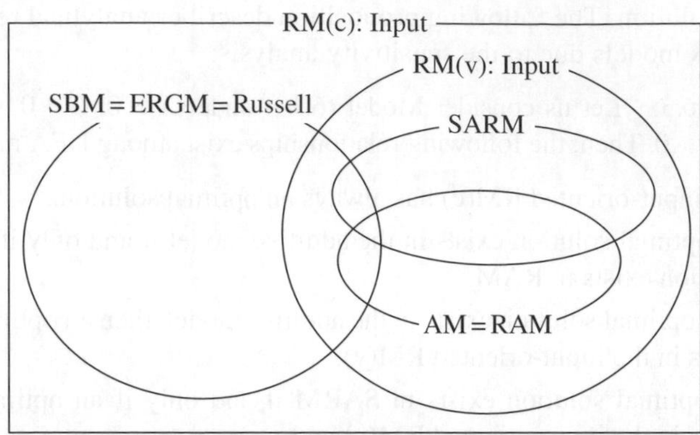

FIGURE 6.1 Sensitivity relationship between seven models (a) Source: Sueyoshi and Sekitani (2009). (b) AM: additive model; ERGM: enhanced Russell graph measure; RAM: range-adjusted measure; RM: radial model; Russell: Russell measure; SARM: slack-adjusted radial measure; and SBM: slack-based measure.

the radial model has $0 \geq 2\lambda_A + \lambda_B + (39/10)\lambda_C$. From the non-negativity of λ_A, λ_B and λ_C, we have $\lambda_A = \lambda_B = \lambda_C = 0$. Hence, we find an infeasible solution in the input-oriented RM(v) if the data is shifted from the current production level of DMU {C} to the other level as listed in Example 6.3.

SBM: If we apply SBM to DMU{C}, then the constraints regarding λ become $\lambda_A + 2\lambda_B + 2\lambda_C \leq (3/2)$, $2\lambda_A + 3\lambda_B + 0.1\lambda_C \geq 4$, and $\lambda_A, \lambda_B, \lambda_C \geq 0$. Since $2\lambda_A + 3\lambda_B + 0.1\lambda_C \geq 4 = (8/3)\lambda_A + (16/3)\lambda_B + (16/3)\lambda_C$, the equation becomes $0 \geq (2/3)\lambda_A + (7/3)\lambda_B + (157/30)\lambda_C$. From the non-negativity of λ_A, λ_B and λ_C, we have $\lambda_A = \lambda_B = \lambda_C = 0$. Hence, an infeasible solution occurs in SBM if the data changes the current production level of DMU {C} to the other level as listed in Example 6.3.

Proposition 6.4: Only the RM(c) can satisfy the property of frontier shift measurability. The other DEA models do not have the property.

[Proof] Along with two examples regarding RM(v) and SBM, both Proposition 6.3 and Figure 6.1 (i.e., the violation of RM(v) and SBM implies the violation of the other models) prove the proposition. Q.E.D.

EXAMPLE 6.3

DMU	Input	Desirable output
{A}	1	2
{B}	2	3
{C}	2 (3/2)	1/10 (4)

(a) Source: Sueyoshi and Sekitani (2009).

Here, it is important to note the following concern on the property of frontier shift measurability. That is, an important feature of RM(c) is that it does not have $e\lambda = 1$ in the formulation so that it can maintain the property. The radial formulation for environmental assessment discussed in Chapter 19, originated from SARM, drops the side constraint so that it can measure the frontier shift in a time horizon. Thus, the radial model for environmental assessment discussed in Chapter 19 does not have $e\lambda = 1$ in the formulation in order to measure a shift of Malmquist index and its related subcomponents in a time horizon.

6.6.3 Invariance to Alternate Optima

The remaining five properties (homogeneity, strict monotonicity, efficiency requirement, unit invariance and translation invariance) depend upon the property of invariance to alternate optima. Therefore, this chapter starts examining the property of invariance to alternate optima by the following proposition:

Proposition 6.5: All DEA models, except the additive model, have the property of invariance to alternate optima.

[Proof] The DEA models, except the additive model, determine these *OE* measures by the optimal objective value $Q(X_k, G_k, \boldsymbol{X}, \boldsymbol{G}, \boldsymbol{A}(\boldsymbol{X}, \boldsymbol{G}), R_h(X_k, G_k))$ of Model (6.1). Therefore, the proposition is valid. Q.E.D.

As documented in Table 6.2, the additive model measures the degree of *OE* by

$$1 - \left(\sum_{i=1}^{m} (x_{ik} - \sum_{j=1}^{n} x_{ij} \lambda_j^*) + \sum_{r=1}^{s} \left(\sum_{j=1}^{n} g_{rj} \lambda_j^* - g_{rk} \right) \right) \tag{6.11}$$

where λ^* is an optimal solution of Model (6.4). The additive model does not use the optimal objective value of Model (6.5) as the degree of *OE* measure. Hence, let Equation (6.11) be $Q(X_k, G_k, \boldsymbol{X}, \boldsymbol{G}, \boldsymbol{A}(\boldsymbol{X}, \boldsymbol{G}), R_h(X_k, G_k))$ for the additive model. Then, it is sufficient to show that Equation (6.11) suffers from an occurrence of multiple optimal solutions in Model (6.4).

The optimal objective value of the additive model (6.4), applied to DMU {C}, becomes –2.9. The optimal solution needs to satisfy $\left\{ \left(\lambda_A^*, \lambda_B^*, \lambda_C^* \right) \;\middle|\; \lambda_A^* + \lambda_B^* = 1,\ \lambda_A^* \geq 0, \lambda_B^* \geq 0, \lambda_C^* = 0 \right\}$.

EXAMPLE 6.4

DMU	Input	Desirable output
{A}	1	2
{B}	2	3
{C}	2	1/10

(a) Source: Sueyoshi and Sekitani (2009).

Hence, Equation (6.11) becomes

$$-\frac{27}{2} \leq 1-\frac{1}{2}\left(\frac{2-\left(\lambda_A^*+2\lambda_B^*\right)}{2}+\frac{\left(2\lambda_A^*+3\lambda_B^*\right)-(1/10)}{(1/10)}\right) \leq -\frac{39}{4}. \quad (6.12)$$

The above inequalities indicate that the additive model applied to DMU {C} suffers from an occurrence of multiple *OE* measures. The upper and lower bounds are specified by Equation (6.12).

6.6.4 Formal Definitions on Other Desirable Properties

The preceding subsections have discussed which model satisfies the three properties (i.e., aggregation, frontier shift measurability and invariance to alternate optima). To examine the other desirable properties, let $\theta[X_k, G_k, \boldsymbol{X}, \boldsymbol{G}, \boldsymbol{A}(\boldsymbol{X}, \boldsymbol{G}), R_h(X_k, G_k)]$ be the *OE* measure of the *k*-th DMU by all DEA models. The *OE* measure (θ) is equivalent to $Q[X_k, G_k, \boldsymbol{X}, \boldsymbol{G}, \boldsymbol{A}(\boldsymbol{X}, \boldsymbol{G}), R_h(X_k, G_k)]$ in all DEA models except RAM and the additive model. In RAM, for example, the *OE* measure is $\theta\left[X_k, G_k, \boldsymbol{X}, \boldsymbol{G}, \boldsymbol{A}(\boldsymbol{X}, \boldsymbol{G}), R_h(X_k, G_k)\right] = 1 - Q\left[X_k, G_k, \boldsymbol{X}, \boldsymbol{G}, \boldsymbol{A}(\boldsymbol{X}, \boldsymbol{G}), R_h(X_k, G_k)\right]$. Using this formulation, this chapter defines the remaining desirable properties as follows:

(a) *Homogeneity (input-oriented)*: For each DMU *k* with $\lambda_k^* = 0$ (i.e., inefficient), every optimal solution λ^* in Model (6.1) needs to satisfy

$$\theta\left[\mu X_k, G_k, \boldsymbol{X}', \boldsymbol{G}, \boldsymbol{A}\left(\boldsymbol{X}', \boldsymbol{G}\right), R_h\left(\mu X_k, G_k\right)\right] = \frac{1}{\mu}\theta\left[X_k, G_k, \boldsymbol{X}, \boldsymbol{G}, \boldsymbol{A}\left(\boldsymbol{X}, \boldsymbol{G}\right), R_h\left(X_k, G_k\right)\right]$$

for all $\mu \geq 1$ where $X'\left[X_1, \ldots, X_{k-1}, \mu X_k, X_{k+1}, \ldots, X_n\right]$.[5]

(b) *Strict Monotonicity*: For all DMUs *k* with $\lambda_k^* = 0$ (so, being inefficient), any optimal solution (λ^*) in Model (6.1) needs to satisfy

$$\theta\left[X_k', G_k', \boldsymbol{X}', \boldsymbol{G}', \boldsymbol{A}\left(\boldsymbol{X}', \boldsymbol{G}'\right), R_h\left(X_k', G_k'\right)\right] < \theta\left[X_k, G_k, \boldsymbol{X}, \boldsymbol{G}, \boldsymbol{A}\left(\boldsymbol{X}, \boldsymbol{G}\right), R_h\left(X_k, G_k\right)\right]$$

for all $\left(-X_k', G_k'\right) \leq \left(-X_k, G_k\right)$ and $\left(-X_k', G_k'\right) \neq \left(-X_k, G_k\right)$.

Here, $\boldsymbol{X}' = \left[X_1, \ldots, X_{k-1}, X_k', X_{k+1}, \ldots, X_n\right]$ and $\boldsymbol{G}' = \left[G_1, \ldots, G_{k-1}, G_k', G_{k+1}, \ldots, G_n\right]$.

[5] The formulation to express the property of homogeneity in this chapter is different from the standard definition of homogeneity discussed by Russell (1985, p. 113 and 1990, p. 258). His definition expresses a general characterization of linear homogeneity. Meanwhile, an important feature of DEA is that it measures the level of *OE* regarding the specific *k*-th DMU. Hence, this chapter needs to modify the property of homogeneity so that it can fit within the framework of DEA-based *OE* measurement. Using the equation related to linear homogeneity, this chapter can examine whether each DEA model can satisfy the property from a unified analytical perspective of *OE* measurement. Thus, this chapter prepares the formulation for homogeneity in terms of DEA-based *OE* measurement.

(c) *Efficiency Requirement*: All DMUs k need to satisfy

$$0 \leq \theta\left[X_k, G_k, \boldsymbol{X}, \boldsymbol{G}, \boldsymbol{A}(\boldsymbol{X}, \boldsymbol{G}), R_h\left(X_k, G_k\right)\right] \leq 1.$$

Even if $\theta\left[X_k, G_k, \boldsymbol{X}, \boldsymbol{G}, \boldsymbol{A}(\boldsymbol{X}, \boldsymbol{G}), R_h(X_k, G_k)\right] = 1$, it may not imply the status of efficiency and the DMU k may not have any feasible solution on λ in Model (6.1) such that $\boldsymbol{X}\lambda \leq X_k$, $\boldsymbol{G}\lambda \geq G_k$ and $(\boldsymbol{X}\lambda, \boldsymbol{G}\lambda) \neq (X_k, G_k)$.

(d) *Unique Projection for Efficiency Comparison*: There exists (X', G') such that $X' = \boldsymbol{X}\lambda^*$ and $G' = \boldsymbol{G}\lambda^*$ for all optimal solutions of λ^* in Model (6.1).

(e) *Unit Invariance*: For any positive diagonal matrix $\boldsymbol{D}^x$ whose order is m and any positive diagonal matrix $\boldsymbol{D}^g$ whose order is s, all DMUs$\{k\}$ need to satisfy

$$\begin{aligned}&\theta\left[\boldsymbol{D}^x X_k, \boldsymbol{D}^g G_k, \boldsymbol{D}^x \boldsymbol{X}, \boldsymbol{D}^g \boldsymbol{G}, \boldsymbol{A}\left(\boldsymbol{D}^x \boldsymbol{X}, \boldsymbol{D}^g \boldsymbol{G}\right), R_h\left(\boldsymbol{D}^x X_k, \boldsymbol{D}^g G_k\right)\right]\\ &\quad = \theta\left[X_k, G_k, \boldsymbol{X}, \boldsymbol{G}, \boldsymbol{A}(\boldsymbol{X}, \boldsymbol{G}), R_h\left(X_k, G_k\right)\right].\end{aligned}$$

(f) *Translation Invariance*: For every pair of the m-dimensional non-negative vector (π_x) and the s-dimensional nonnegative vector (π_g), the following relation:

$$\begin{aligned}&\theta\left[X_k + \pi_x, G_k + \pi_g, \boldsymbol{X} + \Pi_x, \boldsymbol{G} + \Pi_g, \boldsymbol{A}\left(\boldsymbol{X} + \Pi_x, \boldsymbol{G} + \Pi_g\right), R_h\left(X_k + \pi_x, G_k + \pi_g\right)\right]\\ &\quad = \theta\left[X_k, G_k, \boldsymbol{X}, \boldsymbol{G}, \boldsymbol{A}(\boldsymbol{X}, \boldsymbol{G}), R_h\left(X_k, G_k\right)\right]\end{aligned}$$

needs to satisfy for all DMUs$\{k\}$ where $\pi_x = [x, \ldots, x]^{Tr}$ and $\pi_g = [g, \ldots, g]^{Tr}$.

6.6.5 Efficiency Requirement

Proposition 6.1 indicates that SARM may violate the desirable property on efficiency requirement, and Example 6.4 shows that the additive model violates the property. The property on the other DEA models can be summarized as follows:

Proposition 6.6: RAM, SBM and Russell measure satisfy the property of efficiency requirement. The other models [i.e., RM(v), RM(c), an additive model and SARM] violate the property[6].

[Proof] See Appendix of this chapter.

[6]The following two comments are important for understanding Proposition 6.6. First, two input-oriented RMs, SBM and Russell measure determine these *OE* measures on the range of $0 \leq \theta\left[X_k, G_k, \boldsymbol{X}, \boldsymbol{G}, \boldsymbol{A}(\boldsymbol{X}, \boldsymbol{G}), R_h(X_k, G_k)\right] \leq 1$. Second, when we measure the *OE* on DMUs by the two RMs, we need to pay attention to not only these *OE* measures (θ) but also the existence of a slack(s) in these DEA models. Hence, even if the two RMs have $\theta\left[X_k, G_k, \boldsymbol{X}, \boldsymbol{G}, \boldsymbol{A}(\boldsymbol{X}, \boldsymbol{G}), R_h(X_k, G_k)\right] = 1$, we cannot immediately conclude that DMU {k} is fully efficient. Hence, the two RMs do not satisfy the property.

It is true that both RM(v) and RM(c) have the property of $\theta\left[X_k,G_k,\boldsymbol{X},\boldsymbol{G},A(\boldsymbol{X},\boldsymbol{G}),R_h(X_k,G_k)\right]=1$. However, even if these *OE* measures (θ) are unity, we cannot immediately conclude that DMU {*k*} is fully efficient because we need to confirm whether all slacks are zero or not.

6.6.6 Homogeneity

Proposition 6.7: Both input-oriented RM(v) and RM(c) satisfy the property of homogeneity, but the other DEA models violate the property[7]. In the two radial measures, all DMUs with $\lambda_k^*=0$ for every optimal solution λ^* in Model (6.1) satisfy the following condition:

$$\theta\left[\mu X_k,G_k,\boldsymbol{X}',\boldsymbol{G},A(\boldsymbol{X}',\boldsymbol{G}),R_h(\mu X_k,G_k)\right]-\frac{1}{\mu}\theta\left[X_k,G_k,\boldsymbol{X},\boldsymbol{G},A(\boldsymbol{X},\boldsymbol{G}),R_h(X_k,G_k)\right]=0$$

for all $\mu\geq 1$ where $\boldsymbol{X}'=\left[X_1,\ldots,X_{k-1},\mu X_k,X_{k+1},\ldots,X_n\right]$. (6.13)

[Proof] See Appendix of this chapter.

The property of homogeneity should be discussed only on an inefficient DMU (so, $\lambda_k^*=0$). The status change of an efficient DMU due to μ does not have any theoretical implication because the change reorganizes the structure of an efficiency frontier.

The non-radial models violate the property of homogeneity. For instance, this chapter can confirm the violation of RAM by Example 6.5.

In Example 6.5, μ is an arbitrary positive number. If $\mu\geq 1$, then DMU {A} is fully efficient. If $2\geq\mu\geq 1$, then DMU {C} is fully inefficient. If $\mu\geq 2$, then DMU {B} is fully inefficient. The objective of RAM becomes 1–0.5 μ on DMU {B} if $2\geq\mu\geq 1$ and it is 0 if $\mu\geq 2$. When $\mu=1$,the *OE* measure of DMU {B} is 0.5. The homogeneity of RAM for $(x_k,g_k)=(1.5, 0.5)$ has the following relationship:

$$\theta\left[\mu X_k,G_k,\boldsymbol{X}',\boldsymbol{G},A(\boldsymbol{X}',\boldsymbol{G}),R_h(\mu X_k,G_k)\right]-\frac{1}{\mu}\theta\left[X_k,G_k,\boldsymbol{X},\boldsymbol{G},A(\boldsymbol{X},\boldsymbol{G}),R_h(X_k,G_k)\right]$$
$$=1-\frac{1}{2}\mu-\frac{1}{2\mu}=\frac{(\mu-1)^2}{2\mu}>0$$

EXAMPLE 6.5

DMU	Input	Desirable output
{A}	1	2
{B}	μ	1/2
{C}	2	1/2

(a) Source: Sueyoshi and Sekitani (2009).

[7] The proof on Proposition 6.7 is related to only homogeneity of degree minus one in the input quantities. The input-based proof can be applied to examination regarding homogeneity of degree one in the desirable output quantities.

EXAMPLE 6.6

DMU	Input 1	Input 2	Output
{A}	1	2	1
{B}	5/4	1	1
{C}	μ	3μ	1

(a) Source: Sueyoshi and Sekitani (2009).

for all μ such that $1 < \mu \le 2$ and

$$\theta\left[\mu X_k, G_k, \boldsymbol{X}', \boldsymbol{G}, \boldsymbol{A}(\boldsymbol{X}', \boldsymbol{G}), R_h(\mu X_k, G_k)\right] - \frac{1}{\mu}\theta\left[X_k, G_k, \boldsymbol{X}, \boldsymbol{G}, \boldsymbol{A}(\boldsymbol{X}, \boldsymbol{G}), R_h(X_k, G_k)\right] = 0 - \frac{1}{2\mu} < 0$$

for all $\mu \ge 2$. Hence, RAM does not satisfy the property of homogeneity on $(X_k, G_k) = (1.5, 0.5)$.

The following example indicates that SBM and Russell measure violate the property of homogeneity:

In Example 6.6, μ is an arbitrary positive number. If $\mu \ge 1$, then DMUs {A, C} are efficient. If $1 \le \mu \le 5/4$, then the *OE* measure of DMU {B} is $1/\mu - 1/6$. If $\mu = 1$, then the inputs and output of DMU {C} become $X_k = (1,3)$ and $g_k = 1$. In this case, DMU {B} becomes efficient and the *OE* measure of DMU {C} is 5/6. Hence, a DMU {k} with $X_k = (1, 3)$ and $g_k = 1$ has the following OE measure under $1 \le \mu \le 5/4$:

$$\frac{1}{\mu}\theta\left[X_k, G_k, \boldsymbol{X}, \boldsymbol{G}, \boldsymbol{A}(\boldsymbol{X}, \boldsymbol{G}), R_h(X_k, G_k)\right] - \theta \ge \left[\mu X_k, G_k, \boldsymbol{X}', \boldsymbol{G}, \boldsymbol{A}(\boldsymbol{X}', \boldsymbol{G}), R_h(\mu X_k, G_k)\right] = \frac{5}{6\mu} - \frac{1}{\mu} + \frac{1}{6} > 0.$$

Thus, the SBM violates the property of homogeneity.

The Russell measure of DMU {B} is 8/9. Under $1 \le \mu \le 5/4$, the *OE* measure of DMU {B} is $2/9 + 2/(3\mu)$. Thus, the DMU with $X_k = (1, 3)$ and $g_k = 1$ has the following *OE* measure under $1 \le \mu \le 5/4$:

$$\frac{1}{\mu}\theta\left[X_k, G_k, \boldsymbol{X}, \boldsymbol{G}, \boldsymbol{A}(\boldsymbol{X}, \boldsymbol{G}), R_h(X_k, G_k)\right] - \theta\left[\mu X_k, G_k, \boldsymbol{X}', \boldsymbol{G}, \boldsymbol{A}(\boldsymbol{X}', \boldsymbol{G}), R_h(\mu X_k, G_k)\right] = \frac{8}{9\mu} - \frac{2}{9} - \frac{2}{3\mu} = \frac{2}{9}\left(\frac{1}{\mu} - 1\right) < 0.$$

Thus, the Russell measure violates the property of homogeneity.

Note that this chapter does not prove that the additive model and SARM violate the property of homogeneity, but the above proof is applicable to the two DEA models in a similar manner.

6.6.7 Strict Monotonicity

Proposition 6.8: Both SBM and the Russell measure satisfy the property of strict monotonicity, but all other DEA models violate the property. The first assertion on SBM and the Russell measure is proved by examining whether all DMUs $\{k\}$, having $\lambda_k^* = 0$ on every optimal solution λ^* in Model (6.1), satisfy:

$$\theta\left[X_k', G_k', \boldsymbol{X}', \boldsymbol{G}', \boldsymbol{A}(\boldsymbol{X}', \boldsymbol{G}'), R_h(X_k', G_k')\right] < \theta\left[X_k, G_k, \boldsymbol{X}, \boldsymbol{G}, \boldsymbol{A}(\boldsymbol{X}, \boldsymbol{G}), R_h(X_k, G_k)\right]$$

on each DMU {k} which has $(-X_k', G_k') \leq (-X_k, G_k)$ with $(-X_k', G_k') \neq (-X_k, G_k)$. Here, $\boldsymbol{X}' = \left[X_1, \ldots, X_{k-1}, X_k', X_{k+1}, \ldots, X_n\right]$ and $\boldsymbol{G}' = \left[G_1, \ldots, G_{k-1}, G_k', G_{k+1}, \ldots, G_n\right]$.

[Proof] See Appendix of this chapter.

This chapter confirms the second assertion on the violation of RM(c), RM(v) and SARM.

In Example 6.7, μ is an arbitrary positive number less than 2. Let us consider a case in which input-oriented RM(c) is applied to DMU {B}. The constraints related to the desirable outputs become $2\lambda_A + \mu\lambda_B \geq \mu$, $\lambda_A + \lambda_B \geq 1$, $\lambda_A \geq 0$ and $\lambda_B \geq 0$. Since μ is an arbitrary number that is less than 2, $2\lambda_A + \mu\lambda_B \geq \mu$ becomes redundant. The optimal solution $(\lambda_A^*, \lambda_B^*)$ satisfies $\lambda_A^* + \lambda_B^* = 1$. Hence, if $\mu < 2$, then the optimal solution of DMU{B} is $(\lambda_A^*, \lambda_B^*) = (1, 0)$. The optimal objective value is 0.5. The *OE* measure of DMU {B} is invariant even if the first output is reduced. Thus, the RM(c) violates the property of strict monotonicity. The violation of other radial models [e.g., RM(v) and SARM] is found in a similar manner.

In addition to the radial models, it is necessary for us to prove that both RAM and the additive model do not satisfy the property of strict monotonicity. Before examining the property, this chapter reviews the statement prepared by Cooper *et al.* (1999, p. 17) which is written the following: "RAM will also have properties that are referred to as strong monotonicity. We describe this property as follows: Holding all other inputs and outputs constant, an increase in any of its inputs will increase the inefficiency score for an inefficient DMU The same is true for a decrease in any of its outputs."

This chapter considers that the comment on RAM's strict monotonicity is important. However, Example 6.5 shows an example in which the RAM violates the property of strict monotonicity. In Example 6.5, let μ be an arbitrary number which is more than or equal to 1. If $\mu \geq 2$, then DMU {A} is efficient. If $2 > \mu \geq 1$, then DMU {C} is efficient. The efficiency measured by RAM becomes $1–0.5\mu$ if $2 > \mu \geq 1$ and it is 0 if $\mu \geq 2$. This result indicates that if $\mu \geq 2$, then holding the input and desirable output of the other DMUs, an increase in the input of DMU {B} does

EXAMPLE 6.7

DMU	Input	Desirable output 1	Desirable output 2
{A}	1	2	1
{B}	2	μ	1

(a) Source: Sueyoshi and Sekitani (2009).

not change the efficiency measure (= 0: fully inefficient). Meanwhile, if $2 > \mu \geq 1$, the efficiency of DMU {B} depends upon an increase of the input. Thus, the RAM does not maintain the property of strict monotonicity. The violation of RAM is because an increase in the *i*-th input, while holding the others, simultaneously increases the range adjustment (R_i^x). Consequently, the level of RAM efficiency is not changed by the input increase. The discussion here is applicable to the desirable output-based RAM that violates the property of strict monotonicity[8]. Finally, note that this chapter can prove the violation of strict monotonicity by the additive model in a similar manner.

6.6.8 Unique Projection for Efficiency Comparison

Proposition 6.9: All DEA models violate the property of unique projection for efficiency comparison.

The following examples confirm the violation of unique projection for efficiency comparison:

When RAM, SBM and Russell measure are applied to the data set of Example 6.8, the three models have multiple optimal solutions on DMU {C} such as $\{\lambda^* | \ \lambda_A^* + \lambda_B^* = 1, \lambda_C^* = \lambda_D^* = 0, \lambda_A^* \geq 0, \lambda_B^* \geq 0\}$. Multiple projections occur on the following line segment:

$$\left\{ \begin{pmatrix} X\lambda^* \\ G\lambda^* \end{pmatrix} \middle| \lambda_A^* + \lambda_B^* = 1, \lambda_C^* = \lambda_D^* = 0, \lambda_A^* \geq 0, \lambda_B^* \geq 0 \right\}$$

$$= \left\{ \begin{pmatrix} \lambda_A^* + 2\lambda_B^* \\ 2\lambda_A^* + \lambda_B^* \\ 2 \end{pmatrix} \middle| \ \lambda_A^* + \lambda_B^* = 1, \lambda_A^* \geq 0, \lambda_B^* \geq 0 \right\}.$$

EXAMPLE 6.8

DMU	Input 1	Input 2	Desirable output
{A}	1	2	2
{B}	2	1	2
{C}	2	2	2
{D}	6	6	1

(a) Source: Sueyoshi and Sekitani (2009).

[8] In examining the violation of RAM's strict monotonicity, we consider a case where RAM includes zero in the efficiency measure. The zero in RAM's efficiency measure is important because the measure should be ranged between 0 (full inefficiency) and 1 (full efficiency). That is due to the property of efficiency requirement. As discussed by Pastor *et al.* (1999), RAM may have zero as the lower bound of its efficiency measure. In other words, RAM can identify the full inefficiency. The property of RAM's strict monotonicity needs to satisfy within the range [0, 1] under the efficiency requirement. If we omit zero (full inefficiency) in RAM's efficiency measure, it indicates violation of the efficiency requirement. The RAM violates the property of strict monotonicity due to the efficiency requirement. Hence, we must consider zero in the RAM's efficiency measure.

EXAMPLE 6.9

DMU	Input	Desirable output 1	Desirable output 2
{A}	1	2	1
{B}	1	3	1
{C}	1	1	1

(a) Source: Sueyoshi and Sekitani (2009).

When input-oriented RM(c) is applied to DMU{C} of Example 6.9, the DMU has unity in the degree of *OE*. However, the DMU suffers from an occurrence of multiple solutions on the following line segment:

$$\left\{\begin{pmatrix} X\lambda^* \\ G\lambda^* \end{pmatrix} \middle| \lambda_A^* + \lambda_B^* + \lambda_C^* = 1, \lambda_A^* \geq 0, \lambda_B^* \geq 0, \lambda_C^* \geq 0\right\}$$
$$= \left\{\begin{pmatrix} 1 \\ 3\lambda_B^* + \lambda_C^* \\ 1 \end{pmatrix} \middle| \lambda_B^* + \lambda_C^* = 1, \lambda_B^* \geq 0, \lambda_C^* \geq 0\right\}.$$

Thus, the RM(c) suffers from an occurrence of multiple projections. The assertion is applicable to input-oriented RM(v), as well. Therefore, the radial measures violate the property of unique projection.

The optimal solution of SARM applied to DMU {C} of Example 6.10 is on $\left\{\lambda^* \middle| \lambda_A^* + \lambda_B^* = 1, \lambda_C^* = 0, \lambda_A^* \geq 0, \lambda_B^* \geq 0\right\}$. Hence, $\left\{\begin{pmatrix} X\lambda^* \\ G\lambda^* \end{pmatrix} \middle| \lambda_A^* + \lambda_B^* = 1, \lambda_A^* \geq 0, \lambda_B^* \geq 0, \lambda_C^* = 0\right\}$ becomes a projection point set for DMU {C}. In other words, a projection point(s) of the DMU is not unique and the projection(s) may occur on a line segment: $\left\{\begin{pmatrix} 1 \\ \lambda_A^* + 2\lambda_B^* \\ 2\lambda_A^* + \lambda_B^* \end{pmatrix} \middle| \lambda_A^* + \lambda_B^* = 1, \lambda_A^* \geq 0, \lambda_B^* \geq 0\right\}$. The discussion on the above three examples can be applied to other DEA models, including the additive model.

6.6.9 Unit Invariance

Proposition 6.10: All DEA models, except the additive model, satisfy the property of unit invariance. Given any positive diagonal input matrix $\boldsymbol{D}^x$ whose order is *m* and any positive diagonal desirable output matrix $\boldsymbol{D}^g$ whose order is *s*, all DMUs ($k=1,\ldots,n$) satisfy the following condition:

$$\theta\left[\boldsymbol{D}^x X_k, \boldsymbol{D}^g G_k, \boldsymbol{D}^x \boldsymbol{X}, \boldsymbol{D}^g \boldsymbol{G}, \boldsymbol{A}\left(\boldsymbol{D}^x \boldsymbol{X}, \boldsymbol{D}^g \boldsymbol{G}\right), R_h\left(\boldsymbol{D}^x X_k, \boldsymbol{D}^g G_k\right)\right]$$
$$= \theta\left[X_k, G_k, \boldsymbol{X}, \boldsymbol{G}, \boldsymbol{A}\left(\boldsymbol{X}, \boldsymbol{G}\right), R_h\left(X_k, G_k\right)\right].$$

[Proof] See Appendix of this article.

EXAMPLE 6.10

DMU	Input	Desirable output 1	Desirable output 2
{A}	1	1	2
{B}	1	2	1
{C}	2	1	1

(a) Source: Sueyoshi and Sekitani (2009).

EXAMPLE 6.11

DMU	Input	Desirable output
{A}	10 (1/10)	2
{B}	20 (2/10)	3
{C}	20 (2/10)	3/2

(a) Source: Sueyoshi and Sekitani (2009).

When the additive model is applied to the data set listed in Example 6.11, the optimal solution on DMU {C} becomes $\left(\lambda_A^*, \lambda_B^*, \lambda_C^*\right) = (1,0,0)$. The optimal objective value is –10.5. Using Model (6.4), this chapter identifies that the *OE* measure is 3/8. Here, let us consider another data set in which the inputs are divided by 100. The data set is listed within parentheses in Example 6.11. Using the data set, the additive model applied to DMU {C} produces a unique optimal solution: $\left(\lambda_A^*, \lambda_B^*, \lambda_C^*\right) = (0,1,0)$. The optimal objective value is –1.5. By transforming the objective value by Model (6.4), the *OE* measure becomes 0.5. Hence, the additive model violates the property of unit invariance.

6.6.10 Translation Invariance

Proposition 6.11: RAM satisfies the property of translation invariance. The other DEA models violate the property. The first assertion on RAM is proved as follows: Given a pair of an *m*-dimensional non-negative vector (π) and a *s*-dimensional nonnegative vector (ψ), all DMUs satisfy the following condition:

$$\begin{aligned}&\theta\left[X_k+\pi, G_k+\psi, \boldsymbol{X}+\Pi, \boldsymbol{G}+\Psi, \boldsymbol{A}\left(\boldsymbol{X}+\Pi, \boldsymbol{G}+\Psi\right), R_h\left(X_k+\pi, G_k+\psi\right)\right]\\&=\theta\left[X_k, G_k, \boldsymbol{X}, \boldsymbol{G}, \boldsymbol{A}\left(\boldsymbol{X}, \boldsymbol{G}\right), R_h\left(X_k, G_k\right)\right]\end{aligned} \tag{6.14}$$

where $\pi = [\pi_1, \ldots, \pi_m]^{Tr}, \psi = [\psi_1, \ldots, \psi_s]^{Tr}, \Pi = [\pi, \pi, \ldots, \pi]$ and $\Psi = [\psi, \psi, \ldots, \psi]$.

[Proof] See Appendix of this chapter.

The following example confirms the assertion. That is, input-oriented RM(v), RM(c), SARM, SBM and the Russell measure violate the property of translation invariance.

DMU {C} has the optimal solution $\left(\lambda_A^*, \lambda_B^*, \lambda_C^*\right) = (1,0,0)$ under RM(c), RM(v), RAM, SARM, SBM and the Russell measure. The *OE* measure is 3/7 by the two radial models, SARM and SBM. The *OE* of Russell measure is 0.655. To examine whether these DEA models satisfy the property of translation invariance, this

EXAMPLE 6.12

DMU	Input	Desirable output
{A}	3 (4)	2
{B}	4 (5)	1
{C}	7/2 (9/2)	2

(a) Source: Sueyoshi and Sekitani (2009).

EXAMPLE 6.13

DMU	Input	Desirable output
{A}	3 (4)	1
{B}	4 (5)	1

(a) Source: Sueyoshi and Sekitani (2009).

chapter adds one to the three input values, as listed in the second (input) column of Example 6.12. The optimal solution of DMU {C} becomes $\left(\lambda_A^*, \lambda_B^*, \lambda_C^*\right) = (1, 0, 0)$ under the five DEA models, but the optimal objective value, indicating the degree of *OE*, becomes 4/9. The *OE* measure of Russell measure becomes 2/3. Thus, the two radial models, SARM, SBM and the Russell measure do not satisfy the property of translation invariance.

The following example confirms the assertion on the additive model:

The additive model, applied to DMU {B} in Example 6.13, produces $\left(\lambda_A^*, \lambda_B^*\right) = (1, 0)$ as an optimal solution and –1 as an optimal objective value of Model (6.4). The objective value is changed to 7/8 as the *OE* measure. If we add one to the two input values, as listed in the second (input) column of Example 6.13, then the additive model applied to DMU {B} produces $\left(\lambda_A^*, \lambda_B^*\right) = (1, 0)$ as the optimal solution and –1 as the optimal objective value of Model (6.4). However, the *OE* measure becomes 9/10. Hence, the additive model violates the property of translation invariance.

6.7 SUMMARY

This chapter reviewed desirable properties for *OE* in a unified analytical synthesis. Then, we theoretically compared seven radial and non-radial DEA models based upon the *OE* criteria. The findings on *OE* criteria were summarized in Table 6.1. This table provided us with the following guidance in selecting an appropriate DEA model. First, there is no perfect DEA model that can satisfy all desirable *OE* properties. It is necessary for us to develop the perfect DEA model that can satisfy all the desirable *OE* properties. That will be important but being a difficult future research task. Second, all DEA models violate both the property of unique projection for efficiency comparison and the property of aggregation. Cooper *et al.* (2007a) discussed how to deal with the aggregation issue. Third, the additive model violates all desirable *OE* properties. This chapter recommends the use of RAM as an extended formulation of the additive model because RAM satisfies several desirable *OE* properties. Fourth, both the Russell measure and SBM (= ERGM) perform as well as RAM as a non-radial measure. If we are interested in the property of strict monotonicity, the two models (SBM and Russell measure) outperform the other models including RAM. In contrast, if we are interested in the property of translation invariance, RAM outperforms the Russell measure and SBM (= ERGM). In Section II (DEA environmental assessment), the proposed DEA model for non-radial measurement originates from the RAM because it needs to utilize the property of translation invariance. See Chapter 26 which discusses how to handle an occurrence of many zeros in a data set. Fifth, if we are interested in the property of homogeneity, the two RMs are useful for such a purpose. All the non-radial models do not satisfy the desirable property[9]. Sixth, RM(c) is useful in measuring a frontier shift among different periods[10]. See Chapter 19 which describes how to measure a frontier shift among different periods by using the radial measurement for environmental assessment, under constant RTS. Finally, if a data set contains zero and/or negative values, then we need to depend upon the property of "disaggregation" although this chapter has not discussed it. See Chapter 27 for a detailed description on the property of disaggregation. In the case of non-radial measurement, RAM becomes an appropriate model for the data set[11] that contains zero and/or negative values. Meanwhile, in the case of both radial and non-radial measurements, the property of disaggregation is useful in handling such zero and/or negative values.

In addition to the above implications for DEA users, Table 6.1 indicates that all DEA models examined in this chapter suffer from an occurrence of multiple projections. As a result of such a difficulty, all the DEA models cannot uniquely determine the *OE* measure for efficiency comparison. The difficulty is often associated with an occurrence of multiple reference sets and multiple projections. Unfortunately, the research issue has been insufficiently explored in previous DEA studies. To overcome the difficulty, we need to develop a new approach that

[9] If non-radial measures do not satisfy the property of homogeneity, we may consider the property of sub-homogeneity as suggested by Färe *et al.* (1983).

[10] There are several articles in which shifts are appropriately (and probably better) measured and different from RM under constant RTS (e.g., Portela and Thanassoulis, 2006; Grifell-Tatjé *et al.*, 1998; Thrall, 2000).

[11] Several approaches provide us with DEA models to deal with negative data that probably perform better than RAM. For instance, the range directional model (Portela *et al.*, 2004) provides efficiency scores (with a similar interpretation as the traditional efficiency scores of the radial DEA models) that can be readily used without a need to transform the negative data. Its extension to the non-radial case proposed by Pastor and Ruiz (2007) is based on a measure of Cooper *et al.* (1999). Sueyoshi (2001b, 2004, 2006) provided an approach to deal with a negative value(s) without any transformation. His approach applied to DEA-discriminant analysis (DA) can also be directly used for all DEA models without any translation. In this case, the approach will be formulated by an integer programming model. No research has explored the computational issue. See Chapters 26 and 27 that discuss how to handle an occurrence of zero and/or negative values within the computational framework of DEA environmental assessment.

can identify the existence of unique projection for efficiency comparison. The proposed approach combines a DEA model (primal formulation) with its dual model and then adds SCSCs in a combined model. The incorporation of SCSCs restricts projections of an inefficient DMU in a "minimum face" on an efficiency frontier when it changes its status from inefficiency to efficiency. Consequently, the combined primal/dual model with SCSCs can identify a projection set as the minimum face on an efficiency frontier. See Chapter 7 that discusses how to incorporate the SCSCs into DEA assessment.

At the end of this chapter, it is necessary to describe the following four concerns for research extension from conventional DEA to its environmental assessment. First, chapters in Section II need to incorporate undesirable outputs in their computational frameworks. Unification between desirable and undesirable outputs will be very important in terms of DEA environmental assessment. See Chapters 15, 16 and 17 on output unification. Second, a unified efficiency measure for environmental assessment has the assumption that undesirable outputs are "by-products" of desirable outputs. As a result of the assumption, a desirable output vector should have an increasing trend, and an undesirable output vector should have an increasing and decreasing trend along with an increase in an input vector. To identify an occurrence of such an increasing and decreasing trend, DEA environmental assessment needs to incorporate the concept of "congestion." See Chapters 9, 20, 21, 22 and 23. Thus, "strict monotonicity" and "liner homogeneity," for example, are desirable properties for conventional DEA as discussed in this chapter, but are not necessary for environmental assessment. Third, this chapter did not discuss that dual variables, or multipliers, obtained from a radial or non-radial model should be positive in their signs. Otherwise, production factors, corresponding to dual variables with zero, will be not utilized in DEA assessment. RAM and SARM can satisfy the property. Therefore, the two models will be utilized for non-radial and radial models for environmental assessment, respectively. Finally, all models used for environmental assessment should be intuitively and visually discussed. As explored in this chapter, DEA environmental assessment will consider a balance between theory and practicality. These concerns will be explored in chapters in Section II.

APPENDIX

Proof of Proposition 6.1

Any combination between $\theta \geq 1$, $\lambda_k = 1$ and $\lambda_j = 0$ $(j \neq k)$ becomes a feasible solution. The objective function $\sum_{i=1}^{m} R_i^x \sum_{j=1}^{n} x_{ij}\lambda_j - \sum_{r=1}^{s} R_r^g \left(\sum_{j=1}^{n} g_{rj}\lambda_j - g_{rk}\right)$ is bounded, because $\sum_{j=1}^{n} \lambda_j = 1$ and $\lambda_j \geq 0$ $(j = 1, 2, \ldots, n)$. If $\left(1 - \sum_{i=1}^{m} R_i^x x_{ik}\right) < 0$, then SARM does not have an optimal solution. In contrast, if $\left(1 - \sum_{i=1}^{m} R_i^x x_{ik}\right) \geq 0$, then the objective function of SARM has a lower bound because

$$\left(1-\sum_{i=1}^{m}R_i^x x_{ik}\right)\theta+\sum_{i=1}^{m}\left(R_i^x\sum_{j=1}^{n}x_{ij}\lambda_j\right)+\sum_{r=1}^{s}\left(R_r^g\sum_{j=1}^{n}\left(g_{rj}\lambda_j-g_{rk}\right)\right)$$

$$\geq\left(1-\sum_{i=1}^{m}R_i^x x_{ik}\right)\theta-\sum_{r=1}^{s}R_r^g\left(\sum_{j=1}^{n}g_{rj}\lambda_j-g_{rk}\right)\geq-\sum_{r=1}^{s}R_r^g\sum_{j=1}^{n}g_{rj}\lambda_j\geq-\sum_{r=1}^{s}R_r^g\ \underset{j=1,\ldots,n}{\text{maximize}}\ g_{rj}.$$

Let (θ^*,λ^*) be an optimal solution of SARM, then $X\lambda^*\leq\theta^*x_k$, $G\lambda^*\geq g_k$ and $\lambda^*\geq0$. It follows from $x_k>0$ that $\theta^*\geq\underset{i=1,\ldots,m}{\text{maximize}}\frac{\sum_{j=1}^{n}x_{ij}\lambda_j^*}{x_{ik}}\geq\frac{\sum_{j=1}^{n}x_{\ell j}\lambda_j^*}{x_{\ell k}}$, $\forall\ell=1,\ldots,m$. This indicates that $\left(\underset{i=1,\ldots,m}{\text{maximize}}\frac{\sum_{j=1}^{n}x_{ij}\lambda_j^*}{x_{ik}},\lambda^*\right)$ is a feasible solution of SARM and $\left(1-\sum_{i=1}^{m}R_i^x x_{ik}\right)\geq0$. Hence, the following relationship is maintained on optimality:

$$\left(1-\sum_{i=1}^{m}R_i^x x_{ik}\right)\theta^*+\sum_{i=1}^{m}R_i^x\sum_{j=1}^{n}x_{ij}\lambda_j^*-\sum_{r=1}^{s}R_r^g\sum_{j=1}^{n}(g_{rj}\lambda_j^*-g_{rk})$$

$$\geq\left(1-\sum_{i=1}^{m}R_i^x x_{ik}\right)\left(\underset{i=1,\ldots,m}{\text{maximize}}\frac{\sum_{j=1}^{n}x_{ij}\lambda_j^*}{x_{ik}}\right)+\sum_{i=1}^{m}R_i^x\sum_{j=1}^{n}x_{ij}\lambda_j^*-\sum_{r=1}^{s}R_r^g\left(\sum_{j=1}^{n}g_{rj}\lambda_j^*-g_{rk}\right).$$

The above relationship implies that $\left(\underset{i=1,\ldots,m}{\text{maximize}}\frac{\sum_{j=1}^{n}x_{ij}\lambda_j^*}{x_{ik}},\lambda^*\right)$ is on the optimality of SARM. Q.E.D.

Proof of Proposition 6.6

For any $\lambda\geq0$, $f(X_k,G_k,\boldsymbol{X},\boldsymbol{G},\lambda)\geq0$ is maintained in two RMs (input-oriented), SBM and the Russell measure. For a feasible solution of λ, RAM has $f\left(X_k,G_k,\boldsymbol{X},\boldsymbol{G},\lambda\right)\leq0$, $0\leq x_{ik}-\sum_{j=1}^{n}x_{ij}\lambda_j\leq1/[(m+s)R_i^x]$ $(i=1,\ldots,m)$ and $0\leq\sum_{j=1}^{n}g_{rj}\lambda_j-g_{rk}\leq1/[(m+s)R_r^g]\,(r=1,\ldots,s)$. Hence, $-1\leq f(X_k,G_k,\boldsymbol{X},\boldsymbol{G},\lambda)\leq0$. The solution of $\lambda_k=1$ and $\lambda_j=0$ $(j\neq k)$ is feasible in two RMs (input-oriented), SBM and the Russell measure. For a feasible solution of λ, two RMs (input-oriented), SBM and the Russell measure have $f(X_k,G_k,\boldsymbol{X},\boldsymbol{G},\lambda)=1$. Hence, RAM has $-1\leq Q[X_k,G_k,\boldsymbol{X},\boldsymbol{G},\boldsymbol{A}(\boldsymbol{X},\boldsymbol{G}),R_h(X_k,G_k)]\leq0$ and two RMs, SBM and Russell measure have $0\leq Q[X_k,G_k,\boldsymbol{X},\boldsymbol{G},\boldsymbol{A}(\boldsymbol{X},\boldsymbol{G}),R_h(X_k,G_k)]\leq1$. In other words, two RMs (input-oriented), RAM, SBM and Russell measure have $0\leq\theta[X_k,G_k,\boldsymbol{X},\boldsymbol{G},\boldsymbol{A}(\boldsymbol{X},\boldsymbol{G}),R_h(X_k,G_k)]\leq1$.

Next, we prove that if $\theta[X_k,G_k,\boldsymbol{X},\boldsymbol{G},\boldsymbol{A}(\boldsymbol{X},\boldsymbol{G}),R_h(X_k,G_k)]=1$ is identified by SARM, RAM, SBM and Russell measure, then DMU {k} is fully efficient by using contradiction. We assume that there exists a feasible solution $\hat{\lambda}$ of Model (6.1) for DMU $\{k\}$ such that $\boldsymbol{X}\hat{\lambda} \le x_k$, $\boldsymbol{G}\hat{\lambda} \ge G_k$ and $(\boldsymbol{X}\hat{\lambda}, \boldsymbol{G}\hat{\lambda}) \ne (X_k, G_k)$. Since $\boldsymbol{X}>0$ and $\boldsymbol{G}>0$, we have

$$\left(\frac{-\sum_{j=1}^{n} x_{1j}\hat{\lambda}_j}{x_{1k}}, \ldots, \frac{-\sum_{j=1}^{n} x_{mj}\hat{\lambda}_j}{x_{mk}}, \frac{\sum_{j=1}^{n} g_{1j}\hat{\lambda}_j}{g_{1k}}, \ldots, \frac{\sum_{j=1}^{n} g_{sj}\hat{\lambda}_j}{g_{sk}} \right) \ge (-1, \ldots, -1, 1, \ldots, 1)$$

and

$$\left(\frac{-\sum_{j=1}^{n} x_{1j}\hat{\lambda}_j}{x_{1k}}, \ldots, \frac{-\sum_{j=1}^{n} x_{mj}\hat{\lambda}_j}{x_{mk}}, \frac{\sum_{j=1}^{n} g_{1j}\hat{\lambda}_j}{g_{1k}}, \ldots, \frac{\sum_{j=1}^{n} g_{sj}\hat{\lambda}_j}{g_{sk}} \right) \ne (-1, \ldots, -1, 1, \ldots, 1).$$

Let $\lambda_k' = 1$ and $\lambda_j' = 0$ for all $j \ne k$, then we have

$$\begin{aligned} \theta\left[X_k, G_k, \boldsymbol{X}, \boldsymbol{G}, \boldsymbol{A}(\boldsymbol{X}, \boldsymbol{G}), R_h(X_k, G_k)\right] &\le f\left(X_k, G_k, \boldsymbol{X}, \boldsymbol{G}, \hat{\lambda}\right) \\ &< f\left(X_k, G_k, \boldsymbol{X}, \boldsymbol{G}, \lambda'\right) = 1 \end{aligned}$$

for SBM and Russell measure. This is in contradiction to $\theta[X_k, G_k, \boldsymbol{X}, \boldsymbol{G}, \boldsymbol{A}(\boldsymbol{X}, \boldsymbol{G}), R_h(X_k, G_k)] = 1$.

Similarly, we have

$$\begin{aligned} &\left(-R_1^x \sum_{j=1}^{n} x_{1j}\hat{\lambda}_j, \ldots, -R_m^x \sum_{j=1}^{n} x_{mj}\hat{\lambda}_j, R_1^g \sum_{j=1}^{n} g_{1j}\hat{\lambda}_j, \ldots, R_s^g \sum_{j=1}^{n} g_{sj}\hat{\lambda}_j \right) \\ &\ge \left(-R_1^x x_{1k}, \ldots, -R_m^x x_{mk}, R_1^g g_{1k}, \ldots, R_s^g g_{sk} \right) \end{aligned}$$

and

$$\begin{aligned} &\left(-R_1^x \sum_{j=1}^{n} x_{1j}\hat{\lambda}_j, \ldots, -R_m^x \sum_{j=1}^{n} x_{mj}\hat{\lambda}_j, R_1^g \sum_{j=1}^{n} g_{1j}\hat{\lambda}_j, \ldots, R_s^g \sum_{j=1}^{n} g_{sj}\hat{\lambda}_j \right) \\ &\ne \left(-R_1^x x_{1k}, \ldots, -R_m^x x_{mk}, R_1^g g_{1k}, \ldots, R_s^g g_{sk} \right). \end{aligned}$$

Therefore, SARM has

$$\begin{aligned} \theta\left[X_k, G_k, \boldsymbol{X}, \boldsymbol{G}, \boldsymbol{A}(\boldsymbol{X}, \boldsymbol{G}), R_h(X_k, G_k)\right] &\le f\left(X_k, G_k, \boldsymbol{X}, \boldsymbol{G}, \hat{\lambda}\right) \\ &< f\left(X_k, G_k, \boldsymbol{X}, \boldsymbol{G}, \lambda'\right) \le 1 \end{aligned}$$

and RAM has

$$\begin{aligned} 1 &\ge 1 + f\left(X_k, G_k, \boldsymbol{X}, \boldsymbol{G}, \lambda'\right) > 1 + f\left(X_k, G_k, \boldsymbol{X}, \boldsymbol{G}, \hat{\lambda}\right) \\ &\ge 1 + Q\left[X_k, G_k, \boldsymbol{X}, \boldsymbol{G}, \boldsymbol{A}(\boldsymbol{X}, \boldsymbol{G}), R_h(X_k, G_k)\right] \\ &= \theta\left[X_k, G_k, \boldsymbol{X}, \boldsymbol{G}, \boldsymbol{A}(\boldsymbol{X}, \boldsymbol{G}), R_h(X_k, G_k)\right]. \end{aligned}$$

This is in contradiction to $\theta[X_k, G_k, \boldsymbol{X}, \boldsymbol{G}, \boldsymbol{A}(\boldsymbol{X}, \boldsymbol{G}), R_h(X_k, G_k)] = 1$. Q.E.D.

Proof of Proposition 6.7

Both of the two input-oriented RMs have $\boldsymbol{A(X,G)}=\boldsymbol{A(X',G')}$ and $R_h(X_k, G_k) = R_h(\mu X_k, G_k)$. Let λ^* be an optimal solution in Model (6.1). Then we have $\boldsymbol{A}(\boldsymbol{X}',\boldsymbol{G})\lambda^* \le R_h(X_k, G_k)$ and $\lambda^* \ge 0$. It follows from $\lambda_k^* = 0$ and $\mu \ge 1$ that

$$
\begin{aligned}
&\frac{1}{\mu}Q\left[X_k, G_k, \boldsymbol{X}, \boldsymbol{G}, \boldsymbol{A}(\boldsymbol{X}, \boldsymbol{G}), R_h(X_k, G_k)\right] \\
&= \frac{1}{\mu}\underset{i=1,\dots,m}{\text{maximize}} \frac{\sum_{j=1}^{n} x_{ij}\lambda_j^*}{x_{ik}} = \underset{i=1,\dots,m}{\text{maximize}} \frac{\sum_{j=1}^{n} x_{ij}\lambda_j^*}{\mu x_{ik}} \\
&= \underset{i=1,\dots,m}{\text{maximize}} \frac{\sum_{j\ne k} x_{ij}\lambda_j^* + \mu x_{ik}\lambda_k^*}{\mu x_{ik}} \\
&\ge \text{minimize}\left\{ \underset{i=1,\dots,m}{\text{maximize}} \frac{\sum_{j\ne k} x_{ij}\lambda_j + \mu x_{ik}\lambda_k}{\mu x_{ik}} \;\middle|\; \begin{array}{c} \boldsymbol{A}(\boldsymbol{X}', G)\lambda \le R_h(\mu X_k, G_k) \\ \lambda \ge 0 \end{array} \right\} \\
&= Q\left[\mu X_k, G_k, \boldsymbol{X}', \boldsymbol{G}, \boldsymbol{A}(\boldsymbol{X}', \boldsymbol{G}), R_h(\mu X_k, G_k)\right] \\
&\ge \text{minimize}\left\{ \underset{i=1,\dots,m}{\text{maximize}} \frac{\sum_{j\ne k} x_{ij}\lambda_j + x_{ik}\lambda_k}{\mu x_{ik}} \;\middle|\; \begin{array}{c} \boldsymbol{A}(\boldsymbol{X}', \boldsymbol{G})\lambda \le R_h(\mu X_k, G_k) \\ \lambda \ge 0 \end{array} \right\} \\
&\ge \frac{1}{\mu}\text{minimize}\left\{ \underset{i=1,\dots,m}{\text{maximize}} \frac{\sum_{j\ne k} x_{ij}\lambda_j + x_{ik}\lambda_k}{x_{ik}} \;\middle|\; \begin{array}{c} \boldsymbol{A}(\boldsymbol{X}, \boldsymbol{G})\lambda \le R_h(X_k, G_k) \\ \lambda \ge 0 \end{array} \right\} \\
&= \frac{1}{\mu}Q\left[X_k, G_k, \boldsymbol{X}, \boldsymbol{G}, \boldsymbol{A}(\boldsymbol{X}, \boldsymbol{G}), R_h(X_k, G_k)\right].
\end{aligned}
$$

Since $\frac{1}{\mu}Q[X_k, G_k, \boldsymbol{X}, \boldsymbol{G}, \boldsymbol{A}(\boldsymbol{X}, \boldsymbol{G}), R_h(X_k, G_k)] = Q[\mu X_k, G_k, \boldsymbol{X}', \boldsymbol{G}, \boldsymbol{A}(\boldsymbol{X}', \boldsymbol{G}),$ $R_h(\mu X_k, G_k)]$ is maintained in the above equation, we have

$$
\begin{aligned}
&\theta\left[\mu X_k, G_k, \boldsymbol{X}', \boldsymbol{G}, \boldsymbol{A}(\boldsymbol{X}', \boldsymbol{G}), R_h(\mu X_k, G_k)\right] \\
&\quad = \frac{1}{\mu}\theta\left[X_k, G_k, \boldsymbol{X}, \boldsymbol{G}, \boldsymbol{A}(\boldsymbol{X}, \boldsymbol{G}), R_h(X_k, G_k)\right].
\end{aligned}
$$

Q.E.D.

Proof of Proposition 6.8

Let λ^* be an optimal solution of Model (6.1) for SBM, then we have $\boldsymbol{A}(\boldsymbol{X},\boldsymbol{G})\lambda^* \le R_h(X_k,G_k)$. It follows from the definitions of X'_k and G'_k, $(-X'_k,G'_k) \le (-X_k,G_k)$ and $\lambda^*_k = 0$ that $\boldsymbol{A}(\boldsymbol{X}',\boldsymbol{G}')\lambda^* \le R_h(X'_k,G'_k)$. Therefore, λ^* is a feasible solution of

$$\begin{aligned}
&\text{minimize} \quad \left(\frac{1}{m}\sum_{i=1}^{m}\frac{\sum_{j=1}^{n}x'_{ij}\lambda_j}{x'_{ik}}\right)\Bigg/\left(\frac{1}{s}\sum_{r=1}^{s}\frac{\sum_{j=1}^{n}g'_{rj}\lambda_j}{g'_{rk}}\right)\\
&s.t. \quad A\left(X',G'\right)\lambda \le R_h\left(X'_k,G'_k\right),
\end{aligned}$$

where x'_{ij} is an (i,j) component of X' and g'_{rj} is an (r,j) component of G'. Since $(-X'_k,G'_k) \le (-X_k,G_k)$, we have

$$\sum_{i=1}^{m}\frac{\sum_{j=1}^{n}x'_{ij}\lambda^*_j}{x'_{ik}} = \sum_{i=1}^{m}\frac{\sum_{j=1}^{n}x_{ij}\lambda^*_j}{x'_{ik}} \le \sum_{i=1}^{m}\frac{\sum_{j=1}^{n}x_{ij}\lambda^*_j}{x_{ik}}$$

$$\text{and} \quad \sum_{r=1}^{s}\frac{\sum_{j=1}^{n}g'_{rj}\lambda^*_j}{g'_{rk}} = \sum_{r=1}^{s}\frac{\sum_{j=1}^{n}g_{rj}\lambda^*_j}{g'_{rk}} \ge \sum_{r=1}^{s}\frac{\sum_{j=1}^{n}g_{rj}\lambda^*_j}{g_{rk}}.$$

Since $(-X'_k,G'_k) \ne (-X_k,G_k)$, we have either

$$\sum_{i=1}^{m}\frac{\sum_{j=1}^{n}x_{ij}\lambda^*_j}{x_{ik}} > \sum_{i=1}^{m}\frac{\sum_{j=1}^{n}x_{ij}\lambda^*_j}{x'_{ik}} = \sum_{i=1}^{m}\frac{\sum_{j=1}^{n}x'_{ij}\lambda^*_j}{x'_{ik}}$$

or

$$\sum_{r=1}^{s}\frac{\sum_{j=1}^{n}g_{rj}\lambda^*_j}{g_{rk}} < \sum_{r=1}^{s}\frac{\sum_{j=1}^{n}g_{rj}\lambda^*_j}{g'_{rk}} = \sum_{r=1}^{s}\frac{\sum_{j=1}^{n}g'_{rj}\lambda^*_j}{g'_{rk}}.$$

Therefore, it follows that

$$\begin{aligned}
\theta\left(X'_k,G'_k,\boldsymbol{X}',\boldsymbol{G}',\boldsymbol{A}\left(\boldsymbol{X}',\boldsymbol{G}'\right),R_h\left(X'_k,G'_k\right)\right) &\le \sum_{i=1}^{m}\frac{\sum_{j=1}^{n}x'_{ij}\lambda^*_j}{\sum_{j=1}^{n}g'_{rj}\lambda^*_j} < \sum_{i=1}^{m}\frac{\sum_{j=1}^{n}x_{ij}\lambda^*_j}{\sum_{j=1}^{n}g_{rj}\lambda^*_j}\\
&= \theta\left(X_k,G_k,\boldsymbol{X},\boldsymbol{G},\boldsymbol{A}\left(\boldsymbol{X},\boldsymbol{G}\right),R_h\left(X_k,G_k\right)\right).
\end{aligned}$$

The satisfaction of the Russell measure can be proved in a similar manner. Q.E.D.

Proof of Proposition 6.10

$\boldsymbol{A}(\boldsymbol{D}^x\boldsymbol{X},\boldsymbol{D}^g\boldsymbol{G})\lambda \le R_h(\boldsymbol{D}^x X_k,\boldsymbol{D}^g G_k)$ is equivalent to $\boldsymbol{A}(\boldsymbol{X},\boldsymbol{G})\lambda \le R_h(X_k,G_k)$. The objective function of two input-oriented RMs, SBM and the Russell measure consists of $\sum_{j=1}^{n} x_{ij}\lambda_j / x_{ik}$ $(i=1,\ldots,m)$ and $\sum_{j=1}^{n} g_{rj}\lambda_j / g_{rk}$ $(r=1,\ldots,s)$. Let D_i^x be an (i,i) component of $\boldsymbol{D}^x$ and let D_r^g an (r,r) component of $\boldsymbol{D}^g$, then we have $\sum_{j=1}^{n} D_i^x x_{ij}\lambda_j / D_i^x x_{ik} = \sum_{j=1}^{n} x_{ij}\lambda_j / x_{ik}$ $(i=1,\ldots,m)$ and $\sum_{j=1}^{n} D_r^g g_{rj}\lambda_j / D_r^g g_{rk} = \sum_{j=1}^{n} g_{rj}\lambda_j / g_{rk}$ $(r=1,\ldots,s)$. Both $\sum_{j=1}^{n} x_{ij}\lambda_j / x_{ik}$ and $\sum_{j=1}^{n} g_{rj}\lambda_j / g_{rk}$ are unit invariant. Therefore, two input-based RMs, SBM and the Russell measure are unit invariant.

The objective function of RAM and SARM consists of $R_i^x\left(x_{ik} - \sum_{j=1}^{n} x_{ij}\lambda_j\right)$ and $R_r^g\left(\sum_{j=1}^{n} g_{rj}\lambda_j - g_{rk}\right)$. Since R_i^x for $\boldsymbol{D}^x\boldsymbol{X}$ is $D_i^x R_i^x$ $(i=1,\ldots,m)$ and R_r^g for $\boldsymbol{D}^g\boldsymbol{G}$ is $D_r^g R_r^g$, we have $\left(x_{ik} - \sum_{j=1}^{n} x_{ij}\lambda_j\right)\Big/R_i^x = \left(D_i^x x_{ik} - \sum_{j=1}^{n} D_i^x x_{ij}\lambda_j\right)\Big/D_i^x R_i^x$ and $\left(\sum_{j=1}^{n} g_{rj}\lambda_j - g_{rk}\right)\Big/R_r^g = \left(\sum_{j=1}^{n} D_r^g g_{rj}\lambda_j - D_r^g g_{rk}\right)\Big/D_r^g R_r^g$. Therefore, both RAM and SARM are unit invariant. Q.E.D.

Proof of Proposition 6.11

For RAM, $\boldsymbol{A}(\boldsymbol{X},\boldsymbol{G})\lambda \le R_h(X_k,G_k)$ is equivalent to $\boldsymbol{A}(\boldsymbol{X}+\Pi,\boldsymbol{G}+\Psi)\lambda \le R_h(X_k+\pi, G_k+\psi)$. The objective function of RAM consists of

$$x_{ik} - \sum_{j=1}^{n} x_{ij}\lambda_j,\ \sum_{j=1}^{n} g_{rj}\lambda_j - g_{rk},\ R_i^x\left(i=1,\ldots,m\right) \text{ and } R_r^g\left(r=1,\ldots,s\right).$$

Since $\sum_{j=1}^{n} \lambda_j = 1$, we have

$$\begin{aligned} x_{ik} - \sum_{j=1}^{n} x_{ij}\lambda_j &= x_{ik} + \pi_i - \sum_{j=1}^{n} x_{ij}\lambda_j - \pi_i = x_{ik} + \pi_i - \sum_{j=1}^{n} x_{ij}\lambda_j - \pi_i \sum_{j=1}^{n} \lambda_j \\ &= x_{ik} + \pi_i - \sum_{j=1}^{n} x_{ij}\lambda_j - \sum_{j=1}^{n} \lambda_j \pi_i = x_{ik} + \pi_i - \sum_{j=1}^{n} \left(x_{ij} + \pi_i\right)\lambda_j \end{aligned}$$

for $i=1,\ldots,m$, where π_i is the i-th component of a m-dimensional non-negative vector (π).

In the same manner, we can prove $\sum_{j=1}^{n} g_{rj}\lambda_j - g_{rk} = \sum_{j=1}^{n} (g_{rj} + \psi_r)\lambda_j - (g_{rk} + \psi_r)$ for all $r=1,\ldots,s$, where ψ_r is the r-th component of an s-dimensional non-negative vector ψ. From the definitions, the range of R_i^x and the range of R_r^g are equal to those of $\boldsymbol{X}+\Pi$ and $\boldsymbol{G}+\Psi$, respectively. Therefore, we have

$$\begin{aligned} &\theta\left[X_k+\pi, G_k+\psi, \boldsymbol{X}+\Pi, \boldsymbol{G}+\Psi, \boldsymbol{A}\left(\boldsymbol{X}+\Pi,\boldsymbol{G}+\Psi\right), R_h\left(X_k+\pi, G_k+\psi\right)\right] \\ &= \theta\left[X_k, G_k, \boldsymbol{X}, \boldsymbol{G}, \boldsymbol{A}\left(\boldsymbol{X},\boldsymbol{G}\right), R_h\left(X_k, G_k\right)\right] \end{aligned}$$

for every $k=1,\ldots,n$. Q.E.D.

7

STRONG COMPLEMENTARY SLACKNESS CONDITIONS

7.1 INTRODUCTION

This chapter[1] discusses the importance of strong complementary slackness conditions (SCSCs) in DEA. The use of SCSCs will be extended in Chapter 25 for an assessment on energy utility firms. The methodological concern of Chapter 25 is how to reduce the number of efficient organizations by applying both SCSCs and DEA-DA (discriminant analysis). See Chapter 11 on DEA-DA. Thus, this chapter provides a mathematical basis for Chapter 25 (ranking analysis).

To initiate this chapter, we return to Table 6.1 in which at least one DEA model(s) satisfies desirable properties related to *OE*, except unique projection for efficiency comparison and property of aggregation. The aggregation

[1]This chapter is partly based upon the following three articles: (a) Sueyoshi, T and Sekitani, K. (2009) An occurrence of multiple projections in DEA-based measurement of technical efficiency: theoretical comparison among DEA the model in Equations from desirable properties. *European Journal of Operational Research*, 196, 764–794, (b) Sueyoshi, T. and Goto, M. (2012c) DEA radial and non-radial models for unified efficiency under natural and managerial disposability: Theoretical extension by strong complementary slackness conditions. *Energy Economics*, 34, 700–713, and (c) Sueyoshi, T. and Goto, M. (2012i) Returns to scale and damages to scale with strong complementary slackness conditions in DEA assessment: Japanese corporate effort on environment protection. *Energy Economics*, 34, 1422–1434.

Environmental Assessment on Energy and Sustainability by Data Envelopment Analysis,
First Edition. Toshiyuki Sueyoshi and Mika Goto.

property, which not all models satisfy, may not be a serious problem in a practical perspective of DEA. An exception may be found in restructuring of industrial organizations, including merger and acquisition (M & A), where the sum of production factors needs to be examined by comparing it with their original achievements by separated firms. This type of managerial issue is very important from a business perspective, but difficult for us to discuss the issue by DEA because it needs to examine the corporate governance of an aggregated organization in a time horizon. A practical approach for discussing the industrial organization issue can be found in Charnes *et al.* (1988) and Sueyoshi (1991) that discussed how to examine a large-scale industrial organization issue (e.g., the restructuring policy of telecommunication industry in the United States during the 1980s). See also Sueyoshi *et al.* (2010). Furthermore, since there are previous research efforts (e.g., Cooper *et al.*, 2007a) on the property of aggregation, this chapter does not discuss the property of aggregation, rather it is directed toward a description on how to handle the occurrence of multiple projections and multiple reference sets as well as the implications of the non-Archimedean small number (ε_n) from SCSCs that has been not sufficiently examined in previous DEA studies.

Before discussing the implication of SCSCs in DEA, this chapter acknowledges that researchers, including Olesen and Petersen (1996, 2003) and Cooper *et al.* (2007b), clearly understand the existence of such a problem of multiple projections and reference sets in the *OE* measurement. However, no previous DEA research has formally explored how to deal with the occurrence of multiple projections, in particular when we cannot access any prior information to restrict the range of dual variables. In this case, we are unable to use multiplier restriction methods such as the AR and CR methods discussed in Chapter 4. As a theoretical extension of Table 6.1, this chapter examines the importance of SCSCs in handling the occurrence of multiple projections and reference sets in this chapter.

This chapter is organized as follows: Section 7.2 describes the combination of primal and dual models for the development of SCSCs. Section 7.3 summarizes illustrative examples on the use of SCSCs. Section 7.4 discusses theoretical implications on SCSCs. Section 7.5 discusses a guideline on how to extend the use of SCSCs from radial to non-radial measurement. Section 7.6 summarizes this chapter.

7.2 COMBINATION BETWEEN PRIMAL AND DUAL MODELS FOR SCSCs

Using input-oriented RM(v), or Model (4.4), this chapter can slightly modify it by a vector expression for our descriptive convenience. The modified version can be expressed by the following formulation:

$$\begin{aligned}
&\text{Minimize} \quad \theta \\
&\quad s.t. \; -\sum_{j=1}^{n} X_j\lambda_j - d^x \qquad +\theta X_k = 0, \\
&\qquad \sum_{j=1}^{n} G_j\lambda_j \qquad -d^g \qquad = G_k, \\
&\qquad \sum_{j=1}^{n} \lambda_j \qquad\qquad\qquad = 1, \\
&\qquad \lambda_j \ge 0\left(j=1,\ldots,n\right),\, d^x \ge 0, \\
&\qquad d^g \ge 0 \text{ and } \theta : \text{URS.}
\end{aligned} \tag{7.1}$$

The dual formulation of Model (7.1) is formulated as the following vector expression:

$$\begin{aligned}
&\text{Maximize} \; UG_{\text{k}} + \sigma \\
&\quad s.t. \quad -VX_j + UG_j + \sigma \le 0 \;\left(j=1,\ldots,n\right), \\
&\qquad\quad VX_k \qquad\qquad = 1, \\
&\qquad\quad V \ge 0, U \ge 0 \;\text{ and }\; \sigma : \text{URS,}
\end{aligned} \tag{7.2}$$

where $V = (v_1,\ldots,v_m)$ and $U = (u_1,\ldots,u_s)$ are two row vectors of dual variables related to the first and second sets of constraints in Model (7.1). A dual variable (σ), which is unrestricted in the sign, is due to the third constraint of Model (7.1).

Complementary slackness conditions (CSCs) always exist between every optimal solution $(\theta^*, \lambda^*, d^{x*}, d^{g*})$ of Model (7.1) and every optimal solution (V^*, U^*, σ^*) of Model (7.2). They are specified as follows:

$$\lambda_j^*\left(V^*X_j - U^*G_j - \sigma^*\right) = 0\left(j=1,\ldots,n\right), \tag{7.3}$$

$$v_i^* d_i^{x*} = 0\left(i=1,\ldots,m\right) \text{and} \tag{7.4}$$

$$u_r^* d_r^{g*} = 0\left(r=1,\ldots,s\right). \tag{7.5}$$

Equation (7.3) multiples –1 on both sides of the first equations of Model (7.2) to maintain $V^*X_j - U^*G_j - \sigma^* \ge 0$ $(j=1,\ldots,n)$. This is for our descriptive convenience.

Pairing the optimal solution $(\theta^*, \lambda^*, d^{x*}, d^{g*})$ of Model (7.1) with the optimal solution (V^*, U^*, σ^*) of Model (7.2) may satisfy the following additional conditions between them:

$$\lambda_j^* + V^*X_j - U^*G_j - \sigma^* > 0\left(j=1,\ldots,n\right), \tag{7.6}$$

$$v_i^* + d_i^{x*} > 0\,(i = 1,\ldots,m)\text{ and} \tag{7.7}$$

$$u_r^* + d_r^{g*} > 0\,(r = 1,\ldots,s). \tag{7.8}$$

In addition to Equations (7.3) to (7.5), satisfaction of the additional conditions in Equations (7.6) to (7.8) implies SCSCs[2]. Thus, the SCSCs consist of Equations (7.3) to (7.8) in the input-oriented RM(v).

To deal with an occurrence of multiple projections and reference sets, this chapter combines Model (7.1) with Model (7.2) and then adds Equations (7.6) to (7.8). The final model, combining all formulations for SCSCs, becomes as follows:

$$\begin{aligned}
&\text{Maximize } \eta \\
&\quad s.t. \quad \text{all constraints in Models (7.1) and (7.2),} \\
&\qquad\qquad \theta = UG_k + \sigma, \\
&\qquad\qquad \lambda + VX - UG - \sigma e^{Tr} \geq \eta e^{Tr}, \\
&\qquad\qquad V^{Tr} + d^x \geq \eta e^{Tr}, \\
&\qquad\qquad U^{Tr} + d^g \geq \eta e^{Tr} \text{ and} \\
&\qquad\qquad \eta \geq 0.
\end{aligned} \tag{7.9}$$

Here, the superscript *Tr* indicates a vector transpose. The second equation of Model (7.9) indicates that, for optimality, the objective of Model (7.1) should be equivalent to that of Model (7.2). The last three groups of constraints indicate that an optimal solution obtained from Model (7.9) needs to satisfy the additional conditions required to satisfy SCSCs. A new decision variable (η) is unknown and it is incorporated into Model (7.9) in order to maintain the SCSCs on optimality. See Sueyoshi and Sekitani (2007a, c).

Motivation for Model (7.9): An underlying rationale regarding why the development of Model (7.9) is necessary for DEA is because of the following rationale. That is, if we face an occurrence of multiple solutions (e.g., multiple projections and multiple reference sets), it is necessary for us to restrict dual variables (multipliers) by a unique projection from inefficiency to efficiency. However, the multiplier restriction methods (i.e., AR and CR), as discussed in Chapter 4, depend upon the availability of prior information (e.g., previous experience). Furthermore, such multiplier restriction methods do not always guarantee a unique projection and a unique reference set. Therefore, it is necessary for us to consider the multiplier restriction without any prior information, consequently

[2] See Cooper *et al.* (2006, p. 279) for a comment on previous applications of SCSCs. Sueyoshi and Sekitani (2007 a,c) used SCSCs to characterize a supporting hyperplane on a production possibility set.

dealing with an occurrence of multiple projections and multiple reference sets. That is the rationale concerning why we need SCSCs for conventional DEA.

Here, it is important to note that the incorporation of SCSCs into Model (7.9) does not immediately imply that the optimal solution of Model (7.9) always satisfies the conditions for SCSCs. To satisfy the SCSCs by Model (7.9), the proposed combined model needs to incorporate a new variable (η) that is maximized in Model (7.9). The maximization is necessary because we look for a positive value. Consequently, the optimal solution of Model (7.9) satisfies the SCSCs. It is also true that there is a possibility in which η becomes zero on optimality. In this case, Model (7.9) does not work as we expect. An easy approach to handle the problem can be found in Chapters 16 and 17 that discuss how to avoid the use of SCSCs for DEA environmental assessment by a data range adjustment. See Chapter 26, as well, which discusses a methodological problem of SCSCs from ranking analysis on all DMUs.

A unique reference set covering all possible reference sets: A benefit of solving Model (7.9), equipped with SCSCs, is that a reference set measured by the model may be considered to be unique because it contains all possible reference sets generated by Model (7.1). It is true that a single reference set does not solve the occurrence of multiple projections. However, the single reference set measured by Model (7.9) can specify the range of multiple projection sets (due to multiple projections) on an efficiency frontier, because Model (7.9) can identify the interior of the convex hull of efficient DMUs within the reference set that covers all possible projection sets. Thus, it is possible for us to identify an occurrence of all possible multiple projections by examining the dimension of such a projection set determined by SCSCs. Further, by examining the reference set of Model (7.9), we can specify which surface on a production possibility set contains all possible projection sets in the interior region even if multiple projections occur.

Positive dual variable on efficient DMUs: The incorporation of SCSCs into DEA, as formulated by Model (7.9), has two analytical features to be discussed, here. One of the two features is that, if the k-th DMU is efficient, then $V^*X_j - U^*G_j - \sigma^* = 0$ ($j \in RS_k$), where RS_k stands for a reference set of the k-th DMU, is identified for the optimality of Model (7.1). Furthermore, the efficient DMU has $d_i^{x*} = 0$ ($i=1,\ldots,m$) and $d_r^{g*} = 0$ ($r=1,\ldots,s$) for the optimality of Model (7.2). As a result, the efficient DMU has $\lambda_j^* > 0$ ($j \in RS_k$), $v_i^* > 0$ ($i=1,\ldots,m$) and $u_r^* > 0$ ($r=1,\ldots,s$) for optimality. Thus, the positivity on dual variables is identified without any multiplier restriction methods. That is an important feature of SCSCs incorporated into Model (7.9) that is a combined formulation between Models (7.1) and (7.2). Of course, the positivity on dual variables does not guarantee that Model (7.9) always produces a perfectly acceptable solution. Rather, the model is the first step for avoiding an occurrence of zero on dual variables (multipliers) of efficient DMUs. Such an occurrence is clearly inappropriate and unacceptable in many cases. Thus, DEA equipped with SCSCs can provide us with an opportunity to examine if information on all production factors is fully utilized in the computation process of DEA.

The second feature is that the property for positive dual variables is useful on only an efficient DMU. The property of SCSCs does not function on inefficient DMUs, because an inefficient DMU has usually at least one positive slack (i.e., $d_i^{x*} > 0$ for some i and $d_r^{g*} > 0$ for some r)[3]. Consequently, the proposed SCSCs are not useful on an inefficient DMU. However, the inefficient DMU is not important in terms of making an efficiency frontier, so not determining the degree of *OE* on the other DMUs. The elimination of the inefficient DMU does not influence the computational process of other DMU's *OE* measures. Furthermore, the incorporation of SCSCs may reduce the magnitude of *OE* on inefficient DMUs because of the restriction on dual variables (multipliers). See Chapter 10 on the computational concern on an inefficient DMU in DEA.

Weakness: A drawback of Model (7.9) is that it increases the computational burden, compared with Models (7.1) and (7.2), because the former model has more side constraints than the latter ones. The computational issue depends upon a programming capability of a DEA user. Considering the capability of a modern personal computer, such a computational drawback is very minor because Model (7.9) is formulated by linear programming. As long as DEA models are formulated by linear programming, we can easily solve Model (7.9) without any computational difficulty. Such a computational concern belongs to "common sense" for optimization.

Strength: Model (7.9) has a methodological benefit. As mentioned previously, multipliers (V, W and σ) become positive without depending upon multiplier restriction methods (e.g., the AR and CR approach in Chapter 4) which incorporate prior information into DEA computational process. In other words, Model (7.9) can restrict multipliers (i.e., dual variables) to be positive by the incorporation of SCSCs. This chapter does not discuss such a methodological benefit, rather directing toward how to handle an occurrence of multiple projections and multiple reference sets, hereafter.

7.3 THREE ILLUSTRATIVE EXAMPLES

This chapter uses three illustrative examples to discuss the importance of SCSCs in DEA from different computation concerns.

[3] See Chapter 10 which provides a DMU classification into subgroups. One of such groups has unity in an efficiency score but zero in all slacks.

7.3.1 First Example

The first data set contains six DMUs whose production processes use two inputs to yield a single desirable output. The six DMUs are {A, B, C, D, E, F}, whose two inputs (x_1 and x_2) and one desirable output (g) are

{A} = (10, 10, 10), {B} = (10, 10, 15), {C} = (8, 10, 14), {D} = (7, 10, 13), {E} = (6, 10, 12), and {F} = (5, 10, 10).

Figure 7.1 depicts the data set in three-dimensional space.

As depicted in Figure 7.1, the piece-wise linear concur segments ({B}-{C}-{E}-{F}) consist of an efficiency frontier. The five DMUs, except {A}, are all on the efficiency frontier, so being efficient in terms of the *OE* measurement. The remaining DMU {A} is inefficient. When we solve DMU {A} by Model (7.1), the following two optimal solutions become feasible in Model (7.1):

$$\theta^* = 1, \lambda_E^* = 1, d_1^{x*} = 4, d_2^{x*} = 0 \text{ and } d_1^{g*} = 2$$

and

$$\theta^{\#} = 1, \lambda_C^{\#} = 1, d_1^{x\#} = 2, d_2^{x\#} = 0 \text{ and } d_1^{g\#} = 4.$$

The first optimal solution indicates a projection from {A} to {E}, while the second one indicates the projection from {A} to {C}. Both indicate that DMU {A} is inefficient because of positive slacks although the efficiency score is unity. The unity in efficiency is due to the fact that Model (7.1) is an input-oriented model and the second input is 10 for all DMUs. The two results indicate that Model (7.1) produces two different optimal solutions (so, two projections) for DMU {A}. Furthermore, all the points on the line segment between {C} and {E} can become optimal solutions (so, multiple projections) for DMU {A}. Thus, an

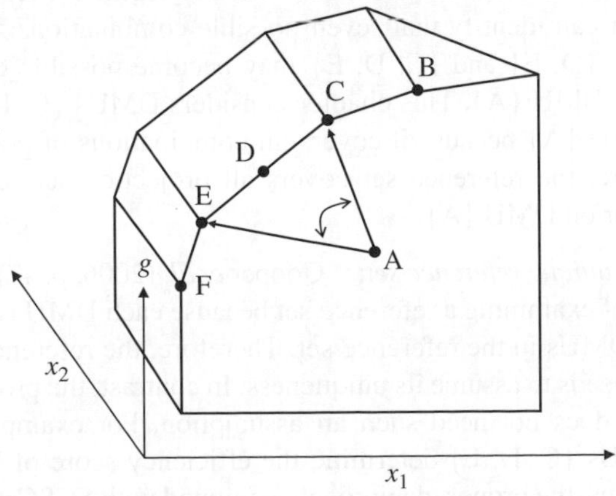

FIGURE 7.1 Occurrence of multiple projections.

infinite number of optimal solutions occur on DMU {A} by Model (7.1). This example of Figure 7.1 visually describes such an occurrence of multiple projections in DEA, or Model (7.1).

What is the difficulty in OE measurement?: An occurrence of multiple projections implies that we must deal with multiple solutions (e.g., multiple comparisons for *OE*) in Model (7.1). When this type of problem occurs in *OE* measurement, we cannot uniquely compare the input vector of a DMU with an efficient one because multiple projections produce multiple efficient input vectors. A difficulty in input vector comparison can be easily applied to the occurrence of multiple projections (so, multiple comparisons) on desirable outputs, as well. Consequently, the *OE* measure of a DMU needs to be determined on an efficiency frontier that contains all possible multiple projections. Furthermore, the occurrence of multiple projections is usually associated with a possible occurrence of multiple reference sets. Such a combined occurrence makes the problem of multiple projections more difficult in DEA performance assessment. It is indeed true that if we pay attention to the level of *OE*, then an occurrence of multiple projections and reference sets does not produce any major difficulty to us. In contrast, if we are interested in determining the type of returns to scale (RTS) as discussed in Chapter 8 and/or the occurrence of congestion as discussed in Chapter 9, then multiple projections and multiple reference sets become serious problems in not only conventional DEA but also its environmental assessment.

To explain the difficulty due to a simultaneous occurrence of multiple projections and multiple reference sets more clearly in the framework of DEA, this chapter returns to the example of Figure 7.1 in which we identify an occurrence of multiple reference sets for an inefficient DMU {A}. The occurrence depends upon which part of an efficiency frontier the inefficient DMU is projected onto. In Figure 7.1, the line segment between {C} and {E} is part of the efficiency frontier ({B}-{C}-{E}-{F}) onto which DMU {A} may be projected. The set is referred to as a "projection set (Ω)," hereafter. On the line segment (i.e., as a projection set), this chapter can identify that seven possible combinations: {C}, {D}, {E}, {C, D}, {C, E}, {D, E} and {C, D, E} may become possible candidates for a reference set for DMU {A}. This chapter considers DMUs {C, D, E} as a reference set for DMU {A} because it covers all combinations of possible reference sets. Furthermore, the reference set covers all projection sets even if multiple projections occur on DMU {A}.

Practicality of a unique reference set: Cooper *et al.* (2006, p. 47) have discussed the importance of examining a reference set because each DMU is relatively evaluated by other DMUs in the reference set. Therefore, the reference set measured by Model (7.1) needs to assume its uniqueness. In contrast, the proposed approach by Model (7.9) does not need such an assumption. For example, returning to Figure 7.1, DMUs {C, D, E} determine the efficiency score of DMU {A} as a reference set. Thus, the proposed approach, equipped with SCSCs, can identify all efficient DMUs that need to be included in the reference set. Such an analytical capability of Model (7.9) can enhance the practicality of DEA in identifying a reference set.

Mathematical expression on a projection set: To discuss more formally a reference set under an occurrence of multiple projections, this chapter proposes the following two conditions:

(a) *Condition 1*: A reference set should cover all possible reference sets on a projection set (Ω).
(b) *Condition 2*: The convex hull of the reference set should contain Ω.

These two conditions guarantee that we can determine an *OE* measure based upon a single reference set even if multiple projections occur on a DMU.

Returning to the example of Figure 7.1, it is necessary for us to describe that the problem, due to multiple projections and multiple reference sets, occurs not only on an inefficient DMU such as {A} but also on another efficient DMU {D}. The efficient DMU has multiple reference sets on the line segment on {C}, {D}, and {E}, so having seven possible reference combinations: {C}, {D}, {E}, {C, D}, {C, E}, {D, E} and {C, D, E}. The proposed model (7.9) applied to {A} and {D} produces the results summarized in Table 7.1. Model (7.9) identifies DMUs {C, D, E} as a reference set for inefficient DMU {A} and efficient DMU {D}, but with different weights on intensity variables (λ_j^*). Thus, it is possible for us to specify a single projection on a single reference set. The *OE* measure of each DMU is determined by comparing its performance with those of efficient DMUs that are identified on a single reference set. As discussed here, Model (7.9) equipped with SCSCs can identify an occurrence of multiple projections by finding a single reference set.

TABLE 7.1 Computational Summary for DMUs {A} and {D} by Model (7.9)

DMU	θ*	λ*	d^{x*}	d^{g*}	Reference Set
{A}	1	$\lambda_C^* = 0.9998$	$d_1^{x*} = 2.0003$	$d^{g*} = 3.9997$	{C, D, E}
		$\lambda_D^* = 0.0001$	$d_2^{x*} = 0$		
		$\lambda_E^* = 0.0001$			
{D}	1	$\lambda_C^* = 0.0587$	$d_1^{x*} = d_2^{x*} = 0$	$d^{g*} = 0$	{C, D, E}
		$\lambda_D^* = 0.8826$			
		$\lambda_E^* = 0.0587$			

(a) Source: Sueyoshi and Sekitani (2009). (b) DMU {A} is inefficient because of positive slacks on the first input and the desirable output. Meanwhile, DMU {D} is efficient because an efficient score is unity and all slacks are zero. (c) The efficiency measure of inefficient DMU {A} is unity because all DMUs have 10, so no difference between them, on their second inputs. (d) Both inefficient DMU {A} and efficient DMU {D} have seven possible reference combinations: {C}, {D}, {E}, {C, D}, {C, E}, {D, E} and {C, D, E}. The reference set should be {C, D, E} because this covers all possible combinations.

7.3.2 Second Example

Returning to Table 2.4 in Chapter 2, this chapter applies four radial measurements: two models: RM(c) and RM(v) × two cases: with and without SCSCs to the data set for our methodological comparison. First, Table 7.2 duplicates the computational results of Table 2.5 that are solved by input-oriented RM(c), or Model (2.5). In other words, it corresponds to Model (7.1) without $\sum_{j=1}^{n} \lambda_j = 1$ and SCSCs. Second, Table 7.3 summarizes computational results that are solved by input-oriented RM(v), or Model (7.1). Third, Table 7.4 corresponds to the result of RM(c) with SCSCs. Finally, Table 7.5 corresponds to the result of RM(v) with SCSCs. Although the analytical capability of RM(c) with SCSCs is not discussed in the first illustrative example, this second example describes such an important feature of SCSCs on both RM(v) and RM(c) in our comparative analysis.

TABLE 7.2 RM(c) without SCSCs

Company	DEA Efficiency	Reference Set	v_1^*	v_2^*	u^*	d_1^{x*}	d_2^{x*}	d_1^{g*}
{A}	0.833	C(λ_C^*=0.333), D(λ_D^*=0.667)	0.167	0.167	0.833	0.000	0.000	0.000
{B}	0.727	D(λ_D^*=0.909), E(λ_E^*=0.091)	0.182	0.091	0.727	0.000	0.000	0.000
{C}	1	C(λ_C^*=1)	0.000	1	1	0.000	0.000	0.000
{D}	1	D(λ_D^*=1)	0.250	0.125	1	0.000	0.000	0.000
{E}	1	E(λ_E^*=1)	0.500	0.000	1	0.000	0.000	0.000
{F}	1	C(λ_C^*=1)	0.000	1	1	2.000	0.000	0.000

TABLE 7.3 RM(v) without SCSCs

Company	DEA Efficiency	Reference Set	v_1^*	v_2^*	u^*	d_1^{x*}	d_2^{x*}	d_1^{g*}
{A}	0.833	C(λ_C^*=0.333), D(λ_D^*=0.667)	0.167	0.167	0.000	0.000	0.000	0.000
{B}	0.727	D(λ_D^*=0.909), E(λ_E^*=0.091)	0.182	0.091	0.000	0.000	0.000	0.000
{C}	1	C(λ_C^*=1)	0.000	1	0.000	0.000	0.000	0.000
{D}	1	D(λ_D^*=1)	0.250	0.125	0.000	0.000	0.000	0.000
{E}	1	E(λ_E^*=1)	0.500	0.000	0.000	0.000	0.000	0.000
{F}	1	C(λ_C^*=1)	0.000	1	0.000	2.000	0.000	0.000

TABLE 7.4 RM(c) with SCSCs

Company	DEA Efficiency	Reference Set	v_1^*	v_2^*	u^*	d_1^{x*}	d_2^{x*}	d_1^{g*}
{A}	0.833	C(λ_C^*=0.333), D(λ_D^*=0.667)	0.167	0.167	0.833	0.000	0.000	0.000
{B}	0.727	D(λ_D^*=0.909), E(λ_E^*=0.091)	0.182	0.091	0.727	0.000	0.000	0.000
{C}	1	C(λ_C^*=1)	0.167	0.333	1	0.000	0.000	0.000
{D}	1	D(λ_D^*=1)	0.231	0.154	1	0.000	0.000	0.000
{E}	1	E(λ_E^*=1)	0.300	0.100	1	0.000	0.000	0.000
{F}	1	C(λ_C^*=0.5), F(λ_F^*=0.5)	0.000	1	1	1.000	0.000	0.000

TABLE 7.5 RM(v): with SCSCs

Company	DEA Efficiency	Reference Set	v_1^*	v_2^*	u^*	d_1^{x*}	d_2^{x*}	d_1^{g*}
{A}	0.833	C(λ_C^*=0.333), D(λ_D^*=0.667)	0.167	0.167	0.167	0.000	0.000	0.000
{B}	0.727	D(λ_D^*=0.909), E(λ_E^*=0.091)	0.182	0.091	0.091	0.000	0.000	0.000
{C}	1	C(λ_C^*=1)	0.167	0.333	0.167	0.000	0.000	0.000
{D}	1	D(λ_D^*=1)	0.231	0.154	0.077	0.000	0.000	0.000
{E}	1	E(λ_E^*=1)	0.300	0.100	0.100	0.000	0.000	0.000
{F}	1	C(λ_C^*=0.5), F(λ_F^*=0.5)	0.000	1	0.500	1.000	0.000	0.000

Tables 7.2 and 7.3 document input-oriented RM(c) and RM(v) without SCSCs, respectively. Three DMUs {C, D, E} are efficient because their efficiency measures are unity and their slacks are all zero. Two DMUs {A, B} are inefficient because their efficiency measures are less than unity. DMU {F} is inefficient because of positivity (2.000) in the first input slack. The *OE* measures are same between the two tables. This is just a coincidence. Radial measures usually produce different results between RM(v) and RM(c).

A problem found in Tables 7.2 and 7.3 is that DMUs {C, D, E} are efficient, but they contain zero in these dual variables. The result indicates that the inputs and desirable outputs, corresponding to the dual variables with zero, are not fully utilized in DEA assessment by RM(c) and RM(v). Thus, an occurrence of zero in dual variables, in particular on efficient DMUs, is problematic in DEA assessment.

To overcome such a difficulty, this chapter incoporates SCSCs into RM(c) and RM(v). The result of incoporating SCSCs in the two radial models is summarized in Tables 7.4 and 7.5, respectively. All the three efficient DMUs {C, D, E} have positive dual variables, as documented in the two tables. DMU {F} has zero in v_1^*, but it is acceptable because it has a positive slack in the first input ($d_1^{x*} = 1$) so that it is inefficient. Thus, such an identification capability cannot be found without SCSCs. Moreover, the reference set of DMU {F} is only DMU {C} in Tables 7.2 and 7.3. However, the incoporation of SCSCs indicates that the refrence set contains two DMUs {C, F} in Tables 7.4 and 7.5. Here, it is easily thought that the incoporation of SCSCs can enhance the quality of DEA performance assessment.

To describe a use of SCSCs in a detailed formulation, this chapter returns to Model (7.9) which is applied to the performance assessment on DMU {A} as an illustrative example. In the case, the formulation with SCSCs is structured as follows:

$$
\begin{aligned}
&\text{Maximize} \quad \eta \\
&\quad s.t. \quad -4\lambda_A - 4\lambda_B - 4\lambda_C - 3\lambda_D - 2\lambda_E - 6\lambda_F - d_1^x + 4\theta = 0, \\
&\qquad\quad -2\lambda_A - 3\lambda_B - \lambda_C - 2\lambda_D - 4\lambda_E - \lambda_F - d_2^x + 2\theta = 0, \\
&\qquad\quad \lambda_A + \lambda_B + \lambda_C + \lambda_D + \lambda_E + \lambda_F - d_1^g = 1, \\
&\qquad\quad \lambda_A + \lambda_B + \lambda_C + \lambda_D + \lambda_E + \lambda_F = 1, \\
&\qquad\quad 4v_1 + 2v_2 = 1, \\
&\qquad\quad \theta - u - \sigma = 0, \\
&\qquad\quad -4v_1 - 2v_2 + u + \sigma \le 0, \\
&\qquad\quad -4v_1 - 3v_2 + u + \sigma \le 0, \\
&\qquad\quad -4v_1 - v_2 + u + \sigma \le 0, \\
&\qquad\quad -3v_1 - 2v_2 + u + \sigma \le 0, \\
&\qquad\quad -2v_1 - 4v_2 + u + \sigma \le 0, \\
&\qquad\quad -6v_1 - v_2 + u + \sigma \le 0, \\
&\qquad\quad \lambda_A + 4v_1 + 2v_2 - u - \sigma - \eta \ge 0, \\
&\qquad\quad \lambda_B + 4v_1 + 3v_2 - u - \sigma - \eta \ge 0, \\
&\qquad\quad \lambda_C + 4v_1 + v_2 - u - \sigma - \eta \ge 0, \\
&\qquad\quad \lambda_D + 3v_1 + 2v_2 - u - \sigma - \eta \ge 0, \\
&\qquad\quad \lambda_E + 2v_1 + 4v_2 - u - \sigma - \eta \ge 0, \\
&\qquad\quad \lambda_F + 6v_1 + 1v_2 - u - \sigma - \eta \ge 0, \\
&\qquad\quad v_1 + d_1^x - \eta \ge 0, \\
&\qquad\quad v_2 + d_2^x - \eta \ge 0, \\
&\qquad\quad u + d_1^g - \eta \ge 0, \\
&\qquad\quad \text{all variables } (\text{except } \theta \text{ and } \sigma) \ge 0, \theta \text{ and } \sigma : \text{URS}.
\end{aligned}
\tag{7.10}
$$

As mentioned previously, the incorporation of SCSCs may slightly increase our computational effort, but this extra effort is easily accommodated by the computational capability of modern personal computers. Furthermore, computational results with SCSCs are more reliable than those without SCSCs that often contain many zeros in dual variables.

7.3.3 Third Example

To highlight the occurrence of multiple reference sets (so, multiple projections) by Model (7.1) and to document how the incorporation of SCSCs in Model (7.9) can produce a unique reference set, this chapter returns to the research of Sueyoshi and Goto (2012 h) that has used the data set summarized in Table 7.6.

Table 7.7 summarizes the computational result obtained from Model (7.1). The table indicates that the reference set of DMU {F} contains two DMUs {A, D}. The result is problematic in identifying the reference set of DMU {F}, measured by RM(v). That is, the reference set should cover all possible reference sets. For example, DMU {F} may have 14 possible reference set combinations such

TABLE 7.6 Third illustrative data

Data	A	B	C	D	E	F
Input 1	1.2500	1	3	5	2	4
Input 2	1.2500	3	1	5	0.6666	4
Output	1.1250	1.5000	1.5000	3	0.5000	1.5000

(a) Source: Sueyoshi and Goto (2012c), who obtained an original data set from Krivonozhko and Forsund (2009), listed in the website: http://www.sv.uio.no/econ/forskning/publikasjoner/memorandum/pdf-filer/2009/Memo-15-2009.pdf. (b) According to Sueyoshi and Goto (2012c), the original data set contained 2/3. None directly handles 2/3 in the computer operation for DEA. This chapter changed it to 0.6666, as found in the second input of DMU {E}. Therefore, there is a possibility on a round-off error in changing 2/3 to 0.6666.

TABLE 7.7 Computational result of input-oriented *RM*(*v*): Model (7.1)

DMU	λ_A^*	λ_B^*	λ_C^*	λ_D^*	λ_E^*	λ_F^*	θ^*
{A}	1						1
{B}		1					1
{C}			1				1
{D}				1			1
{E}					1		1
{F}	0.8			0.2			0.5

(a) Source: Sueyoshi and Goto (2012c). (b) A blank space indicates "zero." The λ^* score indicates an optimal intensity variable. The degree of θ^* indicates an *OE* score. The input-oriented RM(v) model identifies DMUs {A, D} as a reference set of DMU {F}. The other DMUs each have their own reference set.

TABLE 7.8 Computational result of input-oriented RM(v) with SCSCs: Model (7.9)

DMU	λ_A^*	λ_B^*	λ_C^*	λ_D^*	λ_E^*	λ_F^*	θ^*
{A}	1						1
{B}		1					1
{C}			1				1
{D}				1			1
{E}					1		1
{F}	0.6	0.125	0.125	0.15			0.5

(a) Source: Sueyoshi and Goto (2012c). This chapter updates their result. (b) A blank space indicates "zero." The λ^* score indicates an optimal intensity variable. The θ^* is an efficiency score. After incorporating SCSCs, Model (7.9) can identify DMUs {A, B, C, D} as a reference set of DMU {F}. (c) The other DMUs have their own reference set. DMU {E} is on an efficiency frontier, but it does not consist of a reference set of DMU {F}, because DMU {E} is not included in a projection area, or on a minimum face, on the efficiency frontier.

as: {A}, {B}, {C}, {D}, {A, B}, {A, C}, {A, D}, {B, C}, {B, D}, {C, D}, {A, B, C}, {A, B, D}, {A, C, D} and {A, B, C, D}. In these cases, it is reasonable for us to consider DMUs {A, B, C, D} as a reference set of DMU {F} because it covers all possible reference sets. Such a unique reference set consists of "a minimum face" on the efficiency frontier for this example. Thus, the conventional use of RM(v) suffers from an occurrence of multiple reference sets and therefore, the occurrence of multiple projections. Note that DMU {E} is efficient, but it does not consist of a reference set of DMU {F}, because DMU {E} is not included in its projection area, or the minimum face, on the efficiency frontier.

Table 7.8 summarizes the computational result after this example incorporates SCSCs into the radial model. Model (7.9) is used for the computation. After incorporating SCSCs, the reference set of DMU {F} is identified as DMUs {A, B, C, D}, consisting of the minimum face containing the maximum number of efficient DMUs, on the efficiency frontier. Thus, Model (7.9) can uniquely identify a reference set for DMU {F}, so indicating a unique projection of DMU {F} on the efficiency frontier.

The comparison between Tables 7.7 and 7.8 indicates that all radial and non-radial models may suffer from the occurrence of multiple reference sets and multiple projections. As a result of incorporating SCSCs, DEA can identify a single reference set and a unique projection in the assessment.

7.4 THEORETICAL IMPLICATIONS OF SCSCs

To discuss the mathematical implications of SCSCs within the conventional context of radial measurement, this chapter starts by examining the relationship between the non-Archimedean small number (ε_n) and SCSCs in RM(v). This chapter slightly modifies Models (7.1), (7.2) and (7.9) by adding slacks so that we return to these original formulations as discussed in Chapter 4. The modified versions, whose objective function have slacks weighted by ε_n, can be expressed by the following three formulations[4]:

$$\begin{aligned} &\text{Minimize } \theta - \varepsilon_n\left(ed^x + ed^g\right) \\ &s.t. \quad -\sum_{j=1}^{n} X_j\lambda_j - d^x + \theta X_k = 0, \\ &\sum_{j=1}^{n} G_j\lambda_j - d^g = G_k, \\ &\sum_{j=1}^{n} \lambda_j = 1, \\ &\lambda_j \geq 0 \ (j=1,\ldots,n), d^x \geq 0, \\ &d^g \geq 0 \text{ and } \theta : \text{URS}. \end{aligned} \tag{7.11}$$

$$\begin{aligned} &\text{Maximize } UG_k + \sigma \\ &s.t. \quad -VX_j + UG_j + \sigma \leq 0 \ (j=1,\ldots,n), \\ &VX_k = 1, \\ &V \geq \varepsilon_n e, \ U \geq \varepsilon_n e \text{ and } \sigma : \text{URS}, \end{aligned} \tag{7.12}$$

and

$$\begin{aligned} &\text{Maximize } \eta \\ &s.t. \quad \text{all constraints in Models } (7.11) \text{ and } (7.12), \\ &\theta - \varepsilon_n\left(ed^x + ed^g\right) = UG_k + \sigma, \\ &\lambda + VX - UG - \sigma e^{Tr} \geq \eta e^{Tr}, \\ &V^{Tr} + d^x \geq \left(\varepsilon_n + \eta\right)e^{Tr}, \\ &W^{Tr} + d^g \geq \left(\varepsilon_n + \eta\right)e^{Tr} \text{ and} \\ &\eta \geq 0. \end{aligned} \tag{7.13}$$

After setting $\varepsilon_n = 0$ in these three models, they correspond to Models (7.1), (7.2) and (7.9), respectively.

The non-Archimedean small number (ε_n) needs to satisfy the following property:

[4]The very small number (ε_n) works as a lower bound for all multipliers in Model (7.9) so that all multipliers are always positive. However, it is true that the lower bound is very close to zero so that some multipliers become very close to zero. Practically speaking, these multipliers are not fully utilized in DEA performance assessment. Hence, some kind of multiplier restriction is necessary for DEA applications.

Proposition 7.1: DMU{k} has an optimal solution of the model in Model (7.11) if and only if $1 \geq \varepsilon_n \sum_{i=1}^{m} x_{ik}$. Suppose that there is an optimal solution (θ^*, λ^*) of Model (7.11) for the k-th DMU such that $\sum_{i=1}^{m} \sum_{j=1}^{n} \left(\theta^* x_{ik} - x_{ij}\lambda_j^*\right) = \sum_{r=1}^{s} \sum_{j=1}^{n} \left(g_{rj}\lambda_j^* - g_{rk}\right) = 0$, then there exists a positive number ($\bar{\varepsilon}$) such that

$$\theta^* = \text{the optimal value of Model (7.11) for any positive } \varepsilon_n \leq \bar{\varepsilon}. \tag{7.14}$$

[Proof] See the Appendix of this chapter.

In the remaining part of this chapter, we assume that the non-Archimedean small number (ε_n) in Models (7.11), (7.12) and (7.13) satisfies the condition (7.14). Then, the assumption guarantees that both Models (7.11) and (7.12) have an optimal solution. Hence, Model (7.13) has a feasible solution. The relationship can be specified as follows:

Proposition 7.2: Let ε_n be a non-Archimedean small number, then $(\theta^*, \lambda^*, d^{x*}, d^{g*})$ is the optimal solution of Model (7.11) and (V^*, U^*, σ^*) is the optimal solution of Model (7.12) if and only if $(\eta^*, \theta^*, \lambda^*, d^{x*}, d^{g*}, V^*, U^*, \sigma^*)$ is a feasible solution of Model (7.13).

[Proof] The proposition can be proved by the duality theorem of linear programming.

Q.E.D.

Proposition 7.2 indicates that θ obtained from a feasible solution of Model (7.13) always equals the degree of *OE* on DMU {k} measured by Model (7.11). Furthermore, Proposition 7.2 guarantees that we can separate a feasible solution of Model (7.13) into the optimal solutions of Models (7.11) and (7.12). Thus, the optimal solution of Model (7.13) becomes the optimal solution of Models (7.11) and (7.12).

Proposition 7.3: There exists a pair of an optimal solution $(\theta^*, \lambda^*, d^{x*}, d^{g*})$ of Model (7.11) and an optimal solution (V^*, U^*, σ^*) of Model (7.12) such that $(\theta^*, \lambda^*, d^{x*}, d^{g*}, V^*, U^*, \sigma^*)$ satisfies the SCSCs. Therefore, Model (7.13) has a positive optimal objective value.

[Proof] The existence of an optimal solution pair between Models (7.11) and (7.12) satisfies SCSCs.

Q.E.D.

Let $(\bar{\eta}^*, \bar{\theta}^*, \bar{\lambda}^*, \bar{d}^{x*}, \bar{d}^{g*}, \bar{V}^*, \bar{U}^*, \bar{\sigma}^*)$ be the optimal solution of Model (7.13). Then, $\bar{\theta}^*$ becomes the optimal solution of Model (7.11). Hence, we can determine the degree of *OE* measure of DMU {k} by $\bar{\theta}^*$. Furthermore, Conditions 1 and 2, discussed in Section 7.3, satisfy on the optimality of Model (7.13) because $\bar{\eta}^*$ is positive.

The following two propositions guarantee the satisfaction of the two conditions:

Proposition 7.4: The optimal solution $\left(\bar{\eta}^*, \bar{\theta}^*, \bar{\lambda}^*, \bar{d}^{x*}, \bar{d}^{g*}, \bar{V}^*, \bar{U}^*, \bar{\sigma}^*\right)$ of Model (7.13) satisfies

$$\left\{j|\lambda_j^* > 0\right\} \subseteq \left\{j|\bar{\lambda}_j^* > 0\right\} = \left\{j|\bar{V}^* X_j = \bar{U}^* G_j + \bar{\sigma}^*\right\} \subseteq \left\{j|V^* X_j = U^* G_j + \sigma^*\right\} \tag{7.15}$$

for any optimal solution $(\theta^*, \lambda^*, d^{x*}, d^{g*})$ of Model (7.11) and any optimal solution (V^*, U^*, σ^*) of Model (7.12). Let $(\bar{X}, \bar{G}) = (\boldsymbol{X}\bar{\lambda}^*, \boldsymbol{G}\bar{\lambda}^*)$ and $\bar{J} = \{j|\bar{\lambda}_j^* > 0\}$, then

$$\left\{(X,G)\middle|(X,G) = \sum_{j\in\bar{J}}\left(X_j, G_j\right)\lambda_j, \sum_{j\in\bar{J}}\lambda_j = 1, \lambda \geq 0\right\} \tag{7.16}$$

is "a face" of the production possibility set (P) where $P = (G\!: X$ can produce $G)$. The face has the following relationship with any face (F) including $(\bar{X}, \bar{G})$:

$$\left\{(X,G)\middle|(X,G) = \sum_{j\in\bar{J}}\left(X_j, G_j\right)\lambda_j, \sum_{j\in\bar{J}}\lambda_j = 1, \lambda \geq 0\right\} \subseteq F. \tag{7.17}$$

[Proof] See the Appendix of this chapter.

Let $\bar{F}$ be $\{(X,G)\,|(X,G) = \sum_{j\in\bar{J}}(X_j, G_j)\lambda_j, \sum_{j\in\bar{J}}\lambda_j = 1, \lambda \geq 0\}$ of the data domain (7.16) and let the projection set be

$$\Omega_k = \{(X,G)|X = \boldsymbol{X}\lambda^*, G = \boldsymbol{G}\lambda^*,\ (\theta^*, \lambda^*, d^{x^*}, d^{g^*}) \text{ is the optimal solution of Model } (7.11)\}.$$

Then, the explicit expression of the projection set (Ω_k) becomes as follows:

Proposition 7.5: Let $\left(\bar{\eta}^*, \bar{\theta}^*, \bar{\lambda}^*, \bar{d}^{x*}, \bar{d}^{g*}, \bar{V}^*, \bar{U}^*, \bar{\sigma}^*\right)$ be the optimal solution of Model (7.13) and let $(\bar{X}, \bar{G}) = (\boldsymbol{X}\bar{\lambda}^*, \boldsymbol{G}\bar{\lambda}^*)$, then

$$\Omega_k = \bar{F} \cap \left\{(X,G)\middle|\ X \leq \bar{\theta}^* X_k, G \geq G_k\right\} \cap \left\{(X,G)\middle|e\bar{X} + e\bar{G} = eX + eG\right\} \tag{7.18}$$

[Proof] It follows directly from Proposition 7.4 and Model (7.11).

Q.E.D.

Implication (Uniqueness and mathematical expression on the projection set): Condition 1: A reference set should cover all possible reference sets on a projection set (Ω), indicating that, given any optimal solution $(\theta^*, \lambda^*, d^{x*}, d^{g*})$ of

Model (7.11), $\bar{J} \supseteq \{j | \lambda_j^* > 0\}$ indicates a desirable reference set ($\bar{J}$). From Proposition 7.4, $\bar{J} \supseteq \{j | \lambda_j^* > 0\}$ implies that $\bar{J}$ of Model (7.13) contains all possible reference sets identified by solving Model (7.11). Hence, the reference set ($\bar{J}$) identified by solving Model (7.13) is unique.

Meanwhile, Condition 2: the convex hull of the reference set should contain Ω, indicating that the convex hull of $\{(X_j, G_j) | j \in \bar{J}\}$ of Model (7.13) contains a projection set on an efficiency frontier. This result is because of Proposition 7.5. In the occurrence of multiple projections, the projection set is a subset of the convex hull of $\{(X_j, G_j) | j \in \bar{J}\}$ of Model (7.13). Thus, we can specify part of the efficiency frontier on which the inefficient DMU is projected under the occurrence of multiple projections. Consequently, we can determine the face including multiple projections by identifying such a desirable reference set ($\bar{J}$).

The right-hand side of Equation (7.18) is useful in specifying the dimension (dim) of the projection set (Ω_k). For example, if $\bar{F} \subseteq \{(X,G) | \ X \leq \bar{\theta}^* X_k, G \geq G_k\} \cap \{(X,G) | e\bar{X} + e\bar{G} = eX + eG\}$, then $\Omega_k = \bar{F}$. Such a case can be found for DMU {A} in Figure 7.1. Furthermore, we can apply the reverse search method proposed by Avis and Fukuda (1996) in order to identify all vertex points on the production possibility set. Consequently, the projection set (Ω_k) can be expressed by a convex combination of these vertex points. Here, for a convex set S, the dimension of S is denoted by dim S. The equation, dim S=0, means that S is a point and dim S>0 means that the set (S) contains multiple points. Thus, the following proposition determines the uniqueness of the projection point(s):

Proposition 7.6: A projection point of DMU $\{k\}$ is unique if and only if $\dim \Omega_k = 0$. If the optimal solution $(\bar{\eta}^*, \bar{\theta}^*, \bar{\lambda}^*, \bar{d}^{x*}, \bar{d}^{g*}, \bar{V}^*, \bar{U}^*, \bar{\sigma}^*)$ of Model (7.13) satisfies at least one of the following conditions:

(a) $\{j | \bar{\lambda}_j^* > 0\}$ consists of one element and
(b) $\{i | \bar{d}_i^{x*} > 0\} \cup \{r | \bar{d}_r^{g*} > 0\}$ consists of at most one element,

then $\dim \Omega_k = 0$ and the projection is unique.

[Proof] See the Appendix of this chapter.

Implication (Test for sufficient condition on unique projection): The unique projection for efficiency comparison is a desirable property of *OE*. Proposition 7.6 discusses a mathematical condition regarding unique projection in the *OE* measurement. The proposition provides a test for the occurrence of multiple projections by examining whether an optimal solution of Model (7.13) satisfies either condition (a) or condition (b). However, a unique projection may occur even if an optimal solution of Model (7.13) violates the two conditions. Proposition 7.6 is a sufficient condition for such a unique projection. See Sueyoshi and Sekitani (2009) for the necessary and sufficient conditions concerning a unique projection.

7.5 GUIDELINE FOR NON-RADIAL MODELS

In the previous discussion, this chapter used RM(v) to illustrate how to handle the occurrence of multiple projections. The proposed approach can be applied to not only radial models but also non-radial models. To outline how to handle an occurrence of multiple projections in the non-radial models, this chapter describes the following computational process:

Step 1: Combine the primal model of a non-radial model with its dual model as found in Model (7.13). Then, solve the combined model that documents the combined model.

Step 2: Identify a reference set ($\bar{J}$) from the optimal solution of the combined model. The reference set is unique even if multiple projections occur because it contains all possible reference sets generated by the original primal model.

Step 3: Identify a projection set (Ω_k) from $\Omega_k \subseteq \bar{F}$ by Model (7.18). Here, $\bar{F}$ is a convex set that is made from the reference set:

$$\bar{F} = \left\{ (X, G) \middle| (X, G) = \sum_{j \in \bar{J}} \left(X_j, G_j \right) \lambda_{j \in \bar{J}}, \sum_{j \in \bar{J}} \lambda_j = 1, \lambda \geq 0 \right\}.$$

Step 4: Check the dimension of the projection set (Ω_k). If the dimension is zero, then the projection is unique. In contrast, if the dimension is not zero (≥1), then it indicates an occurrence of multiple projections.

The following two comments are important in applying the proposed computation to non-radial models. One of these comments is that, if multiple projections occur on a DMU, mainly on an inefficient DMU, then the proposed approach can identify a projection set on an efficiency frontier. A single reference set determined by the proposed approach uniquely determines the projection set. Thus, the property of unique efficiency comparison is attained by the proposed approach even under an occurrence of multiple projections because the reference set is unique. The other comment is that our outline, discussed in this chapter, is useful for all non-radial models except the Russell measure, which is a non-linear programming formulation.

7.6 SUMMARY

It is widely known that DEA has three difficulties in the applications. First, we may face the occurrence of multiple projections. Second, we need to consider the occurrence of multiple references. These two difficulties were discussed in Chapters 5 and 6. For the latter difficulty, dual variables (multipliers) are often zero. This is also problematic because related productions factors are not fully utilized in DEA performance assessment. The first two difficulties are known from previous studies, but they have not been sufficiently explored to obtain their uniqueness on projection and a reference set. The latter difficulty of an occurrence of zero on dual variables has been discussed by multiplier restriction methods (e.g., AR and CR). However, no previous study discussed how to handle the difficulty when we cannot access prior information. Furthermore, these approaches often produce an infeasible solution. To deal with all three types of difficulty in a simultaneous manner, this chapter discusses the use of SCSCs. A shortcut to solve the third difficulty (i.e., positive dual variables) is that we incorporate the non-Archimedean small number (ε_n) in the objective function of RMs.

Finally, it is important to note that the incorporation of SCSCs in DEA has still a difficulty in determining the level of *OE* on all DMUs. Even if we apply the SCSCs to any radial and non-radial models, we must face the occurrence of many efficient DMUs in DEA assessment. In this case, DEA equipped with SCSCs is not a perfect methodology. To reduce the number of efficient DMUs, Chapter 25 will discuss how to handle such a DMU ranking problem by combining SCSCs with DEA-DA (discriminant analysis).

APPENDIX

Proof of Proposition 7.1

Since Model (6.1) in Chapter 6, expressing a general DEA model, has a feasible solution such that $\theta = 1$, $\lambda_k = 1$, $d^x = 0$ and $d^g = 0$, it is sufficient for us to show the two following conditions:

(a) The objective function, or $\theta - \varepsilon_n\left(\sum_{i=1}^{m} d_i^x + \sum_{r=1}^{s} d_r^g\right)$ of Model (7.13), has a lower bound if $1 \geq \varepsilon \sum_{i=1}^{m} x_{ik}$ and
(b) Model (7.13) is infeasible if $1 < \varepsilon \sum_{i=1}^{m} x_{ik}$.

The objective function of Model (7.13) is reduced to the following equivalent formulation:

$$\begin{aligned}
\theta - \varepsilon_n\left(\sum_{i=1}^{m} d_i^x + \sum_{r=1}^{s} d_r^g\right) &= \theta - \varepsilon_n\left[\sum_{i=1}^{m}\left(\theta x_{ik} - \sum_{j=1}^{n} x_{ij}\lambda_j\right) + \sum_{r=1}^{s}\left(\sum_{j=1}^{n} g_{rj}\lambda_j - g_{rk}\right)\right] \\
&= \theta - \varepsilon_n\sum_{i=1}^{m}\theta x_{ik} + \varepsilon_n\sum_{i=1}^{m}\sum_{j=1}^{n} x_{ij}\lambda_j - \varepsilon_n\sum_{r=1}^{s}\sum_{j=1}^{n} g_{rj}\lambda_j + \varepsilon_n\sum_{r=1}^{s} g_{rk} \\
&= \theta\left(1 - \varepsilon_n\sum_{i=1}^{m} x_{ik}\right) + \varepsilon_n\sum_{j=1}^{n}\left(\sum_{i=1}^{m} x_{ij} - \sum_{r=1}^{s} g_{rj}\right)\lambda_j + \varepsilon_n\sum_{r=1}^{s} g_{rk}.
\end{aligned} \tag{7.A1}$$

Since $\left\{\lambda \middle| \sum_{j=1}^{n} \lambda_j = 1, \lambda \geq 0\right\}$ is compact, $\sum_{j=1}^{n}\left(\sum_{i=1}^{m} x_{ij} - \sum_{r=1}^{s} g_{rj}\right)\lambda_j$ has the minimum value on $\left\{\lambda \middle| \sum_{j=1}^{n} \lambda_j = 1, \lambda \geq 0\right\}$. The minimum value is denoted by ϖ.

Suppose that, if $\varepsilon_n \sum_{i=1}^{m} x_{ik} \leq 1$, then $\theta\left(1-\varepsilon_n \sum_{i=1}^{m} x_{ik}\right) + \varepsilon_n \sum_{j=1}^{n}\left(\sum_{i=1}^{m} x_{ij} - \sum_{r=1}^{s} g_{rj}\right)\lambda_j + \varepsilon_n \sum_{r=1}^{s} g_{rk} \geq \varepsilon_n\left(\varpi + \sum_{r=1}^{s} g_{rk}\right)$. Equation (7.A1) indicates that the objective function of Model (7.13) has a lower bound. Therefore, Model (7.13) has an optimal solution. Conversely, assume that, if $\varepsilon_n \sum_{i=1}^{m} x_{ik} > 1$, then $V \geq \varepsilon_n e$ implies $Vx_k \geq \varepsilon_n e x_k = \varepsilon_n \sum_{i=1}^{m} x_{ik} > 1$. Hence, Model (7.13) is infeasible. Therefore, if the dual problem of Model (7.13) is feasible, Model (7.13) has an optimal solution so that we have $\varepsilon_n \sum_{i=1}^{m} x_{ik} \leq 1$.

Suppose that there exists an optimal solution (θ^*, λ^*) of Model (6.2) of Chapter 6 such that $\sum_{i=1}^{m}\sum_{j=1}^{n}(\theta^* x_{ik} - x_{ij}\lambda_j^*) = \sum_{r=1}^{s}\sum_{j=1}^{n}(g_{rj}\lambda_j^* - g_{rk}) = 0$, then we have $\theta^* x_{ik} = \sum_{j=1}^{n} x_{ij}\lambda_j^*$ for all $i = 1,\ldots,m$ and $\sum_{j=1}^{n} g_{rj}\lambda_j^* = g_{rk}$ for all $r = 1,\ldots,s$. Since Model (6.2) becomes Model (7.11) with $\varepsilon_n = 0$, Model (7.12) with $\varepsilon_n = 0$ corresponds to the dual problem of Model (6.2). Hence, SCSCs imply that there exists a dual optimal solution (V^*, U^*, σ^*) such that $V^* > 0$ and $U^* > 0$. Furthermore, the duality theorem of linear programming implies $\theta^* = U^* G_k + \sigma^*$. Here, let $\bar{\varepsilon} = \text{minimize}\left\{\underset{i=1,\ldots,m}{\text{minimize}}\, v_i^*, \underset{r=1,\ldots,s}{\text{minimize}}\, u_r^*\right\}$, then $V^* \geq \varepsilon_n e$ and $U^* \geq \varepsilon_n e$ for any positive $\varepsilon_n \leq \bar{\varepsilon}$. This indicates that (V^*, U^*, σ^*) is also an optimal solution of Model (7.13) with any positive $\varepsilon_n \leq \bar{\varepsilon}$ and $\theta^* = U^* G_k + \sigma^*$ is the optimal value of Model (7.13) with any positive $\varepsilon_n \leq \bar{\varepsilon}$. Since Model (7.13) is the dual problem of Model (7.11), the optimal value of Model (7.13) is $\theta^* = U^* G_k + \sigma^*$ for any positive $\varepsilon_n \leq \bar{\varepsilon}$.

Q.E.D.

Proof of Proposition 7.4

From Proposition 7.3, the optimal solution $(\bar{\eta}^*, \bar{\theta}^*, \bar{\lambda}^*, \bar{d}^{x*}, \bar{d}^{g*}, \bar{V}^*, \bar{U}^*, \bar{\sigma}^*)$ of Model (7.13) satisfies $\{j|\bar{\lambda}_j^* > 0\} = \{j|\bar{V}^* X_j = \bar{U}^* G_j + \bar{\sigma}^*\}$. Any optimal solution $(\theta^*, \lambda^*, d^{x*}, d^{g*})$ of Model (7.11) satisfies $\bar{V}^* X_j = \bar{U}^* G_j + \bar{\sigma}^*$ for all $j \in \{j\ |\lambda_j^* > 0\}$. Hence, $\{j|\lambda_j^* > 0\} \subseteq \{j|\bar{\lambda}_j^* > 0\} = \{j|\bar{V}^* X_j = \bar{U}^* G_j + \bar{\sigma}^*\}$. If $\bar{\lambda}_j^* > 0$, then it implies that $V^* X_j = U^* G_j + \sigma^*$ is satisfied for any optimal solution (V^*, U^*, σ^*) of Model (7.12). This means $\{j|\bar{\lambda}_j^* > 0\} \subseteq \{j|V^* X_j = U^* G_j + \sigma^*\}$ and $\{j|\lambda_j^* > 0\} \subseteq \{j|\bar{\lambda}_j^* > 0\} = \{j|\bar{V}^* X_j = \bar{U}^* G_j + \bar{\sigma}^*\} \subseteq \{j|V^* X_j = U^* G_j + \sigma^*\}$.

Equation (7.16) indicates the face of the production possibility set (P) if and only if:

$$\left\{(X,G)\in P\middle|\bar{V}*X=\bar{U}*G+\bar{\sigma}*\right\}=\left\{(X,G)\middle|(X,G)=\sum_{j\in\bar{J}}\left(X_j,G_j\right)\lambda_j,\sum_{j\in\bar{J}}\lambda_j=1,\lambda\geq 0\right\}. \tag{7.A2}$$

Choose $(X,G)\in\{(X,G)\in P|\bar{V}*X=\bar{U}*G+\bar{\sigma}*\}$ arbitrarily, then both $\bar{V}*\geq\varepsilon_n e$ and $\bar{U}*\geq\varepsilon_n e$ implies that (X,G) is on the efficient frontier of P. Therefore, there exists a convex coefficient vector λ (for intensity variables) such that $(X,G)=\sum_{j=1}^{n}(X_j,G_j)\lambda_j$. Since $\bar{V}*X_j-\bar{U}*G_j-\bar{\sigma}*\leq 0$ for all $j=1,\ldots,n$ and $\bar{V}*X-\bar{U}*G-\bar{\sigma}*=\sum_{j=1}^{n}(\bar{V}*X_j-\bar{U}*G_j-\bar{\sigma}*)\lambda_j=0$, $\lambda_j>0$ implies $\bar{V}*X_j=\bar{U}*G_j+\bar{\sigma}*$. Hence, we have $\{j|\lambda_j>0\}\subseteq\bar{J}$ and $(X,G)=\sum_{j\in\bar{J}}(X_j,G_j)\lambda_j$. This result indicates the following relationship:

$$\left\{(X,Y)\in P\middle|\bar{V}*X=\bar{U}*G+\bar{\sigma}*\right\}\subseteq\left\{(X,G)\middle|(X,G)=\sum_{j\in\bar{J}}\left(X_j,G_j\right)\lambda_j,\sum_{j\in\bar{J}}\lambda_j=1,\lambda\geq 0\right\}. \tag{7.A3}$$

Conversely, consider $(X,G)=\sum_{j\in\bar{J}}(X_j,G_j)\lambda_j$ such that $\sum_{j\in\bar{J}}\lambda_j=1$ and $\lambda\geq 0$, then it follows from $\bar{J}=\{j|\bar{V}*X_j=\bar{U}*G_j+\bar{\sigma}*\}$ that $\bar{V}*X-\bar{U}*G-\bar{\sigma}*=\sum_{j\in\bar{J}}(\bar{V}*X_j-\bar{U}*G_j-\bar{\sigma}*)\lambda_j=0$. The result implies from Equation (7.A3) that Equation (7.A2) is satisfied.

Consider any supporting hyperplane H of P on $(\bar{X},\bar{G})$, then there exists an optimal solution of Model (7.11) such that $H=\{(X,G)|\,V*X=U*G+\sigma*\}$. Since $\{j|\bar{V}*X_j=\bar{U}*G_j+\bar{\sigma}*\}\subseteq\{j|V*X_j=U*G_j+\sigma*\}$,

$$\left\{(X,G)\in P\middle|\bar{V}*X=\bar{U}*G+\bar{\sigma}*\right\}=\left\{(X,G)\middle|\bar{V}*X=\bar{U}*G+\bar{\sigma}*\right\}\cap$$
$$P\subseteq\left\{(X,G)\middle|V*X=U*G+\sigma*\right\}\cap P=H\cap P.$$

For any face F including $(\bar{X},\bar{G})$, we have $\{(X,G)|(X,G)=\sum_{j\in\bar{J}}(X_j,G_j)\lambda_j$, $\sum_{j\in\bar{J}}\lambda_j=1,\lambda\geq 0\}\subseteq F$.

Q.E.D.

Proof of Proposition 7.6

The set Ω_k consists of a single projection point if and only if $\dim\Omega_k=0$. Therefore, a projection point is unique if and only if $\dim\Omega_k=0$. Suppose that $\{j|\bar{\lambda}_j^*>0\}$ consists of one element, then $\{j|\lambda_j^*>0\}\subseteq\{j|\bar{\lambda}_j^*>0\}$ of Equation (7.15) implies $\dim\Omega_k=0$. Suppose that $\{i|\bar{d}_i^{x*}>0\}\cup\{r|\bar{d}_r^{g*}>0\}$ consists of at most one element, then we have the following two cases:

(a) Case 1: $\{i|\bar{d}_i^{x*}>0\}\cup\{r|\bar{d}_r^{g*}>0\}=\varnothing$
The case implies that any projection point is on the efficient frontier on P and it is $(\bar{\theta}*X_k,G_k)$. Therefore, $\dim\Omega_k=0$.

(b) Case 2: $\{i|\bar{d}_i^{x*}>0\}\cup\{r|\bar{d}_r^{g*}>0\}\neq\varnothing$

Without loss of generality, this study assumes $\bar{d}_1^{x*} > 0$ and $\bar{d}_i^{x*} = 0$ for all $i \neq 1$ and $\bar{d}_r^{g*} = 0$ for all r. Any optimal solution $(\theta^*, \lambda^*, d^{x*}, d^{g*})$ of Model (7.11) also satisfies $d_1^{x*} > 0$ and $d_i^{x*} = 0$ for all $i \neq 1$ and $d_r^{g*} = 0$ for all r. Model (7.17) indicates

$$\Omega_k \subseteq \left\{(X,G) \in P \middle| \, X \leq \bar{\theta} * X_k, G \geq G_k\right\} \cap \left\{(X,G) \middle| e\bar{X} + e\bar{G} = eX + eG\right\}.$$

Here, let e_1 be a unit vector whose first component is 1, then we have

$$\left\{(X,G) \in P \middle| \, X \leq \bar{\theta} * X_k, G \geq G_k\right\} = \left\{(X,G) \in P \middle| \, X = \bar{\theta} * X_k - \mu e_1, G = G_{k,} \, \mu \geq 0\right\}.$$

Since we have $\{(X,G) \in P | \, X = \bar{\theta} * X_k - \mu e_1, G = G_{k,} \, \mu \geq 0\} \not\subset \{(X,G) | e\bar{X} + e\bar{G} = eX + eG\}$, the intersection $\{(X,G) \subset P | \, X = \bar{\theta}^* X_k - \mu e_1, G = G_k, \mu \geq 0\} \cap \{(X,G) | e\bar{X} + e\bar{G} = eX + eG\}$ consists of a single point. Therefore, we have dim $\Omega_k = 0$ that indicates a unique projection.

Q.E.D.

8

RETURNS TO SCALE

8.1 INTRODUCTION

This chapter[1] discusses the concept and type of returns to scale (RTS) that serve as a methodological basis for examining useful scale-related measures by DEA. The measure provides us with information on not only a change of desirable outputs due to a unit increase of inputs but also the shape of an efficiency frontier. The first research effort on RTS was discussed by Banker (1984) and then followed by many other studies (e.g., Chang and Guh, 1991). At an early stage of this type of research, the previous efforts were interested in how to measure the type of RTS on the specific k-th DMU. In a conventional framework of DEA, the measurement looks for the type of "local" RTS. Hereafter, this chapter and the others related to scale measures do not use the term "local" because we clearly understand the implication.

After the research of Banker and Thrall (1992), the central theme moved to the RTS measurement under an occurrence of multiple solutions on an intercept (σ) of a supporting hyperplane although their measurements need to assume a unique projection and a unique reference set. They discussed how to handle multiple solutions on σ, but not considering the other types of multiple solutions such as

[1] This chapter is partly based upon the article: Sueyoshi, T. (1999) DEA duality on returns to scale (RTS) in production and cost analyses: an occurrence of multiple solutions and differences between production-based and cost-based RTS estimates. *Management Science*, **45**, 1593–1608.

Environmental Assessment on Energy and Sustainability by Data Envelopment Analysis, First Edition. Toshiyuki Sueyoshi and Mika Goto.

multiple projections and multiple reference sets. See Chapter 7 on such an occurrence of multiple projections and multiple reference sets.

To intuitively describe the concept and type of RTS, this chapter starts with a description on the measurement on scale elasticity, which is an economic concept between a desirable output and an input, following the research effort of Sueyoshi (1999). The proposed approach is extended into the measurement of RTS in the framework of multiple desirable outputs and inputs in this chapter. Chapters 20–23 will further extend his research effort into RTS, damages to scales (DTS), returns to damage (RTD), damages to return (DTR), rate of substitution (RSU) and marginal rate of transformation (MRT) for DEA environmental assessment. An important extension is that these measures will be discussed under an occurrence of congestion. See Chapter 9 on the occurrence of congestion within a conventional use of DEA. Thus, this chapter provides a fundamental strategy to examine how to obtain these scale measures by DEA, all of which are very important for discussing sustainability in energy sectors. It is often misunderstood among researchers and users that DEA is only for efficiency measurement. We disagree with the widely believed common opinion. The methodology should be discussed from not only efficiency assessment but also various scale measurements on production activities of organizations. Thus, this chapter provides us with an underlying concepts and related implications on such scale measures by starting a description on RTS.

The remainder of this chapter is organized as follows: Section 8.2 reviews underlying concepts. Section 8.3 discusses production-based RTS measurement. Section 8.4 discusses cost-based RTS measurement. Section 8.5 describes scale efficiencies and economies of scale. Section 8.6 summarizes this chapter.

8.2 UNDERLYING CONCEPTS

First, let us consider that there is a functional form $[g = f(x)]$ between a desirable output (g) and an input (x). Then, we measure the degree of desirable output-based scale elasticity (e_g) as follows:

$$e_g = \frac{\text{marginal product}}{\text{average product}} = \frac{\mathrm{d}g/\mathrm{d}x}{g/x}. \tag{8.1}$$

In a similar manner, this chapter considers that cost (c) is expressed by a function of a desirable output (g). The degree of cost-based scale elasticity (e_c) is measured by

$$e_c = \frac{\text{marginal cost}}{\text{average cost}} = \frac{\mathrm{d}c/\mathrm{d}g}{c/g}. \tag{8.2}$$

Using the two scale elasticity measures, each DMU is classified by the following rule:

$$
\begin{aligned}
&e_g > 1(e_c < 1) \leftrightarrow \text{increasing RTS,} \\
&e_g = 1(e_c = 1) \leftrightarrow \text{constant RTS and} \\
&e_g < 1(e_c > 1) \leftrightarrow \text{Decreasing RTS.}
\end{aligned} \quad (8.3)
$$

On the other hand, the input-based scale elasticity measures are classified by the following rule:

$$
\begin{aligned}
&e_g < 1 \leftrightarrow \text{increasing RTS,} \\
&e_g = 1 \leftrightarrow \text{constant RTS and} \\
&e_g > 1 \leftrightarrow \text{decreasing RTS.}
\end{aligned} \quad (8.4)
$$

In the above case, the relationship between the degree of scale elasticity and the type of RTS is same as the cost-based classification of e_c.

Figure 8.1 visually classifies the three types (i.e., increasing, constant, decreasing) of RTS by considering a desirable output-based model. The figure has two coordinates for x and g. Such a visual description is our convenience. A counter line, connecting {A}–{B}–{C}–{D}, depicts an efficiency frontier. A production possibility set locates within the south-east region from the efficiency

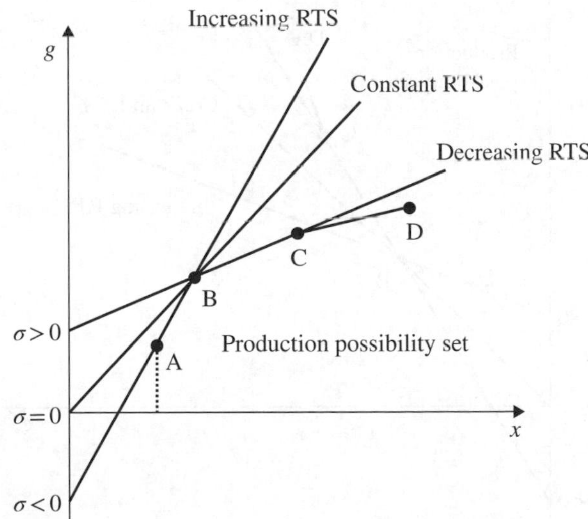

FIGURE 8.1 Three types of RTS and supporting hyperplanes

frontier. Paying attention to DMU {B}, for example, Figure 8.1 depicts the three types of slopes and intercepts of a supporting hyperplane.

Accordingly, Figure 8.1 classifies the three types of desirable output-based RTS on DMU {B} by the sign of an intercept (σ) as follows:

$$\begin{aligned} &\text{Increasing RTS} \leftrightarrow \sigma < 0, \\ &\text{Constant RTS} \leftrightarrow \sigma = 0 \text{ and} \\ &\text{Decreasing RTS} \leftrightarrow \sigma > 0. \end{aligned} \tag{8.5}$$

In a similar manner, Figure 8.2 depicts the three types of RTS that are classified by the cost model. As depicted in the figure, the three types of RTS are characterized by the sign of the intercept (σ) of a supporting hyperplane on a production possibility set that locates within the north-west region from an efficiency frontier that consists of the four DMUs {A, B, C, D}. The classification becomes opposite to the case of production-based RTS. The classification is summarized as follows:

$$\begin{aligned} &\text{Increasing RTS} \leftrightarrow \sigma > 0, \\ &\text{Constant RTS} \leftrightarrow \sigma = 0 \text{ and} \\ &\text{Decreasing RTS} \leftrightarrow \sigma < 0. \end{aligned} \tag{8.6}$$

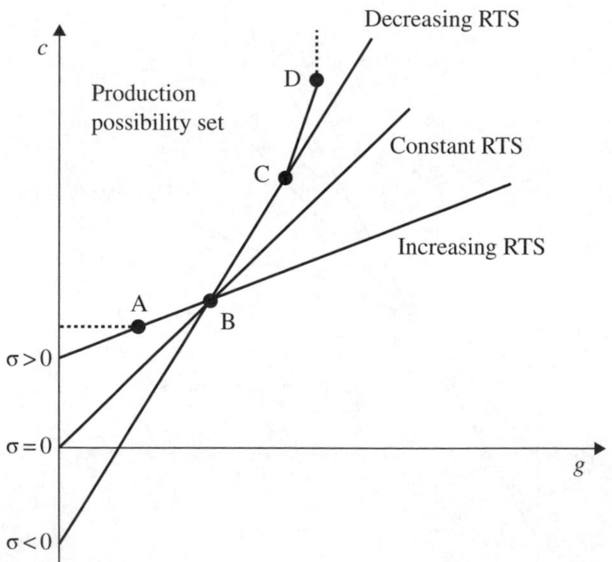

FIGURE 8.2 Type of cost-based RTS and supporting hyperplane

Here, it is important to describe four concerns. First, the sign of scale elasticity and the sign of the intercept are based on an input-oriented RM(v) that is equivalent to the case of the cost model, where RM(v) stands for a radial model under variable RTS. See Chapter 4. Second, RM(c) can assume constant RTS by setting $\sigma = 0$ in the formulation where RM(c) stands for a radial model under the assumption. Third, the scale elasticity is defined by a single desirable output in production economics. DEA can easily extend it to the case of multiple desirable outputs. In the case, the concept, corresponding to "scale elasticity," is referred to as "scale economies." Finally, production economics assumes that all DMUs are efficient so that the scale elasticity is measured on an efficiency frontier. In contrast, DEA does not need such an assumption because inefficient DMUs are projected onto the efficiency frontier. Hence, a slack adjustment is necessary for determining the type and magnitude of RTS.

A difficulty associated with the proposed DEA-based RTS measurement is that it contains a possibility in which an infinite number of σ may occur at efficient DMUs such as {A, B, C, D}, all of which consist of vertexes of an efficiency frontier. Such an occurrence is depicted in Figure 8.3. The figure visually describes an occurrence of an infinite number of supporting hyperplanes on DMU {B} and these related intercepts in the case of cost-based RTS. Therefore, it is necessary for us to examine an upper bound ($\bar{\sigma}$) and a lower bound ($\underline{\sigma}$) to determine the scale elasticity measures (e_g and e_c) and to determine the type of RTS by conventional DEA.

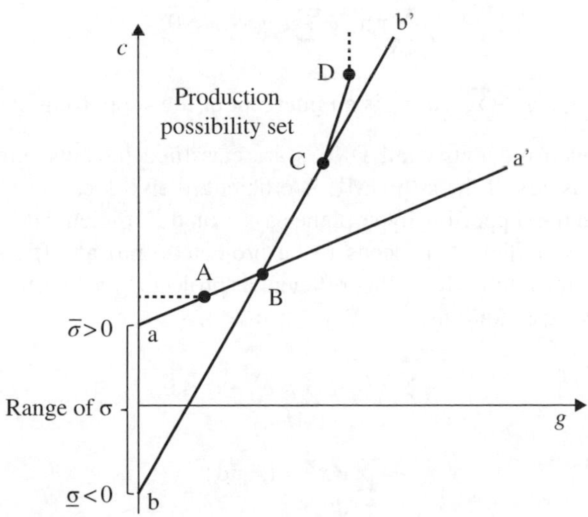

FIGURE 8.3 Occurrence of multiple supporting hyperplanes

8.3 PRODUCTION-BASED RTS MEASUREMENT

As discussed in Section 8.2, the location of a supporting hyperplane determines the type of RTS. Therefore, it is necessary for us to characterize the supporting hyperplane in a data space of multiple inputs and desirable outputs.

Proposition 8.1: In an input-oriented RM(v), or Model (4.5), a supporting hyperplane of the k-th DMU is expressed by $-\sum_{i=1}^{m} v_i x_i + \sum_{r=1}^{s} u_r g_r + \sigma = 0$, where v_i (i =1,...,m) and u_r (r=1,...,s) are unknown and non-negative variables to express a directional vector of a supporting hyperplane. An intercept is expressed by an unrestricted decision variable (σ). These unknown variables are characterized and determined by

$$-\sum_{i=1}^{m} v_i x_{ij} + \sum_{r=1}^{s} u_r g_{rj} + \sigma = 0 \ \left(j \in RS_k\right) \text{ and } \sum_{i=1}^{m} v_i x_{ik} = 1.$$

Here, $RS_k = \{j | \lambda_j^* > 0, j \in J\}$ is a reference set of the k-th DMU (Tone, 1996), measured by Model (4.4). The set (J) stands for a set of all DMUs.

[Proof] In the $m+s$ dimensional space, a supporting hyperplane of the k-th DMU is characterized as

$$-\sum_{i=1}^{m} v_i \left(x_i - x_{ik}\right) + \sum_{r=1}^{s} u_r \left(g_r - g_{rk}\right) = 0. \tag{8.7}$$

Equation (8.7) becomes

$$-\sum_{i=1}^{m} v_i x_i + \sum_{r=1}^{s} u_r g_r + \sigma = 0, \tag{8.8}$$

where $\sigma = \sum_{i=1}^{m} v_i x_{ik} - \sum_{r=1}^{s} u_r g_{rk}$ is an intercept of the supporting hyperplane.

As mentioned previously, all DMUs are classified into two groups by their efficiency measures. If the k-th DMU is efficient, then it locates on an efficiency frontier so that the supporting hyperplane passes on the efficient DMU. In contrast, if the DMU is inefficient, it needs to be projected onto an efficiency frontier. Consequently, the k-th DMU, either efficient or projected on an efficiency frontier, has the following conditions:

$$-\sum_{i=1}^{m} v_i x_{ik}^* + \sum_{r=1}^{s} u_r g_{rk}^* + \sigma \leq 0, \tag{8.9}$$

$$\sum_{i=1}^{m} v_i x_{ik}^* = 1, \text{ and} \tag{8.10}$$

$$\sum_{r=1}^{s} u_r g_{rk}^* + \sigma = 1. \tag{8.11}$$

Here, the symbol (*) implies that the k-th DMU is on an efficiency frontier.

Equations (8.9) and (8.10) are obtained from the first and second constraints of Model (4.5), respectively, under the condition that production factors of the k-th DMU are on an efficiency frontier. Equation (8.11) is because the objective value of Model (4.5) becomes unity if the k-th DMU is on an efficiency frontier.

As mentioned above, x_{ik} and g_{rk} are the observed values of two production factors regarding the k-th DMU. Meanwhile, x_{ik}^{*} and g_{rk}^{*} are these projected values on an efficiency frontier. Thus, the superscript (*) implies the projection onto an efficiency frontier, so simultaneously indicating the status of optimality.

The location of the k-th (inefficient) DMU on an efficiency frontier needs the following projections by eliminating slacks.

$$x_{ik}^{*} = \theta^{*} x_{ik} - d_{i}^{x*} = \sum_{j \in RS_k} x_{ij} \lambda_{j}^{*} \text{ and} \tag{8.12}$$

$$g_{rk}^{*} = g_{rk} + d_{r}^{g*} = \sum_{j \in RS_k} g_{rj} \lambda_{j}^{*}. \tag{8.13}$$

The above formulations are obtained from Model (4.4).

By incorporating Equations (8.10) and (8.11), Equation (8.9) becomes as follows:

$$-\sum_{i=1}^{m} v_i x_{ik}^{*} + \sum_{r=1}^{s} u_r g_{rk}^{*} + \sigma = 0. \tag{8.14}$$

Using $\sum_{j \in RS_k} \lambda_j^{*} = 1$ and Equations (8.12) and (8.13), Equation (8.14) becomes as follows:

$$-\sum_{i=1}^{m} v_i \left(\sum_{j \in RS_k} x_{ij} \lambda_j^{*} \right) + \sum_{r-1}^{s} u_r \left(\sum_{j \in RS_k} g_{rj} \lambda_j^{*} \right) + \sum_{j \in RS_k} \sigma \lambda_j^{*} = 0. \tag{8.15}$$

In other words, it becomes:

$$\sum_{j \in RS_k} \left(-\sum_{i=1}^{m} v_i x_{ij} + \sum_{r=1}^{s} u_r g_{rj} + \sigma \right) \lambda_j^{*} = 0. \tag{8.16}$$

Since $\lambda_j^{*} > 0$ holds for all $j (j \in RS_k)$ on an efficiency frontier, the supporting hyperplane of the k-th DMU is determined by

$$-\sum_{i=1}^{m} v_i x_{ij} + \sum_{r=1}^{s} u_r g_{rj} + \sigma = 0 \quad \left(j \in RS_k \right). \tag{8.17}$$

The second part (i.e., $\sum_{i=1}^{m} v_i x_{ik} = 1$) is due to Equation (8.10).

Q.E.D.

As mentioned previously, the RTS measurement is determined by the location of a supporting hyperplane. As a result, we must face an occurrence of multiple supporting hyperplanes, as depicted in Figure 8.3. In the case of RM(v), the upper and lower bounds of the intercept is determined by

$$\begin{aligned} &\text{Maximize/Mininimize} \quad \sigma \\ &\quad s.t. \quad \text{all constraints in Models (4.4) and (4.5) and} \\ &\quad \sum_{r=1}^{s} u_r g_{rk} + \sigma = \theta. \end{aligned} \tag{8.18}$$

Model (8.18) can be specifically expressed as follows:

$$\begin{aligned} &\text{Maximize/Mininimize} \quad \sigma \\ &s.t. \quad -\sum_{j=1}^{n} x_{ij}\lambda_j - d_i^x + \theta x_{ik} = 0 \quad (i = 1,\ldots,m), \\ &\quad \sum_{j=1}^{n} g_{rj}\lambda_j - d_r^g = g_{rk} \quad (r = 1,\ldots,s), \\ &\quad \sum_{j=1}^{n} \lambda_j = 1, \\ &\quad -\sum_{i=1}^{m} v_i x_{ij} + \sum_{r=1}^{s} u_r g_{rj} + \sigma \le 0 \quad (j=1,\ldots,n), \\ &\quad \sum_{i=1}^{m} v_i x_{ik} = 1, \\ &\quad \sum_{r=1}^{s} u_r g_{rk} + \sigma = \theta, \\ &\quad \theta : \text{URS}, \ \lambda_j \ge 0 \ (j = 1,\ldots,n), \\ &\quad d_i^x \ge 0 \ (i = 1,\ldots,m), d_r^g \ge 0 \ (r = 1,\ldots,s), \\ &\quad v_i \ge 0, u_r \ge 0 \text{ and } \sigma : \text{URS}. \end{aligned}$$

Both the primal model (4.4) and the dual model (4.5) are incorporated together in Model (8.18), along with the condition $\left(\sum_{r=1}^{s} u_r g_{rk} + \sigma = \theta\right)$ on optimality. The condition is necessary because the objective value of the primal formulation should equal that of the dual model on optimality. Therefore, the condition needs to be incorporated into Model (8.18).

The upper ($\bar{\sigma}^*$) and lower ($\underline{\sigma}^*$) bounds of σ is identified by maximization or minimization of Model (8.18). It is true that the measurement of upper and lower bounds by Model (8.18) has more constraints than Model (4.5), so being computationally more time-consuming than Model (4.5). However, the measurement by Model (8.18) is more reliable than that of Model (4.5) in terms of determining the upper ($\bar{\sigma}^*$) and lower ($\underline{\sigma}^*$) bounds. Thus, there is a tradeoff between them.

To reduce the computational burden, this chapter may approximate Model (8.18) as follows:

$$\begin{aligned}
&\text{Maximize/Mininimize} \quad \sigma \\
&s.t. \quad -\sum_{i=1}^{m} v_i x_{ij} + \sum_{r=1}^{s} u_r g_{rj} + \sigma \le 0 \quad (j \in J - IE) \\
&\qquad \sum_{i=1}^{m} v_i x_{ik} = 1, \\
&\qquad \sum_{r=1}^{s} u_r g_{rk} + \sigma = \theta_k^*, \\
&\qquad v_i \ge 0, u_r \ge 0 \text{ and } \sigma : \text{URS.}
\end{aligned} \tag{8.19}$$

Model (8.19) is formulated in the manner that the degree of operational efficiency (*OE*: θ_k^*) on the *k*-th DMU, which is determined by Model (4.4), is incorporated in the dual model, or Model (4.5). The upper (under maximization) and lower (under minimization) bounds of σ are obtained after the degree of *OE* is incorporated into Model (8.19). Thus, Model (8.19) consists of two stages for computation for Model (4.4) and (8.19). Here, *IE* represents a group of inefficient DMUs that are not on the efficiency frontier. The DMU group (*J*) stands for all DMUs to be measured by DEA. The group of *J-IE* contains DMUs in which clearly inefficient (dominated) DMUs are eliminated from all ones (*J*). See Chapter 5 for a mathematical definition on *J* and that of *IE*.

Model (8.19) has an important feature to be noted here. That is, the dual-related constraints are also dropped from Model (8.19). It is easily imagined that the computational time to solve Model (8.19) is much less than Model (8.18) because of both less constraints related to inefficient DMUs and no constraints related to the primal and dual formulations. However, the reduced computational effort provides only an approximated range on the intercept, so having a limited classification capability in determining the type of RTS.

After obtaining $\bar{\sigma}^*$ and $\underline{\sigma}^*$ from Model (8.18) or (8.19), the type of input-based RTS is classified by the following rule:

$$\begin{aligned}
&\text{Increasing RTS} \leftrightarrow \underline{\sigma}^* > 0, \\
&\text{Constant RTS} \leftrightarrow \bar{\sigma}^* \ge 0 \ge \underline{\sigma}^* \text{ and} \\
&\text{Decreasing RTS} \leftrightarrow 0 > \bar{\sigma}^*.
\end{aligned} \tag{8.20}$$

As specified in the classification rule (8.20), the input-oriented RM(v) exhibits increasing RTS on the k-th DMU if the upper and lower bounds of the intercept are both positive in these signs. In contrast, if they are negative, then the DMU exhibits decreasing RTS. The constant RTS is found on the DMU if the upper bound is positive and the lower bound is negative. The definition on constant RTS is different from production economics because it uses a smooth curve for a production function. In contrast, DEA does not use such a smooth production curve. Such is indeed a strength (i.e., no assumption on a production function) but simultaneously a drawback (i.e. the upper and lower bounds). Thus, DEA is not associated with uniqueness, so suffering from multiple solutions on RTS.

Here, it is necessary to mention that Equations (8.20) is for the input-based RM(v). The classification becomes opposite in the case when the desirable output-based RM(v) is employed by DEA. That is,

$$\begin{aligned}&\text{Increasing RTS} \leftrightarrow 0 > \bar{\sigma}^*,\\&\text{Constant RTS} \leftrightarrow \bar{\sigma}^* \geq 0 \geq \underline{\sigma}^* \text{ and}\\&\text{Decreasing RTS} \leftrightarrow \underline{\sigma}^* > 0.\end{aligned} \tag{8.21}$$

Note that if we drop the assumption on uniqueness from an intercept (σ) from Equations (8.5), it becomes Equations (8.21). Similarly, the uniqueness changes from Equations (8.6) to Equations (8.21).

8.4 COST-BASED RTS MEASUREMENT

By shifting our interest from the production-based RTS measurement to the cost-based RTS measurement, this chapter starts with Proposition 8.2 which characterizes the location of a supporting hyperplane in a data domain on cost and desirable outputs.

Proposition 8.2: A supporting hyperplane on the k-th DMU that is measured by a cost model, or Model (4.34), is specified as $c = \sum_{r=1}^{s} u_r g_r + \sigma$. Here, u_r $(r=1,...,s)$ and σ are unknown variables to represent a directional vector and an intercept of a supporting hyperplane. Given p_{ik} $(i=1,...,m)$ of the k-th DMU, the supporting hyperplane is obtained by

$$-\sum_{i=1}^{m} p_{ik} x_{ij} + \sum_{r=1}^{s} u_r g_{rj} + \sigma = 0 \left(j \in RS_k \right),$$

where $RS_k = \{ j | \lambda_j^* > 0, j \in J \}$ stands for a reference set of the k-th DMU that is measured by Model (4.34).

[Proof] In the $s+1$ dimensional space, a supporting hyperplane of the k-th DMU (c_k, G_k) becomes

$$-\left(c - c_k\right) + \sum_{r=1}^{s} u_r \left(g_r - g_{rk}\right) = 0. \tag{8.22}$$

Equation (8.22) is restructured as follows:

$$c = \sum_{r=1}^{s} u_r g_r + \left(c_k - \sum_{r=1}^{s} u_r g_{rk} \right) = \sum_{r=1}^{s} u_r g_r + \sigma. \tag{8.23}$$

In Equation (8.23), we replace $c_k - \sum_{r=1}^{s} u_r g_{rk}$ by σ.

If the k-th DMU is efficient, then the supporting hyperplane passes on the DMU that locates on an efficiency frontier. In contrast, if the DMU is inefficient, it needs to be projected on an efficiency frontier to attain the status of efficiency. The k-th DMU, projected on the efficiency frontier, satisfies the following conditions via Models (4-34) and (4-36):

$$-\sum_{i=1}^{m} p_{ik} x_i^* + \sum_{r=1}^{s} u_r g_{rk}^* + \sigma \leq 0, \tag{8.24}$$

$$\sum_{i=1}^{m} p_{ik} x_i^* = c_k^* \text{ and} \tag{8.25}$$

$$\sum_{r=1}^{s} u_r g_{rk}^* + \sigma = c_k^*. \tag{8.26}$$

Equation (8.24) originates from the first constraints of Model (4-36). Equations (8.25) and (8.26) are because of $\sum_{i=1}^{m} p_{ik} x_i^* = \sum_{r=1}^{s} u_r^* g_{rk} + \sigma^* = c_k^*$ on the optimality of Models (4.34) and (4.36). Here, c_k and g_{rk} are the observed values of cost and desirable outputs regarding the k-th DMU. Meanwhile, c_k^* and g_{rk}^* are these projected values on an efficiency frontier. As mentioned previously, the superscript (*) implies the projection onto an efficiency frontier, so simultaneously indicating the status of optimality.

The k-th DUM has the following projection on an efficiency frontier:

$$x_i^* = \sum_{j \in RS_k} x_{ij} \lambda_j^* \tag{8.27}$$

and

$$g_{rk}^* = g_{rk} + d_r^{g*} = \sum_{j \in RS_k} g_{rj} \lambda_j^* \tag{8.28}$$

Equations (8.25) and (8.26) change Equation (8.24) by the following equality:

$$-\sum_{i=1}^{m} p_{ik} x_i^* + \sum_{r=1}^{s} u_r g_{rk}^* + \sigma = 0. \tag{8.29}$$

By incorporating Equations (8.27) and (8.28) into Equation (8.24), it becomes as follows:

$$-\sum_{i=1}^{m} p_{ik} \left(\sum_{j \in RS_k} x_{ij} \lambda_j^* \right) + \sum_{r=1}^{s} u_r \left(\sum_{j \in RS_k} g_{rj} \lambda_j^* \right) + \sum_{j \in RS_k} \sigma \lambda_j^* = 0 \tag{8.30}$$

with $\sum_{j\in RS_k}\lambda_j^* = 1$. Consequently, Equation (8.30) becomes

$$\sum_{j\in RS_k}\left(-\sum_{i=1}^{m}p_{ik}x_{ij}+\sum_{r=1}^{s}u_r g_{rj}+\sigma\right)\lambda_j^* = 0. \tag{8.31}$$

Since $\lambda_j^* > 0$ holds for all $j \in RS_k$ in Model (4-34), unknown variables (u_r for all r and σ) of a supporting hyperplane are determined by

$$-\sum_{i=1}^{m}p_{ik}x_{ij}+\sum_{r=1}^{s}u_r g_{rj}+\sigma = 0 \quad \left(j \in RS_k\right). \tag{8.32}$$

Q.E.D.

To determine the type of cost-based RTS, this chapter measures the upper and lower bounds of σ by the following model:

$$\begin{aligned}
&\text{Maximize/Minimize } \sigma \\
&\quad s.t. \quad \text{all constraints in Models } (4.34) \text{ and } (4.36) \text{ and} \\
&\qquad\qquad -\sum_{i=1}^{m}p_{ik}x_i+\sum_{r=1}^{s}u_r g_{rk}+\sigma = 0.
\end{aligned} \tag{8.33}$$

More specifically, Model (8.33) is rewritten by the following formulation:

$$\begin{aligned}
&\text{Maximize / Minimize } \sigma \\
&s.t. \quad -\sum_{j=1}^{n}x_{ij}\lambda_j + x_i \ge 0 \qquad (i=1,\dots,m), \\
&\qquad \sum_{j=1}^{n}g_{rj}\lambda_j \ge g_{rk} \qquad (r=1,\dots,s), \\
&\qquad \sum_{j=1}^{n}\lambda_j = 1, \\
&\qquad -\sum_{i=1}^{m}v_i x_{ij}+\sum_{r=1}^{s}u_r g_{rj}+\sigma \le 0 \quad (j=1,\dots,n), \\
&\qquad v_i \le p_{ik} \ (i=1,\dots,m), \\
&\qquad \sum_{r=1}^{s}u_r g_{rk}+\sigma = \sum_{i=1}^{m}p_{ik}x_i, \\
&\qquad \lambda_j \ge 0\ (j=1,\dots,n),\ x_i \ge 0\ (i=1,\dots,m), \\
&\qquad v_i \ge 0\ (i=1,\dots,m),\ u_r \ge 0\ (r=1,\dots,s) \text{ and} \\
&\qquad \sigma : \text{URS}.
\end{aligned}$$

To reduce our computational burden, this chapter approximates Model (8.33) by the following formulation:

$$\begin{aligned} &\text{Maximize / Minimize } \sigma \\ &s.t. \quad \sum_{r=1}^{s} u_r g_{rj} + \sigma \le \sum_{i=1}^{m} p_{ik} x_{ij} \quad (j \in J - IE), \\ &\sum_{r=1}^{s} u_r g_{rk} + \sigma = c_k^*, \\ &u_r \ge 0 \text{ and } \sigma : \text{URS}. \end{aligned} \tag{8.34}$$

The set (IE) is a group of operationally inefficient DMUs. The minimum cost (c_k^*) of the k-th DMU, measured by Model (4.34), is incorporated in Model (8.34). The model determines the upper ($\bar{\sigma}^*$) and lower ($\underline{\sigma}^*$) bounds of σ on the estimated cost minimum. As discussed in Section 8.3, Model (8.34) also provides an approximation method to determine the type of RTS.

The type of RTS, using the two bounds, is classified by the following rule:

$$\begin{aligned} &\text{Increasing RTS} \leftrightarrow \underline{\sigma}^* > 0, \\ &\text{Constant RTS} \leftrightarrow \bar{\sigma}^* \ge 0 \ge \underline{\sigma}^* \text{ and} \\ &\text{Decreasing RTS} \leftrightarrow 0 > \bar{\sigma}^*. \end{aligned} \tag{8.35}$$

8.5 SCALE EFFICIENCIES AND SCALE ECONOMIES

After returning to the scale economies discussed in Chapter 4, this chapter discusses analytical relationships between production and cost-based scale efficiencies and scale economies (i.e., RTS).

Proposition 8.3: If $AE_v = AE_c$, then it leads to $CSE = PSE$. In contrast, if $AE_v \ne AE_c$, then it becomes $CSE = (AE_c/AE_v)PSE$.

[Proof] The proof is because

$$CSE = CE_c/CE_v = (AE_c OE_c)/(AE_v/OE_v) = (AE_c)/(AE_v)PSE.$$

Q.E.D.

Proposition 8.3 indicates that a ratio between AE_c (allocative and scale efficiency) and AE_v (allocative efficiency) indicates the relationship between PSE (production-based scale efficiency) and CSE (cost-based scale efficiency). A difficulty in empirical studies is that PSE and CSE can provide qualitative information on scale efficiency measures. However, they cannot provide any quantitative information on scale economies, or a magnitude of RTS.

Hereafter, it is necessary for us to measure the degree of scale economies in multiple components of the two production factors. Here, let e_g be the elasticity on production in an input-oriented model and e_c be the elasticity on cost. The two elasticity measures in the case of multiple components are defined as follows:

$$e_g = \frac{\sum_{i=1}^{m} v_i^{g*} x_{ik}}{\sum_{r=1}^{s} u_r^{g*} g_{rk}} = \frac{1}{\sum_{r=1}^{s} u_r^{g*} g_{rk}} \text{ and} \tag{8.36}$$

$$e_c = \frac{c_k^*}{\sum_{r=1}^{s} u_r^{c*} g_{rk}}. \tag{8.37}$$

In the above two equations, the subscripts g and c are added to each dual variable to distinguish between variables obtained from the production model and those from the cost model.

The theoretical relationship between Equations (8.36) and (8.37) is summarized as follows:

Proposition 8.4: Under the assumption on uniqueness on an optimal solution, if the k-th DMU has both $OE_V = 1$ and $\sigma^{g*} = \sigma^{c*}/c_k^*$, then the DMU has $e_c = e_g$. In the other cases,

$$\frac{e_c}{e_g} = \frac{OE_v - \sigma^{g*}}{1 - \sigma^{c*}/c_k^*}. \tag{8.38}$$

[Proof] Equation (8.36) becomes

$$e_g = \frac{1}{\sum_{r=1}^{s} u_r^{g*} g_{rk}} = \frac{1}{\left(\theta^* - \sigma^{g*}\right)} = \frac{1}{\left(OE_v - \sigma^{g*}\right)}, \tag{8.39}$$

because of Equation (4.6). In a similar manner, Equation (8.37) becomes

$$e_c = \frac{c_k^*}{\sum_{r=1}^{s} u_r^{c*} g_{rk}} = \frac{c_k^*}{c_k^* - \sigma^{c*}} = \frac{1}{1 - \sigma^{c*}/c_k^*} \tag{8.40}$$

because of Equation (8.26). Dividing Equation (8.40) by Equation (8.39) produces

$$\frac{e_c}{e_g} = \frac{OE_v - \sigma^{g*}}{1 - \sigma^{c*}/c_k^*}.$$

Q.E.D.

As discussed previously, it is possible to drop the assumption on a unique optimal solution on σ^{g*}. In the case, let the upper and lower bounds of σ^{g*} be $\bar{\sigma}^{g*}$ and $\underline{\sigma}^{g*}$, then Equation (8.39) can be specified in terms of RTS as follows:

$$\begin{aligned}
&\text{Increasing RTS}: \frac{1}{OE_v - \bar{\sigma}^{g*}} > e_g > \frac{1}{OE_v - \underline{\sigma}^{g*}} > 1,\\
&\text{Constant RTS}: \frac{1}{OE_v - \bar{\sigma}^{g*}} \geq e_g(=1) \geq \frac{1}{OE_v - \underline{\sigma}^{g*}} \text{ and}\\
&\text{Decreasing RTS}: 1 > \frac{1}{OE_v - \bar{\sigma}^{g*}} > e_g > \frac{1}{OE_v - \underline{\sigma}^{g*}}.
\end{aligned} \tag{8.41}$$

In a similar manner, Equation (8.40) is used for the cost-based RTS as follows:

$$\begin{aligned}
&\text{Increasing RTS}: \frac{1}{1 - \bar{\sigma}^{c*}/c_k^*} > e_c > \frac{1}{1 - \underline{\sigma}^{c*}/c_k^*} > 1,\\
&\text{Constant RTS}: \frac{1}{1 - \bar{\sigma}^{c*}/c_k^*} \geq e_c(=1) \geq \frac{1}{1 - \underline{\sigma}^{c*}/c_k^*} \text{ and}\\
&\text{Decreasing RTS}: 1 > \frac{1}{1 - \bar{\sigma}^{c*}/c_k^*} > e_c > \frac{1}{1 - \underline{\sigma}^{c*}/c_k^*}.
\end{aligned} \tag{8.42}$$

Proposition 8.5: When we face an occurrence of multiple solutions, the upper and lower bounds of e_c/e_g are described as follows:

$$\frac{OE_v - \underline{\sigma}^{g*}}{1 - \bar{\sigma}^{c*}/c_k^*} \geq \frac{e_c}{e_g} \geq \frac{OE_v - \bar{\sigma}^{g*}}{1 - \underline{\sigma}^{c*}/c_k^*}. \tag{8.43}$$

[Proof] Dividing the upper bound of Equations (8.42) by the lower bound of Equations (8.41) produces the upper bound of e_c/e_g. In contrast, dividing the lower bound of Equations (8.42) by the upper bound of Equations (8.41) produces the lower bound of e_c/e_g.

Q.E.D.

Proposition 8.6: In measuring the economies of scale by e_g and e_c, increasing RTS and decreasing RTS are mutually exclusive in the DEA assessment.

[Proof] Since $e_g \geq 0$ and $e_c \geq 0$, the ratio becomes $e_c/e_g \geq 0$. If one of the two is increasing RTS ($\bar{\sigma}^{g*} \geq \underline{\sigma}^{g*} > 0$ or $\bar{\sigma}^{c*} \geq \underline{\sigma}^{c*} > 0$) and the other is decreasing RTS ($0 > \bar{\sigma}^{g*} \geq \underline{\sigma}^{g*}$ or $0 > \bar{\sigma}^{c*} \geq \underline{\sigma}^{c*}$), then we have $e_c/e_g < 0$. This is inconsistent with the non-negative condition on the ratio between the two economies of scale.

Q.E.D.

Applicability: This chapter does not document an illustrative application of the proposed approach because Chapter 20 will discuss how to measure the type and magnitude of RTS for DEA environmental assessment.

8.6 SUMMARY

This chapter discussed how to measure the type and degree of RTS in a conventional framework of DEA production and cost analyses. The measurement on RTS has often suffered from an occurrence of the three types of multiple solutions. The first one is an occurrence of multiple projections and the second one is an occurrence of multiple reference sets. See Chapter 7, which discusses an occurrence of such multiple solutions. The last one, discussed in this chapter, is an occurrence of multiple intercepts (σ). This chapter considered a range of multiple intercepts by assuming both a unique projection and a unique reference set.

Dual relationships between production-based RTS and cost-based RTS were also discussed in this chapter. In Section II, Chapter 20 will extend the concept of RTS into damages to scale (DTS) and Chapter 23 will further extend it to returns to damage (RTD) and damages to return (DTR). Thus, this chapter may serve as conceptual and methodological bases for such a new type of scale economies for DEA environmental assessment.

At the end of this chapter, it is important for us to note that DEA does not assume any functional form to express a production function or a cost function. If we are interested in measuring only a level of efficiency, the unique feature may become a methodological benefit, indeed. However, if we pay attention to the measurement of RTS, the unique feature often becomes a methodological drawback. For example, scale elasticity and scale economies often become very large at the level of infinite. Thus, as discussed in Chapters 22 and 23, DEA environmental assessment needs to restrict multipliers (dual variables) without any prior information. Such a new approach will be discussed as "explorative analysis" in those two chapters.

9

CONGESTION

9.1 INTRODUCTION

This chapter[1] discusses the economic concept of "congestion."[2] It is a widely observed phenomenon that identifies inefficiency in such a manner that any reduction in an input(s) results in an increase in a desirable output(s) without worsening other inputs and desirable outputs (Cooper *et al.*, 2004). Conversely, congestion implies that any increase in an input(s) decreases the desirable output(s) without worsening other production factors. This type of inefficiency is clearly different from the concept of "operational inefficiency" discussed in previous chapters, which indicates the existence of an excess amount of input(s)

[1] This chapter is partly based upon the articles: (a) Sueyoshi, T. and Sekitani, K. (2008) DEA congestion and returns to scale under an occurrence of multiple optimal projections. *European Journal of Operational Research*, **194**, 592–607 and (b) Sueyoshi, T. and Goto, M. (2016) Undesirable congestion under natural disposability and desirable congestion under managerial disposability in U.S. electric power industry measured by DEA environmental assessment. *Energy Economics*, **55**, 173–188.

[2] The inefficiency associated with congestion has not received much attention in economics literature (Brockett *et al.*, 2004, p. 208 and Cooper *et al.* 2001a, p. 63). For example, well-known economists such as Leibenstein (1966, 1976) and Stigler (1976) argued that the issue and its related topics did not constitute proper subjects for study in economics. However, the situation was changed after Färe and Svensson (1980) explored the congestion issue within a mathematical framework of linear programming. See debates between Cooper *et al.* (2001c) and Cherchye *et al.* (2001).

Environmental Assessment on Energy and Sustainability by Data Envelopment Analysis, First Edition. Toshiyuki Sueyoshi and Mika Goto.

and/or a shortfall of desirable output(s). A typical example can be found in the line capability limit of electric power transmission. Another example is the transportation capacity limit on a pipeline network between upstream and downstream in a petroleum industry.

Several previous studies (e.g., Färe *et al.*, 1985) paid attention to the occurrence of congestion in DEA. The concept is defined as "evidence of congestion is present when reductions in one or more inputs can be associated with increases in one or more desirable outputs, or in reverse, when increases in one or more inputs can be associated with decreases in one or more desirable outputs, without worsening any other input or desirable output." See Cooper *et al.* (2001b) on the definition on congestion within a conventional framework of DEA.

Two groups of researchers have argued a possible occurrence of congestion in previous DEA studies. One of the two groups includes production economists (i.e., Färe and Svensson, 1980; Cherchye *et al.*, 2001), who have proposed the measurement of congestion by a radial model, or RM(v). See Chapter 4 on the radial measurement. Meanwhile, Cooper and his associates (Brockett *et al.* 1998, 2004; Cooper *et al.*, 2001a, b, c) proposed an alternative DEA approach which was based upon non-radial models in which slacks were used for efficiency measurement. See Chapter 5 on the non-radial measurement.

It is indeed true that the measurement of congestion is important for DEA-based performance evaluation. In this book, the concept of congestion will serve as the basis for developing two disposability concepts, referred to as "undesirable congestion under natural disposability" and "desirable congestion under managerial disposability," for environmental assessment. See Chapter 21. Both undesirable and desirable congestions are very important concepts in discussing DEA environmental assessment in energy sectors. Before discussing the concept of congestion from the perspective of environmental assessment, this chapter discusses conventional implications on congestion and then discusses these analytical implications from primal and dual formulations. Such descriptions provide us with mathematical understanding related to the occurrence of congestion and further a mathematical linkage to returns to scale (RTS).

The purpose of this chapter is to document how to formulate primal and dual models related to the possible occurrence of congestion within a conventional DEA framework and then to discuss its mathematical properties and economic implications, all of which can be derived from these dual formulations. See Chapters 21, 22 and 23 which extend the concept of congestion, discussed in this chapter, into DEA environmental assessment.

The remainder of this chapter is organized as follows: Section 9.2 shows an illustrative example on congestion. Section 9.3 provides foundational discussions. Section 9.4 describes a supporting hyperplane. Section 9.5 discusses how to identify an occurrence of congestion. Section 9.6 shows the relationship between congestion and RTS. Section 9.7 provides a computational process on how to measure the degree of congestion. Section 9.8 describes economic implications. Section 9.9 summarizes this chapter.

9.2 AN ILLUSTRATIVE EXAMPLE

For our illustrative purpose, this chapter visually discusses the possible occurrence of congestion in a transmission grid line for connecting generators and end users. The visual illustration is important in understanding what type of problem is due to congestion. No previous study clearly discussed such a real business example, so that the concept of congestion often invited many misunderstandings in the DEA community. Therefore, this chapter starts from an illustrative example, obtained from the research effort of Sueyoshi and Goto (2016).

First, let us consider a wholesale power exchange market that is separated into multiple zones based on the geographic location of nodes and transmission grid structure. Figure 9.1 depicts a power market that consists of three zones (I, II and III). The figure depicts only three zones for our descriptive convenience. Each zone consists of several generators and loads, or demands. There are two types of transmission connections in the power market: intra-zonal links and inter-zonal links. Intra-zonal links are connections that exist among generators and wholesalers within a zone. Inter-zonal links are connections between zones. A common market clearing price (MCP) may exist if these zones are linked together with

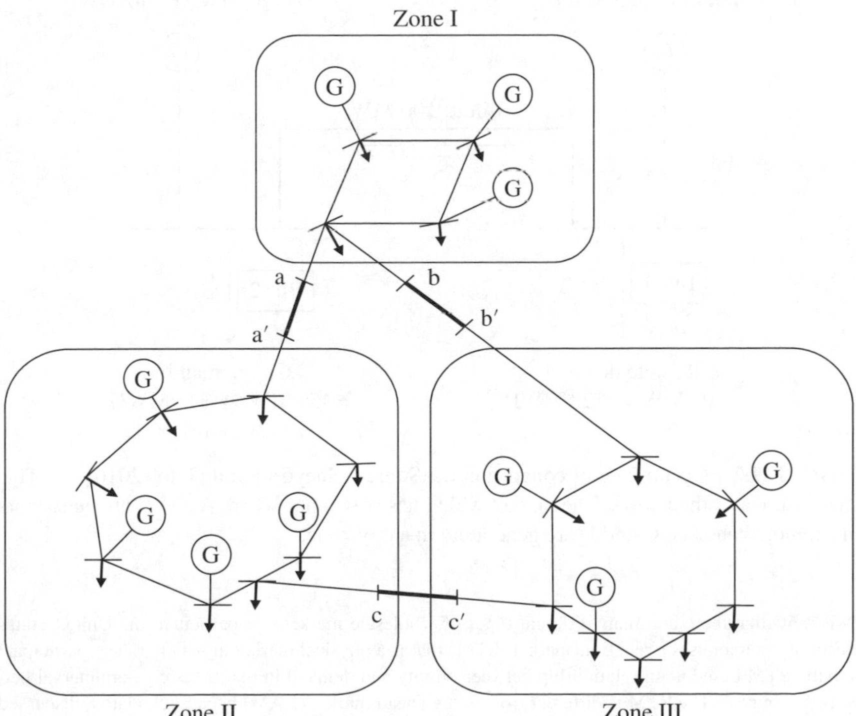

FIGURE 9.1 Congestion among three zones (a) Source: Sueyoshi and Goto (2016).

each other. However, if these zones are functionally separated by a capacity limit on an interconnection line, so indicating an occurrence of congestion, then each zone has a different locational marginal price (LMP) as their MCP.

Figure 9.1 assumes that congestion occurs in the inter-zonal links (a–a′, b–b′ and c–c′). This chapter excludes an occurrence of congestion on intra-zonal link because the congestion can be handled by obtaining electricity from a generator(s) either in a same zone or in another zone. All further references to the word “link” mean an inter-zonal link. A capacity limit on the link usually increases MCP, because it limits the ability of each zone to obtain electricity from an inexpensive generator(s) in another zone(s). For example, let us consider a case where Zone I supplies electricity to Zone II and congestion may occur on the link (a–a′). The MCP of Zone II depends upon whether it can access a power supply from inexpensive generators in Zone III or access Zone I’s excess capacity through Zone III. In the latter case, the MCP increases because of the use of an expensive generator(s).

To describe how a line limit increases MCP, this chapter considers a wholesale market that consists only of a day ahead market (DAM)[3]. Figure 9.2 depicts

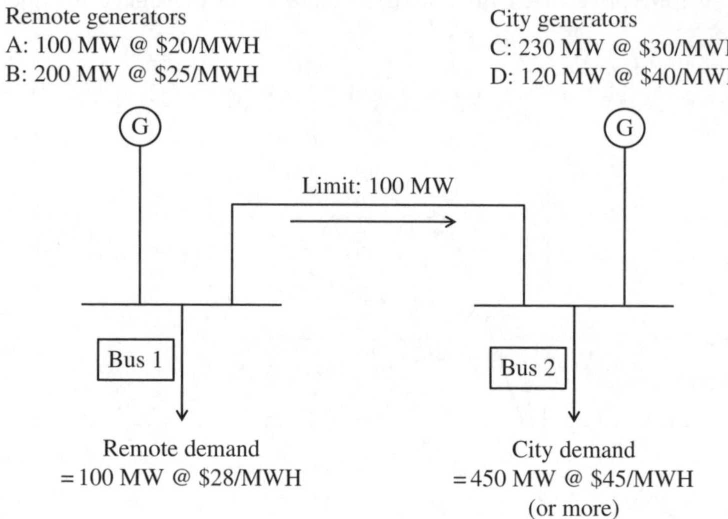

FIGURE 9.2 Occurrence of congestion (a) Source: Sueyoshi and Goto (2016). (b) This chapter updates their work. The market which has two zones where A and B are generators in a remote zone and C and D are generators in a city zone.

[3] We know that there are many different types of wholesale markets. In particular, the United States electricity market uses a real time market (RTM) that is a physical market in which traders participate into the RTM based upon relationship between supply and demand in real time (e.g., an interval between 5 minutes). The RTM is different from a day ahead market (DAM) because the latter, discussed in this chapter, is a financial market in which traders participate into the markets, opened one day before real electricity delivery, to arrange a delivery schedule and generate a profit by their bids.

a DAM which has two zones where A and B are generators in a remote zone and C and D are generators in a city zone. The market in the remote zone is self-sufficient and can be cleared because the total supply (300 MW = 100 MW + 200 MW) is larger than total demand (100 MW). The MCP on the equilibrium point (EP) becomes US$ 20/MWH (megawatt hour) and the load is 100 MW. Meanwhile, the market for city zone cannot be cleared, because the total supply (350 MW = 230 MW + 120 MW) is less than the total demand (450 MW) or more.

To clear the market for the city zone, the independent system operator (ISO) controls transmission in such a manner that it arranges for electricity to be transmitted from the remote to the city in a wholesale market. If there is a capacity limit on the link between them, then the limit influences the MCP. To explain this influence, this chapter prepares two scenarios: (a) absence of a capacity limit and (b) the opposite case.

One of these two scenarios is depicted in Figure 9.3 that explains a market price clearing mechanism without a capacity limit on the link. The horizontal axis indicates the amount of generation (or load) and the vertical axis indicates the market price. The supply side is depicted by an increasing step function, while the demand side is depicted by a decreasing step function. An intersection between supply and demand indicates the equilibrium point (EP) at which a MCP is determined as US$ 30/MWH and a total amount of power supply is 550 MW (100 MW + 450 MW). In this case, inexpensive generators (A and B) in

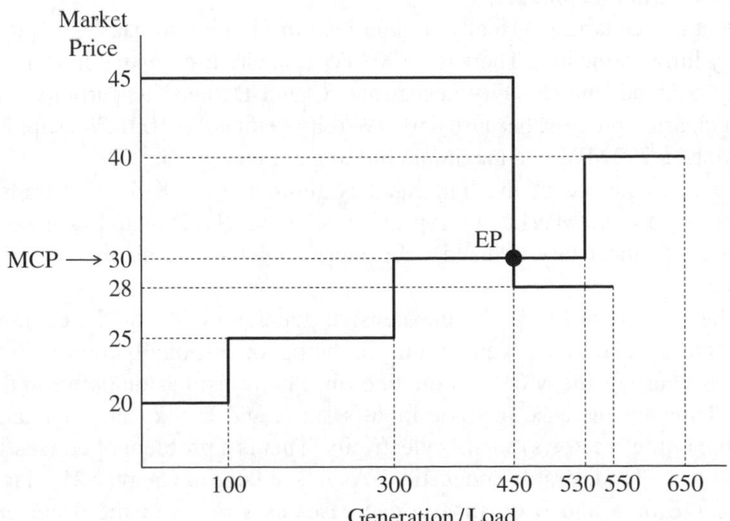

FIGURE 9.3 Market clearing price: no line limit (a) Source: Sueyoshi and Goto (2016). This chapter updates their work.

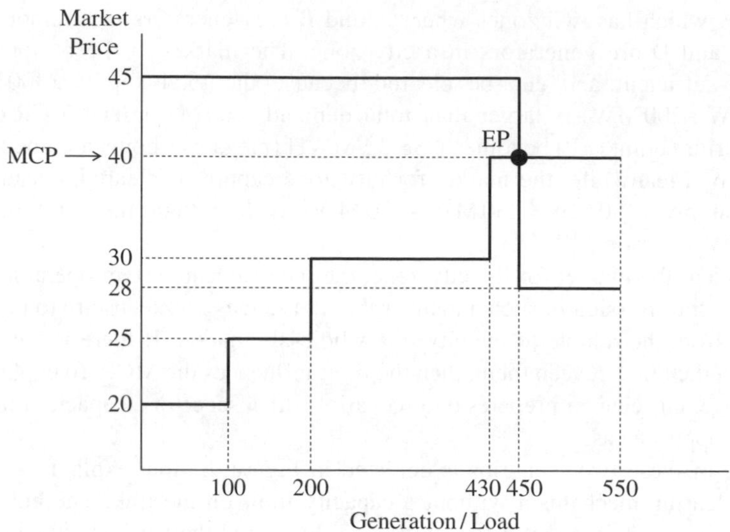

FIGURE 9.4 Market clearing price (line limit) (a) Source: Sueyoshi and Goto (2016). This chapter updates their work.

the remote are fully utilized to satisfy not only remote demand (100 MW) but also part of the city demand (200 out of 450 MW). The 200 MW of electricity flows through the link from the remote to the city. The expensive generator (D) is excluded from the market.

The other scenario is visually summarized in Figure 9.4 that depicts it with a capacity limit on the link. There is a 100 MW capacity limit on the link. To satisfy the city demand, the two city generators (C and D) need to participate in the market clearing process, because 450 MW (city demand) – 100 MW (supply from the remote) > 230 MW (the maximum bidding amount of C).

As a consequence of the link capacity limit, the MCP for this transaction increases to US$ 40/MWH. The expensive generator (D) is included in the power generation to meet city demand, thus increasing the overall market price for the zone.

In the context of DEA, the inexpensive generators (A and B) cannot fully participate in electricity generation, so being operationally inefficient as a whole, because of the MCP and the line limit on transmission between the two zones. Thus, an increase in some input(s) in A and B does not increase their desirable output vector(s) such as electricity. That is a problem of congestion, or referred to as "undesirable congestion" (e.g., see later in Chapter 21). Thus, the level of *OE* in A and B drastically decreases as a result of the occurrence of congestion.

9.3 FUNDAMENTAL DISCUSSIONS

Many previous studies in production economics (e.g., Färe *et al.*, 1985; Färe and Svensson, 1980) have suggested that a possible occurrence of congestion, later referred to as an "undesirable congestion", is identified by comparing the following two input-oriented models:

Input-oriented RM(v)

$$
\begin{aligned}
&\text{Mimimize} \quad \theta \\
&s.t. \quad -\sum_{j=1}^{n} x_{ij}\lambda_j + \theta x_{ik} \geq 0 \quad (i=1,\ldots,m), \\
&\qquad \sum_{j=1}^{n} g_{rj}\lambda_j \geq g_{rk} \quad (r=1,\ldots,s), \\
&\qquad \sum_{j=1}^{n} \lambda_j = 1, \\
&\qquad \theta\text{: URS and } \lambda_j \geq 0 \ (j=1,\ldots,n).
\end{aligned}
\tag{9.1}
$$

and

Input-oriented congestion

$$
\begin{aligned}
&\text{Minimize} \quad \alpha \\
&s.t. \quad -\sum_{j=1}^{n} x_{ij}\lambda_j + \alpha x_{ik} = 0 \quad (i=1,\ldots,m), \\
&\qquad \sum_{j=1}^{n} g_{rj}\lambda_j \geq g_{rk} \quad (r=1,\ldots,s), \\
&\qquad \sum_{j=1}^{n} \lambda_j = 1, \\
&\qquad \alpha\text{: URS and } \lambda_j \geq 0 \ (j=1,\ldots,n).
\end{aligned}
\tag{9.2}
$$

Models (9.1) and (9.2) drop a slack-related efficiency adjustment, or $\varepsilon_n\left(\sum_{i=1}^{m} d_i^x + \sum_{r=1}^{s} d_r^g\right)$, from these formulations in order to simplify our description. They are input-oriented RM(v) and its modification, which are formulated with and without a possible occurrence of congestion, respectively, by assigning equality constraints on inputs, not desirable outputs. It is important to note that the equality assignment is a major research component for identifying a possible occurrence of congestion in DEA. The problem is which production factors are expressed by equality constraints.

Model (9.1) provides a radial measure for OE_v. An important feature of the efficiency measure is that it avoids the assumption on constant RTS, so being formulated under variable RTS, because the assumption of constant RTS is inconsistent with identifying a possible occurrence of congestion.

The two variables (θ and α) in the objective function of the two RMs represent, respectively, a level of OE_v under two different production technologies. These are unrestricted URS in Models (9.1) and (9.2), where each of the two variables can take any sign. Comparison between the two models, as mentioned above, indicates that the first set of constraints on inputs is formulated by inequality in Model (9.1) and by equality in Model (9.2). Such a difference seems minor, but being very important in congestion identification. See Chapters 21, 22 and 23 that discuss such a difference from DEA environmental assessment in which equality is given to desirable or undesirable outputs under two different conditions (i.e., natural or managerial disposability). We compare this chapter with those ones, later.

To describe the implication of congestion, this chapter needs to discuss the dual formulations of Models (9.1) and (9.2), both of which have the following structures, respectively:

Input-oriented RM(v)*: Dual formulation*

$$
\begin{aligned}
\text{Maximize} \quad & \sum_{r=1}^{s} u_r g_{rk} + \sigma \\
s.t. \quad & -\sum_{i=1}^{m} v_i x_{ij} + \sum_{r=1}^{s} u_r g_{rj} + \sigma \le 0 \quad (j = 1,\ldots,n), \\
& \sum_{i=1}^{m} v_i x_{ik} = 1, \\
& v_i \ge 0 \; (i = 1,\ldots,m), \\
& u_r \ge 0 \; (r = 1,\ldots,s) \text{ and } \sigma\text{: URS.}
\end{aligned}
\tag{9.3}
$$

Input-oriented congestion: Dual formulation

$$
\begin{aligned}
\text{Maximize} \quad & \sum_{r=1}^{s} u_r g_{rk} + \sigma \\
s.t. \quad & -\sum_{i=1}^{m} v_i x_{ij} + \sum_{r=1}^{s} u_r g_{rj} + \sigma \le 0 \quad (j = 1,\ldots,n), \\
& \sum_{i=1}^{m} v_i x_{ik} = 1, \\
& v_i\text{: URS } (i = 1,\ldots,m), \\
& u_r \ge 0 \; (r = 1,\ldots,s) \text{ and } \sigma\text{: URS.}
\end{aligned}
\tag{9.4}
$$

Here, both v_i $(i=1,\ldots,m)$ and u_r $(r=1,\ldots,s)$ are dual variables, often referred to as "multipliers", which are derived from the first and second constraint sets in Models (9.1) and (9.2), respectively. Similarly, the unknown variable (σ) is a dual variable derived from the last constraint of the two models. Almost no difference is identified between Models (9.3) and (9.4). However, a careful examination between them indicates that v_i, or the dual variable related to the i-th input, is non-negative in Model (9.3) but it is unrestricted in Model (9.4).

Following the conventional description on DEA, let the optimal objective value of Model (9.1) be θ^* and that of Model (9.2) be α^*, where the superscript (*) indicates the status of optimality. Then, the input-oriented congestion (IC) is measured by the following ratio:

$$\text{Input-oriented congestion: IC}\left(\theta^*, \alpha^*\right) = \theta^* / \alpha^*. \tag{9.5}$$

If IC $(\theta^*, \alpha^*) < 1$, then the congestion occurs on the k-th DMU. Meanwhile, if IC $(\theta^*, \alpha^*) = 1$, then the congestion does not occur on the k-th DMU.

Intuitively speaking, the optimal objective value of Model (9.1) is more freely determined than that of Model (9.2) in terms of the minimization. Therefore, $\theta^* \leq \alpha^*$. Consequently, IC $(\theta^*, \alpha^*) = \theta^* / \alpha^* < 1$ indicates an occurrence of congestion while IC $(\theta^*, \alpha^*) = \theta^* / \alpha^* = 1$ implies no occurrence of congestion.

Shifting our description from the input-oriented to output-oriented measurement, this chapter reorganizes Models (9.1) and (9.2) by the following formulations:

Desirable output-oriented RM(v)

$$\begin{aligned}
&\text{Maximize} \quad \tau \\
&s.t. \quad \sum_{j=1}^{n} x_{ij}\lambda_j \qquad \leq x_{ik} \quad (i=1,\ldots,m), \\
&\qquad -\sum_{j=1}^{n} g_{rj} + \tau g_{rk} \leq 0 \qquad (r=1,\ldots,s), \\
&\qquad \sum_{j=1}^{n} \lambda_j \qquad = 1, \\
&\qquad \tau: \text{URS and } \lambda_j \geq 0 \ (j=1,\ldots,n).
\end{aligned} \tag{9.6}$$

As discussed in Chapter 4, the output-oriented RM(v) uses τ to express the level of operational efficiency by $OE^* = 1/\tau^*$.

Here, it is important to note two concerns. One of the two concerns is that this chapter uses only the term "output-oriented congestion" since there is no distinction between the two output categories (i.e., desirable and undesirable outputs). This type of congestion on desirable outputs will be discussed later as an occurrence of "undesirable congestion" in Chapter 21.

The occurrence of congestion is measured by

Desirable output-oriented congestion

$$
\begin{aligned}
&\text{Maximize} \quad \beta \\
&s.t. \quad \sum_{j=1}^{n} x_{ij}\lambda_j = x_{ik} \left(i=1,\ldots,m\right), \\
&\qquad -\sum_{j=1}^{n} g_{rj} + \beta g_{rk} \le 0 \ \left(r=1,\ldots,s\right), \\
&\qquad \sum_{j=1}^{n} \lambda_j = 1, \\
&\qquad \beta\text{: URS and } \lambda_j \ge 0 \ \left(j=1,\ldots,n\right).
\end{aligned} \tag{9.7}
$$

Models (9.6) and (9.7) have the following dual formulations, respectively:

Desirable output-oriented RM(v)*: Dual formulation*

$$
\begin{aligned}
&\text{Minimize} \ \sum_{i=1}^{m} v_i x_{ik} + \sigma \\
&s.t. \quad \sum_{i=1}^{m} v_i x_{ij} - \sum_{r=1}^{s} u_r g_{rj} + \sigma \ge 0 \quad \left(j=1,\ldots,n\right), \\
&\qquad \sum_{r=1}^{s} u_r g_{rk} = 1, \\
&\qquad v_i \ge 0 \ \left(i=1,\ldots,m\right), \ u_r \ge 0 \ \left(r=1,\ldots,s\right) \text{ and} \\
&\qquad \sigma\text{: URS.}
\end{aligned} \tag{9.8}
$$

Desirable output-oriented congestion: Dual formulation

$$
\begin{aligned}
&\text{Minimize} \ \sum_{i=1}^{m} v_i x_{ik} + \sigma \\
&s.t. \quad \sum_{i=1}^{m} v_i x_{ij} - \sum_{r=1}^{s} u_r g_{rj} + \sigma \ \ge 0 \quad \left(j=1,\ldots,n\right), \\
&\qquad \sum_{r=1}^{s} u_r g_{rk} = 1, \\
&\qquad v_i\text{: URS } \left(i=1,\ldots,m\right), \ u_r \ge 0 \left(r=1,\ldots,s\right) \text{ and} \\
&\qquad \sigma\text{: URS.}
\end{aligned} \tag{9.9}
$$

Let the optimal objective value of Model (9.6) be τ^* and that of Model (9.7) be β^*. Then, the desirable output-oriented congestion (OC) is measured by the following ratio (Färe *et al.*, 1985):

$$\text{Output-oriented congestion: OC}\left(\tau^*, \beta^*\right) = \tau^* / \beta^*. \tag{9.10}$$

If OC $(\tau^*, \beta^*) > 1$, then the congestion occurs on the k-th DMU. The sign is the opposite of the input-oriented case. Meanwhile, if OC $(\tau^*, \beta^*) = 1$, then there is no congestion on the DMU.

In the desirable output-oriented case, the objective value of Model (9.6) is more freely determined than that of Model (9.7) in the maximization, so having $\tau^* \geq \beta^*$. Therefore, OC $(\tau^*, \beta^*) > 1$ indicates an occurrence of congestion. Meanwhile, OC $(\tau^*, \beta^*) = \tau^* / \beta^* = 1$ indicates no occurrence of congestion.

A methodological benefit of the conventional approach, discussed here, is that it is possible to measure the occurrence of congestion even if the proposed DEA models suffer from an occurrence of multiple solutions. Meanwhile, a drawback of the approach is that it cannot identify which input or desirable output is related to such an occurrence of congestion. So, it is impossible for us to identify how and why the congestion occurs in part or whole of a production facility.

Departing from the conventional discussion in DEA, hereafter, this chapter needs to discuss economic implications derived from dual variables obtained from the congestion-related formulations. These implications are summarized as follows:

Proposition 9.1: If $v_i^* \geq 0$ for $i = 1, \ldots, m$ are identified on the optimality of Models (9.4) and (9.9), then the congestion does not occur on the k-th DMU. In contrast, if $v_i^* < 0$ is found on at least one input (for some i), then congestion[4] occurs on the k-th DMU.

[Proof] If $v_i^* \geq 0$ is found on all the inputs on the optimal value of Model (9.4), then the objective value of Model (9.3) equals that of Model (9.4) on optimality. Hence, the k-th DMU has $\theta^* = \alpha^*$ on the optimality of the two input-oriented models. Thus, we have IC $(\theta^*, \alpha^*) = 1$, indicating no occurrence of congestion. In contrast, if $v_i^* < 0$ is found on at least one input on the optimality of Model (9.4), then the optimal objective value of Model (9.4) is not equivalent to that of Model (9.3). This indicates that $\theta^* < \alpha^*$, implying the occurrence of congestion along with IC $(\theta^*, \alpha^*) < 1$. Q.E.D.

Proposition 9.1 indicates the computational benefit that we can determine the possible occurrence of congestion by examining only dual variables derived from Models (9.4) and (9.9). The ratio used for IC(θ^*, α^*) and that of OC (τ^*, β^*) are not necessary in identifying an occurrence of congestion so that we can avoid the additional computation burden on Models (9.1), (9.2), (9.6) and (9.7). Furthermore, the dual variable examination can easily identify which input causes such an occurrence of congestion.

[4] This type of congestion is considered as the status of "strong" congestion in this chapter. See also Chapters 21, 22 and 23 that have fully utilized the type of congestion to identify an occurrence of eco-technology innovation or its related managerial challenges.

In addition to the computational benefit, it is necessary to mention that two drawbacks are associated with the dual variable examination. One of the two drawbacks is that the proposed dual formulations may suffer from an occurrence of multiple solutions (e.g., multiple projection and multiple reference sets). See Chapter 7 that describes an occurrence of multiple solutions in DEA. The other drawback is that "zero" may occur in these dual variables. If the dual variable (v_i^*) has zero, then it indicates an initial stage of congestion, or "weak congestion", which is very different from an occurrence of "strong congestion" with at least one negative dual variable, or $v_i^* < 0$ on some i. Such a separation between the two types (weak and strong) of congestion is important in the classification, but such a drawback is not discussed in Proposition 9.1. Another serious difficulty may occur when u_r^* becomes zero on some r. This type of difficulty will be discussed later in this chapter.

9.4 SUPPORTING HYPERPLANE

9.4.1 Location of Supporting Hyperplane

To explore further economic implications of a negative dual variable(s) on any component of an input vector, this chapter needs to characterize the supporting hyperplane on the k-th DMU that is measured by Model (9.4) with the possible occurrence of input-oriented congestion. The mathematical characterization is summarized by the following proposition:

Proposition 9.2: A supporting hyperplane on the k-th DMU is mathematically expressed by $-\sum_{i=1}^{m} v_i x_i + \sum_{r=1}^{s} u_r g_r + \sigma = 0$, where v_i $(i=1,\ldots,m)$ are unrestricted and u_r $(r=1,\ldots,s)$ are non-negative in their signs. The two groups of parameters indicate a vector direction of the supporting hyperplane in the $(m+s)$ dimensional $(X,\ G)$ space. The variable (σ) is unrestricted and it indicates an intercept of the supporting hyperplane. The supporting hyperplane is characterized by the following equations:

$$-\sum_{i=1}^{m} v_i x_{ij} + \sum_{r=1}^{s} u_r g_{rj} + \sigma = 0,\ j \in RS_k \quad \text{and} \quad \sum_{i=1}^{m} v_i x_{ik} = 1, \tag{9.11}$$

where $RS_k = \{j \mid \lambda_j^* > 0,\ j \in J\}$ is a reference set of the k-th DMU measured by Model (9.2). The set (J) indicates all DMUs.

[Proof] The complementary slackness conditions between Models (9.2) and (9.4) provide

$$\left(-\sum_{i=1}^{m} v_i x_{ij} + \sum_{r=1}^{s} u_r g_{rj} + \sigma\right)\lambda_j = 0\ (j=1,\ldots,n). \tag{9.12}$$

Since λ_j is positive in a reference set (RS_k) of the k-th DMU, Equation (9.12) immediately indicates the first part: $-\sum_{i=1}^{m} v_i x_{ij} + \sum_{r=1}^{s} u_r g_{rj} + \sigma = 0,\ j \in RS_k$. We can also obtain the second part of Equation (9.11), or $\sum_{i=1}^{m} v_i x_{ik} = 1$, from the second group of the constraints in Model (9.4).

Q.E.D.

Proposition 8.1 in Chapter 8 provides a detailed proof on the location of a supporting hyperplane. Proposition 9.2 documents a short cut for the proof to avoid our mathematical duplication. It is indeed true that there is almost no difference between Propositions 8.1 and 9.2. However, there is a very important difference between them. That is, the input-related dual variables (v_i for $i = 1,\ldots,m$) are all unrestricted (URS) in Proposition 9.2, but they are non-negative in Proposition 8.1. Such a seemingly minor difference produces a major difference in RTS measurement and congestion identification, as later discussed in this chapter (see the figure in Section 9.6; i.e., Figure 9.7). Moreover, such a difference becomes very important in DEA environmental assessments as documented in Chapters 21, 22 and 23.

Shifting to the output-oriented formulation (9.9), this chapter can identify a supporting hyperplane that is mathematically expressed by $\sum_{i=1}^{m} v_i x_i - \sum_{r=1}^{s} u_r g_r + \sigma$. The supporting hyperplane is characterized by the following equations:

$$\sum_{i=1}^{m} v_i x_{ij} - \sum_{r=1}^{s} u_r g_{rj} + \sigma = 0,\ j \in RS_k \quad \text{and} \quad \sum_{r=1}^{s} u_r g_{rk} = 1,$$

where the reference set (RS_k) is measured by Model (9.7). The proof on the output-oriented case is directly obtained from extending the input-oriented case. Therefore, this chapter does not discuss it further, here.

9.4.2 Visual Description of Congestion and RTS

Figure 9.5 depicts a possible occurrence of congestion in the x-g space. For our visual convenience, the two production factors have a single component (x and g). It is not difficult to theoretically extend to the case of their multiple components in the framework of DEA. The piece-wise linear contour line indicates an efficiency frontier in the x–g space. An occurrence of congestion can be found on the right-hand side of Figure 9.5, on {F} for example, on which an increase in the input (x) decreases the desirable output (g). As depicted on the left-hand side of Figure 9.5, an input increase (x) enhances the desirable output (g) until the input attains the amount specified by {A}. However, when the input becomes more than {A}, the input increase is not associated with an increase of a desirable output anymore, as

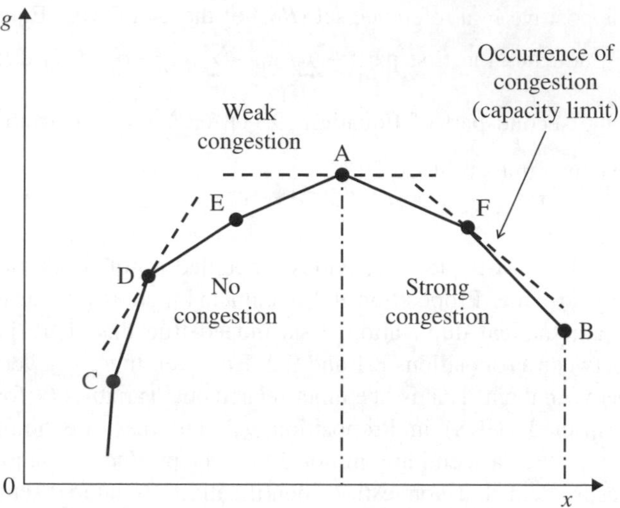

FIGURE 9.5 Occurrence of congestion (a) A possible occurrence of congestion is measured by allocating equality to an input vector (X) as found in Models (9.2) and (9.7). The analytical feature will be further explored and changed in Chapter 21. (b) RM(c) cannot measure an occurrence of congestion because $\sigma = 0$ under constant RTS. Thus, it is necessary for us to use RM(v) for the identification.

found in the linear contour line segments between {A} and {B}. This situation is considered as the phenomenon of congestion.

An implication of Figure 9.5 is that the possible occurrence of congestion can be separated into three categories, including "no congestion" found on {D}, "weak congestion" found on {A} and "strong congestion" found on {F}. The phenomenon of no congestion implies that an increase of some input(s) leads to that of some desirable output(s), without worsening the other components of the two production factor vectors. The weak congestion implies that an increase of some input(s) does not change some desirable output(s), without worsening the other components. The phenomenon of weak congestion indicates an initial stage of strong congestion, or a capacity limit on part or all of a production facility.

Returning to the classification of RTS in Chapter 8, Figure 9.6 visually describes the relationship between the five types of RTS (increasing, constant, decreasing, zero and negative) and their related supporting hyperplanes, along with a possible occurrence of congestion. For our visual convenience, Figure 9.6 considers a single component of the two production factors. As depicted on the right-hand side of the figure, the strong congestion corresponds to "negative RTS" and the weak congestion corresponds to "no RTS." The two types of RTS were not discussed in Chapter 8 because these types should be discussed only under no occurrence of congestion. Therefore, this chapter needs to discuss the two types of RTS (i.e., negative and no), with any occurrence of congestion as an additional phenomenon. Meanwhile, under no occurrence of congestion on the left-hand

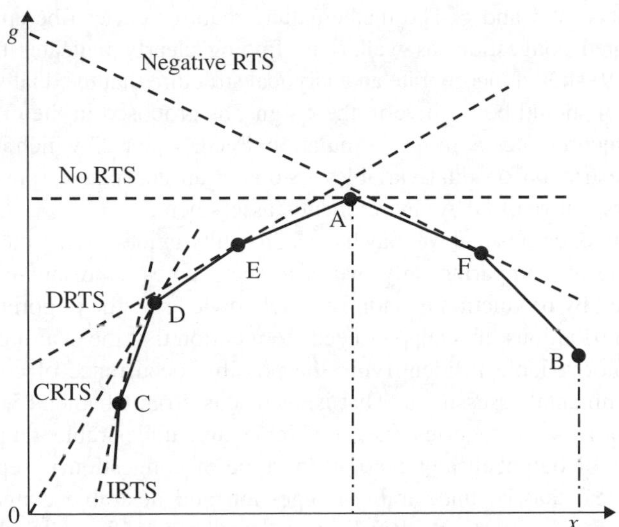

FIGURE 9.6 Five types of returns to scale (a) Source: Sueyoshi and Goto (2016b). This chapter reorganizes their work. (b) RTS: Returns to scale; IRTS: increasing RTS; CRTS: constant RTS; DRTS: decreasing RTS. (c) Positive RTS under no congestion is further separated into IRTS, CRTS and DRTS. (d) Negative RTS and No RTS correspond to strong congestion and weak congestion if two production factors have only a single component. If they have multiple components, then these concepts do not link to each other. (e) RM(c) cannot measure an occurrence of congestion because $\sigma = 0$ under constant RTS. Thus, it is necessary for us to use only RM(v) for this type of identification.

side of Figure 9.6, the measurement of RTS, whose supporting hyperplanes have positive slopes, are classified into three different non-congestion cases: increasing RTS (IRTS) as found on {C}< constant RTS (CRTS) as found on {D} and decreasing RTS (DRTS) as found on {E}.

In the occurrence of congestion on {A} and {F}, Figure 9.6 visually specifies the two types of RTS: (a) no RTS on {A} and (b) negative RTS on {F}. The conventional use of point elasticity (e_p) for RTS classification indicates that *dg/dx* becomes zero on {A}. Moreover, *dg/dx* becomes negative on {F}. Hence, e_p becomes negative on {F}. Thus, Figure 9.6 visually indicates a theoretical linkage among the occurrence of congestion, the slope of a supporting hyperplane and the type of RTS.

Here, it is important to note two concerns on a possible occurrence of congestion. One of these concerns is that in the simple case where an input (x) is used to produce a desirable output (g), a supporting hyperplane is specified by $-vx+ug+\sigma=0$, or $g=(v/u)x-\sigma/u$ in the input-oriented measure. A mathematical requirement for identifying the supporting hyperplane is that u should be positive in the sign. Both v and σ are unrestricted in their signs. If u is zero, it is very difficult for us to determine the location of the supporting hyperplane because v/u and σ/u become infinite. Thus, the dual variable related to a desirable output should be strictly positive in determining a supporting hyperplane in a data space of x and g. The mathematical requirement can be applied to the output-oriented congestion, as well. This finding clearly indicates that Models (9.4) and (9.9) should incorporate an analytical structure that the dual variables (u_r for $r=1,\ldots,s$) should be positive in their signs, as proposed in the original DEA model that incorporates ε_n in the formulation. See Chapter 23 which incorporates multiplier restriction on dual variables to obtain an acceptable (positive) range from the perspective of DEA environmental assessment.

The other concern is that we may have a difficulty in identifying the occurrence of congestion by comparing only two efficiency scores, as discussed in IC and OC, proposed by production economists. Acknowledging the importance of such previous contributions, this chapter needs to mention that the examination for IC and OC is not reliable in identifying the possible occurrence of congestion in DEA environmental assessment. That is, as discussed in Chapter 15, outputs are classified into two categories (i.e., desirable and undesirable outputs). As a consequence of output disaggregation, the type of congestion is separated into two categories, that is, undesirable congestion and desirable congestion. The efficiency-based congestion identification cannot specify on which outputs become a source of the occurrence of congestion. That is problematic. Therefore, we will not utilize the conventional approach for congestion identification in DEA environmental assessment. See, for example, Chapter 21 that discusses how to identify the occurrence of these two types of congestion.

9.5 CONGESTION IDENTIFICATION

9.5.1 Slack Adjustment for Projection

As mentioned in Section 9.4, the identification process of congestion by examining $v_i^* < 0$ for some i, measured by Models (9.4) and (9.9), may have a difficulty in finding the occurrence because the dual variables related to desirable outputs may take $u_r = 0$ for some r. In contrast, if $u_r > 0$ are observed for all $r=1,\ldots,s$, then the identification has no problem.

Let us consider output-oriented congestion, hereafter, because it is directly linked to a possible occurrence of congestion in environmental assessment, later in Section II. See, for example, Chapter 21 which considers two types of congestion by considering desirable and undesirable outputs. It is easily imagined that our discussion here is easily extendable to the input-oriented congestion.

The production possibility set of Model (9.6) and that of Model (9.7) are specified as

$$P = \left\{ (X,G) \left| \begin{array}{l} x \geq \sum_{j=1}^{n} x_j \lambda_j,\ g \leq \sum_{j=1}^{n} g_j \lambda_j, \\ \sum_{j=1}^{n} \lambda_j = 1,\ \lambda_j \geq 0,\ j=1,\ldots,n \end{array} \right. \right\}$$

and

$$P_{congestion} = \left\{ (X,G) \left| \begin{array}{l} x = \sum_{j=1}^{n} x_j \lambda_j, \; g \le \sum_{j=1}^{n} g_j \lambda_j, \\ \sum_{j=1}^{n} \lambda_j = 1, \lambda_j \ge 0, \; j = 1,\ldots,n \end{array} \right. \right\},$$

respectively. The two types of RM(v) can be expressed by maximize $\{\tau | (x_k, \tau g_k) \in P\}$ and maximize $\{\beta | (x_k, \beta g_k) \in P_{\text{congestion}}\}$ on the two production possibility sets.

Previous research extended the concept of congestion further by linking it to other economic concepts related to efficiency. For example, Brockett *et al.* (2004, p. 216) and Tone and Sahoo (2004, p. 757)[5] proposed to measure the concept of "strong efficiency" that may be identified on DMUs {A, B, C, D, E, F} in Figures 9.5 and 9.6 and then discussed the relationship between congestion and strong efficiency. Strong efficiency is defined as follows:

Definition 9.1: Let τ^* be the optimal objective value of Model (9.6) and let $(d^{x*}, d^{g*}, \lambda^*)$ be an optimal solution of the following non-radial problem (Sueyoshi and Sekitani, 2008):

$$\begin{aligned} &\text{Maximize} \quad \sum_{i=1}^{m} d_i^x + \sum_{r=1}^{s} d_r^g \\ &\text{s.t.} \quad \sum_{j=1}^{n} x_{ij}\lambda_j + d_i^x = x_{ik} \quad (i=1,\ldots,m), \\ &\qquad \sum_{j=1}^{n} g_{rj}\lambda_j - d_r^g = \tau^* g_{rk} \; (r=1,\ldots,s), \\ &\qquad \sum_{j=1}^{n} \lambda_j = 1, \\ &\qquad \lambda_j \ge 0 \; (j=1,\ldots,n), \; d_i^x \ge 0 \; (i=1,\ldots,m) \text{ and} \\ &\qquad d_r^g \ge 0 \; (r=1,\ldots,s). \end{aligned} \tag{9.13}$$

The status of *OE*, where $OE = 1/\tau^*$, is identified within *P* if and only if $\tau^* = 1$, $d^{x*} = 0$ and $d^{g*} = 0$. DMUs in this type are on the vertexes of an efficiency

[5] Tone (1996) formally characterized the relationship between a supporting hyperplane on a production possibility set and RTS in DEA production analysis. Sueyoshi (1999) extended his characterization into DEA cost analysis and discussed the relationship between production-based and cost-based RTS measures. See Chapter 8. Tone and Sahoo (2005) explored cost efficiency and scale efficiency from the perspective of India's life insurance company. Tone and Sahoo (2006) also discussed their investigation on cost efficiency and scale elasticity by comparing their results with Sueyoshi's approach (1999) in terms of cost-based RTS measurement. Unfortunately, all the previous research efforts suffered from an occurrence of multiple solutions in their production-based and cost-based RTS models.

frontier. Similarly, let β^* be the optimal value of Model (9.7) and let (d^{g*}, λ^*) be the optimal solution of the following problem:

$$
\begin{aligned}
&\text{Maximize} \quad \sum_{r=1}^{s} d_r^g \\
&s.t. \quad \sum_{j=1}^{n} x_{ij}\lambda_j \qquad = x_{ik} \quad (i=1,\ldots,m), \\
&\qquad \sum_{j=1}^{n} g_{rj}\lambda_j - d_r^g = \beta^* g_{rk} \, (r=1,\ldots,s), \\
&\qquad \sum_{j=1}^{n} \lambda_j \qquad = 1, \\
&\qquad \lambda_j \geq 0 \, (j=1,\ldots,n) \text{ and } d_r^g \geq 0 \, (r=1,\ldots,s).
\end{aligned}
\tag{9.14}
$$

The status of "strong *OE*" is identified within $P_{\text{congestion}}$ if and only if $\beta^* = 1$ and $d_r^g * = 0$.

Definition 9.2: A DMU is "congested" if it is strongly *OE* with respect to $P_{\text{congestion}}$ and there exists a production activity in $P_{\text{congestion}}$ that uses less resource in one or more inputs for making more products in one or more desirable outputs.

Definition 9.2 assumes that a DMU is strongly *OE* with respect to $P_{\text{congestion}}$. Hence, the approach projects all DMUs onto an efficiency frontier of $P_{\text{congestion}}$ and then identifies whether each DMU is congested or not under strong *OE* with respect to $P_{\text{congestion}}$. To identify whether the *k*-th DMU is congested, the proposed approach needs to solve both Models (9.7) and (9.14). The approach for identifying a projected point (X_k', G_k') of a specific *k*-th DMU is as follows:

Step1: Determine β^* by solving Model (9.7).

Step2: Given β^* and (X_k, G_k), obtain an optimal solution (λ^*, d_r^{g*}) of Model (9.14). Then, we project the observed production factors of the *k*-th DMU by the following manner:

$$X_k' \leftarrow X_k (\text{unchanged}) \text{ and } G_k' \leftarrow \beta^* G_k + d_r^g *.$$

The projected point (X_k', G_k') is strong *OE* with respect to $P_{\text{congestion}}$.

Note that the proposed approach is effective under two conditions: unique projection and unique reference set. Under an occurrence of multiple projections and/or multiple reference sets, the proposed approach may produce different results on congestion identification. See Chapter 7.

9.5.2 Congestion Identification on Projected Point

Proposition 9.3: The projected (X_k', G_k') is on the boundary of $P_{\text{congestion}}$ and the optimal objective value of Model (9.7), applied on the projected DMU, is $\beta^* = 1$.

Then, the k-th DMU is congested if and only if any optimal solution (V^*, U^*, σ^*) of Model (9.9) on the projected DMU satisfies either

(a) $v_i^* < 0$ for at least one $i \in \{1, \ldots, m\}$ or

(b) $V^* \geq 0$, $v_i^* = 0$ for at least one $i \in \{1, \ldots, m\}$ and $u_r^* = 0$ for at least one $r \in \{1, \ldots, s\}$.

[Proof] Sueyoshi and Sekitani (2008) have provided a mathematical proof on the proposition in their Theorem 3.

Q.E.D.

See Figure 9.5 for a visual description of Proposition 9.3. The visual description proves the validity of Proposition 9.3. In addition, condition (a) indicates an occurrence of strong congestion. Meanwhile, condition (b) indicates the occurrence of weak congestion.

9.6 THEORETICAL LINKAGE BETWEEN CONGESTION AND RTS

In the optimal solution (V^*, U^*, σ^*) of Model (9.8), each V^* is considered as a virtual input price vector of X_k. Similarly, U^* is considered as a virtual desirable output price vector of G_k. Each component of these vectors indicates a slope of a supporting hyperplane(s) on a production possibility set. The supporting hyperplane(s) is closely related to RTS in such a manner that RTS is determined by examining a sign of the dual variable (σ^*).

In changing Model (9.8) to Model (9.9) for congestion identification, this chapter finds that there is almost no difference between them, as mentioned previously. An exception can be found in the fact that the vector (V^*) may be a negative component(s) because the vector is unrestricted in the sign in Model (9.9). A negative component of V^* influences the sign of σ^*, so influencing the type of RTS. This concern indicates that it is necessary for us to characterize both RTS and congestion from the structure of a supporting hyperplane(s)[5]. The following theorem initiates the research task:

Proposition 9.4: Consider the k-th DMU and let β^* be the optimal value of Model (9.7). For every optimal solution (V^*, U^*, σ^*) of Model (9.9), $H = \left\{ (X, G) \middle| \sum_{i=1}^{m} v_i^* x_i - \sum_{r=1}^{s} u_r^* g_r + \sigma^* = 0 \right\}$ is a supporting hyperplane of $P_{\text{congestion}}$ on $(X_k, \beta^* G_k)$. Hence, there exists a supporting hyperplane of $P_{\text{congestion}}$ on $(X_k, \beta^* G_k)$ such that (V^*, U^*, σ^*) is the optimal solution of Model (9.9), where v_i $(i = 1, 2, \ldots, m)$ is unrestricted in their signs.

[Proof] The proof is trivial from the previous discussion. Q.E.D

As visually described in Figure 9.6, a marginal rate of transformation indicates the slope of the supporting hyperplane. Hence, this chapter needs to formally specify the marginal rate of transformation to characterize RTS under a possible occurrence of congestion. Let dual variables obtained from Model (9.9) be v_i^* $(i=1, 2,\ldots,m)$, u_r^* $(r=1, 2,\ldots,s)$ and σ^* for the k-th DMU. Proposition 9.4 indicates that Equation (9.15) expresses an estimated supporting hyperplane on the k-th DMU by

$$\sum_{r=1}^{s} u_r^* g_r = \sum_{i=1}^{m} v_i^* x_i + \sigma^*. \tag{9.15}$$

Let $\kappa = \sum_{r=1}^{s} u_r^* g_r$, then the derivative of κ produces the following results from Equation (9.15):

$$\frac{\partial \kappa}{\partial x_i} = v_i^*\left(i=1,\ldots,m\right) \quad \text{and} \quad \frac{\partial \kappa}{\partial g_r} = u_r^*\left(r=1,\ldots,s\right).$$

Hence,

$$\frac{\partial g_r}{\partial x_i} = \left(\frac{\partial \kappa}{\partial x_i}\right) \bigg/ \left(\frac{\partial \kappa}{\partial g_r}\right) = v_i^* / u_r^*. \tag{9.16}$$

The above equation provides an economic implication: if v_i^* is negative, the marginal rate of transformation $(\partial g_r / \partial x_i)$ becomes negative under the condition that all u_r^* $(r=1,\ldots,s)$ are assumed to be positive. If v_i^* is positive, then the marginal rate is positive. If the dual variable is zero, then the marginal rate becomes zero. The negative marginal rate indicates that an increase in the i-th input reduces the amount of the r-th desirable output under an occurrence of strong congestion. The case corresponds to the right-hand side of Figure 9.6, thus implying that RTS is "negative." The negative RTS visually implies the supporting hyperplane on {F} in Figure 9.6. Meanwhile, if v_i^* is zero, the marginal rate of transformation $(\partial g_r / \partial x_i)$ becomes zero. An increase in the i-th input does not change the amount of the r-th desirable output under weak congestion, so implying the situation of no RTS. The supporting hyperplane on {A} in Figure 9.6 depicts such a case of no RTS.

The above discussion provides us with an economic implication on congestion based upon Proposition 9.4. As mentioned previously, the optimal solution (V^*, U^*, σ^*) of Model (9.8) consists of a supporting hyperplane on $(X_k, \tau^* G_k)$. An intercept of the supporting hyperplane is expressed by σ^* and a type of RTS is determined by the intercept. Proposition 9.4 indicates that an optimal solution of Model (9.9) consists of a supporting hyperplane on $(X_k, \beta^* G_k)$ of $P_{\text{congestion}}$. Furthermore, $P_{\text{congestion}} \subseteq P$. If $\tau^* = \beta^*$, then the optimal solution of Model (9.8) is equivalent to Model (9.9). Using the optimal solution of Model (9.9), the following relationship is identified between RTS and congestion on the k-th DMU under the assumption of a unique optimal solution (e.g., a unique projection and a unique reference set):

(a) Increasing RTS (no congestion) $\leftrightarrow$ Any optimal solution (V^*, U^*, σ^*) of Model (9.9) satisfies $v_i^* \geq 0$ $(i = 1,\ldots,m)$ and $\sigma^* < 0$,

(b) Constant RTS (no congestion) $\leftrightarrow$ There exists an optimal solution (V^*, U^*, σ^*) of Model (9.9) that satisfies $v_i^* \geq 0$ $(i = 1,\ldots,m)$ and $\sigma^* = 0$,

(c) Decreasing RTS (no congestion) $\leftrightarrow$ Any optimal solution (V^*, U^*, σ^*) of Model (9.9) satisfies $v_i^* \geq 0$ $(i = 1,\ldots,m)$ and $\sigma^* > 0$,

(d) Negative RTS (strong congestion) $\leftrightarrow$ Any optimal solution (V^*, U^*, σ^*) of Model (9.9) satisfies $v_i^* < 0$ for at least one $i \in \{1,\ldots,m\}$ and

(e) No RTS (weak congestion) $\leftrightarrow$ All other cases, including $V^* \geq 0$ and $v_i^* = 0$ for at least one component for $i \in \{1,\ldots,m\}$.

It is important to note four concerns on the above classification on RTS and congestion. First, if $v_i^* < 0$ for some i and $v_{i'}^* = 0$ for the other i', then both strong and weak congestion phenomena coexist on the k-th DMU. In the case, we consider it as an occurrence of the strong congestion. In addition, if the assumption on unique projection is violated on the k-th DMU, then the proposed RTS classification is ineffective. See Chapter 7 on the use of SCSCs. See also Chapters 21 and 23 on a possible occurrence of congestion in DEA environmental assessment. Second, as discussed in Chapter 8, it is necessary for the RTS classification to measure the upper and lower bounds on the intercept (σ^*). This chapter does not discuss such an occurrence of multiple intercepts to avoid an increase in our computational tractability. Third, as a basis for making a theoretical linkage to Chapters 21, 22 and 23, where dual variables related to desirable outputs are all positive, this chapter drops $u_r^* = 0$ from the proposed classification. The need for such examination was discussed in Proposition 9.3. Finally, Figure 9.7 visually specifies the classification of RTS under a possible occurrence of congestion.

9.7 DEGREE OF CONGESTION

To measure the degree of congestion, this chapter needs to discuss the marginal rate of transformation by measuring $\partial g_r / \partial x_i$ for a combination between r and i. If the marginal rate becomes zero or negative, then the k-th DMU is weakly or strongly congested, respectively. Here, let us pay attention to the k-th DMU with $\beta^* = 1$ in Model (9.7). The ratio $e_p(i,r)$ of the marginal rate of transformation to the average product is expressed by $e_p = (\partial g_r / \partial x_i)/(g_r / x_i)$ for the i-th input and the r-th desirable output. Note that if $\beta^* > 1$, then it is necessary to project an observation of the k-th DMU onto an efficiency frontier.

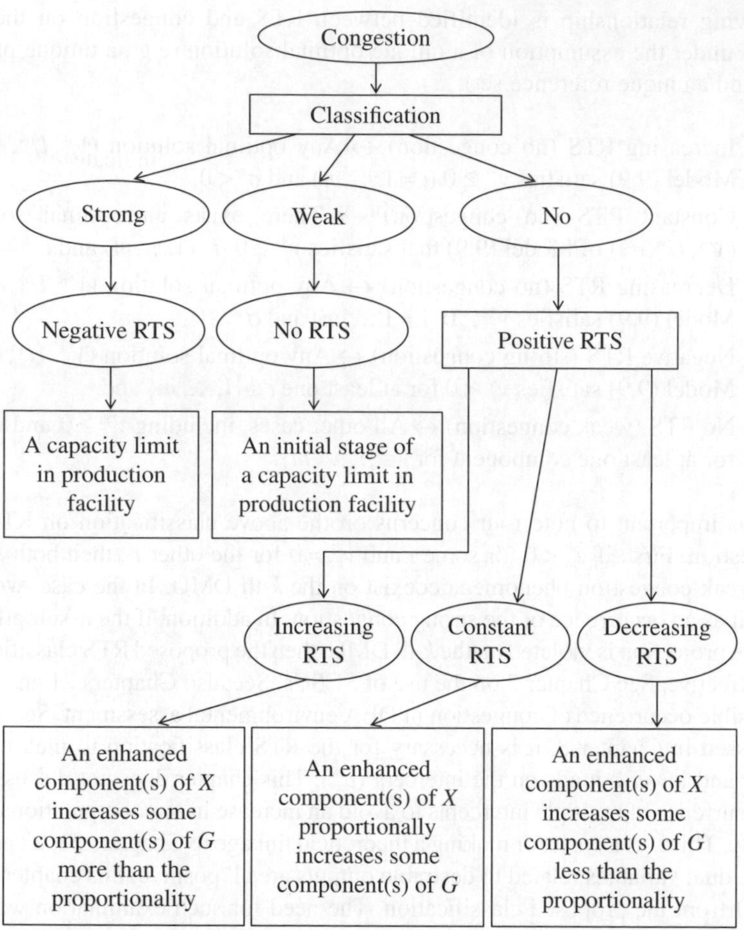

FIGURE 9.7 Classification of returns to scale under congestion

To simplify our discussion, this chapter assumes that there is a unique supporting hyperplane $H = \left\{ (X,G) \middle| \sum_{i=1}^{m} v_i^* x_i - \sum_{r=1}^{s} u_r^* g_r + \sigma^* = 0 \right\}$ on (X_k, G_k) in $P_{\text{congestion}}$. Model (9.9) has a unique optimal solution (V^*, U^*, σ^*) and $e_p(i,r)$ is expressed for the k-th DMU as follows:

$$e_p = \frac{dg_r/dx_i}{g_r/x_i} = \frac{v_i^*/u_r^*}{g_{rk}/x_{ik}} = \frac{v_i^* x_{ik}}{u_r^* g_{rk}}. \tag{9.17}$$

Since (V^*, U^*, σ^*) is an optimal solution of Model (9.9), the sum of the above ratios in terms of desirable outputs becomes

$$\sum_{r=1}^{s} e_p = \sum_{r=1}^{s} \frac{v_i^* x_{ik}}{u_r^* g_{rk}} = v_i^* x_{ik}. \tag{9.18}$$

See Model (9.9) that contains $\sum_{r=1}^{s} u_r g_{rk} = 1$. The value of $v_i^* x_{ik}$ indicates the ratio of the marginal rate of transformation to the average product with respect to the i-th input. Thus, this chapter considers $v_i^* x_{ik}$ as the degree of congestion.

In the above discussion, this chapter can consider DMU with $\beta^* = 1$ in Model (9.7). In a case when $\beta^* > 1$ is identified on the k-th DMU, it is necessary for us to project the DMU on an efficiency frontier of $P_{\text{congestion}}$. Then, the degree of congestion is measured on the projected point as discussed in Section 9.5.2.

Given a projected point $(X_k, \beta^* G_k)$ of the k-th DMU, the degree of scale elasticity (for a single desirable output) and the degree of scale economies (for multiple desirable outputs) are defined as follows:

$$e_p = \frac{dg_r / dx_i}{g_r / x_i} = \frac{\beta^* v_i^* / u_r^*}{g_{rk} / x_{ik}} = \frac{\beta^* v_i^* x_{ik}}{u_r^* g_{rk}} \text{ and} \tag{9.19}$$

$$\sum_{r=1}^{s} e_p = \sum_{r=1}^{s} \frac{\beta^* v_i^* x_{ik}}{u_r^* g_{rk}} = \beta^* v_i^* x_{ik}. \tag{9.20}$$

Applicability: This chapter does not document an application of the proposed approach because Chapters 21, 22 and 23 will discuss how to attain the status of sustainability under a possible occurrence of congestion.

9.8 ECONOMIC IMPLICATIONS

The occurrence of congestion has two implications in DEA environmental assessment. One of the two implications is a capacity limit on part or all of a production system as discussed in Section 9.2. This type of implication can be found in a conventional discussion on DEA. The other implication includes a possible occurrence of "economic limit."

To discuss differences between the two types of implication on congestion, Figure 9.8 visually specifies such differences. The left-hand side of Figure 9.8 discusses a case where the horizontal coordinate indicates the number of employees who work for a power plant. The vertical coordinate indicates an amount of electricity. A capacity limit on transmission is found in an inter-zonal link between A and B. Meanwhile, the right-hand side of Figure 9.8 visually indicates another type of congestion in which a possible occurrence of economic limit may be found on the horizontal coordinate for the number of people and the vertical coordinate for gross domestic product (GDP). In the case, DEA is applied

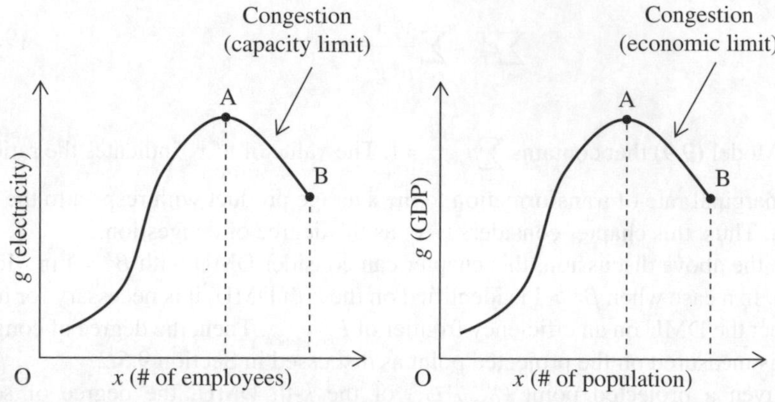

FIGURE 9.8 Implications of congestion

to DMUs (e.g., states, prefectures and provinces) at the level of local and central governments, so not at a corporate level. Such an example can be found in Chapter 23 on regional planning in China. The occurrence of economic difficulty is identified at the level of regions where an amount of GDP decreases along with an increase in the number of people, for example. This indicates an economic difficulty of each region.

9.9 SUMMARY

The phenomenon of congestion was explored by a series of research efforts by Cooper and his associates (e.g., Brockett *et al.*, 1998, 2004; Cooper *et al.*, 2000, 2001a, b, c). They made a major contribution on the DEA-based congestion issue. Unfortunately, their research efforts did not pay attention to a theoretical linkage between congestion and RTS, even if the two economic concepts were closely related to each other in terms of production economics.

Applicability: Chapter 21 documents an occurrence of congestion in United States coal-fired power plants. The purpose of this chapter is to cover fundamental knowledge on congestion within the conventional framework of DEA. Therefore, this chapter does not describe such an application, rather noting the following two concerns:

Two Important Concerns: This chapter has two concerns that we need to keep in mind for our research on DEA. One of the two concerns is that equality constraints are assigned to an input vector (X) in identifying a possible occurrence of congestion. This analytical feature is a source of misunderstanding in the identification of eco-technology innovation by DEA environmental assessment, as discussed in Chapter 21. As a straightforward extension of the conventional congestion as summarized in this chapter, many preceding works assign the equality on components of an undesirable output vector. See Chapter 28 on the previous research efforts on "weak disposability" and "strong disposability." The concept on weak and strong disposability combination has dominated previous works on pollution prevention. Chapters 15–27 extend the concept of congestion discussed in this chapter further by considering two cases: (a) equality on undesirable outputs and (b) equality on desirable outputs. The other concern is that an occurrence of congestion is identified by $v_i^* < 0$ for some i. However, the type of RTS is determined by the intercept of the hyperplane(s), or σ^*. See Chapter 8. Thus, there is a difference between an occurrence of congestion and the type identification of RTS.

Finally, the phenomenon of congestion is very important in identifying not only a capacity limit on a production facility but also eco-technology innovation to reduce an amount of various industrial pollutions, as discussed in Chapters 21, 22 and 23. The gist of DEA environmental assessment is the phenomenon of congestion, later referred to as "desirable congestion" in Chapter 21. Section II will discuss how to reduce an amount of pollution by eco-technology innovation which is identified by examining an occurrence of desirable congestion.

10

NETWORK COMPUTING

10.1 INTRODUCTION

This chapter[1] proposes a computing architecture that is computationally structured by multi-stage parallel processes to enhance its algorithmic efficiency in solving various DEA problems. Most of previous computations were traditionally performed by a computer with a single processor without much thought for computational efficiency and time savings. The computational approach proposed in this chapter may not produce a computational benefit for small DEA problems or one-time applications. However, when a problem becomes very large (e.g., more than a few thousand DMUs), the traditional approach for computation becomes painfully insufficient and very time-consuming[2]. The proposed computational

[1] This chapter is partly based upon the article: Sueyoshi, T. and Honma, T. (2003) DEA network computing in multi-stage parallel processes. *International Transactions in Operational Research*, **10**, 217–244.

[2] To overcome this computational difficulty on DEA, we address the following two research concerns: One of the two concerns is that previous studies, such as Ali and Seiford (1993), Charnes *et al*. (1986, 1991), Gonzalez-Lema *et al*. (1996), Phillips *et al*. (1990), Sueyoshi and Chang (1989), and Sueyoshi (1990, 1992), explored computational aspects of DEA, fully utilizing the technology of a single processing computer. Acknowledging the contributions of these previous studies, we need to mention that incredibly rapid developments are now observed in network technology. The new technology makes it possible to synchronously operate multiple PCs. Previous DEA studies did not sufficiently

Environmental Assessment on Energy and Sustainability by Data Envelopment Analysis,
First Edition. Toshiyuki Sueyoshi and Mika Goto.

scheme is useful in dealing with such large DEA problems, including DEA environmental assessment.

The purpose of this chapter is to document how network computing technology can be linked to DEA performance evaluation on spreadsheet software. It is expected that the proposed network computing can solve large problems with a considerably reduced central processing unit (CPU) time. It is true that we do not have any question on the modern use of personal computers (PCs) for DEA. Their technology development is amazing and very effective. However, it is important for us to discuss how to connect multiple PCs for handling a large problem such as a bootstrap approach that needs many re-sampling processes in order to outline the shape of an efficiency frontier and statistical inferences. It is easily envisioned that the proposed DEA network computing can add a new computational perspective to DEA theory and practicality.

The remainder of this chapter is organized as follows: Section 10.2 describes the proposed DEA computer architecture for network computing. An input-oriented RM(v) is used for our description of this chapter. All descriptions of RM(v) are applicable to other radial and non-radial models discussed in previous chapters. Section 10.3 discusses algorithmic strategies that are incorporated into the proposed network computing. The computational performance is examined by a large-scale simulation study. Section 10.4 summarizes such simulation results obtained from previous research effort (Sueyoshi and Honma, 2003). A concluding summary is discussed in Section 10.5.

10.2 NETWORK COMPUTING ARCHITECTURE

To develop the DEA network computing approach in a distributed computer environment, this chapter discusses (a) organizing a group of PCs as a network computing system located within an organization and (b) cloud computing[3] in the

explore such a simultaneous use of multiple PCs, or so-called "computer network and distributed processing." It is easily expected that an algorithmic design on the network technology can enhance our computing capability to solve large DEA problems, because DEA requires the solution of many, very similar, independent, linear programming problems. Thus, DEA is well suited to parallel computing and there is potential for speed-ups and computational gains. Hence, a clear linkage between DEA and network technology needs to be explored in this study. Note that Phillips *et al.* (1990) first discussed a simultaneous operation of multiple PCs. Barr and Durchholz (1997) considered parallel implementation, using a shared memory, closed coupled, multi-processor machine. They reported nearly linear speed-ups, using 15 processors.

[3] Cloud computing, known as "on demand computing," is a kind of internet-based computing, where shared resources and information are provided to computers and other devices on demand. It is a model for enabling ubiquitous, on demand access to a shared pool of configurable computing resources. Cloud computing and storage solutions provide users and enterprises with various capabilities to store and process their data in third-party resources. It relies on the sharing of resources to achieve coherence and economies of scale. The foundation of cloud computing is a broader concept of converged infrastructure and shared services. See https://en.wikipedia.org/wiki/Cloud_computing.

larger context through Internet where we can access various network technologies for interconnecting PCs. The proposed network computing approach can be easily extended into a geographically much larger computer network via the Internet. See Hayashi and Sueyoshi (1994) for a historical perspective regarding the history of Internet development.

Client–Server Environment: Many computations are usually implemented in a client–server environment in modern computer technology. A typical example is a super computer, in which the proposed algorithmic approach can easily fit with its computation scheme. In the client-server computing paradigm, a computing entity, called a "client," requests services (e.g., instructions regarding which DMU is to be evaluated and how to solve DEA problems). There are usually multiple clients in a network paradigm. Another computing entity called a "server," running on some other computer, provides the various services requested by many clients. The communication network provides a means of conveying information back and forth between multiple clients and the server. Figure 10.1 illustrates such a client–server environment.

Network Environment: Each PC is connected to a hub that generally supports multiple PCs in a star configuration, as depicted in Figure 10.1. DEA results are transmitted through transmission control protocol/Internet protocol (TCP/IP). The TCP/IP allows interconnected PCs to be able to exchange information freely (e.g., DMUs to be evaluated, their inputs, desirable outputs and efficiency results) with each other as if these PCs were all directly conncted together to each other. See Martin (1994) and Taniguchi (1998) that provided a detailed technical description on TCP/IP.

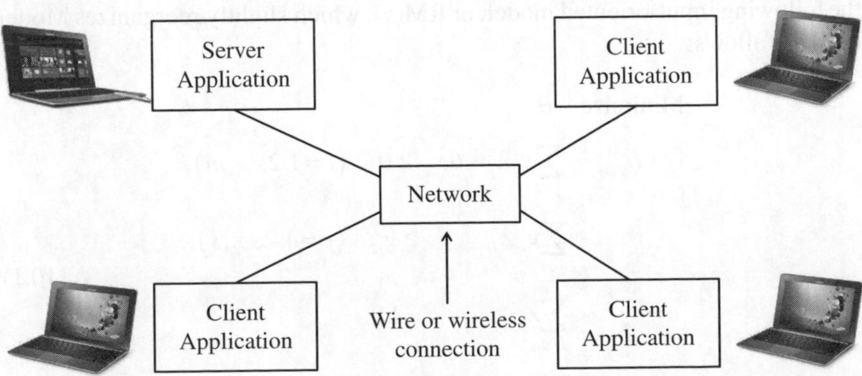

FIGURE 10.1 Client–server environment (star configuration)

The architecture of TCP/IP has multiple layers. An application (upper) layer, in which the server prepares each data set, transforms the data set into digital signals, usually called "packets." All the packets are shifted to a transport layer that provides a data delivery service. The application (bottom) layer exchanges messages through the transport layer that examines whether the data set is properly transformed into carrying packets. The TCP adds a header (i.e., information regarding what is being transmitted to the destination PC) to each packet at the transport layer.

These packets with TCP headers are shifted to the network layer where an IP header (i.e., the IP address of the source PC and that of the destination PC) is added to each carrying packet. Here, the IP header indicates a destination address that recognizes and specifies where each data set should be delivered in a computer network. The data set with TCP/IP headers is then shifted to the data link layer that controls all devices for network connection. The data link layer functions at the hub that transmits packets through the computer network. Thus, a hierarchical process of TCP/IP is established for network computing. See the research presented by Sueyoshi and Honma (2003) that visually discussed the hierarchical process of TCP/IP for network computing.

10.3 NETWORK COMPUTING FOR MULTI-STAGE PARALLEL PROCESSES

10.3.1 Theoretical Preliminary

To describe the proposed network computing, this chapter starts with a description of the analytical structure, assuming that there are n DMUs. They are denoted by $j \in J = \{j \mid j = 1,2,\ldots,n\}$ in this chapter. It is also assumed that no rank deficit occurs in the data domain for inputs and desirable outputs, as discussed in Chapter 2. Under these assumptions, DEA measures the level of efficiency on a specific DMU $\{k\}$, whose performance (X_k, G_k) is measured by the following input-oriented model, or RM(v), which slightly reorganizes Model (4.4) as follows:

$$
\begin{aligned}
&\text{Minimize} \quad \theta \\
&s.t. \quad -\sum_{j\in J} x_{ij}\lambda_j + \theta x_{ik} \geq 0 \quad (i = 1,2,\ldots,m), \\
&\qquad \sum_{j\in j} g_{rj}\lambda_j \geq g_{rk} \quad (r = 1,2,\ldots,s), \\
&\qquad \sum_{j\in j} \lambda_j = 1, \\
&\qquad \lambda_j \geq 0 \; (j \in J) \text{ and } \theta : \text{URS}.
\end{aligned}
\tag{10.1}
$$

A difference between Models (4.4) and (10.1) is that the latter expresses a set of all DMUs by J for our descriptive convenience. As discussed in Chapter 4, the optimal θ^* score of Model (10.1) indicates the level of operational efficiency (OE_v) under variable RTS. In Model (10.1), $\theta^* = 1$ and zero in all slacks indicates the state of full efficiency. In the other case, (e.g., $\theta^* < 1$ and at least one positive slack) represents some level of inefficiency. For our descriptive convenience, Model (10.1) drops slacks from the objective function. As discussed in Chapters 4 and 5, there are seemingly efficient DMUs with $\theta^* = 1$, but these DMUs belong to an inefficient DMU set (later referred to as IF and IF' in this chapter) that contains a positive slack(s) on optimality. Chapter 4 considers that such a DMU is inefficient because of the existence of a positive slack(s).

The development of network computing fully utilizes previous research results on the DEA special algorithm, first documented in Sueyoshi (1990, 1992). To briefly explain his algorithm, before extending it into the proposed network computing, this chapter starts with a description of the DMU classification, returning to Chapter 5. Here, it is important to note that Chapter 5 uses the DMU classification to describe differences between the Debreu–Farrell (radial) and the Pareto–Koopmans (non-radial) measurements. Meanwhile, this chapter uses the DMU classification to describe an algorithm development on network computing.

According to the DMU classification discussed in Chapter 5, the whole set (J), representing all the DMUs, is first separated into the following two subsets J_d and J_n:

$$\left[x_{1j},\ldots,x_{mj},-g_{1j},\ldots,-g_{sj}\right]^{Tr} \geq \left[x_{1j'},\ldots,x_{mj'},-g_{1j'},\ldots,-g_{sj'}\right]^{Tr}. \tag{10.2}$$

Equation (10.2) indicates that DMU j' clearly outperforms DMU j where at least one inequality strictly satisfies the concept of "dominance." A rationale for supporting Equation (10.2) is that the DMU j becomes clearly inefficient since the omission of such a clearly inefficient DMU group does not influence the whole assessment process of DEA. Thus, it is easily imagined that the inefficient DMU(s) should be eliminated at the beginning of DEA algorithm to reduce the whole computational burden. Of course, the performance of the clearly inefficient DMU(s) is evaluated, after all the other DMUs are evaluated, in the proposed algorithm.

Using Equation (10.2), the whole DMU set is separated into the following two DMU subsets:

$$J_d = \left\{ j \in J \mid \text{there exists some DMU}_{j'} \text{ which satisfies Equation} (10.2)\right\} \text{ and}$$
$$J_n = J - J_d.$$

Here, J_d is considered as a "dominated DMU set," while J_n is the complement set for representing a "non-dominated DMU set" ($J = J_d \cup J_n$). As discussed in Chapter 5, the two DMU subsets are further classified as follows:

$$J_n = E \cup E' \cup IF' \cup IE' \text{ and } J_d = IF \cup IE. \tag{10.3}$$

where

$$E = \left\{k \in J_n \middle| \theta_k^* = 1, \lambda_k^* = 1, \lambda_j^* = 0 \left(\forall j \neq k \in J_n\right),\right.$$
$$\left.\text{all zero slacks and unique solution}\right\},$$

$$E' = \left\{k \in J_n \middle| \theta_k^* = 1, \lambda_k^* \leq 1, \lambda_j^* > 0 \left(\exists j \in J_n\right),\right.$$
$$\left.\text{all zero slacks and multiple solutions}\right\},$$

$$IF' = \left\{k \in J_n \middle| \theta_k^* = 1, \text{at least one positive slack and multiple solutions}\right\},$$

$$IE' = \left\{k \in J_n \middle| \theta_k^* < 1, \lambda_k^* = 0 \,\&\, \lambda_j^* > 0 \left(\exists j \in E\right)\right\},$$

$$IF = \left\{k \in J_d \middle| \theta_k^* = 1 \text{ and at least one positive slack}\right\} \text{ and}$$

$$IE = \left\{k \in J_d \middle\| \theta_k^* < 1, \lambda_k^* = 0 \,\&\, \lambda_j^* > 0 \left(\exists j \in E\right)\right\},$$

where the symbols ($\forall$ and $\exists$) means "all" and "some," respectively.

Before describing the proposed network computing, the following conditions are necessary for the above DMU classification:

(a) First, all DMUs in E, E' and IF' have $\theta^* = 1$ on optimality.

(b) Second, differences among the three groups can be specified as follows: An occurrence of alternate optima (multiple solutions) may be found in E' and IF'. In contrast, an optimal solution is unique for E.

(c) Third, the DMU(s) in IF' has at least one positive slack on optimality, while the DMUs in E and E' have no slack. Hence, a conclusive classification can be obtained by both searching for alternate optima and applying the second-stage test using the additive model.

(d) Finally, the following computational processes are necessary for separation among E, E' and IF':

Step 1: Solve Model (10.1) to obtain the optimal solution of DMU $\{k\}$. Then, the additive model is applied to DMU $\{k\}$. If the objective of the additive model is positive, then the DMU belongs to IF'. Meanwhile, if the objective is zero (so, all slacks are zero), the DMU belongs to either E or E' and go to Step 2.

Step 2: If λ_k^* is a non-basic variable and its reduced cost is zero in the final tableau, then DMU $\{k\}$ has alternate optima (so, multiple solution). In the case, it belongs to E'. Meanwhile, if λ_k^* is a basic variable, then the optimal solution of DMU $\{k\}$ is unique and therefore, it belongs to E.

All the DMUs in IE and IF are "clearly inefficient" in terms of the dominance and non-dominance relationship. So, the two groups consist of the dominated DMU set (J_d). Meanwhile, IE' indicates a group of inefficient but non-dominated DMUs. Alternate optima may occur on the three DMU groups (IF, IE and IE').

Here, it is important to note that Sueyoshi and Chang (1989, p. 207) and Sueyoshi (1990, p. 250; 1992, p. 146) used the concept of dominance and non-dominance in their algorithmic developments in order to identify the clearly inefficient (i.e., dominated) group of DMU(s) in the framework of radial and non-radial models.

10.3.2 Computational Strategy for Network Computing

The development of the proposed network computing uses the following three major computation strategies:

(a) *Strategy 1*: When a data set becomes large (e.g., more than 1000 DMUs), the formulation can be expressed by the revised simplex tableau for Model (10.1) that consists of a long column (related to DMUs) and a relatively short row (related to inputs and desirable outputs). Therefore, it is easily imagined that the column reduction technique becomes an effective computation strategy[4].

(b) *Strategy 2*: The DEA efficiency measurement of each DMU is determined by comparing its production achievement with an efficiency frontier (comprising efficient DMUs in $E \cup E'$). Therefore, an effective DEA algorithm can identify the efficiency frontier in the early stage.

(c) *Strategy 3*: The computational task is achieved in a distributed computing environment. The network computing reduces the computational load of each PC, and consequently reduces the total computation time to solve all DMUs. However, the use of network computing has a negative effect on the total computation time. That is, the total number of information exchanges (regarding optimal solution and DMUs to be solved) among PCs needs to be reduced because each information exchange needs a preparatory time to write an address to which information is transmitted. Many information exchanges among PCs may result in a considerable time loss.

10.3.3 Network Computing in Multi-Stage Parallel Processes

Figure 10.2, obtained from Sueyoshi and Honma (2003), visually describes an algorithmic flow of the proposed network computing architecture. As depicted in the figure, the proposed computation is broken down into three stages. To explain the figure, we consider a computing environment in which a single server and z clients ($c = 1, \ldots, z$) are connected to each other through the network, where c stands for the c-th client PC.

[4] The proposed strategy is not recommended when the number of DMUs is small. For example, if the data set contains only ten DMUs whose performances are measured by these production relationships between two inputs and one output, the tableau of the model in Equation (10.1) consists of 10 columns related to λ_j (plus one column for θ) and four rows. In this case, the column reduction technique cannot drastically reduce the computational burden.

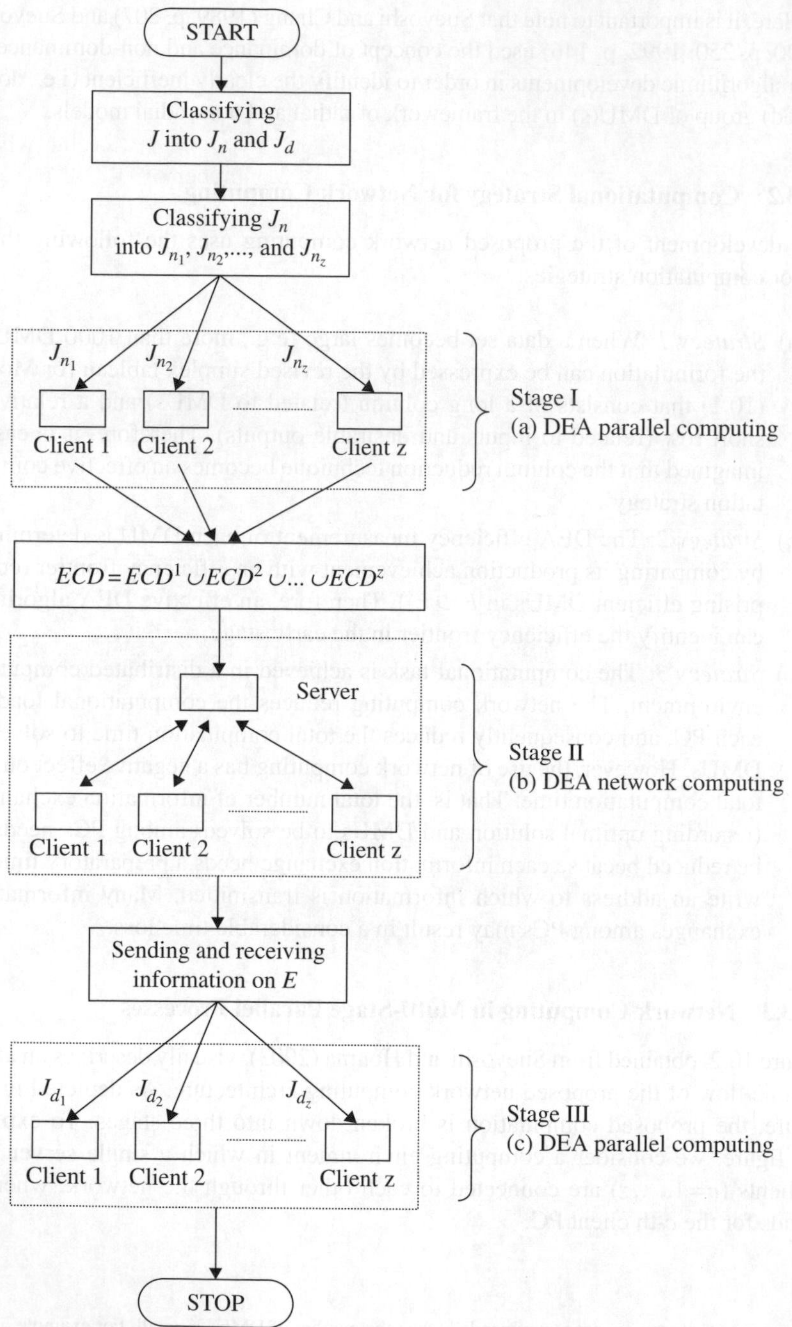

FIGURE 10.2 Computational strategy for network computing (a) Source: Sueyoshi and Honma (2003). (b) This chapter reorganizes it for our description. ECD stands for efficiency candidates, which are a group of DMUs that may belong to $E \cup E'$.

Preparatory: Using Equation (10.2), a server classifies J (a whole DMU set) into J_n (a non-dominated DMU set) and J_d (a dominated DMU set).

Stage I (Parallel computing for J_n): A server randomly separates J_n into its subgroups ($J_n = J_{n_1} \cup J_{n_2} \cup \ldots \cup J_{n_z}$) and transmits information regarding which DMUs (J_{n_c}) should be solved by the c-th client. The number of DMUs in J_{n_c} is referred to as a "block." After evaluating all the DMUs in J_{n_c}, the c-th client informs the server which DMUs become E-candidates (ECD), implying that these DMUs may belong to $E \cup E'$. The client solves only a subset of J_n. Therefore, the proposed algorithm is currently unable to make any decision regarding whether these E-candidates are really efficient or inefficient because such are determined by the final evaluation. Here, the following subset (ECD^c) is defined:

$$ECD^c = \left\{k \in J_{n_c} \middle| \theta_k^* = 1, \lambda_k^* = 1, \lambda_j^* = 0 \left(\text{all } j \neq k \in J_{n_c}\right), \text{and all slacks are zero}\right\},$$

which indicates a set of the E-candidates identified by the c-th client. At the end of Stage I, the server can identify a whole set of E-candidates as follows:

$$ECD = ECD^1 \cup ECD^2 \cup \ldots \cup ECD^z.$$

The ECD maintains the relationship such as $E \cup E' \subseteq ECD \subseteq J_n$. The CPU time of the c-th client depends upon the number of DMUs in each subset (J_{n_c}). The smaller the number becomes, the shorter the CPU time becomes at the c-th client.

Stage II (Network computing for J_n): The server sends all the clients information regarding which DMUs belong to ECD. Each client (e.g., the c-th client) measures the efficiency measure of DMUs, belonging to J_{n_c}, by using ECD. In this case, the simplex tableau contains the λ columns related to DMUs in ECD, not J (the whole DMU set). The columns related to inefficient DMUs are eliminated from the tableau. The set (E) is identified at the end of Stage II. Such DEA results are all transmitted to the server.

Stage III (Parallel computing for J_d): The sever sends all the clients information regarding which DMUs belong to E. The server separates J_d into its subgroups ($J_d = J_{d_1} \cup J_{d_2} \cup \ldots \cup J_{d_z}$) and then transmits information regarding which DMUs (J_{d_c}) should be solved by the c-th client ($c = 1, \ldots, z$). The client returns these DEA results to the server.

A computational flow of DEA network computing depicted in Figure 10.2 can be found in Barr and Durchholtz (1997). The approach is referred to as "hierarchical decomposition (HD)" that can be considered as three-staged, parallel processes for partitioning a whole data set into blocks that are processed independently for DEA efficiency measurement. The premise is that if a DMU is inefficient within a block, then it is inefficient for any superset, including the whole DMU set. The inefficient DMUs are excluded from technology matrixes at the next stage. The process proceeds to aggregate blocks that are used for identification of efficient DMUs. A major difference between theirs (i.e., Barr and Durchholtz, 1997) and the proposed approach can be found in the fact that the latter is implemented in a client–server parallel system on modern computer technology.

10.3.3.1 Stage I (Parallel Computing for J_n) Figure 10.3, obtained from Sueyoshi and Honma (2003), visually provides a detailed outline regarding Stage I. After receiving information on J_{n_c} from the server, the c-th client ($c=1,\ldots,z$) starts the following algorithmic process.

Step 1: Set $k=1$, where "k" is the number of steps and $k=1$ implies the initial step. Note that the k-th step has a one to one correspondence to the k-th DMU to be currently evaluated by DEA. Thus, "k" indicates not only the step but also the order of DMU (J). Set $NY_1^c = J_{n_c}$, $RS_1^c = \phi$ and $ECD_1^c = \phi$, where the symbol "ϕ" stands for an empty set. The superscript "c" indicates the c-th client, as mentioned previously, and the subscript "1" stands for the first step.

Step k: Assuming that client c currently evaluates the k-th DMU (so, the k-th step), this study defines the following three subsets:

$$ECD_k^c = \left\{ j \in J_{n_c} \middle| j \in ECD^c \text{ and the } j\text{-th DMU in } J_{n_c} \text{ has been examined by the } c\text{-th client} \right\},$$

$$NY_k^c = \left\{ j \in ECD_k^c \right\} \cup \left\{ j \in J_{n_c} \middle| \text{the } j\text{-th DMU in } J_{n_c} \text{ has not yet been examined by the } c\text{-th client} \right\},$$

$$RS_k^c = \left\{ j \in ECD^c \middle| \lambda_j^* > 0 \text{ and the } j\text{-th DMU in } J_{n_c} \text{ has not yet been examined by the } c\text{-th client} \right\}.$$

The first subset (ECD_k^c) is a subset of the E-candidate set (ECD^c) at the k-th step of the c-th client. The second subset (NY_k^c) indicates ECD_k^c plus a group of DMUs whose DEA efficiencies are "not yet" examined at the k-th step of the c-th client. The third set (RS_k^c) indicates a reference set for J_{n_c}. Before evaluating DMUs in RS_k^c, the proposed algorithm can predict that they belong to ECD^c. Therefore, these DEA solutions are expressed by $\theta^* = 1$, all slacks are zero and they are unique.

Here, set $k=j$ ($\forall j \in NY_k^c - ECD_k^c$), where $NY_k^c - ECD_k^c$ is a subset whose DMUs have not yet been examined by the c-th client. Then, the OE_V of the k-th DMU

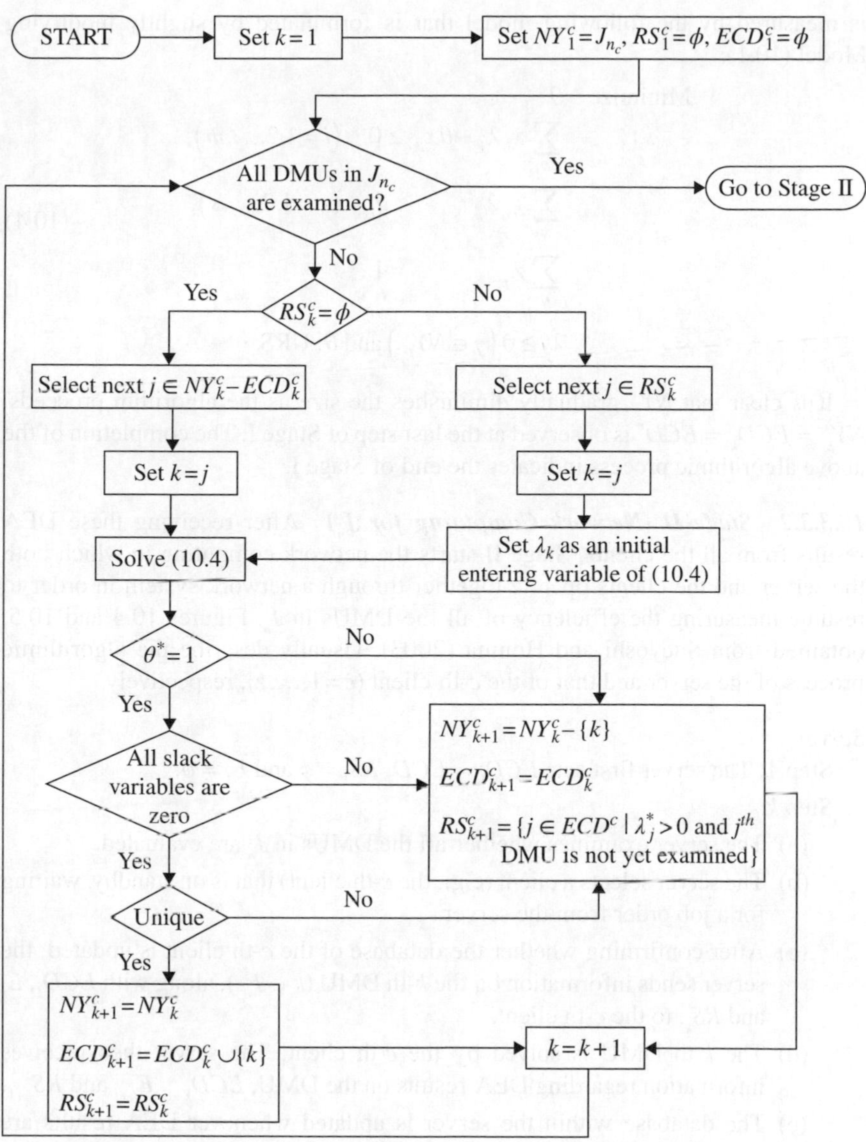

FIGURE 10.3 Stage I (parallel computing for J_n (a) Source: Sueyoshi and Honma (2003)). This chapter reorganizes it for our description. (b) *NY*: a group of DMUs which are "not yet" examined at the current stage.

is measured by the following model that is formulated by slightly modifying Model (10.1):

$$
\begin{aligned}
&\text{Minimize} \quad \theta \\
&\text{s.t} \quad -\sum_{j \in NY_k^c} x_{ij}\lambda_j + \theta x_{ik} \geq 0 \quad (i = 1,2,\ldots,m), \\
&\qquad \sum_{j \in NY_k^c} g_{rj}\lambda_j \geq g_{rk} \quad (r = 1,2,\ldots,s), \\
&\qquad \sum_{j \in NY_k^c} \lambda_j = 1, \\
&\qquad \lambda_j \geq 0 \left(j \in NY_k^c\right) \text{ and } \theta\text{: URS.}
\end{aligned}
\tag{10.4}
$$

It is clear that NY_k^c gradually diminishes the size as the algorithm proceeds. $NY_k^c = ECD_k^c = ECD^c$ is observed at the last step of Stage I. The completion of the above algorithmic process indicates the end of Stage I.

10.3.3.2 Stage II (Network Computing for J_n) After receiving these DEA results from all the clients, Stage II starts the network computing in which both the server and the clients operate together through a network system in order to resume measuring the efficiency of all the DMUs in J_n. Figures 10.4 and 10.5, obtained from Sueyoshi and Honma (2003), visually describe the algorithmic process of the server and that of the c-th client ($c = 1,\ldots, z$), respectively.

Server:

Step 1: The server first sets $ECD_1 = ECD$, $RS_1 = \phi$ and $E_1 = \phi$.

Step k:

(a) The server examines whether all the DMUs in J_n are evaluated.

(b) The server selects a client (e.g., the c-th client) that is on standby, waiting for a job order from the server.

(c) After confirming whether the database of the c-th client is updated, the server sends information on the k-th DMU ($k \in J_{n_c}$), along with ECD_k, E_k and RS_k, to the c-th client.

(d) The k-th DMU is solved by the c-th client. The server then receives information regarding DEA results on the DMU, ECD_{k+1}, E_{k+1} and RS_{k+1}.

(e) The database within the server is updated whenever DEA results are transmitted from the c-th client.

The above computational process of the server is repeated until all the DMUs in $J_n (= J_{n_1} \cup J_{n_2} \cup \ldots \cup J_{n_z})$ are evaluated by DEA.

The c-th *Client* ($c = 1, 2,\ldots, z$):

Step 1: The c-th client receives information on the k-th DMU along with $ECD_1 = ECD$, $RS_1 = \phi$ and $E_1 = \phi$ from the server and updates its database.

Step k: The c-th client measures the efficiency of the k-th DMU ($k \in J_{n_c}$) by

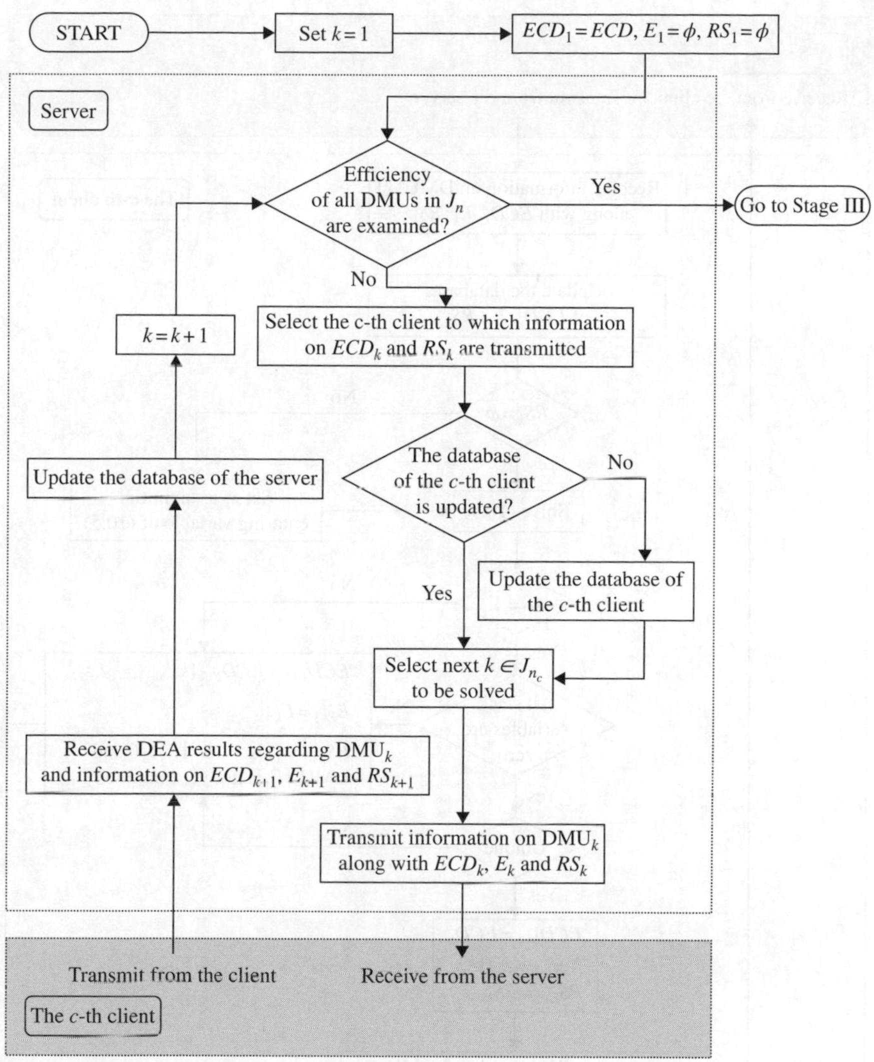

FIGURE 10.4 Stage II (network computing for J_n (a) Source: Sueyoshi and Honma (2003)). This chapter reorganizes it for our description.

$$
\begin{aligned}
&\text{Minimize} \quad \theta \\
&s.t. \quad -\sum_{j \in ECD_k} x_{ij}\lambda_j + \theta x_{ik} \geq 0 \quad (i = 1, 2, \ldots, m), \\
&\qquad \sum_{j \in ECD_k} g_{rj}\lambda_j \qquad \geq g_{rk} \quad (r = 1, 2, \ldots, s), \\
&\qquad \sum_{j \in ECD_k} \lambda_j \qquad = 1, \\
&\qquad \lambda_j \geq 0 \left(j \in ECD_k \right) \text{ and } \theta\text{: URS.}
\end{aligned}
\tag{10.5}
$$

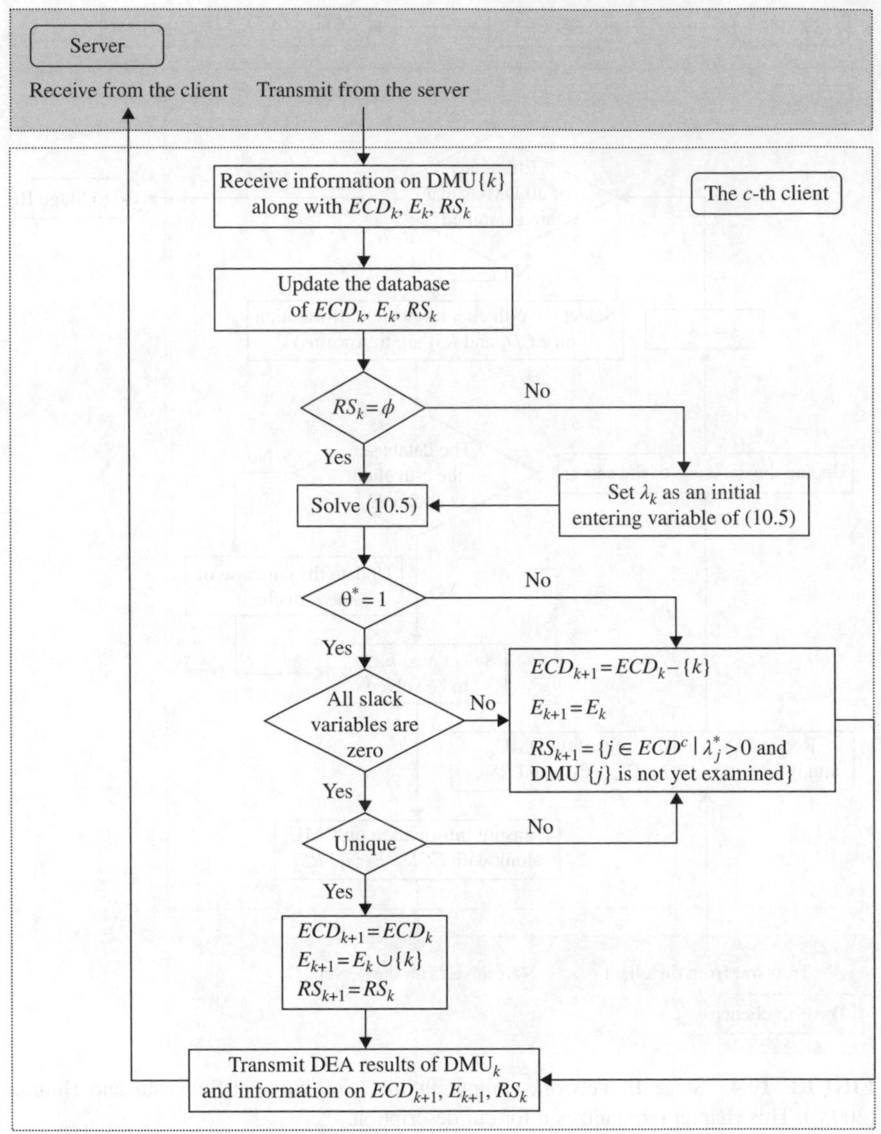

FIGURE 10.5 Network computing at the c-th client (a) Source: Sueyoshi and Honma (2003). This chapter reorganizes it for our description.

The size of ECD_k gradually diminishes as the algorithm proceeds and $ECD_k = E_k = E$ is observed at the last step of Stage II.

10.3.3.3 Stage III (Parallel Computing for J_d) Figure 10.6 depicts the algorithmic process of Stage III. The server first separates J_d into z subsets such as

$J_d = J_{d_1} \cup J_{d_2} \cup \ldots \cup J_{d_z}$. The server transmits the c-th client about information regarding which DMU $\{k\}$ should be solved by the client.

The c-th client measures the efficiency of DMU $\{k\}$ ($k \in J_{d_c}$) by

$$
\begin{aligned}
\text{Minimize} \quad & \theta \\
s.t. \quad & -\sum_{j \in E} x_{ij}\lambda_j + \theta x_{ik} \geq 0 \quad (i = 1, 2, \ldots, m), \\
& \sum_{j \in E} g_{rj}\lambda_j \geq g_{rk} \quad (r = 1, 2, \ldots, s), \\
& \sum_{j \in E} \lambda_j = 1, \\
& \lambda_j \geq 0 \, (j \in E) \text{ and } \theta\text{: URS.}
\end{aligned}
\tag{10.6}
$$

Since the revised simplex tableau has columns related to λ_j ($j \in E$) in Model (10.6), it is easily expected that the CPU time drastically reduces in particular when the number of inefficient DMUs is large because λ_j related to these DMUs are all omitted from Model (10.6). Moreover, z clients operate simultaneously under control of the server, so the total CPU time to solve all DMUs in J_d reduces as the number of clients increases. After each client solves Model (10.6), DEA results are transmitted to the server through the network. The completion of Stage III indicates the end of the whole computation of the proposed DEA network computing approach.

10.4 SIMULATION STUDY

A simulation study was applied to the proposed network computing, using artificially generated data sets. Four PCs were used for the simulation study. See Sueyoshi and Honma (2003) for the network configuration. Following the research efforts of Sueyoshi (1990, 1992) and Sueyoshi and Chang (1989), this chapter classified all DMUs into six different groups for the simulation study, which are specified as follows:

(a) First, efficient DMUs in E are designed to have the input vector X and the desirable output vector G whose components are generated by:

x_1 and x_2 are selected from the uniform distribution of (3, 7),

$x_3 = 10 - \left[100 - (x_1 - 10)^2 - (x_2 - 10)^2\right]^{1/2}$,

$g_1 = x_1^{0.4378} x_2^{0.3926} x_3^{0.1696}$, $g_2 = x_1^{0.1696} x_2^{0.3926} x_3^{0.4378}$ and $g_3 = \left(100 - g_1^2 - g_2^2\right)^{1/2}$.

(b) Second, efficient DMUs in E' are generated as follows: two adjacent efficient DMUs, denoted by $(X, G) = \left[(x_1, x_2, x_3)^{Tr}, (g_1, g_2, g_3)^{Tr}\right]$ and $X', G' = \left[(x_1', x_2', x_3')^{Tr}, (g_1', g_2', g_3')^{Tr}\right]$, are selected from the explicitly

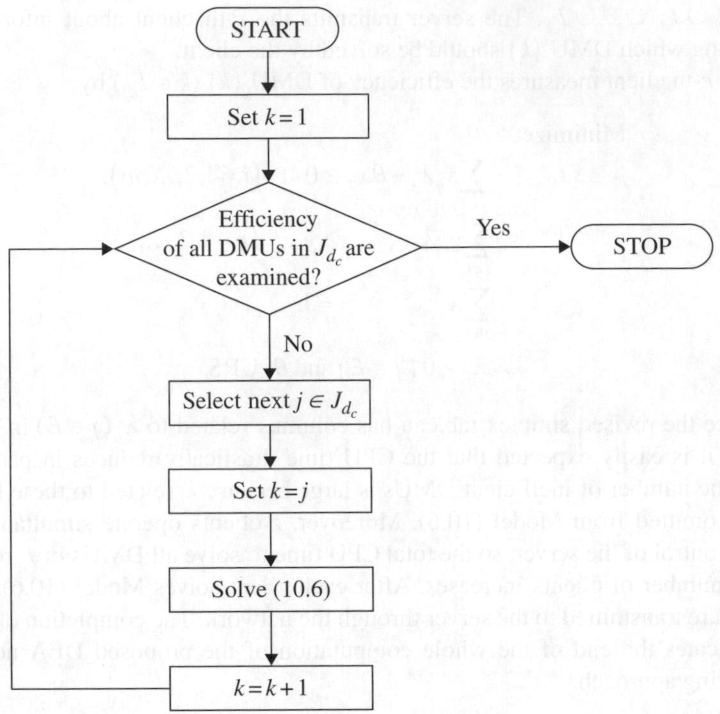

FIGURE 10.6 Stage III (parallel computing for J_d at the c-th client (a) Source: Sueyoshi and Honma (2003). This chapter reorganizes it for our description.

efficient DMU set (E) and then another type of efficient DMU in E' is determined as a convex combination between (X,G) and (X',G') such as $\tilde{X}=\omega X+(1-\omega)X'$ and $\tilde{G}=\omega G+(1-\omega)G'$, where ω has values of (0, 1).

(c) Third, inefficient DMUs in IE are generated as follows: the two adjacent efficient DMUs, or (X,G) and (X',G'), are selected from E and then an input vector $(\tilde{X})$ of an inefficient DMU in IE is determined by $\tilde{X}=\omega X+(\varepsilon-\omega)X'$, where ε is selected in a way that satisfies the following conditions: $\omega x_i+(\varepsilon-\omega)x_i' > max(x_i,x_i')$ for $i=1$, 2 and 3. Furthermore, three components of its desirable output vector are determined by

$$\tilde{g}_1=min(g_1,g_1'), \tilde{g}_2=min(g_2,g_2') \text{ and } \tilde{g}_3=min(g_3,g_3').$$

(d) Fourth, inefficient DMUs in IE' are generated as follows: two efficient DMUs, (X,G) and (X',G'), are selected from E and then an input vector $(\tilde{X})$ of an inefficient DMU in IE' is determined by $\tilde{X}=\omega X+(\varepsilon-\omega)X'$, where ε is selected in a way that satisfies the following conditions: $1<\omega x_i+(\varepsilon-\omega)x_i' < max(x_i,x_i')$, $i=1$, 2 and 3. Furthermore, as specified for IE, the three components of the desirable output vector are determined by

$$\tilde{g}_1=min(g_1,g_1'), \tilde{g}_2=min(g_2,g_2') \text{ and } \tilde{g}_3=min(g_3,g_3').$$

(e) Fifth, after identifying an efficient DMU, denoted by (X,G), from E, we increase one component of the input vector to make an inefficient DMU in IF in either $\tilde{X} = (\tilde{x}_1, x_2, x_3)$ or $\tilde{X} = (x_1, \tilde{x}_2, x_3)$, where $\tilde{x}_1 > x_1$ and $\tilde{x}_2 > x_2$. Furthermore, $\tilde{G} = G$ is maintained for the desirable output vector.

(f) Finally, two adjacent efficient DMUs, (X,G) and (X',G'), are selected from E and then a DMU (X'',G'') is determined by $X'' = \omega X + (1-\omega)X'$ and $G'' = \omega G + (1-\omega)G'$. The input–output vector $(\tilde{X}, \tilde{G})$ of a DMU in IF' is determined by $\tilde{X} = X''$ and $\tilde{G} = \left(\tilde{g}_1, g''_2, g''_3\right)$. Here, $\tilde{g}_1 < g''_1$ is maintained.

The above six DMU generations were repeated until we obtained the number of each DMU group needed for our simulation study. In this chapter, 3000 DMUs were generated for the first simulation and 10000 DMUs were generated for the second simulation.

The following additional two concerns are important in characterizing the proposed data generation process. One of the two concerns is that, as easily found in the structure of the proposed simulation, the number of input components is restricted to three and that of desirable output components are restricted to three (so, the number of these total dimensions is six) for generating the six different groups of DMUs. Of course, it is possible for us to generate these data sets by increasing the number of inputs and that of desirable outputs. However, in the case (more than three dimensions in an input vector or a desirable output vector), we cannot control the type of generated DMUs, as required by the proposed group classification. Consequently, it is difficult for us to examine how the percentage of efficient DMUs influences a speedup of the proposed network computing. Therefore, the proposed simulation study restricts the dimension of inputs plus desirable outputs to six. The other concern is that the proposed data generation started with the group (E) and then yielded the other groups of DMUs by extending E. To examine whether the data generation really produced such different groups, we reconfirmed all the generated DMUs by both searching for alternate optima and applying the second-stage test by using the additive model.

The first simulation has examined how the CPU time of the proposed network computing approach is affected by a block size, a size of DMUs and a percentage of efficient DMUs. The structure of our simulation study can be summarized as follows:

(a) Number of DMUs: $n = 1000, 1500, 2000, 2500$ and 3000.
(b) Block size: $J_{n_c} = 50, 100, 150, 200, 250, 300, 350, 400, 450$ and 500.
(c) Percentage of efficient DMUs: pe = 10, 20, 30, 40 and 50%.

As indicated above, the structure of the first simulation study was a $5 \times 10 \times 5$ factorial experiment in which each treatment had 200 replications, so totally 50000 ($5 \times 10 \times 5 \times 200$) DEA evaluations were in the simulation study. Each replication produced a slightly different CPU time. Thus, we measured an average of these 200 replications on each treatment.

Table 10.1 compares CPU times of the proposed DEA network computing approach when the number of efficient DMUs ($E \cup E'$) is 20% of a whole DMU sample (J). The table lists CPU times used for the preparatory treatment for classifying J into J_n and J_d, Stage I (DEA parallel computing for J_n), Stage II (DEA network computing for J_n), and Stage III (DEA parallel computing for J_d), along with these total CPU times and average times (per DMU). Each average CPU time is listed within pointed parentheses (i.e., < >), next to each total CPU time. The unit of CPU time to be used for the proposed simulation study is the millisecond (MS), corresponding to 10^{-3} second. For example, see the two numbers 146817 and <48.94> listed in the row label “Total” and a block size = 50 at the right-hand side of Table 10.1. The two numbers indicate that 146817 MS, or 146.817 seconds, is necessary to complete a DEA efficiency measurement of 3000 DMUs on average. This implies that the proposed network computing approach needs, on average, 48.94 MS (=146 817/3000), or 0.04894 seconds to evaluate each DMU.

The finding of Table 10.1 is that the total CPU time is mutually affected by the number of DMUs. That is, the number of DMUs influences the total CPU time of the proposed network computing. CPU time is also influenced by a block size. This finding can be confirmed in the average CPU time, as well. For example, the average CPU time increases along with the number of DMUs, until the block size reaches 350. Similar CPU times are observed when the block size is 400. When the block size becomes more than 450, the average CPU time to measure 1000 DMUs is longer than that for 3000 DMUs. This result indicates that there may be an optimal combination between the block size and the number of DMUs in terms of minimizing the average CPU time.

Table 10.2 extends the simulation results of Table 10.1 by incorporating different percentages of efficient DMUs (pe = 10, 20, 30, 40 and 50%). The two numbers listed within curved, (), and square, [], brackets indicate the percentage of DMUs in E and E', respectively. The finding of Table 10.2 is that the total CPU time reduces as the percentage of efficient DMUs decreases. The reduction of total CPU time occurs at any block size. This result is because the computational strategy incorporated into the network computing is effective when the number of efficient DMUs is small.

Next, we examined how the number of PCs (clients) influences the total CPU time. The structure of this simulation study may be summarized as follows:

(a) Number of DMUs: n = 1000, 1500, 2000, 2500, 3000, 5000, 7500, 10000
(b) Number of PCs: z = 1, 2, 3 and 4
(c) Percentage of efficient DMUs: pe = 10, 20, 30, 40 and 50%

Thus, the second simulation had the structure of an 8 × 4 × 5 factorial experiment in which each treatment had 200 replications, so that a total of 32000 (8 × 4 × 5 × 200) DEA evaluations were conducted for the whole simulation study.

TABLE 10.1 Comparison between CPU time and network computing

Block Size	Time \ Number of DMUs	1000	1500	2000	2500	3000
50	Treatment for Dominance	165 < 0.17 >	365 < 0.24 >	648 < 0.32 >	1001 < 0.40 >	1431 < 0.48 >
	Stage I	1701 < 1.70 >	2814 < 1.88 >	3251 < 1.63 >	4197 < 1.68 >	5109 < 1.70 >
	Stage II	12210 < 12.21 >	25392 < 16.93 >	41527 < 20.76 >	70126 < 28.05 >	93124 < 31.04 >
	Stage III	6705 < 6.71 >	14131 < 9.42 >	21986 < 10.99 >	34134 < 13.65 >	47153 < 15.72 >
	Total	20782 < 20.78 >	42701 < 28.47 >	67411 < 33.71 >	109457 < 43.78 >	146817 < 48.94 >
100	Treatment for Dominance	165 < 0.17 >	366 < 0.24 >	646 < 0.32 >	1002 < 0.40 >	1432 < 0.48 >
	Stage I	3387 < 3.39 >	4327 < 2.88 >	5472 < 2.74 >	7311 < 2.92 >	8262 < 2.75 >
	Stage II	11122 < 11.12 >	21190 < 14.13 >	36353 < 18.18 >	61308 < 24.52 >	82358 < 27.45 >
	Stage III	6615 < 6.62 >	13755 < 9.17 >	22102 < 11.05 >	34024 < 13.61 >	45304 < 15.10 >
	Total	21290 < 21.29 >	39637 < 26.42 >	64573 < 32.29 >	103645 < 41.46 >	137356 < 45.79 >
150	Treatment for Dominance	164 < 0.16 >	360 < 0.24 >	644 < 0.32 >	1002 < 0.40 >	1431 < 0.48 >
	Stage I	3903 < 3.90 >	7151 < 4.77 >	7779 < 3.89 >	10385 < 4.15 >	11548 < 3.85 >
	Stage II	9598 < 9.60 >	21276 < 14.18 >	34027 < 17.01 >	59396 < 23.76 >	76927 < 25.64 >
	Stage III	6663 < 6.66 >	13529 < 9.02 >	22598 < 11.30 >	32382 < 12.95 >	44945 < 14.98 >
	Total	20327 < 20.33 >	42316 < 28.21 >	65048 < 32.52 >	103164 < 41.27 >	134850 < 44.95 >
200	Treatment for Dominance	165 < 0.17 >	361 < 0.24 >	643 < 0.32 >	997 < 0.40 >	1433 < 0.48 >
	Stage I	6618 < 6.62 >	6585 < 4.39 >	12290 < 6.15 >	12739 < 5.10 >	13872 < 4.62 >
	Stage II	10091 < 10.09 >	18722 < 12.48 >	35413 < 17.71 >	54243 < 21.70 >	74094 < 24.70 >
	Stage III	6730 < 6.73 >	13705 < 9.14 >	22603 < 11.30 >	35231 < 14.09 >	45809 < 15.27 >
	Total	23604 < 23.60 >	39372 < 26.25 >	70948 < 35.47 >	103209 < 41.28 >	135208 < 45.07 >
250	Treatment for Dominance	166 < 0.17 >	366 < 0.24 >	644 < 0.32 >	1002 < 0.40 >	1430 < 0.48 >
	Stage I	9624 < 9.62 >	10050 < 6.70 >	11994 < 6.00 >	17675 < 7.07 >	19188 < 6.40 >
	Stage II	11467 < 11.47 >	19303 < 12.87 >	33447 < 16.72 >	55555 < 22.22 >	73472 < 24.49 >
	Stage III	6631 < 6.63 >	13900 < 9.27 >	22464 < 11.23 >	33779 < 13.51 >	46200 < 15.40 >
	Total	27888 < 27.89 >	43618 < 29.08 >	68550 < 34.28 >	108011 < 43.20 >	140290 < 46.76 >

(*Continued*)

TABLE 10.1 (Continued)

Block Size	Number of DMUs / Time	1000	1500	2000	2500	3000
300	Treatment for Dominance	165 < 0.17 >	366 < 0.24 >	645 < 0.32 >	996 < 0.40 >	1432 < 0.48 >
	Stage I	13100 < 13.10 >	14396 < 9.60 >	14435 < 7.22 >	19754 < 7.90 >	26445 < 8.82 >
	Stage II	11467 < 11.47 >	18742 < 12.49 >	30546 < 15.27 >	54854 < 21.94 >	76757 < 25.59 >
	Stage III	6676 < 6.68 >	13926 < 9.28 >	22547 < 11.27 >	33309 < 13.32 >	46460 < 15.49 >
	Total	31408 < 31.41 >	47429 < 31.62 >	68172 < 34.09 >	108912 < 43.56 >	151094 < 50.36 >
350	Treatment for Dominance	167 < 0.17 >	361 < 0.24 >	646 < 0.32 >	1002 < 0.40 >	1433 < 0.48 >
	Stage I	18072 < 18.07 >	18717 < 12.48 >	19246 < 9.62 >	20315 < 8.13 >	27869 < 9.29 >
	Stage II	12869 < 12.87 >	21376 < 14.25 >	32420 < 16.21 >	47268 < 18.91 >	73757 < 24.59 >
	Stage III	6618 < 6.62 >	13900 < 9.27 >	22480 < 11.24 >	33764 < 13.51 >	46037 < 15.35 >
	Total	37726 < 37.73 >	54353 < 36.24 >	74792 < 37.40 >	102347 < 40.94 >	149096 < 49.70 >
400	Treatment for Dominance	163 < 0.16 >	361 < 0.24 >	646 < 0.32 >	996 < 0.40 >	1429 < 0.48 >
	Stage I	22412 < 22.41 >	25817 < 17.21 >	24850 < 12.43 >	25432 < 10.17 >	24893 < 8.30 >
	Stage II	14548 < 14.55 >	22868 < 15.25 >	33068 < 16.53 >	50082 < 20.03 >	66811 < 22.27 >
	Stage III	6694 < 6.69 >	13614 < 9.08 >	22387 < 11.19 >	34560 < 13.82 >	46450 < 15.48 >
	Total	43816 < 43.82 >	62660 < 41.77 >	80950 < 40.48 >	111070 < 44.43 >	139583 < 46.53 >
450	Treatment for Dominance	163 < 0.16 >	361 < 0.24 >	644 < 0.32 >	1001 < 0.40 >	1431 < 0.48 >
	Stage I	28047 < 28.05 >	29918 < 19.95 >	31851 < 15.93 >	32577 < 13.03 >	30207 < 10.07 >
	Stage II	15911 < 15.91 >	22548 < 15.03 >	35365 < 17.68 >	52005 < 20.80 >	64694 < 21.56 >
	Stage III	6739 < 6.74 >	13705 < 9.14 >	22297 < 11.15 >	34690 < 13.88 >	45536 < 15.18 >
	Total	50860 < 50.86 >	66531 < 44.35 >	90157 < 45.08 >	120273 < 48.11 >	141867 < 47.29 >
500	Treatment for Dominance	168 < 0.17 >	361 < 0.24 >	646 < 0.32 >	997 < 0.40 >	1431 < 0.48 >
	Stage I	36387 < 36.39 >	35897 < 23.93 >	38710 < 19.36 >	38466 < 15.39 >	37995 < 12.67 >
	Stage II	17439 < 17.44 >	25562 < 17.04 >	37963 < 18.98 >	53838 < 21.54 >	67848 < 22.62 >
	Stage III	6687 < 6.69 >	13880 < 9.25 >	22372 < 11.19 >	36322 < 14.53 >	45549 < 15.18 >
	Total	60680 < 60.68 >	75699 < 50.47 >	99690 < 49.85 >	129621 < 51.85 >	152823 < 50.94 >

(a) Source: Sueyoshi and Honma (2003).

TABLE 10.2 Comparison between CPU time and network computing

Efficient DMUs	Block Size	Number of DMUs 1000	1500	2000	2500	3000
10% (9%) [1%]	50	14848 < 14.85 >	30429 < 20.29 >	47911 < 23.96 >	67923 < 27.17 >	96802 < 32.27 >
	100	15030 < 15.03 >	28116 < 18.74 >	45364 < 22.68 >	64748 < 25.90 >	87146 < 29.05 >
	150	14556 < 14.56 >	31381 < 20.92 >	44313 < 22.16 >	61994 < 24.80 >	87549 < 29.18 >
	200	16971 < 16.97 >	28146 < 18.76 >	49270 < 24.64 >	65970 < 26.39 >	84309 < 28.10 >
	250	20302 < 20.30 >	32542 < 21.69 >	43480 < 21.74 >	73151 < 29.26 >	91180 < 30.39 >
	300	24233 < 24.23 >	36988 < 24.66 >	47954 < 23.98 >	61569 < 24.63 >	98952 < 32.98 >
	350	29503 < 29.50 >	41510 < 27.67 >	51680 < 25.84 >	67282 < 26.91 >	87083 < 29.03 >
	400	35222 < 35.22 >	45481 < 30.32 >	59735 < 29.87 >	73746 < 29.50 >	88829 < 29.61 >
	450	41475 < 41.48 >	52826 < 35.22 >	65888 < 32.94 >	78328 < 31.33 >	96826 < 32.28 >
	500	48486 < 48.49 >	60332 < 40.22 >	73503 < 36.75 >	84437 < 33.77 >	103138 < 34.38 >
20% (18%) [2%]	50	20782 < 20.78 >	42701 < 28.47 >	67411 < 33.71 >	109457 < 43.78 >	146817 < 48.94 >
	100	21290 < 21.29 >	39637 < 26.42 >	67573 < 33.79 >	103645 < 41.46 >	137356 < 45.79 >
	150	20327 < 20.33 >	42316 < 28.21 >	65048 < 32.52 >	103164 < 41.27 >	134850 < 44.95 >
	200	23604 < 23.60 >	39372 < 26.25 >	70948 < 35.47 >	103209 < 41.28 >	135208 < 45.07 >
	250	27888 < 27.89 >	43618 < 29.08 >	68550 < 34.28 >	108011 < 43.20 >	140290 < 46.76 >
	300	31408 < 31.41 >	47429 < 31.62 >	68172 < 34.09 >	108912 < 43.56 >	151094 < 50.36 >
	350	37726 < 37.73 >	54353 < 36.24 >	74792 < 37.40 >	102347 < 40.94 >	149096 < 49.70 >
	400	43816 < 43.82 >	62660 < 41.77 >	80950 < 40.48 >	111070 < 44.43 >	139583 < 46.53 >
	450	50860 < 50.86 >	66531 < 44.35 >	90157 < 45.08 >	120273 < 48.11 >	141867 < 47.29 >
	500	60680 < 60.68 >	75699 < 50.47 >	99690 < 49.85 >	129621 < 51.85 >	152823 < 50.94 >
30% (27%) [3%]	50	25483 < 25.48 >	52486 < 34.99 >	90653 < 45.33 >	128280 < 51.31 >	191660 < 63.89 >
	100	27180 < 27.18 >	51700 < 34.47 >	87560 < 43.78 >	121265 < 48.51 >	186092 < 62.03 >
	150	25951 < 25.95 >	52756 < 35.17 >	87069 < 43.53 >	125601 < 50.24 >	184184 < 61.39 >
	200	29265 < 29.27 >	53777 < 35.85 >	94845 < 47.42 >	124980 < 49.99 >	190842 < 63.61 >

(Continued)

TABLE 10.2 (Continued)

Efficient DMUs	Block Size	Number of DMUs 1000	1500	2000	2500	3000
	250	32758 < 32.76 >	53532 < 35.69 >	99627 < 49.81 >	131770 < 52.71 >	188523 < 62.84 >
	300	39562 < 39.56 >	61934 < 41.29 >	92144 < 46.07 >	145174 < 58.07 >	199530 < 66.51 >
	350	44296 < 44.30 >	66396 < 44.26 >	98644 < 49.32 >	128710 < 51.48 >	214637 < 71.55 >
	400	51410 < 51.41 >	76415 < 50.94 >	105775 < 52.89 >	131985 < 52.79 >	196409 < 65.47 >
	450	60300 < 60.30 >	86044 < 57.36 >	116584 < 58.29 >	143271 < 57.31 >	191095 < 63.70 >
	500	68605 < 68.61 >	90255 < 60.17 >	125068 < 62.53 >	150331 < 60.13 >	206448 < 68.82 >
40% (36%) [4%]	50	32781 < 32.78 >	63527 < 42.35 >	112737 < 56.37 >	161993 < 64.80 >	239979 < 79.99 >
	100	33019 < 33.02 >	63632 < 42.42 >	114637 < 57.32 >	162870 < 65.15 >	235540 < 78.51 >
	150	33792 < 33.79 >	69645 < 46.43 >	116347 < 58.17 >	158784 < 63.51 >	242999 < 81.00 >
	200	36272 < 36.27 >	76660 < 51.11 >	115289 < 57.64 >	162134 < 64.85 >	243524 < 81.17 >
	250	40143 < 40.14 >	67753 < 45.17 >	125584 < 62.79 >	172203 < 68.88 >	233544 < 77.85 >
	300	46545 < 46.55 >	76330 < 50.89 >	118810 < 59.41 >	186018 < 74.41 >	248437 < 82.81 >
	350	52378 < 52.38 >	81988 < 54.66 >	119761 < 59.88 >	183770 < 73.51 >	263458 < 87.82 >
	400	59099 < 59.10 >	92479 < 61.65 >	129997 < 65.00 >	174947 < 69.98 >	271361 < 90.45 >
	450	67397 < 67.40 >	100384 < 66.92 >	134476 < 67.24 >	184471 < 73.79 >	253894 < 84.63 >
	500	77365 < 77.37 >	111245 < 74.16 >	145868 < 72.93 >	191445 < 76.58 >	248793 < 82.93 >
50% (45%) [5%]	50	38501 < 38.50 >	85288 < 56.86 >	140295 < 70.15 >	209541 < 83.82 >	300725 < 100.24 >
	100	39884 < 39.88 >	85568 < 57.05 >	136631 < 68.32 >	205000 < 82.00 >	304783 < 101.59 >
	150	41898 < 41.90 >	86034 < 57.36 >	141340 < 70.67 >	202782 < 81.11 >	302465 < 100.82 >
	200	41677 < 41.68 >	91336 < 60.89 >	142500 < 71.25 >	199022 < 79.61 >	304713 < 101.57 >
	250	46185 < 46.19 >	87041 < 58.03 >	152233 < 76.12 >	211760 < 84.70 >	306125 < 102.04 >
	300	53072 < 53.07 >	89509 < 59.67 >	153432 < 76.72 >	220517 < 88.21 >	312769 < 104.26 >
	350	59667 < 59.67 >	98272 < 65.51 >	147141 < 73.57 >	234573 < 93.83 >	330784 < 110.26 >
	400	64717 < 64.72 >	107035 < 71.36 >	155468 < 77.73 >	216236 < 86.49 >	347891 < 115.96 >
	450	73847 < 73.85 >	113568 < 75.71 >	168643 < 84.32 >	216536 < 86.61 >	336072 < 112.02 >
	500	83768 < 83.77 >	126302 < 84.20 >	173310 < 86.66 >	227507 < 91.00 >	323348 < 107.78 >

(a) Source: Sueyoshi and Honma (2003).

The computational results of the second simulation study are documented in both Tables 10.3 and 10.4. Table 10.3 compares the CPU times of the proposed DEA network for computing when the block size is 50 and the percentage of efficient DMUs ($E \cup E'$) is 20%. Table 10.3 indicates that the total and average CPU times have an increasing trend, corresponding to the number of DMUs. However, the rate of the increasing trend diminishes as the number of PCs increases. For example, the average CPU time is 39.00 MS when two PCs are used for network computing and 1000 DMUs are evaluated by DEA. The corresponding average CPU time becomes 20.78 MS when the number of PCs increases from two to four. Such a change in the number of PCs reduces the average CPU time by 47% [= (1 – 20.78)/39.00]. The decrease in the total and average CPU times can be found in the DEA evaluation of all the numbers of DMUs listed in Table 10.3.

Based upon this result, we may predict an extended case in which the total CPU time, incorporating more than four PCs in the network, becomes less than that of the four PCs. However, the reduction rate is less than that which occurs when the number of PCs is changed from three to four. Thus, it is expected that there is an optimal number of PCs for the proposed network computing that is determined by a combination of the decrease in CPU time within the PCs and the increase in network communication time between them. Network computing may not lessen the total CPU time if the number of PCs is beyond the optimal number.

Each PC in network computing fully utilizes the proposed algorithmic strategy, but it has a limitation on computational improvement. For example, a single PC operation to solve more than 2500 DMUs needs a long computational time (for example, more than 10 hours). So, our simulation terminates the CPU time measurement of single PC operation for more than 2000 DMUs. The proposed network implementation can easily overcome such a limitation in terms of solving large-scale DEA problems by fully utilizing network technology (e.g., a client–server function) to synchronize multiple PCs.

Table 10.4 generalizes the simulation results of Table 10.3 by incorporating different percentages (pe = 10, 20, 30, 40 and 50%) of efficient DMUs. As summarized in Table 10.4, the average CPU time shows an increasing trend matching the number of DMUs. The increasing rate is affected by not only the number of PCs but also the percentage of efficient DMUs. For example, the combination between $z=2$ and pe = 50% produces the highest (or the most time-consuming) average CPU time among the six combinations. The second highest is the combination between $z=2$ and pe = 30%. However, the third highest time-consuming combination is the one for $z=4$ and pe = 50%. This is a surprising result. Thus, the average CPU time and related increasing rate are affected by both the number of PCs and the percentage of efficient DMUs.

TABLE 10.3 Comparison between CPU time and network computing

# of PCs	Treatment	Number of DMUs							
		1000	1500	2000	2500	3000	5000	7500	10000
1	Dominance Computation	1731	2592	4506	Unable to measure				
		65306	147293	253747					
	Total	67037	149885	258253					
	Average (per DMU)	67.04	99.92	129.13					
2	Dominance	160	360	661	991	1502	3425	7401	13049
	Stage I	5428	7731	10315	12428	16644	14821	17545	27579
	Stage II	20068	37645	70821	110298	163505	425342	731999	1652507
	Stage III	13340	24154	29938	68899	78213	279081	520041	127200
	Total	38996	69890	111735	192616	259864	722669	1276986	1820335
	Average (per DMU)	39.00	46.59	55.87	77.05	86.62	144.53	170.26	182.03
3	Dominance	161	360	661	992	1502	3425	7401	12859
	Stage I	3615	5288	7031	9223	11046	9915	12037	18707
	Stage II	13679	26168	46787	76650	110539	284999	534725	1102184
	Stage III	9504	17054	27770	47068	53106	188431	358580	874748
	Total	26959	48870	82249	133933	176193	486770	912743	2008498
	Average (per DMU)	26.96	32.58	41.12	53.57	58.73	97.35	121.70	200.85
4	Dominance	165	365	648	1001	1431	3415	7411	12978
	Stage I	1701	2814	3251	4197	5109	7792	9704	14481
	Stage II	12210	25392	41527	70126	93124	217572	437154	821922
	Stage III	6705	14131	21986	34134	47153	134143	249633	609286
	Total	20781	42702	67412	109458	146817	362922	703902	1458667
	Average (per DMU)	20.78	28.47	33.71	43.78	48.94	72.58	93.85	145.87

(a) Source: Sueyoshi and Honma (2003). (b) It was impossible to measure a CPU time of DEA problems that contained more than 2000 DMUs by a single PC.

TABLE 10.4 Comparison between CPU time and network computing

# of PCs	Efficient DMUs	Number of DMUs 1000	1500	2000	2500	3000	5000	7500	10000
1	10%(9%)[1%]	52981 < 52.98 >	111793 < 74.53 >	173579 < 86.79 >			Unable to measure		
	20%(18%)[2%]	67037 < 67.04 >	149885 < 99.92 >	258253 < 129.13 >					
	30%(27%)[3%]	86165 < 86.17 >	190187 < 126.79 >	345827 < 172.91 >					
	40%(36%)[4%]	113827 < 113.83 >	241089 < 160.73 >	403125 < 201.56 >					
	50%(45%)[5%]	124577 < 124.58 >	279396 < 186.26 >	496588 < 248.29 >					
2	10%(9%)[1%]	25477 < 25.48 >	51494 < 34.33 >	82378 < 41.19 >	120704 < 48.28 >	170625 < 56.88 >	503604 < 100.72 >	836894 < 111.59 >	1723428 < 172.34 >
	20%(18%)[2%]	38996 < 39.00 >	69890 < 46.59 >	111735 < 55.87 >	192616 < 77.05 >	259864 < 86.62 >	722669 < 144.53 >	1276986 < 170.26 >	2965835 < 296.58 >
	30%(27%)[3%]	49712 < 49.71 >	91792 < 61.19 >	179208 < 89.60 >	248467 < 99.39 >	330075 < 110.03 >	977576 < 195.52 >	1962943 < 261.73 >	3775659 < 377.57 >
	40%(36%)[4%]	59175 < 59.18 >	121164 < 80.78 >	243560 < 121.78 >	348752 < 139.50 >	514349 < 171.45 >	1175961 < 235.19 >	2418247 < 322.43 >	5016553 < 501.66 >
	50%(45%)[5%]	78283 < 78.28 >	180830 < 120.55 >	274695 < 137.35 >	392314 < 156.93 >	612591 < 204.20 >	1633409 < 326.68 >	3120748 < 416.10 >	6201888 < 620.19 >

(Continued)

TABLE 10.4 (Continued)

# of PCs	Efficient DMUs	Number of DMUs 1000	1500	2000	2500	3000	5000	7500	10000
3	10%(9%)[1%]	17816 < 17.82 >	38005 < 25.34 >	56241 < 28.12 >	84942 < 33.98 >	113693 < 37.90 >	343554 < 68.71 >	599482 < 79.93 >	1150714 < 115.07 >
	20%(18%)[2%]	26959 < 26.96 >	48870 < 32.58 >	82249 < 41.12 >	133933 < 53.57 >	176193 < 58.73 >	486770 < 97.35 >	912743 < 121.70 >	2008498 < 200.85 >
	30%(27%)[3%]	36592 < 36.59 >	62339 < 41.56 >	120724 < 60.36 >	167281 < 66.91 >	220928 < 73.64 >	652588 < 130.52 >	1284968 < 171.33 >	2261692 < 226.17 >
	40%(36%)[4%]	42260 < 42.26 >	84030 < 56.02 >	163195 < 81.60 >	233926 < 93.57 >	345557 < 115.19 >	786301 < 157.26 >	1617486 < 215.66 >	3202955 < 320.30 >
	50%(45%)[5%]	53617 < 53.62 >	132356 < 88.24 >	183553 < 91.78 >	269397 < 107.76 >	407455 < 135.82 >	1081005 < 216.20 >	2073321 < 276.44 >	4171137 < 417.11 >
4	10%(9%)[1%]	14848 < 14.85 >	30429 < 20.29 >	47911 < 23.96 >	67923 < 27.17 >	96802 < 32.27 >	271420 < 54.28 >	473561 < 63.14 >	898112 < 89.81 >
	20%(18%)[2%]	20782 < 20.78 >	42701 < 28.47 >	67411 < 33.71 >	109457 < 43.78 >	146817 < 48.94 >	362922 < 72.58 >	703902 < 93.85 >	1458667 < 145.87 >
	30%(27%)[3%]	25483 < 25.48 >	52486 < 34.99 >	90653 < 45.33 >	128280 < 51.31 >	191660 < 63.89 >	484437 < 96.89 >	1033556 < 137.81 >	1936935 < 193.69 >
	40%(36%)[4%]	32781 < 32.78 >	63527 < 42.35 >	112737 < 56.37 >	161993 < 64.80 >	239979 < 79.99 >	602186 < 120.44 >	1225402 < 163.39 >	2509158 < 250.92 >
	50%(45%)[5%]	38501 < 38.50 >	85288 < 56.86 >	140295 < 70.15 >	209541 < 83.82 >	300725 < 100.24 >	824235 < 164.85 >	1470154 < 196.02 >	3110954 < 311.10 >

(a) Source: Sueyoshi and Honma (2003).

10.5 SUMMARY

To deal with large DEA problems, this chapter proposed the architecture of network computing that was designed to synchronize multiple PCs in a client–server parallel implementation. Each PC in the network managed its own memory in order to evaluate the performance of DMUs on Microsoft Excel. An important feature of this network computing was that it was computationally structured in multi-stage parallel processes: (a) preparatory treatment for classifying all the DMUs into two subsets (J_n: a non-dominated DMU set; J_d: a dominated DMU set), (b) Stage I (DEA parallel computing for J_n), (c) Stage II (DEA network computing for J_n) and (d) Stage III (DEA parallel computing for J_d). Thus, network computing, which was designed to fully utilize both modern network technology and special algorithmic strategies, enhanced the algorithmic efficiency to solve various large problems. The performance of the proposed network computing was examined in a large simulation study that compared the CPU times by controlling the number of DMUs, the percentage of efficient DMUs, the size of block and the number of PCs.

Here, it is necessary to mention that the computational results were listed in a previous research effort (Sueyoshi and Honma, 2003). It is indeed true that their simulation results were based upon an old type of PC configuration that might be obsolete under modern PC technology. However, the proposed architecture for network computing is still useful and computationally efficient in dealing with a large data set on DEA even now.

Paying attention to modern computer technology, this chapter needs to discuss four problems. First, this chapter has discussed only local area network computing. The computational feature needs to be extended into wide area network computing and further, "cloud computing" and a super computer, which make it possible that we can simultaneously operate computing devices all over the world. Parallel processing, using networks of PCs, is the current form of distributed computing and it is a realistic computing environment for DEA applications. Hence, such a physical network expansion is an important future research task. Second, as explored in Gonzalez–Lima *et al.* (1996), it is possible for us to use a modern algorithm (e.g., the interior point method) to solve various DEA performance assessments. Third, the proposed approach did not explore the analytical relationship between problem size (i.e., the number of DMUs) and computation time. How large is the best block size? No reply was discussed in this chapter. Finally, the proposed network computing approach can be useful in various DEA applications. The proposed approach does not document such empirical results, in particular DEA environmental assessment. That will be an important research task.

11

DEA-DISCRIMINANT ANALYSIS

11.1 INTRODUCTION

This chapter discusses discriminant analysis (DA)[1]. DA[2] is a classification method that can predict the group membership of a newly sampled observation. In the use of DA, a group of observations, whose memberships are already identified, are

[1]This chapter is partly based upon the article: Sueyoshi, T. (2006) DEA-discriminant analysis: methodological comparison among eight discriminant analysis approaches. *European Journal of Operational Research*, **169**, 247-272.

[2]The previous research efforts on DA were methodologically classified into three groups. The first group was interested in "statistics" on DA. The contribution may be dating back to Fisher (1936) and Smith (1947). See, for instance, Kendal *et al.* (1983) and McLachlan (1992) in which previous contributions of statistical DA were compiled. The conventional statistical DA methods usually depended upon underlying assumptions on a group distribution. For example, two groups come from normal populations with different means, but a same covariance matrix, all of which should be known to us before computation. Under these assumptions, the statistical methods provided a theoretical basis for conducting various statistical inferences and tests. Furthermore, an ordinary least squares method was usually used to obtain coefficient estimates of a linear discriminant function. Thus, there is a computational simplicity in the statistical DA methods. Those are methodological strength and contribution, indeed. However, it is also true that many real data sets do not satisfy such underlying assumptions. The second group was "econometrics." If independent variables are normally distributed, the statistical DA estimator is a true maximum-likelihood estimator and therefore asymptotically more efficient than other DA methods. However, the assumption on normality is not satisfied in many real data sets. The last group attempted to overcome the shortcoming related to the statistical DA. Econometricians have

Environmental Assessment on Energy and Sustainability by Data Envelopment Analysis,
First Edition. Toshiyuki Sueyoshi and Mika Goto.

used for the estimation of weights (or parameters)[3] of a discriminant function by some criterion such as minimization of misclassifications or maximization of correct classifications. A new sample is classified into one of several groups by the DA results.

In the last two decades, Sueyoshi (1999a, 2001b, 2004, 2005a, b, 2006), Sueyoshi and Kirihara (1998), Sueyoshi and Hwang (2004a), Sueyoshi and Goto (2009a,b,c, 2012c, 2013b)[4] and Goto (2012) proposed a new type of non-parametric DA approach that provides a set of weights of a linear discriminant function(s), consequently yielding an evaluation score(s) for determining group membership. The new non-parametric DA is referred to as "DEA-DA" because it maintains discriminant capabilities by incorporating the non-parametric feature of

developed other several DA methods that were closely linked to the theory of probabilistic choice discussed by psychologists. The most well-known research effort in this area may be due to McFadden (1973, 1976, 1980) who has investigated logit and probit models. The two models are usually solved by maximum-likelihood methods. An important feature of logit and probit analyses is that they provide the conditional probability of an observation belonging to a certain class, given its independent variables. Both are based on a cumulative probability function and neither requires that independent variables are multivariate normal nor that groups have equal covariance matrices, unlike the requirements of statistical DA. Furthermore, these approaches are closely linked to statistical inferences and various tests.

[3] DA determines unknown parameter estimates on a discriminant function. Meanwhile, DEA determines unknown multipliers (i.e., dual variables), or weights on inputs, desirable and undesirable outputs. DEA-DA combines both methodologies so that it determines unknown weights for a discriminant function.

[4] In reviewing their research works, the first research effort (Sueyoshi, 1999a) proposed a GP approach for DA and then compared it with an additive model of DEA from the perspective of GP. The comparison provided similarities and differences between the proposed GP model and DEA. Then, the name DEA-DA was used to distinguish it from other DA and DEA approaches. Sueyoshi and Kirihara (1998) documented how to incorporate prior information into the evaluation process of DEA-DA. The original GP version of DEA-DA had a major drawback. It could not deal with a negative value in data. Consequently, its classification applicability was very limited on a data set comprising only non-negative observations. To overcome the methodological shortcoming, Sueyoshi (2001) extended it further in the manner that the proposed approach can deal with the negative value in a data set. The refined approach was referred to as "extended DEA-DA." An important feature of the extended DEA-DA approach is that it is designed to minimize the total distance of misclassified observations. Furthermore, the proposed approach is formulated by two-stage DEA-DA formulations and can be solved by any linear programming software. The first stage is used to identify an existence of an overlap between two groups of observations. The second stage classifies observations belonging to the overlap. Although the extended version can handle the issue of a negative data value, Sueyoshi (2004a) has found another methodological issue. That is, the total number of observations classified correctly is usually used to examine the performance of DA methods, not the total distance. This finding indicates a need of the methodological extension in which the extended approach is further reformulated by mixed integer programming (MIP) models, because it needs to incorporate a binary variable to count the number of misclassified observations. Thus, Sueyoshi (2004a) proposed a MIP version of DEA-DA and methodologically compared it with the extended GP model. Sueyoshi (2004b) and Sueyoshi and Hwang (2004) dropped the first stage of the two-stage MIP approach in order to simplify the estimation process of the two-stage MIP approach. The formulation for the second stage (Sueyoshi, 2004a) was used as a MIP model for the standard MIP approach after a slight modification. The standard MIP approach is useful when the overlap between two groups is not a serious problem. In addition to their simplification in the MIP formulation, they discussed how to make a link to a statistical inference and a statistical test from the perspective of practical applications. The comparison between DEA-DA and DEA was discussed in terms of an application perspective (Sueyoshi and Hwang, 2004).

DEA into DA. An important feature of DEA-DA[5] is that it can be considered as an extension of frontier analysis, using the L1-regression analysis discussed in Chapter 3. Moreover, a series of research efforts has attempted to formulate DEA-DA in the manner that they can handle a data set that contains zero and/or negative values. Thus, the development of DEA-DA in this chapter is different from that of DEA discussed in previous chapters[6]. See Chapters 26 and 27 that discuss how to handle zero and/or negative values in DEA environmental assessment.

The remainder of this chapter is organized as follows: Section 11.2 describes two mixed integer programming (MIP) formulations for DEA-DA. Section 11.3 extends the proposed DEA-DA approach to the classification of multiple groups. Methodological comparison by using illustrative data sets is documented in Section 11.4. Section 11.5 reorganizes the proposed approach from frontier analysis so that this chapter discusses an analytical linkage between DEA-DA and DEA-based frontier analysis. Concluding comments are summarized in Section 11.6.

11.2 TWO MIP APPROACHES FOR DEA-DA

11.2.1 Standard MIP Approach

Consider a classification problem in which there are two groups (G_1 and G_2). Here, G_1 is a set of observations in the first group above an estimated discriminant function and G_2 is a set of observations in the second group below that estimated discriminant function. The sum of the two groups contains n observations ($j = 1,\ldots,n$). Each observation is characterized by m independent factors ($i = 1,\ldots,m$), denoted by z_{ij}. Note that there is no distinction between inputs and outputs in DEA-DA, as found in the previous chapters on DEA. The group membership of each

[5] DEA-DA belongs to "mathematical programming (MP)" approaches that have been proposed for solving various DA problems. The first contribution of this group was due to Charnes *et al.* (1955) which documented how to formulate L_1 regression analysis by a GP model and how to solve the problem by linear programming algorithm. A popularity of MP-based DA among OR&MS researchers occurred after the research effort of Freed and Glover (1981a,b). They presented how a DA problem could be formulated by GP. Based upon these optimization techniques, this group of DA studies was further classified into (a) linear programming methods (e.g., Markowski and Markowski, 1987; Lam and Moy, 1997; Mangasarian, 1999), (b) nonlinear programming methods (e.g., Cavalier *et al.*, 1989; Stam and Joachimsthaler, 1989; Glover, 1990; Duarte Silva and Stam, 1994; Falk and Karlov, 2001) and (c) MIP methods (e.g., Bajgier and Hill, 1982; Rubin, 1990; Abad and Banks, 1993; Wilson, 1996; Yanev and Balev, 1999). A comprehensive review on the MP-based DA was found in Stam (1997), Doumpos *et al.* (2001) and Zopounidis and Doumpos (2002). A methodological benefit of the research group is that the MP-based DA methods do not need any assumption on a group distribution. Meanwhile, a shortcoming of the MP-based DA is that statistical inferences and tests are not yet well established at the level of the statistical and econometric DA approaches.

[6] As discussed in Chapter 2, DEA allocates weights to observations, or DMUs, while DEA-DA allocates them to parameters of a discriminant function. Such a distinct feature in their analytical structures makes the two approaches being different in terms of their applications. Acknowledging these differences, this chapter needs to describe that we can apply the two approaches for performance assessment.

observation should be known before the DA computation. The observation (z_{ij}) may consist of a data set that contains zero, positive and/or negative values.

The single-stage approach, often referred to as a "standard MIP approach," is originally formulated as follows:

$$\begin{aligned}
&\text{Minimize} && \sum_{j \in G_1} \mu_j + \sum_{j \in G_2} \mu_j \\
&s.t. && \sum_{i=1}^{m} \lambda_i z_{ij} - cf + M\mu_j \geq 0, && (j \in G_1), \\
& && \sum_{i=1}^{m} \lambda_i z_{ij} - cf - M\mu_j \leq -\varepsilon, && (j \in G_2), \\
& && \sum_{i=1}^{m} |\lambda_i| = 1, \\
& && \lambda_i : \text{URS}, \ \ cf : \text{URS and } \mu_j : \text{binary},
\end{aligned} \tag{11.1}$$

where M is a given large number (e.g., 10000) and ε is a given small number (e.g., 1). Both should be prescribed before the computation of Model (11.1). The objective function minimizes the total number of incorrectly classified observations by counting a binary variable (μ_j). The binary variable (0 or 1) expresses correct or incorrect classification. A discriminant score for cutting-off between two groups is expressed by cf ($j \in G_1$) and cf-ε ($j \in G_2$), respectively. The small number (ε) is incorporated into Model (11.1) in order to avoid a case where an observation(s) exists on an estimated discriminant function. Thus, the small number is for perfect classification. Furthermore, a weight estimate regarding the i-th factor is expressed by λ_i. All the observations (z_{ij} for all i and j) are connected by a weighted linear combination, or $\sum \lambda_i z_{ij}$, that consists of a discriminant function. The weights on factors are restricted in such a manner that the sum of absolute values of λ_i ($i = 1,..., m$) is unity in Model (11.1). The treatment is because we need to deal with zero and/or negative values in an observed data set. It is easily thought that λ_i is a weight on each factor, not each observation (i.e., DMU), as found in a conventional use of DA. The weight simply expresses the importance of each factor in terms of estimating a discriminant function. The importance among weight estimates is specified by a percentage expression.

It is possible for us to formulate Model (11.1) by maximizing the number of correct classifications. Both incorrect minimization and correct maximization does not produce a major difference between them. Because of different MIP formulations and different computations (i.e., minimization or maximization), the two approaches may produce different but similar results in terms of their classification capabilities.

In examining Model (11.1), we understand that it is difficult for us to directly solve the model because $\sum |\lambda_i| = 1$ is incorporated in the formulation. Hence, Model (11.1) needs to be reformulated by separating each weight $\lambda_i (= \lambda_i^+ - \lambda_i^-)$ for

$i = 1,\ldots,m$. Here, $\lambda_i^+ = \left(|\lambda_i| + \lambda_i\right)/2$ and $\lambda_i^- = \left(|\lambda_i| - \lambda_i\right)/2$ indicate the positive and negative parts of the i-th weight, respectively.

Following the research effort of Sueyoshi (2006), this chapter reformulates Model (11.1) as follows:

$$
\begin{aligned}
\text{Minimize} \quad & \sum_{j \in G_1} \mu_j + \sum_{j \in G_2} \mu_j \\
\text{s.t.} \quad & \sum_{i=1}^{m} \left(\lambda_i^+ - \lambda_i^-\right) z_{ij} - cf + M\mu_j \geq 0 \qquad (j \in G_1), \\
& \sum_{i=1}^{m} \left(\lambda_i^+ - \lambda_i^-\right) z_{ij} - cf - M\mu_j \leq -\varepsilon \quad \left(j \in G_2\right), \\
& \sum_{i=1}^{m} \left(\lambda_i^+ - \lambda_i^-\right) z_{ij} = 1, \qquad\qquad\qquad\qquad (11.2) \\
& \upsilon_i^+ \geq \lambda_i^+ \geq \varepsilon \upsilon_i^+ \text{ and } \upsilon_i^- \geq \lambda_i^- \geq \varepsilon \upsilon_i^- \quad \left(i = 1,\ldots,m\right), \\
& \upsilon_i^+ + \upsilon_i^- \leq 1 \qquad\qquad\qquad\qquad \left(i = 1,\ldots,m\right), \\
& \sum_{i=1}^{m} \left(\upsilon_i^+ + \upsilon_i^-\right) = m, \\
& cf : \text{URS}, \upsilon_i^+, \upsilon_i^-, \mu_j : \text{binary and all other variables} \geq 0.
\end{aligned}
$$

In Model (11.2), binary variables (υ_i^+ and υ_i^- for $i = 1,\ldots,m$) are incorporated in the formulation. The first two groups of constraints in Model (11.2) are originated from those of Model (11.1). The third constraint expresses $|\lambda_i|$ by $\lambda_i^+ + \lambda_i^-$ for $i = 1,\ldots,m$. The third one is often referred to as "normalization" in the DA community. The fourth group of constraints indicate the upper and lower bounds of λ_i^+ and those of λ_i^-, respectively. The fifth group of constraints implies that the sum of binary variables (υ_i^+ and υ_i^- for all i) is less than or equal to unity. The restriction avoids a simultaneous occurrence of $\upsilon_i^+ > 0$ and $\upsilon_i^- > 0$. The last constraints are incorporated into Model (11.2) in order to utilize all weights (so, being a full model for a discriminant function).

All observations, or $Z_j = (z_{1j}, \ldots, z_{mj})^{Tr}$ for all j, are classified by the following rule:

(a) If $\sum_{i=1}^{m} \lambda_i^* z_{ij} \geq cf^*$, then the j-th observation belongs to G_1, and
(b) If $\sum_{i=1}^{m} \lambda_i^* z_{ij} \leq cf^* - \varepsilon$, then the j-th observation belongs to G_2.

Here, $\lambda_i^*(i = 1,\ldots,m)$ and cf^* are obtained from the optimality of Model (11.2). The classification of a new sample, or $Z_k = (z_{1k}, \ldots, z_{mk})^{Tr}$ for the specific k, is determined by the same rule by replacing the subscript j by k in the classification rule.

A methodological strength of the standard MIP approach is that we can utilize a single classification (λ_i^* for all $i=1,\ldots,m$ and cf^*). A drawback of the approach is that the computation effort on Model (11.2) is more complex and more tedious (i.e., more time-consuming) than that of the two-stage MIP approach, discussed in Section 11.2.2. The computational concern seems inconsistent with our expectation. The rationale concerning why the computational efficiency occurs in the two-stage MIP approach will be discussed in the next subsection.

Although the standard MIP approach is more time-consuming than the two-stage MIP approach in solving various DA problems, the former has an analytical benefit that should be noted here. That is, it is possible for us to reformulate Model (11.2) by linear programing which minimizes the total amount of slacks for expressing distances between observations of independent factors and an estimated discriminant function. It is trivial that a linear programming formulation is more efficient than the proposed MIP formulation, or Model (11.2). This chapter, along with Chapter 25, does not describe the linear programming formulation, because the performance of DA is usually evaluated by a classification rate, as documented in this chapter. Of course, we clearly understand that if we discuss frontier analysis, a linear programming formulation is sufficient enough to develop an efficiency frontier for measuring a degree of *OE*. See Chapter 3 on frontier analysis.

11.2.2 Two-stage MIP Approach

It is widely known that the main source of misclassification in DA is due to the existence of an overlap between two groups in a data set. This type of problem does not occur in DEA. If we expect to increase the number of observations classified correctly, then DA approaches must consider how to handle an existence of such an overlap. If there is no overlap, most DA approaches can produce a perfect, or almost perfect, classification. However, if there is an overlap in the observations, an additional computation process is usually necessary to deal with this overlap. Thus, it is expected at the initial stage of DA that there is a tradeoff between computational effort (time) and level of classification capability.

In this chapter, a two-stage approach is proposed to handle such an overlap and thus increase the classification capability of DA. Surprisingly, this two-stage approach may decrease the computational burden regarding DA, depending upon the size of the overlap. The proposed computation process of the two-stage approach consists of classification and overlap identification, along with handling the overlap.

The first stage of the two-stage approach, discussed by Sueyoshi (2006), is formulated as follows:

Stage 1: Classification and Overlap Identification (COI):

$$\begin{aligned}
&\text{Minimize} \quad \rho \\
&s.t. \quad \sum_{i=1}^{m}\left(\lambda_i^+ - \lambda_i^-\right) z_{ij} - cf + \rho \ge 0 \qquad (j \in G_1), \\
&\sum_{i=1}^{m}\left(\lambda_i^+ - \lambda_i^-\right) z_{ij} - cf - \rho \le 0 \qquad (j \in G_2), \\
&\sum_{i=1}^{m}\left(\lambda_i^+ - \lambda_i^-\right) z_{ij} = 1, \qquad (11.3) \\
&\upsilon_i^+ \ge \lambda_i^+ \ge \varepsilon \upsilon_i^+, \ \upsilon_i^- \ge \lambda_i^- \ge \varepsilon \upsilon_i^+ \qquad (i-1,\ldots,m), \\
&\upsilon_i^+ + \upsilon_i^- \le 1 \qquad (i=1,\ldots,m), \\
&\sum_{i=1}^{m}\left(\upsilon_i^+ + \upsilon_i^-\right) = m, \\
&cf \text{ and } \rho: \text{URS}, \ \upsilon_i^+, \ \upsilon_i^-: \text{binary and all other variables} \ge 0.
\end{aligned}$$

In Model (11.3), the unrestricted variable "ρ" indicates the absolute distance between the discriminant score and a discriminant function.

Let $\lambda_i^*(=\lambda_i^{+*} - \lambda_i^{-*})$, cf^* and ρ^* be the optimal solution of the first stage (COI), or Model (11.3). On the optimality of Model (11.3), all observations on independent factors, or $Z_j = \left(z_{1j},\ldots,z_{mj}\right)^{Tr}$, for $j=1,\ldots,n$ are classified by the following rule:

(a) If $\sum_{i=1}^{m} \lambda_i^* z_{ij} > cf^* + \rho^*$, then the j-th observation belongs to G_1 (= C_1),
(b) If $\sum_{i=1}^{m} \lambda_i^* z_{ij} < cf^* - \rho^*$, then the observation belongs to G_2 (= C_2) and
(c) If $cf^* - \rho^* \le \sum_{i=1}^{m} \lambda_i^* z_{ij} \le cf^* + \rho^*$, then the observation belongs to an overlap.

Based upon the classification rule, the original sample data set (G) is classified into the following four subsets ($G = G_1 \cup G_2 = C_1 \cup D_1 \cup C_2 \cup D_2$):

(a) $C_1 = \left\{ j \in G_1 \middle| \sum_{i=1}^{m} \lambda_i^* z_{ij} > cf^* + \rho^* \right\}$,
(b) $C_2 = \left\{ j \in G_2 \middle| \sum_{i=1}^{m} \lambda_i^* z_{ij} < cf^* - \rho^* \right\}$,
(c) $D_1 = G_1 - C_1$ and
(d) $D_2 = G_2 - C_2$.

In the classification, C_1 and C_2 indicate a clearly above group and a clearly below group, respectively. Here, C_1 is a partial group of G_1 which are clearly above the estimated discriminant function in Stage 1 and C_2 is a partial group of

G_2 which are clearly below the estimated discriminant function in Stage 1. The remaining two groups ($D_1 \cup D_2$) belong to the overlap in which D_1 is a part of G_1 and D_2 is a part of G_2. A new sample (Z_k) is also classified by the above rules by replacing subscript j by subscript k.

Stage 2: Handling Overlap (HO):

The existence of an overlap is identified by $\rho^* \geq 0$ at the first stage for overlap identification. In contrast, $\rho^* < 0$ indicates no overlap between two groups. In the former case when the overlap is found between them, the second stage needs to reclassify all the observations, belonging to the overlap ($D_1 \cup D_2$), because the group membership of these observations is still unknown and not yet determined. Mathematically, the second (HO) stage is formulated as follows:

$$
\begin{aligned}
\text{Minimize}\quad & \sum_{j\in D_1}\mu_j + \sum_{j\in D_2}\mu_j \\
\text{s.t.}\quad & \sum_{i=1}^{m}\left(\lambda_i^+ - \lambda_i^-\right)z_{ij} - cf + M\mu_j \geq 0 \qquad (j \in D_1), \\
& \sum_{i=1}^{m}\left(\lambda_i^+ - \lambda_i^-\right)z_{ij} - cf - M\mu_j \leq -\varepsilon \qquad (j \in D_2), \\
& \sum_{i=1}^{m}\left(\lambda_i^+ + \lambda_i^-\right) = 1, \\
& \upsilon_i^+ \geq \lambda_i^+ \geq \varepsilon\upsilon_i^+ \text{ and } \upsilon_i^- \geq \lambda_i^- \geq \varepsilon\upsilon_i^- \quad (i = 1,\ldots,m), \\
& \upsilon_i^+ + \upsilon_i^- \leq 1 \qquad (i = 1,\ldots,m), \\
& \sum_{i=1}^{m}\left(\upsilon_i^+ + \upsilon_i^-\right) = p, \\
& cf: \text{URS},\ \upsilon_i^+,\ \upsilon_i^-,\ \mu_j: \text{binary and all other variables} \geq 0.
\end{aligned} \tag{11.4}
$$

In Model (11.4), the binary variable (μ_j) counts the number of observations classified incorrectly. These observations to be examined at the second stage are reduced from $G_1 \cup G_2$ to $D_1 \cup D_2$ as a result of the first stage. If the number of observations belonging to $D_1 \cup D_2$ is larger than m (the number of weights to be estimated), then $\sum_{i=1}^{m}\left(\upsilon_i^+ + \upsilon_i^-\right) = m$ is used in Model (11.4). However, if the number is less than m, then $\sum_{i=1}^{m}\left(\upsilon_i^+ + \upsilon_i^-\right) = p$ is used in Model (11.4) where the number "p" ($< m$) indicates a prescribed number in such a manner that p is less than or equal to the number of observations in $D_1 \cup D_2$. Thus, the degree of freedom needs to be considered for the selection of p. In reality, prior information is necessary for the selection of p.

Observations belonging to the overlap ($D_1 \cup D_2$) are classified at the second (HO) stage by the following rule:

(a) If $\sum_{i=1}^{m}\lambda_i^* z_{ij} \geq cf^*$, then the j-th observation in the overlap belongs to G_1 and
(b) If $\sum_{i=1}^{m}\lambda_i^* z_{ij} \leq cf^* - \varepsilon$, then the j-th observation belongs to G_2.

A new sample (Z_k) is also classified by the above rule by replacing the subscript (j) by the one (k).

Figures 11.1 to 11.4 visually describe the classification processes of the two-stage approach. For our visual convenience, Figure 11.1 has coordinates for z_1 and z_2, the two factors which depict observations belonging to two groups G_1 and G_2. Stage 1 is used to identify an occurrence of an overlap between two groups by two lines. Thus, Stage 1 classifies all observations into these three groups as follows: a part of G_1, a part of G_2 and an overlap group. In the classification, the clearly above group (C_1) is part of G_1 and the clearly below group (C_2) is part of G_2. See Figure 11.3 for a visual description of C_1 and C_2. The remaining group belongs to the overlap. Stage 1 cannot classify observations belonging to the overlap.

Figure 11.2 visually describes how Stage 1 determines the size of an overlap between two groups by using a distance measure (ρ). All the observations are classified into these three groups: C_1 (clearly above), C_2 (clearly below) and an

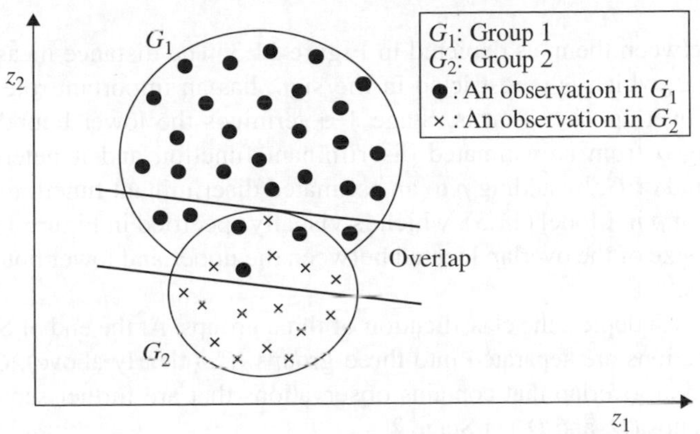

FIGURE 11.1 Classification and overlap identification at Stage 1

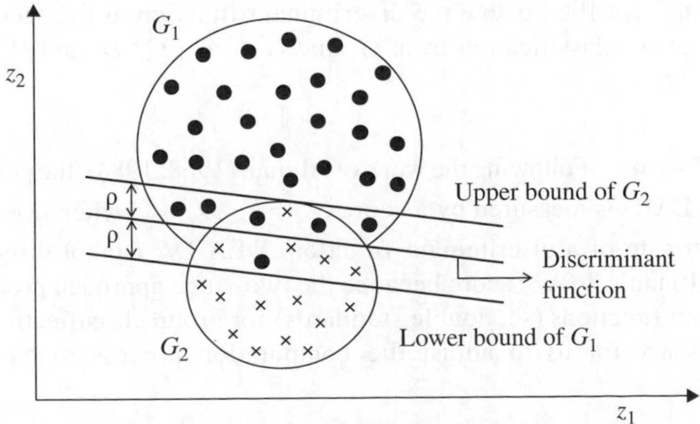

FIGURE 11.2 Overlap identification at Stage 1

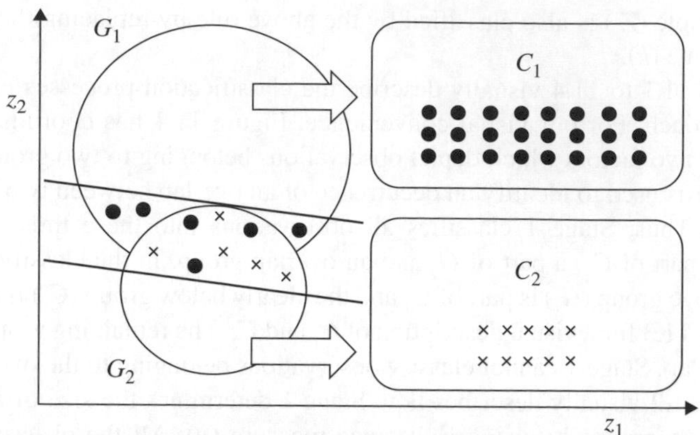

FIGURE 11.3 Classification at Stage 1

overlap between them, as depicted in Figure 11.3. The distance measure (ρ) in Figure 11.2, which is unrestricted in the sign, has an important role in identifying the overlap size. That is, Stage 1 determines the lower bound of G_1 by subtracting ρ from an estimated discriminant function and it determines the upper bound of G_2 by adding ρ to an estimated discriminant function. The minimization of ρ in Model (11.3), which is visually specified in Figure 11.2, determines the size of the overlap located between the upper and lower bounds of the two groups.

Figure 11.3 depicts the classification of three groups. At the end of Stage 1, all the observations are separated into three groups: C_1 (clearly above), C_2 (clearly below) and an overlap that contains observations that are further separated into two subgroups (D_1 and D_2) at Stage 2.

Figure 11.4 visually describes the function of Stage 2 that separates observations belonging to the overlap into D_1 and D_2. D_1 is the group of observations classified above the discriminant function and D_2 is the group of observations classified below the discriminant function at the second stage. Thus, the group classification from G_1 and G_2 to C_1, C_2, D_1 and D_2 indicates the end of Stage 2.

Altman's Z score: Following the work of Altman (1968, 1984), the performance of the *j*-th DMU is measured by a score ($Z_j = \Sigma_{i=1} \lambda_i * z_{ij}$) where λ_i* is the *i*-th weight estimate of a discriminant function. DEA-DA cannot directly compute the Altman's Z (AZ) score because the two-stage approach produces two discriminant functions (so, double standards) for group classification. Hence, it is necessary for us to adjust the computation process to measure the AZ score.

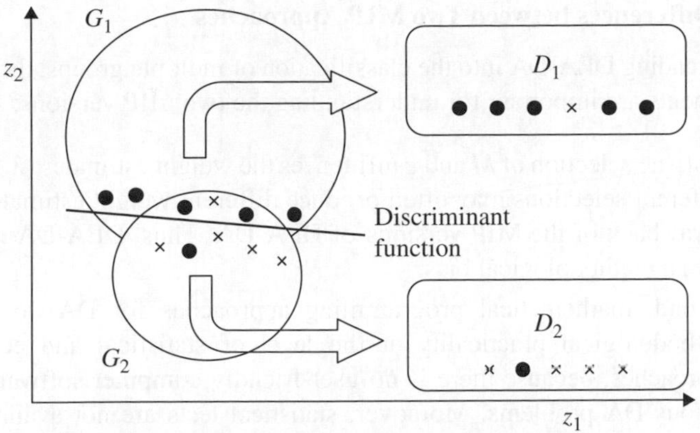

FIGURE 11.4 Final classification at Stage 2

To adjust the problem of double standards, this chapter returns to the proposed approach in which the first stage estimates a discriminant function $\left(\sum_{i=1}^{m} \lambda_i^{1*} z_i\right)$ and the second stage estimates a discriminant function $\left(\sum_{i=1}^{m} \lambda_i^{2*} z_i\right)$. The AZ score for the j-th DMU ($j = 1,\ldots,n$) is defined as follows:

$$\begin{aligned} AZ_j &= \frac{n}{n+\#(\text{overlap})}\sum_{i=1}^{m}\lambda_i^{1*} z_{ij} + \frac{\#(\text{overlap})}{n+\#(\text{overlap})}\sum_{i=1}^{m}\lambda_i^{2*} z_{ij} \\ &= \sum_{i=1}^{m}\left(\frac{n}{n+\#(\text{overlap})}\lambda_i^{1*} + \frac{\#(\text{overlap})}{n+\#(\text{overlap})}\lambda_i^{2*}\right) z_{ij}. \end{aligned} \quad (11.5)$$

This chapter refers to Equation (11.5) as "Altman's Z score," or simply "AZ score." Exactly speaking, the equation is slightly different from the equation for the original AZ score proposed by Altman (1968). However, this chapter uses the term to honor his research work.

The proposed AZ score consists of weight estimates that are adjusted by the number of observations used in the first and second stages of DEA-DA. The adjustment is due to the computational process of DEA-DA in which the first stage (for overlap identification) estimates weights of a discriminant function by using all the observations. The total sample size is "n" for the first stage. Meanwhile, the second stage utilizes only observations in the overlap. The sample size used in the second stage is #(overlap). Thus, this chapter combines two groups of weight estimates measured by DEA-DA into a single set of weight estimates in order to compute the AZ score. Using the AZ score, DEA-DA can determine the financial performance of each DMU. See Chapter 25 on the selection of common multipliers.

11.2.3 Differences between Two MIP Approaches

Before extending DEA-DA into the classification of multiple groups, the following five comments are important for understanding the two MIP versions:

(a) First, the selection of M and ε influences the weight estimates (λ_i* and cf*). Different selections may often produce different weight estimates. That is a drawback of the MIP versions of DEA-DA. Thus, DEA-DA may suffer from a methodological bias.

(b) Second, mathematical programming approaches for DA do not attain methodological practicality at the level of statistical and econometric approaches, because there is no user-friendly computer software to solve various DA problems. Moreover, statistical tests are not available to us, because asymptotic theory is not yet developed to the level of these conventional (statistical and econometrics) approaches.

(c) Third, the two-stage MIP approach is effective only when many observations belong to an overlap. If the number of observations in the overlap is small, the performance of the standard MIP approach is similar to that of the two-stage MIP approach in terms of classification performance.

(d) Fourth, the standard MIP approach may be practical in persuading corporate leaders and policy makers, because it produces a single separation function. A problem of the standard MIP approach is that it is more time-consuming than the two-stage MIP approach. The rationale on why the latter approach is more efficient than the former in terms of their computational times is because of the analytical structure of the two-stage approach. That is, the number of binary variables at the first stage is much lower than that of the second stage. The second stage is similar to the standard approach in their analytical structures. A difference between the two is that the second stage has binary variables used to count the number of incorrectly classified observations in $D_1 \cup D_2$. The number is less than that of observations $G_1 \cup G_2$ in the standard approach under $D_1 \cup D_2 \subset G_1 \cup G_2$. Thus, the two-stage approach is computationally more efficient than the standard approach. The former approach can handle much larger data sets than the latter approach.

(e) Finally, although this chapter does not discuss it in detail, all DA research work must examine whether the sign of each weight is consistent with our prior information, such as previous experience and theoretical requirement. For example, when non-default (G_1) and default (G_2) firms are separated by several financial factors (e.g., profit and return on equity; ROE), the sign of profit is expected to be positive. However, it may be negative due to some structural difficulty in a data set (e.g., multi-collinearity). In that case, it is necessary for us to incorporate an additional side constraint(s), representing prior information, into the proposed DA formulations. See Sueyoshi and Kirihara (1998) for a detailed discussion on how to incorporate such prior information into DEA-DA.

11.2.4 Differences between DEA and DEA-DA

Table 11.1 compares DEA and DEA-DA to summarize their differences. It is clearly identified that both DEA and DEA-DA originate from GP, as discussed in Chapter 3. In their conventional uses, however, DEA is used for performance assessment and DEA-DA is used for group classification.

11.3 CLASSIFYING MULTIPLE GROUPS

A shortcoming of the two MIP approaches (standard and two-stage) is that they cannot handle the classification of more than two groups. To extend them to deal with more than two groups, this chapter considers the classification of three groups by using the standard MIP approach. Then, the formulation is further restructured into the classification of q groups ($\varsigma = 1,\ldots,q$).

The classification of three groups can be formulated by slightly modifying the model in Equation (11.2) as follows:

$$
\begin{aligned}
\text{Minimize} \quad & \sum_{j \in G_1} \mu_j + \sum_{j \in G_2} \mu_j + \sum_{j \in G_3} \mu_j \\
s.t. \quad & \sum_{i=1}^{m} \left(\lambda_i^+ - \lambda_i^-\right) z_{ij} - cf_1 + M\mu_j \ge 0 && \left(j \in G_1\right), \\
& \sum_{i-1}^{m} \left(\lambda_i^+ - \lambda_i^-\right) z_{ij} - cf_1 - M\mu_j \le -\varepsilon && \left(j \in G_2\right), \\
& \sum_{i=1}^{m} \left(\lambda_i^+ - \lambda_i^-\right) z_{ij} - cf_2 + M\mu_j \ge 0 && \left(j \in G_2\right), \\
& \sum_{i=1}^{m} \left(\lambda_i^+ - \lambda_i^-\right) z_{ij} - cf_2 - M\mu_j \le -\varepsilon && \left(j \in G_3\right), \qquad (11.6) \\
& \sum_{i=1}^{m} \left(\lambda_i^+ + \lambda_i^-\right) = 1, \\
& \upsilon_1^+ \ge \lambda_1^+ \ge \varepsilon\upsilon_1^+,\ \upsilon_1^- \ge \lambda_1^- \ge \varepsilon\upsilon_1^- && \left(i = 1,\ldots,m\right), \\
& \upsilon_1^+ + \upsilon_1^- \le 1 && \left(i = 1,\ldots,m\right), \\
& \sum_{i=1}^{m} \left(\upsilon_i^+ + \upsilon_i^-\right) = m, \\
& cf_1, cf_2 : \text{URS},\ \upsilon_i^+,\ \upsilon_i^-,\ \mu_j : \text{binary and all other variables} \ge 0.
\end{aligned}
$$

The modification of the above formulation is that it incorporates two discriminant scores (cf_1 and cf_2) in the model in Equation (11.6). The first discriminant score (cf_1) for a cut-off is used to separate between G_1 and G_2. The second one (cf_2) is for the classification between G_2 and G_3.

TABLE 11.1 Differences between DEA and DEA-DA

Analytical features	DEA	DEA-DA
Input and output classification and group classification	(a) DEA needs to classify all variables into either outputs or inputs. (b) DEA does not need to prescribe group classification.	(a) DEA-DA does not need any classification on factor variables. (b) DEA-DA needs to prescribe group classification before the computation.
Non-parametric	The j-th weight estimate (multiplier) obtained by DEA indicates the importance of the j-th observation or DMU in determining an efficiency frontier of the specific observation.	The i-th weight estimate obtained by DEA-DA indicates the importance of the i-th factor variable for comprising a discriminant function to separate between the two groups.
Firm-specific weights vs group-wide common weights	The result of DEA is DMU-specific so that the efficiency measure depends upon each DMU.	The result of DEA-DA is group-widely applicable because it provides "common weights" for all observations.
Distribution-free	DEA does not have any assumption on how all observations locate below an efficiency frontier.	DEA-DA does not need any assumption on a group distribution above and below a discriminant function.
Rank-sum tests	No conventional statistical tests are available. Rank-sum tests are available for DEA.	No conventional statistical tests are available. Rank-sum tests are available for DEA-DA.
Data imbalance	The data imbalance in DEA implies that a large input or output dominates the others. DEA can solve the problem of data imbalance by adjusting all inputs and outputs by these averages.	The data imbalance in DEA-DA implies different sample sizes between two groups. DEA-DA can deal with the problem of date imbalance by putting more weight on one of two groups.
Multiple solutions and multiple criteria	Multiple solutions may occur on DEA results. Efficiency of a DMU is determined by comparing its performance with an efficiency frontier. Thus, DEA consists of a single criterion.	An occurrence of multiple solutions in MIP approaches for DEA-DA is not important. The standard approach has a single criterion for group classification, but the two-stage approach has double standards for group classification.
Optimality	Since DEA is solved by linear programming, it can guarantee optimality. An exception is the Russell measure.	Since DEA-DA is solved by MIP, it cannot guarantee optimality.

All observations $Z_j = \left(z_{1j},\cdots,z_{mj}\right)^{Tr}$ for $j = 1,\ldots,n$, are classified as follows:

(a) If $\sum_{i=1}^{m} \lambda_i^* z_{ij} \geq cf_1^*$, then the j-th observation belongs to G_1,

(b) If $cf_1^* - \varepsilon \geq \sum_{i=1}^{m} \lambda_i^* z_{ij} \geq cf_2^*$, then the observation belongs to G_2 and

(c) If $cf_2^* - \varepsilon \geq \sum_{i=1}^{m} \lambda_i^* z_{ij}$, then the observation belongs to G_3.

A newly sampled k-th observation, $Z_k = \left(z_{1k},\ldots,z_{mk}\right)^{Tr}$, is classified by the above rule by changing the subscript (j) by the one (k).

More Than Three Groups: The model in Equation (11.6) is further reformulated for classification of more than three groups. The following formulation provides the classification of multiple groups ($\varsigma = 1,\ldots,q$):

$$
\begin{aligned}
&\text{Minimize} \quad \sum_{\varsigma=1}^{q}\sum_{j\in G_\varsigma} \mu_j \\
&s.t. \quad \sum_{i=1}^{m}\left(\lambda_i^+ - \lambda_i^-\right)z_{ij} - cf_\varsigma + M\mu_j \geq 0, \quad \left(j \in G_\varsigma,\ \varsigma = 1,\ldots,q-1\right), \\
&\sum_{i=1}^{m}\left(\lambda_i^+ - \lambda_i^-\right)z_{ij} - cf_\varsigma - M\mu_j \leq -\varepsilon, \quad \left(j \in G_{\varsigma-1},\ \varsigma = 1,\ldots,q-1\right), \\
&\sum_{i=1}^{m}\left(\lambda_i^+ + \lambda_i^-\right) = 1, \qquad (11.7) \\
&\upsilon_i^+ \geq \lambda_i^+ \geq \varepsilon\upsilon_i^+,\ \upsilon_i^- \geq \lambda_i^- \geq \varepsilon\upsilon_i \quad \left(i = 1,\ldots,m\right), \\
&\upsilon_i^+ + \upsilon_i^- \leq 1 \quad \left(i = 1,\ldots,m\right), \\
&\sum_{i=1}^{m}\left(\upsilon_i^+ + \upsilon_i^-\right) = m, \\
&cf_\varsigma : \text{URS}, \upsilon_i^+,\ \upsilon_i^-,\ \mu_j; \text{binary and all other variables} \geq 0.
\end{aligned}
$$

Here, G_ς is a set of observations in the ς-th group (for $\varsigma = 1,\ldots,q-1$), In the above case, multiple groups are separated by cf_ς^* ($\varsigma = 1,\ldots,q-1$) and $\lambda_i^* = \lambda_i^{+*} - \lambda_i^{-*}$ ($i = 1,\ldots,m$), all of which are obtained from the optimality of the model in Equation (11.7).

All observations $Z_j = \left(z_{1j},\ldots,z_{mj}\right)^{Tr}$ for $j = 1,\ldots,$n are classified as follows:

(a) If $\sum_{i=1}^{m} \lambda_i^* z_{ij} \geq cf_1^*$, then the observation belongs to G_1,

(b) If $cf_{\varsigma-1}^* - \varepsilon \geq \sum_{i=1}^{m} \lambda_i^* z_{ij} \geq cf_\varsigma^*$, then the observation belongs to G_ς (ς = 2,…, q–1) and

(c) If $cf_{q-1}^* - \varepsilon \geq \sum_{i=1}^{m} \lambda_i^* z_{ij}$, then the observation belongs to G_q.

A newly sampled observation, or $Z_k = (z_{1k},\ldots,z_{mk})^{Tr}$, is classified by the above rule by changing the subscript (*j*) by the one (*k*).

At the end of this subsection, the following three comments are necessary for using Model (11.7):

(a) First, a unique feature of Model (11.7) is that it produces *q*–1 different discriminant scores (cf_1^* to cf_{q-1}^*) in order to classify *q* groups. Furthermore, it maintains the same weight scores (λ_i^*, $i=1,\ldots,m$). Figure 11.5 visually describes the classification of multiple groups.

(b) Second, when solving Model (11.7), $cf_1^* > cf_2^* > \ldots > cf_{q-1}^*$ is required on optimality. If such a requirement is not satisfied on the optimality, then additional side constraints: $cf_1 \geq cf_2 + \varepsilon, cf_2 \geq cf_3 + \varepsilon, \ldots,$ and $cf_{q-2} \geq cf_{q-1} + \varepsilon$ must be incorporated into Model (11.7). The additional capability is due to methodological flexibility of the MIP-based approach.

(c) Finally, the proposed MIP approach can solve only a specific type of multiple group classification, where the "specific" implies that a whole data set can be arranged in a particular ordering, as depicted in Figure 11.5. In other words, the particular ordering data implies the one that is classified into multiple groups by several separation functions whose slopes (i.e., weights) are same but having different intercepts (i.e., classification scores). If a date set does not have such a special ordering structure, Model (11.7) may produce an "infeasible" solution or a low classification rate. Thus, Model (11.7) has a limited capability for group classification.

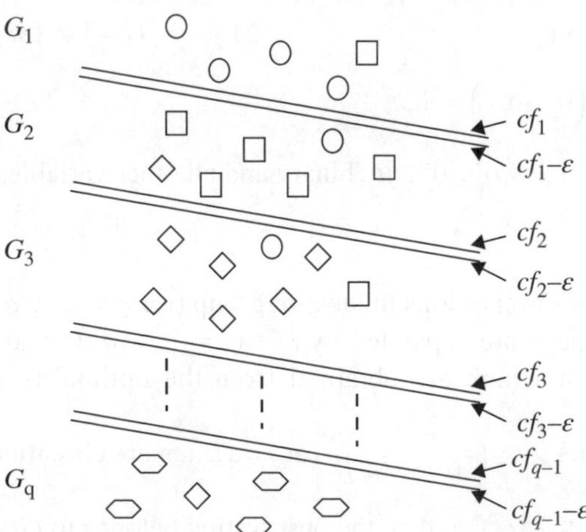

FIGURE 11.5 Classification of multiple groups (a) Source: Sueyoshi (2006). (b) This type of multiple group classification by DEA-DA has a limited classification capability because it produces same cutting off hyperplanes except an intercept. There is a possibility that DEA-DA produces an infeasible solution when such hyperplanes cannot separate multiple groups

11.4 ILLUSTRATIVE EXAMPLES

11.4.1 First Example

This chapter uses two illustrative data sets to document the discriminant capabilities of the proposed DEA-DA approaches. The first data set, listed in Table 11.2, contains 31 observations which are separated into two groups (G_1: the first 20 observations and G_2: the remaining second observations). Each observation is characterized by four factors.

The bottom of Table 11.2 summarizes two hit rates (apparent and leave one out: LOO) for six different DA methods. Here, "apparent" indicates the number of correctly classified observations in a training sample. In measuring the apparent of the first data set, all 31 observations become the training sample. In contrast, the LOO measurement drops each observation from the training sample and then applies a DA method to the remaining observations. LOO examines whether the omitted observation is correctly classified. Thus, the LOO measurement needs to be repeated until all the observations are examined.

Figure 11.6 depicts such a computation process to measure apparent and LOO. In the figure, the apparent measurement uses the whole data set as both a training sample set and a validation sample set. Meanwhile, the LOO measurement uses a single observation omitted from the whole data set as a test sample. The remaining data set is used as a training sample set.

As summarized in Table 11.2, the two-stage MIP approach is the best performer in the apparent and LOO measurements (90.32 and 83.33%, respectively). Two econometric (logit and probit) approaches and Fisher's linear DA perform insufficiently because the data set does not satisfy the underlying assumptions that are necessary for applying those DA methods. In contrast, a shortcoming of the MIP approaches is that they need computational times which are much longer than those of the other DA methods.

11.4.2 Second Example

The second data set, originally obtained from Sueyoshi (2001b, pp. 337–338), is related to Japanese banks that contains 100 observations. All the observations in the data set are listed by their corporate ranks. Therefore, G_1 contains 50 banks ranked from the first to the 50th. Meanwhile, G_2 contains the remaining 50 banks ranked from the 51st to the 100th. This data set has been extensively used by many studies on DA.

Data Accessibility: The data set is documented in Sueyoshi (2006).

In this chapter, the whole data set is used to apply four different grouping cases (two, three, four and five groups). To make each group, all 100 observations are classified by these ranks. For example, in the case of four groups, G_1 consists of observations whose ranks are from the first to the 25th. G_2 is a set of observations

TABLE 11.2 Illustrative data set and hit rates

Observation (j)	Factor 1	Factor 2	Factor 3	Factor 4
1	0.03	3.00	2.40	9.20
2	2.13	2.00	2.00	0.60
3	8.48	2.50	2.20	1.50
4	0.76	1.10	0.90	3.20
5	0.05	2.10	2.00	5.40
6	1.65	1.30	0.90	5.60
7	1.09	0.80	0.30	11.00
8	0.08	3.20	2.90	9.40
9	0.10	1.20	0.80	8.50
10	0.91	0.90	0.30	15.80
11	1.25	0.80	0.20	23.60
12	1.23	0.80	0.50	7.50
13	0.06	1.60	0.80	51.80
14	2.10	0.90	0.10	77.80
15	0.29	1.10	0.70	9.00
16	0.93	1.00	0.50	13.40
17	0.16	1.40	1.10	5.70
18	0.10	1.80	1.20	6.70
19	0.22	1.30	0.90	10.70
20	0.63	1.40	0.90	6.70
21	5.40	1.04	0.43	4.33
22	1.69	2.27	0.98	5.18
23	0.87	2.20	1.11	7.67
24	0.74	0.58	0.45	6.08
25	2.71	1.46	1.23	3.25
26	2.60	0.43	0.35	8.18
27	0.00	0.09	0.04	13.46
28	0.38	1.47	1.30	4.81
29	0.20	1.61	0.16	450.56
30	51.89	0.37	0.14	7.63
31	1.35	0.92	0.42	16.60

Method	Hit rate (%)	
	Apparent	LOO
Standard MIP approach	80.65	77.42
Two-stage MIP approach	90.32	83.33
Logit	70.97	61.29
Probit	70.97	61.29
Fisher's linear DA	67.70	51.60
Smith's quadratic DA	80.60	67.70

(a) Source: Sueyoshi (2006).

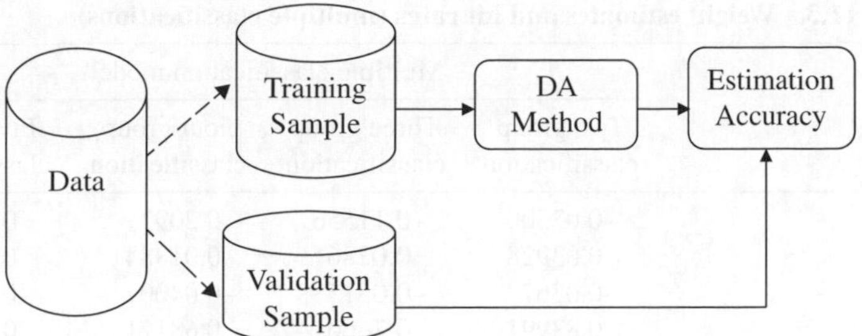

FIGURE 11.6 Structure of performance measurement (a) Source: Sueyoshi (2006)

whose ranks are from the 26th to the 50th. Similarly, G_3 and G_4 are selected from the 51st to the 75th and the 76th to the 100th, respectively. Table 11.3 summarizes the weight estimates of Model (11.7) and two hit rates, all of which arc obtained from Model (11.7) applied to the four different grouping cases.

11.5 FRONTIER ANALYSIS

It is possible for us to apply DEA-DA for frontier analysis, as found in the L1 regression analysis of Chapter 3. See Chapter 25, as well. Since no research has explored that research issue, this chapter describes the analytical link between DEA-DA and frontier analysis.

In applying DEA-DA to frontier analysis, we can easily imagine two types of frontiers. One type is that all observations belong to G_1, but G_2 is empty. An estimated frontier function locates below all observations. The first case is formulated as follows:

$$
\begin{aligned}
\text{Minimize}\quad & \sum_{j\in G_1}\mu_j \\
\text{s.t.}\quad & \sum_{i=1}^{m}\left(\lambda_i^{+}-\lambda_i^{-}\right)z_{ij}-cf+M\mu_j\ge 0, \qquad \left(j\in G_1\right),\\
& \sum_{i=1}^{m}\left(\lambda_i^{+}+\lambda_i^{-}\right)=1,\\
& \upsilon_i^{+}\ge\lambda_i^{+}\ge\varepsilon\upsilon_i^{+},\ \upsilon_i^{-}\ge\lambda_i^{-}\ge\varepsilon\upsilon_i^{-} \qquad (i,\ldots,m),\\
& \upsilon_i^{+}+\upsilon_i^{-}\le 1 \qquad (i,\ldots,m),\\
& \sum_{i=1}^{m}\left(\upsilon_i^{+}+\upsilon_i^{-}\right)=m,\\
& cf:\text{URS},\ \upsilon_i^{+},\ \upsilon_i^{-},\ \mu_{j:}:\text{binary and all other variables}\ge 0.
\end{aligned}
\qquad (11.8)
$$

TABLE 11.3 Weight estimates and hit rates (multiple classifications)

Weight estimates		Multiple classification model			
		Two group classification	Three group classification	Four group classification	Five group classification
λ_1^*		−0.03300	−0.11556	0.20921	0.02930
λ_2^*		0.03928	0.02861	0.01884	0.03663
λ_3^*		−0.03677	−0.03153	−0.04001	−0.03241
λ_4^*		0.83991	0.76601	0.68171	0.85212
λ_5^*		−0.04369	−0.05120	−0.02948	−0.04316
λ_6^*		0.00260	0.00616	0.00574	0.00488
λ_7^*		−0.00475	−0.00094	−0.01502	−0.00149
cf_1^*		−1.76496	−1.25426	−1.80618	−0.95670
cf_2^*		—	−1.44013	−2.06284	−1.17478
cf_3^*		—	—	−2.21214	−1.33047
cf_4^*		—	—	—	−1.48890
Hit rate (%)	Apparent	96	95	97	94
	LOO	90	90	87	87

(a) Source: Sueyoshi (2006).

The other case, where the frontier function locates above all observations, is formulated as follows:

$$\begin{aligned}
&\text{Minimize} \quad \sum_{j\in G_2}\mu_j \\
&s.t. \quad \sum_{i=1}^{m}\left(\lambda_i^+ - \lambda_i^-\right)z_{ij} - cf - M\mu_j \le 0 \qquad \left(j\in G_2\right), \\
&\sum_{i=1}^{m}\left(\lambda_i^+ + \lambda_i^-\right) = 1, \\
&\upsilon_i^+ \ge \lambda_i^+ \ge \varepsilon\upsilon_i^+,\ \upsilon_i^- \ge \lambda_i^- \ge \varepsilon\upsilon_i^- \qquad \left(i=1,\ldots,m\right), \\
&\upsilon_i^+ + \lambda_i^+ \le 1 \qquad \left(i=1,\ldots,m\right), \\
&\sum_{i=1}^{m}\left(\upsilon_i^+ + \upsilon_i^-\right) = m, \\
&cf : \text{URS},\ \upsilon_i^+,\ \upsilon_i^-,\ \mu_j : \text{binary and all other variables} \ge 0.
\end{aligned} \tag{11.9}$$

Models (11.8) and (11.9) drop the small number (ε) from the right-hand side because frontier analysis accepts that a few observations exist on the frontier. Of course, the small number (ε) in the two models can be used to maintain positive weights.

Applicability: Chapter 25 discusses how to apply the proposed DEA-DA to the performance assessment of Japanese energy utility firms. Hence, this chapter does not document the applicability.

11.6 SUMMARY

This chapter discussed two MIP approaches for DEA-DA and compared them with other DA models (e.g., Fisher, Smith, logit and probit), all of which were extensively used in conventional DA on statistics and econometrics. In comparison with these DA methods, this chapter confirmed that the MIP approaches for DEA-DA performed at least as well as the other well known DA methods. Moreover, the standard MIP approach was extendable into the classification of multiple (more than two) groups in the data set, as depicted in Figure 11.5, along with a comment that such an extension has a limited classification capability.

It is true that there is no perfect methodology for DA. Any DA method, including the proposed two types of MIP approaches, has methodological strengths and shortcomings. The methodological drawbacks of the proposed MIP approaches are summarized by the following three concerns. First, the MIP approaches need asymptotic theory, based upon which we can derive statistical tests on DA. The statistical and econometric approaches can provide us with various statistical tests incorporated in prevalent computer software tools. The software, including such traditional approaches, is usually inexpensive. In many cases, users can freely access such DA methods. The availability of user-friendly software including many statistical tests really enhances the practicality of the proposed MIP approaches. Second, a special computer algorithm must be developed for the proposed MIP approaches. The computational time of the proposed MIP approaches should be more drastically reduced at the level of statistical and econometric DA approaches. The methodological shortcoming is because the MIP formulations usually need a long computational time and some effort. In particular, the single-stage approach is much more time-consuming than the two-stage approach. Finally, the selection of M and ε influences the sign and magnitude of weight estimates (λ_i^* and cf^*). Different selections of such pre-specified numbers may produce different weight estimates. That is another shortcoming of the MIP versions of DEA-DA.

At the end of this concluding section, it is important to note that this chapter pays attention to only methodological comparison based upon the two hit rates (i.e., apparent and LOO) in the illustrative examples. However, it is widely known that such a comparison is one of many practical and empirical criteria. Hence, our methodological comparison may be limited from a perspective regarding which DA method is more appropriate in dealing with group classification. It is an important research task to improve currently existing methods and/or develop a new methodology. Moreover, in using any method for applications, we must understand that there is a methodological bias in most empirical studies. Said simply, different methods produce different results. Thus, it is necessary for us to understand the existence of such a methodological bias in deriving any scientific or empirical conclusion.

12

LITERATURE STUDY FOR SECTION I

12.1 INTRODUCTION

This chapter consists of three parts related to Section I (i.e., Chapters 1–11). The first part summarizes available computer codes that can be applied for the conventional use of DEA, which has been discussed in chapters of Section I. The second part discusses the methodological link between DEA and its applications for environmental assessment from a pedagogical perspective. The last part lists the previous research efforts cited in the chapters of Section I.

12.2 COMPUTER CODES

The available computer codes are classified into general and DEA-specific codes. Most of models discussed in the chapters of Section I can be solved by the specific codes. Table 12.1 summarizes the two types of computer codes. The general codes, including GAMS, MATLAB and LINDO, can be applied to all the models discussed in the chapters of Section I.

As a pedagogical tool, it is useful to use DEA-specific codes for students at undergraduate and master levels. A problem associated with the DEA-specific codes is that we have a difficulty in directly applying them for environmental assessment, which will be discussed in the chapters of Section II. Moreover, they

Environmental Assessment on Energy and Sustainability by Data Envelopment Analysis, First Edition. Toshiyuki Sueyoshi and Mika Goto.

TABLE 12.1 Available Computer Codes for DEA

Group	Software	Source	Feature	URL
General modeling software	GAMS	GAMS Development Corporation	The general algebraic modeling system (GAMS) is specifically designed for modeling linear, non-linear and mixed integer optimization problems. The system is especially useful with large, complex problems. GAMS is available for use on personal computers, workstations, mainframes and supercomputers.	https://www.gams.com/
	MATLAB	MathWorks	MATLAB is a high-level language and interactive environment used by millions of engineers and scientists worldwide. It lets you explore and visualize ideas and collaborate across disciplines including signal and image processing, communications, control systems, and computational finance.	http://www.mathworks.com/index.html?s_tid=gn_logo
	LINDO	LINDO Systems Inc	LINDO™ linear, nonlinear, integer, stochastic and global programming solvers have been used by thousands of companies worldwide to maximize profit and minimize cost on decisions involving production planning, transportation, finance, portfolio allocation, capital budgeting, blending, scheduling, inventory, resource allocation and more.	http://www.lindo.com/

DEA specialized software	DEAP	Tim Coelli, CEPA Centre for Efficiency and Productivity Analysis, School of Economics, The University of Queensland		http://www.uq.edu.au/economics/cepa/deap.php
	EMS	University of Dortmund	Efficiency measurement system	http://www.holger-scheel.de/ems/
	PIM-DEA	DEAzone.com		http://deazone.com/en/software
	Frontier Analyst	Banxia Software Ltd		http://www.banxia.com/
	DEAFrontier	Joe Zhu, Foisie School of Business, Worcester Polytechnic Institute		http://www.deafrontier.net/deasoftware.html
	OSDEA GUI	Open Source DEA		http://www.opensourcedea.org/index.php?title=Open_Source_DEA
	Max DEA	CHENG Gang and Beijing Realworld Research and Consultation Company Ltd.		http://www.maxdea.cn/MaxDEA.html
	FEAR	Paul W. Wilson, Department of Economics and School of Computing, Clemson University	A software package for frontier efficiency analysis with R	http://www.clemson.edu/economics/faculty/wilson/Software/FEAR/fear.html
	Benchmarking	Peter Bogetoft, Department of Economics, Copenhagen Business School and Lars Otto	This is "GNU S," a freely available language and environment for statistical computing and graphics which provides a wide variety of statistical and graphical techniques: linear and non-linear modelling, statistical tests, time series analysis, classification, clustering and so on. Package "benchmarking" is for DEA and SFA.	https://cran.r-project.org/web/packages/Benchmarking/

do not have a computational capability to handle a large data set. See Chapter 10 on network computing for dealing with a large data set on DEA. It is important for an instructor to discuss that DEA formulations cannot directly handle zero and negative values. See Chapters 26 and 27 for a description on how to handle such a data set with zero and negative values.

12.3 PEDAGOGICAL LINKAGE FROM CONVENTIONAL USE TO ENVIRONMENTAL ASSESSMENT

It may be possible for us to use a conventional DEA model [e.g., an input-oriented Radial Measure, or RM(v)] under variable RTS, for environmental assessment as a pedagogical tool for undergraduate and master students, although not for research purposes at the level of Ph.D. students and faculty members. To explain how to use DEA as a pedagogical tool, let us consider the original description in Chapter 4 that specifies the definition of operational efficiency by

$$\text{Operational Efficiency}\,(OE) = (\Sigma\text{weighted outputs})/(\Sigma\text{weighted inputs}). \tag{12.1}$$

The symbol Σ indicates the sum of components. The weights are considered as "multipliers" or "dual variables" discussed in Chapter 4. See also Model (2.2) in Chapter 2, which mathematically describes the efficiency specification.

An important difference between a conventional use of DEA and its applications for environmental assessment is that outputs are separated into desirable and undesirable output categories in the latter applications. Thus, a methodological difficulty is often associated with how to unify the two types of outputs for DEA environmental assessment. A straightforward reply to the inquiry is that it is possible for us to reorganize the above definition as follows:

$$\begin{aligned} OE = &\left(\Sigma\text{weighted desirable outputs}\right)/ \\ &\left[\left(\Sigma\text{weighted inputs}\right)+\left(\Sigma\text{weighted undesirable outputs}\right)\right]. \end{aligned} \tag{12.2}$$

The rationale on the above reformulation is very simple. That is, the less undesirable outputs are the better so that undesirable outputs are treated like inputs in the above equation. Many previous research works summarized in Chapter 28 have reformulated their radial (e.g., RMs) and non-radial models (e.g, SBM) by incorporating the simple rationale in their formulations and apply them for environmental assessment.

Acknowledging the pedagogical importance at the level of undergraduate and master education, however, the simple justification does not work for research at the professional academic level because environmental assessment needs to consider output classification into desirable and undesirable output categories.

The latter needs to be treated differently from inputs. Thus, previous research efforts which have incorporated the simple justification have often misguided business strategy and/or environmental policy by providing inappropriate empirical results by DEA. See Chapter 15 that explains how DEA environmental assessment is different from DEA. A methodological difficulty in DEA environmental assessment will be more clearly discussed in the chapters of Section II.

Before proceeding to Section II on DEA environmental assessment, it is necessary to describe the following four research concerns. First, the gist of environmental assessment is that DEA needs to incorporate eco-technology innovation into the proposed formulations. It is believed that green technology innovation and its related managerial challenges are essential in reducing a level of various pollutions. It is very difficult for conventional DEA frameworks to incorporate the eco-technology progress on environmental assessment into the proposed simple modification, or Model (12.2), although such a research direction can be found in many previous DEA studies on environmental assessment. Second, the directional vector of undesirable outputs is different from that of desirable outputs. Both have opposite directions for their betterments in environmental assessment. Therefore, the output unification should be discussed within the conceptual and computational frameworks. Third, the input vector decreases its components in a conventional use of DEA. An opposite case (i.e., an increase of an input vector) is excluded from a conventional framework of DEA as discussed in the chapters of Section I. However, DEA environmental assessment must consider the input increase as long as a desirable output vector can increase and an undesirable output vector can decrease for sustainability enhancement. Finally, the conventional use of DEA implicitly looks for total cost minimization. See Chapter 4. However, DEA environmental assessment needs to consider marginal or average cost in understanding modern business strategy and industrial policy. Furthermore, DEA environmental assessment needs to mention that consumers do not purchase products from dirty-imaged firms. Rather, they buy products from clean-imaged firms even if their prices are higher than those of dirty-imaged firms. Moreover, the occurrence of "opportunity cost" is much larger than any other cost components in considering environmental protection and assessment. A typical example can be found in the disaster of the Fukushima Daiichi nuclear power plant occurred on 11 March 2011. See Chapter 26 about the disaster of Fukushima Daiichi. The chapter examines whether the new fuel mix policy, proposed by the Japanese government, is reasonable or not.

It is easily imagined that many other problems are associated with the simple modification, or Model (12.2), all of which will be overcome by DEA environmental assessment. Thus, the simple modification should be restricted within an educational scope of undergraduate and master levels, but not for advanced research concerning environmental assessment. Of course, we clearly acknowledge the contribution of previous research efforts which utilizes Model (12.2) for environmental assessment. We also know that there is an exceptional case such as performance assessment on renewable energies where the output classification is not so important. For example, solar generation does not produce CO_2 emission and the other type of greenhouse gas (GHG) emissions. In the case, it is possible for us to use the conventional framework of DEA for environmental assessment. See, for instance, Sueyoshi and Goto (2014) that have internationally compared the performance of the United States and Germany in terms of their photovoltaic power stations.

REFERENCES FOR SECTION I

The previous works on DEA and other research efforts used in Section I are summarized by the following list of references:

Adad, P.L. and Banks, W.J. (1993) New LP-based heuristics for the classification problem. *European Journal of Operational Research*, **67**, 88–100.

Ahn, T.S., Charnes, A. and Cooper W.W. (1988) Some statistical and DEA evaluations of relative efficiencies of public and private institutions of higher learning *Socio-Economic Planning Sciences*, **22**, 259–269.

Appa, G. and Smith, C. (1973) On L_1 and Chebychev estimation. Mathematical Programming, **5**, 73–87.

Aida, K., Cooper, W.W., Pastor, J.T. and Sueyoshi, T. (1998) Evaluating water supply services in Japan with RAM: A range-adjusted measure of inefficiency. *Omega*, **26**, 207–232.

Ali, I. and Seiford, L. M. (1990) Translation invariance in data envelopment analysis. *Operations Research Letters*, **9**, 403–405.

Altman, E.I. (1968) Financial ratios, discriminant analysis and the prediction of corporate bankruptcy. *Journal of Finance*, **23**, 589–609.

Altman, E.I. (1984) A further empirical investigation of the bankruptcy cost question. *Journal of Finance*, **39**, 1067–1089.

Ali, I. and Seiford, L. M. (1993) Computational accuracy and infinitesimals in data envelopment analysis. *Information Systems and Operational Research*, **31**, 290–297.

Armstrong, R.D. and Kung, M.T. (1980) An Algorithm for a least absolute value regression problem with bounds on the parameters. *Applied Mathematics and Computers*, **7**, 267–279.

Arthanari, T.S. and Dodge, Y. (1981) *Mathematical Programming in Statistics*. John Wiley & Sons, Inc., New York.

Avis, D. and Fukuda, K. (1996) Reverse search for enumeration. *Discrete Applied Mathematics*, **65**, 21–46.

Bajgier, S.M. and Hill, A.V. (1982) A comparison of statistical and linear programming approaches to the discriminant problem. *Decision Sciences*, **13**, 604–618.

Banker, R.D. (1984) Estimating most productive scale size using data envelopment analysis, *European Journal of Operational Research*, **17**, 35–44.

Banker, R.D., Charnes, A. and Cooper, W.W. (1984) Some models for estimating technical and scale inefficiencies in data envelopment analysis, *Management Science*, **30**, 1078–1092.

Banker, R.D. and Thrall, R.M. (1992) Estimating returns to scale using data envelopment analysis, *European Journal of Operational Research*, **62**, 74–84.

Barr, R.S. and Durchholz, M.L. (1997) Parallel and hierarchical decomposition approaches for solving large-scale data envelopment analysis models. *Annals of Operational Research*, **73**, 339–372.

Bardhan, I., Bowlin, W.F., Cooper, W.W. and Sueyoshi, T. (1996a) Models and measures for efficiency dominance in DEA. Part I: Additive models and MED measures, *Journal of the Operations Research Society of Japan*, **39**, 322–332.

Bardhan, I., Bowlin, W.F., Cooper, W.W. and Sueyoshi, T. (1996b) Models and measures for efficiency dominance in DEA. Part II: Free disposal Hulls and Russell measures, *Journal of the Operations Research Society of Japan*, **39**, 333–344.

Barrodele, I. and Robert, F.D.K. (1978) An efficient algorithm for discrite L1 linear approximation with linear constraints. *SIAM Journal of Numerical Analysis*, **15**, 603–611.

Bassett, G. and Koenker, R. (1978) Asymptotic theory of least absolute error regression. *Journal of American Statistical Association*, **73**, 618–622.

Bassett, G. and Koenker, R. (1982) An empirical quartile function for linear models with IID errors. *Journal of American Statistical Association*, **77**, 407–415.

Bjurek, H. (1996) The Malmquist total factor productivity index. *Scandinavian Journal of Economics*, **98**, 303–313.

Blackorby, C. and Russell, R.R. (1999) Aggregation of efficiency indices. *Journal of Productivity Analysis*, **12**, 5–20.

Bres, E., Burns, D., Charnes, A. and Cooper, W.W. (1980) A goal programming model for planning officer accessions, *Management Science*, **26**, 73–783.

Brockett, P.L., Charnes, A., Cooper, W.W., Huang, Z. and Sun, D.B. (1997) Data transportation in DEA Cone ratio envelopment approaches for monitoring bank performance. *European Journal of Operational Research*, **98**, 250–265.

Brockett, P.L., Charnes, A., Cooper, W.W., Kwon, K.H. and Ruefli, T.W. (1992) Chance constrained programming approaches for empirical analyses of mutual fund investment strategies, *Decision Sciences*, **23**, 385–408.

Brockett, P.L., Cooper, W.W., Deng, H., Golden, L.L and Ruefli, T.W. (2004) Using DEA to identify and manage congestion. *Journal of Productivity Analysis*, **22**, 207–226.

Brockett, P. L., Cooper, W. W., Shin, H. C. and Wang, Y. (1998) Inefficiency and congestion in Chinese production before and after the 1978 economic reforms. *Socio-Economic Planning Science*, **32**, 1–20.

Bryne, R.F., Charnes, A., Cooper, W.W. and Kortanek, K. (1967) A chance-constrained approach to capital budgeting with portfolio type payback and liquidity constraints and horizon type posture controls, *Journal of Financial and Quantitative Analysis*, **2**, 339–364.

Cavalier, T.M., Ignizio, J.P. and Soyster, A.L. (1989). Discriminant analysis via mathematical programming: certain problems and their causes. *Computers and Operations Research*, **16**, 353–362.

Chang, K.P. and Guh, Y.Y. (1991) Linear production functions and data envelopment analysis, *European Journal of Operational Research*, **52**, 215–223.

Charnes, A. and Cooper, W.W. (1954) Linear programming, *Scientific American*, **191**, 21–23.

Charnes, A. and Cooper, W.W. (1957) Management models and industrial applications of linear programming, *Management Science*, **4**, 38–91.

Charnes, A. and Cooper, W.W. (1959) Chance constrained programming, *Management Science*, **6**, 73–79.

Charnes, A. and Cooper, W.W. (1961) *Management Models and Industrial Applications of Linear Programming*, John Wiley & Sons, Inc., New York.

Charnes, A. and Cooper, W.W. (1962) Normal deviates and chance constraints, *Journal of the American Statistical Association*, **57**, 134–148.

Charnes, A. and Cooper, W.W. (1962) On some works of Kantorovich Koopmans and others, *Management Science*, **8**, 246–263.

Charnes, A. and Cooper, W.W. (1962) Programming with linear fractional functional, *Naval Research Logistics Quarterly*, **9**, 181–186.

Charnes, A. and Cooper, W.W. (1963) Deterministic equivalents for optimizing and satisficing under chance constraints, *Operations Research*, **11**, 18–39.

Charnes, A. and Cooper, W.W. (1973) An explicit general solution in linear fractional programming, *Naval Research Logistics Quarterly*, **20**, 449–468.

Charnes, A. and Cooper, W.W. (1975) Goal programming and constrained regressions, *Omega*, **3**, 403–409.

Charnes, A. and Cooper, W.W. (1977) Goal programming and multiple objective optimizations Part I. *European Journal of Operational Research*, **1**, 39–54.

Charnes, A., Cooper, W.W. and Ferguson, R.O. (1955) Optimal estimation of executive compensation by linear programming, *Management Science*, **1**, 138–151.

Charnes, A., Cooper, W.W., Golany, B., Seiford, L. and Stutz, J. (1985) Foundations of data envelopment analysis for Pareto–Koopmans efficient empirical production functions, *Journal of Econometrics*, **30**, 91–107.

Charnes, A., Cooper, W.W. and Henderson, A. (1959) *Introduction to Linear Programming*, John Wiley & Sons, Inc., New York.

Charnes, A., Cooper, W.W. and Ijiri, Y. (1963) Breakeven budgeting and programming to goals. *Journal of Accounting Research*, **1**, 16–43.

Charnes, A., Cooper, W.W., Lewin, A.Y. and Seiford, L. (1994) *Data Envelopment Analysis: Theory, Methodology and Applications*, Kluwer Academic Publishers, New York.

Charnes, A., Cooper, W.W. and Rhodes, E. (1978) Measuring the efficiency of decision making units, *European Journal of Operational Research*, **6**, 429–444.

Charnes, A., Cooper, W.W. and Rhodes, E. (1981) Data envelopment analysis as an approach for evaluating program and managerial efficiency–with an illustrative application to the program follow through experiment in U.S. public school education, *Management Science*, **27**, 668–697.

Charnes, A., Cooper, W.W., Seiford, L. and Stutz, J. (1982) A multiplicative model for efficiency analysis, *Socio-Economic Planning Sciences*, **16**, 223–224.

Charnes, A., Cooper, W.W., Seiford, L. and Stutz, J. (1983) Invariant multiplicative efficiency and piecewise Cobb–Douglas envelopments, *Operations Research Letters*, **2**, 101–103.

Charnes, A., Cooper, W.W., Wei, Q.L. and Huang, Z.M. (1989) Cone ratio data environmental analysis and multi-objective programming. *International Journal of System Science*, **20**, 1099–1118.

Charnes, A., Cooper, W.W. and Sueyoshi, T. (1986) Least squares/ridge regressions and goal programming/constrained regression alternatives, *European Journal of Operational Research*, **27**, 146–157.

Charnes, A., Cooper, W.W. and Sueyoshi, T. (1988) A goal programming/constrained regression review of the Bell system breakup, *Management Science*, **34**, 1–26.

Charnes, A., Cooper, W.W. and Thrall, R.M. (1986) Classifying and characterizing efficiencies and inefficiencies in data envelopment analysis. *Operations Research Letters*, **5**, 105–110.

Charnes, A., Cooper, W.W. and Thrall, R.M. (1991) A structure for classifying and charactering efficiency and inefficiency in data envelopment analysis. *Journal of Productivity Analysis*, **2**, 197–237.

Charnes, A. and Neralic, L. (1990) Sensitivity analysis of the additive model in data envelopment analysis. *European Journal of Operational Research*, **48**, 332–341.

Charnes, A., Rousseau, J.J. and Semple, J.H. (1996) Sensitivity and stability of efficiency classifications in data envelopment analysis. *Journal of Productivity Analysis*, **7**, 5–18.

Cherchye, L., Kuosmanen, T. and Post, T. (2001) Alternative treatment of congestion in DEA: a rejoinder to Cooper, Gu, and Li. *European Journal of Operational Research*, **132**, 75–80.

Cooper, W.W. (2005) Origins, uses of, and relations between goal programming and data envelopment analysis. *Journal of Multi-Criteria Decision Analysis*, **15**, 3–11.

Cooper, W. W., Bisheng, G. and Li, S. (2001a) Comparisons and evaluations of alternative approaches to the treatment of congestion in DEA. *European Journal of Operational Research*, **132**, 62–74.

Cooper, W. W., Deng, H., Gu, B., Li, S. and Thrall, R. M. (2001b) Using DEA to improve the management of congestion in Chinese industries (1981–1997), *Socio-Economic Planning Sciences*, **35**, 1–16.

Cooper, W.W., Deng, H., Huang, Z. and Li, S.X. (2002) Chance constrained programming approaches to technical efficiencies and inefficiencies in stochastic data envelopment analysis. *Journal of the Operational Research Society*, **53**, 1347–1356.

Cooper, W. W., Gu, B. and Li, S. (2001). Note: alternative treatments of congestion in DEA – a response to the Cherchye, Kuosmanen and Post critique. *European Journal of Operational Research*, **132**, 81–87.

Cooper, W.W., Huang, Z. Li, S., Parker, B.R. and Pastor, J.T. (2007) Efficiency aggregation with enhanced Russell measures in data envelopment analysis. *Socio-Economic Planning Sciences*, **41**, 1–21.

Cooper, W.W., Park, K.S. and Pastor, J.T. (1999) RAM: A range adjusted measure of inefficiency for use with additive models, and relations to other models and measures in DEA. *Journal of Productivity Analysis*, **11**, 5–42.

Cooper, W.W., Park, K.S. and Pastor, J.T. (2000) Marginal rates and elasticities of substitution with additive models in DEA. *Journal of Productivity Analysis*, **13**, 105–123.

Cooper, W.W., Park, K.S. and Pastor, J.T. (2001) The range adjusted measure (RAM) in DEA: A response to the comments by Steinmann and Zweifl. *Journal of Productivity Analysis*, **15**, 145–152.

Cooper, W.W., Park, K.S. and Yu, G. (1999) IDEA and AR-IDEA: models for dealing with imprecise data in DEA. *Management Science*, **45**, 597–607.

Cooper, W.W., Ruiz, J.L and Sirvent, I. (2007). Choosing weights from alternative optimal solutions of dual multiplier models in DEA. *European Journal of Operational Research*, **180**, 443–458.

Cooper, W.W., Seiford, L. and Tone, K. (2000) *Data Envelopment Analysis: A Comprehensive Text with Models, Applications, References and DEA-Solver Software*, Kluwer Academic Publishers, Norwell.

Cooper, W.W., Seiford, L.M. and Tone, K. (2006) *Introduction to Data Envelopment Analysis and Its Uses, with DEA Solver Software and References*, Springer Science and Business Media, Inc., New York.

Cooper, W.W., Seiford, L. and Zhu, J (2005) *Handbook on Data Envelopment Analysis*, Kluwer Academic Publishers, Norwell.

Cooper, W.W., Seiford, L.M. and Zhu, J. (2000) A unified additive model approach for evaluating inefficiency and congestion with associated measures in DEA. *Socio-Economic Planning Science*, **34**, 1–25.

Debreu, G. (1951) The Coefficient of Resource Utilization, *Econometrica*, **19**, 273–292.

Debreu, G. (1959) *Theory of Value*, Colonial Press, Cambridge, Mass.

Dmitruk, A.V. and Koshevoy, G. A. (1991) On the existence of a technical efficiency criterion. *Journal of Economic Theory*, **55**, 121–144.

Dielman, T.E. and Pfaffenberger, R.C. (1984) Computational algorithm for calculating least value and Chebychev estimates for multiple regression. *American Journal of Mathematical and Management Science*, **4**, 169–197.

Doyle, J. and Rodney Green, R. (1994) Efficiency and cross-efficiency in DEA: derivations, meanings and uses. *The Journal of the Operational Research Society*, **45**, 567–578.

Doumpos, M., Zanakis, S.H. and Zopounidis, C. (2001) Multicriteria preference disaggregation for classification problems with an application to global investing risk. *Decision Sciences*, **32**, 333–385.

Duarte Silva, A.P. and Stam, A. (1994) Second-order mathematical programming formulations for discriminant analysis. *European Journal of Operational Research*, **74**, 4–22.

Dyson, R.G. and Thanassoulis, E. (1988) Reducing weight flexibility in data environmental analysis, *Journal of the Operational Research Society*, **39**, 563–576.

Eisenhart, C. (1964) The meaning of 'least' in least squares. *Journal of the Washington Academy of Science*, **54**, 24–33.

Emrouznejad, A. and Amin, G.R. (2009) DEA models for a ratio data: convexity consideration. *Applied Mathematical Modelling*, **33**, 486–498.

Falk, J.E. and Karlov, V.E. (2001) Robust separation of finite sets via quadratics. *Computers and Operations Research*, **28**, 537–561.

Färe, R., Grosskopf, S. and Lovell, C.A.K. (1983) Measuring the Technical Efficiency of Multiple Outputs Technologies. *Quantitative Studies on Production and Prices*, W. Eichhorn, R. Henn, R.K. Neumann and R.W. Shephard, (eds), Physica-Verlag, Würzburg and Viena.

Färe, R., Grosskopf, S. and Lovell, C.A.K. (1994a) *Production Frontiers*, Cambridge University Press, Cambridge.

Färe, R., Grosskoph, S. and Lovell, C.A.K. (1985) *The Measurement of Efficiency of Production*, Kluwer-Nijhoff Publishing, Boston.

Färe, R., Grosskopf, S., Norris, M. and Zhang, Z. (1994b) Productivity growth, technical progress, and efficiency change in industrialized countries. *American Economic Review*, **84**, 66–83.

Färe, R. and Svensson, L (1980) Congestion of production factors. *Econometrica*, **48**, 1745–1753.

Farrell, M.J. (1954) An application of activity analysis to the theory of the firm. *Econometrica*, **22**, 291–302.

Farrell, M.J. (1957) The measurement of productive efficiency. *Journal of the Royal Statistical Society*, Series A, **120** (Part 3) 253–290.

Farrell, M.J. and Fieldhouse, M. (1962) Production functions under increasing returns to scale. *Journal of the Royal Statistical Society, Series A (General)*, **125**, 252–267.

Fisher, R.A. (1936). The use of multiple measurements in taxonomy problems. *Annals of Eugenics*, **7**, 179–188.

Freed, N. and Glover, F. (1981a) Simple but powerful goal programming models for discriminant problems. *European Journal of Operational Journal*, **7**, 44–60.

Freed, N. and Glover, F. (1981b) A linear programming approach to the discriminant problem. *Decision Sciences*, **12**, 68–74.

Fukuyama, H. (2000) Returns to scale and scale elasticity in data envelopment analysis. *European Journal of Operational Research*, **130**, 498–509.

Gonzalez-Lima, M.D., Tapia, R.A. and Thrall, R.M. (1996) On the construction of strong complementary slackness solutions for DEA linear programming problems using a primal–dual interior-point method. *Annals of Operations Research*, **66**, 139–162.

Green, R., Doyle, J.R. and Cook, W.D. (1996) Preference voting and project ranking using DEA and cross-evaluation. *European Journal of Operational Research*, **90**, 461–472.

Grifell-Tatjé, E., Lovell, C.A.K. and Pastor, J.T., (1998) A quasi-Malmquist productivity index. *Journal of Productivity Analysis*, **10**, 7–20.

Glover, F. (1990) Improved linear programming models for discriminant analysis. *Decision Sciences*, **21**, 771–785.

Glover, F. and Sueyoshi, T. (2009) Contributions of Professor William W. Cooper in operations research and management science. *European Journal of Operational Research*, **197**, 1–16.

Harter, H.L. (1974) The method of least squares and some alternatives – Part I and Part II. *International Statistical Review*, **42**, 147–174 and 235–264.

Harter, H.L. (1975) The method of least squares and some alternatives – Part III and Part IV. *International Statistical Review*, **43**, 1–44 and 125–190.

Hayashi, K. and Sueyoshi, T. (1994) Information infrastructure development: international comparison between the United States and Japan. *Telematics and Informatics*, **11**, 153–166.

Huber P.J. (1981) *Robust Regression*, John Wiley & Sons, Inc., New York.

Ijiri, Y. and Sueyoshi, T. (2010) Accounting essays by Professor William W. Cooper: revisiting in commemoration of his 95th birthday. *ABACUS: A Journal of Accounting, Finance and Business Studies*, **46**, 464–505.

Kendal, S.M., Stuart, A. and Ord, J.K. (1983) *The Advanced Theory of Statistics*, Vol. 3. Charles Grifin, London.

Koenker, R. and Bassett, G. (1978) Regression quantiles. *Econometrica*, **46**, 33–50.

Koopmans, T.C. (1951) Analysis of production as an efficient combination of activities. *Activity Analysis of Production and Allocation*, Edited by T.C. Koopmans. John Wiley & Sons, Inc., New York.

Lam, K.F. and Moy, J.W. (1997) An experimental comparison of some recently developed linear programming approaches to the discriminant problem. *Computers and Operations Research*, **24**, 593–599.

Lambert, D. (1999) Scale and the Malmquist productivity index. *Applied Economic Letters*, **6**, 593–596.

Leibenstein, H. (1966) Allocative efficiency vs. X-efficiency. *American Economic Review*, **56**, 392–415.

Leibenstein, H. (1976) *Beyond Economic Man*, Harvard University Press, Cambridge, Mass.

Lovell, C.A.K. and Pastor, J.T. (1995) Units invariant and translations invariant DEA. *Operations Research Letters*, **18**, 147–151.

Martin, J. (1994) *Local Area Networks*. Prentice Hall, New Jersey.

Mangasarian, O.L. (1999) Arbitrary-norm separating plane. *Operations Research Letters*, **24**, 15–23.

Markowski, C.A. and Markowski, E.P. (1987). An experimental comparison of several approaches to the discriminant problem with both qualitative and quantitative variables. *European Journal of Operational Research*, **28**, 74–78.

McFadden, D. (1973) Conditional logit analysis of qualitative choice behavior. In P. Zarembka (ed.), Frontier in Econometrics. Academic Press, New York.

McFadden, D. (1976) A comment on discriminant analysis versus logit analysis. *Annals of Economics and Social Measurement*, **5**, 511–523.

McFadden, D. (1980) Econometric models for probabilistic choice. *Journal of Business*, **53**, 513–529.

McLachlan, G.L. (1992) *Discriminant Analysis and Statistical Pattern Recognition*. John Wiley & Sons, Inc., New York.

Narula, S.C. and Wellington, J.F. (1982) The minimum sum of absolute errors regression: a state of the art survey. *International Statistical Review*, **50**, 317–326.

Norton, R.M. (1984) The double exponential distribution: Using calculus to find a maximum likelihood estimator. *The American Statistician*, **38**, 135–136.

Olesen, O. and Petersen, N. (1996) Indicators of ill-conditioned data sets and model misspecification in data envelopment analysis: an extended facet approach. *Management Science*, **42**, 205–219.

Olesen, O. and Petersen, N. (2003) Identification and use of efficient faces and facets in DEA. *Journal of Productivity Analysis*, **20**, 323–360.

Pastor, J.T. (1996) Translation invariance in DEA: a generalization, *Annals of Operations Research*, **66**, 93–102.

Pastor, J.T. and Ruiz J.L. (2007) Variables with negative values in DEA. *Modeling Data Irregularities and Structural Complexities in Data Envelopment Analysis*, J. Zhu and W.D. Cook (eds), pp. 63–84, Springer, New York.

Paster, J.T., Ruiz, J.L. and Sirvent, I. (1999) A statistical test for detecting influential observations in DEA. *European Journal of Operational Research*, **115**, 542–554.

Phillips, F.Y., Parsons, R.G. and Donoho, A. (1990) Parallel microcomputing for data envelopment analysis. *Computer, Environment and Urban Systems*, **115**, 542–554.

Robinson, A. (1966) *Nonstandard Standard Analysis:Studies in the Foyundations of Mathematics*, North Holland, Amsterdam.

Portela A.S. and Thanassoulis, E. (2006) Malmquist indexes using a geometric distance function (GDF). application to a sample of Portuguese bank branches. *Journal of Productivity Analysis*, **25**, 25–41.

Premachandra, I.M., Bhabra, G.S. and Sueyoshi, T. (2008) DEA as a tool for bankruptcy assessment: a comparative study with logistic regression technique. *European Journal of Operational Research*, **193**, 412–424.

Rubin, P.A. (1990). Heuristic solution procedures for a mixed-integer programming discriminant model. *Managerial and Decision Economics*, **11**, 255–266.

Russell, R.R. (1985) Measures of technical efficiency. *Journal of Economic Theory*, **35**, 109–126.

Russell, R.R. (1990) Continuity of measures of technical Efficiency *Journal of Economic Theory*, **51**, 255–267.

Schaible S. (1996) Fractional programming. *Encyclopedia of Operations Research and Management Science*, Edited by S.I Gass and C.M. Harris, Kluwer Academic Publishers, Norwell.

Smith, C.A.B. (1947) Some examples of discrimination. *Annals of Eugenics*, **13**, 272–282.

Sueyoshi, T. (1990) Special algorithm for an additive model in data envelopment analysis. *Journal of the Operational Research Society*, **41**, 249–257.

Sueyoshi, T. (1991) Estimation of stochastic frontier cost function using data envelopment analysis: an application to AT&T divestiture. *Journal of the Operational Research Society*, **42**, 463–477.

Sueyoshi, T. (1992) Measuring technical, allocative and overall efficiencies using DEA algorithm. *Journal of the Operational Research Society*, **43**, 141–155.

Sueyoshi, T. (1996) Divestiture of Nippon Telegraph and Telephone. *Management Science*, **42**,1326–1351.

Sueyoshi, T. (1997) Measuring scale efficiencies and returns to scale of Nippon Telegraph and Telephone in production and cost analyses. *Management Science*, **43**, 779–796.

Sueyoshi, T. (1999a) DEA-discriminant analysis in the view of goal programming. *European Journal of Operational Journal*, **3**, 564–582.

Sueyoshi, T. (1999b) DEA duality on returns to scale (RTS) in production and cost analyses: an occurrence of multiple solutions and differences between production-based and cost-based RTS estimates. *Management Science*, **45**, 1593–1608.

Sueyoshi, T. (2001a) *Data Envelopment Analysis* (in Japanese), Asakura-Shoten Publishing Inc., Tokyo.

Sueyoshi, T. (2001b) Extended DEA-discriminant analysis. *European Journal of Operational Research*, **131**, 324–351.

Sueyoshi, T. (2004) Mixed integer programming approach of extended DEA-discriminant analysis. *European Journal of Operational Research*, **152**, 45–55.

Sueyoshi, T. (2005a) Financial ratio analysis of the electric power industry. *Asia-Pacific Journal of Operational Research*, **22**, 349–376.

Sueyoshi, T. (2005b) A Comparison between standard and two stage mixed integer approaches for discriminant analysis. *Asia-Pacific Journal of Operational Research*, **22**, 513–528.

Sueyoshi, T. (2006) DEA-discriminant analysis: methodological comparison among eight discriminant analysis approaches. *European Journal of Operational Research*, **169**, 247–272.

Sueyoshi, T. and Aoki, S. (2001) A use of a nonparametric statistic for DEA frontier shift: the Kruskal and Wallis Rank test. *OMEGA*, **29**, 1–18.

Sueyoshi, T. and Chang, Y. (1989) Goal programming approach for regression median. *Decision Sciences*, **20**, 4.

Sueyoshi, T. and Chang, Y. (1989) Efficient algorithm for the additive and multiplicative models in data envelopment analysis. *Operations Research Letters*, **8**, 205–213.

Sueyoshi, T. and Goto, M. (2009a) Can RandD expenditure spending avoid corporate bankruptcy? comparison between Japanese machinery and electric equipment industries using DEA-discriminant analysis. *European Journal of Operational Research*, **196**, 289–311.

Sueyoshi, T. and Goto, M. (2009b) DEA-DA for bankruptcy-based performance assessment: Misclassification nalysis of the Japanese construction industry. *European Journal of Operational Research*, **199**, 576–594.

Sueyoshi, T. and Goto, M. (2009c) Methodological comparison between DEA (data envelopment analysis) and DEA-DA (discriminant analysis) from the perspective of bankruptcy assessment. *European Journal of Operational Research*, **199**, 561–575.

Sueyoshi, T. and Goto, T. (2012a) DEA radial and non-radial models for unified efficiency under natural and managerial disposability: theoretical extension by strong complementary slackness conditions. *Energy Economics*, **34**, 700–713.

Sueyoshi, T. and Goto, M. (2012b) Returns to scale and damages to scale with strong complementary slackness conditions in DEA environmental assessment: Japanese corporate effort on environmental protection. *Energy Economics*, **34**, 1422–1434.

Sueyoshi, T. and Goto, M. (2012c) Efficiency-based rank assessment for electric power industry: a combined use of data envelopment analysis (DEA) and DEA-discriminant analysis (DA). *Energy Economics*, **34**, 634–644.

Sueyoshi, T. and Goto, M. (2013a) Pitfalls and remedies in DEA applications: how to handle an occurrence of zero in multipliers by strong complementary slackness conditions. *Engineering*, **5**, 29–34.

Sueyoshi, T. and Goto, M. (2013b) A use of DEA-DA to measure importance of R&D expenditure in Japanese information technology industry. *Decision Support Systems*, **54**, 941–952.

Sueyoshi, T. and Goto, M. (2014) Photovoltaic power stations in Germany and the United States: A comparative study by data envelopment analysis *Energy Economics*, **42**, 271–288.

Sueyoshi, T., Goto, M. and Omi, Y. (2010) Corporate governance and firm performance: Evidence from Japanese manufacturing industries after the lost decade *European Journal of Operational Research*, **203**,724–736.

Sueyoshi, T. and Hwang, S.N. (2004) A Use of nonparametric test for DEA-DA: a methodological comparison. *Asia-Pacific Journal of Operational Research*, **21**, 179–197.

Sueyoshi, T. and Honma, T. (2003) DEA network computing in multi-stage parallel processes. *International Transactions in Operational Research*, **10**, 217–244.

Sueyoshi, T and Kirihara, Y. (1998) Efficiency measurement and strategic classification of Japanese banking institutions. *International Journal of Systems Science*, **29**, 1249–1263.

Sueyoshi, T. and Sekitani, K. (1998a) Mathematical properties of least absolute value estimation with serial correlation, *Asia-Pacific Journal of Operational Research*, **15**, 75–92.

Sueyoshi, T. and Sekitani, K. (1998b) Goal programming regression with serial correlation: policy implications for Japanese information infrastructure development, *OMEGA*, **26**,195–205.

Sueyoshi, T. and Sekitani, K. (2005) Returns to scale in dynamic DEA. *European Journal of Operational Research*, **161**, 536–544.

Sueyoshi, T. and Sekitani, K. (2007a) Measurement of returns to scale using a non-radial DEA model: A range-adjusted measure approach. *European Journal of Operational Research*, **176**, 1918–1946.

Sueyoshi, T. and Sekitani, K. (2007b) Computational strategy for Russell measure in DEA: second-order cone programming. *European Journal of Operational Research*, **180**, 459–471.

Sueyoshi, T. and Sekitani, K. (2007c) The Measurement of returns to scale under a simultaneous occurrence of multiple solutions in a reference set and a supporting hyperplane. *European Journal of Operational Research*, **181**, 549–570.

Sueyoshi, T. and Sekitani, K. (2008) DEA congestion and returns to scale under an occurrence of multiple optimal projections. *European Journal of Operational Research*, **194**, 592–607.

Sueyoshi, T. and Sekitani, K. (2009) An occurrence of multiple projections in DEA-based measurement of technical efficiency: theoretical comparison among DEA models from desirable properties. *European Journal of Operational Research*, **196**, 764–794.

Stam, A. (1997). Nontraditional approaches to statistical classification: some perspectives on L_p-norm methods. *Annals of Operations Research*, **74**, 1–36.

Stam, A. and Joachimsthaler, E.A. (1989) Solving the classification problem via linear and nonlinear programming methods. *Decision Sciences*, **20**, 285–293.

Stigler, S.M. (1973) Studies in the history of probability and statistics. XXXII. *Biometrika*, **60**, 439–445.

Stigler, G. S. (1976) The X-istence of X-efficiency. *American Economic Review*, **66**, 213–216.

Taniguchi, I. (1998) Dictionary of Network Computing (in Japanese). Nippon Jitsugiyou Shupan, Tokyo, Japan.

Thompson, R.G., Singleton, F.D., Thrall, R.M. and Smith, B.A. (1986) Comparative site evaluation for locating a high-energy physics lab in Texas. *Interface*, **16**, 35–49.

Thrall, R. (2000) Measures in DEA with an application to the Malmquist index. *Journal of Productivity Analysis*, **13**, 125–137.

Tone, K. (1993) *Measurement and Improvement on Management Efficiency* (in Japanese), Asakura Publishing Inc.

Tone, K. (2001) A Slack-based measure of efficiency in data environment analysis. *European Journal of Operational Research*, **130**, 498–509.

Tone, K. (1996) A simple characterization of returns to scale. *Journal of Operations Research Society of Japan*, **39**, 604–613.

Tone, K. and Sahoo, B.K. (2004) Degree of scale economies and congestion: a unified DEA approach. *European Journal of Operational Research*, **158**, 755–772.

Tone, K. and Sahoo, B.K. (2005) Evaluating cost efficiency and returns to scale in the life insurance corporation of India using data envelopment analysis. *Socio-Economic Planning Sciences*, **39**, 261–285.

Tone, K. and Sahoo, B.K. (2006) Re-examining scale elasticity in DEA. *Annals of Operations Research*, **145**, 69–87.

Tran, N.A., Shively, G. and Preckel, P. (2008) A new method for detecting outliers in data envelopment analysis. *Applied Economics Letters*, **15**, 1–4.

Wagner, H.M. (1959) Linear programming techniques for regression analysis. *Journal of the American Statistical Association*, **54**, 206–212.

Wei, Q. and Yu, G. (1997) Analyzing properties of K-cones in the generalized data environment analysis model. *Journal of Econometrics*, **80**, 63–84.

Wilson, J.M. (1996). Integer programming formulation of statistical classification problems. *Omega: International Journal of Management Science*, **24**, 681–688.

Wong, Y.B. and Beasley, J.E. (1990) Restricting weight flexibility in data environment analysis. *Journal of the Operational Research Society*, **41**, 829–835.

Yanev, N. and Balev, S. (1999) A combinatorial approach to the classification problem. *European Journal of Operational Research*, **115**, 339–350.

Yu, G., Wei, Q. and Brockett, P. (1996) A generalized data envelopment analysis model: a unification and extension of existing models for efficiency analysis of decision making units. *Annals of Operations Research*, **66**, 47–89.

Yu, G., Wei, Q., Brockett, P. and Zhou, L. (1996) Construction of all DEA efficient surfaces of the production possibility set under the generalized data envelopment analysis model. *European Journal of Operational Research*, **95**, 491–510.

Zopounidis, C. and Doumpos, M. (2002) Multicriteria classification and sorting methods: a literature review. *European Journal of Operational Research*, **138**, 229–246.

SECTION II

DEA ENVIRONMENTAL ASSESSMENT

13

WORLD ENERGY

13.1 INTRODUCTION

This chapter describes a recent energy trend in the world. Energy is separated into primary and secondary categories. Primary energy is further classified into fossil fuels and non-fossil fuels. Fossil fuels include oil, natural gas, and coal, while non-fossil fuels include nuclear and renewable energies (e.g., solar, water, wind, biomass and other energy resources). Energy consumption is essential for developing economic prosperity in all nations. This chapter considers electricity as a representative of secondary energy because it is produced through the use of primary energy sources.

As the initial step for the DEA environmental assessment to be discussed in Section II, this chapter summarizes a general trend in energy whose consumption has been increasing along with economic development and population increase in the world. The purpose of this chapter is to describe the research necessity of DEA environmental assessment from the perspective of supply and demand on energy, along with population increase, in the world. Based upon our observation on energy, this chapter conveys an academic rationale regarding why DEA is important in investigating various issues on energy, environment and sustainability.

The remainder of this chapter is organized as follows: Section 13.2 describes the world energy trend. Section 13.3 describes primary energy

Environmental Assessment on Energy and Sustainability by Data Envelopment Analysis,
First Edition. Toshiyuki Sueyoshi and Mika Goto.

sources. Section 13.4 discusses electricity as secondary energy. Section 13.5 discusses the current situation on petroleum and world trade. Section 13.6 explains energy economics by using an illustrative example of power trading. Section 13.7 summarizes this chapter.

13.2 GENERAL TREND

World primary energy consumption continues to increase, along with economic growth. From 1965 to 2013, it increased from 3.8 to 12.7 billion tons of oil equivalent, indicating an average annual growth rate of 2.6%. The growth of energy consumption varies, depending on the region and its regional industrialization. For example, advanced countries such as OECD nations (where OECD stands for the Organization for Economic Co-operation and Development) had lower growth rates. In contrast, developing countries (e.g., non-OECD countries) had higher growth rates than the more advanced nations. One rationale is that the advanced counties have already attained a high level of industrial infrastructures so that they are sufficient in maintaining moderate growth rates in their economies and populations. Furthermore, they have improved the efficiency level of energy consumption equipment over the past due to technology development. In contrast, energy consumption has been steadily increasing in developing countries. In particular, a significant increase in the world energy consumption can be found in the Asia-Pacific region. Under such an energy consumption trend, the share of OECD countries in energy consumption has decreased from 70% in 1965 to 43% in 2013, as depicted in Figure 13.1, where the left vertical axis indicates the amount of energy consumption, measured by million tons of oil equivalent and the right vertical axis indicates the percentage of the OECD share. The annual period is shown on the horizontal axis.

Figure 13.2 visually describes world primary energy consumption by each energy source. In the figure, the vertical axis indicates the amount of energy consumption by each energy source and the horizontal axis indicates the annual period. Oil has been a major source of primary energy consumption, accounting for the largest share of total energy consumption with 31.53% in 2012, particularly supported by a steady increase in the usage of a transport sector. The average annual growth rate was 1.2% over the period from 1971 to 2012.

The consumption of coal and natural gas has grown faster than oil over the observed annual periods. Coal consumption has increased for electricity generation, particularly in Asian counties such as China, where coal is very popular as an inexpensive generation fuel. Natural gas consumption is favored in developed countries not only for generation fuel but also for city gas demand because they are required to cope with the global warming and climate change problem.

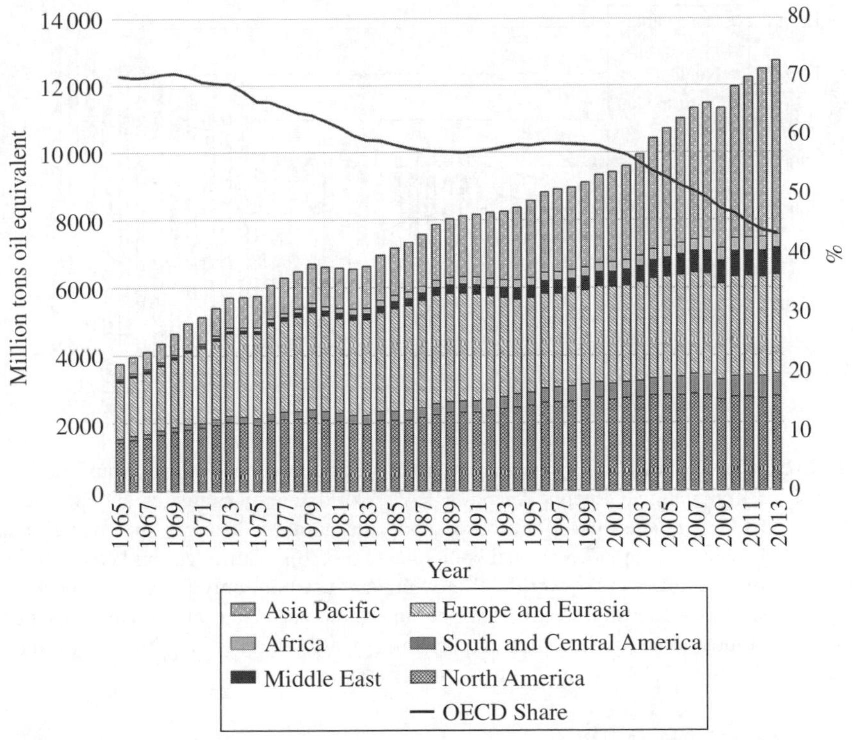

FIGURE 13.1 Trend of world primary energy consumption (a) Source: BP Statistical Review of World Energy (2014)[1]. (b) We prepared the figure based upon the numbers listed in the data source. (c) A large increase in energy consumption can be found in the Asia Pacific region during the observed decades. An increase can be found in Europe and Eurasia, as well. A rapid economic development has been accomplished in the two regions.

The fastest growing energy sources during the observed annual periods (1971–2012) were nuclear and renewable energies (including geothermal, solar, wind, biofuels and waste, and excluding hydro) whose annual average growth rates in consumption were 7.9% and 2.1%, respectively. The rationale for this rapid growth of nuclear and renewable energies includes the necessity of diversified energy supply capabilities, directing toward a low carbon society. However, it is important to mention that their shares in primary energy consumption in total were not high enough, with approximately 5% and 11%, respectively, in 2012. Thus, even now, it is almost impossible for most nations to use their renewable energies as their main energy sources. Of course, it is hoped that their shares will be able to increase more in future. See Chapter 26 for our concern on nuclear energy.

[1] http://www.bp.com/en/global/corporate/about-bp/energy-economics/statistical-review-of-world-energy/statistical-review-downloads.html.

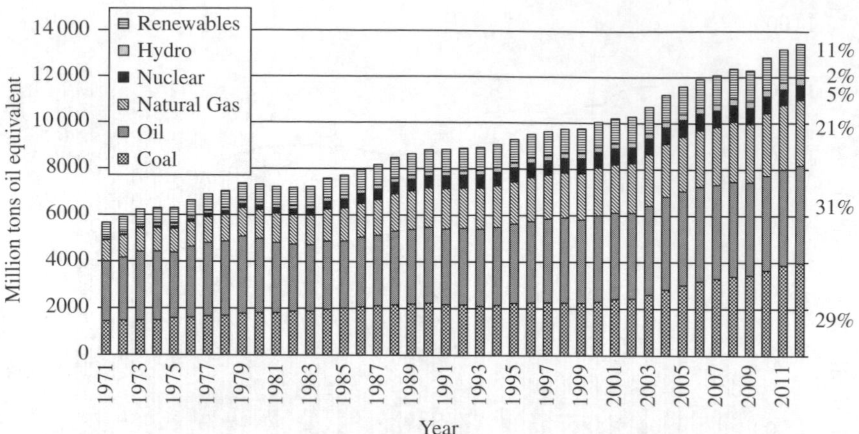

FIGURE 13.2 World energy consumption by energy sources in the annual periods (a) Source: International Energy Agency (IEA), World Energy Statistics and Balances, OECD[2]. (b) We prepared the figure based upon the numbers listed in the source data. (c) The figure indicates that fossil fuel components (i.e., oil, natural gas and coal) are the major energy resources in the world, all of which are used not only for modern business (e.g., transport and industry sectors) and the household sector but also for military purposes. (d) Fluctuations in the oil price may influence the market condition of business because oil is the most important primary energy resource.

13.3 PRIMARY ENERGY

13.3.1 Fossil Fuel Energy

13.3.1.1 Oil

The amount of proved oil reserve as of 2012 was one trillion and 687 billion barrels, after excluding Canadian oil sands and Venezuela's Orinoco Belt, as depicted in Figure 13.3, where the oil reserve of each nation is expressed as a part of the pie chart. The reserve-production (RP) ratio calculated from the numbers is 52.9 years. The RP ratio has remained almost constant during the four decades after the 1980s due to an improvement in resource recovery technology as well as newly detected and confirmed oil resources, although oil resource depletion was a serious problem after the oil shocks in the 1970s. In particular, recently, the RP ratio rather increased because of an increase in the heavy oil reserve in Venezuela and Canada.

In 2012, the country with the largest proved reserves was Venezuela, although Saudi Arabia had been in first position for a long time before it became second in 2010. The share of the proved reserve of Venezuela was 18% with 297.6 billion barrels, followed by 16% with 265.9 billion barrels in Saudi Arabia, and 10% with 174.3 billion barrels in Canada. These were followed by Iran (9%), Iraq (9%),

[2]Organization for Economic Co-operation and Development, Paris, France, OECDiLibrary http://www.oecd-ilibrary.org/energy/data/iea-world-energy-statistics-and-balances_enestats-data-en.

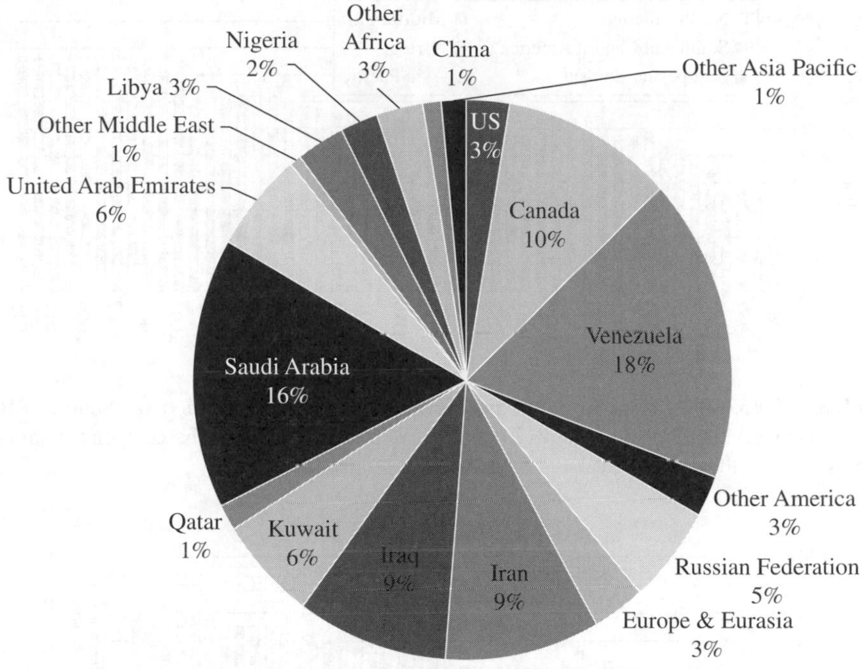

FIGURE 13.3 World oil proved reserves as of 2012 (a) Source: BP Statistical Review of World Energy (2014)[3]. (b) We prepared the figure based upon the numbers listed in the data source. (c) The figure excludes an amount of Canadian oil sands and Venezuela's Orinoco Belt. The shale oil reserve is also excluded from the figure. The largest shale oil reserve exists in China and the second is the United States.

Kuwait (6%) and United Arab Emirates (6%). Middle East countries accounted for approximately half of the total share of proved oil reserves in the world.

Figure 13.4 illustrates that world oil production, on the vertical axis, has increased from 53.66 to 86.75 million barrels per day from 1972 to 2013, so becoming approximately 1.6 times larger than the level of 1972 over the past four decades. Since 2000, European countries decreased in their amount of oil production, while Asia-Pacific region, Africa, and Latin America remained almost constant in their oil production. Production from Russia and the Middle East steadily increased during the observed annual periods. As depicted in the figure, the world has a large amount of supply capability to satisfy demand.

As depicted in Figure 13.5, oil production in the OPEC countries, on the vertical axis, decreased in the early 1980s after a large increase in the 1970s, but

[3] http://www.bp.com/en/global/corporate/about-bp/energy-economics/statistical-review-of-world-energy/statistical-review-downloads.html.

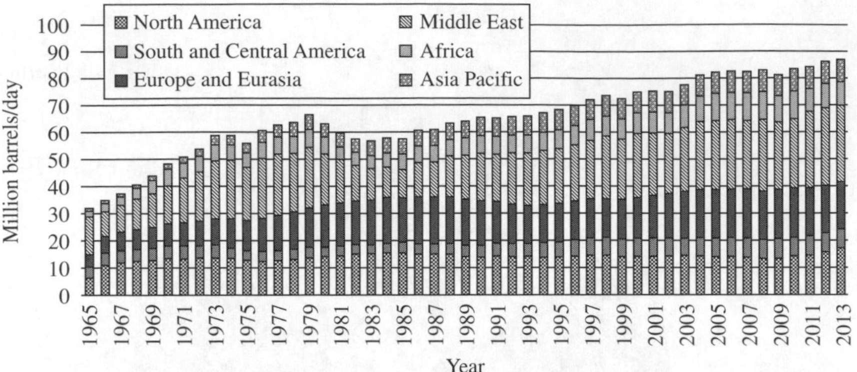

FIGURE 13.4 Trend of world oil production: regional classification (a) Source: BP Statistical Review of World Energy (2014)[4]. (b) We prepared the figure based upon the numbers listed in the source data.

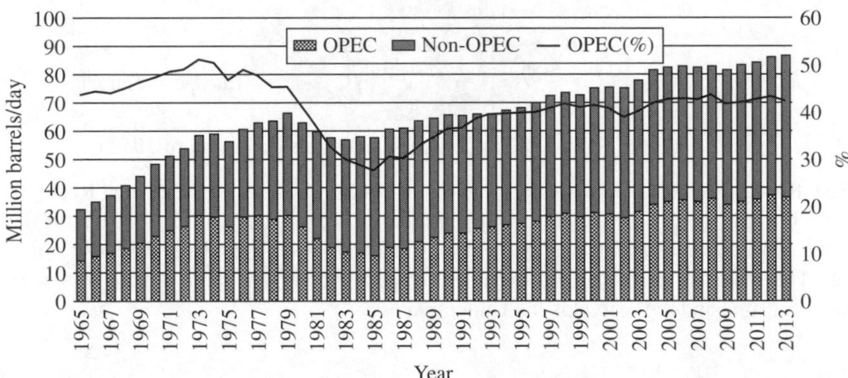

FIGURE 13.5 Trend of world oil production by OPEC and non-OPEC nations (a) Source: BP Statistical Review of World Energy (2014)[5]. (b) We prepared the figure based upon the numbers listed in the source data.

the amount of production gradually recovered in the late 1980s. Here, OPEC stands for the Organization of Petroleum Exporting Counties.

The decreasing oil production trend of OPEC nations in the early 1980s was because of both a production increase from non-OPEC countries, looking for a high oil price, and lower oil consumption in the world. Consequently, the production share of OPEC countries decreased from more than 50% in the early 1970s to

[4] http://www.bp.com/en/global/corporate/about-bp/energy-economics/statistical-review-of-world-energy/statistical-review-downloads.html.

[5] http://www.bp.com/en/global/corporate/about-bp/energy-economics/statistical-review-of-world-energy/statistical-review-downloads.html.

a level of less than 30% in the middle of 1980s. However, it increased again to a level of low 40% in the 2000s.

Oil production in non-OPEC countries, including the former republics of the Soviet Union, the United States, Mexico, Canada, the United Kingdom, Norway, China and Malaysia, has steadily grown from 17.88 to 49.93 million barrels per day from 1965 to 2013. In recent years, oil production in the United States has received a major attention from the world, because their production grew rapidly due to the shale oil and gas revolution. A problem with shale oil production is that the production cost is high (e.g., US$ 50 per barrel) due to the technical difficulty of water cracking, so that many US oil companies, dependent upon shale oil production, may have a financial problem because of the recent low oil price. See Chapter 24 for a detailed description of the United States petroleum industry. See also Chapter 16 which compares international petroleum companies with national ones.

It is important to add, here, that recent shale technology has replaced the hydro-cracking approach by using "supercritical CO_2." Carbon dioxide is a gas in air at standard temperature and pressure. It becomes a supercritical fluid below the critical condition on temperature and pressure. The new cracking technology uses the liquid stage of CO_2 to extract oil and gas from underground wells. The new approach can keep a large amount of CO_2 in drilling wells so that it can reduce the amount in the air, so being ecological but costly. See https://en.wikipedia.org/wiki/Supercritical_carbon_dioxide.

Figure 13.6 depicts that world oil consumption, on the vertical axis, has grown from 55.56 to 91.33 million barrels per day from 1973 to 2013. The annual average growth rate was 1.3%. In OECD countries, oil consumption increased during the late 1970s, from 41.32 million barrels per day in 1973, and then decreased in the beginning of the 1980s because an economic recession occurred after the two oil shocks. Energy sources, such as nuclear and natural gas, were proposed as an alternative to oil, as well. Along with an expansion of the economy after the late 1980s, oil consumption slowly increased, but it stagnated after 2005 because of the improved fuel efficiency of vehicles and a rising oil price.

In contrast, non-OECD countries exhibited a large amount of oil consumption in recent years. The increase was supported by their economic growth. For example, the consumption increased from 14.25 to 45.77 million barrels per day from 1973 to 2013, indicating an increase of 3.0% as an average annual growth rate. As a result, the share of oil consumption in non-OECD counties increased from 26% in 1973 to 50% in 2013, whereas developed countries decreased their shares of consumption from 74% to 50% during the same annual periods.

World oil trading has steadily increased along with the increase in oil consumption. The volume of total oil trade has reached 55.67 million barrels per day in 2013. The 50% of the total volume of oil imports was occupied by the three large markets, including Japan, the United States and European nations. Meanwhile, the Middle East occupied the largest share of the total volume of exports with 35% share in 2013. In addition, the 10% of the total volume of

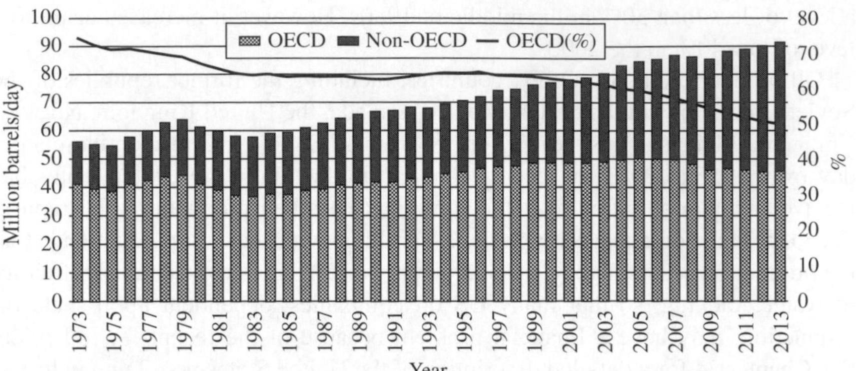

FIGURE 13.6 Trend of world oil consumption by OECD and non-OECD nations (a) Source: BP Statistical Review of World Energy (2014)[6]. (b) We prepared the figure based upon the numbers listed in the data source. (c) There was no major increase in OECD nations, but there was an increasing trend in the non-OECD nations. The result indicates that industrial nations in the OECD attained the almost maximum limit on oil consumption. In contrast, non-OECD nations increased consumption for their industrial developments along with their population increase.

exports from the Middle East (2.01 million barrels per day) was delivered to the United States, 11% (2.07 million barrels per day) to Europe and 76% (14.74 million barrels per day) to the Asia-Pacific region. The evidence confirms that the Asia-Pacific region is the largest sales channel of oil from the Middle East. This regional oil dependency on the Middle East remained higher in Asian countries than in Europe and the United States over the 1990s in order to support their rapid economic growth, in particular China and Japan.

13.3.1.2 Natural Gas

As visually summarized in Figure 13.7, the world gas reserve was 185.7 trillion cubic meters as of at the end of 2013. The Middle East occupied the largest share of gas reserve with 43%, followed by Europe, Russia and the former republics of the Soviet Union with 31% of the total share. Differing from the high level of regional concentration of oil reserves in the Middle East, natural gas has a low regional concentration. Natural gas production was 3.4 trillion cubic meters in 2013. The average annual growth rate was 2.5% between 2003 and 2013, which was higher than the growth rate (1.1%) of oil during the same annual periods. Two large regions of natural gas production in 2013 were North America (with 27% share) and Europe, Russia and the former republics of the Soviet Union (with 31% share).

Although the amount of natural gas reserve in the Middle East was 43%, the production share was only 17%. This gap between reserve and production

[6] http://www.bp.com/en/global/corporate/about-bp/energy-economics/statistical-review-of-world-energy/statistical-review-downloads.html.

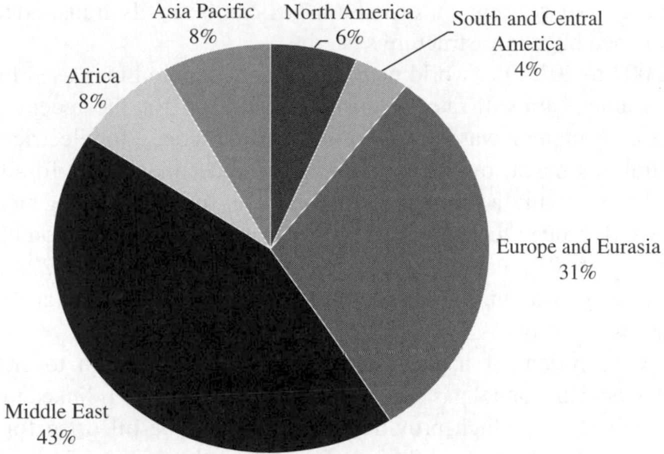

FIGURE 13.7 World natural gas: proved reserves as of 2013 (a) Source: BP Statistical Review of World Energy (2014)[7]. (b) We prepared the figure based upon the numbers listed in the source data.

occurred because of two business rationales. One of these rationales was that a very large amount of investment was necessary to transport natural gas through a huge pipeline network. The other rationale was that investment for natural gas production was relatively small because gas was usually produced alongside oil. The oil price (per unit) was much higher than the gas price. Thus, no gas pipeline network was constructed from the Middle East to the large consumption areas in the world. The situation was different between Russia and Western Europe where a gas pipeline network already existed between them. Most of the natural gas produced in Middle East countries was consumed by themselves and the remaining was liquefied and exported as liquefied natural gas (LNG).

Responding to increasing natural gas consumption in the world, the major oil companies in Europe and the United States developed large natural gas plants. In particular, new LNG projects have been planned and prepared to increase the amount of LNG. In addition, new technologies such as gas to liquids (GTL) and di-methyl ether (DME) have been applied to natural gas production. Part of these has been already commercialized for gas production.

About 60% of the world natural gas consumption arises from North America, Europe, Russia and the former republics of the Soviet Union. There are two rationales for the large share in those regions. One of these rationales is that they produce an abundant amount of natural gas and have promoted the usage of natural gas. The other rationale is that these areas have already well developed pipeline

[7] http://www.bp.com/en/global/corporate/about-bp/energy-economics/statistical-review-of-world-energy/statistical-review-downloads.html.

infrastructures. A large amount of natural gas can be easily transported through their established huge infrastructure systems.

From 2003 to 2013, the world natural gas consumption increased by 2.6% as an average annual growth rate. A business rationale for the recent growth of natural gas consumption was because of a demand increase for electricity generation. Natural gas has a lower environmental impact than other fossil fuels. In addition, the economic advantage of natural gas for electricity generation has been increased through technological progress using gas turbine combined cycle generation. As of 2013, natural gas accounted for 30%, 32% and 22% in the total primary energy consumption in the United States, the OECD nations in Europe, and Japan, respectively.

The pricing system of natural gas varies from one region to another. For example, the price of natural gas (i.e., LNG) exported to Japan is linked to the Japan crude cocktail (JCC), which provides an average crude oil price for Japanese imports. The pricing formula is designed to reduce the degree of variation in the natural gas price. Meanwhile, the gas price in the United States and north-western Europe such as the United Kingdom is determined by the relationship between demand and supply in each gas market. In the other countries in continental Europe, the natural gas price is linked to the prices of alternative fuels such as petroleum products or crude oil.

13.3.1.3 Coal

Confirmed coal resources were 891.5 billion tons at the end of 2013, most of which were reserved in the United States (26.6%), Russia (17.6%) and China (12.8%), respectively. Bituminous coal, referred to as "black coal," amounted to 403.2 billion tons. Subbituminous coal, referred to as "brown coal," amounted to 488.3 billion tons. The advantage of coal is its lower regional concentration than oil and natural gas. Coal reserves are widely distributed over the world. Besides, according to BP statistics in 2014, the RP ratio was 113 years, considerably longer than the ratio for other energy resources, such as oil.

World coal production in 2013 was estimated as 7.896 billion tons. Among the total coal production in 2013, the sum from China (47%) and the United States (13%) provided more than half of the total, followed by Australia, Indonesia, India and Russia, whose production sum became 84% of the total coal production.

China has been increasing coal production since 2001 in order to cope with rapidly expanding domestic energy consumption that is mainly used for electricity generation. In the United States, coal has been long positioned as an important energy resource, followed by oil. Coal-fired power generation had more than 50% share of electricity generation until the early 2000s. However, because of increased social consciousness on various air pollution problems and increased natural gas generation, the number of coal-fired power plants has gradually decreased so that the share of electricity generation in the United States became about 43% in 2013. Another rationale for explaining the share decrease was because of a large price decrease in natural gas, caused by the recent development of shale gas. The decrease of coal-fired power generation reduced the consumption and production of coal.

The total coal consumption in the world was estimated at 7.697 billion tons in 2012, implying a growth of 2.3% over the previous year. The two largest coal consumption countries were China (48%) and the United States (11%), whose sum accounted for approximately 60% of the world consumption in 2012.

The total coal export in the world was estimated at 1.255 billion tons in 2012. The largest exporter of coal was Indonesia, occupying 30.5% of the world total. The second was Australia with a share of 24.0%, followed by Russia (10.7%). China was the second largest exporter of coal in 2003. However, the amount of Chinese export drastically decreased after 2004 because of its rapid expansion in domestic coal consumption. In recent years, Asian countries such as China and India increased their coal consumption for electricity generation at many coal-fired power plants to satisfy an increase in electricity consumption. In 2012, the total sum of coal imports by Asian countries, including Japan, China, Korea, India and Taiwan, was estimated to be 0.822 billion tons, or 64.4% of the world total coal import. In particular, China's import of coal exceeded 0.1 billion tons in 2009 and China became a pure importer of coal as a result of a drastic increase in coal consumption.

Note that Chapter 17 discusses coal-fired power plants in the United States. These have made a fuel mix shift from coal to other types of fuels (e.g., natural gas, renewable and nuclear energies) under strict regulation on air pollution. Chapter 18 confirms that United States coal-fired power plants with bituminous coal have outperformed those of subbituminous coal in terms of their efficiency measures.

13.3.2 Non-fossil Energy

13.3.2.1 Nuclear

After the world's first nuclear power generator began operation in 1951 in the United States, many other countries have actively promoted the development of nuclear power generation. However, from the late 1980s, the growth of a nuclear power generation capacity became steady across the world, as visually summarized in Figure 13.8. The figure depicts the nuclear generation capacity of three groups (i.e., Europe, Asia and America) of OECD nations.

Many nations have paid serious attention to nuclear power generation both to alleviate global competition for fossil-fuel energy resources and to tackle global warming and climate change. As a result, the total nuclear power generation capacity has steadily increased in the world, as depicted on the vertical axis of Figure 13.8.

Figure 13.9 visually describes the amount of nuclear power generation in these three groups of OECD nations. The United States and Europe have constructed only a limited number of new nuclear power plants. However, during the observed annual periods, the amount of nuclear power generation indicated an increasing trend because of its enhanced generation capacity and improved utilization factor. For example, the utilization factor was approximately 90% in the United States as

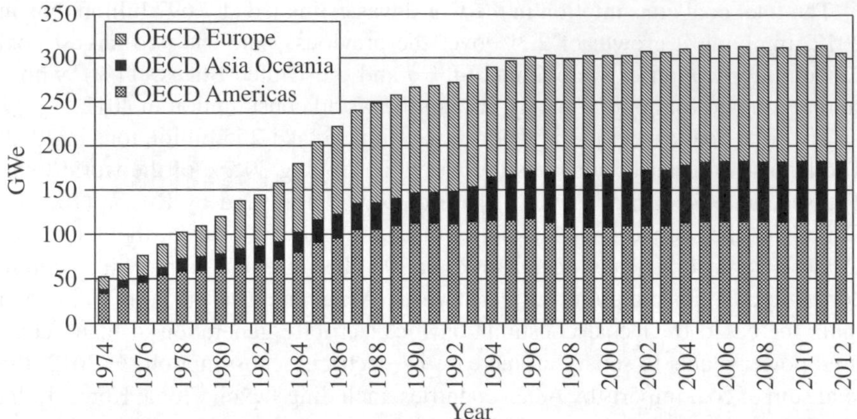

FIGURE 13.8 Nuclear generation capacity in three groups of OECD nations (a) Source: IEA Electricity Information Statistics[8]. (b) We prepared the figure based upon the numbers listed in the source data. (c) GWe stands for Gigawatt-electrical.

a result of efforts for high operational efficiency since the accident of Three Mile Island. Meanwhile, after the disaster of the Fukushima Daiichi nuclear power plant on 11 March 2011, the amount of nuclear power generation decreased in Japan and Asian regions, but it has not largely changed in the other regions. See Sueyoshi and Goto (2015c) and Chapter 26 for a detailed description of Japanese fuel mix strategy after the disaster at the Fukushima Daiichi's nuclear power plant. See also Goto and Sueyoshi (2016)[9].

Uranium resources are widely distributed in the world. As of 2012, Canada, Australia and Kazakhstan ranked high in terms of their amount of uranium reserve and production. The uranium price in the spot market fluctuated with nuclear power plant constructions. The price was also influenced by other difficulties such as oil shocks and accidents at nuclear power plants. In 2007, the price once rose to US$ 136 per pound of U_3O_8 and it remained above US$ 60 per pound of U_3O_8 until March 2011 before the disaster of Fukushima Daiichi's nuclear power plant. Here, U_3O_8 stands for triuranium octoxide. After the disaster, the uranium price slightly dropped and it has remained at a relatively stable level because of tight conditions on supply and demand as well as the influence of speculation money.

[8] Organization for Economic Co-operation and Development, Paris, France, OECDiLibrary http://www.oecd-ilibrary.org/energy/data/iea-electricity-information-statistics_elect-data-en.

[9] Goto, M. and Sueyoshi, T. (2016) Electricity market reform in Japan after Fukushima: Implications for climate change policy and energy market developments. *Economics of Energy and Environmental Policy*, **5**, 15–30.

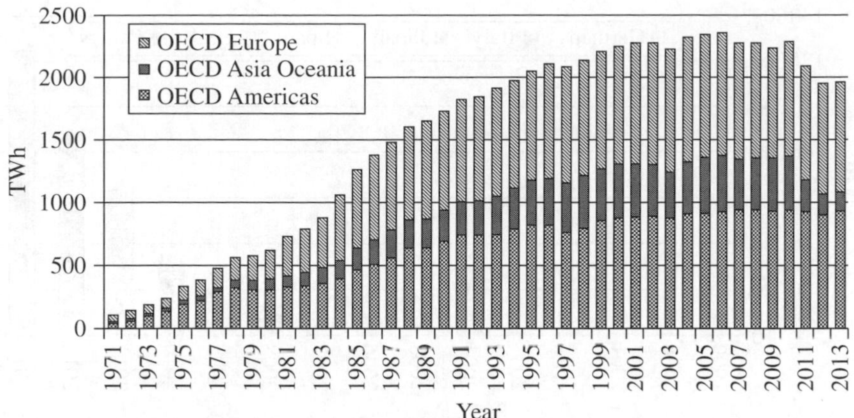

FIGURE 13.9 Amount of nuclear power generation in three groups of OECD nations (a) Source: IEA Electricity Information Statistics[10]. (b) We prepared the figure based upon the numbers listed in the data source. (c) TWh stands for Terawatt hour.

13.3.2.2 Renewable Energy

Solar Photovoltaic Power Generation: According to the statistics of the International Energy Agency, Photovoltaic Power Systems Program (IEA PVPS, 2014), the total installed capacity of solar photovoltaic (PV) power generation was 125 GW in 2013 over IEA countries.

Figure 13.10 visually describes the amount of cumulative installed PV power from 1999 to 2013. Although Japan was the largest installer of the PV capacity before 2004, the installation grew faster in Germany and Spain because the two nations adopted a feed-in tariff (FIT) to support the high cost of PV. See Sueyoshi and Goto (2014d) about their positive and negative concerns on FIT.

Wind Power Generation: The world installed capacity of wind power generation rapidly increased in recent years, reaching a level of 369.55 GW in 2014. Figure 13.11 depicts the global cumulative installed wind power generation capacity between 1997 and 2014. As of 2015, the new installation of wind power generation capacity was 30753 MW (48.5%) in China and was 8598 MW (13.5%) in the United States, together accounting for 62.0% of world capacity. In addition, off-shore wind power generation has been rapidly expanding, reaching a cumulative capacity of 12.1 GW by 2015. In particular, the United Kingdom focuses on the off-shore wind power generation, accounting for 41.8% of the accumulated installed capacity in the world in 2015.

Biomass: In 2012, biomass supplied approximately 10% of the world's primary energy. In particular, biomass accounts for an average of 4.8% of the primary

[10] Organization for Economic Co-operation and Development, Paris, France, OECDiLibrary http://www.oecd-ilibrary.org/energy/data/iea-electricity-information-statistics_elect-data-en.

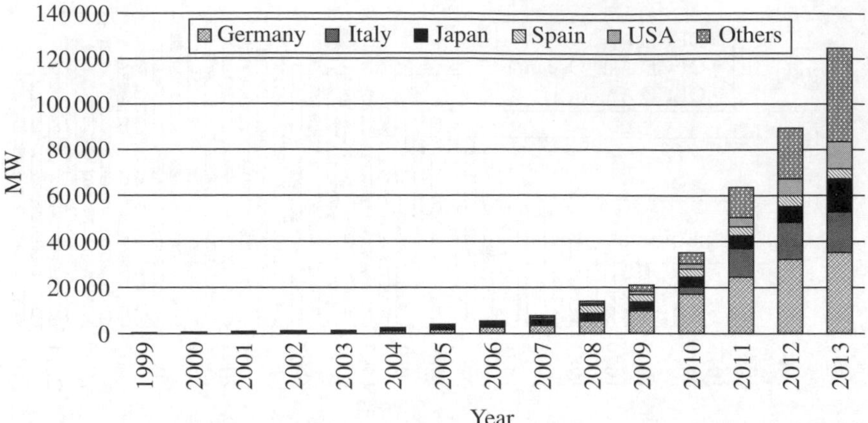

FIGURE 13.10 Cumulative installed photovoltaic power from 1999 to 2013 (a) Source: IEA Photovoltaic Power Systems Program, Trends 2014 in Photovoltaic Applications (IEA-PVPS T1-25, 2014)[11]. (b) We prepared the figure based upon the numbers listed in the data source. (c) MW stands for Megawatt.

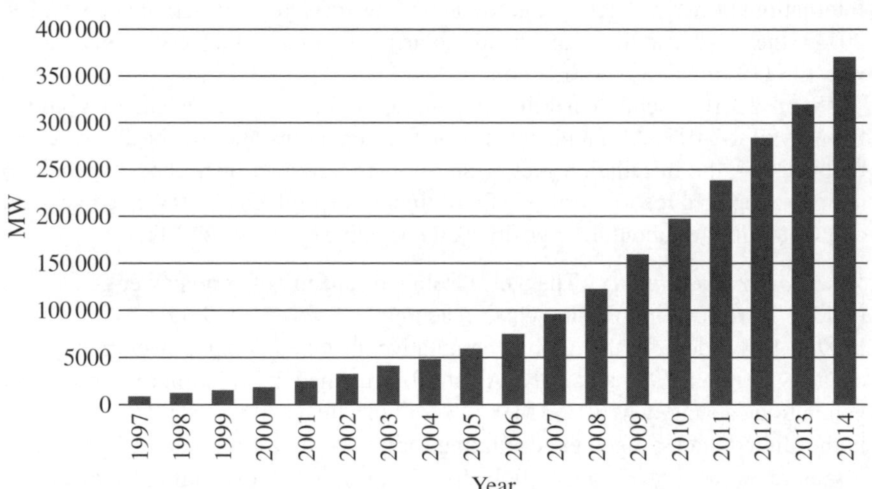

FIGURE 13.11 Global cumulative installed wind power generation capacity from 1997 to 2014 (a) Source: Global Wind Energy Council[12]. (b) We prepared the figure based upon the numbers listed in the data source. (c) MW stands for Megawatt.

[11] http://www.iea-pvps.org/fileadmin/dam/public/report/statistics/IEA_PVPS_Trends_2014_in_PV_Applications_-_lr.pdf.

[12] http://www.gwec.net/global-figures/graphs/.

energy supply in OECD countries, while it is 13.6% in non-OECD countries. The OECD countries, such as the United States and European nations, have been promoting biomass generation through their energy policies in the context of a countermeasure against global warming and climate change.

To enhance biomass usage, many countries have been developing various energy policies that attempt to reduce oil dependency in their transportation sector and reduce their GHG emissions. Meanwhile, there are social concerns about the rapid increase in biomass usage. For example, the use of biomass seriously influences a steep rise in food prices and it invites cutting down rain forests to convert them into farm land. Thus, to reduce the impact of biomass usage upon the natural environment and food markets, international conferences are open to discuss how to construct global sustainability standards on biomass. In addition, research has been promoted to produce biofuel from non-food materials such as straw, timber and algae. International major oil companies are currently focusing on new research and development on the next-generation biofuel.

Hydro: The capacity of hydro power generation amounted to 1010 GW, which was approximately 20% of the total generation capacity in the world as of 2012. Countries with large hydro power generation capacities include China, the United States, Canada and Japan.

Geothermal: The installed capacity of geothermal power generation amounted to 11.7 GW as of 2013. Countries with large geothermal power generation include the United States, Philippine and Indonesia. In 2013, they had a generation capacity of approximately 3.4, 1.9 and 1.3 GW, respectively.

13.4 SECONDARY ENERGY (ELECTRICITY)

As depicted in Figure 13.12, world electricity consumption has increased constantly until today. In the 1970s, the annual growth rate remained high at 5.3% on average, although there was a temporary stagnation in growth after the oil shock. The growth rate gradually decreased to 3.6% in the 1980s and 2.5% in the 1990s, but it recovered to attain a steady growth at 3.1% in the 2000s.

The electrification rate in the world increased from 12.2% in 1980 to 18.1% in 2012, exhibiting an increase by 5.9% during the observed periods. A rationale for the increase was the rapid and widespread growth of the use of electric appliances in the world. Generation capacity in the world continuously increased and reached 5680 GW in 2012.

The average annual growth rate of a total generation capacity was 3.5% in the 1980s, decreasing to 2.2% in the 1990s but increasing to 3.9% in the 2000s. In the world, China's growth forecast will be tremendous in future. According to the Chinese government's official announcement in the 12th version of their five-year energy development plan from 2011 to 2015, China has set a new policy goal on generation capacity to increase from 970 to 1490 GW, so indicating an increase of 9% as an average annual growth.

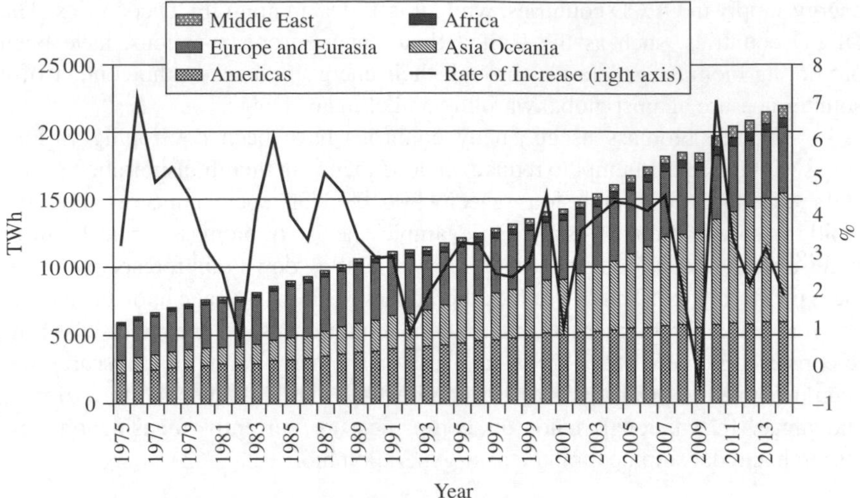

FIGURE 13.12 Trend in world electricity consumption (a) Source: IEA World Energy Statistics and Balances[13]. (b) We prepared the figure based upon the numbers listed in the data source. (c) TWh stands for Terawatt hour.

Considering the world's generation capacity in 2013, we find that steam-power generations by fossil fuels were major energy sources, accounting for 64.5% share of the total generation capacity. However, since the oil shocks in the 1970's, it became necessary for many counties to develop alternative energy sources to oil. Nuclear power generation was promoted for such an industrial goal. Consequently, the nuclear power generation capacity had 9.6% as an average annual growth rate in the 1980s. However, the growth of nuclear power generation decreased in developed countries. The annual growth rate stagnated at 0.5% on average in the 1990s and remained 0.8% in the 2000s. In similar manner, hydro power generation capacity had a problem in identifying new sites for construction, so that its growth rate was low as a result of growing capacity in the 1990s.

The world's electricity generation continuously increased and produced 23.3 million GWh in 2013. The average annual growth rate of generation capacity was 3.5% in the 1980s and 2.2% in the 1990s, whereas the growth rate of generation amount was larger than that of generation capacity, exhibiting 3.8% in the 1980s and 2.5% in the 1990s, respectively. The average annual growth rates in capacity and generation indicated that the utilization rate of generation plants increased during the observed periods. However, the average annual growth rate of generation was 3.0% on average in the 2000s, which was lower than that of the

[13] Organization for Economic Co-operation and Development, Paris, France, OECDiLibrary http://www.oecd-ilibrary.org/energy/data/iea-world-energy-statistics-and-balances_enestats-data-en.

generation capacity with 3.8%. This was due to the influence of world-wide economic recession after the financial crisis in the fall of 2008.

Among the fossil fuels, coal-fired power generation increased its share from 37% in 1975 to 41% in 2013, indicating that coal-fired power generation increased faster than total power generation. The amount of oil-fired power generation steadily increased at an average annual growth rate of 5.7% in the 1970s. However, as a result of the shift from oil to alternative energy sources because of the influence of oil shocks, the annual average growth rate became constantly negative: –2.3% in the 1980s, –0.8% in the 1990s and –2.2% in the 2000s. In contrast, the annual growth rate of gas-fired power generation was 4.1% on average in the 1970s, thus exhibiting an increasing trend. The growth rate of gas-fired power generation was 5.4% in the 1980s, 4.4% in the 1990s and 5.4% in the 2000s, which was larger than the growth rate of total generation. Thus, it is easily thought that gas-fired power generation served as an alternative energy to coal-fired and/or oil-fired ones.

13.5 PETROLEUM PRICE AND WORLD TRADE

Oil is the most important primary energy source in terms of trade volume and wide applicability in electricity generation, chemical products, automobile fuels and military fuels for operating armed forces. Oil is clearly a strategic energy source for business and military.

According to the US Energy Information Administration (2016)[14], the price fluctuation may be separated into five periods after 2004. The first period was from 2004 to 2008, when the oil price sharply increased during all five years. After 2008, corresponding to the financial crisis, the crude oil price in the second period decreased to less than US$ 62 per barrel until 2009. The third period was from 2009 to 2011. The crude oil price returned to the price level held before the crisis. The fourth period was from 2011 to 2014. The oil price was mostly unchanged, compared to the other periods. Finally, the oil price after 2014 dropped below US$ 60 per barrel and remained at this low price until recently.

The decline in oil price after 2014 was because of several compound reasons on oil production. Five concerns need to be discussed here as examples. First, there was economic recession in China. According to the Energy Information Administration (EIA, 2014), China was the world's largest importer of oil in 2014. There was a sharp drop in oil demand within the global oil market partly because of China's demand drop. Second, Saudi Arabia, which had been a market power, could no longer control the market price. See Chapter 16 for a description on the size of Saudi Arabia's oil production. There is only a single company, named "Saudi Aramco," that is 100% owned by the King of Saudi Arabia. It may be a good strategic policy direction for Saudi Arabia to privatize the company so that privatization may generate more benefits for the nation. For example, the

[14] https://research.stlouisfed.org/fred2/series/DCOILWTICO/downloaddata

government of Saudi Arabia can obtain huge capital by selling its equity. Third, there was speculation on a rising crude oil production from Iran, after the economic sanction from Western nations was lifted. The nation is the second largest OPEC supplier regarding oil production capacity. Fourth, the oil production nations (i.e., OPEC and non-OPEC) had a difficulty in consensus building among themselves. Finally, there was an increasing supply from shale oil and gas production in the United States. The US shale revolution may considerably change the relationship between supply and demand in the global oil market.

Acknowledging the existence of an oil price fluctuation, the world economy has been increasing from 2004 to 2016. During the annual periods, world trade has exhibited a steady growth. It is important to note that the amount of world trade in 2008 and 2009 temporarily dropped below the volume of 2005 under the impact of a financial crisis. During 2009–2011, recovery came up to the same volume as before the crisis. After 2014, the volume of world trade in terms of the US dollar decreased despite the growing volume of world trade. A possible rationale for such a decline in the dollar amount was perhaps the decreasing price of oil. See CPB World Trade Monitor (2016)[15].

Implications for DEA: Energy resources serve as inputs to produce desirable outputs (e.g, electricity). The amount of all energy resources has increased in most periods, as found in the figures summarized in this chapter. There are two implications for DEA and its environmental assessment. The first implication is that the world population has increased and it is expected to reach 11.2 billion in the year 2100[16]. Thus, along with the population increase, DEA environmental assessment needs to incorporate an increasing direction into an input vector, or energy resources, until the increase can reach an efficiency frontier shaped by undesirable outputs. The frontier may serve as an upper limit on the increase of an input vector. The methodological implication is inconsistent with a conventional use of DEA, discussed in the chapters of Section I, where an input vector should decrease or maintain a current level for efficiency enhancement. See Chapter 4. The other implication is that science development and technology innovation make it possible to increase the world population. Thus, economic growth, supported by science and technology, is essential for sustainability development in the world. Therefore, it is necessary for DEA environmental assessment to consider such technology innovation in the proposed performance assessment.

13.6 ENERGY ECONOMICS

It is necessary for us to discuss DEA from the perspective of energy economics, which can be considered as an auction market in this chapter. Our description is important in understanding the underlying concepts and frameworks to develop DEA environmental assessment. To simplify our discussion, let us consider a

[15] http://www.cpb.nl/en/number/cpb-world-trade-monitor-january-2016

[16] https://en.wikipedia.org/wiki/World_population

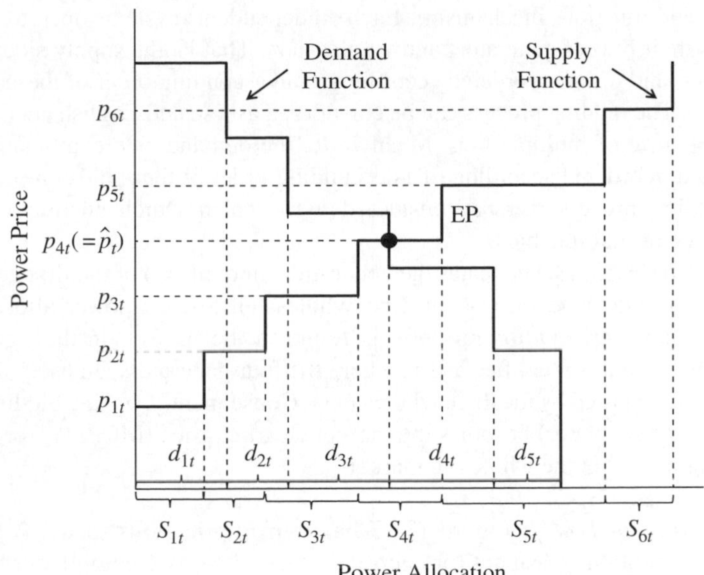

FIGURE 13.13 An equilibrium point in power trading market (a) Source: Sueyoshi and Tadiparthi (2008). (b) The market clearing price (MCP) is found on the equilibrium point (EP).

power trading market, as an illustrative example, in the United States, where generators and wholesalers participate by their bids based upon generation capabilities and demand forecasts on electricity consumption. See, for example, Sueyoshi and Tadiparthi (2007)[17] for a detailed description on the US wholesale market for power trading. See also Sueyoshi and Tadiparthi (2008)[18].

Figure 13.13 visually describes the market coordination mechanism of power trading. In the market, the i-th generator ($i=1, \ldots, 6$) bid on an amount of generation (s_{it}) and its bidding price (p_{it}) for the t-th period. Similarly, the j-th wholesaler ($j=1, \ldots, 5$) bid an amount of demand (d_{jt}) and its bidding price (p_{jt}) in the wholesale market. The number of generators and wholesalers can easily increase to any realistic specification (e.g., 10,000) in a computer simulator.

The bidding price of generators is usually determined by a "marginal cost" for generation[19]. Considering the bidding price, the figure visually describes the

[17] Sueyoshi, T and Tadiparthi, G.R. (2007) An Agent-based approach to handle business complexity in US wholesale power trading. *IEEE Transactions on Power Systems*, **22**, 532–543.

[18] Sueyoshi, T. and Tadiparthi, G.R. (2008) An agent-based decision support system for wholesale electricity market. *Decision Support Systems*, **44**, 425–446.

[19] The bidding prices of generators are determined by "marginal cost" of their generations, not "total cost," as discussed in Chapter 4. Thus, a conventional framework of DEA does not fit with the business reality of US energy sectors. This view can be applied to other sectors in the world. Thus, it is necessary for us to reorganize the analytical structure of DEA in applying it for environmental assessment.

market coordination mechanism. Each independent system operator (ISO) reorders their bids of generators and wholesalers. That is, the supply side combinations (s_{it} and p_{it}) are reordered according to an ascending order of these bids on price (p_{it}). The bidding process can be considered as a sealed English auction with the acceptance of multiple bids. Meanwhile, the demand side combinations (d_{jt} and p_{jt}) are reordered according to a descending order of these bids on price (p_{jt}). The bidding process can be considered as a sealed Dutch auction with the acceptance of multiple bids.

In Figure 13.13, ISO allocates the generation amount (s_{1t}) of the first generator to satisfy the demand (d_{1t}) of the first wholesaler. Such a power allocation is continued until an equilibrium point is found in the market. In the figure, the equilibrium point is identified as EP, where the four generators are used to satisfy the demand required by the three wholesalers. Consequently, p_{4t} (the bidding price of the fourth generator) becomes the market clearing price (MCP: $\hat{P}_t$) for all participating traders in the wholesale market.

Implications for DEA: Figure 13.13 has four implications for DEA applied to environmental assessment for energy sectors. First, two smooth curves usually express a supply and demand relationship, as found in any textbook of economics. Such smooth curves are due to conceptual description and computational tractability. However, the relationship should be expressed by two step functions as depicted in Figure 13.13. It can be easily imagined that the supply and demand relationship is not differentiable, rather being in reality expressed by step functions. Thus, the DEA models proposed in this book, which do not need differentiability, can easily fit into an investigation of the relationship between supply and demand. Second, as discussed in Chapter 9, if congestion occurs in a grid system, then the market clearing mechanism does not function as depicted in Figure 13.13. Therefore, DEA needs to consider such a possible occurrence of congestion in the performance assessment. See, for example, Chapters 9 and 21. Third, the amount of a desirable output (e.g., electricity) is influenced by a market clearing price on EP. If there are many players, they can reduce their bidding prices in such a manner that they have an opportunity to generate electricity in the wholesale power market. In DEA, the performance of a DMU (e.g., a generator and a wholesaler) should be relatively compared by the others. Therefore, it is necessary for DEA to consider how to measure the effectiveness of the strategies of many DMUs (e.g., generators and wholesalers) in a market dynamics process with a time horizon. See Chapter 19. Finally, Figure 13.13 does not incorporate undesirable outputs. The existence of undesirable outputs may be not important in this illustrative example on energy economics. However, the undesirable outputs are very important production factors in discussing various issues on environmental protections. Thus, the DEA environmental assessment needs to incorporate undesirable outputs in the mathematical formulations, as discussed in Chapters from 14 to 27.

13.7 SUMMARY

This chapter described a recent energy trend in the world. Energy was separated into primary and secondary categories. Primary energy was further classified into fossil and non-fossil fuels. Fossil fuels included oil, natural gas and coal, while non-fossil fuels included nuclear and renewable energies (e.g., solar, wind, biomass and others). This chapter also discussed a fuel mix issue on electricity generation as a representative of the secondary energy.

It is easily imagined that energy consumption is essential for the development of economic prosperity in all nations. However, the use of various energy sources usually produces many different types of pollution (e.g., air, soil and water pollution) on the earth, so resulting in great damage on our society, economics and human health. See Chapter 14 for a detailed description on environmental issues. Thus, it is essential for us to understand a general trend of world energy, as discussed in this chapter.

After reviewing the conventional use of DEA in Section I, this chapter discussed its important methodological implications in various applications for environmental assessment as a holistic methodology. One such methodological implication is that DEA applications to energy and environmental assessment need to pay attention to the fact that total energy consumption, along with an increase in the world population, has been increasing in most annual periods after World War II, except some periods under economic recession. Energy resources are usually employed as inputs in DEA performance analysis. So, is it possible for conventional DEA to find an efficiency frontier that locates above an observed input vector? The answer is "no" because the conventional DEA excludes such an efficiency frontier required for environmental assessment. See Chapters 4 and 5 for a description of DEA radial and non-radial models in their conventional frameworks. As discussed in these chapters, all the components of an input vector need to decrease under input-oriented measurement in a conventional framework of DEA. Meanwhile, they should maintain the current observed values under the measurement. An opposite case is observed under output-oriented measurement. Thus the conventional use of DEA is useful in conducting the performance assessment of many different organizations under economic stability, but not under economic growth for sustainability. As a result, all the previous discussions in Section I reflect part of business reality and economic phenomena, but most previous DEA studies had limited practicality in modern business and economy where environmental concerns are essential. Of course, we clearly acknowledge their contributions that have advanced our knowledge dissemination on DEA-based performance assessment.

In addition to the above methodological implication, this chapter needs to mention that it is not easy for us to maintain a social balance (so, sustainability) between economic development and environmental protection. DEA, as discussed in Section II, may serve as a holistic methodology to make such a balance by identifying the source(s) of efficiency and inefficiency, later referred to as unified (operational and environmental) efficiency and inefficiency. The unification process to attain a status of sustainability needs a new approach to combine desirable outputs (e.g., electricity) and undesirable outputs (e.g., GHG emission) in the performance assessment. The existence of undesirable outputs was not considered in the conventional assessment by DEA. Consequently, the conventional DEA has a limited capability in environmental assessment. An exception can be found in Chapter 25. See Chapter 12.

As the initial step for DEA environmental assessment, this chapter starts proposing a new research direction of DEA by considering the current trend of energy in the world. The methodological importance and practicality of DEA environmental assessment will be discussed in the remaining chapters of Section II.

14

ENVIRONMENTAL PROTECTION

14.1 INTRODUCTION

After describing the world energy trend in Chapter 13, this chapter rediscusses previous policy efforts for environmental protection in four industrial regions. This retreat is important for understanding why we are now facing different types of industrial pollution (e.g., air, water, soil and others) and serious environmental issues (e.g., global warming and climate change) alongside our industrial development and economic growth in the world. The four industrial regions to be discussed in this chapter include Europe, Japan, China and the United States, all of which are selected because they are the regions whose environmental achievements are measured by the proposed DEA approaches in Section II. This chapter also discusses a deregulation trend on energy sectors, in particular the electric power industry. This deregulation is important for sustainability development and enhancement. Thus, this chapter discusses environmental issues and a future direction on deregulation, consequently enhancing the status of sustainability in the four regions.

The remainder of this chapter is organized as follows: Section 14.2 discusses environmental protection in the European Union (EU). Section 14.3 describes Japanese industrial pollution and prevention efforts. Section 14.4 examines Chinese pollution and governmental efforts. Chapter 14.5 discusses environmental issues in the United States. Section 14.6 summarizes this chapter.

Environmental Assessment on Energy and Sustainability by Data Envelopment Analysis, First Edition. Toshiyuki Sueyoshi and Mika Goto.

14.2 EUROPEAN UNION

14.2.1 General Description

Industrial development originated in the United Kingdom (UK) in the middle of the eighteenth century. Under the industrial revolution, technology innovation occurred in the textile industry that used coal to operate a steam engine. The change of industrial technology resulted in rapid urbanization of the UK population. Workers moved to settle around increasingly mechanized industrial centers. The industrial revolution later shifted to France, Germany, other European countries and the United States. From the end of the nineteenth century to the beginning of the twentieth century, technology innovation shifted from the textile industry to the metal and chemical industries that heavily used oil and electricity.

Economic prosperity was tempered by heavy social and environmental impacts. A famous archaic example is "Great Smog of 1952" in London[1], which killed more than 10,000 people due to SO_x and other air pollution substances emitted from coal combustion in industries and households. This disaster posed an opportunity for people to think seriously about the problem of air pollution, and it led to the establishment of the City of London Act in 1954 and the Clean Air Act (CAA) in 1956 and 1968.

The middle of the twentieth century was the birth of modern environmentalism and the development of public environmental conscientiousness among people. Understanding climate change showed the limit of a growth-centered economic model and various environmental issues (e.g., air, water and soil pollutions). This limitation led to regulation at the national level, along with the development of a comprehensive body for environmental laws. Environmental law was rooted in a response to industrialization.

According to the *EU Environmental Policy Handbook*[2], edited by Stefan Scheuer, and the environmental policy of the European Union[3], Europe's environmental policy started in 1973, following the United Nations (UN) Conference on the Environment in 1972 in Stockholm. The conference addressed public and scientific evidences about "limits to growth." After the conference, along with growing concern about the limits to growth, western societies increasingly acknowledged the importance of ecological awareness and an industrial shift towards sustainable development. Environmental conscientiousness was expressed by international laws concerning biodiversity,

[1] Great Smog. See hhps://Wikipedia.org/wiki/Great_Smog.

[2] *EU Environmental Policy Handbook: A Critical Analysis of EU Environmental Legislation* (2005) Stefan Scheuer (ed.), European Environmental Bureau, Brussels.

[3] See https://en.wikipedia.org/wiki/Environmental_policy_of_the_European_Union.

atmospheric pollution and climate change, which enshrined the "polluter pays" principle.

In addition to the UN Conference on the Environment in June 1972, the Paris Summit meeting, opened in October 1972 by the heads of state and government in the European Community (EC), was considered as the beginning of the EU's environmental policy. A declaration on environmental and consumer policy was adopted at the summit which requested the European Commission to draw up an action program for environmental protection. The first environmental action program (EAP) was adopted in July 1973 that represented the EU's first environmental policy. The task force within the European Commission, which drew up this action program, eventually led to the formation of a Directorate General for the Environment (DG Environment).

The primary reason at that time for introducing a common environmental policy was due to the concern, which diversified environmental standards, might result in trade barriers and competitive distortions in a common market within the EU. Different national standards for particular products, such as limitations on vehicle emissions, posed significant barriers on free trade within the EU. Additional motivations driving the EU's emerging environmental policy were increasing international conscientiousness on environmental problems and growing recognition from the beginning of the 1970s that environmental pollution did not stop at national borders. Rather, it had to be addressed by cross-border measures. The agreement was not finalized until the middle of the 1980s. After signing the Single European Act (SEA) in 1986, economic and ecological objectives were put on more equal footing within the EU community.

EU environmental policy was shaped by a variety of players, including all the main EU institutions as well as lobby groups, which made up the wider Brussels policy-making community. Member states shaped EU environmental policy by working within the Council of Ministers. The Council is now a central player in decision making within the EU, which shares a policy-making power with the European Parliament under an ordinary legislative procedure. There are different Council formations that specify ministers who are responsible for particular policy areas. One of them is the Environment Council. The number of Environment Council meetings has significantly increased over time.

The European Commission not only has the exclusive right to propose new environmental policy, but it has also a responsibility to ensure the implementation of environmental rules. Since its creation in the 1950s, the European Commission has served as the center of the EU's policy making. However, it did not set up a policy system dedicated to environmental issues until the 1970s, and the full DG Environment was not set up until 1981. The DG Environment was first perceived as a relatively weak DG. However, it has gradually become more assertive through the development of technical and political expertise.

14.2.2 Environmental Action Program

Environmental Action Programs (EAPs)[4] provide a general framework for the EU's environment policy. These programs define the most important medium- and long-term goals. They also determine basic strategy with appropriate concrete measures. The EAPs may date back to the UN Conference on the Environment held in 1972. That conference had agreed that a common community environmental policy was essential and the European Commission developed EAPs. The Treaty of Maastricht in 1992 created a contractual basis for EAPs. When the Treaty of Lisbon entered into force, this contractual basis was determined in Article 192(3) of the Treaty on the Functioning of the EU. Under this provision, EAPs are issued on the proposal of the Commission by the European Parliament and the Council in an ordinary legislative procedure, and are thus formal legislative acts.

Seven EAPs have been adopted so far, with durations ranging from three to 10 years: the first EAP (1973–1976), the second (1977–1981), the third (1982–1986), the fourth (1987–1992), the fifth (1993–2000), the sixth (2002–2012) and the seventh (2014–2020).

The first EAP emphasized the need for a comprehensive assessment on impacts of other policies in an effort to avoid damaging activities. In this way, the first EAP already contained many of the ideas behind "sustainable development." The first EAP devoted most of its attention to water protection and waste. It also contained a sectorial approach, with special reference to agriculture and spatial planning. It also mentioned preparatory activities for emissions control.

The second EAP was a follow-up to the first one in terms of approach and objective, along with simply a greater range of problems to be dealt with. Nature protection received special attention in the second EAP. The first and second EAPs determined a detailed list of actions to control a broad range of pollution issues. Eleven principles were listed and the priorities for environmental policies remained valid in subsequent action programs.

The third EAP tried to provide an overall strategy for protecting the environment and natural resources in the European Community. It shifted its emphasis from pollution control to pollution prevention. It also broadened the concept of environmental protection, including land use planning and the integration of environmental concerns into other European Community policies.

The fourth EAP gave substance to new obligations for integrating the environmental dimension into other Community policies by emphasizing four areas of activity. The four areas were: (a) effective implementation of existing Community legislation, (b) regulation of all environmental impacts of 'substances' and 'sources' of pollution, (c) increased public access and (d) dissemination of information and

[4]Web page of Federal Ministry for the Environment, Nature Conservation, Building and Nuclear Safety. See http://www.bmub.bund.de/en/topics/sustainability-international/europe-and environment/environment-action-programmes/.

job creation. This was an initial commitment for strategic reorientation of environmental policies in the European Community. "Sustainable development" gradually became a normative reference for environmental policy in the EU from the beginning of the 1990s until now.

Strategic reorientation became visible at the end of the fourth EAP. Then, it was explicitly formulated under the fifth EAP. The general approach and strategy of the fifth EAP differed from previous programs. As indicated by the title, "Towards Sustainability," the fifth one implied the program related to longer-term objectives and focused on a more global approach.

The fifth EAP legislation in the late 1990s was impressive, including new complex and holistic framework legislation, such as the Ambient Air Quality Directive and the Water Framework Directive. They formulated an ambitious work program for several decades. Furthermore, policy preparation at the EU level became much more participatory. Policy preparation invited environmental non-government organizations (NGOs) to play an important role in committees, expert networks and numerous consultation processes. This new movement might be slightly counterbalanced by influential industries that have lobby activities at all levels of the Commission.

The sixth EAP was a decision of the European Parliament adopted by the Council on 22 July 2002. It determined the framework for environmental policy-making within the EU for the annual periods 2002–2012 and outlined actions that needed to be taken to achieve them. The sixth EAP identified four priority areas: (a) climate change, (b) nature and biodiversity, (c) environment and health and (d) natural resources and wastes. The sixth EAP promoted the full integration of environmental protection requirements into all Community policies and actions. It also provided environmental components of the Community's strategy for sustainable development.

The objective of the seventh EAP was summarized in the title "Living well, within the limits of our planet." EU's environment policy had to steer a course between mankind's justified desires for well-being and limits set by the environment. This was underpinned by a vision for 2050 in the following manner:

> "In 2050, we live well, within the planet's ecological limits. Our prosperity and healthy environment stem from innovative circular economy where nothing is wasted and where natural resources are managed sustainably, and biodiversity is protected, valued and restored in ways that enhance our society's resilience. Our low-carbon growth has long been decoupled from resource use, setting the pace for a safe and sustainable global society."

To realize the vision, priority objectives to be achieved by 2020 have been identified for three areas: (a) natural capital, (b) resource-efficient, green and competitive low-carbon economy and (c) environment and health. These achievements are measured by four aspects: (a) implementation, (b) knowledge base, (c) environmental externalities and (d) coherence. They also have spatial dimensions: (a) sustainable cities and (b) international environmental protection.

14.3 JAPAN

As discussed by Sueyoshi and Goto (2014b), Japanese environmental pollution may date back to the industrial revolution in the late 1880s when pollution became very serious as the result of industrial development. Along with the progress of industrialization and urbanization, Japanese environmental pollution was considered as the deterioration of public health such as air pollution, water pollution, noise, vibration and bad odor. Such historic public concern about industrial development is the same as that of today. See a detailed description on Japanese industrial development and pollution that can be found in the research by the Society for Environmental Economics and Policy Studies (2006)[5].

In Japan, environmental pollution problems before World War II occurred in mining industries. During the Meiji Period (1868–1912), Japan owned and operated the world's top class copper mines. The production and export of copper constituted a large share of the Japanese economy. At that period, Japanese copper mines contained 40% sulfur, producing major damage to people's health and the environment by sulfurous acid gas.

An initial case of Japan's environmental pollution could be found in the Ashio mineral poison case, in which disposed mineral ore badly influenced human life. The mineral ore flowed out to the Watarase River, killing fishes and reducing the rice harvest by contaminating the rice paddies. Water contamination continued to be trouble until after World War II. Meanwhile, there were cases of air pollution and poisoning by sulfurous acid gas from minerals at Bessi, Hitachi and Kosaka. In these cases, the injured farmers negotiated with the mineral mine companies. The companies' engineers made efforts to improve their pollution problems by environmental technologies, such as the world's highest chimney at that time and advanced equipment for the desulfurization of flue gas.

After World War II, Japan experienced a rapid economic growth during the 20 years since 1955 so that it could facilitate heavy chemical industrialization and nationwide urbanization. Along with such economic growth, many companies produced various types of industrial pollution, because they prioritized their economic success over environmental safety. Environmental regulation by central and local governments was not effective in guiding Japanese corporations toward environmental safety. The two most well-known environmental pollution cases, "Minamata" disease (using the name of the polluted region) and "Itai-itai" disease (implying a very serious pain in Japanese), occurred due to water pollution. Unfortunately, these serious pollution problems could not be solved immediately, rather they were gradually changed, with support for injured people from antipollution public opinion and lawsuits against the firms that caused the industrial pollution. People's conscientiousness about the environment gradually influenced

[5] *Environmental Economics and Policy Studies: Basic Facts and Concepts* (in Japanese) (2006) Society for Environmental Economics and Policy Studies, Yuhikaku, Tokyo.

Japanese industrial policy in such a manner that firms increased the amount of investment for pollution prevention and public agencies increased the number of officials for environmental protection.

An important feature of Japanese environmental policy was that it was not primarily guided by public opinion, rather it was influenced by international pressure on environmental issues. The pressure was usually initiated by the progress of internationalization of environmental pollution such as global environmental issues and cross-border pollution problems. For example, to attain "sustainable development" proposed by the UN Conference on Environment and Development held in 1992, Japan promoted not only a pollution prevention policy but also other related issues that transformed Japan toward sustainable development as required by international organizations. Under such international pressure, the Japanese government began challenging the issues of pollution and environment protection by the Basic Environment Law enacted in 1993, the Act on the Promotion of Global Warming Countermeasures enacted in 1998, the Basic Law for Establishing the Recycling-based Society enacted in 2000 and the Basic Act on Biodiversity enacted in 2008. In 2001, the Environment Agency was upgraded to the Ministry of the Environment under central government reorganization. The mission of the Ministry of the Environment includes protecting the global environment, preventing pollution, protecting the natural and other environments and ensuring safety in nuclear research, development and use.

Japan is now directing toward a sustainable society where economic success can coexist together with environmental safety. To attain the goal of Japanese industrial policy on environmental protection, all public and private entities have started paying more serious attention to industrial pollution issues than before. In particular, during the past decade, firms faced increasing pressures to enhance "corporate sustainability" beyond their financial performance measures. Demands for corporate sustainability, as part of global sustainability in the world, are motivated by various business factors, including external ones such as risk of regulatory mandates, fear of loss of sales, potential decline in reputation and internal factors such as expectations for potential improvement in productivity through eco-technology innovation for environmental protection. Among such sustainability challenges faced by many companies, the control of greenhouse gas (GHG) emission is one of the most imminent tasks for their business survivability.

14.4 CHINA

According to the historical description of Chinese environmental protection, prepared by Sueyoshi and Yuan (2015a, 2016b), Chinese representatives attended the first UN Conference held in Sweden on the Human Environment in 1972. On 5 August 1973, the Environmental Protection Leadership Commission was first established at the National Environmental Protection Conference of China. The members of the commission were leaders of more than 20 related national departments. By then, the country's environment was already in a serious situation that was further exacerbated by economic reforms of the late 1970s. Under the leadership of Deng Xiaoping in 1978, the reforms boosted China's industrial outputs at an average annual growth rate of more than 11.4%[6]. In the same year, China has amended the constitution to add the following statement: "The state protects the environment and natural resources. It also prevents and controls pollution and other public hazards." The amendment formed the constitutional foundation for the country's administrative framework for environment protection. In 1979, China passed the Environmental Protection Law for trial implementation. This law included a chapter on the governmental structure and responsibilities, which required all related departments under the State Council, as well as provincial and municipal governments, to set up specialized environmental protection and supervision institutes. China's environmental protection framework was thereby formally enacted into law.

China has conducted the first national administrative reform after the transformation for a market mechanism in the early 1980s. As part of the reform, the Environmental Protection Leadership Commission was dissolved and the Environmental Protection Agency was set up as part of the newly formed Ministry of Urban Construction and Environmental Protection[7]. Thus, environmental protection belonged to the category of urban construction, specifying the government's responsibility to protect the environment as secondary. In 1983, the Chinese government announced that environmental protection became a state policy. The Air Pollution Prevention and Control Law was enacted in 1987. In 1989, the Standing Committee of the National People's Congress set up the administrative framework by making the 1979 Environmental Protection Law permanent. The decision strengthened the governmental responsibility and authority over environmental protection. Later, more laws were enacted, such as the Energy Conservation Law of 1997, and several important international agreements, such as the Kyoto and Montreal protocols.

The legacy of decentralization characterized by Deng's reforms remained at the heart of China's environmental struggles[8]. The reforms diffused authority to the provinces, creating a proliferation of Township and Village Enterprises (TVEs) to encourage development in rural industries. In 1997, TVEs generated almost one-third of the national Gross Domestic Product (GDP). However, local governments had difficulty monitoring their performance and therefore seldom upheld environmental standards. Today, environmental policies remain difficult in these

[6] History of Pollution: economics, politics, and Deng XiaoPing. See https://edblogs.columbia.edu/scppx3335-001-2014-1/2014/03/12/history-of-pollution-economics-politics-and-deng-xiaoping/.

[7] Qiu, X. and Li, H. (2009) China's environmental super ministry reform: background, challenges, and the future. *Environmental Law Reporter*, **39**, 10152–10163.

[8] History of Pollution: economics, politics, and Deng XiaoPing. See https://edblogs.columbia.edu/scppx3335-001-2014-1/2014/03/12/history-of-pollution-economics-politics-and-deng-xiaoping/.

enforcements at a local level, where officials often retain their economic incentives and ignore environment protection.

In 1998, the Chinese government upgraded the Leading Group to a ministry-level agency, which became the State Environmental Protection Administration (SEPA). However, although SEPA was directly under the State Council's control, it was still not at the cabinet level and did not have a voting power in the Council's decisions. The Clean Production Promotion Law, enacted in June 2002, established demonstration programs for pollution regulation in 10 major Chinese cities and designated several river valleys as priority areas for the cleaning-up of pollution.

From 2001 to 2005, Chinese environmental authorities received more than 2.53 million letters and 430,000 visits by 597,000 petitioners who requested environmental redress. The number of mass protests caused by concerns over environmental issues grew steadily from 2001 to 2007. The increased attention on environmental matters caused the Chinese government to display an increased level of concern towards environmental issues and subsequently it implemented strict environmental regulations. Consequently, subsidies for some polluting industries were cancelled, while other polluting industries were shut down. However, many internal environmental targets were still missed in China's industrial development[9,10].

After 2007, the influence of corruption was a hindrance to effective enforcement, because local authorities often ignored orders and hampered the effectiveness of central decisions. In response, the Communist Party of China (CPC) implemented the "Green" project, where China's GDP was adjusted to compensate for negative environmental effects. However, the program quickly lost official influence due to unfavorable data.

On 15 March 2008, China's Eleventh National People's Congress passed the Super Ministry Reform (SMR) motion, which was proposed by the State Council, and created five "super ministries," mostly combinations of two or more previous ministries or departments. The main purpose of the SMR was to avoid overlapping governmental responsibilities by combining departments with similar authority and closely related administrative functions. One of the highlights was the elevation of SEPA to the Ministry of Environmental Protection (MEP), which is also referred to as the environmental SMR. In the SMR, MEP was upgraded and was the only department to retain its organizational structure and governmental responsibilities. Since its status as a cabinet member is protected by law and cannot be changed by the State Council, the establishment of MEP demonstrates the strong political direction and commitment of China's central government to environmental protection.

[9] McCubbin, P.R. (2009) China and climate change: domestic environmental needs, differentiated international responsibilities, and rule of law weaknesses. *Environmental and Energy Law and Policy Journal*, **3**, 200–235.

[10] Joseph Kahn and Jim Yardley (2007) As China roars, pollution reaches deadly extremes. *The New York Times*, August 26, New York.

Citizen activism in government decisions increased in the 2010s[11] and more than 50,000 environmental protests occurred in China during 2012. In response to increasing air pollution problems, the Chinese government announced a US$ 277 billion plan for five years to address the issue in 2013. According to the Xinhua News Agency report, a policy effort including an investment of 950 million Yuan was made by the Chinese government to build an information broadcast system, referred to as the "National Environmental Air Quality Monitoring Platform," which started to operate on 1 January 2013. Since then, the real-time air quality index, including PM 2.5, PM 10, SO_2, NO_2 and CO_2, could be retrieved from the platform (http://113.108.142.147:20035/emcpublish/). Northern China will receive particular attention, as the government aims to reduce air emissions by 25% by 2017, compared with 2012 levels[12].

In March 2014, the CPC "declared war" on pollution during the opening of the National People's Congress. The parliament approved new environmental law in April 2014 and the new environmental protection provisions went into effect in January 2015[13]. The new law empowered environmental enforcement agencies with great punitive power, defined areas which require extra protection and gave independent environmental groups more ability to operate in the country[14]. Companies that break the law will be "named and shamed," with company executives subject to prison sentences of 15 days. There will be no upper limit on fines. In all, the new law has 70 provisions, compared to the 47 of the existing law. More than 300 different groups will be able to sue on behalf of Chinese people harmed by pollution[15].

Under the new law, local governments will be subject to discipline for enforcing environmental laws. Regions will no longer be judged solely on their economic progress, but instead must balance progress with environmental protection[16]. Additionally, local governments will be required to disclose environmental information to the public. Individuals are encouraged by the law to "adopt a low-carbon and frugal lifestyle and perform environmental protection duties" such as recycling their garbage.

Starting from April 2014 and passing in October 2015, China has proposed the "13th five-year" plan (2016–2020). It is hoped that the green economy will be the most important part of the 13th five-year plan. As the world's largest carbon

[11] Bradsher, K. (2012) Bolder protects against pollution win project's defeat in China. *The New York Times,* July 4, New York.

[12] Upton, J. (2013) "China to spend big to clean up its air." *Grist.* July 25, Grist Magazine, Inc, New York.

[13] Jabeen, A., Sheng, H.X. and Aamir, M. (2015) Environmental stability still in danger, loopholes of new environmental protection law in China. *US-China Law Review*, **12**, 951–964.

[14] Valli, R. (2014) China revises environmental law. *Voice of America,* April 25, New York.

[15] Environmental policy in China. See https://en.wikipedia.org/wiki/Environmental_policy_in_China.

[16] Environmental policy in China. See https://en.wikipedia.org/wiki/Environmental_policy_in_China.

emitter, China's transition to the green economy will be important from an international perspective. Currently, China no longer relies solely on GDP for assessing officials' performance. If environmental protection is not ensured, officials are classed as not up to standard. The CPC's chairman, or Xi Jinping, has discussed the importance of green mountains and clean waters in his major speech. This may be a good signal for environmental protection in China. The energy plan will build safe, clean, efficient and sustainable modern energy systems that have proposed strategic tasks for increasing domestic energy productions. These strategies include (a) promotion of energy conservation and efficiency, (b) optimization of energy structure, (c) vigorous developments of hydro, nuclear, wind, solar and geothermal energies, (d) encouragement of international cooperation, (e) promotion of energy technologies and institutional innovation, and (f) strengthening energy regulation.

The goal is energy security to find a balance between operational efficiency and cleanness so that China can achieve the sustainable development of energy[17].

14.5 THE UNITED STATES OF AMERICA

14.5.1 General Description

The Fish and Wildlife Coordination Act was the first law to protect the environment of the United States (US). It was enacted on 10 March 1934 to protect fish and wildlife when federal actions resulted in control or modification of a natural stream or body of water. The Act provided basic authority for the involvement of the US Fish and Wildlife Service in evaluating impacts on fish and wildlife from proposed water resource development projects.

On 25 June 1947, the Federal Insecticide, Fungicide, and Rodenticide Act was created which was another US federal law that was set up to the basic system of pesticide regulation to protect applicators, consumers, and environment. Later, the Federal Water Pollution Control Act was enacted in 1948 and the Air Pollution Control Act was applied in 1955.

Beginning in the late 1950s and through the 1960s, Congress reacted to increasing public concern about the impact that human activity could have on the environment. A key legislative option to address the concern was the declaration of a national environmental policy[18]. Advocates argued that without specific

[17] See Zhang Chun and Xu Nan, on 9 March 2014: "Green sector will boom during China's 13th five-year plan." See https://www.chinadialogue.net/article/show/single/en/7280-Green-sector-will-boom-during-China-s-13th-Five-Year-Plan.

[18] Luther, L. (2008) *The National Environmental Policy Act: Background and Implementation*. Congressional Research Service, Washington D.C.

policy, federal agencies were neither able nor inclined to consider environmental impacts of their actions in fulfilling the agency's mission. The statute that ultimately addressed the issue was the National Environmental Policy Act (NEPA) in 1969[19].

The law was signed by President Nixon on 1 January 1970. NEPA was the first of several major environmental laws passed in the 1970s. It declared a national policy to protect the environment and created a Council on Environmental Quality (CEQ) in the Executive Office of the President. To implement national policy, NEPA required that a detailed statement of environmental impacts should be prepared for all major federal actions significantly affecting the environment. The "detailed statement" would ultimately be referred to as an "Environmental Impact Statement (EIS)."

In 1970, President Richard Nixon proposed an executive reorganization that consolidated many of federal government's environmental responsibilities under one agency, the new Environmental Protection Agency (EPA). The reorganization proposal was reviewed and passed by the House and Senate[20]. It was created for the purpose of protecting human health and the environment by writing and enforcing regulations based on laws passed by Congress and began operation on 2 December 1970. The agency was led by its administrator, who was appointed by the President and approved by Congress.

For at least 10 years before NEPA was enacted, Congress debated issues that the act would ultimately address[21]. The act was modeled in the Resources and Conservation Act of 1959. The bill would have established an environmental advisory council in the office of the President, declared national environmental policy and required the preparation of an annual environmental report. In the years, following the introduction of Senator Murray's bill, similar bills were introduced and hearings were held to discuss the state of the environment and Congress's potential responses to perceived problems. In 1968, a joint House–Senate colloquium was convened by the chairmen of the Senate Committee on Interior and Insular Affairs and the House Committee on Science and Astronautics to discuss the need for and potential means of implementing national environmental policy. In the colloquium, some members of Congress expressed a continuing concern over federal agency actions affecting the environment[22].

The EPA conducts environmental assessment, research and education. It has the responsibility of maintaining and enforcing national standards under a variety

[19] The Guardian: Origins of the EPA. *EPA History webpage*. US Environmental Protection Agency website. 1992.

[20] US Environmental Protection Agency. *When and How Was EPA Created*?

[21] Jackson, H.M. "Scoop" (1969-07-09). Senate Report No. 91-296-National Environmental Policy Act of 1969. *Calendar 287*, U.S. Congress, p. 11.

[22] Luther, L. (2008) *The National Environmental Policy Act: Background and Implementation*. Congressional Research Service, Washington D.C.

of environmental laws, in consultation with state, tribal and local governments. It delegates some permit-granting, monitoring and enforcement responsibility to US states and federally recognized tribes. EPA enforcement powers include fines, sanctions and other measures. The agency also works with industries and all levels of government in a wide variety of voluntary pollution prevention programs and energy conservation efforts.

The EPA began regulating GHG emissions from mobile and stationary sources of air pollution under the CAA for the first time on 2 January 2011. Standards for mobile sources were established pursuant to Section 202 of the CAA, and GHG emissions from stationary sources were controlled under the authority of Part C of Title I of the Act.

14.5.2 Regional Comparison between PJM and California ISO

After a general description on US environmental protection efforts, this chapter will describe regional differences in terms of pollution prevention, focusing upon the electric power industry as a representative secondary energy, because the United States have environmental regulation policies that are different among regions at the level of states. See Chapter 13 about energy classification. This subsection compares East and West regions for our illustrative purpose. It is easily imagined that our description on regulation differences in the electric power industry is applicable to other industries in the United States.

As discussed by previous research efforts such as Sueyoshi and Goto (2010b, 2012b,e,f,g, 2015b, 2016), US power plants are a major source of air pollution. In particular, in the United States, roughly two-thirds of all SO_2 and one-quarter of all NO_x come from electric power generation that relies on burning fossil fuels[23]. Coal-fired power plants account for a large share of these numbers. The power plants are also the largest polluter of toxic mercury and the largest contributor of hazardous air toxics. Although coal-fired power plants produced approximately 30% of electricity in 2016 in the United States[24], they were responsible for over 68% of the carbon dioxide (CO_2) pollution from the electric power sector.

In North America, there are Independent System Operators (ISOs) and Regional Transmission Organizations (RTOs). PJM Interconnection, which is the most well-known among them, operates a wholesale electricity market in the north-east part of the United States. The organization manages a high-voltage electricity grid network to ensure reliability for more than 60 million people in all or parts of 13 states (Delaware, Illinois, Indiana, Kentucky, Maryland, Michigan, New Jersey, North Carolina, Ohio, Pennsylvania, Tennessee, Virginia, West Virginia) and the District of Columbia. PJM's utility members, more than 750, include

[23] See https://www.epa.gov/acidrain/what-acid-rain.

[24] See https://www.eia.gov/tools/faqs/

power generators, transmission owners and electricity distributors. The headquarters is located at Valley Forge, Pennsylvania. While PJM began in 1927 as a power pool, today the organization operates the region's power grid and wholesale electricity market as a RTO.

PJM's function, as a federally regulated RTO, is to act independently in managing the region's transmission system and the wholesale electricity market. PJM ensures the reliability of the largest centrally dispatched grid in North America. The voltage in PJM's transmission grid is monitored 24 hours a day and seven days a week in order to keep electricity supply and demand in balance by directing power generators as to how much energy needs to be produced. The company dispatches about 185,600 MW (megawatt) of generating capacity over 62,591 miles of transmission lines. In addition, PJM has system operators that conduct dispatch operations and monitor the status of each grid, using telemetered data from virtually 74,000 points across the whole grid system.

California ISO was established in 1996 to support the electric power industry. California ISO manages the flow of electricity across 80% of California's power grid and serves a population of nearly 30 million people. The company was formed under the direction of FERC to operate a robust and reliable power system that balances between high transmission reliability and low cost. The company is now a key player in working with the state of California to achieve its clean energy goals. As an objective grid operator, California ISO ensures open access to its wholesale power market that is designed to diversify resources and improve operational efficiency. The organization gives equal access to approximately 25,865 circuit-miles of power lines and reduces barriers to diverse resources which compete to bring power to customers. The organization operates independently to manage the power grid traffic and to ensure an amount of electricity that is safely and reliably delivered to utilities and consumers.

14.5.3 Federal Regulation on PJM and California ISO

The two organizations are regulated by EPA. Under the CAA, there are three major environmental requirements:[25] (a) Mercury and Air Toxics Standards (MATS), (b) Reciprocating Internal Combustion Engines and National Emission Standards for Hazardous Air Pollutants (RICE NESHAP) and (c) Greenhouse Gas Tailoring Rule (GGTR). Besides the federal requirements, PJM and California ISO are also bound by different state and regional air quality regulations.

As the first requirement, MATS applies the CAA Maximum Achievable Control Technology (MACT) requirement to new or modified sources of emissions regarding mercury, arsenic, acid gas, nickel, selenium and cyanide.

[25] See https://www.epa.gov/clean-air-act-overview.

Emissions standards set under MATS indicate federal air pollution limits that individual facilities must meet by a set date. The EPA must set emission standards for existing sources in an appropriate category that are at least as stringent as the emission reductions achieved by the average of the top 12% best controlled sources. These rules set technology-based emission limitation standards for mercury and other toxic air pollutants, reflecting levels achieved by the best-performing sources currently in operation. Thus, the rules set standards for all hazardous air pollutants (HAPs)[26] emitted by coal- and oil-fired electric generating units (EGUs) with a capacity of 25 MW or greater.

Regarding the second requirement, RICE NESHAP sets rules for regulating owners and operators of RICE, including stationary electrical generation facilities, and applies it for reducing emissions of toxic air pollutants such as formaldehyde, acetaldehyde, methanol, CO_2, NO_x, volatile organic compounds and other particulates[27]. It is implemented in February 2010 and issued by the EPA. The standard applies to existing equipment for non-emergency applications and does not apply to equipment used only in emergency applications. As a minimum, the affected stationary diesel engines must comply with CO_2 emission limits or must be fitted with emission control equipment, such as diesel oxidation catalysts, to reduce CO_2 emissions by 70%. The regulation includes a number of other provisions, including work practices for engine operators. While the regulation does not mandate the emission control technology, the EPA has designed the standards based on the capabilities of the diesel oxidation catalyst (DOC)[28].

Finally, the GGTR was issued in May 2010 to regulate CO_2 and other GHG emissions under the existing framework of New Source Review and Prevention of Significant Deterioration. Thus, the EPA provides guidance on new or modified units to install or implement the best available control technology (BACT), which gives a way to state environmental regulators that determine BACT on a project by project basis. This rule is policy aimed at shielding small polluters from strict requirements permitted by the CAA[29].

14.5.4 Local Regulation on PJM

The PJM not only follows federal regulations but also follows state and regional regulations. Notably, New Jersey, Connecticut, Delaware, Maine, Maryland, Massachusetts, New Hampshire, New York, Rhode Island and Vermont have separate state regulation requirements. Similarly, the Cross State Air Pollution Rule (CSAPR) regulates the region encompassing PJM.

[26] See https://www.epa.gov/matsl.

[27] See http://www.monitoringanalytics.com/reports/PJM_State_of_the_Market/2017.shtml.

[28] See http://www.monitoringanalytics.com/reports/PJM_State_of_the_Market/2017.shtml.

[29] See http://www.monitoringanalytics.com/reports/PJM_State_of_the_Market/2017.shtml.

The Regional Greenhouse Gas Initiative (RGGI) is the nation's first mandatory cap and trade program for GHG emissions. RGGI involves nine states: Connecticut, Delaware, Maine, Maryland, Massachusetts, New Hampshire, New York, Rhode Island and Vermont. The RGGI cap and trade system applies only to carbon dioxide emissions from electric power plants with capacities to generate 25 MW or more – approximately 168 facilities. The RGGI emissions cap took effect on 1 January 2009, based on an agreement signed by RGGI governors in 2005[30]. New Jersey was an initiation member of RGGI, but it withdrew from RGGI due to the Governor's decision, who said the program was not effective in reducing GHGs and the whole system was not working as intended.

The CSAPR was finalized by the EPA under a provision of the CAA to reduce transported pollution. CSAPR defined upwind state obligations to reduce pollution significantly contributing to downwind non-attainment and maintenance areas based on the magnitude of a state's contribution, cost of controlling pollution from various sources and air quality impact of reductions. The intent of CSAPR is to reduce emissions of SO_2 and NO_x from power plants in the eastern half of the United States. This rule seeks to reduce air pollution that damages the ozone and results in the emission of fine particles, and put in place a framework to address pollution that affects air quality in downwind states. The EPA anticipates improved efficiency at existing sources, improved performance of pollution control equipment, a load shift to existing cleaner units, the use of lower sulfur coal and a switch among fuel mixes[31].

14.5.5 Local Regulation on California ISO

The California ISO is regulated by two different organizations. The first is the Federal Energy Regulatory Commission (FERC) which is an independent federal agency that regulates the interstate transmission of electricity, natural gas and oil. It is also regulated by the California Public Utilities Commission (CPUC), which regulates investor-owned utilities that operate in the ISO balancing authority area. The two organizations provide guidance not only on rates but also on monitoring GHG emissions produced from the operation of utility facilities[32].

The FERC is the leading agency under the California Environmental Quality Act (CEQA) that requires state and local agencies to identify the significant environmental impacts of their actions and to avoid or mitigate those impacts, for all thermal power plants (50 MW or more) that are proposed for construction and operation in the state. The Energy Commission's licensing process is the equivalent of the CEQA environmental impact report process, which is certified as an

[30] Ramseur, J.L. (2016) The regional greenhouse gas initiative: Lessons learned and issues for congress. *Congressional Research Service*, **7**-5700, R41836.

[31] Ramseur, J.L. (2016) The regional greenhouse gas initiative: Lessons learned and issues for congress. *Congressional Research Service*, **7**-5700, R41836.

[32] See https://www.ferc.gov/.

equivalent regulatory program. The FERC responsibilities that directly affect California ISO are its approval of rates for wholesale power and transmission in interstate commerce for jurisdictional utilities, power marketers, power pools, power exchanges and independent system operators; review of rates set by the federal power marketing administrations; and certification of qualifying small power production and cogeneration facilities. However, the FERC's main responsibility is to determine whether new and ongoing projects have a significant adverse environmental impact resulting from GHG emissions, and if needed, to mitigate those impacts[33].

Secondary to the FERC, the CPUC regulates investor-owned electric and natural gas utilities operating in California. The CPUC is guided by California's Air Resources Board (CARB), the "clean air agency" in the government of California. California is the only state that is permitted to have such a regulatory agency, as it was the only state to have one before the federal CAA. As with any other air-quality institution, the goals of the CARB include maintaining healthy air quality, protecting the public from exposure to toxic air contaminants and providing innovative approaches for complying with air pollution rules and regulations. The CPUC, through its oversight over these utilities, plays a key role in improving California's energy-related initiatives designed to benefit consumers, the environment and the economy. Under the CARB, California ISO must follow the California Global Warming Solutions Act of 2006, which is a comprehensive and multi-year program to reduce GHG emission levels, by 2020, to the same levels as those experienced in 1990. Among the reduction programs, the CARB notably implements measures to reduce sulfur hexafluoride emission. This is a very potent GHG, with a global warming potential that is approximately 23,000 times more powerful than CO_2. The CARB staff has included the reduction of sulfur hexafluoride from gas insulated switchgear as a possible emission reduction measure within its plan[34].

Table 14.1 summarizes different air quality requirements on PJM and California ISO. The PJM follows air quality policies as mandated by EPA as well as RGGI that put a cap on carbon dioxide emissions produced by power generation. Since PJM primarily uses coal for electricity generation, this cap has a negative impact on electricity generation, which helps to promote power plants to use newer equipment and processes to generate electricity. In addition, PJM has policies like the CSAPR that monitors and regulates air pollution produced from electricity generation and transferred from state to state.

Meanwhile, the California ISO follows policies set forth by the EPA as well as the CEQA that monitors the impact of new and ongoing projects, particularly thermal electricity generation of 50 MW or more. In addition, the California ISO has an ongoing goal to reach 1990 GHG emission levels by the year 2020 as well as monitoring and regulating sulfur hexafluoride.

[33] See https://www.ferc.gov/.

[34] See https://www.arb.ca.gov/homepage.htm.

TABLE 14.1 PJM and California ISO on air quality requirements

	PJM	California ISO
Follows EPA air policies:	X	X
MATS (monitoring hazardous air pollutants emitted by coal- and oil-fired electric generating units – capacity of 25 MW or greater)		
RICE NESHAP (reduction of air pollutants caused by stationary electrical generation facilities)		
Greenhouse Gas Tailoring Rule (implementation of the best available control technology)		
Follows policies set by California Environmental Quality Act (environmental impact of new and ongoing projects, e.g., thermal power plants – 50 MW or more)		X
Follows guidance set by California's Air Resources Board (a multi-year program to reduce GHG emissions to the 1990 level by 2020 and monitoring levels of sulfur hexafluoride)		X
Follows policies set by Regional Greenhouse Gas Initiative (an effort to cap CO_2 emissions produced by power generation facilities)	X	
Follows guidance under Cross State Air Pollution Rule (a provision to reduce transported pollution – downwind attainment and maintenance areas)	X	

(a) Source: Sueyoshi and Goto, (2013a). This chapter updates their work for our description.

14.6 SUMMARY

Before discussing how to apply DEA for environmental assessment, this chapter provided a historical view on ongoing policy making efforts for environmental protection in four industrial regions in the world. This retreat is important in understanding why we are now facing various industrial pollutions (e.g., global warming and climate change), along with their industrial developments and economic growths. This chapter has reviewed such previous and current efforts for environmental protections in Europe, Japan, China and the United States as our descriptive examples.

Implications for DEA: A lesson from this chapter is that the environmental assessment needs to incorporate not only desirable outputs (e.g., GDP and electricity), related to activities in industry and whole economy, but also undesirable outputs (e.g., CO_2, GHG emissions and water contamination), related to various environmental pollutions, in order to measure a level of sustainability. The first component is for industrial deregulation and international competition, while the second has been under various governmental regulations, depending upon each nation's industrial structure and development. As mentioned previously, the conventional use of DEA was based on economic theory that did not consider the existence of undesirable outputs in performance assessment, as discussed in Section I of this book. Thus, a straightforward use of DEA provides limited practical and theoretical implications for the development of a sustainable society. See Chapter 12 on our concerns for the conventional use of DEA. A new use of the proposed DEA environmental assessment in Section II maximizes the level of operational efficiency for economic success, but simultaneously minimizes the level of environmental pollution. The two components should be carefully balanced to attain a high level of sustainability. See Chapter 15 on the definition on sustainability within the framework of DEA environmental assessment proposed in this book.

Finally, the DEA approach may be not the best methodology in terms of environmental assessment and protection. There are many methodological alternatives. However, DEA has a high level of analytical capability and practical potential to assess economic success and industrial pollution prevention achievements in a unified manner. The assessment, as an initial step for sustainability development, is important in attaining a balanced development between economic prosperity and pollution prevention. The remaining chapters in Section II will provide a detailed description on DEA environmental assessment on many different sectors (e.g., industries and nations).

15

CONCEPTS

15.1 INTRODUCTION

The Intergovernmental Panel on Climate Change (IPCC; http://ipcc.ch/index.htm), established by the United Nations (UN) environmental program, reported the policy suggestion in April of 2014 that it is necessary for us to reduce the amount of greenhouse gas (GHG) emissions, in particular CO_2, by 40–70% (compared with 2010) until 2050 and to reduce the level to almost zero by the end of the twentyfirst century via shifting our current systems to energy-efficient ones. Otherwise, it warns that global warming and climate change will destroy our natural and socio-economic systems. Consequently, we will have to face various risks (e.g., heat waves, droughts, floods and food crises, as well as damage to human, social and economic systems) on the earth. Consequently, we will lose our social sustainability whose current achievement requires managing economic developments and environmental protection in a similar way to synchronized swimming.

Although it is ideal to establish ecological societies without producing any GHG emissions, global warming and climate change have been influencing not only our social systems but also corporate behavior and operations in real business, because every private sector needs to change their business strategies to adapt various regulation changes on preventing industrial pollution.

Environmental Assessment on Energy and Sustainability by Data Envelopment Analysis,
First Edition. Toshiyuki Sueyoshi and Mika Goto.

It is easily imagined that such an environmental implication, as discussed above, is immediately applicable to organizations in public sectors, because they have social responsibility and accountability to people for whom they provide various services. The environmental implication becomes more seriously applicable to private sectors. For example, the more environment-conscientious consumers refuse to purchase products and services from dirty-imaged companies even if their prices are much lower than those of green-imaged companies. The conventional business philosophy and practice (e.g., less expensive and higher quality) may not function anymore in modern business because all firms belong to part of a world-wide trend toward the development of a sustainable society. Thus, all the organizations in private sectors cannot survive without paying attention to various pollution issues as going concerns that have been operating in ecologically conscientious markets.

The benefits from adapting GHG reduction technologies in private sectors range from intangible ones, such as improved public image as a green corporate citizen, to measurable ones such as lower direct and indirect emission levels. Unfortunately, while acknowledging the importance of reducing the amount of GHG emissions, many organizations often misunderstand the strategic linkages between cost of employing GHG reduction technologies and enhancement of their overall performance and business opportunities. It may be true in a myopic horizon that environmental protection needs a large amount of investment for the reduction of GHG emissions. It is believed that the investment that does not produce any direct benefit to firms.

However, such a business concern may be different in a long-term horizon. As discussed by Porter and van der Linde (1995)[1], environmental regulation does not jeopardize corporate performance, rather it provides firms with a new business opportunity to improve efficiency and competitiveness through environmental innovation in their manufacturing processes and products. Such examples can be easily observed in many new products such as hybrid cars, electric vehicles and clean coal technology. In modern business reality, some companies clearly understand the synergy effect between their investments for low GHG emissions, including low-carbon technologies, and enhancements in their operational performance. It is easily thought that organizations with green corporate images become more competitive in today's environment-conscientious markets because they are more efficient in reaping benefits from their sustainability by making effective investment in technology for low GHG emissions. This view clearly indicates that modern corporations in all industrial sectors need to consider their eco-technology investments on environmental protection and corporate performance enhancement from the perspective of corporate sustainability in short and long-term horizons.

A difficulty associated with attaining a high level of sustainability is that policy makers, corporate leaders and academia do not have a common practical

[1] Porter, M. and van der Linde, C., 1995. Toward a new conception of the environment–competitiveness relationship. *Journal of Economic Perspectives*, **9**, 97–118.

methodology to assess the performance of organizations in terms of their operational and environmental achievements. Moreover, there is no appropriate methodology that can assist them in preparing their investment strategies to attain a high level of sustainability. As mentioned in Chapter 14, DEA is such a methodology to assess the environmental performance and more broadly the level of sustainability. As an initial step for the methodology development, this chapter is concerned with a description on strategic concepts incorporated in the proposed DEA environmental assessment.

The remainder of this chapter is organized as follows: Section 15.2 describes the role of DEA in measuring the unified (operational and environmental) performance of organizations. Simply, this section explains a rationale regrading why DEA is important in environmental assessment. Section 15.3 discusses differences between social and corporate sustainability concepts. Section 15.4 describes strategic adaptation to various environmental changes. Section 15.5 extends them to natural and managerial disposability, both of which serve as a conceptual basis for the proposed DEA environmental assessment. Section 15.6 discusses a unification process between disposability concepts. Section 15.7 visually describes a reason concerning why DEA environmental assessment is much more difficult than the conventional use of DEA discussed in Section I. Section 15.8 describes a possible occurrence of desirable congestion, or eco-technology innovation, for reducing the amount of undesirable outputs. Section 15.9 compares the proposed ones with the previous concepts that have been widely used for DEA. Section 15.10 summarizes this chapter.

15.2 ROLE OF DEA IN MEASURING UNIFIED PERFORMANCE

As discussed in Chapter 14, it is indeed true that many nations have been working on the reduction of various industrial pollutions and climate change issues. However, most of their policy efforts were not as successful as expected. A rationale was that the previous policy efforts did not have an appropriate methodology to holistically evaluate all economic activities regarding both economic success and environmental protection. To overcome such a methodological difficulty, we propose a new use of DEA for environmental assessment, hereafter, in the proceeding chapters of Section II.

Before discussing the computational framework of DEA environmental assessment, this chapter needs to describe a role of DEA in measuring unified (operational and environmental) performance. Figure 15.1 visually describes the importance of DEA as part of "social intelligence" for both mitigating the level of environmental pollutions and enhancing the status of economic prosperity. Each organization in public and private sectors needs to have all such intelligence capabilities to increase the level of social sustainability. A methodology selection (e.g., DEA in this book) is determined within such an intelligence capability, where "intelligence" implies a social capability by which an organization can handle

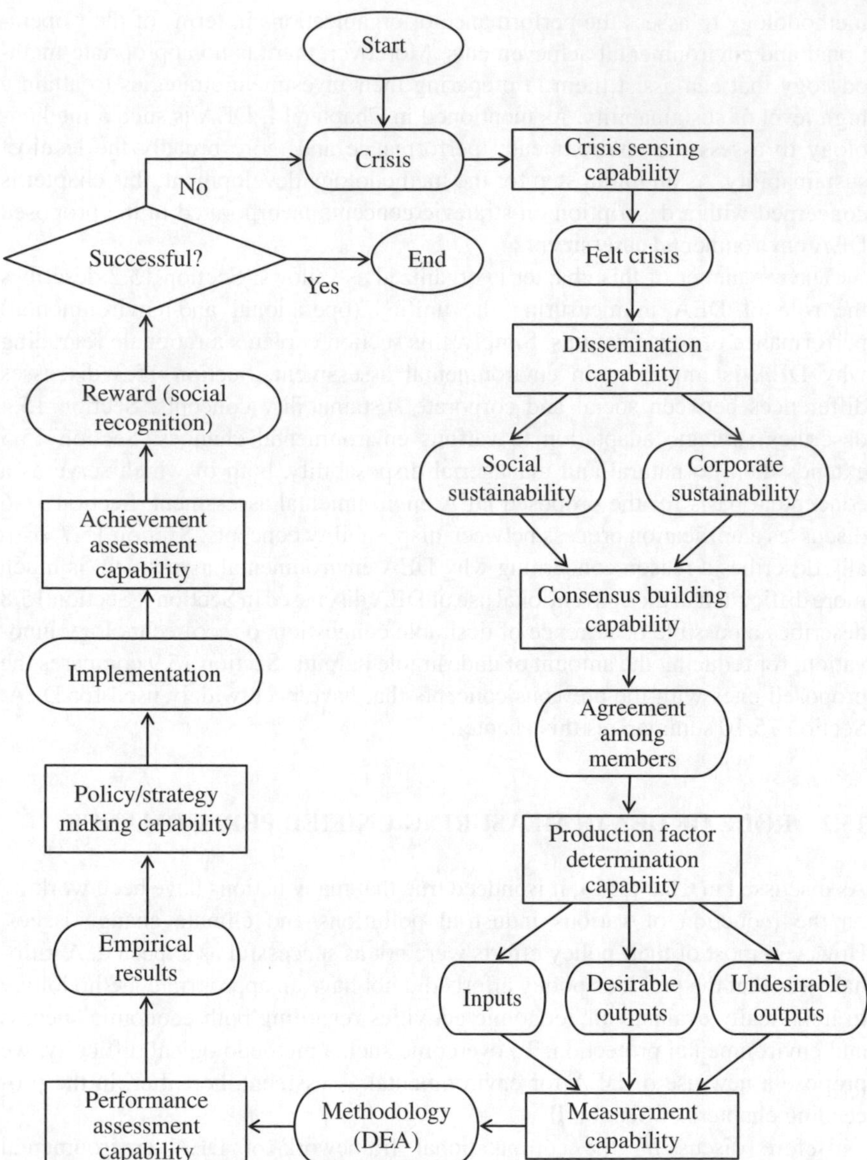

FIGURE 15.1 Position of DEA in environmental assessment (a) This figure was first proposed by Sueyoshi and Goto (2015a). This chapter reorganizes it by incorporating a consensus building capability into the original figure.

various decisional issues. The social capability may function for systematic adjustment to various changes (e.g., a regulation change on industrial pollution), so not directly linking to any conventional engineering context (e.g., artificial intelligence in computer science). It is true that such a social capability depends upon the utilization of a data processing capability equipped in a modern personal computer and other types of computing devices that can simultaneously deal with a large database, often called "big data," related to economy and environment, both of which exist in our business and society. See Chapter 10 for a description on network computing that can handle a large database to operate the computation of DEA.

As an initial step in Figure 15.1, an organization, or a decision making unit (DMU: a firm or a public agency), needs to identify that it faces a serious crisis due to industrial pollution (e.g., air, soil or water pollution). As a result of such recognition on the existence of pollution, consumers do not purchase products from dirty-imaged firms. They purchase products from green-imaged companies even if their prices are much higher than those of dirty-imaged companies. More seriously, a very large opportunity cost may often occur, along with industrial pollution, which creates huge damage to each firm. For example, the disaster of the Fukushima Daiichi nuclear power plant, caused by the Great East Japan Earthquake occurred on 11 March in 2011, was an example of such an opportunity cost. See Chapter 25 which discusses the Japanese fuel mix plan after the disaster of the Fukushima Daiichi nuclear power plant. The opportunity cost is much larger than any other cost component in the operation of modern business and public administration.

To avoid the problem, each organization needs to incorporate "a sensing capability" by which the organization can identify an existence of such a serious crisis due to industrial pollution. The sensing capability may be often found in an individual member from the top to the bottom or in a group of members within the organization. A result of the sensing capability, incorporated inside an organization, is referred to as a "felt crisis". For example, Tokyo Electric Power Company did not have such a sensing capability, so that it did not prepare for multiple electric power sources to reduce the temperature of nuclear power reactors at Fukushima Daiichi for "just in case" (e.g., a "tsunami" due to a very large earthquake). The lack of such a sensing capability within the electric utility company (in particular, within the executive board members) produced the disaster on 11 March 2011. That was clearly a corporate mistake under the Japanese incompetent industrial policy on nuclear power generation.

The felt crisis is extended to "social sustainability" in the public sector or "corporate sustainability" in the private sector. The concept of "social sustainability," supported by many international organizations such as UN, is related to global social responsibility and accountability for preventing the global warming and climate change that have been gradually changing our ecological systems. The concept of social sustainability contains world-wide implications in a long-term horizon. In contrast, the concept of corporate sustainability has other implications, different from social sustainability, all of which are related to modern business concerns. That is, corporate sustainability is closely related to business survivability via a public image as a good corporate citizen. Moreover, as mentioned previously, the opportunity cost (e.g., damage due to industrial pollution) is much larger than any other production-related cost component and may terminate the existence as a going concern. Thus, it is important for us to clearly distinguish between the two sustainability concepts in discussing world-wide policy issues (e.g., global warming and climate change) and modern corporate behavior for preventing industrial pollution.

After identifying an organizational objective (e.g., social or corporate sustainability), each organization needs to begin "consensus building" among its members. In the consensus building process, the necessity of sustainability is "disseminated" among all members. For example, see the web page of the 2015 United Nations Climate Change Conference, the so-called "Conference of the Parties (COP) 21," held in Paris, France[2]. The conference, as a consensus building process, negotiated the Paris Agreement, or the global agreement on reduction of climate change, which represented conscientiousness among the representatives of 197 parties who attended the conference. The agreement entered into force on 4 November 2016. As of August 2017, 159 parties have ratified of 197 parties to the convention. According to the organizing committee, the Paris Agreement's objective is to strengthen the global response to the threat of climate change by keeping a global temperature rise in this century well below 2 degrees Celsius above the pre-industrial level and to pursue efforts to limit the temperature increase even further to 1.5 degrees Celsius. The agreement calls for zero net GHG emissions to be reached during the second half of the 21st century. The conference can be considered an example of such a consensus building process that shares the necessity of social sustainability in the world.

After establishing the consensus building among members, the organization needs to determine its production factors (i.e., inputs, desirable and undesirable outputs) for sustainability assessment. This type of social intelligence is referred to as "a production factor determination capability." The DEA, as a holistic approach, is used in this stage as one of many methodological alternatives for assessing the level of sustainability. Some member(s) should have an analytical capability for

[2] The current policy direction of the Republican (Trump) government in the United States is the withdrawal from the agreement. The selection of President Trump indicates that "Democracy is now dead in the United States." Thus, it is very difficult for us to imagine that the United States will follow the agreement. China may be slightly better than the United States, but the central government does not have any monitoring capability on the performance of local governments. All statistics, originally prepared by local governments and summarized by the central government, are not reliable and artificially created to provide incorrect information on an amount of GHG emissions. Thus, the agreement will be not satisfied as we expect because the two large emission pollutants will not satisfy it. Another rationale regrading why the agreement will be not satisfied is because the incompetent UN cannot control member nations, in particular permanent Security Council members (i.e., China, France, Russian Federation, the United Kingdom, and the United States) in all aspects on international affairs including collaboration to reduce the amount of GHG emissions in the world. A major problem on the global warming and climate change is partly because of inadequate and inefficient governance of UN. Maybe, "the universe may have a limit and a shape, but the stupidity of the current UN governance is limitless and shapeless." Now, it is the time to change the governance of UN. That is the first step to combat various climate changes in the world.

operating DEA assessment in this stage. This type of social intelligence is referred to as "a measurement capability."

After applying DEA for performance assessment, the organization obtains empirical results, indicating the level of sustainability, which is identified by relatively comparing its performance with others. This is referred to as "a performance assessment capability." Based upon the empirical results, the organization needs to prepare policy or corporate strategy for pollution prevention. This type of intelligence is referred to as "a policy/strategy making capability." The organization needs to implement the policy or strategy for sustainability enhancement. This type of intelligence is referred to as "an achievement assessment capability."

In the DEA assessment, if the implementation is successful, then a reward or a social recognition should be given to the organization. In contrast, if it is not successful, the result implies a necessity of felt crisis for the organization by returning to the initial stage of the proposed social intelligence. Thus, a gradual improvement process will be necessary and repeated many times until each organization can attain eliminating the crisis-related environmental pollution.

Finally, it is important to note that, if one of the intelligence capacities depicted in Figure 15.1 does not exist in an organization, then it will be unable to survive as a going concern in our modern business and society.

15.3 SOCIAL SUSTAINABILITY VERSUS CORPORATE SUSTAINABILITY

DEA environmental assessment has underlying concepts related to sustainability. The concept of sustainability is indeed important in terms of attaining economic prosperity and pollution reduction, both of which need to be carefully synchronized together in these simultaneous developments.

As the initial step for a description on DEA environmental assessment, this chapter separates the concept of sustainability into social and corporate categories (Sueyoshi and Goto, 2014b; Wang *et al.*, 2014), as mentioned previously. The concept of "social (public) sustainability," supported by many international organizations such as UN, is related to global social responsibility and accountability in preventing the global warming and climate change that has been gradually destroying our ecological and economic systems. The concept of social sustainability may contain world-wide implications in a long-term horizon.

In contrast, "corporate (private) sustainability" has other type of implications, different from social sustainability, all of which are closely related to modern business perspectives. As mentioned previously, consumers try to avoid purchasing products from dirty-imaged firms that produce acid rain gases and emit GHG, even if the prices of their products are much lower than those of green companies. It is easily imagined that corporate sustainability is closely related to survivability via public image as a good corporate citizen. Moreover, as discussed in Figure 15.1, the opportunity cost (e.g., damage due to industrial pollution) is much larger than any production-related cost item in modern business. Therefore, it is important for us to clearly distinguish between "social" and "corporate" sustainability concepts in discussing the global warming and climate change and modem corporate behavior for preventing environmental pollution as a good image for enhancing marketing strategy.

15.3.1 Why Is Social Sustainability Important?

The global environmental problem includes various ecological issues such as

(a) Stratospheric ozone depletion,
(b) Global warming and climate change (because of an increase of greenhouse gases – CO_2: carbon dioxide; CH_4: methane; N_2O: nitrous oxide; and others),
(c) Air pollution such as PM 2.5 (where PM stands for "particulate matter" and 2.5 indicates the size of both solid particles and liquid droplets found in the air),
(d) Soil and water pollutions (e.g., polluted water, due to shale gas and oil production in horizontal wells, reaches underground water sources used for drinking) and
(e) Acid precipitation problems (e.g., NO_x and SO_2) on the Earth's surface.

Such ecological problems need to be solved by not only natural science and engineering efforts but also international cooperation among all nations in the world.

In discussing these environmental pollution issues, our discussion often pays attention to only the contribution of natural science, often neglecting the importance of social science such as industrial policy, energy policy, environmental policy and ecological economics in public sectors and corporate strategy and eco-technology innovation in private sectors, all of which must necessarily be carefully coordinated together to prepare policy and business rationales for guiding multi-dimensional aspects on environmental pollution prevention.

To describe the importance of social sustainability further by using a policy example, let us consider the resource allocation by a government that has often unequally allocated energy and business resources (e.g., capital, education and people) to different regions when preparing industrial and economic policy agendas. The equality in resource allocation is a fundamental principle of social stability. In contrast, if there is "regional imbalance," it leads to "income imbalance" among people in different regions. The income imbalance often produces "social instability" among different regions in many nations. Therefore, the concept of social sustainability, developed with social stability, is important in future industrial and economic planning of each nation. See, for example, Sueyoshi and Yuan (2015a, 2016b) for a detailed description on Chinese regional planning and sustainability. See also Chapter 23 on Chinese regional planning.

15.3.2 Why Is Corporate Sustainability Important?

According to Sarkis *et al.* (2010)[3], followed by Wang *et al.* (2014) and Sueyoshi and Wang (2014a), green management is classified into eight research groups such as (a) business complexity, (b) ecological modernization, (c) information utilization, (d) institutional externality, (e) resource-based view, (f) resource dependency, (g) social network and (h) stakeholder involvement. This chapter follows research effort of Sueyoshi and Wang (2014a) for describing a literature review on green supply chains in private sectors.

(a) *Business Complexity*: Business complexity is defined through heterogeneity or diversity in factors such as consumers, governmental regulations and technological advancements. As their corporate size increases, firms have difficulty in planning and predicting their business actions, including product return, recycling, product inspection and quality check (Chakravarthy, 1997[4]; Vachon and Klassen, 2006[5]). This type of concern was also discussed by Sueyoshi and Goto (2010a). Utilizing a data set on 220 Japanese manufacturing firms from 2004 to 2007, a total of 853 observations, their study concluded that the Japanese firms tried to accumulate capital and then used it to prevent industrial pollution. As a result, the size of firms was a very important component in understanding their corporate behavior toward environmental protection. Acknowledging the influence of regulation, their study also discussed that firms could not make any strategic step toward corporate sustainability without capital accumulation even under strict regulation.

(b) *Ecological Modernization*: Through eco-innovation, firms are geared toward achieving industrial development and environmental protection through technology innovation. Regulation has influenced corporate efforts on environmental innovation (Murphy and Gouldson, 2000[6]). The study also suggested that manufacturing firms could overcome barriers to innovation and gain operational opportunities for performance improvement. However, as reported by Revell (2007)[7], the corporate efforts did not bring financial benefits. Therefore, it is necessary to develop a diffusion mechanism in which the

[3] Sarkis, J., Gonzalez-Torre, P. and Adenso-Diaz, B., 2010. Stakeholder pressure and the adoption of environmental practices: the mediating effect of training. *Journal of Operations Management*, **28**, 163–176.

[4] Chakravarthy, B., 1997. A new strategy framework for coping with turbulence. *Sloan Management Review*, **38**, 69–82.

[5] Vachon, S. and Klassen, R.D., 2006. Extending green practices across the supply chain: the impact of upstream and downstream integration. *International Journal of Operations and Production Management*, **26**, 795–821.

[6] Murphy, J. and Gouldson, A., 2000. Environmental policy and industrial innovation: integrating environment and economy through ecological modernization. *Geoforum*, **31**, 33–44.

[7] Revell, A., 2007. The ecological modernisation of SMEs in the UK's construction industry. *Geoforum*, **38**, 114–126.

core large firms motivate toward environmental innovation and then diffuse innovation technology to small and medium firms (Hall, 2001)[8]. See Sueyoshi and Goto (2010a) for a description on the diffusion process in Japanese firms.

(c) *Information Utilization*: Firms need to report their environmental performance to outside stakeholders (e.g., customers and equity holders), but they lack full knowledge on products, processes and material flows. Consequently, information asymmetry has occurred between firms and outside stakeholders (Simpson, 2010)[9]. Information sharing is critical for coordinating between firms and stakeholders in terms of enhancing corporate image and satisfying regulation requirements (Wong *et al.*, 2009)[10]. A difficulty associated with information asymmetry and sharing is that all stakeholders need a brief summary which they can easily understand, not detailed and complicated information on the operational and environmental performance of firms.

(d) *Institution Externality*: External pressures have influenced a firm to adapt organizational practice (Lai *et al.*, 2006)[11]. The concept is used to examine how a firm addresses green issues because of external pressures (Jennings and Zandbergen, 1995)[12]. The concept of externality indicates the importance of a research direction to explain environmental practices. Public agencies are an example of powerful institutions that influence the business practice of an organization (Rivera, 2004)[13]. Moreover, according to Carter *et al.* (2000)[14], 75% of United States consumers made their purchasing decisions after taking the firm's environmental reputation into consideration, and 80% of them were willing to pay more for environmentally friendly products. Consumers now have an increasingly heightened environmental awareness and purchase green products. See Harris (2006)[15].

(e) *Resource-based View*: The competitive advantage of firms is sustained by harnessing resources (e.g., trust from customers) that are valuable, rare,

[8] Hall, J., 2001. Environmental supply-chain innovations. *Greener Management International*, **35**, 105–119.

[9] Simpson, D., 2010. Use of supply relationships to recycle secondary materials. *International Journal of Production Research*, **48**, 227–249.

[10] Wong, C.W.Y., Lai, K.H. and Cheng, T.C.E., 2009. Complementarities and alignment of information systems management and supply chain management. *International Journal of Shipping and Transport Logistics*, **1**, 156–171.

[11] Lai, K.H., Wong, C.W.Y. and Cheng, T.C.E., 2006. Institutional isomorphism and the adoption of information technology for supply chain management. *Computers in Industry*, **57**, 93–98.

[12] Jennings, P.D. and Zandbergen, P.A., 1995. Ecologically sustainable organizations: an institutional approach. *The Academy of Management Review*, **20**, 1015–1052.

[13] Rivera, J., 2004. Institutional pressures and voluntary environmental behavior in developing countries: evidence from the Costa Rican hotel industry. *Society and Natural Resources*, **17**, 779–797.

[14] Carter, C.R., Kale, R. and Grimm, C.M., 2000. Environmental purchasing and firm performance: an empirical investigation. *Transportation Research Part E: Logistics and Transportation Review*, **36**, 219–228.

[15] Harris, P.G., 2006. Environmental perspectives and behavior in China: Synopsis and bibliography. *Environment and Behavior*, **38**, 5–21.

imperfectly imitable and non-substitutable (Barney, 1991)[16]. Firm resources are often defined as assets, production and sale capabilities, firm attributes and history, all of which need to be utilized for improving their efficiency, effectiveness and competitiveness (Barney, 1991; Russo and Fouts, 1997)[17].

(f) *Resource Dependency*: Firms should seek high performance gains in a long-term horizon instead of pursuing short-term benefits at the expense of others. Such a long-term perspective is essential for the development of corporate sustainability. See Paloviita and Luoma-aho (2010)[18]. To attain the long-term perspective, firms need to manage internal and external coordination to attain such performance goals where partners' coordination and resource sharing are beneficial for operational and environmental improvements (Yang *et al.*, 2008)[19].

(g) *Social Network*: The concept discusses how to develop corporate sustainability by establishing social relationships inside and outside a firm. A firm makes decisions according to information obtained through its social network (Wuyts *et al.*, 2004)[20]. For example, informal human relationships consist of an internal social network and relationships with customers are an external social network. To enhance the social network development, it is important for firms to share information such as their new recyclable products, clean process and corporate efforts for environmental protection (Walton *et al.*, 1998[21]; Theycl, 2001[22]).

(h) *Stakeholders*: This group of individuals can affect or be affected by the achievement of a firm's objectives (Freeman, 1984)[23]. All firms have externalities that influence a variety of stakeholders both internal and external to the firm. The scope of environmental issues is widely diverse because stakeholders consist of many different types of individual. An important group of stakeholders are the stock holders who pay attention to both a return to their investments in a short term horizon and the status of a going concern in a long-term horizon (Jacobs *et al.*, 2010)[24]. All firms always face a huge risk originating from environmental issues because every violation obliges them to pay a large opportunity cost, often jeopardizing the status

[16] Barney, J.B., 1991. Firm resources and sustained competitive advantage. *Journal of Management*, **17**, 99–120.

[17] Russo, M.V. and Fouts P.A., 1997. A Resource-based perspective on corporate environmental performance and profitability. *Academy of Management Journal*, **40**, 534–559.

[18] Paloviita, A. and Luoma-aho, V., 2010. Recognizing definitive stakeholders in corporate environmental management. *Management Research Review*, **33**, 306–316.

[19] Yang, J., Wang, J., Wong, C.W.Y. and Lai, K.H., 2008. Relational stability and alliance performance in supply chain. *Omega*, **36**, 600–608.

[20] Wuyts, S., Stremersch, S., Van Den Bulte, C. and Franses, P.H., 2004. Vertical marketing systems for complex products: a triadic perspective. *Journal of Marketing Research*, **41**, 479–487.

[21] Walton, S.V., Handfield, R.B. and Melnyk, S.A., 1998. The green supply chain: integrating suppliers into environmental management processes. *The Journal of Supply Chain Management*, **34**, 2–11.

[22] Theyel, G., 2001. Customer and supplier relations for environmental performance. *Greener Management International*, **35**, 61–69.

[23] Freeman, R.E., 1984. *Strategic Management: A Stakeholder Approach*. Pittman, Marshfield.

[24] Jacobs, B.W., Singhal, V.R. and Subramanian, R., 2010. An empirical investigation of environmental performance and the market value of the firm. *Journal of Operations Management*, **28**, 430–441.

of a going concern. Consumers are another type of stakeholder who recently begin to think of environmental protection as their first priority in their decision making on purchases. It is almost impossible for modern corporations to survive if they cannot sell their products because of their dirty-image. The reality may be often different, but the image of a green company is essential for modern business. The emission scandal of Volkswagen on 18 September 2015 was such an example (https://en.wikipedia.org/wiki/Volkswagen_emissions_scandal) of the insincere green image. The last stakeholder group consists of employees who take pride in the fact that they are working for a green company. An important research issue in the area is that it needs empirical evidence to confirm a positive relationship among stakeholders in terms of environmental protection (Tate *et al.*, 2010[25]; Vachon, 2007[26]).

15.4 STRATEGIC ADAPTATION

An important feature of conventional DEA is that it can measure the relationship between desirable outputs and inputs without specifying any production function between them. Meanwhile, DEA environmental assessment extends DEA by incorporating undesirable outputs in addition to the two groups of production factors. Thus, it is necessary for us to consider the triangle relationship between desirable and undesirable outputs, both of which are produced by input resources.

Figure 15.2 visually describes the triangle between the three types of production factors. The figure incorporates the three types of performance evaluation which are incorporated in DEA environmental assessment: (a) operational evaluation where the first priority is operational performance and the second priority is environmental performance, (b) environmental evaluation where the first priority is environmental performance and the second priority is operational performance and (c) unified (operational and environmental) evaluation where both operational and environmental performance measures are treated equally, but depending upon a data structure on production factors of all DMUs.

The figure has two important concerns to be discussed here. One of the two concerns is that an increase in some input(s) usually leads to an increase in some desirable and/or undesirable output(s). However, eco-technology development can change such a situation, so being able to increase some desirable output(s) but simultaneously decrease some undesirable output(s). The other concern is an assumption that undesirable outputs are "byproducts" of desirable outputs.

[25]Tate, W.L., Ellram, L.M. and Kirchoff, J.F., 2010. Corporate social responsibility reports: A thematic analysis related to supply chain management. *The Journal of Supply Chain Management*, **46**, 19–44.
[26]Vachon, S., 2007. Green supply chain practices and the selection of environmental technologies. *International Journal of Production Research*, **45**, 4357–4379.

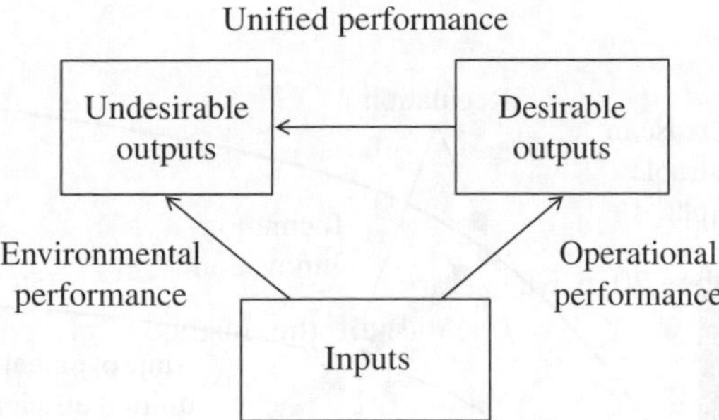

FIGURE 15.2 Triangular relationship between three production factors (a) Undesirable outputs are byproducts of desirable outputs, both of which are produced by inputs. The more desirable outputs and less undesirable outputs produce the better performance in a DEA environmental assessment.

Conventionally, it is not important for DEA to consider the assumption because only the existence of desirable outputs, as discussed in chapters of Section I, is incorporated in the conventional use. However, the prevention of undesirable outputs is equally as important as the production of desirable outputs under recent environmental regulation and consumer preference on green products.

Many corporate leaders often believe that environmental regulation may jeopardize their performance, so claiming the necessity of deregulation in their industries. We support a deregulation direction on desirable outputs because many firms are competing to attain their corporate goals (e.g., economic success and job security). However, their view is not always true in modern business reality. To visually explain our rationale on why regulation is important on reduction of undesirable outputs in modern business, where eco-technology innovation provides us with a new business strategy, Figure 15.3 depicts the relationship between desirable and undesirable outputs under two adaptation strategies to a regulation change. The figure will be later discussed as the concept of managerial disposability in this chapter.

For our visual convenience, Figure 15.3 depicts a case where each of the three production factors has only a single component. The observation (g_n, b_n, x_n) indicates the current achievement of an organization under no regulation, where the desirable and undesirable outputs are listed, but not the input in the figure. The subscript (n) indicates no regulation. The figure assumes two smooth curves to express the functional relationships between the two types of output.

The conventional strategy to adapt a regulation change can be considered in such a manner that an input decrease leads to a decrease in an undesirable output so that it can satisfy the regulation change by natural reduction. The amount of an undesirable output is shifted from b_n to b_r where the subscript (r) indicates regulation. However, it may simultaneously increase the desirable output from g_n to g_r by a managerial effort so that the organization can attain an efficiency

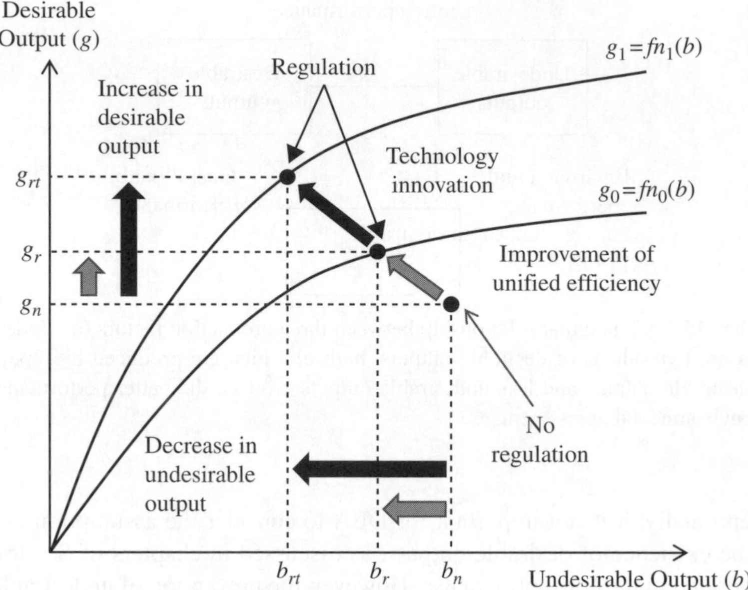

FIGURE 15.3 Influence of regulation and eco-technology innovation (a) Source: Sueyoshi *et al.* (2013b). (b) The amount of an input is not listed in the figure for our descriptive convenience. The figure implicitly considers an input increase or decrease. (c) An important feature of this figure is that an input change causes shifts in a desirable output and an undesirable output as depicted in the figure. The first shift is due to a change from non-regulation to regulation. The second shift is due to eco-technology innovation. Such a change implies "desirable congestion" due to a regulation change and an occurrence of eco-technology innovation. See Chapter 21 for a description on desirable congestion. (d) This type of strategy is referred to as "managerial disposability", later in this chapter.

frontier, or a smooth curve, shaped by $g_0 = fn_0(b)$ in Figure 15.3, where *fn* stands for a production function. Of course, if the firm does not make a managerial effort in this case, the desirable output may reduce the magnitude below the curve and the undesirable output may fall at a level that satisfies the governmental regulation.

The second strategy for adaptation is considered to utilize eco-technology innovation, usually along with a managerial challenge, that shifts the functional form from g_0 to $g_1 = fn_1(b)$. The operation of the organization is improved by its engineering capability that is often given by investment on new eco-technology. In Figure 15.3, the adaptive strategy makes it possible for the organization to increase the input to attain a more desirable combination (g_{rt}, b_{rt}) by both increasing the desirable output and decreasing the undesirable output from the previous combination (g_r, b_r). The subscript (t) stands for technology innovation.

Considering a case where the three production factors have multiple components, this chapter can find an important concern from Figure 15.3. That is, it is possible for the organization to increase the directional vector of inputs in order to decrease that of undesirable outputs by eco-technology innovation (e.g., clean coal technology in an electricity industry) and/or a managerial challenge (e.g., the use of coal combustion with lower CO_2 emission) under a regulation change. The enhanced input vector increases the directional vector of desirable outputs as much as possible under the new eco-technology. Thus, the regulation may serve as a source of eco-technology innovation for increasing the vector of desirable outputs and preventing that of undesirable outputs. This new aspect related to technology, found in Figure 15.3, has been not sufficiently incorporated in a conventional framework of DEA because an increased input vector is usually expected to increase both desirable and undesirable outputs. Thus, inputs are expected to decrease or maintain the directional vector in the conventional framework. Later, Chapters 21 and 23 discuss the second adaptation strategy from an occurrence of "desirable congestion," or eco-technology innovation, under managerial disposability.

At the end of this section, note that this chapter does not clearly address the minimization of production cost in Figure 15.3. However, it is easily imagined that technological innovation and environmental protection are associated with long-term cost-saving through increased competitiveness and successful adaptation to the regulation change.

15.5 TWO DISPOSABILITY CONCEPTS

In order to shift the conceptual discussion of the preceding sections to mathematical formulations for DEA environmental assessment, this chapter needs to introduce concepts concerning "natural disposability" and "managerial disposability". To describe them more clearly, let us consider $X \in R_+^m$ as an input vector with m components, $G \in R_+^s$ as a desirable output vector with s components and $B \in R_+^h$ as an undesirable output vector with h components. In these column vectors, the subscript j is used to stand for the j-th DMU, whose vector components are strictly positive.

Using an axiomatic expression, this chapter specifies a unified (operational and environmental) production possibility set to express natural (N) and managerial (M) disposability by the two types of output vectors and an input vector, respectively, as follows:

$$P_v^N(X)=\left\{(G,B): G \le \sum_{j=1}^{n} G_j\lambda_j, B \ge \sum_{j=1}^{n} B_j\lambda_j, X \ge \sum_{j=1}^{n} X_j\lambda_j, \sum_{j=1}^{n}\lambda_j = 1, \lambda_j \ge 0\,(j=1,\ldots,n)\right\},$$

$$P_v^M(X)=\left\{(G,B): G \le \sum_{j=1}^{n} G_j\lambda_j, B \ge \sum_{j=1}^{n} B_j\lambda_j, X \le \sum_{j=1}^{n} X_j\lambda_j, \sum_{j=1}^{n}\lambda_j = 1, \lambda_j \ge 0\,(j=1,\ldots,n)\right\}. \tag{15.1}$$

$P_v^N(X)$ stands for a production and pollution possibility set under natural (N) disposability and $P_v^M(X)$ stands for that of managerial disposability. The subscript v stands for "variable" RTS because the constraint $\left(\sum_{j=1}^{n}\lambda_j = 1\right)$ is incorporated into the two axiomatic expressions. See Chapter 8 for a detailed description of RTS. See also Chapter 20 on the linkage between RTS and DTS where DTS stands for damage to scale on inputs and undesirable outputs.

The difference between the two disposability concepts is that the production technology under natural disposability, or $P_v^N(X)$, has $X \geq \sum_{j=1}^{n} X_j\lambda_j$ in Equation 15.1, implying that a DMU can attain an efficiency frontier by reducing a directional vector of inputs. In contrast, that of the managerial disposability, or $P_v^M(X)$, has $X \leq \sum_{j=1}^{n} X_j\lambda_j$ in Equation 15.1, implying that a DMU can attain a status of an efficiency frontier by increasing a directional vector of inputs. Meanwhile, a common feature of the two disposability concepts is that both have $G \leq \sum_{j=1}^{n} G_j\lambda_j$ and $B \geq \sum_{j=1}^{n} B_j\lambda_j$ in their axiomatic expressions. These conditions intuitively appeal to us because an efficiency frontier for desirable outputs should locate above or on all observations, while that of undesirable outputs should locate below or on these observations. The first part of Equation (15.1) on natural disposability is originated from a conventional framework of DEA, while the second one is originated from the adaptive strategy depicted in Figure 15.3.

By changing these axiomatic structures to constant RTS and DTS, the production and pollution possibility set becomes as follows:

$$P_c^N(X) = \left\{ (G,B) : G \leq \sum_{j=1}^{n} G_j\lambda_j, B \geq \sum_{j=1}^{n} B_j\lambda_j, X \geq \sum_{j=1}^{n} X_j\lambda_j, \lambda_j \geq 0\,(j=1,\ldots,n) \right\},$$
$$P_c^M(X) = \left\{ (G,B) : G \leq \sum_{j=1}^{n} G_j\lambda_j, B \geq \sum_{j=1}^{n} B_j\lambda_j, X \leq \sum_{j=1}^{n} X_j\lambda_j, \lambda_j \geq 0\,(j=1,\ldots,n) \right\}, \quad (15.2)$$

where the two equations drop $\sum_{j=1}^{n}\lambda_j = 1$ from their axiomatic structures by assuming constant RTS and DTS. The subscript c is used to express constant RTS and DTS.

Here, it is necessary to discuss that an input vector is usually assumed to project toward a decreasing direction in the previous research efforts on DEA, as discussed in Section I. The assumption is often inconsistent with the reality related to environmental protection. For example, many governments and firms consider an increase in input resources to yield an annual "growth" of a desirable output. Thus, the conventional framework of DEA is not consistent with the economic expansion, or "economic growth," because the previous DEA studies have implicitly assumed the minimization of total production cost. See Chapter 4. The cost concept may be acceptable for performance analysis under "economic recession" or "stagnation," but not in many cases where industrial planning and corporate strategy are based upon their economic growths. Thus, it is easily imagined that DEA environmental assessment, as discussed in Section II, is conceptually different from the conventional use of DEA. The cost concept for guiding public and private entities in their strategy development is average cost (under constant RTS) or marginal cost (under variable RTS) but not the total cost, anymore. Furthermore, an opportunity cost, originated from business risk due to industrial pollutions and the other types of various environmental problems (e.g., the nuclear power plant accident at Fukushima Daiichi in Japan), has a major role in modern corporate governance issues. Such cost concepts for current policy making and modern business are implicitly incorporated in formulating the two disposability concepts, in particular in managerial disposability.

15.6 UNIFIED EFFICIENCY UNDER NATURAL AND MANAGERIAL DISPOSABILITY

The natural and managerial disposability concepts can be unified together into a new type of efficiency measure, or unified efficiency (UE), that measures the operational and environmental performance of DMUs in a combined form. Figure 15.4 visually describes such a computational process to measure the level of UE. Sueyoshi and Goto (2012a) have first unified desirable and undesirable outputs to measure the UE of a DMU by relatively comparing its performance with those of the other DMUs.

As depicted at the top of Figure 15.4, the proposed DEA environmental assessment unifies operational and environmental performance in such a manner that it can determine the level of UE of DMUs. Then, the unification is separated in natural (N) and managerial (M) disposability concepts, referred to as UEN and UEM, respectively. See Chapters 16 and 17 for a description on UEN and UEM. The conceptual framework of Figure 15.4 is supported by Porter's hypothesis, implying that "investment on environmental protection does not jeopardize their performance, rather providing an opportunity to enhance the performance." It is easily imagined that such a business direction does not immediately produce any improvement in a short-term horizon. Rather, the concept is related to a long-term horizon. It is important for policy makers and business leaders to understand the importance of gradual improvement on unified efficiency. Finally, the two unified efficiency measures may be combined together into unified efficiency under natural and managerial disposability (UENM). See Chapter 24 for a detailed description on UENM.

Figure 15.5 depicts a "gradual improvement process" for performance unification, as discussed by Sueyoshi and Goto (2010). DMUs (e.g., manufacturing firms) cannot survive without a good corporate image because consumers do not purchase products from firms that produce industrial pollution. Therefore, modern

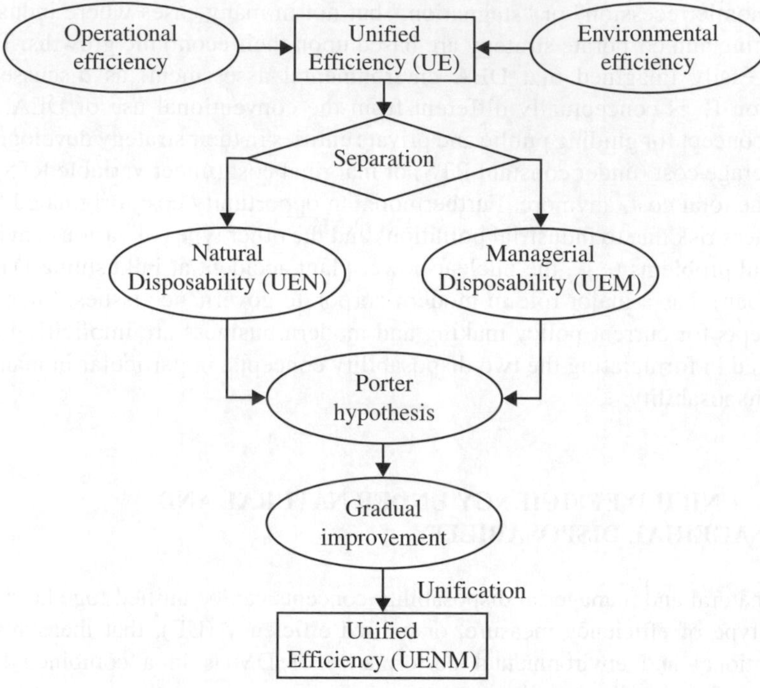

FIGURE 15.4 Computational flow for unified efficiency

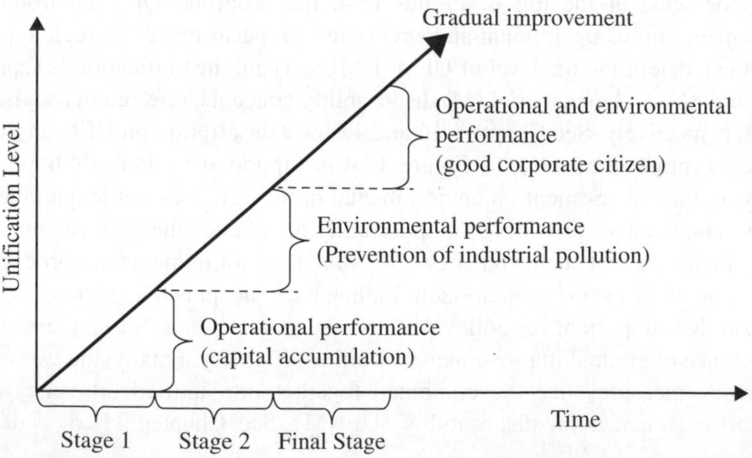

FIGURE 15.5 Gradual improvement for performance unification (a) For our visual convenience, this chapter depicts the relationship between a unification level and a time horizon by a straight line. The reality is much different from the figure. Therefore, it may be characterized by a nonlinear relationship between them.

business needs to pay attention to how to prevent industrial pollution. However, it can be easily imagined that they have a difficulty in implementing environmental protection without "capital accumulation." Thus, the unification process can be considered as a gradual process for improving their operational and environmental performance whose results are usually observed in a long-term horizon.

15.7 DIFFICULTY IN DEA ENVIRONMENTAL ASSESSMENT

As mentioned previously, an important aspect of DEA environmental assessment is that it separates outputs into desirable and undesirable categories. The output separation cannot be found in the conventional framework of DEA. The two types of outputs are different in terms of their analytical structures (e.g., the direction of these vectors). A problem to be overcome after the output separation is that DEA must consider a conceptual framework and its related formulations for unifying desirable and undesirable outputs. Figure 15.6 depicts the unification process from Stages I to III, all of which are incorporated in the proposed DEA formulations. For our visual description, the figure depicts a case of a single component of three production factors.

The first stage (I) is separated into two sub-stages: (A) and (B). Sub-stage (A) of stage (I) indicates the production relationship between an input (x) and a desirable output (g) under the assumption that all DMUs produce a same amount of undesirable output (b). A production possibility set (PrPS) locates below a piece-wise linear convex curve, depicting an efficiency frontier (EF_g) in the x–g space. Stage (I) also has sub-stage (B). A pollution possibility set (PoPS) locates above the piece-wise linear concave curve that expresses an efficiency frontier (EF_b) in the x–b space under the assumption that they produce a same amount of a desirable output (g). An important feature of stage (I) is that the production possibility set (A) is independent from the pollution possibility set (B). This assumption is slightly unrealistic. See stage III in Figure 15.6, which changes the assumption.

To unify the two sets related to stage (I), stage (II) combines them in a unified framework. The horizontal and vertical coordinates in the right-upper corner of Figure 15.6 indicate x and $g\&b$, respectively. The unification process makes it possible to identify a production and pollution possibility set (Pr&PoPS) between PrPS and PoPS, locating between the two efficiency frontiers (EF_g and EF_b). All DMUs locate within Pr&PoPS, which is a product set, so that they are between the production and pollution possibility sets depicted in stage (I).

Stage (III) is the final unification process which incorporates an assumption that "an undesirable output is a byproduct of a desirable output." The assumption seems trivial, but it drastically changes the structure of DEA environmental assessment. That is, the assumption changes the two efficiency frontiers (EF_g and EF_b) to be shaped by two convex forms, as depicted in Stage III (A) of Figure 15.6. Here, it is important to note that the efficiency frontier (EF_g) should have an increasing trend along with an input increase. However, the efficiency frontier

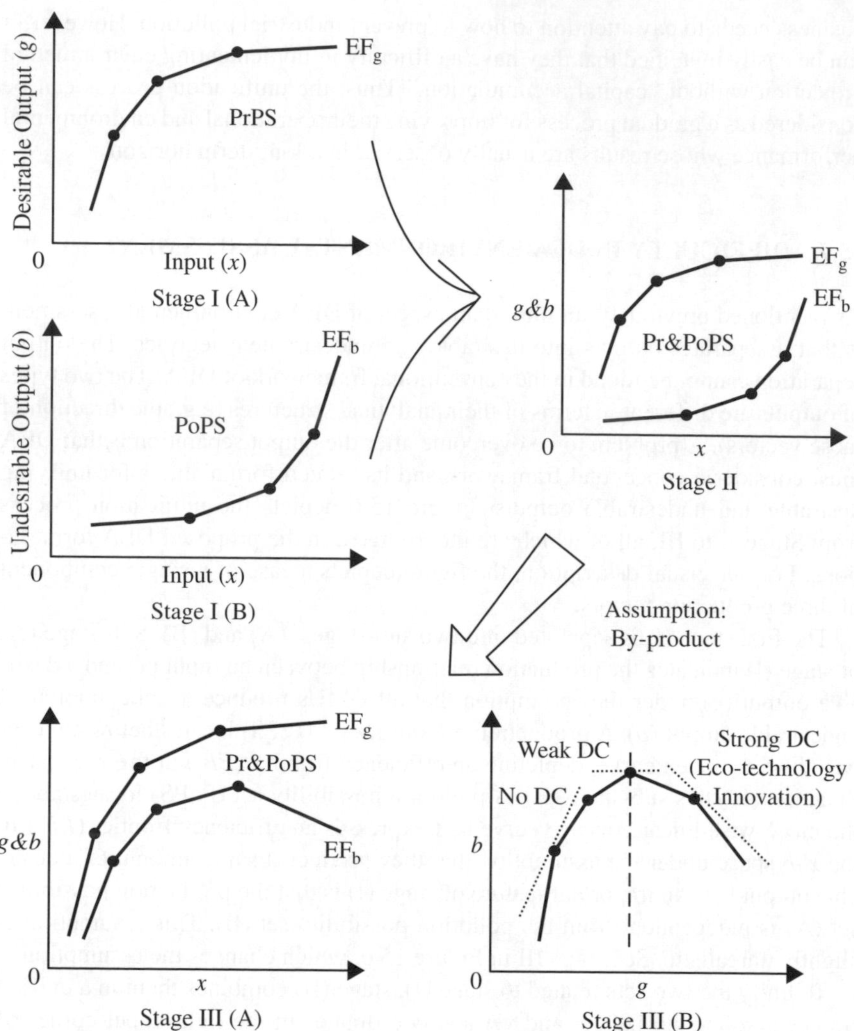

FIGURE 15.6 Unification process between desirable and undesirable outputs (a) EF stands for efficiency frontier. (b) PrPS, PoPS and Pr&PoPS indicate production possibility set, pollution possibility set, and production and pollution possibility set, respectively. (c) Stage (III) is the final unification process which incorporates an assumption that an undesirable output is a byproduct of a desirable output. The figure incorporates an occurrence of desirable congestion.

(EF_b) should have an increasing and decreasing trend because we are interested in reducing an amount of pollution (b). Both curves should have piece-wise linear convex forms because of the byproduct assumption. Thus, they are structurally different from those of the first and second stages. In DEA environmental assessment, the increasing and decreasing trend is due to an occurrence of eco-technology innovation for pollution prevention. Stage III (B) depicts such a relationship between g on the horizontal axis and b on the vertical axis. The relationship is necessary because we are interested in production activities with an increase in a desirable output and a decrease in an undesirable output. That is the purpose of environmental assessment. Thus, Figure 15.6 visually describes a rationale regarding why DEA environmental assessment is more complicated and more difficult than a conventional use of DEA that does not have any separation between desirable and undesirable outputs.

15.8 UNDESIRABLE CONGESTION AND DESIRABLE CONGESTION

Stage III (B) of Figure 15.6 visually specifies the importance of congestion on undesirable outputs in DEA environmental assessment. The concept of congestion is separated into undesirable congestion (UC) and desirable congestion (DC) in the proposed environmental assessment. Figure 15.7 depicts differences between UC and DC. The left-hand side of Figure 15.7 exhibits three types of UC in a space of an undesirable output (b) on the horizontal axis and a desirable output (g) on the vertical axis. The right-hand side of Figure 15.7 exhibits three types of DC in a space of a desirable output (g) on a horizontal axis and an undesirable output (b) on the vertical axis.

The analytical importance of such an occurrence of UC is that the occurrence is characterized by a supporting hyperplane. For example, as depicted in the left-hand side of Figure 15.7, the negative slope of a supporting line indicates an occurrence of "strong UC." For example, the occurrence indicates a capacity limit on part or whole of a production facility. In contrast, a positive slope implies an opposite case (i.e., no occurrence of UC), so being

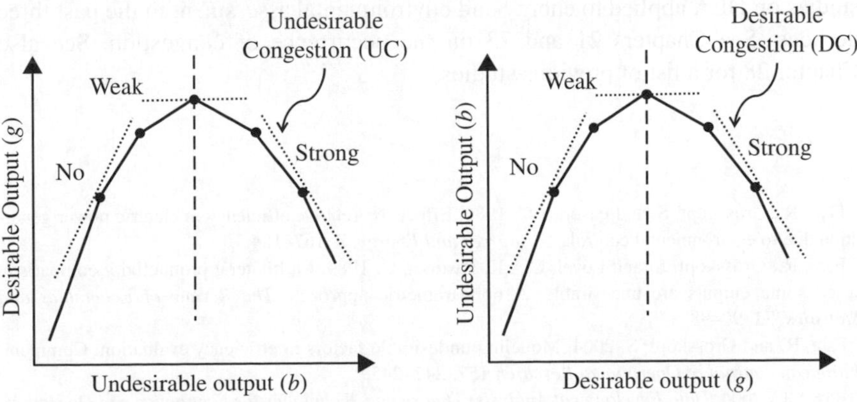

FIGURE 15.7 Undesirable congestion and desirable congestion

"no UC." An occurrence of "weak UC" is identified between strong UC and no UC, so being "weak UC."

The occurrence of UC is problematic in many industries, as discussed in many conventional DEA studies. See Chapter 9 for an illustrative example of UC in the electric power industry. Acknowledging the importance of conventional research on UC, however, this chapter is interested in sustainability development by enhancing the level of economic prosperity and reducing the amount of industrial pollution. Therefore, this chapter pays attention to a possible occurrence of DC, hereafter, because the occurrence indicates eco-technology innovation so that it can enhance the level of social or corporate sustainability.

The right-hand side of Figure 15.7 exhibits such a possible occurrence of DC. The negative slope of a supporting hyperplane indicates an occurrence of "strong DC," or eco-technology innovation for reducing an undesirable output (e.g., industrial pollution). In contrast, a positive slope implies an opposite case (i.e., no occurrence of DC), so being "no DC." An occurrence of "weak DC" is identified between strong and no DC, so being "weak DC."

15.9 COMPARISON WITH PREVIOUS DISPOSABILITY CONCEPTS

In reviewing previous research efforts on DEA environmental assessment, we find very important work by Färe *et al.* (1986[27], 1989[28]) that developed the economic concept of weak and strong disposability. See Chapter 28 for a literature study on previous works. As suggested by Färe and Grosskopf (2004)[29] and Ray (2004, p. 170)[30]: "the weak disposability can identify an occurrence of congestion, while the strong disposability cannot." After their discovery on the relationship between the two disposability concepts and a possible occurrence of congestion (i.e., UC in this chapter), their studies conceptually and methodologically dominated most studies on DEA applied to energy and environmental assessment in the past three decades. See Chapters 21 and 23 on the occurrence of congestion. See also Chapter 28 for a list of previous studies.

[27] Färe, R., Grosskopf, S. and Pasurka, C. 1986. Effects on relative efficiency in electric power generation due to environmental controls. *Resources and Energy*, **8**, 167–184.

[28] Färe, R., Grosskopf, S. and Lovel, C.A.K, Pasurka, C. 1989. Multilateral productivity comparison when some outputs are undesirable: A nonparametric approach. *The Review of Economics and Statistics*, **71**, 90–98.

[29] Färe, R. and Grosskopf, S. 2004. Modeling undesirable factors in efficiency evaluation: Comment. *European Journal of Operational Research* **157**, 242–245.

[30] Ray, S.C. 2004. *Data Envelopment Analysis: Theory and Techniques for Economics and Operations Research*, Cambridge University Press, Cambridge, UK.

15.9.1 Weak and Strong Disposability

Following Färe *et al.* (1989, pp. 91–92), production technology (*P*) can be considered as a mapping $P: X \in R_{+}^{m} \rightarrow P(X) \subseteq R_{+}^{s}$, where the output set *P(X)* denotes a set of all desirable output vectors that are producible by the input vector *X*.

Considering both undesirable and desirable output vectors, the study by Färe *et al.* (1989, p. 92) has specified an output vector as (*G*, *B*). Then, the weak disposability (*W*) is axiomatically specified by the following vector notation on the two types of output vectors:

$$P_{v}^{w}(X)=\left\{(G,B): G \leq \sum_{j=1}^{n} G_{j}\lambda_{j}, B=\sum_{j=1}^{n} B_{j}\lambda_{j}, X \geq \sum_{j=1}^{n} X_{j}\lambda_{j}, \sum_{j=1}^{n}\lambda_{j}=1, \lambda_{j} \geq 0 \ (j=1,\ldots,n)\right\}. \tag{15.3}$$

The equality constraints $\left(B=\sum_{j=1}^{n} B_{j}\lambda_{j}\right)$ are given to undesirable outputs. This equality assignment is a straightforward extension of the conventional definition on congestion (UC) on inputs. See Chapter 9 for the description on a possible occurrence of UC.

The strong disposability (*S*) is specified by the following vector notation on the two output vectors:

$$P_{v}^{s}(X)=\left\{(G,B): G \leq \sum_{j=1}^{n} G_{j}\lambda_{j}, B \leq \sum_{j=1}^{n} B_{j}\lambda_{j}, X \geq \sum_{j=1}^{n} X_{j}\lambda_{j}, \sum_{j=1}^{n}\lambda_{j}=1, \lambda_{j} \geq 0 \ (j=1,\ldots,n)\right\}. \tag{15.4}$$

The inequality constraints $\left(B \leq \sum_{j=1}^{n} B_{j}\lambda_{j}\right)$ allow for strong disposability on undesirable outputs.

15.9.2 Null-joint Relationship (Assumption on "Byproducts")

In addition to the definition on weak and strong disposability, Färe and Grosskopf (2004) discussed a "null-joint relationship" between desirable and undesirable outputs. The relationship is very important in considering DEA environmental assessment, as depicted in Figure 15.6. The relationship indicates that "undesirable outputs can be produced only if desirable outputs are produced." Thus, undesirable outputs are the byproducts of desirable outputs. An important question, discussed in this chapter, is how to mathematically combine desirable and undesirable outputs under such an assumption. Previously, undesirable outputs were not important in a conventional framework of DEA because they are just the byproducts of desirable outputs. However, they are now as important as desirable outputs in modern business and economics. Consumers prefer purchasing products from clean-imaged companies even if their prices are more expensive than those of dirty-imaged ones in modern business.

Figure 15.8 visually compares differences between natural and managerial disposability and strong and weak disposability. The figure visually indicates two important concerns. One of these concerns is that the weak disposability expressed by Equation (15.3) is equivalent to the natural disposability with equality on undesirable outputs (*B*). The other concern is that the concept of weak and strong disposability has been discussed on a possible occurrence of UC, implying a capacity limit on part or whole of a production facility. In contrast, the managerial disposability implies a possible occurrence of DC, or

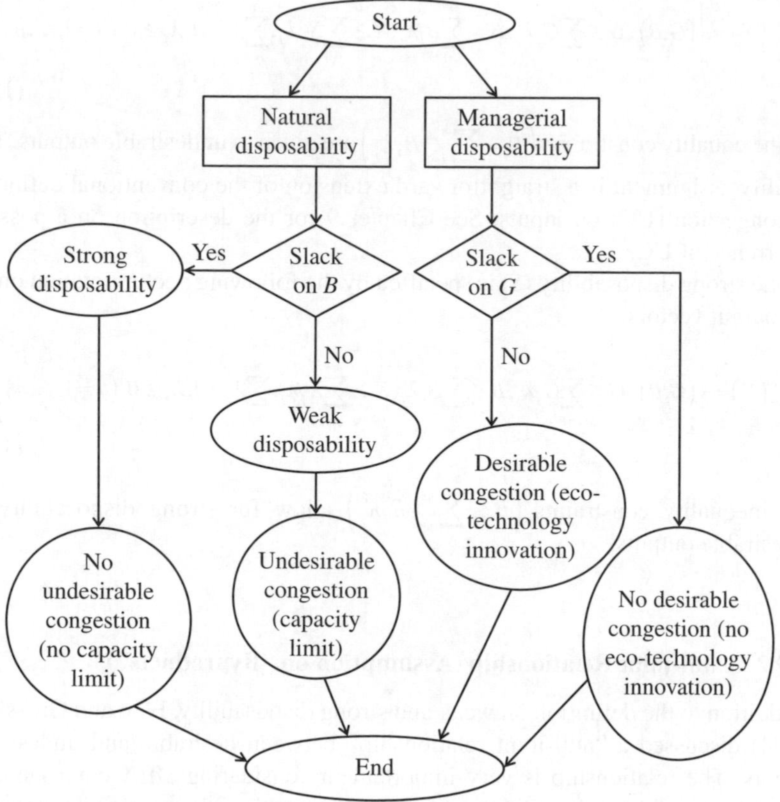

FIGURE 15.8 Natural and managerial disposability and strong and weak disposability (a) An existence of "slack" means that constraints are considered as "inequality" so that corresponding dual variables are positive or zero in their signs. In contrast, "no slack" implies that they are considered as "equality". The corresponding dual variables are unrestricted so that they may take positive, zero or negative in their signs. Thus, we can identify a possible occurrence of capacity limit under natural disposability or a possible occurrence of eco-technology innovation under managerial disposability. (b) The conventional framework of DEA incorporates only the left-hand side of the above four flows. Meanwhile DEA environmental assessment incorporates all of the four methodological flows. Thus, there are analytical differences between DEA and its environmental assessment.

eco-technology innovation by assigning equality on desirable outputs (G) under the condition of an increase in inputs (X).

As mentioned previously, any possible occurrence of congestion in DEA environmental assessment should be separated into the two categories. One of these categories is an occurrence of UC, or the conventional type of congestion, which has been long discussed in previous DEA studies. See Chapter 9 for a detailed description regarding the conventional type of congestion. The other type of congestion indicates an occurrence of DC, indicating an existence of eco-technology innovation and/or its related managerial challenge, applied for mitigation of undesirable outputs.

It is indeed true that the identification and separation between UC and DC are important from the perspective of operating energy sectors because the performance of energy firms is often influenced by both UC and DC. We consider that an occurrence of UC is important in the energy sectors, but DC is more important than UC in terms of environmental assessment to overcome the global warming and climate change by eco-technology innovation. Therefore, the technology development should be combined with managerial and economic challenges.

Finally, Table 15.1 summarizes two structural differences between UC and DC. An important question is why the inequality on undesirable outputs is expressed by $B \le \sum_{j=1}^{n} B_j \lambda_j$ under strong disposability. The concept implies that an efficiency frontier locates above or on observed undesirable outputs. Thus, DMUs which produce more undesirable outputs (e.g., GHG emission and other types of industrial pollutions) are evaluated as better performers. It is indeed not intuitive to us. To overcome such a problem, Färe *et al.* (1989) assigned the equality on undesirable outputs under the "null-joint relationship (assumption of byproduct)" between G and B with the assumption of congestion (i.e., UC). However, their allocation of equality on B under weak disposability has limited

TABLE 15.1 Structural differences between two disposability combinations

	Weak and strong disposability	Natural and managerial disposability
Undesirable outputs	$B = \sum_{j=1}^{n} B_j \lambda_j$ for weak disposability and $B \le \sum_{j=1}^{n} B_j \lambda_j$ for strong disposability.	$B \ge \sum_{j=1}^{n} B_j \lambda_j$ for both natural and managerial disposability.
Desirable outputs	$G \le \sum_{j=1}^{n} G_j \lambda_j$ for both weak and strong disposability.	$G \le \sum_{j=1}^{n} G_j \lambda_j$ for both natural and managerial disposability.
Inputs	$X \ge \sum_{j=1}^{n} X_j \lambda_j$ for both weak and strong disposability.	$X \ge \sum_{j=1}^{n} X_j \lambda_j$ for natural disposability and $X \le \sum_{j=1}^{n} X_j \lambda_j$ for managerial disposability.

implication for eco-technology innovation for pollution mitigation. To express the eco-technology development, this chapter allocated the equality on G, not B. An occurrence of congestion becomes DC, not UC anymore. Of course, we clearly acknowledge their contributions and the importance of UC in the conventional framework of DEA. However, it is very difficult for us to accept the previous research direction because we are interested in the reduction of various pollutions and discuss how to make a balance between economic success and environmental protection. Thus, DEA environmental assessment looks for the status of improvement regarding sustainability, not a capacity limit on part or whole production system, discussed in the weak disposability. The concept of congestion is separated into UC (i.e., a capacity limit) under natural disposability and DC (i.e., eco-technology innovation) under managerial disposability. The weak disposability corresponds to the possible occurrence of UC. Again, such an occurrence of UC is important in DEA, but DC is more important than UC in terms of environmental assessment. See Sueyoshi and Goto (2016) for a visual description of UC.

15.10 SUMMARY

This chapter described conceptual frameworks that serve as a basis for developing mathematical formulations of DEA applied to energy and environmental assessment. It is true that we are now facing various risks on the earth (e.g., heat waves, droughts, floods and food crises as well as damage to human, social and economic systems). To handle part of this very large-scale problem, the concept of natural and managerial disposability were first introduced and further extended to other economic and strategic concerns in this chapter. The two disposability concepts are very important in terms of energy and environment assessment because they are linked to an operational difficulty in energy sectors and eco-technology innovation on undesirable outputs such as GHG emissions, soil and water pollutions.

An underlying premise of this chapter is that eco-technology innovation, implying engineering development along with a managerial challenge in a broad sense, may combat the global warming and climate change as well as other types of pollution problem which have been currently destroying our natural and socio-economic systems in the world. Departing from a conventional use of DEA, as discussed in the chapters of Section I, this chapter discussed how to reorganize the computational structure of DEA to establish the status of sustainability. The concept of sustainability, implying economic prosperity and environmental protection, was separated into social and corporate sustainability in this chapter. The new use of DEA, as will be analytically discussed in Section II, can serve as a holistic methodology that makes a linkage between engineering and economics/business for further sustainable development. It is easily envisioned that engineering developments on environmental protection should be guided by social science, including business strategy as well as economic and environmental policies.

16

NON-RADIAL APPROACH FOR UNIFIED EFFICIENCY MEASURES

16.1 INTRODUCTION

This chapter[1] discusses the three types of non-radial models for DEA environmental assessment, all of which originate from the range-adjusted measure (RAM) discussed in Chapter 5. See the properties of RAM in Chapter 6. This study utilizes RAM as a basic model because it produces positive multipliers, or dual variables, related to all production factors. The analytical feature implies that all production factors are fully utilized in RAM-based assessment. Furthermore, as discussed in Chapter 6, the non-radial model has the property of "translation invariance." This property indicates that an efficiency measure should be not influenced even if inputs and/or outputs are shifted toward the same direction by adding or subtracting a specific real number. The desirable property makes it possible to evaluate the performance of various organizations (i.e., DMUs), whose production factors contain zeros and negative values in a data set. See Chapter 26 for a detailed description on how to use the desirable property for DEA environmental assessment. The two unique features indicate the important

[1]This chapter is partly based upon the article: Sueyoshi, T. and Goto, M. (2012a) Data envelopment analysis for environmental assessment: Comparison between public and private ownership in petroleum industry. *European Journal of Operational Research*, **216**, 668–678.

Environmental Assessment on Energy and Sustainability by Data Envelopment Analysis, First Edition. Toshiyuki Sueyoshi and Mika Goto.

methodological benefits of RAM. However, it is necessary for this chapter to reorganize RAM for environmental assessment because it cannot directly handle the existence of undesirable outputs in the original formulation, as discussed in Chapter 5.

To measure the performance of DMUs whose production activities are characterized by three types of production factors, this chapter incorporates the concept of natural and/or managerial disposability into the proposed non-radial approach. The underlying premise is that DMUs attempt to adapt a regulation change on undesirable outputs by considering two disposability concepts: natural and/or managerial disposability. See Chapter 15 for a description of the disposability concepts.

As an illustrative example to document the practicality, this chapter applies the proposed non-radial approach to compare national petroleum firms under public ownership with international petroleum firms under private ownership in terms of their efficiency measures under these disposability concepts.

This chapter has two research interests on the petroleum industry. One of the two interests is whether national petroleum firms outperform international ones under natural disposability. In contrast, international petroleum firms outperform national ones under managerial disposability. The two research concerns will provide the petroleum industry with business guidance for efficient operation around the world.

The remainder of this chapter is organized as follows: Section 16.2 introduces the proposed non-radial models to measure the level of unified (operational and environmental) efficiency. Section 16.3 describes the unified efficiency (*UE*) under natural disposability. Section 16.4 extends it to the measurement of *UE* under managerial disposability. Section 16.5 discusses the desirable properties related to the three *UE* measures. Section 16.6 applies the proposed approach for comparison between national and international petroleum companies. Section 16.7 summarizes this chapter.

16.2 UNIFIED EFFICIENCY

16.2.1 Formulation

A non-radial approach to measure the level of unified (operational and environmental) efficiency (UE_v^{NR}) for the k-th DMU can be reformulated under natural and/or managerial disposability concepts. The superscript NR stands for non-radial measure. The subscript v indicates variable RTS and DTS. See Chapter 20 for a description on RTS and DTS. In the proposed approach, the performance of the specific k-th DMU (x_{ik}, g_{rk}, b_{fk}) is relatively compared by n DMUs, including itself, which are expressed by x_{ij}, g_{rj}, b_{fj} for $j=1,\ldots,n$, where x, g and b stand for the input, desirable output and undesirable output, respectively. The subscriptions $i=1,\ldots,m$, $r=1,\ldots,s$ and $f=1,\ldots,h$ are for the three production factors of the j-th DMU.

The magnitude of UE_v^{NR} is measured by the following Model (16.1) (Sueyoshi and Goto, 2012a):

$$\begin{aligned}
&\text{Maximize} \sum_{i=1}^{m} R_i^x \left(d_i^{x+} + d_i^{x-}\right) + \sum_{r=1}^{s} R_r^g d_r^g + \sum_{f=1}^{h} R_f^b d_f^b \\
&s.t. \quad \sum_{j=1}^{n} x_{ij}\lambda_j - d_i^{x+} + d_i^{x-} = x_{ik} \ (i=1,\ldots,m), \\
&\sum_{j=1}^{n} g_{rj}\lambda_j \quad - d_r^g = g_{rk} \ (r=1,\ldots,s), \qquad (16.1) \\
&\sum_{j=1}^{n} b_{fj}\lambda_j \quad + d_f^b = b_{fk} \ (f=1,\ldots,h), \\
&\sum_{j=1}^{n} \lambda_j = 1, \\
&\lambda_j \geq 0 \ (j=1,\ldots,n),\ d_i^{x+} \geq 0 \ (i=1,\ldots,m), d_i^{x-} \geq 0 \ (i=1,\ldots,m), \\
&d_r^g \geq 0 \ (r=1,\ldots,s) \text{ and } d_f^b \geq 0 \ (f=1,\ldots,h).
\end{aligned}$$

Here, the slack variable (d_i^x) related to the i-th input is separated into positive and negative parts (d_i^{x+} and d_i^{x-}) to express the two types of disposability. The slack is measured by a distance of the observed value from an efficiency frontier. They are incorporated together into the first group of constrains of Model (16.1). The column vector $\lambda = \left(\lambda_1,\ldots,\lambda_n\right)^{Tr}$, often referred to as the intensity or structural vector, is used for connecting the input, desirable and undesirable output vectors by a convex combination between them. Since the sum of these intensity variables is restricted to unity in Model (16.1), a production and pollution possibility set for Model (16.1) is structured under variable RTS and DTS. See Chapter 20 for a detailed description of these concepts in the context of DEA environmental assessment. See Chapter 1, as well. The conventional use of DEA does not have the concept of DTS.

An important feature of Model (16.1) is that production factors are adjusted by the data ranges in the objective function. The data range adjustments are determined by the upper and lower bounds on inputs and those of desirable and undesirable outputs in the following manner:

(a) $R_i^x = (m+s+h)^{-1}\left(\max\left\{x_{ij} \ \middle| j=1,\ldots,n\right\} - \min\left\{x_{ij} \ \middle| j=1,\ldots,n\right\}\right)^{-1}$: a data range adjustment related to the i-th input ($i=1,...,m$),

(b) $R_r^g = (m+s+h)^{-1}\left(\max\left\{g_{rj} \ \middle| j=1,\ldots,n\right\} - \min\left\{g_{rj} \ \middle| j=1,\ldots,n\right\}\right)^{-1}$: a data range adjustment related to the r-th desirable output ($r=1,...,s$) and

(c) $R_f^b = (m+s+h)^{-1}\left(\max\left\{b_{fj} \ \middle| j=1,\ldots,n\right\} - \min\left\{b_{fj} \ \middle| j=1,\ldots,n\right\}\right)^{-1}$: a data range adjustment related to the f-th undesirable output ($f=1,...,h$).

A computational benefit of the data range adjustment is that it can control the magnitude of the *UE* measure between full efficiency and full inefficiency. Without the data range adjustment, Model (16.1) may produce a *UE* whose magnitude is very close to unity in all DMUs, as found in the additive model (see Chapter 6). It is indeed true that such a result is mathematically acceptable, but practically unacceptable, because we usually look for a wide range of an efficiency distribution for our performance assessments. See Chapter 25 for the use of common multipliers to adjust the efficiency range. Furthermore, the data range adjustment produces positive dual variables, so being able to avoid an occurrence of zero in them. This analytical feature is very important in measuring scale measures, such as RTS and DTS in Chapter 20. See Chapters 21, 22 and 23 as well. All the methodologies related to DEA environmental assessment in the three chapters fundamentally need positive dual variables, with a few exceptions, to measure the occurrence of congestion (i.e., undesirable congestion under natural disposability and desirable congestion under managerial disposability).

Computational Difficulty: Mathematically, the two slacks related to the *i*-th input are defined as $d_i^{x+} = \left(\left|d_i^x\right| + d_i^x\right)/2$ and $d_i^{x-} = \left(\left|d_i^x\right| - d_i^x\right)/2$. The parentheses in the objective function of Model (16.1) indicate $d_i^{x+} + d_i^{x-} = \left|d_i^x\right|$. The input slacks in the first group of constraints in Model (16.1) indicate $d_i^{x+} - d_i^{x-} = d_i^x$. The variable transformation of input slacks needs the non-linear conditions: $d_i^{x+} d_i^{x-} = 0$ $(i = 1, \ldots, m)$, implying that the two slack variables are mutually exclusive. Consequently, the simultaneous occurrence of both $d_i^{x+} > 0$ and $d_i^{x-} > 0$ $(i = 1, \ldots, m)$ needs to be excluded from the optimal solution of Model (16.1).

Unbounded Solution: When Model (16.1) has such a simultaneous occurrence on positive and negative slacks regarding inputs, a computer code may produce "an unbounded solution" because of non-linear condition violations. To make Model (16.1) satisfy the non-linear conditions, previous studies (e.g., Sueyoshi and Goto, 2012a) suggested the following two computational alternatives:

(a) One of the two alternations is that Model (16.1) incorporates the non-linear conditions into the formulation as side constraints and then we solve Model (16.1) with $d_i^{x+} d_i^{x-} = 0$ $(i = 1, \ldots, m)$ as a non-linear programming problem.

(b) The other alternative is that Model (16.1) incorporates the following side constraints: $d_i^{x+} \le M\gamma_i^+$, $d_i^{x-} \le M\gamma_i^-$, $\gamma_i^+ + \gamma_i^- \le 1$, γ_i^+ and γ_i^-: binary $(i = 1, \ldots, m)$ into the formulation and then we solve Model (16.1) with the side constraints as a mixed integer programming problem. Here, *M* stands for a very large number that we need to prescribe before the computational operation.

Infeasibility in Dual Formulation of Model (16.1): Model (16.1) may have a feasible solution. However, it often produces "an unbounded solution," as mentioned above, because the dual formulation is "infeasible." In this case, we must depend upon one of the above two alternatives. Both usually produce a similar optimal solution. See Table 3 of Sueyoshi and Goto (2011c) for an illustrative example where the two approaches produce a similar result. It is important to note that the unbounded solution implies the violation of non-linear conditions, not a conventional unbounded solution (where an optimal objective value becomes infinite) in linear programming. The violation, indicated by an unbounded solution, implies a simultaneous occurrence of $d_i^{x+} > 0$ and $d_i^{x-} > 0$ for some i on optimality of Model (16.1).

Unified Efficiency (UE_v^{NR}): After obtaining an optimal solution from Model (16.1) with non-linear conditions, this chapter determines the *UE* score (θ^*) on optimality as follows:

$$UE_v^{NR} = \theta^* = 1 - \left[\sum_{i=1}^{m} R_i^x \left(d_i^{x+*} + d_i^{x-*} \right) + \sum_{r=1}^{s} R_r^g d_r^{g*} + \sum_{f=1}^{h} R_f^b d_f^{b*} \right]. \quad (16.2)$$

Here, all slack variables represent the level of inefficiency in Equation (16.2) and they are determined on the optimality of Model (16.1). The *UE* is determined by subtracting the level of inefficiency from unity. The asterisk (*) indicates optimality.

The degree of unified (operational and environmental) efficiency of the k-th DMU, measured by Equation (16.2), is evaluated by both natural and managerial disposability. Thus, Equation (16.2) measures a *UE* score of a DMU when two different disposability concepts coexist in the DEA assessment. This aspect is different from these separate treatments on disposability, discussed later in this chapter.

Overcoming Computational Difficulty in Handling Model (16.1): The two alternatives (i.e., non-linear and mixed integer programming) incorporating non-linear conditions can overcome the methodological problem of Model (16.1). Therefore, this chapter discusses the importance of incorporating non-linear conditions in Model (16.1) so that the model no longer has any computational problem due to the unbounded solution.

A computational alternative, to be solved by linear programming, is that we reformulate Model (16.1) by slightly changing the objective function as follows:

$$\begin{aligned} &\text{Maximize} \sum_{i=1}^{m} R_i^x \left(d_i^{x+} - d_i^{x-} \right) + \sum_{r=1}^{s} R_r^g d_r^g + \sum_{f=1}^{h} R_f^b d_f^b \\ &s.t. \quad \text{same constraints in Model } (16.1). \end{aligned} \quad (16.3)$$

The difference between Models (16.1) and (16.3) is that Model (16.3) changes the sign of the input slack ($+d_i^{x-}$) from positive to negative ($-d_i^{x-}$). This is the only difference between the two models.

Paying attention to only the slack variables on inputs, this chapter can intuitively assert that Model (16.3) maximizes the positive slack (d_i^{x+}), simultaneously minimizing the negative slack (d_i^{x-}). Consequently, it is expected that Model (16.3) has $d_i^{x+*} > 0$ and $d_i^{x-*} = 0$ ($i = 1,\ldots,m$) on optimality.

The level of *UE* is measured by Equation (16.2). That is,

$$UE_v^{NR} = 1 - \left[\sum_{i=1}^{m} R_i^x \left(d_i^{x+*} + d_i^{x-*} \right) + \sum_{r=1}^{s} R_r^g d_r^{g*} + \sum_{f=1}^{h} R_f^b d_f^{b*} \right],$$

where all slacks are measured on the optimality of Model (16.3). Here, it is important to note that $\left(d_i^{x+} - d_i^{x-}\right)$ is used in the objective function of Model (16.3) and the sum of input slacks $\left(d_i^{x+*} + d_i^{x-*}\right)$ for all $i = 1,\dots,m$ is used to determine the degree of *UE*. Moreover, Model (16.3) is different from Model (16.1) in terms of these objective functions because of the computational tractability of Model (16.3). However, the level of inefficiency is determined by the total amount of slacks so that Equation (16.2) is used for identifying the level of *UE*.

The dual formulation of Model (16.3) becomes as follows:

$$\begin{aligned}
&\text{Minimize} \quad -\sum_{i=1}^{m} v_i x_{ik} - \sum_{r=1}^{s} u_r g_{rk} + \sum_{f=1}^{h} w_f b_{fk} + \sigma \\
&s.t. \quad -\sum_{i=1}^{m} v_i x_{ij} - \sum_{r=1}^{s} u_r g_{rj} + \sum_{f=1}^{h} w_f b_{fj} + \sigma \ \geq \ 0 \qquad (j = 1,\dots,n), \\
&\qquad v_i = R_i^x \qquad (i = 1,\dots,m), \\
&\qquad u_r \geq R_r^g \qquad (r = 1,\dots,s), \\
&\qquad w_f \geq R_f^b \qquad (f = 1,\dots,h), \\
&\qquad \sigma : \text{URS}.
\end{aligned} \tag{16.4}$$

where v_i $(i = 1,\dots,m)$, u_r $(r = 1,\dots,s)$, w_f $(f = 1,\dots,h)$ are all dual variables related to the first, second and third groups of constraints in Model (16.3). The dual variable σ is obtained from the fourth constraint of Model (16.3). The symbol URS indicates that a variable is "unrestricted."

Model (16.3) shows the primal feasibility and Model (16.4) shows the dual feasibility. Consequently, we can solve the two models by linear programming. This chapter does not explore Models (16.3) and (16.4) further because if we change $v_i = R_i^x$ to $v_i \geq R_i^x$ in Model (16.4), then it corresponds to a *UE* measurement under managerial disposability, whose formulation will be discussed later in this chapter. The proposed model for managerial disposability can be solved by linear programming and it has more practical implications than Models (16.3) and (16.4) in terms of DEA environmental assessment.

Finally, a similar description of Model (16.3) can be applied for changing the objective function of Model (16.1) as follows:

$$\begin{aligned}
&\text{Maximize} \sum_{i=1}^{m} R_i^x \left(-d_i^{x+} + d_i^{x-} \right) + \sum_{r=1}^{s} R_r^g d_r^g + \sum_{f=1}^{h} R_f^b d_f^b \\
&s.t. \quad \text{same constraints in Model } (16.1).
\end{aligned} \tag{16.5}$$

The formulation corresponds to the *UE* measurement under natural disposability in this chapter.

As mentioned previously, this chapter will not utilize Models (16.3) and (16.5) further, because the proposed non-radial models under natural or managerial disposability concept, discussed later, have more practical and theoretical implications than these formulations. Furthermore, the proposed models can have a direct linkage to the concepts discussed in Chapter 15.

16.2.2 Visual Implications of UE

An important feature of Model (16.1) is that it can express the two efficiency frontiers by a single set of intensity variables $\left(\lambda_j \geq 0;\ j=1,\ldots,n\right)$. The directional vector of possible projections in Model (16.1) includes both natural and managerial disposability for environmental assessment.

Figure 16.1 depicts the analytical structure for measuring *UE*. For our visual convenience, Figure 16.1 considers {K}, the achievement of a DMU that uses a single component of an input vector (*x* in the horizontal axis) to produce both a single component of the desirable output vector (*g* on the vertical axis) and a single component of the undesirable output vector (*b* on the vertical axis). The description can be easily extendable to the case of multiple components of three vectors in the proposed formulations.

Here, let us consider that the achievement of the *k*-th DMU is specified as $\left(x_{ik}, g_{rk}, b_{fk}\right)$ in Figure 16.1. To measure the level of *UE*, Model (16.1) needs to incorporate the two efficiency frontiers into the proposed analytical structure, as visually described in Figure 16.1. One of the two frontiers is an efficiency frontier for the desirable output. This frontier is listed as the top frontier in Figure 16.1. The top frontier corresponds to Stage I (A) of Figure 15.6. The other (a bottom frontier) is an efficiency frontier for the undesirable output. The bottom frontier corresponds to Stage I (B) of Figure 15.6. The efficiency frontier on the undesirable output is separated into the frontier under natural disposability (the left-hand side of the efficiency frontier for the undesirable output in Figure 16.1) and the frontier under managerial disposability (the right-hand side of the efficiency frontier for the undesirable output). The combined two efficiency frontiers in Figure 16.1 correspond to Stage II of Figure 15.6. As depicted in Figure 16.1, the DEA formulations proposed in this chapter do not incorporate the assumption on the byproduct status of undesirable outputs.

Possible performance enhancements, depicted in Figure 16.1, are classified by the following four types of projections.

(a) NW (North-West) projections to enhance operational performance: If the *k*-th DMU is inefficient in operational performance, then it needs to improve the operational efficiency by increasing the amount of a desirable output and/or decreasing the amount of an input. A possible projection can be found on {B} as such an example. A piece-wise linear contour line ({A}–{B}–{C}) indicates an efficiency frontier where {K} needs to be projected. The direction of such

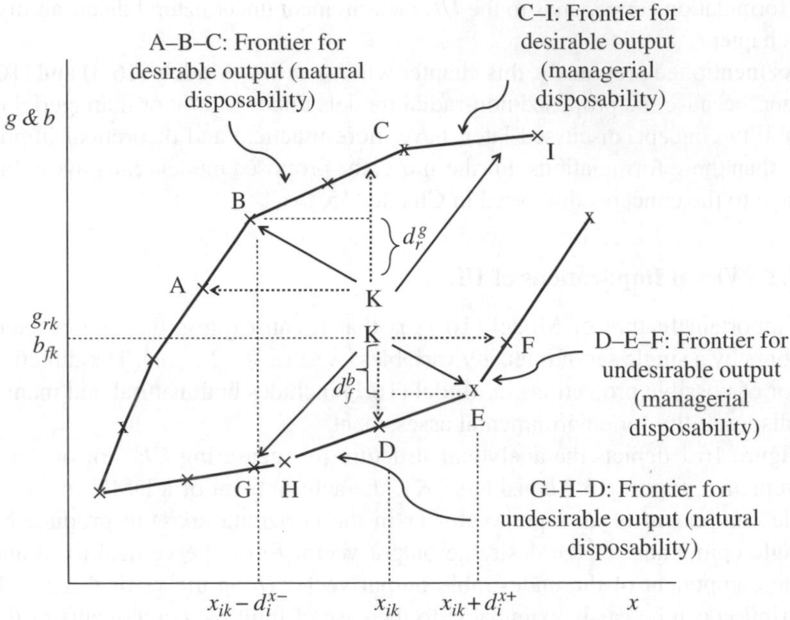

FIGURE 16.1 Two efficiency frontiers for measuring unified efficiency (a) The direction of NW (north-west) indicates a projection of {K} toward an efficiency frontier ({A}–{B}–{C}) that is incorporated into the conventional framework of DEA. (b) DEA environmental assessment incorporates the projection of SW (south-west) from {K} toward an efficiency frontier ({G}–{H}–{D}) under natural disposability and the projection of SE (south-east) from {K} toward an efficiency frontier ({D}–{E}–{F}) under managerial disposability. The projection of NW serves a basis for the performance assessment of conventional DEA. The projection of NE (north-east) toward an efficient frontier ({C}–{I}) has been excluded from the conventional use. The proposed environmental assessment incorporates the projection toward NE along with an increase in components of an input vector. (c) The figure corresponds to Stages I (A) and (B) of Figure 15.6. As depicted in Figure 15.6, the proposed DEA environmental assessment is much more completed than the one described in this figure. (d) NW + SW: natural disposability and NE + SE: managerial disposability. Both are incorporated in the proposed DEA environmental assessment.

A possible projection toward {A}–{B}–{C} is expressed by both the sign of an input-related slack and that of a desirable output-related slack, along with the direction of a unified inefficiency measure in the radial case, as depicted in Figure 16.1. The non-radial measurement does not have such an inefficiency score. This type of projections belongs to "NW" which indicates the enhancement of operational performance. Such projections serve a conceptual basis for enhancing the level of operational efficiency within conventional DEA which is based upon modern production economics. The projection looks for the best strategy to increase a desirable output under "economic stagnation" expressed by the input reduction. The reduction naturally implies a decrease of an undesirable output (i.e., industrial pollution), so indicating the concept of "natural disposability".

(b) NE (North-East) projections to enhance operational performance: The projection of NE has been excluded from the conventional use of DEA. The proposed DEA assessment incorporates the type of projections toward NE on the piece-wise linear contour line ({C}–{I}), along with an increase in both an input and a desirable output, as depicted in the right hand side of Figure 16.1. The type of projection for enhancing the level of operational performance implies "economic growth" of the *k*-th DMU. The economic growth is usually associated with various industrial pollutions. Therefore, this study must consider how to make an industrial balance between them. We clearly understand that it is not easily attainable such a balance because it needs to make a considerable effort, so being referred to as "managerial disposability". This type of performance enhancement has been never discussed in conventional DEA. The contribution of DEA environmental assessment is that it incorporates the type of projections, or corporate efforts, toward NE.

(c) SW (South-West) projections to enhance environmental performance under natural disposability: Shifting our interest from operational to environmental performance enhancement, Figure 16.1 visually describes a projection to enhance the performance in the input (on a horizontal coordinate) and the undesirable output (on a vertical coordinate) whose type of projections direct toward SW along with an input decrease. One of such directions is the projection from {K} to {G} in Figure 16.1. Here, {G} stands for the performance on an efficiency frontier for an undesirable output and an input. The piece-wise linear contour line ({G}-{H}-{D}) indicates such a frontier for unified efficiency. This type of projections, occurring with an input decrease, naturally decreases an amount of the undesirable output. Thus, this study considers the projection as "natural disposability" to enhance environmental performance.

(d) SE (South-East) projections to enhance environmental performance under managerial disposability: The type of projections from {K} to {E} indicates an opposite case of the SW projection, but indicating an enhancement of environmental performance. In the case, an efficiency frontier for the *k*-th DMU can be found on the piece-wise linear contour line ({D}-{E}-{F}) where the DMU needs to reduce the amount of the undesirable output by eco-technology and/or managerial effort, along with an input increase. This type of projections, under managerial disposability, has been never explored in the previous DEA studies.

Lower and Upper Bounds on Input Vector: As depicted in Figure 16.1, it is possible for each DMU to reduce the amount of an input until it can reach an efficiency frontier on a desirable output and an input. Thus, the efficiency frontier indicates the lower bound for the input reduction. In contrast, the DMU can increase an input until it can reach an efficiency frontier on an undesirable output and an input. Here, the efficiency frontier on the bottom line at the right hand indicates the upper bound for an input increase. This upper bound indicates the limit of "economic growth" under current eco-technology. The advancement of green technology shifts the upper bound for the right hand side in the manner that we can attain a maximum level of input increase, so indicating the maximum enhancement of economic growth.

The importance of green technology innovation is that it can shift the upper bound in the manner that an input increase can enhances the amount of a desirable output with a reduced undesirable output (i.e., less pollution). Thus, the eco-technology makes it possible that we can attain the status of sustainability. In contrast, the conventional use of DEA based upon modern economics has never considered such an analytical scheme that is necessary for sustainability development by economic growth and pollution reduction. As depicted in Figure 16.1, the conventional use of DEA, expressed only by the projections toward NW, does not have any practicality on sustainability development. See Chapter 14 that discusses the importance of a change in an input vector to attain the status of sustainability in energy sectors.

16.3 UNIFIED EFFICIENCY UNDER NATURAL DISPOSABILITY

The unified (operational and environment) efficiency, measured by Model (16.1), can be separated into two different non-radial models. One of the two models is formulated under natural disposability. The other is formulated under managerial disposability.

The non-radial model under natural disposability (N) to measure the unified efficiency (*UEN*) combining the desirable and undesirable outputs of the *k*-th DMU is formulated by the following augmented model (Sueyoshi and Goto, 2012a):

$$
\begin{aligned}
\text{Maximize} \quad & \sum_{i=1}^{m} R_i^x d_i^{x-} + \sum_{r=1}^{s} R_r^g d_r^g + \sum_{f=1}^{h} R_f^b d_f^b \\
s.t. \quad & \sum_{j=1}^{n} x_{ij}\lambda_j + d_i^{x-} = x_{ik} \quad (i=1,\ldots,m), \\
& \sum_{j=1}^{n} g_{rj}\lambda_j - d_r^g = g_{rk} \quad (r=1,\ldots,s), \\
& \sum_{j=1}^{n} b_{fj}\lambda_j + d_f^b = b_{fk} \quad (f=1,\ldots,h), \\
& \sum_{j=1}^{n} \lambda_j = 1, \\
& \lambda_j \geq 0 \ (j=1,\ldots,n), \ d_i^{x-} \geq 0 \ (i=1,\ldots,m), \\
& d_r^g \geq 0 \ (r=1,\ldots,s) \text{ and } d_f^b \geq 0 \ (f=1,\ldots,h).
\end{aligned}
\tag{16.6}
$$

Model (16.6) considers only deviations $+d_i^{x-}$ $(i=1,\ldots,m)$ related to inputs in order to attain the strategy under natural disposability.

The unified efficiency (UEN_v^{NR}) of the *k*-th DMU under natural disposability and variable RTS is measured by

$$UEN_v^{NR} = 1 - \left(\sum_{i=1}^{m} R_i^x d_i^{x-*} + \sum_{r=1}^{s} R_r^g d_r^{g*} + \sum_{f=1}^{h} R_f^b d_f^{b*} \right) \quad (16.7)$$

where all slack variables are determined on the optimality of Model (16.6). Thus, the equation within the parenthesis, obtained from the optimality of Model (16.6), indicates the level of unified inefficiency under natural disposability. The unified efficiency is obtained by subtracting the level of inefficiency from unity, as formulated in Equation (16.7). All the slacks are measured by

$$d_i^{x-*} = x_{ik} - \sum_{j=1}^{n} x_{ij} \lambda_j^*, \; d_r^{g*} = \sum_{j=1}^{n} g_{rj} \lambda_j^* - g_{rk} \text{ and } d_f^{b*} = b_{fk} - \sum_{j=1}^{n} b_{fj} \lambda_j^* \quad (16.8)$$

for the optimality of Model (16.6).

Model (16.6) has the following dual formulation:

$$\begin{aligned} \text{Minimize } & \sum_{i=1}^{m} v_i x_{ik} - \sum_{r=1}^{s} u_r g_{rk} + \sum_{f=1}^{h} w_f b_{fk} + \sigma \\ s.t. \quad & \sum_{i=1}^{m} v_i x_{ij} - \sum_{r=1}^{s} u_r g_{rj} + \sum_{f=1}^{h} w_f b_{fj} + \sigma \geq 0 \quad (j = 1,\ldots,n), \\ & v_i \geq R_i^x \quad (i = 1,\ldots,m), \\ & u_r \geq R_r^g \quad (r = 1,\ldots,s), \\ & w_f \geq R_f^b \quad (f = 1,\ldots,h), \\ & \sigma : \text{URS}, \end{aligned} \quad (16.9)$$

where v_i $(i=1,\ldots,m)$, u_r $(r=1,\ldots,s)$ and w_f $(f=1,\ldots,h)$ are all dual variables related to the first, second and third groups of constraints in Model (16.6). The dual variable σ is obtained from the fourth line in Model (16.6). An important feature of Model (16.9) is that all dual variables are positive so that all the production factors are fully utilized in the proposed environmental assessment.

Models (16.6) and (16.9) have three concerns to be noted. First, the two models have

$$\sum_{i=1}^{m} R_i^x d_i^{x-*} + \sum_{r=1}^{s} R_r^g d_r^{g*} + \sum_{f=1}^{h} R_f^b d_f^{b*} = \sum_{i=1}^{m} v_i^* x_{ik} - \sum_{r=1}^{s} u_r^* g_{rk} + \sum_{f=1}^{h} w_f^* b_{fk} + \sigma^* \quad (16.10)$$

on optimality. Equation (16.10) indicates the level of unified inefficiency, measured by Model (16.9). The unified efficiency is determined by the following equation:

$$UEN_v^{NR} = 1 - \left(\sum_{i=1}^{m} v_i^* x_{ik} - \sum_{r=1}^{s} u_r^* g_{rk} + \sum_{f=1}^{h} w_f^* b_{fk} + \sigma^* \right). \quad (16.11)$$

Equation (16.11) indicates that each dual variable specifies a change in the degree of unified inefficiency, due to a unit increase in the corresponding production factor, under natural disposability.

Second, it is possible for us to reformulate Model (16.6) with the input constraints $\left(\sum_{j=1}^{n} x_{ij}\lambda_j \le x_{ik}\right)$ for $i=1,\ldots,m$ by eliminating slacks in the formulation without the proposed range adjustment on inputs. In this case, the reformulation changes the dual variables to be $v_i \ge 0$ for $i=1,\ldots,m$ in Model (16.9). The change implies that these variables (v_i) may become zero, so indicating that the *i*-th input may not be fully utilized in the performance assessment. In contrast, Model (16.6) can always produce positive multipliers (i.e., dual variables) as long as the data ranges are positive, as mentioned previously. Thus, information on inputs as well as desirable and undesirable outputs is fully utilized in the assessment process of Model (16.6). Such an observation is confirmed by examining the dual formulation. This concern can be applied to other production factors (i.e., desirable and undesirable outputs).

Finally, Model (16.6) incorporates three types of slack variables in the formulation. All the constraints are considered as non-equality because of the existence of slack variables. This type of formulation is referred to as an "augmented form" in linear programming.

16.4 UNIFIED EFFICIENCY UNDER MANAGERIAL DISPOSABILITY

Shifting our research interest from natural disposability to managerial disposability, this chapter can measure *UE* under managerial disposability (M) where environmental performance is the first priority and operational performance is the second priority. The priority of managerial disposability is the opposite to the priority of natural disposability.

To measure the unified efficiency (UEM_v^{NR}) of the *k*-th DMU under managerial disposability, the research of Sueyoshi and Goto (2012a) proposed the following non-radial model in an augmented form:

$$
\begin{aligned}
\text{Maximize} \; & \sum_{i=1}^{m} R_i^x d_i^{x+} + \sum_{r=1}^{s} R_r^g d_r^g + \sum_{f=1}^{h} R_f^b d_f^b \\
s.t. \quad & \sum_{j=1}^{n} x_{ij}\lambda_j - d_i^{x+} = x_{ik} \quad (i=1,\ldots,m), \\
& \sum_{j=1}^{n} g_{rj}\lambda_j - d_r^g = g_{rk} \quad (r=1,\ldots,s), \\
& \sum_{j=1}^{n} b_{fj}\lambda_j + d_f^b = b_{fk} \quad (f=1,\ldots,h), \\
& \sum_{j=1}^{n} \lambda_j = 1, \\
& \lambda_j \ge 0 \; (j=1,\ldots,n), \; d_i^{x+} \ge 0 \; (i=1,\ldots,m), \\
& d_r^g \ge 0 \; (r=1,\ldots,s) \text{ and } d_f^b \ge 0 \; (f=1,\ldots,h).
\end{aligned}
\tag{16.12}
$$

Model (16.12) considers only deviations $-d_i^{x+}$ $(i=1,\ldots,m)$ related to inputs in order to attain the status of managerial disposability.

The unified efficiency (UEM_v^{NR}) of the k-th DMU under managerial disposability and variable DTS is measured by

$$UEM_v^{NR} = 1 - \left(\sum_{i=1}^{m} R_i^x d_i^{x+*} + \sum_{r=1}^{s} R_r^g d_r^{g*} + \sum_{f=1}^{h} R_f^b d_f^{b*} \right) \tag{16.13}$$

where all slack variables are determined on the optimality of Model (16.12). The equation within the parenthesis, obtained from the optimality of Model (16.12), indicates the level of unified inefficiency under managerial disposability. The unified efficiency is obtained by subtracting the level of inefficiency from unity, as formulated in Equation (16.13). All the slacks are measured by

$$d_i^{x+*} = \sum_{j=1}^{n} x_{ij}\lambda_j^* - x_{ik},\ d_r^{g*} = \sum_{j=1}^{n} g_{rj}\lambda_j^* - g_{rk} \text{ and } d_f^{b*} = b_{fk} - \sum_{j=1}^{n} b_{fj}\lambda_j^* \tag{16.14}$$

on the optimality of Model (16.12).

Model (16.12) has the following dual formulation:

$$\begin{aligned}
&\text{Minimize } -\sum_{i=1}^{m} v_i x_{ik} - \sum_{r=1}^{s} u_r g_{rk} + \sum_{f=1}^{h} w_f b_{fk} + \sigma \\
&s.t. \quad -\sum_{i=1}^{m} v_i x_{ij} - \sum_{r=1}^{s} u_r g_{rj} + \sum_{f=1}^{h} w_f b_{fj} + \sigma \ \geq 0 && (j=1,\ldots,n), \\
&\qquad v_i \geq R_i^x && (i=1,\ldots,m), \\
&\qquad u_r \geq R_r^g && (r=1,\ldots,s), \\
&\qquad w_f \geq R_f^b && (f=1,\ldots,h), \\
&\qquad \sigma : \text{URS},
\end{aligned} \tag{16.15}$$

where v_i $(i=1,\ldots,m)$, u_r $(r=1,\ldots,s)$ and w_f $(f=1,\ldots,h)$ are all dual variables related to the first, second and third groups of constraints in Model (16.12). The dual variable σ is obtained from the fourth equation of Model (16.12).

Models (16.12) and (16.15) measure the level of unified inefficiency under managerial disposability. The following condition:

$$\sum_{i=1}^{m} R_i^x d_i^{x+*} + \sum_{r=1}^{s} R_r^g d_r^{g*} + \sum_{f=1}^{h} R_f^b d_f^{b*} = -\sum_{i=1}^{m} v_i^* x_{ik} - \sum_{r=1}^{s} u_r^* g_{rk} + \sum_{f=1}^{h} w_f^* b_{fk} + \sigma^* \tag{16.16}$$

satisfies the optimality of the primal and dual models.

The level of *UEM*, measured from Model (16.16), is determined by

$$UEM_v^{NR} = 1 - \left(-\sum_{i=1}^{m} v_i^* x_{ik} - \sum_{r=1}^{s} u_r^* g_{rk} + \sum_{f=1}^{h} w_f^* b_{fk} + \sigma^* \right). \quad (16.17)$$

Equation (16.17) indicates that each dual variable specifies a change in the degree of unified inefficiency, due to one unit increase in each production factor, under managerial disposability.

16.5 PROPERTIES OF NON-RADIAL APPROACH

The proposed non-radial approach, originated from the range-adjusted measure (RAM), has two useful properties as a DEA environmental assessment measure. See Chapter 5 for a description of RAM. One of the two properties is "efficiency requirement," indicating that a *UE* measure exists between zero and one, where "zero" implies full inefficiency and "one" implies full efficiency. The other property is "translation invariance," indicating that *UE* should be not influenced even if production factors are shifted toward the same direction by adding or subtracting a specific real number. See Chapter 6 on these two desirable properties of conventional DEA. See also Chapter 26 that describes the property of translation invariance with the framework of DEA environmental assessment. Therefore, this chapter does not discuss it here, rather focusing upon the property of efficiency requirement in Model (16.6). The description can be applied to Model (16.12), as well.

Efficiency Requirement ($0 \leq UEN_v^{NR^*} \leq 1$): The objective function of Model (16.6) is restructured by

$$\sum_{i=1}^{m} R_i^x \left(x_{ik} - \sum_{j=1}^{n} x_{ij} \lambda_j \right) + \sum_{r=1}^{s} R_r^g \left(\sum_{j=1}^{n} g_{rj} \lambda_j - g_{rk} \right) + \sum_{f=1}^{h} R_f^b \left(b_{fk} - \sum_{j=1}^{n} b_{fj} \lambda_j \right), \quad (16.18)$$

where all slacks are replaced by three groups of production factors. Since all the slacks are non-negative, the objective function is equal to or more than zero. Therefore, it indicates $UEN_v^{NR^*} \leq 1$. Next, each term of the objective function is expressed by

$$\left(x_{ik} - \sum_{j=1}^{n} x_{ij} \lambda_j^* \right) \geq 1/R_i^x, \left(\sum_{j=1}^{n} g_{rj} \lambda_j^* - g_{rk} \right) \geq 1/R_r^g \text{ and } \left(b_{fk} - \sum_{j=1}^{n} b_{fj} \lambda_j^* \right) \geq 1/R_f^b. \quad (16.19)$$

Hence, the objective function is equal to or less than unity, so that it indicates $UEN_v^{NR*} \geq 0$. Consequently, $0 \leq UEN_v^{NR*} \leq 1$ is identified by Model (16.6). In a similar manner, this chapter can easily prove $0 \leq UEM_v^{NR*} \leq 1$.

Magnitude Adjustment of UE: As discussed above, the three *UE* measures satisfy the property of efficiency requirement. A problem is that the three *UE* measures often produce many efficient DMUs and inefficient ones whose efficiency scores become very close to unity (e.g., 0.999). The result is mathematically acceptable, but not practical, because DEA studies usually look for a wide efficiency distribution for DMU classification.

To overcome the drawback of the proposed non-radial approach, this study needs to adjust the objective function of Model (16.6) for natural disposability and that of Model (16.12) for managerial disposability as follows:

$$\begin{aligned} &\text{Maximize} \quad \varepsilon\left(\sum_{i=1}^{m} R_i^x d_i^{x-} + \sum_{r=1}^{s} R_r^g d_r^g + \sum_{f=1}^{h} R_f^b d_f^b\right) \\ &s.t. \quad \text{same constraints in Model } (16.6), \end{aligned} \tag{16.20}$$

and

$$\begin{aligned} &\text{Maximize} \quad \varepsilon\left(\sum_{i=1}^{m} R_i^x d_i^{x+} + \sum_{r=1}^{s} R_r^g d_r^g + \sum_{f=1}^{h} R_f^b d_f^b\right) \\ &s.t. \quad \text{same constraints in Model } (16.12). \end{aligned} \tag{16.21}$$

Here, the prescribed small number (ε), indicating a "magnitude control factor (MCF)," is incorporated in Models (16.20) and (16.21) for *UE* measures. The small number should be selected in the manner that a UE measure can satisfy the property of efficiency requirement (i.e., $0 \le UEN_v^{NR} \le 1$ and $0 \le UEM_v^{NR} \le 1$).

It is important to note that the small number is not the non-Archimedean small number (ε_n), as discussed in Section I, which is widely used in the conventional use of DEA. Rather, the small number (ε) is a specific number (e.g., 1.0, 0.5 or 0.1) for controlling the magnitude of *UE* measures. It is true that none knows what the non-Archimedean small number is in reality and it is different from the MCF utilized in the DEA environmental assessment. A problem of MCF is that it may include our subjectivity in the selection. So, if we cannot determine the degree of MCF, then this study recommends the selection of $\varepsilon = 1$, as formulated in Models (16.6) and (16.12).

Finally, it is possible for us to incorporate ε into Model (16.1) for *UE* measurement. As discussed in Chapter 24, *UEN* and *UEM* are unified to measure a unified efficiency (*UENM*) under natural and managerial disposability. In the combined model, the small number is used to control the magnitude of *UENM* for another purpose. Therefore, this chapter does not describe further the use of ε in *UE* measurement, here.

16.6 NATIONAL AND INTERNATIONAL FIRMS IN PETROLEUM INDUSTRY

Hereafter, this chapter documents the practicality of the proposed approach by comparing national and international petroleum companies.

Data Accessibility: The original data set used in this chapter can be found in the supplemental section of Sueyoshi and Goto (2015a).

Readers with an interest in the topic can use the data set from that study on Excel for academic purposes. They used the data set for a time series analysis of petroleum companies. This chapter is not interested in the time-series aspect, rather treating the date set as pooled data. In contrast, Sueyoshi and Goto (2015a) used the same data set to investigate the components of a frontier shift in a time horizon.

This chapter examines the three hypotheses that can be summarized as follows. The first null hypothesis is whether firms under public ownership perform as well as those under private ownership in the petroleum industry in terms of unified (operational and environmental) efficiency. Firms under public ownership usually look for natural disposability, where operational performance is the first priority and environmental performance is the second priority. Regulation on undesirable outputs and monitoring for pollution reduction are limiting on public firms because public companies and their governments are usually connected to each other in many aspects on their operations. Furthermore, they usually operate within their own nations and they can easily attain capital from their governmental supports for technology development. A problem faced by these national companies is that they use the capital for production technology of oil and gas (e.g, drilling technology) and only allocate a limited amount for eco-technology because their operations are within their own counties under lower regulation standards. See Chapter 23 for a description of Chinese energy companies, most of which are in public ownership and have been operating under local government control. In contrast, privately owned firms may look for managerial disposability because their operations do not have any national boundary. Therefore, they need to operate under international standards whose regulation is much stricter than their domestic standards.

The proposed hypothesis has two sub-hypotheses. One of these is whether firms under public ownership perform as well as those under private ownership in terms of UE under natural disposability. The other sub-hypothesis is the same, but in terms of UE under managerial disposability.

16.6.1 Business Structure

The petroleum industry is usually separated into two business functions. One function is related to "upstream" (e.g., exploration, development and production of crude oil or natural gas). The other function is related to "downstream" (e.g., oil tankers, refineries, storages and retails). This chapter measures only the performance of petroleum companies in the upstream. It is widely known that international oil companies (IOCs) dominate the downstream market so that it is not necessary to measure the performance of petroleum firms in the downstream.

16.6.2 National and International Oil Companies

To compare the performance of petroleum firms in public ownership with that of private petroleum firms, this chapter selects 14 firms whose equities are partially or completely owned by their governments. This type of petroleum firm is considered a national oil company (NOC) because their operations are under governmental influence. This chapter also selects five private petroleum firms whose governmental ownership is zero. All the firms are well known as IOCs, often referred to as "Major."

The NOC firms are as follows: (a) China National Offshore Oil Corporation in China, (b) Eni in Italy, (c) Gazprom in Russia, (d) Industrija Nafte in Croatia, (e) MOL Hungarian Oil and Gas Company in Hungary, (f) Österreichische Mineralölverwaltung in Austria, (g) Petroleum Development Oman in Oman, (h) Petroleos Mexicanos in Mexico, (i) Petrobras Brazil in Brazil, (j) PetroChina Company Limited in China, (k) Rosneft in Russia, (l) Saudi Aramco in Saudi Arabia, (m) China Petroleum and Chemical Corporation in China and (n) Statoil in Norway.

The IOC firms under private ownership are as follows: (a) British Petroleum in the United Kingdom, (b) Chevron in the United States, (c) Total in France, (d) ExxonMobil in the United States and (e) Shell in the Netherlands.

The proposed approach evaluates the 19 (=14 national + 5 international) petroleum firms based upon their four inputs (i.e., total oil reserve, total gas reserve, total operating cost, number of employees), two desirable outputs (total oil production, total gas production) and one undesirable output (total CO_2 emission). The observed annual periods are from 2005 to 2009. The data set used in this chapter is obtained from the annual reports (2005–2009) of 19 petroleum firms. In addition, this chapter conducted a telephone survey on the CO_2 emissions of these 19 petroleum firms. Descriptive statistics are listed in Sueyoshi and Goto (2012a, 2015a).

16.6.3 UE Measures

Table 16.1 summarizes the *UE* of the 19 petroleum firms measured by Model (16.1) equipped with non-linear conditions. As summarized in Table 16.1, NOCs outperformed IOCs from 2005 to 2008 on average. See the bottom two rows of Table 16.1. The opposite case was found in 2009. Since the total average (0.804) of NOCs is larger than that (0.770) of IOCs in terms of unified (operational and environmental) efficiency, we consider that NOCs under public ownership outperform IOCs under private ownership in the petroleum industry. Another

TABLE 16.1 Unified efficiency (2005–2009)

No.	Company and Country (Government Ownership)	2005	2006	2007	2008	2009	Average
1	China National Offshore Oil Corporation (CNOOC): China (71%)	0.831	0.833	0.835	0.844	0.839	0.836
2	Eni: Italy (30%)	0.628	0.632	0.613	0.626	0.653	0.630
3	Gazprom: Russia (51%)	0.862	0.861	0.865	0.873	0.869	0.866
4	Industrija Nafte (INA): Croatia (75%)	0.789	0.784	0.794	0.804	0.794	0.793
5	Hungarian Oil and Gas Company (MOL): Hungary (25%)	0.832	0.833	0.832	0.789	0.818	0.821
6	Österreichische Mineralölverwaltung (OMV): Austria (32%)	0.809	0.702	0.716	0.711	0.716	0.731
7	Petroleum Development Oman (PDO): Oman (60%)	0.794	0.836	0.841	0.841	0.835	0.829
8	Petroleos Mexicanos (Pemex): Mexico (100%)	0.585	0.569	0.565	0.541	0.532	0.558
9	Petrobras Brazil (Petrobras): Brazil (32%)	0.685	0.662	0.670	0.701	0.713	0.686
10	PetroChina Company Limited (PetroChina): China (90%)	0.945	0.917	0.944	0.947	1.000	0.951
11	Rosneft: Russia (100%)	0.962	0.965	1.000	1.000	1.000	0.985
12	Saudi Aramco: Saudi Arabia (100%)	1.000	1.000	0.957	1.000	1.000	0.991
13	China Petroleum & Chemical Corporation (Sinopec): China (57%)	0.807	0.803	0.800	0.776	0.785	0.794
14	Statoil: Norway (71%)	0.711	0.731	0.809	0.812	0.818	0.776
15	British Petroleum (BP): UK (0%)	0.727	0.924	0.825	0.910	0.895	0.856
16	Chevron: USA (0%)	0.625	0.625	0.627	0.617	0.654	0.630
17	Total: France (0%)	0.653	0.643	0.639	0.629	0.661	0.645
18	ExxonMobil: USA (0%)	0.927	0.972	1.000	0.808	1.000	0.941
19	Shell: Netherland (0%)	0.580	0.681	0.687	0.928	1.000	0.775
	Average of NOCs (1–14)	0.803	0.795	0.803	0.805	0.812	0.804
	Average of IOCs (15–19)	0.702	0.769	0.756	0.778	0.842	0.770

(a) Source: Sueyoshi and Goto, (2012a). (b) The number within brackets for each firm indicates the level of governmental ownership.

important finding is that IOCs exhibited an increasing trend from 2005 to 2009. Meanwhile, NOCs did not exhibit any increasing or decreasing trend in the observed annual periods.

16.6.4 UE Measures under Natural Disposability

Table 16.2 documents *UEN* for 19 petroleum firms, as measured by Model (16.6). The bottom two rows of Table 16.2 indicate the annual average *UEN* scores of NOCs and IOCs, measured under natural disposability. Table 16.2 indicates that NOCs outperformed IOCs on average in the observed annual periods.

Many NOCs such as firms {1,3,4,7,8 and 11} exhibited unity (100%) on average in their *UEN*. The other NOC firms also exhibited a high level of *UEN*. In contrast, IOCs exhibited some level of unified inefficiency that was larger than NOCs. Such a difference between NOCs and IOCs can be confirmed in Table 16.2.

As discussed by Sueyoshi and Goto (2012a), the operation cost of NOCs is less than that of IOCs on average because they can easily access various types of public funds. Acknowledging that there is a scale difference between NOCs and IOCs, the data set used in this chapter shows that the average operation cost of NOCs is 34.84 (US$ billion), while that of IOCs is 249.44 (US$ billion). Further, it is easily expected that each government allocates productive oil and gas wells to NOCs. This explains why NOCs outperform IOCs, as found in Table 16.2. Thus, public ownership outperforms private ownership in the petroleum industry in terms of *UEN*.

16.6.5 UE Measures under Managerial Disposability

Table 16.3 documents *UEM* for 19 petroleum firms, as measured by Model (16.12). The bottom two rows summarize the annual average of NOCs under public ownership and that of IOCs under private ownership. Table 16.3 indicates that IOCs outperformed NOCs on average in the observed annual periods. The finding obtained from Table 16.3 is completely opposite to that of Table 16.2.

Business Implication: The three tables suggest an important business implication, that is, NOCs attempt to enhance their operational performance, while simultaneously paying attention to the level of CO_2 emissions that satisfy only the national standard where each company operates. An interesting exception can be found in Saudi Aramco (Firm 12) that exhibited a high level of *UEM*. The company used to serve as a market power that could control oil price in the petroleum market because of its production scale. As a result, Saudi Aramco had a large amount of capital accumulation which could be allocated to pollution prevention equipment. The firm was an exceptional case among the NOCs.

TABLE 16.2 Unified efficiency under natural disposability (2005–2009)

No.	Company and Country (Government Ownership)	2005	2006	2007	2008	2009	Average
1	China National Offshore Oil Corporation (CNOOC): China (71%)	1.000	1.000	1.000	1.000	1.000	1.000
2	Eni: Italy (30%)	0.935	0.942	0.933	0.941	0.955	0.941
3	Gazprom: Russia (51%)	1.000	1.000	1.000	1.000	1.000	1.000
4	Industrija Nafte (INA): Croatia (75%)	1.000	1.000	1.000	1.000	1.000	1.000
5	Hungarian Oil and Gas Company (MOL): Hungary (25%)	1.000	1.000	1.000	0.992	0.989	0.996
6	Österreichische Mineralölverwaltung (OMV): Austria (32%)	1.000	0.997	1.000	0.998	1.000	0.999
7	Petroleum Development Oman (PDO): Oman (60%)	1.000	1.000	1.000	1.000	1.000	1.000
8	Petroleos Mexicanos (Pemex): Mexico (100%)	1.000	1.000	1.000	1.000	1.000	1.000
9	Petrobras Brazil (Petrobras): Brazil (32%)	1.000	0.991	0.983	0.972	1.000	0.989
10	PetroChina Company Limited (PetroChina): China (90%)	1.000	0.947	0.948	1.000	1.000	0.979
11	Rosneft: Russia (100%)	1.000	1.000	1.000	1.000	1.000	1.000
12	Saudi Aramco: Saudi Arabia (100%)	1.000	1.000	0.977	1.000	1.000	0.995
13	China Petroleum & Chemical Corporation (Sinopec): China (57%)	1.000	1.000	0.966	0.971	1.000	0.987
14	Statoil: Norway (71%)	1.000	1.000	0.995	1.000	1.000	0.999
15	British Petroleum (BP): UK (0%)	1.000	1.000	0.975	1.000	1.000	0.995
16	Chevron: USA (0%)	0.937	1.000	1.000	0.937	0.973	0.970
17	Total: France (0%)	0.939	0.926	0.923	0.906	0.932	0.925
18	ExxonMobil: USA (0%)	0.989	1.000	1.000	0.945	1.000	0.987
19	Shell: Netherland (0%)	1.000	1.000	0.996	1.000	1.000	0.999
	Average of NOCs (1–14)	0.995	0.991	0.986	0.991	0.996	0.992
	Average of IOCs (15–19)	0.973	0.985	0.979	0.958	0.981	0.975

(a) Source: Sueyoshi and Goto, (2012a). (b) The number within brackets for each company indicates the level of governmental ownership.

TABLE 16.3 Unified efficiency under managerial disposability (2005–2009)

No.	Company and Country (Government Ownership)	2005	2006	2007	2008	2009	Average
1	China National Offshore Oil Corporation (CNOOC): China (71%)	0.831	0.833	0.835	0.844	0.839	0.836
2	Eni: Italy (30%)	0.651	0.669	0.661	0.696	0.684	0.672
3	Gazprom: Russia (51%)	0.862	0.861	0.865	0.873	0.869	0.866
4	Industrija Nafte (INA): Croatia (75%)	0.789	0.784	0.794	0.804	0.794	0.793
5	Hungarian Oil and Gas Company (MOL): Hungary (25%)	0.832	0.833	0.832	0.789	0.818	0.821
6	Österreichische Mineralölverwaltung (OMV): Austria (32%)	0.809	0.702	0.716	0.711	0.716	0.731
7	Petroleum Development Oman (PDO): Oman (60%)	0.794	0.836	0.841	0.841	0.835	0.829
8	Petroleos Mexicanos (Pemex): Mexico (100%)	0.667	0.654	0.654	0.651	0.628	0.651
9	Petrobras Brazil (Petrobras): Brazil (32%)	0.652	0.662	0.670	0.663	0.657	0.661
10	PetroChina Company Limited (PetroChina): China (90%)	0.896	0.919	0.946	0.931	1.000	0.938
11	Rosneft: Russia (100%)	0.962	0.965	1.000	1.000	1.000	0.985
12	Saudi Aramco: Saudi Arabia (100%)	1.000	1.000	0.988	1.000	1.000	0.998
13	China Petroleum & Chemical Corporation (Sinopec): China (57%)	0.958	0.953	0.965	1.000	1.000	0.975
14	Statoil: Norway (71%)	0.711	0.731	0.729	0.731	0.736	0.728
15	British Petroleum (BP): UK (0%)	0.849	0.950	0.901	1.000	0.920	0.924
16	Chevron: USA (0%)	0.754	0.765	0.777	0.812	0.758	0.773
17	Total: France (0%)	0.747	0.759	0.777	0.818	0.769	0.774
18	ExxonMobil: USA (0%)	0.969	0.978	1.000	1.000	1.000	0.989
19	Shell: Netherland (0%)	0.868	0.884	0.907	1.000	1.000	0.932
	Average of NOCs (1–14)	0.815	0.815	0.821	0.824	0.827	0.820
	Average of IOCs (15–19)	0.837	0.867	0.872	0.926	0.889	0.878

(a) Source: Sueyoshi and Goto, (2012a). (b) The number within brackets for each company indicates the level of governmental ownership.

See Chapter 13 for a description of the petroleum industry. In contrast, IOCs operate internationally and they need to pay serious attention to not only enhancing their oil and gas productions but also reducing their CO_2 emissions using advanced eco-technology. Their operations attempt to satisfy the international standard on CO_2 emissions that is much stricter than the national standard of any country. Thus, private ownership outperforms public ownership in terms of *UEM*.

Implication for Sustainability: Figure 16.2 visually describes a causal relationship between "capital accumulation" and "social sustainability enhancement." As depicted in Figure 16.2, the initial concern is whether or not an organization or a nation can have enough capital to establish petroleum business. The concern is important in the

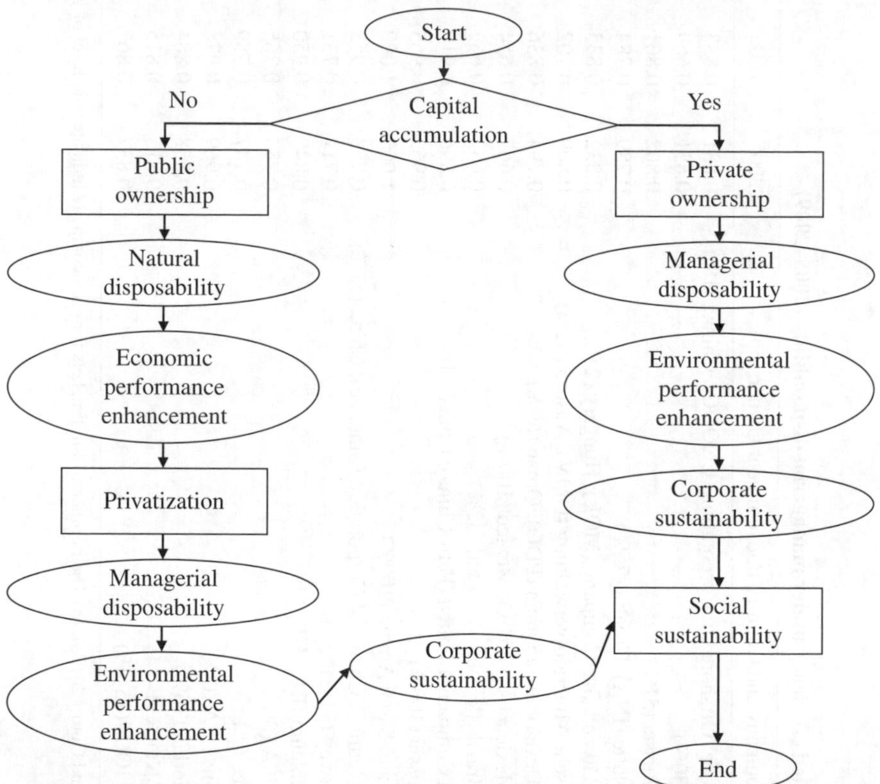

FIGURE 16.2 Sustainability development (a) Energy firms under public ownership can more easily access capital (so attaining a large amount of capital accumulation) than those under private ownership. As the result of such easy capital accessibility, public firms cannot make corporate efforts toward their growth processes. Public ownership often invites serious corruption between governments and energy firms, as found in many developing countries. Privatization will be necessary for most energy firms under public ownership.

initial stage because this industry is a very risky business. A large investment is always necessary for drilling oil and gas fields. There is no guarantee of successful drilling for production. As the result of such business risk, many petroleum companies operate under the structure of public ownership, so becoming NOCs. In contrast, many petroleum companies in industrial nations (e.g., the United States) operate under private ownership, so being IOCs. See Chapter 24 for a detailed description of the petroleum industry in the United States. NOCs under public ownership may have multiple business goals. For example, they need to increase the number of employees involved in the production of oil and gas. NOCs also increase the amount of revenue for their nation, to attain its economic success and welfare. Thus, it is easily thought that their operations are under natural disposability, as depicted in the left-hand side of Figure 16.2. The operation of NOCs may generate capital accumulation for their governments. Under international pressure, as discussed in Chapter 14, they must try to prevent air and water pollution, consequently enhancing their level of corporate sustainability and more globally their level of social sustainability. In contrast, IOCs have social responsibility and accountability in their international operations under the strict regulations of many nations. Their operations enhance the level of corporate sustainability, so increasing the status of social sustainability as well, as depicted in the right-hand side of Figure 16.2.

Privatization: At the end of the visual description, Figure 16.2 incorporates the implication of "privatization" on NOCs, which can shift them from public to private ownership. As a consequence of privatization, they increase their operational independence from their governments and, as private entities, they pay attention to global business concerns. See Chapter 13 for a detailed description of the current situation of the petroleum market. Oil is traded in a commodity market, being a symbol of "capitalism," which may have difficulty coexisting with public ownership. It is true that energy companies need a very large initial investment. To avoid business risk, public ownership may function at the initial stage of corporation establishment. NOCs under public ownership can provide people with highly paid jobs under government control. However, the relationship between government and business under public ownership becomes a source of inefficiency (e.g., corruption) in business, so producing an income imbalance among their people. Moreover, most NOCs cannot prepare any effective business strategy to adjust oil price fluctuations, as discussed in Chapter 13, under incompetent governmental energy policy. Thus, privatization will be essential for the corporate sustainability of all NOCs.

16.7 SUMMARY

This chapter discussed non-radial models for DEA environmental assessment. An important feature of this chapter was that it formulated mathematical models under natural or managerial disposability to adapt to a regulation change on undesirable outputs, following the conceptual discussion in Chapter 15.

This chapter applied the proposed approach to compare the performance of private and publicly owned petroleum firms. This chapter identified two important findings for the petroleum industry. One of these was that NOCs under public ownership outperformed IOCs under private ownership in both *UE* and *UEN*. However, the performance of IOCs exhibited an increasing trend in the *UEN*. The other finding was that IOCs outperformed NOCs in *UEM*. A rationale to support these two implications is that NOCs need to satisfy only the environmental standards of their own country. Hence, they may choose a strategy to focus on oil and gas production to improve their operational efficiency. In contrast, IOCs need to compete with each other across the world and need to satisfy international standards. As a consequence, IOCs need to consider their environmental performance more seriously than NOCs while enhancing their operational efficiency. Therefore, they accumulate capital and invest it into new eco-technology to satisfy the international standard on CO_2 emission reduction. Consequently, IOCs outperform NOCs in their *UEM* measures.

Finally, the empirical findings in this chapter indicate that the corporate sustainability of petroleum firms is influenced by their public or private ownership. The concern will be re-examined in Chapter 24 that will discuss the performance of US petroleum firms, all of which are under private ownership. The chapter will use a new non-radial approach for combining natural and managerial disposability concepts for the proposed DEA environmental assessment.

17

RADIAL APPROACH FOR UNIFIED EFFICIENCY MEASURES

17.1 INTRODUCTION

As a methodological alternative to the non-radial approach for DEA environmental assessment discussed in Chapter 16, this chapter introduces a radial approach to determine the level of unified efficiency measures. By shifting the previous formulations to the radial measurement, this chapter[1] discusses a new type of DEA environmental assessment to determine the three unified efficiency measures (i.e., *UE*, *UEN* and *UEM*) that may serve as our empirical basis for enhancing the level of social and corporate sustainability.

An important feature of the radial measurement is that it incorporates an inefficiency score to determine the level of unified efficiency measures, each of which is measured by subtracting the inefficiency score from unity. The analytical feature is different from the non-radial measurement discussed in Chapter 16, which determines the level of three unified efficiencies by the total amount of slacks. Furthermore, the radial approach incorporates a very small number (ε_s) that indicates the relationship between the inefficiency score and a total amount of all slacks. The incorporation of such a small number can reduce the influence of slacks in the unified efficiency measurement and it can control the magnitude

[1]This chapter is partly based upon the article: Sueyoshi, T. and Goto, M. (2012e) Environmental assessment by DEA radial measurement: U.S. coal-fired power plants in ISO (independent system operator) and RTO (regional transmission organization). *Energy Economics*, **34**, 663–676.

Environmental Assessment on Energy and Sustainability by Data Envelopment Analysis, First Edition. Toshiyuki Sueyoshi and Mika Goto.

of all dual variables to be positive so that all production factors are fully utilized in the proposed environmental assessment.

To document the practicality of the proposed radial approach, this chapter applies it to compare the performance of coal-fired power plants, which have been operating under an independent system operator (ISO) and a regional transmission organization (RTO), with that of the other coal-fired power plants in the United States, not belonging to any ISO and RTO. The comparison serves as an empirical basis to understand the operational benefit of ISO and RTO in terms of coal-fired power plant operations for corporate sustainability enhancement. See Chapter 14 that explains the differences between California ISO and PJM, which is the most well-known RTO, in the United States.

The remainder of this chapter is organized as follows: Section 17.2 introduces a radial model to measure a level of unified efficiency. Section 17.3 describes how to unify desirable and undesirable outputs within the framework of this radial measurement. Section 17.4 describes unified efficiency under natural disposability. Section 17.5 shifts it to managerial disposability. Section 17.6 applies the proposed approach to compare two groups of coal-fired power plants in the United States (i.e., ISO and RTO vs the others). Section 17.7 summarizes this chapter.

17.2 UNIFIED EFFICIENCY

According to Sueyoshi and Goto (2012b), this chapter uses the following radial model, corresponding to Model (16.1) in the case of non-radial measurement, in order to determine the level of unified efficiency (UE_v^R) concerning the k-th DMU:

$$
\begin{aligned}
&\text{Maximize } \xi + \varepsilon_s \left[\sum_{i=1}^{m} R_i^x \left(d_i^{x+} + d_i^{x-} \right) + \sum_{r=1}^{s} R_r^g d_r^g + \sum_{f=1}^{h} R_f^b d_f^b \right] \\
&s.t. \quad \sum_{j=1}^{n} x_{ij}\lambda_j - d_i^{x+} + d_i^{x-} = x_{ik} \; (i=1,\ldots,m), \\
&\quad \sum_{j=1}^{n} g_{rj}\lambda_j - d_r^g - \xi g_{rk} = g_{rk} \; (r=1,\ldots,s), \\
&\quad \sum_{j=1}^{n} b_{fj}\lambda_j + d_f^b + \xi b_{fk} = b_{fk} \; (f=1,\ldots,h), \qquad (17.1) \\
&\quad \sum_{j=1}^{n} \lambda_j = 1, \\
&\quad \lambda_j \geq 0 \; (j=1,\ldots,n), \; \xi : \text{URS}, \\
&\quad d_i^{x+} \geq 0 \; (i=1,\ldots,m), \; d_i^{x-} \geq 0 \; (i=1,\ldots,m), \\
&\quad d_r^g \geq 0 \; (r=1,\ldots,s) \text{ and } d_f^b \geq 0 \; (f=1,\ldots,h).
\end{aligned}
$$

where ξ is an unknown inefficiency score, indicating the distance between the efficiency frontier and an observed vector of desirable and undesirable outputs. The superscript R indicates radial measurement. A very small number (ε_s) is also incorporated into Model (17.1) to reduce the influence of slacks in the objective function. Both ξ and ε_s cannot be found in the non-radial approach discussed in Chapter 16.

A problem of the very small number is that a subjective decision may be included in the selection of ε_s. The specification difficulty is a drawback of Model (17.1). Historically, the number (ε_s) was considered as a "non-Archimedean small number (ε_n)" in the DEA community that was due to our mathematical convenience in the manner that all dual variables (multipliers) are positive in their signs. It is indeed true that the analytical aspect is important for DEA. A problem is that none knows what it is in reality. Therefore, this chapter proposes the use of the very small number in such a manner that we can control the relationship between an inefficiency score and the total sum of slacks. The subjectivity issue on ε_s still exists in the selection process, as well.

In avoiding such a specification difficulty, it is possible for us to set $\varepsilon_s = 0$ in Model (17.1). However, in such a case, dual variables (multipliers) may become zero in some production factors so that information on the corresponding production factors is not fully utilized in Model (17.1). The result is problematic and unacceptable. Such an unacceptable result has been often found in many DEA performance assessments. For example, let us consider that an environmental study uses the amount of CO_2 as an undesirable output, but the factor is not fully utilized in many previous DEA studies because its corresponding dual variable (i.e., multiplier) is zero. In this case, the environmental assessment has very limited practicality. Therefore, this chapter uses the very small number ε_s in Model (17.1). See Chapters 7 and 23 that use strong complementary slackness conditions (SCSCs) to avoid the use of ε_s.

As discussed in Chapter 16 (for non-radial measure), Model (17.1) is also formulated as a non-linear programming problem so that it is difficult for us to directly solve the model because an unbounded solution may occur in Model (17.1). To solve Model (17.1), Sueyoshi and Goto (2012b) proposed the following two computational alternatives, as suggested in Chapter 16:

(a) One of the two alternatives is that Model (17.1) incorporates non-linear conditions, or $d_i^{x+} d_i^{x-} = 0$ (i = 1,…,m), into the formulation as side constraints and then we solve it as a non-linear programming problem.

(b) The other alternative is that Model (17.1) incorporates the following side constraints: $d_i^{x+} \le M\gamma_i^+$, $d_i^{x-} \le M\gamma_i^-$, $\gamma_i^+ + \gamma_i^- \le 1$, γ_i^+ and γ_i^-: binary (i = 1,…,m) into the formulation and then we solve Model (17.1) with the side constraints as a mixed integer programming problem. Here, M stands for a very large number that we need to prescribe before our computation.

Unified Efficiency (UE_v^R): The incorporation of the non-linear conditions makes it possible to compute Model (17.1). The level of unified efficiency (UE_v^R) regarding the *k*-th DMU is determined by

$$UE_v^R = 1 - \left[\xi^* + \varepsilon_s \left(\sum_{i=1}^{m} R_i^x \left\{ d_i^{x+*} + d_i^{x-*} \right\} + \sum_{r=1}^{s} R_r^g d_r^{g*} + \sum_{f=1}^{h} R_f^b d_f^{b*} \right) \right], \quad (17.2)$$

where the inefficiency score and all the slacks within the parentheses are obtained from the optimality of Model (17.1).

17.3 RADIAL UNIFICATION BETWEEN DESIRABLE AND UNDESIRABLE OUTPUTS

An important research inquiry is why the radial measurement proposed in this chapter can unify desirable and undesirable outputs, both of which have opposite directional vectors for performance betterment.

To visually reply to the inquiry, this chapter prepares Figure 17.1, consisting of a single input (x) and two desirable outputs (g_1 and g_2). The horizontal axis of Figure 17.1 is for g_1/x and the vertical axis is for g_2/x. In this case, it is assumed that all DMUs produce the same amount of an undesirable output (b). Meanwhile, Figure 17.2 extends it to the case of two undesirable outputs (b_1 and b_2). The

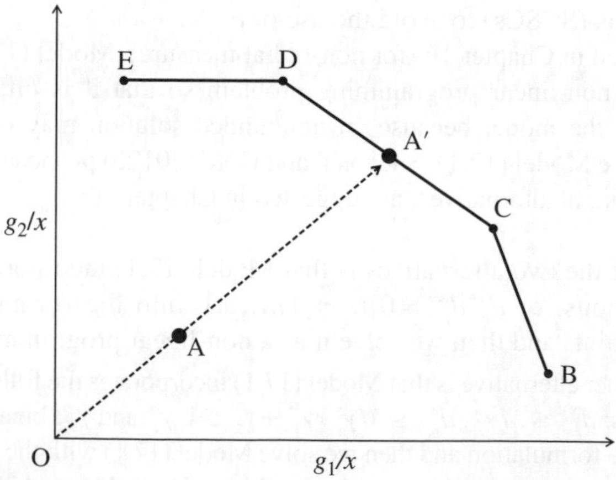

FIGURE 17.1 Projection onto efficiency frontier for desirable outputs (a) Source: Sueyoshi and Goto (2014a).

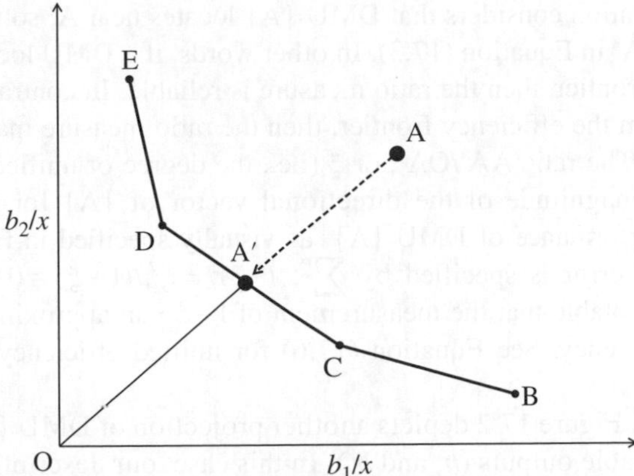

FIGURE 17.2 Projection onto efficiency frontier for undesirable outputs (a) Source: Sueyoshi and Goto (2014a).

horizontal axis of Figure 17.2 is for b_1/x and the vertical axis is for b_2/x under the assumption that all DMUs produce the same amount of a desirable output (g). In the two figures, {A}, {B}, {C}, {D} and {E} indicate the performance of five DMUs. A piece-wise linear contour line ({B}–{C}–{D}–{E}) indicates the efficiency frontier in each figure.

In the two figures, the inefficient DMU {A} directs the location toward the projected point A′ to enhance the joint performance between operational and environmental achievements. The directional vector of the desirable outputs is identified as (g_1^A, g_2^A) in Figure 17.1 and that of the undesirable outputs becomes $(-b_1^A, -b_2^A)$ in Figure 17.2. The two vector directions $(g_1^A, g_2^A, -b_1^A, -b_2^A)$ are given in examining the performance enhancement. However, the magnitude of the two directional vectors is unknown because none knows a distance between {A} and A′. It is true that such a magnitude can be visually identified in the two figures. However, if the number of coordinates is more than three, then none knows how to determine the magnitude of a directional vector for performance betterment. One purpose, not all, of DEA environmental assessment is to identify the distance between the two points ({A} and A′) for its relative comparison.

To visually explain the distance measurement, let us consider the ratio (OA)/(OA′) in Figure 17.1 that indicates the magnitude of unified efficiency concerning DMU {A}. The ratio is expressed by the following equation:

$$\frac{OA}{OA'} = \frac{OA}{OA + AA'} = \frac{1}{1 + \dfrac{AA'}{OA}} \cong \frac{1}{1 + \dfrac{AA'}{OA'}} = \frac{1}{1+\xi} = 1 - \xi + \xi^2 - \xi^3 + \cdots \cong 1 - \xi. \quad (17.3)$$

The above equation considers that DMU {A} locates near A′ so that OA can be replaced by OA′ in Equation (17.3). In other words, if a DMU locates on or near an efficiency frontier, then the ratio measure is reliable. In contrast, if the DMU locates far from the efficiency frontier, then the ratio measure may approximate the exact one. The ratio AA′/OA′, or ξ (i.e., the degree of unified inefficiency), indicates the magnitude of the directional vector of {A} for improving the operational performance of DMU {A}, as visually specified in Figure 17.1. An approximation error is specified by $\sum_{z=2}^{\infty}(-\xi)^z = \xi^2/(1+\xi) \cong 0$ because $\xi \leq 1$. Thus, it is acceptable that the measurement of $1-\xi$ can approximate the degree of unified efficiency. See Equation (17.6) for unified efficiency under natural disposability.

Meanwhile, Figure 17.2 depicts another projection of DMU {A} in the case of two undesirable outputs (b_1 and b_2). In this case, our description is straightforward. That is, the ratio OA′/OA indicates the degree of unified efficiency of {A} by

$$\frac{OA'}{OA} = \frac{OA - AA'}{OA} = 1 - \xi. \tag{17.4}$$

The ratio AA′/OA, or ξ, indicates a magnitude of the directional vector of {A} to improve the environmental performance in the two undesirable outputs. Thus, it is possible for us to unify desirable and undesirable outputs and to measure the magnitude of unified efficiency by $1-\xi$, as visually discussed in Figure 17.2. See Equation (17.11) for unified efficiency under managerial disposability.

Finally, this chapter needs to specify two analytical concerns. One of the two concerns is that the essence of DEA environmental assessment by radial measurement is to measure the magnitude of unified inefficiency (ξ), which is "unrestricted" in the sign. Such an analytical feature will be discussed in Chapter 19, which discusses how to measure a frontier shift among different periods. It was often misunderstood in previous DEA studies that the unified efficiency score should be "non-negative" in sign. Such a belief may be acceptable in a cross-sectional data analysis, but it is unacceptable in a time series analysis. See a detailed description on DEA time-series analysis in Chapter 19. The other concern is that Figures 17.1 and 17.2 visually describe how to unify desirable and undesirable outputs in a single unified framework even if they have opposite directional vectors for performance betterment. The unified measure identifies the level of unified (operational and environmental) inefficiency. This analytical feature is very different from the conventional use of DEA radial measurement that is designed to measure only the degree of operational efficiency (*OE*) as discussed in Chapters in Section I.

17.4 UNIFIED EFFICIENCY UNDER NATURAL DISPOSABILITY

The previous research of Sueyoshi and Goto (2012b) proposed the following radial model to measure the level of unified efficiency on the k-th DMU under natural disposability (N):

$$\begin{aligned}
&\text{Maximize } \xi+\varepsilon_s\left(\sum_{i=1}^{m}R_i^x d_i^{x-}+\sum_{r=1}^{s}R_r^g d_r^g+\sum_{f=1}^{h}R_f^b d_f^b\right)\\
&s.t. \quad \sum_{j=1}^{n}x_{ij}\lambda_j+d_i^{x-} = x_{ik}\ (i=1,\ldots,m),\\
&\qquad \sum_{j=1}^{n}g_{rj}\lambda_j - d_r^g - \xi g_{rk} = g_{rk}\ (r=1,\ldots,s),\\
&\qquad \sum_{j=1}^{n}b_{fj}\lambda_j + d_f^b+\xi b_{fk} = b_{fk}\ (f=1,\ldots,h),\\
&\qquad \sum_{j=1}^{n}\lambda_j=1,\\
&\qquad \lambda_j\ge 0\ (j=1,\ldots,n),\ \xi:\text{URS},\ d_i^{x-}\ge 0\ (i=1,\ldots,m),\\
&\qquad d_r^g\ge 0\ (r=1,\ldots,s) \text{ and } d_f^b\ge 0\ (f=1,\ldots,h).
\end{aligned} \tag{17.5}$$

The unified efficiency score (UEN_v^R) of the k-th DMU under natural disposability is measured by

$$UEN_v^R = 1-\left[\xi^*+\varepsilon_s\left(\sum_{i=1}^{m}R_i^x d_i^{x-*}+\sum_{r=1}^{s}R_r^g d_r^{g*}+\sum_{f=1}^{h}R_f^b d_f^{b*}\right)\right], \tag{17.6}$$

where the inefficiency score and all slack variables are determined on the optimality of Model (17.5). Thus, the equation within the parentheses is obtained from the optimality of Model (17.5). The unified efficiency under natural disposability is obtained by subtracting the level of inefficiency from unity. All the slacks are measured by

$$d_i^{x-*}=x_{ik}-\sum_{j=1}^{n}x_{ij}\lambda_j^*,\ d_r^{g*}=\sum_{j=1}^{n}g_{rj}\lambda_j^*-(1+\xi^*)g_{rk} \text{ and } d_f^{b*}=(1-\xi^*)b_{fk}-\sum_{j=1}^{n}b_{fj}\lambda_j^*$$

on the optimality of Model (17.5).

Model (17.5) has the following dual formulation:

$$
\begin{aligned}
&\text{Minimize } \sum_{i=1}^{m} v_i x_{ik} - \sum_{r=1}^{s} u_r g_{rk} + \sum_{f=1}^{h} w_f b_{fk} + \sigma \\
&s.t. \quad \sum_{i=1}^{m} v_i x_{ij} - \sum_{r=1}^{s} u_r g_{rj} + \sum_{f=1}^{h} w_f b_{fj} + \sigma \geq 0 \qquad (j = 1,\ldots,n), \\
&\sum_{r=1}^{s} u_r g_{rk} + \sum_{f=1}^{h} w_f b_{fk} = 1, \\
&v_i \geq \varepsilon_s R_i^x \qquad (i = 1,\ldots,m), \\
&u_r \geq \varepsilon_s R_r^g \qquad (r = 1,\ldots,s), \\
&w_f \geq \varepsilon_s R_f^b \qquad (f = 1,\ldots,h), \\
&\sigma : \text{URS}.
\end{aligned}
\tag{17.7}
$$

where v_i $(i = 1,\ldots,m)$, u_r $(r = 1,\ldots,s)$ and w_f $(f = 1,\ldots,h)$ are all positive dual variables (i.e., multipliers) related to the first, second and third groups of constraints in Model (17.5). The dual variable (σ) is obtained from the fourth constraint of Model (17.5).

The comparison between Models (17.5) and (17.7) provides us with the following three concerns: First, the objective value of Model (17.5) equals that of Model (17.7) on optimality. Thus, we have the following relationship:

$$
\xi^* + \varepsilon_s \left(\sum_{i=1}^{m} R_i^x d_i^{x-*} + \sum_{r=1}^{s} R_r^g d_r^{g*} + \sum_{f=1}^{h} R_f^b d_f^{b*} \right) = \sum_{i=1}^{m} v_i^* x_{ik} - \sum_{r=1}^{s} u_r^* g_{rk} + \sum_{f=1}^{h} w_f^* b_{fk} + \sigma^* \tag{17.8}
$$

on optimality. Using the dual formulation, the unified efficiency score (UEN_v^R) of the k-th DMU under natural disposability is measured by

$$
UEN_v^R = 1 - \left(\sum_{i=1}^{m} v_i^* x_{ik} - \sum_{r=1}^{s} u_r^* g_{rk} + \sum_{f=1}^{h} w_f^* b_{fk} + \sigma^* \right). \tag{17.9}
$$

Second, an important feature of Model (17.5) is that all dual variables are positive as formulated in Model (17.7). Thus, information on the three groups of production factors is fully utilized in Model (17.5). Finally, each dual variable indicates an amount of change in unified inefficiency under natural disposability due to a unit change in the corresponding production factor.

17.5 UNIFIED EFFICIENCY UNDER MANAGERIAL DISPOSABILITY

Shifting our research interest from natural disposability to managerial disposability (M), where the first priority is environmental performance and the second priority is operational performance, this chapter utilizes the following radial model that measures the unified efficiency of the k-th DMU under managerial disposability (Sueyoshi and Goto, 2012b):

$$\begin{aligned}
&\text{Maximize } \xi + \varepsilon_s \left(\sum_{i=1}^{m} R_i^x d_i^{x+} + \sum_{r=1}^{s} R_r^g d_r^g + \sum_{f=1}^{h} R_f^b d_f^b \right) \\
&s.t. \quad \sum_{j=1}^{n} x_{ij}\lambda_j - d_i^{x+} = x_{ik} \ (i=1,\dots,m), \\
&\qquad \sum_{j=1}^{n} g_{rj}\lambda_j - d_r^g - \xi g_{rk} = g_{rk} \ (r=1,\dots,s), \\
&\qquad \sum_{j=1}^{n} b_{fj}\lambda_j + d_f^b + \xi b_{fk} = b_{fk} \ (f=1,\dots,h), \qquad (17.10) \\
&\qquad \sum_{j=1}^{n} \lambda_j = 1 , \\
&\qquad \lambda_j \ge 0 \ (j=1,\dots,n), \xi: \text{URS}, d_i^{x+} \ge 0 \ (i=1,\dots,m), \\
&\qquad d_r^g \ge 0 \ (r=1,\dots,s) \text{ and } d_f^b \ge 0 \ (f=1,\dots,h).
\end{aligned}$$

An important feature of Model (17.10) is that it changes $+d_i^{x-}$ of Model (17.5) to $-d_i^{x+}$ in order to attain the status of managerial disposability. A unified efficiency score (UEM_v^R) of the k-th DMU under managerial disposability is measured by

$$UEM_v^R = 1 - \left[\xi^* + \varepsilon \left(\sum_{i=1}^{m} R_i^x d_i^{x+*} + \sum_{r=1}^{s} R_r^g d_r^{g*} + \sum_{f=1}^{h} R_f^b d_f^{b*} \right) \right], \qquad (17.11)$$

where the inefficiency score and all slacks are determined on the optimality of Model (17.10). Thus, the equation within the parentheses, obtained from the optimality of Model (17.10), indicates the level of unified inefficiency under managerial disposability. The unified efficiency is obtained by subtracting the level of inefficiency from unity.

Model (17.10) has the following dual formulation:

$$
\begin{aligned}
\text{Minimize} \; & -\sum_{i=1}^{m} v_i x_{ik} - \sum_{r=1}^{s} u_r g_{rk} + \sum_{f=1}^{h} w_f b_{fk} + \sigma \\
s.t. \quad & -\sum_{i=1}^{m} v_i x_{ij} - \sum_{r=1}^{s} u_r g_{rj} + \sum_{f=1}^{h} w_f b_{fj} + \sigma \geq 0 \qquad (j=1,\ldots,n), \\
& \sum_{r=1}^{s} u_r g_{rk} + \sum_{f=1}^{h} w_f b_{fk} = 1, \\
& v_i \geq \varepsilon_s R_i^x \qquad (i=1,\ldots,m), \\
& u_r \geq \varepsilon_s R_r^g \qquad (r=1,\ldots,s), \\
& w_f \geq \varepsilon_s R_f^b \qquad (f=1,\ldots,h), \\
& \sigma : \text{URS},
\end{aligned}
\tag{17.12}
$$

where v_i ($i = 1,\ldots,m$), u_r ($r = 1,\ldots,s$) and w_f ($f = 1,\ldots,h$) are all dual variables related to the first, second and third groups of constraints in Model (17.10). The dual variable (σ) is obtained from the fourth constraint of Model (17.10). The objective value of Model (17.10) equals that of Model (17.12) on optimality. Consequently, we have the following relationship:

$$
\xi^* + \varepsilon_s \left(\sum_{i=1}^{m} R_i^x d_i^{x+*} + \sum_{r=1}^{s} R_r^g d_r^{g*} + \sum_{f=1}^{h} R_f^b d_f^{b*} \right) = -\sum_{i=1}^{m} v_i^* x_{ik} - \sum_{r=1}^{s} u_r^* g_{rk} + \sum_{f=1}^{h} w_f^* b_{fk} + \sigma^*
\tag{17.13}
$$

on optimality. A unified efficiency score (UEM_v^R) of the k-th DMU under managerial disposability is measured by

$$
UEM_v^R = 1 - \left(-\sum_{i=1}^{m} v_i^* x_{ik} - \sum_{r=1}^{s} u_r^* g_{rk} + \sum_{f=1}^{h} w_f^* b_{fk} + \sigma^* \right). \tag{17.14}
$$

The difference between Models (17.7) and (17.12) is that the former model has $\sum_{i=1}^{m} v_i x_{ik}$ in the objective function and $\sum_{i=1}^{m} v_i x_{ij}$ for $j = 1,\ldots,n$ in the constraints, while the latter model has $-\sum_{i=1}^{m} v_i x_{ik}$ in the objective function and $-\sum_{i=1}^{m} v_i x_{ij}$ for $j = 1,\ldots,n$ in the constraints. No other difference exists between them. The previous description on dual variables obtained from Model (17.7) can be applied to those of Model (17.12).

17.6 COAL-FIRED POWER PLANTS IN THE UNITED STATES

Power plants are a major source of air pollution. In particular, in the United States, roughly two-thirds of all SO_2 and one-quarter of all NO_x come from electric power generation that relies on the burning of fossil fuels[2]. Coal-fired power plants account for a large share of these numbers. These power plants are also the largest polluter for toxic mercury and the largest contributor of hazardous air toxics. Although coal-fired power plants produced approximately 30% of electricity in the United States[3], they were responsible for over 68% of the carbon dioxide (CO_2) pollution from the electric power sector in 2016. See Chapter 13 for the detailed structure of the primary energy sources used to generate electricity as secondary energy.

Because of such various pollution problems, as discussed in Chapter 14, this chapter is interested in how DEA evaluates the unified performance of US coal-fired power plants. The rationale regarding why this chapter examines both their environmental and operational performance is that their operation is under the control of their transmission system operators. Therefore, the operation of a transmission grid system is as important as their environmental performance for power plants to effectively comply with environmental regulation. See Chapter 9 that visually describes the possible occurrence of congestion in a transmission grid system.

This chapter is fully aware of the fact that the United States had approximately 22.1% of the world's total proved reserves of coal at the end of 2016[4]. The macro aspect of coal-fired power plants explains a rationale regarding why the United States depends upon them as a principal fuel for electricity generation, even if they produce pollution problems.

17.6.1 ISO and RTO

In the United States, ISO (e.g., California ISO and New York ISO) and RTO (e.g., PJM Interconnection and New England ISO) control each regional electricity grid system, serve the wholesale power market as a market operator and provide reliability for region's electric power system. Meanwhile, there are some regions in the United States where ISO and RTO do not exist (e.g., State of New Mexico). In such regions, energy utilities are vertically integrated and control their own transmission networks. They conform to the open access rule in their transmission networks, as used by all market participants. Furthermore, power trades among energy utilities are mostly facilitated through bilateral contracts and power purchase agreements in their regions.

[2] https://www.epa.gov/acidrain/what-acid-rain

[3] https://www.eia.gov/tools/faqs/

[4] British Petroleum (2017) *BP Statistical Review of World Energy* (June 2017), British Petroleum, London.

In considering the practical importance of ISO and RTO, this chapter first pays attention to the opinion prepared by the Electric Power Supply Association[5] on 1 June 2011. The association has described that "ISO and RTO serve as a third party independent operator of a transmission system. However, there is an inherent conflict of interest when a single company owns all of transmission and distribution system and some of generation. The ISO and RTO, as the third-party independent operators, ensure that no preference should be given in the dispatch of a utility-owned generator." The association adds that "ISO and RTO provide fair transmission access to facilitate competition for the benefit of consumers. They provide transaction support as part of their market duties and engage in regional planning to ensure that the right infrastructure is built in the right place at the right time."

This chapter clearly acknowledges the fact that ISO and RTO, in particular PJM, are highly regarded by policy makers, corporate leaders and many individuals who are involved in the electric power industry. No question about it. However, no previous study has empirically confirmed the effectiveness of ISO and RTO from the perspective of efficient electric power plant operations. Thus, this chapter has a business inquiry regarding whether ISO and RTO really contribute to the enhancement of efficient market coordination and system operation through the effective dispatch of electricity generated by power plants, as we expect.

To investigate this research concern, this chapter compares the performance of coal-fired power plants under the control of ISO and RTO with that of other coal-fired power plants in the regions where such organizations do not exist. The criteria for our comparison are three unified efficiency measures under natural or managerial disposability as well as both combined disposability, all of which are measured by the proposed radial approach.

This chapter has two research concerns, or research hypotheses, to be examined by the proposed DEA approach. One of the two concerns, or the first hypothesis, is whether "coal-fired power plants under the control of ISO and RTO perform as well as those which do not belong to such organizations in the three efficiency measures (i.e., *UE*, *UEN* and *UEM*)." The first hypothesis is supported by professional organizations like the California ISO[6]. The ISO mentions several rationales regarding why ISOs and RTOs are important for the electric power industry of North America. One of such rationales is "efficient grid dispatch." The function of an efficient grid dispatch is summarized by the California ISO as follows: "Through the use of advanced technologies and market-driven incentives, the performance of power plants with regional markets tends to be better than performance in areas under monopoly control. There are lower power outage rates within competitive market regions because generation owners are motivated to keep plants online, especially during peak periods, to maximize their revenues."

[5] https://epsa.org/

[6] https://www.caiso.com/Documents/CompanyInformation_Facts.pdf

The other concern, or the second hypothesis, examined in this chapter is related to technology advancement in coal-fired power plants. That is, the second hypothesis is whether "US coal-fired power plants have improved their amounts of generation and their capabilities for environmental protection by technology innovation during the observed annual periods (1995–2007)."

As discussed by the California ISO, it may be true that coal-fired power plants in ISO and RTO control areas pay attention to technology advancement. To empirically confirm the claim, this chapter is interested in whether coal-fired power plants in ISO and RTO utilize their production capabilities for environmental protection to increase the amount of generation and to reduce the amount of CO_2 emission and other types of pollution.

Finally, this chapter summarizes the Mann–Whitney rank sum test to examine statistically a null hypothesis (i.e., no difference between two groups of observations) and the Kruskal–Wallis rank sum test to examine statistically a null hypothesis (i.e., no difference between multiple periods). See the Appendix for a detailed description on the Mann–Whitney and Kruskal–Wallis rank sum tests.

17.6.2 Data

Table 17.1 summarizes the number of coal-fired power plants in ISO and RTO as well as that of coal-fired power plants not belonging to these organizations from 1995 to 2007. This chapter pools all annual data sets and computes their efficiency scores to examine their dynamic changes during the observed annual periods.

The generation of each power plant is characterized by a single desirable output, four inputs and three undesirable outputs. The desirable output is measured by the annual amount of net generation (MWh: megawatt hours). The four

TABLE 17.1 Number of coal-fired power plants

	Total	ISO and RTO	Others
1995	172	97	75
1996	176	98	78
1997	176	97	79
1998	177	98	79
1999	170	96	74
2000	177	98	79
2001	172	95	77
2002	174	96	78
2003	173	93	80
2004	174	95	79
2005	173	93	80
2006	173	95	78
2007	136	76	60

(a) Source: Sueyoshi and Goto (2012e).

inputs are measured by: (a) the number of employees at each power plant, (b) the sum of all the investment costs at each plant level (the sum represents a running total of all additions, retirements and adjustments at the plant level), (c) the total non-fuel operating and maintenance (O&M) cost that stands for the sum of the total non-fuel operating and maintenance expenses at each plant level, and (d) the fuel consumption that stands for fuel consumption (i.e., coal in this chapter) of each plant, measured by weight. The undesirable outputs are measured in the following manner: (a) the total amount of SO_2 released each year (in tons), based on the emission rate per hour and operating hours for the year, (b) the tons of NO_x emitted each year by measuring (or estimating) NO_x emissions as pounds per mmBtu (million British thermal unit) and then calculating its annual total, based on heat input (mmBtu), (c) the annual amount of CO_2 emission, as computed by the amount of CO_2 emitted each hour times the operation hours of each plant. Note that SO_2 and NO_x are acid rain gases and CO_2 is greenhouse gas (GHG). Descriptive statistics for the data set can be found in Sueyoshi and Goto (2012e).

Data Accessibility: The data set on the desirable output and the four inputs was obtained from Powerdat, published by Platts (part of McGraw-Hill), and the electric section of the Federal Energy Regulatory Commission (FERC)[7]. The three undesirable outputs were obtained from the US Environmental Protection Agency[8].

17.6.3 Unified Efficiency

Table 17.2 summarizes the unified (operational and environmental) efficiency measures of coal-fired power plants, all of which are measured by Model (17.1). The row on Avg. in the table lists the annual average of unified efficiency measures regarding the two groups of coal-fired power plants. One of the two groups consists of coal-fired power plants in ISO and RTO. The other group consists of coal-fired power plants not belonging to these organizations.

An important finding in Table 17.2 is that coal-fired power plants in the second group has outperformed those of the first (ISO and RTO) group during almost all of the observed annual periods. This is the opposite result to the first hypothesis, assuming that the two groups operate at the same efficiency level. To confirm this claim statistically, this chapter examines the first hypothesis (i.e., the first group performs as well as the second group) by a rank sum test. For this statistical test, this chapter applies the Mann–Whitney rank sum test to the ranks of the two groups of coal-fired power plants.

The Mann–Whitney statistic becomes –3.10, implying the rejection of the first hypothesis at the 1% level of significance because 3.10 > 2.58 (the critical value of the standard normal distribution). The Mann–Whitney rank sum test indicates

[7] https://www.ferc.gov/docs-filing/forms.asp

[8] https://www.epa.gov/energy/emissions-generation-resource-integrated-database-egrid

TABLE 17.2 Unified efficiency (1995–2007)

		1995	1996	1997	1998	1999	2000	2001
ISO & RTO	Avg.	0.732	0.717	0.718	0.716	0.723	0.748	0.753
	S.D.	0.108	0.107	0.112	0.106	0.113	0.099	0.095
	Max.	1.000	1.000	1.000	0.979	0.953	0.968	0.988
	Min.	0.526	0.350	0.159	0.256	0.184	0.541	0.546
Others	Avg.	0.722	0.732	0.737	0.746	0.753	0.765	0.771
	S.D.	0.092	0.085	0.090	0.090	0.085	0.090	0.088
	Max.	1.000	0.901	1.000	0.967	0.967	0.994	0.982
	Min.	0.542	0.555	0.568	0.555	0.556	0.553	0.561
Total	Avg.	0.727	0.724	0.727	0.729	0.736	0.755	0.761
	S.D.	0.101	0.098	0.103	0.100	0.103	0.095	0.092
	Max.	1.000	1.000	1.000	0.979	0.967	0.994	0.988
	Min.	0.526	0.350	0.159	0.256	0.184	0.541	0.546

		2002	2003	2004	2005	2006	2007
ISO & RTO	Avg.	0.757	0.771	0.787	0.796	0.807	0.799
	S.D.	0.101	0.103	0.106	0.107	0.108	0.115
	Max.	0.993	0.986	1.000	1.000	1.000	1.000
	Min.	0.528	0.544	0.529	0.560	0.577	0.592
Others	Avg.	0.773	0.770	0.794	0.794	0.790	0.806
	S.D.	0.094	0.116	0.091	0.091	0.087	0.092
	Max.	0.975	0.989	1.000	1.000	1.000	1.000
	Min.	0.537	0.155	0.627	0.624	0.605	0.639
Total	Avg.	0.764	0.771	0.790	0.795	0.800	0.802
	S.D.	0.098	0.109	0.099	0.099	0.099	0.105
	Max.	0.993	0.989	1.000	1.000	1.000	1.000
	Min.	0.528	0.155	0.529	0.560	0.577	0.592

(a) Source: Sueyoshi and Goto (2012e). (b) Avg.: Average, S.D.: Standard Deviation, Max.: Maximum and Min.: Minimum.

that the second group (i.e., coal-fired power plants which do not belong to ISO and RTO) has outperformed the first group (i.e., coal-fired power plants belonging to ISO and RTO) in terms of their unified efficiency measures. This empirical result is very surprising to us and is inconsistent with the opinion of many well-known organizations (e.g., California ISO and FERC[9]). It is also important to note that the two groups of coal-fired power plants have operated at almost the same level of unified efficiency after 2003. This indicates that there was no major difference between the two groups in their recent performance on unified efficiency. This is partly consistent with the first hypothesis.

Besides the comparison between the two groups, this chapter examines whether a technological progress has existed in US coal-fired power plants during the

[9] https://www.ferc.gov/industries/electric/indus-act/rto.asp

observed annual periods. The Kruskal–Wallis statistic is 166.27, implying the rejection of the null hypothesis of no technological progress at the 1% level of significance because 166.27 > 26.11 (the critical value of the χ^2 distribution). Consequently, the Kruskal–Wallis rank sum test indicates that we cannot reject the second hypothesis. Thus, the US coal-fired power plants have showed technological improvement in their unified performance.

17.6.4 Unified Efficiency under Natural Disposability

Table 17.3 summarizes the unified efficiency measures of US coal-fired power plants, all of which are measured by Model (17.5) under natural disposability. The Avg. row in the table lists the annual average of the unified efficiency regarding the two groups of coal-fired power plants. An important finding of Table 17.3 is that

TABLE 17.3 Unified efficiency under natural disposability (1995–2007)

		1995	1996	1997	1998	1999	2000	2001
ISO & RTO	Avg.	0.818	0.805	0.809	0.810	0.813	0.834	0.824
	S.D.	0.102	0.104	0.115	0.104	0.116	0.099	0.095
	Max.	1.000	1.000	1.000	1.000	1.000	1.000	0.994
	Min.	0.537	0.435	0.174	0.312	0.199	0.575	0.561
Others	Avg.	0.833	0.837	0.840	0.842	0.847	0.857	0.849
	S.D.	0.095	0.085	0.090	0.089	0.087	0.087	0.086
	Max.	1.000	1.000	1.000	1.000	1.000	1.000	1.000
	Min.	0.622	0.609	0.601	0.645	0.656	0.644	0.666
Total	Avg.	0.825	0.819	0.823	0.824	0.828	0.844	0.836
	S.D.	0.099	0.097	0.105	0.098	0.105	0.094	0.091
	Max.	1.000	1.000	1.000	1.000	1.000	1.000	1.000
	Min.	0.537	0.435	0.174	0.312	0.199	0.575	0.561

		2002	2003	2004	2005	2006	2007
ISO & RTO	Avg.	0.824	0.835	0.841	0.845	0.850	0.847
	S.D.	0.101	0.099	0.098	0.102	0.100	0.104
	Max.	1.000	0.998	1.000	1.000	1.000	1.000
	Min.	0.535	0.553	0.554	0.565	0.594	0.634
Others	Avg.	0.854	0.843	0.859	0.857	0.844	0.859
	S.D.	0.088	0.119	0.091	0.090	0.088	0.088
	Max.	1.000	1.000	1.000	1.000	1.000	1.000
	Min.	0.661	0.155	0.649	0.683	0.637	0.685
Total	Avg.	0.838	0.839	0.849	0.850	0.847	0.852
	S.D.	0.096	0.109	0.095	0.097	0.095	0.097
	Max.	1.000	1.000	1.000	1.000	1.000	1.000
	Min.	0.535	0.155	0.554	0.565	0.594	0.634

(a) Source: Sueyoshi and Goto (2012e). (b) Avg.: Average, S.D.: Standard Deviation, Max.: Maximum and Min.: Minimum.

coal-fired power plants in the second group have outperformed those of the first group (ISO and RTO) during all observed annual periods with the exception of 2006. This is the opposite result to the first hypothesis. To confirm the claim statistically, this chapter examines the first hypothesis (i.e., the first group performs as well as the second group) by applying the Mann–Whitney rank sum test to the ranks of the two groups of coal-fired power plants.

The Mann–Whitney statistic is –4.49, implying the rejection of the first hypothesis at the 1% level of significance because 4.49 > 2.58 (the critical value of the standard normal distribution). The rank sum test indicates that the second group (i.e., coal-fired power plants which do not belong to ISO and RTO) has outperformed the first group (coal-fired power plants belonging to ISO and RTO) in unified efficiency under natural disposability. This empirical result is inconsistent with the opinion of many professional organizations (e.g., California ISO and FERC).

This chapter also examines whether a technological progress has existed in coal-fired power plants during the observed annual periods. The Kruskal–Wallis statistic is 22.96, implying the rejection of the null hypothesis (i.e., no technology progress) at the 5% level of significance because 22.96 > 21.03 (the critical value of the χ^2 distribution). The rank sum test implies that the coal-fired power plants have exhibited technological improvement in their operational performance.

17.6.5 Unified Efficiency under Managerial Disposability

Table 17.4 summarizes the unified efficiency measures of coal-fired power plants, all of which are measured by Model (17.10) under managerial disposability. The Avg. row in the table lists the annual average of unified efficiency score regarding the two groups of coal-fired power plants.

An important finding in Table 17.4 is that coal-fired power plants in the second group have outperformed those of the first (ISO and RTO) group during most of the observed annual periods. This is the opposite result to the first hypothesis. To confirm the claim statistically, this chapter examines the first hypothesis (i.e., the first group performs as well as the second group). The Mann–Whitney rank sum test is applied to the ranks of the two groups of coal-fired power plants.

The Mann–Whitney statistic is –2.08, implying the rejection of the first hypothesis at the 5% level of significance because 2.08 > 1.96 (the critical value of the standard normal distribution). The rank sum test indicates that the second group (i.e., coal-fired power plants not belonging to the group of ISO and RTO) has outperformed the first group (coal-fired power plants belonging to these well-known organizations) in their unified efficiency under managerial disposability. This empirical result is inconsistent with the opinion of many professional organizations (e.g., California ISO and FERC). It is important to note that the trend has been reversed in the last three annual periods (2005–2007) and the two groups seem to be in an almost same level of unified efficiency under managerial disposability. Thus, this is partly consistent with the first hypothesis.

TABLE 17.4 Unified efficiency under managerial disposability (1995–2007)

		1995	1996	1997	1998	1999	2000	2001
ISO & RTO	Avg.	0.781	0.751	0.751	0.741	0.751	0.773	0.780
	S.D.	0.124	0.122	0.122	0.107	0.111	0.106	0.103
	Max.	1.000	1.000	1.000	0.982	1.000	1.000	1.000
	Min.	0.545	0.400	0.332	0.501	0.529	0.544	0.549
Others	Avg.	0.761	0.754	0.765	0.772	0.779	0.788	0.796
	S.D.	0.105	0.096	0.101	0.101	0.094	0.095	0.089
	Max.	1.000	1.000	1.000	1.000	1.000	1.000	1.000
	Min.	0.567	0.560	0.569	0.555	0.556	0.553	0.561
Total	Avg.	0.772	0.753	0.757	0.755	0.763	0.779	0.787
	S.D.	0.116	0.111	0.113	0.105	0.104	0.101	0.097
	Max.	1.000	1.000	1.000	1.000	1.000	1.000	1.000
	Min.	0.545	0.400	0.332	0.501	0.529	0.544	0.549

		2002	2003	2004	2005	2006	2007
ISO & RTO	Avg.	0.783	0.799	0.816	0.827	0.844	0.838
	S.D.	0.107	0.109	0.109	0.112	0.109	0.112
	Max.	1.000	1.000	1.000	1.000	1.000	1.000
	Min.	0.531	0.551	0.533	0.570	0.591	0.595
Others	Avg.	0.802	0.808	0.822	0.822	0.825	0.836
	S.D.	0.097	0.099	0.090	0.092	0.094	0.096
	Max.	0.993	1.000	1.000	1.000	1.000	1.000
	Min.	0.537	0.560	0.652	0.624	0.605	0.645
Total	Avg.	0.792	0.803	0.819	0.825	0.835	0.837
	S.D.	0.103	0.104	0.101	0.103	0.103	0.105
	Max.	1.000	1.000	1.000	1.000	1.000	1.000
	Min.	0.531	0.551	0.533	0.570	0.591	0.595

(a) Source: Sueyoshi and Goto (2012e). (b) Avg.: Average, S.D.: Standard Deviation, Max.: Maximum and Min.: Minimum.

This chapter also examines whether a technological progress has existed in coal-fired power plants during the observed annual periods. The Kruskal–Wallis statistic is 159.05, implying the rejection of the null hypothesis at the 1% level of significance because 159.05 > 26.22 (the critical value of the χ^2 distribution). The rank sum test indicates that US coal-fired power plants have exhibited technological improvement in terms of their environmental performance.

17.7 SUMMARY

As an alternative to the non-radial approach, discussed in Chapter 16, for DEA environmental assessment, this chapter proposed a radial approach to determine the three unified efficiency measures that may serve as an empirical basis for developing corporate sustainability.

An important feature of the proposed radial measurement is that it incorporates an inefficiency score to determine the level of unified efficiency measures, each of which is measured by subtracting the inefficiency score from unity. Furthermore, the proposed approach incorporates a prescribed very small number (ε_s) that indicates the relationship between the inefficiency score and the total amount of slacks. This incorporation reduces the influence of slacks in the measurement of unified efficiency and makes all dual variables positive so that all production factors are fully utilized in the proposed DEA environmental assessment. The analytical feature on the inefficiency score cannot be found in the non-radial measurement discussed in Chapter 16.

To document the practicality of the proposed measurement, this chapter applied three radial models to compare the performance of US coal-fired power plants under ISO and RTO with that of power plants not belonging to any organization. This chapter identified that the latter group of coal-fired power plants outperformed the group of power plants belonging to ISO and RTO in the three efficiency measures (i.e., unified efficiency, unified efficiency under natural disposability and unified efficiency under managerial disposability). This result implies that the transmission network coordination in ISO and RTO, which requires transmission unbundling of electric power firms, does not always outperform transmission network operation under vertically integrated utilities, in particular the operation of coal-fired power plants. The empirical result may be very different from the previous opinion (i.e., the benefit of ISO and RTO) that has been long supported by many professional organizations such as the California ISO and FERC in the United States. This chapter considers that the empirical result is due to the fact that US power plants have been gradually shifting from coal combustion to gas, nuclear and renewable energies. Consequently, coal-fired electricity generation has been less important in power plants in ISO and RTO regions under strict regulation on air pollution. See Chapter 14 for a description of air pollution regulation in the United States. See also Chapter 20 for the fuel mix shift in US power plants. In addition, this chapter found that there was technological progress in coal-fired power plants from 1995 to 2007. The technology innovation (e.g., clean coal technology) improved their operational, environmental and both unified performance measures during the observed annual periods. As a result of technology dissemination in the US coal-fired power plants, the two groups exhibited close performance measures in the recent annual periods.

APPENDIX

Mann–Whitney Rank Sum Test: To examine statistically a null hypothesis (i.e., no difference between two groups of observations), this chapter applies the Mann–Whitney rank sum test to the rank of two groups of observations (e.g, coal-fired power plants in this chapter). In the rank sum test, all power plants are ranked by their unified efficiency scores from the greatest to the least in most of cases although the ranking order may be changed by the purpose of its application. The formulation used for the rank sum test is as follows:

$$U = \left(n_1 \times n_2\right) + \frac{n_1\left(n_1+1\right)}{2} - \sum R_1 \quad \text{or} \quad U = \left(n_1 \times n_2\right) + \frac{n_2\left(n_2+1\right)}{2} - \sum R_2 \qquad (17.\text{A1})$$

Here, n_1 and n_2, respectively, represent the number of observations in the first group (G_1) and that in the second group (G_2). $\sum R_1$ and $\sum R_2$ represent the sum of the ranks of each group, respectively. Note that the two parts of Equation (17.A1) may produce two different values, but the two always produce the same statistic.

It is mathematically approximated that each group follows a normal distribution that has a mean expressed by $n_1 n_2/2$ and a variance expressed by $n_1 n_2\left(n_1+n_2+1\right)/12$ if a sample size is more than 20. See Mann and Whitney (1947). Hence, the statistic, or

$$\mu = \left[U - n_1 n_2/2\right] \Big/ \sqrt{n_1 n_2\left(n_1+n_2+1\right)/12} \qquad (17.\text{A2})$$

follows a standard normal distribution $N(0,1)$. See Hollander and Wolfe (1999)[10] for a description regarding how to deal with a case where many firms are on the same rank.

Kruskal–Wallis Rank Sum Test: To examine the null hypothesis (a group of observations distributes randomly among multiple periods), this chapter uses the Kruskal–Wallis rank sum test. The entire observed period is separated into T periods. To compute the Kruskal–Wallis statistic (H), this chapter combines all observations $\left(n = \sum_{t=1}^{T} n_t\right)$ in T periods. Then, we rank them from the greatest to the least by these efficiency scores over the entire period. Let R_{jt} denote the rank of the j-th plant in the t-th period. The rank sum of all plants in the t-th period is $R_t = \sum_{j=1}^{n_t} R_{jt}$. Then, the Kruskal–Wallis statistic (H) is determined by

$$H = \frac{12}{n(n+1)} \sum_{t=1}^{T} \frac{R_t^2}{n_t} - 3(n+1). \qquad (17.\text{A3})$$

The statistic (H) follows the χ^2 distribution with a degree of freedom (df = $T - 1$). See Hollander and Wolfe (1999) for a detailed description of both the H statistic and how to deal with many firms on the same rank.

[10] Hollander, M., Wolfe, D.A. (1999) *Nonparametric Statistical Methods*, John Wiley and Sons, Inc., New York.

18

SCALE EFFICIENCY

18.1 INTRODUCTION

This chapter[1] describes how to measure the degree of scale efficiency under natural or managerial disposability. Scale efficiency indicates a managerial level regarding how each DMU can control its operational size in such a manner that it can enhance the level of unified efficiency.

To document the practicality of the proposed scale efficiency measurement, this chapter applies it to examine the performance of coal-fired power plants in the north-east region of the United States. This chapter is also interested in examining whether there is a significant difference between the two types (BIT: bituminous coal and SUB: subbituminous coal) of coal-fired power plants in their unified efficiency measures, including their scale efficiency measures, under natural and managerial disposability concepts, respectively. Chapter 17 has discussed the coal-fired power plants in ISO and RTO. Meanwhile, this chapter is not interested in such regional differences in power plant control, rather being concerned with the type of coal combustion in the power plants.

[1] This chapter is partly based upon the article: Sueyoshi, T. and Goto, M. (2015b), Environmental assessment on coal-fired power plants in U.S. north-east region by DEA non-radial measurement. *Energy Economics*, **50**, 125–139.

Environmental Assessment on Energy and Sustainability by Data Envelopment Analysis, First Edition. Toshiyuki Sueyoshi and Mika Goto.

As discussed in Chapter 14, the north-east region (e.g., West Virginia) is controlled by PJM. The PJM, as the most well-known Regional Transmission Organization (RTO) in the United States, covers areas near the Appalachian Mountains which produce a large amount of SUB. The operation of PJM mainly uses SUB combustion in its coal-fired power plants. Consequently, the research concern of this chapter is whether such a fuel structure has a difficulty from the perspective of environmental assessment. This chapter is also interested in measuring the scale efficiency of coal-fired power plants, along with unified efficiency under natural or managerial disposability.

The remainder of this chapter is organized by the following structure. Section 18.2 discusses how to measure the degree of scale efficiency by non-radial models under natural disposability. Section 18.3 shifts it to managerial disposability. Section 18.4 discusses how to measure the degree of scale efficiency by radial models under natural disposability. Section 18.5 changes it to managerial disposability. Section 18.6 measures the performance of coal-fired power plants in the United States. Section 18.7 summarizes this chapter.

18.2 SCALE EFFICIENCY UNDER NATURAL DISPOSABILITY: NON-RADIAL APPROACH

To examine how each DMU carefully manages the operational size under natural disposability, this chapter measures the degree of scale efficiency by

$$SEN^{NR} = UEN_c^{NR} / UEN_v^{NR} \tag{18.1}$$

where *SEN* is scale efficiency under natural disposability which is measured by a non-radial model (NR). UEN_v^{NR} is unified efficiency under natural disposability that is measured by Model (16.20). UEN_c^{NR} is the unified efficiency under natural disposability that is measured by Model (16.20) without $\sum_{j=1}^{n} \lambda_j = 1$. Since $UEN_c^{NR} \le UEN_v^{NR}$, SEN^{NR} is less than or equals unity in the degree of scale efficiency. The higher score in *SEN* indicates the better scale management under natural disposability.

As mentioned above, this chapter uses Model (16.20) for the proposed empirical study that is formulated by the following non-radial model under natural disposability:

$$\begin{aligned}
\text{Maximize} \quad & \varepsilon \left(\sum_{i=1}^{m} R_i^x d_i^{x-} + \sum_{r=1}^{s} R_r^g d_r^g + \sum_{f=1}^{h} R_f^b d_f^b \right) \\
\text{s.t.} \quad & \sum_{j=1}^{n} x_{ij} \lambda_j + d_i^{x-} = x_{ik} \quad (i = 1, \ldots, m), \\
& \sum_{j=1}^{n} g_{rj} \lambda_j - d_r^g = g_{rk} \quad (r = 1, \ldots, s),
\end{aligned}$$

$$\begin{aligned}
&\sum_{j=1}^{n} b_{fj}\lambda_j \qquad\qquad\qquad\qquad + d_f^b = b_{fk}\ (f=1,\ldots,h),\\
&\sum_{j=1}^{n} \lambda_j = 1, \qquad\qquad\qquad\qquad\qquad\qquad\qquad\qquad (18.2)\\
&\lambda_j \ge 0\ (j=1,\ldots,n),\ d_i^{x-} \ge 0\ (i=1,\ldots,m),\\
&d_r^g \ge 0\ (r=1,\ldots,s) \text{ and } d_f^b \ge 0\ (f=1,\ldots,h).
\end{aligned}$$

As discussed in Chapter 16, a drawback of the non-radial approach is that it produces efficient and inefficient DMUs whose unified efficiency measures are very close to unity. To overcome the problem, this chapter adds a small number (ε) to control the magnitude of unified efficiency under natural disposability. The small number is referred to as the "magnitude control factor (MCF)" in Chapter 16.

The unified efficiency (UEN_v^{NR}) of the k-th DMU under natural disposability and variable RTS is measured by

$$UEN_v^{NR} = 1 - \varepsilon\left(\sum_{i=1}^{m} R_i^x d_i^{x-*} + \sum_{r=1}^{s} R_r^g d_r^{g*} + \sum_{f=1}^{h} R_f^b d_f^{b*}\right) \qquad (18.3)$$

where all slack variables are determined on the optimality of Model (18.2). The equation within the parenthesis, obtained from the optimality of Model (18.2), indicates the level of unified inefficiency under natural disposability. The unified efficiency is obtained by subtracting the level of inefficiency from unity, as formulated in Equation (18.3).

Meanwhile, the unified efficiency (UEN_c^{NR}) of thc k-th DMU, incorporating constant RTS, is measured by

$$\begin{aligned}
&\text{Maximize} \quad \varepsilon\left(\sum_{i=1}^{m} R_i^x d_i^{x-} + \sum_{r=1}^{s} R_r^g d_r^g + \sum_{f=1}^{h} R_f^b d_f^b\right)\\
&\text{s.t.} \quad \sum_{j=1}^{n} x_{ij}\lambda_j + d_i^{x-} \qquad\qquad\qquad = x_{ik}\ (i=1,\ldots,m),\\
&\qquad\ \ \sum_{j=1}^{n} g_{rj}\lambda_j \qquad\qquad - d_r^g \qquad = g_{rk}\ (r=1,\ldots,s), \qquad (18.4)\\
&\qquad\ \ \sum_{j=1}^{n} b_{fj}\lambda_j \qquad\qquad\qquad + d_f^b = b_{fk}\ (f=1,\ldots,h),\\
&\qquad\ \ \lambda_j \ge 0\ (j=1,\ldots,n),\ d_i^{x-} \ge 0\ (i=1,\ldots,m),\\
&\qquad\ \ d_r^g \ge 0\ (r=1,\ldots,s) \text{ and } d_f^b \ge 0\ (f=1,\ldots,h).
\end{aligned}$$

Model (18.4) drops $\sum_{j=1}^{n}\lambda_j = 1$ from Model (18.2), so incorporating constant RTS. See Chapter 20 for a description of RTS. The degree of UEN_c^{NR} is determined by Equation (18.3). Thus, we can identify the degree of scale efficiency by

Equation (18.1), or $SEN^{NR} = UEN_c^{NR} / UEN_v^{NR}$, by using the proposed two non-radial models, or Models (18.2) and (18.4).

Figure 18.1 visually describes two efficiency frontiers under variable and constant RTS in the horizontal axis (an input) and the vertical axis (a desirable output). Figure 18.1, corresponding to Stage I (A) of Figure 15.6, assumes that all DMUs use the same amount of an undesirable output, so being not listed in the *x–g* dimension.

In Figure 18.1, the straight line from the origin to a projected production point, or G (passing DMU {B}), is an efficiency frontier under constant RTS, while DMUs {A, B, C, D, E} consist of the frontier under variable RTS. A production possibility set under variable RTS consists of the region Ω_1, while that of constant RTS comprises all three regions $(\Omega_1 \cup \Omega_2 \cup \Omega_3)$. DMU {F} is inefficient. The DMU needs to shift the performance from {F} to {C} under variable RTS and to G under constant RTS. This chapter measures the degree of UEN_v^{NR} by the distance {F} to H divided by the distance {C} to H. Meanwhile, the degree of UEN_c^{NR} is measured by the distance {F} to H divided by that of G to H. Here, the location of H is for an artificially created production point which uses an amount of input but has no production on the desirable output. Consequently, we have the following visual relationship in Figure 18.1:

$$SEN^{NR} = UEN_c^{NR} / UEN_v^{NR} = (\text{Distance C to H})/(\text{Distance G to H}).$$

The degree of SEN^{NR} is less than or equals unity. The status of unity in SEN^{NR} implies that a DMU operates under constant RTS in terms of desirable outputs.

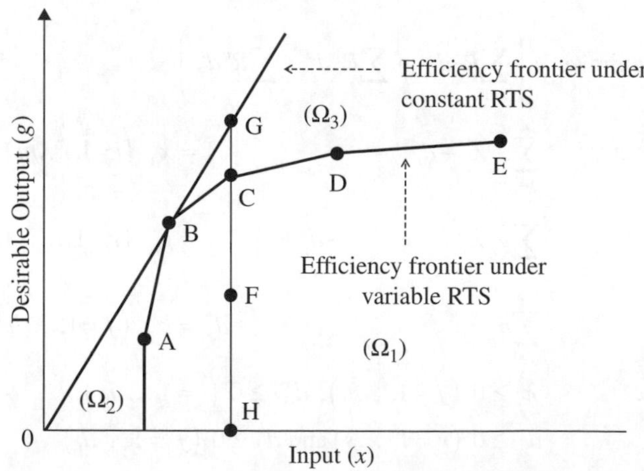

FIGURE 18.1 Two efficiency frontiers under different RTS (a) Source: Sueyoshi and Goto (2015b).This study updates the original work. (b) See Figure 15.6. This figure corresponds to Stage I (A) of Figure 15.6. (c) This figure is prepared for a desirable output-oriented case. (d) RTS stands for returns to scale.

18.3 SCALE EFFICIENCY UNDER MANAGERIAL DISPOSABILITY: NON-RADIAL APPROACH

To measure how each DMU carefully manages the operational size under managerial disposability, this chapter measures the degree of scale efficiency by

$$SEM^{NR} = UEM_c^{NR} / UEM_v^{NR}, \tag{18.5}$$

where SEM^{NR} is scale efficiency under managerial disposability. UEM_v^{NR} is unified efficiency under managerial disposability as measured by Model (16.21). UEM_c^{NR} is the unified efficiency under managerial disposability, measured by Model (16.21) without $\sum_{j=1}^{n} \lambda_j = 1$, which incorporates constant DTS. See Chapter 20 for a detailed description on DTS. Since $UEM_c^{NR} \leq UEMk_v^{NR}$, SEM^{NR} is less than or equal to unity. The higher score in SEM^{NR} indicates the better scale management under managerial disposability.

This chapter uses Model (16.21) for the proposed empirical study that is formulated by the following non-radial model under managerial disposability:

$$\begin{aligned}
&\text{Maximize} \quad \varepsilon\left(\sum_{i=1}^{m} R_i^x d_i^{x+} + \sum_{r=1}^{s} R_r^g d_r^g + \sum_{f=1}^{h} R_f^b d_f^b\right) \\
&\text{s.t.} \quad \sum_{j=1}^{n} x_{ij}\lambda_j - d_i^{x+} = x_{ik} \quad (i=1,\ldots,m), \\
&\qquad \sum_{j=1}^{n} g_{rj}\lambda_j - d_r^g = g_{rk} \quad (r=1,\ldots,s), \\
&\qquad \sum_{j=1}^{n} b_{fj}\lambda_j + d_f^b = b_{fk} \quad (f=1,\ldots,h), \\
&\qquad \sum_{j=1}^{n} \lambda_j = 1, \\
&\qquad \lambda_j \geq 0 \ (j=1,\ldots,n), \ d_i^{x+} \geq 0 \ (i=1,\ldots,m), \\
&\qquad d_r^g \geq 0 \ (r=1,\ldots,s) \text{ and } d_f^b \geq 0 \ (f=1,\ldots,h),
\end{aligned} \tag{18.6}$$

where a prescribed small number (ε) is incorporated into Model (18.6) to control the magnitude of unified efficiency under managerial disposability.

The unified efficiency (UEM_v^{NR}) of the k-th DMU under managerial disposability and variable DTS is measured by

$$UEM_v^{NR} = 1 - \varepsilon\left(\sum_{i=1}^{m} R_i^x d_i^{x+*} + \sum_{r=1}^{s} R_r^g d_r^{g*} + \sum_{f=1}^{h} R_f^b d_f^{b*}\right) \tag{18.7}$$

where all slack variables are determined on the optimality of Model (18.6). The equation within the parentheses, obtained from the optimality of Model (18.6), indicates the level of unified inefficiency under managerial disposability. The unified efficiency is obtained by subtracting the level of inefficiency from unity, as formulated in Equation (18.7).

Meanwhile, the unified efficiency (UEM_c^{NR}) of the k-th DMU, incorporating constant DTS, is measured by

$$\begin{aligned}
\text{Maximize} \quad & \varepsilon\left(\sum_{i=1}^{m} R_i^x d_i^{x+} + \sum_{r=1}^{s} R_r^g d_r^g + \sum_{f=1}^{h} R_f^b d_f^b\right) \\
\text{s.t.} \quad & \sum_{j=1}^{n} x_{ij}\lambda_j - d_i^{x+} = x_{ik} \ (i=1,\ldots,m), \\
& \sum_{j=1}^{n} g_{rj}\lambda_j \quad - d_r^g = g_{rk} \ (r=1,\ldots,s), \\
& \sum_{j=1}^{n} b_{fj}\lambda_j \quad + d_f^b = b_{fk} \ (f=1,\ldots,h), \\
& \lambda_j \geq 0 \ (j=1,\ldots,n),\ d_i^{x+} \geq 0 \ (i=1,\ldots,m), \\
& d_r^g \geq 0 \ (r=1,\ldots,s) \text{ and } d_f^b \geq 0 \ (f=1,\ldots,h).
\end{aligned} \tag{18.8}$$

Model (18.8) drops $\sum_{j=1}^{n} \lambda_j = 1$ from Model (18.6), so incorporating constant DTS. The degree of UEM_c^{NR} is determined by Equation (18.7). Thus, this chapter can identify the degree of scale efficiency by Equation (18.5), or $SEM^{NR} = UEM_c^{NR} / UEM_v^{NR}$.

Figure 18.2, corresponding to Stage I (B) in Figure 15.6, visually describes two efficiency frontiers under constant and variable DTS in the horizontal axis (input : x) and the vertical axis (undesirable output : b). The figure assumes that all DMUs produce the same amount of a desirable output (g), so being not listed in the x–b dimension. The straight line from the origin to a projected point (G, passing on {B}) is an efficiency frontier under constant DTS, while DMUs {A, B, C, D, E} consist of the frontier under variable DTS. A pollution possibility set under variable DTS consists of the region Ω_1, while that of constant DTS comprises all three regions ($\Omega_1 \cup \Omega_2 \cup \Omega_3$). DMU {F} is inefficient. The DMU needs to shift the performance from {F} to {C} under variable DTS and to G under constant DTS. This chapter measures the degree of UEM_v^{NR} by the distance {C} to H divided by the distance {F} to H. Meanwhile, the degree of UEM_c^{NR} is measured by the distance G to H divided by that of {F} to H. The H is an artificially created production point which uses an amount of input (x) but it does not produce any pollution. Consequently, this chapter has the following relationship in Figure 18.2: $SEM^{NR} = \text{UEM}_c^{NR} / UEM_v^{NR}$ = (Distance G to H)/(Distance C to H). Here, it is important to note that such a specification on SEM^{NR} in Figure 18.2 is visually opposite to that of SEN^{NR} = (Distance C to H)/(Distance G to H) in Figure 18.1.

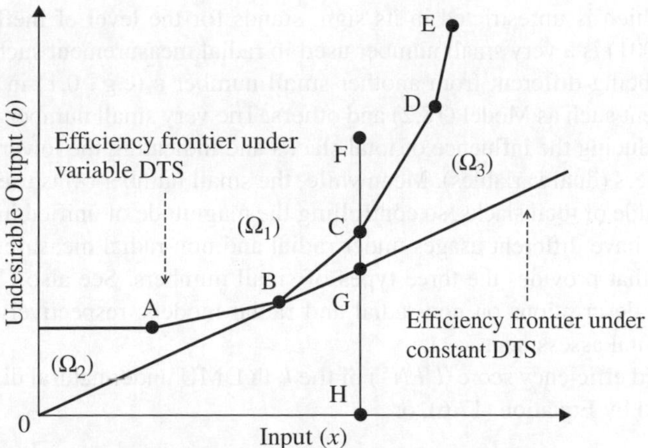

FIGURE 18.2 Two efficiency frontiers under different DTS (a) Source: Sueyoshi and Goto (2015b). This study updates the original work. (b) See Figure 15.6. This figure corresponds to Stage I (B) of Figure 15.6. (c) This figure is prepared for an undesirable output-oriented case. (d) DTS stands for damages to scale.

18.4 SCALE EFFICIENCY UNDER NATURAL DISPOSABILITY: RADIAL APPROACH

The computational procedure of a radial approach is conceptually similar to that of the non-radial approach in Sections 18.2 and 18.3, but using Model (17.5) in Chapter 17. Returning to the model, this chapter measures the unified efficiency of the k-th DMU under natural disposability by duplicating the following formulation:

$$\begin{aligned}
&\text{Maximize } \xi + \varepsilon_s \left(\sum_{i=1}^{m} R_i^x d_i^{x-} + \sum_{r=1}^{s} R_r^g d_r^g + \sum_{f=1}^{h} R_f^b d_f^b \right) \\
&\text{s.t.} \quad \sum_{j=1}^{n} x_{ij}\lambda_j + d_i^{x-} = x_{ik} \ (i=1,\dots,m), \\
&\qquad \sum_{j=1}^{n} g_{rj}\lambda_j - d_r^g - \xi g_{rk} = g_{rk} \ (r=1,\dots,s), \\
&\qquad \sum_{j=1}^{n} b_{fj}\lambda_j + d_f^b + \xi b_{fk} = b_{fk} \ (f=1,\dots,h), \\
&\qquad \sum_{j=1}^{n} \lambda_j = 1, \\
&\qquad \lambda_j \ge 0 \ (j=1,\dots,n),\ \xi : \text{URS},\ d_i^{x-} \ge 0 \ (i=1,\dots,m), \\
&\qquad d_r^g \ge 0 \ (r=1,\dots,s) \text{ and } d_f^b \ge 0 \ (f=1,\dots,h).
\end{aligned} \tag{18.9}$$

Here, ξ, which is unrestricted in its sign, stands for the level of inefficiency. ε_s (e.g., 0.00001) is a very small number used in radial measurement such as Model (18.9), so being different from another small number ε (e.g., 0.1) in non-radial measurement such as Model (18.2) and others. The very small number (ε_s) is used for both reducing the influence of total slacks and indicating the lower bounds of all multipliers (dual variables). Meanwhile, the small number (ε) is used to adjust the magnitude of total slacks, so controlling the magnitude of unified inefficiency. Thus, both have different usages under radial and non-radial measurements. See Chapter 1 that provides the three types of small numbers. See also Chapters 16 and 17 for descriptions on non-radial and radial models, respectively, for DEA environmental assessment.

A unified efficiency score (UEN_v^R) of the k-th DMU under natural disposability is measured by Equation (17.6), or

$$UEN_v^R = 1 - \left[\xi^* + \varepsilon_s\left(\sum_{i=1}^{m} R_i^x d_i^{x-*} + \sum_{r=1}^{s} R_r^g d_r^{g*} + \sum_{f=1}^{h} R_f^b d_f^{b*}\right)\right], \quad (18.10)$$

where the inefficiency score (ξ) and all slack variables are determined on the optimality of Model (18.9). Thus, the equation within the parenthesis is obtained from the optimality of Model (18.9). The unified efficiency under natural disposability is obtained by subtracting the level of inefficiency from unity. The superscript R stands for radial measurement.

Meanwhile, the unified efficiency (UEN_c^R) of the k-th DMU is measured by

$$\begin{aligned}
\text{Maximize} \quad & \xi + \varepsilon_s\left(\sum_{i=1}^{m} R_i^x d_i^{x-} + \sum_{r=1}^{s} R_r^g d_r^g + \sum_{f=1}^{h} R_f^b d_f^b\right) \\
\text{s.t.} \quad & \sum_{j=1}^{n} x_{ij}\lambda_j + d_i^{x-} = x_{ik} \ (i=1,\dots,m), \\
& \sum_{j=1}^{n} g_{rj}\lambda_j - d_r^g - \xi g_{rk} = g_{rk} \ (r=1,\dots,s), \\
& \sum_{j=1}^{n} b_{fj}\lambda_j + d_f^b + \xi b_{fk} = b_{fk} \ (f=1,\dots,h), \\
& \lambda_j \geq 0 \ (j=1,\dots,n), \ \xi : \text{URS}, \ d_i^{x-} \geq 0 \ (i=1,\dots,m), \\
& d_r^g \geq 0 \ (r=1,\dots,s) \text{ and } d_f^b \geq 0 \ (f=1,\dots,h).
\end{aligned} \quad (18.11)$$

Model (18.11) drops $\sum_{j=1}^{n} \lambda_j = 1$ from Model (18.9), so incorporating constant DTS. The degree of UEN_c^R is determined by $UEN_c^R = 1 - \left[\xi^* + \varepsilon_s\left(\sum_{i=1}^{m} R_i^x d_i^{x-*} + \sum_{r=1}^{s} R_r^g d_r^{g*} + \sum_{f=1}^{h} R_f^b d_f^{b*}\right)\right]$, where the inefficiency score and all the slacks are obtained from the optimality of Model (18.11). Thus, we can identify the degree of scale efficiency by

$$SEN^R = UEN_c^R / UEN_v^R, \tag{18.12}$$

where SEN^R is scale efficiency under natural disposability. UEN_v^R is unified efficiency under natural disposability that is measured by Model (18.9) and UEN_c^R is the unified efficiency under natural disposability that is measured by Model (18.11) with constant RTS.

18.5 SCALE EFFICIENCY UNDER MANAGERIAL DISPOSABILITY: RADIAL APPROACH

The computational procedure of a radial approach under managerial disposability conceptually follows that of the non-radial approach, but using Model (17.10) in Chapter 17. Returning to the radial model, this chapter measures the unified efficiency of the k-th DMU under managerial disposability by duplicating the following formulation:

$$\begin{aligned}
&\text{Maximize } \xi + \varepsilon_s \left(\sum_{i=1}^{m} R_i^x d_i^{x+} + \sum_{r=1}^{s} R_r^g d_r^g + \sum_{f=1}^{h} R_f^b d_f^b \right) \\
&\text{s.t.} \quad \sum_{j=1}^{n} x_{ij}\lambda_j - d_i^{x+} = x_{ik} \ (i=1,\dots,m), \\
&\qquad \sum_{j=1}^{n} g_{rj}\lambda_j - d_r^g - \xi g_{rk} = g_{rk} \ (r=1,\dots,s), \\
&\qquad \sum_{j=1}^{n} b_{fj}\lambda_j + d_f^b + \xi b_{fk} = b_{fk} \ (f=1,\dots,h), \\
&\qquad \sum_{j=1}^{n} \lambda_j = 1, \\
&\qquad \lambda_j \ge 0 \ (j=1,\dots,n), \ \xi : \text{URS}, \ d_i^{x+} \ge 0 \ (i=1,\dots,m), \\
&\qquad d_r^g \ge 0 \ (r=1,\dots,s) \text{ and } d_f^b \ge 0 \ (f=1,\dots,h).
\end{aligned} \tag{18.13}$$

A unified efficiency score (UEM_v^R) of the k-th DMU under managerial disposability is measured by Equation (17.11), or

$$UEM_v^R = 1 - \left[\xi^* + \varepsilon_s \left(\sum_{i=1}^{m} R_i^x d_i^{x+*} + \sum_{r=1}^{s} R_r^g d_r^{g*} + \sum_{f=1}^{h} R_f^b d_f^{b*} \right) \right], \tag{18.14}$$

where the inefficiency score and all slack variables are determined on the optimality of Model (18.13). Thus, the equation within the parenthesis is obtained from the optimality of Model (18.13). The unified efficiency under managerial disposability is obtained by subtracting the level of inefficiency from unity.

Meanwhile, the unified efficiency (UEM_c^R) of the k-th DMU, incorporating constant DTS, is measured by

$$\begin{aligned}
&\text{Maximize } \xi + \varepsilon_s \left(\sum_{i=1}^{m} R_i^x d_i^{x+} + \sum_{r=1}^{s} R_r^g d_r^g + \sum_{f=1}^{h} R_f^b d_f^b \right) \\
&\text{s.t.} \quad \sum_{j=1}^{n} x_{ij}\lambda_j - d_i^{x+} = x_{ik} \ (i=1,\ldots,m), \\
&\qquad \sum_{j=1}^{n} g_{rj}\lambda_j - d_r^g - \xi g_{rk} = g_{rk} \ (r=1,\ldots,s), \qquad (18.15) \\
&\qquad \sum_{j=1}^{n} b_{fj}\lambda_j + d_f^b + \xi b_{fk} = b_{fk} \ (f=1,\ldots,h), \\
&\qquad \lambda_j \geq 0 \ (j=1,\ldots,n),\ \xi : \text{URS},\ d_i^{x+} \geq 0 \ (i=1,\ldots,m), \\
&\qquad d_r^g \geq 0 \ (r=1,\ldots,s) \text{ and } d_f^b \geq 0 \ (f=1,\ldots,h).
\end{aligned}$$

Model (18.15) drops $\sum_{j=1}^{n} \lambda_j = 1$ from Model (18.13), so incorporating constant DTS in the formulation. The degree of UEM_c^R is measured by Equation (18.14), or $UEM_c^R = 1 - \left[\xi^* + \varepsilon_s \left(\sum_{i=1}^{m} R_i^x d_i^{x+*} + \sum_{r=1}^{s} R_r^g d_r^{g*} + \sum_{f=1}^{h} R_f^b d_f^{b*} \right) \right]$. Thus, we can identify the degree of scale efficiency by

$$SEM^R = UEM_c^R / UEM_v^R \qquad (18.16)$$

where SEM^R is scale efficiency under managerial disposability. UEM_v^R is unified efficiency under managerial disposability that is measured by Model (18.13) and UEM_c^R is the unified efficiency under managerial disposability that is measured by Model (18.15) with constant DTS.

18.6 UNITED STATES COAL-FIRED POWER PLANTS

18.6.1 The Clean Air Act

Returning to Table 14.1, Chapter 14 has described the history of the Clean Air Act (CAA) that contains a chronological description on the law. The table summarizes legal actions concerning CAA and their related descriptions from the perspective of energy policy and regulation in the United States. Acknowledging the policy-making effort of US governments, this chapter needs to mention that it is necessary for us to examine whether the policy implementation of CAA has been effective or not over the pollution control of electric utility firms and their power plant operations. Thus, focusing on the research interest at a plant level, this chapter examines whether CAA implementation really influences the operation of coal-fired power plants because they are a major group of air polluters in the world.

In addition to air regulation, market liberalization[2] is also important for power plant operations. In the United States, the Federal Government (e.g., Federal Energy Regulatory Commission; FERC) has the authority to control both competition at the wholesale level and the promotion of equal access to transmission grids. Meanwhile, Public Utility Commission (PUC) at each state is responsible for restructuring its retail electricity market. In particular, the year (1996) was significantly important in the history of US electric utility regulation because FERC issued Orders 888 and 889 that enabled open access to transmission lines for electricity market participants. Some states (e.g., Rhode Island and California) passed the restructuring legislation on the electric utility industry in 1996. The liberalization policy was due to the economic assertion that competition in the electric power industry would enhance the managerial effort of utility firms to improve their business efficiencies, to reduce consumers' financial burden and to increase economic prosperity. Thus, electric power firms faced liberalization at federal and state level as well as regulation on their air pollution under the CAA. Figure 18.3 visually summarizes such policy components of the current situation surrounding coal-fired power plants in the United States.

Acknowledging the importance of liberalization as a general business trend in the US electric power industry, this chapter needs to mention the existence of an opposite regulatory trend where many states have not yet accepted full retail competition in electricity because of various reasons, such as the California electricity crisis[3] and the protection of utility industries and consumers within their states. As mentioned previously, the policy decision regarding electricity restructuring is separated at federal and state levels. Thus, the level of market liberalization depends upon each state in the United States. Consequently, the operation of coal-fired power plants is classified into either plant operation in liberalized and deregulated states or plant operation in regulated states. Therefore, this chapter focuses upon coal-fired power plants in PJM (the highly regarded RTO), operating

[2]"Liberalization" is a process of market reforms to introduce competition and a less restrictive regulation framework for companies in the electricity industry. Meanwhile, "deregulation" is the modification or repeal of existing regulation or the removal of state control and the introduction of a more formal regulatory framework. See Rousaki, K. (1999) Liberalization of electricity markets and coal use. *Energeia*, **10**, 1–6.

[3]During the summer in 2000, wholesale electricity prices in California were approximately 500% higher than those during the same months in 1998–1999. The price hike was unexpected by many policy makers and individuals who were involved in the electric utility industry. Since then, they have been long wondering whether the US electricity deregulation policy produced benefits from competition promised to consumers. See Sueyoshi, T. and Tadiparthi, G.R. (2008) Why did California electricity crisis occur?: A numerical analysis using multiagent intelligent simulator. *IEEE Transactions on Systems, Man and Cybernetics Part C: Applications and Reviews*, **38**, 779–790. See also Sueyoshi, T. (2010) Beyond economics for guiding large public policy issue: Lessons from the Bell system divestiture and the California electricity crisis. *Decision Support Systems*, **48**, 457–469.

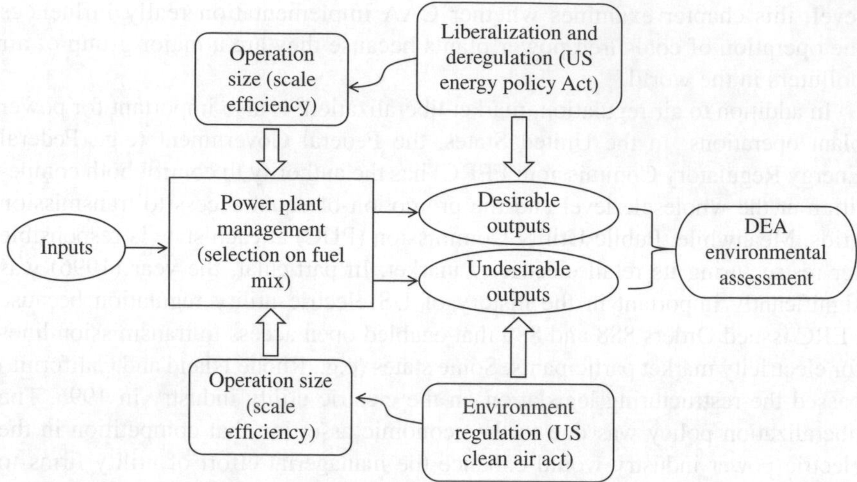

FIGURE 18.3 Power plant management under different regulation environments (a) Source: Sueyoshi and Goto (2015b). (b) The "liberalization" is a process of market reforms to introduce competition and a less restrictive regulation framework for companies in the electricity industry. (c) The "deregulation" is the modification or repeal of existing regulation or the removal of state control and the introduction of a more formal regulatory framework.

in a liberalized market, which supplies electricity to the north-east part of the United States. See Chapter 14 which describes the region where PJM provides electricity. Furthermore, an important concern to be examined in this chapter is whether a coal-fired power plant operates with scale efficiency because the electricity is generated in a large process facility. Thus, proper scale management is important as one of the efficiency enhancement components.

At the end of this subsection, it is important to add that liberalization is a process of market reforms to introduce competition and a less restrictive regulation framework for companies in the electricity industry. In contrast, deregulation is the modification or repeal of existing regulation or the removal of state control and the introduction of a more formal regulatory framework. In the United States, "liberalization" is widely used in descriptions of the electric power industry. Thus, this chapter uses the term hereafter.

18.6.2 Production Factors

In this chapter, the generation of each power plant is characterized by two inputs, a single desirable output and five undesirable outputs. They are considered as production factors. The two inputs are measured by the nameplate capacity (MW: Megawatt) and the amount of annual heat input (mmBtu). The desirable output is measured by the amount of annual net generation (MWh: Megawatt hours). The undesirable outputs are measured by five items: the annual amount of NO_x emission (tons), the annual amount of SO_2 emission (tons), the annual amount of CO_2 emission (tons), the annual amount of CH4 emission (pounds; lbs) and the annual amount of N_2O emission (lbs) at each coal-fired power plant.

Data Accessibility: The supplementary section of Sueyoshi and Goto (2015) lists the original data set.

The data source is the database[4] of EPA's "eGRID year 2010." The data set consists of 68 coal-fired power plants that are operating under the PJM Interconnection. This chapter excludes power plants used for combined heat and power (CHP) and those under mixed combustion with biomass fuel. Descriptive statistics are listed in the research of Sueyoshi and Goto (2015).

In discussing greenhouse gas (GHG) control and performance assessment, this chapter needs to specify the type of GHG emissions, including CO_2, methane (CH_4), nitrous oxide (N_2O), hydrofluorocarbons (HFCs), perfluorocarbons (PFCs) and sulfur hexafluoride (SF_6). The SO_2 and NO_x emissions do not belong to the GHG, but both belong to the acid rain gases that produce major damage on the surface of the earth and in our lives. Thus, it is important for us to pay attention to not only GHG but also acid rain gases whose damage is much more serious than those of the former emissions.

As discussed in Chapter 2, for our computational convenience, each production factor is divided by each factor average for research in this chapter. Such a data treatment is necessary for us in order to avoid the situation where a large production factor dominates the others in the computational process of the proposed DEA environmental assessment. Thus, all production factors are unit-less in the proposed research. See Chapters 1, 26 and 27 on a description on the data adjustment.

18.6.3 Research Concerns

Table 18.1 summarizes the type of plant primary fuel and the plant capacity factor of 68 power plants. Bituminous coal (BIT)[5] is relatively soft coal that contains a tarlike substance, called "bitumen." Subbituminous coal (SUB) is a type of coal whose properties range from those of lignite to those of bituminous coal. Coal is used primarily as fuel for steam-electric power generation. The capacity factor of each power plant is the ratio of the actual output over a period to its potential output, if it were possible for it to operate at full nameplate capacity continuously over the same period. The computation of plant capacity factor is defined by [the amount of annual generation (KWh/year)/the number of annual hours (356 days × 24 hours)] × plant capacity (KW).

[4] https://www.epa.gov/energy

[5] https://www.britannica.com/science/bituminous-coal and https://en.wikipedia.org/wiki/Sub-bituminous_coal

TABLE 18.1 Plant primary fuel and capacity factor

No.	Plant name	Plant primary fuel	Plant capacity factor
1	Shelby Municipal Light Plant	BIT	0.1928
2	Dover	BIT	0.2003
3	Painesville	BIT	0.1769
4	Orrville	BIT	0.4562
5	Whitewater Valley	BIT	0.2569
6	Picway	BIT	0.0676
7	FirstEnergy Rivesville	BIT	0.0060
8	FirstEnergy Willow Island	SUB	0.0708
9	Bremo Bluff	BIT	0.4421
10	FirstEnergy Lake Shore	SUB	0.3362
11	Titus	BIT	0.3239
12	FirstEnergy Albright	BIT	0.2388
13	Niles	BIT	0.1655
14	FirstEnergy Armstrong Power Stat	BIT	0.5862
15	New Castle Plant	BIT	0.2350
16	Joliet 9	SUB	0.3997
17	Kanawha River	BIT	0.2964
18	FirstEnergy Ashtabula	BIT	0.2284
19	FirstEnergy Mitchell Power Station	BIT	0.2450
20	Sunbury Generation LP	BIT	0.3976
21	FirstEnergy R E Burger	BIT	0.2236
22	Crawford	SUB	0.4639
23	State Line Energy	SUB	0.6239
24	Portland	BIT	0.3139
25	Shawville	BIT	0.4679
26	Cheswick Power Plant	BIT	0.3193
27	Fisk Street	SUB	0.2669
28	East Bend	BIT	0.7539
29	Killen Station	BIT	0.6713
30	Clinch River	BIT	0.2405
31	Kammer	BIT	0.2401
32	Waukegan	SUB	0.5541
33	Avon Lake	BIT	0.4035
34	Indian River Generating Station	BIT	0.3275
35	Clover	BIT	0.8338
36	Big Sandy	BIT	0.6820
37	Tanners Creek	BIT	0.4050
38	Philip Sporn	BIT	0.2549
39	FirstEnergy Fort Martin Power Station	BIT	0.6278
40	Will County	SUB	0.5206
41	FirstEnergy Eastlake	BIT	0.5698
42	Mountaineer	BIT	0.7090
43	Kincaid Generation LLC	SUB	0.5322
44	FirstEnergy Pleasants Power Station	BIT	0.6916
45	Brandon Shores	BIT	0.5061

TABLE 18.1 (Continued)

No.	Plant name	Plant primary fuel	Plant capacity factor
46	W H Zimmer	BIT	0.7768
47	Walter C Beckjord	BIT	0.3005
48	Miami Fort	BIT	0.6769
49	Muskingum River	BIT	0.5002
50	Morgantown Generating Plant	BIT	0.5282
51	PPL Brunner Island	BIT	0.7413
52	Mitchell	BIT	0.7161
53	PPL Montour	BIT	0.6675
54	Mt Storm	BIT	0.7015
55	Hatfields Ferry Power Station	BIT	0.5929
56	Powerton	SUB	0.5762
57	Cardinal	BIT	0.6106
58	Conemaugh	BIT	0.7372
59	Keystone	BIT	0.8212
60	Conesville	BIT	0.3900
61	Homer City Station	BIT	0.6245
62	FirstEnergy Harrison Power Station	BIT	0.7001
63	J M Stuart	BIT	0.6268
64	FirstEnergy W H Sammis	BIT	0.5707
65	Rockport	SUB	0.7741
66	General James M Gavin	BIT	0.8292
67	FirstEnergy Bruce Mansfield	BIT	0.7524
68	John E Amos	BIT	0.5988

(a) Source: Sueyoshi and Goto (2015b). (b) BIT stands for bituminous coal and SUB stands for subbituminous coal.

The capacity factors of PJM's 68 coal-fired power plants, summarized in Table 18.1, indicate two research concerns, summarized by null hypotheses, on the operation of these power plants. One of these research concerns is whether BIT outperforms SUB in terms of coal quality, so producing a lower amount of GHG emission. Therefore, we are interested in examining whether coal-fired power plants, utilizing BIT, outperform those of SUB in terms of their unified efficiency measures. That is the first hypothesis concerning whether "there is no difference between coal-fired power plants with BIT and SUB in terms of their unified efficiency measures."

The other concern is that large coal-fired power plants may serve as "base-load," but small ones cannot be the base-load power plants in PJM. The base-load plants provide part of the minimum level of demand on an electrical supply system over 24 hours. Consequently, regulatory agencies (e.g., public utility commission and Environmental Protection Agency; EPA) pay attention to the operation and environmental protection of large power plants, but give limited attention to the small ones. Furthermore, small power plants have more freedom in their bidding strategies than large ones in PJM's wholesale power market because they are not base-load power plants. See Chapter 13 which describes a bidding process for wholesale power trading. The concern is summarized by the following second hypothesis on whether there is no difference between coal-fired power plants with small operation (less than 50% in plant capacity factor) and large operation (more than 50% in plant capacity factor) in their unified efficiency measures. It is generally acceptable to consider that there is a correlation between an operation size and a plant capacity factor, depending upon the type of fuel. The correlation may be found in coal-fired power generation because they can serve as both base and non-base load plants.

18.6.4 Unified Efficiency Measures of Power Plants

Table 18.2 summarizes the unified efficiency measures of PJM's 68 coal fired power plants in 2010. Model (18.2) under natural disposability and Model (18.6) under managerial disposability are used to measure their unified efficiency scores along with scale efficiency measures. The scale efficiency measures are determined by Equation (18.1) under natural disposability and Equation (18.5) under managerial disposability.

A computational result is summarized in Table 18.2 where S.D. stands for standard deviation. Max. and Min. indicate maximum and minimum unified efficiencies, respectively. In Table 18.2, the coal-fired power plants under natural and managerial disposability exhibit unity or close to unity (at the level of more than 90%) in their unified efficiency scores. One rationale is that the desirable output (i.e., annual net generation) of power plants is controlled by PJM and undesirable outputs (i.e., five GHGs) are regulated by governmental agencies (e.g., EPA), so that they do not have any major difference in their unified (operational and environmental) performance measures. For example, their unified efficiency measures under natural disposability are 0.979 and 0.976 on average, respectively. These average scores slightly change to 0.992 and 0.962, respectively, when they are measured under managerial disposability. See the bottom of Table 18.2. These unified efficiency measures determine the degree of scale efficiencies as 0.997 and 0.970, on average, under natural and managerial disposability.

18.6.5 Mean Tests

Table 18.3 summarizes the mean tests on unified efficiency measures. The table contains two statistical findings. One of the two findings is that this chapter cannot find any statistical difference between small and large operations in unified efficiencies under natural and managerial disposability concepts. An exception can be found in the unified efficiency under managerial disposability with variable DTS. This is because their plant operations are frequently monitored by US regulatory agencies regardless of their capacity factors or sizes. Thus, this chapter cannot statistically reject the second hypothesis.

TABLE 18.2 Unified efficiencies on power plants: non-radial approach

Plant No.	*UENv*	*UENc*	*SEN*	*UEMv*	*UEMc*	*SEM*
1	1.000	0.996	0.996	0.998	0.996	0.998
2	1.000	0.996	0.996	0.997	0.992	0.995
3	0.999	0.996	0.997	1.000	1.000	1.000
4	0.987	0.983	0.997	1.000	0.971	0.971
5	0.997	0.994	0.997	1.000	1.000	1.000
6	0.999	0.996	0.997	0.999	0.997	0.998
7	1.000	0.997	0.997	1.000	1.000	1.000
8	0.996	0.993	0.997	1.000	1.000	1.000
9	0.989	0.987	0.998	0.990	0.977	0.987
10	0.995	0.992	0.997	0.997	0.989	0.992
11	0.990	0.988	0.997	0.993	0.980	0.987
12	0.989	0.986	0.997	0.994	0.974	0.979
13	0.985	0.982	0.997	1.000	0.981	0.981
14	0.979	0.976	0.997	0.981	0.971	0.990
15	0.986	0.984	0.997	0.995	0.976	0.981
16	0.987	0.985	0.998	0.992	0.965	0.972
17	1.000	1.000	1.000	1.000	1.000	1.000
18	0.986	0.984	0.997	0.997	0.970	0.973
19	0.990	0.987	0.998	1.000	1.000	1.000
20	0.964	0.961	0.998	0.975	0.899	0.922
21	0.980	0.977	0.998	0.994	0.962	0.967
22	0.986	0.984	0.998	0.995	0.965	0.970
23	0.971	0.970	0.998	0.973	0.946	0.971
24	0.976	0.973	0.998	1.000	0.986	0.986
25	0.959	0.957	0.998	0.967	0.937	0.968
26	0.979	0.977	0.998	1.000	1.000	1.000
27	0.986	0.983	0.998	1.000	0.981	0.981
28	1.000	1.000	1.000	1.000	1.000	1.000
29	0.982	0.981	0.998	0.987	0.954	0.966
30	0.981	0.979	0.998	1.000	0.976	0.976
31	0.978	0.975	0.997	1.000	0.989	0.989
32	0.979	0.978	0.999	0.988	0.934	0.946
33	0.971	0.967	0.996	1.000	1.000	1.000
34	0.976	0.974	0.998	1.000	0.971	0.971
35	1.000	1.000	1.000	0.970	0.967	0.997
36	0.968	0.968	0.999	0.970	0.953	0.983
37	0.965	0.964	0.999	0.986	0.950	0.963
38	0.968	0.966	0.998	1.000	0.976	0.976
39	0.960	0.959	0.999	0.970	0.958	0.988
40	0.961	0.960	0.999	0.981	0.877	0.893
41	0.950	0.949	0.999	0.961	0.925	0.962
42	1.000	1.000	1.000	1.000	1.000	1.000
43	0.932	0.932	0.999	0.954	0.835	0.875

(*Continued*)

TABLE 18.2 (Continued)

Plant No.	*UENv*	*UENc*	*SEN*	*UEMv*	*UEMc*	*SEM*
44	0.980	0.980	1.000	0.993	0.980	0.987
45	1.000	1.000	1.000	1.000	1.000	1.000
46	1.000	1.000	1.000	0.991	0.988	0.997
47	0.917	0.916	0.998	1.000	0.861	0.861
48	0.971	0.971	1.000	0.984	0.956	0.971
49	0.917	0.917	0.999	1.000	0.909	0.909
50	0.975	0.975	1.000	1.000	1.000	1.000
51	0.952	0.952	1.000	0.964	0.924	0.958
52	1.000	1.000	1.000	1.000	1.000	1.000
53	1.000	1.000	1.000	1.000	1.000	1.000
54	1.000	0.988	0.988	1.000	1.000	1.000
55	1.000	1.000	1.000	1.000	1.000	1.000
56	0.930	0.930	1.000	0.958	0.784	0.819
57	1.000	1.000	1.000	1.000	1.000	1.000
58	1.000	1.000	1.000	1.000	1.000	1.000
59	0.982	0.981	0.999	1.000	1.000	1.000
60	0.947	0.947	0.999	1.000	0.901	0.901
61	0.912	0.911	0.999	1.000	0.897	0.897
62	1.000	1.000	1.000	1.000	1.000	1.000
63	1.000	0.959	0.959	1.000	0.952	0.952
64	0.950	0.949	0.999	1.000	0.856	0.856
65	0.925	0.923	0.998	0.940	0.767	0.815
66	1.000	1.000	1.000	1.000	0.973	0.973
67	1.000	0.972	0.972	1.000	1.000	1.000
68	1.000	0.971	0.971	1.000	1.000	1.000
Total Mean	0.979	0.976	0.997	0.992	0.962	0.970
Total S.D.	0.023	0.023	0.007	0.014	0.051	0.044
Total Max.	1.000	1.000	1.000	1.000	1.000	1.000
Total Min.	0.912	0.911	0.959	0.940	0.767	0.815

(a) Source: Sueyoshi and Goto (2015b). (b) The plant names are listed in Table 18.1. (c) *UENv* and *UENc* stand for unified efficiencies, i.e., UEN_v^{NR} and UEN_c^{NR} under natural disposability. *SEN* stands for SEN^{NR} under natural disposability. The lowercase v and c stand for variable and constant RTS, respectively. (d) *UEMv* and *UEMc* stand for unified efficiencies, i.e., UEM_v^{NR} and UEM_c^{NR} under managerial disposability. *SEM* stands for SEM^{NR} under managerial disposability. The lowercase v and c stand for variable and constant DTS, respectively. (e) All unified efficiency scores are measured under non-radial models with $\varepsilon = 0.5$.

The other finding is that Table 18.3 statistically confirms that coal-fired power plants using BIT outperformed the other plants using SUB in the three unified efficiency measures under managerial disposability. This is because the disposability concept has the first priority to environmental performance, so that *UEM* and *SEM* exhibit differences depending on the type of the coal combustion. The result

TABLE 18.3 Mean tests on unified efficiencies: non-radial approach

	UENv		*UENc*		*SEN*		*UEMv*		*UEMc*		*SEM*	
Statistics	Small	Large	Small	Large	Small	Large	Small	Large	Small	Large	Small	Large
Avg.	0.982	0.976	0.980	0.973	0.998	0.996	0.996	0.988	0.973	0.952	0.977	0.963
St.Dev.	0.017	0.028	0.017	0.027	0.001	0.009	0.007	0.017	0.032	0.063	0.030	0.054
t-score	1.018		1.262		0.708		2.702*		1.820		1.351	
p-value	0.313		0.212		0.484		0.010		0.074		0.182	
Statistics	BIT	SUB	BIT	SUB	BIT	SUB	BIT	SUB	BIT	SUB	BIT	SUB
Avg.	0.981	0.968	0.978	0.966	0.997	0.998	0.994	0.980	0.972	0.913	0.977	0.931
St.Dev.	0.022	0.027	0.022	0.026	0.007	0.001	0.011	0.021	0.036	0.084	0.034	0.068
t-score	1.544		1.385		−1.634		2.209*		2.280*		2.226*	
p-value	0.147		0.189		0.107		0.049		0.044		0.048	

(a) Source: Sueyoshi and Goto (2015b). (b) See the notes in the header of Table 18.2. (c) All unified efficiency scores are measured under non-radial models with $\varepsilon = 0.5$. (d) An asterisk indicates the significance at the 10% level.

also confirms that the quality difference of coal influences the environmental performance of these power plants in PJM. Thus, this chapter can statistically reject the first hypothesis under managerial disposability.

18.7 SUMMARY

This chapter described how to measure the degree of scale efficiency under natural and managerial disposability concepts by using both non-radial and radial measurements. To document the practicality of scale efficiency, this chapter applied the proposed approach to examine the performance of coal-fired power plants in the north-east region of the United States. This chapter confirmed that there was a significant difference between the two types (BIT and SUB) of coal-fired power plants in their unified efficiency measures, including their scale efficiencies, under the concept of managerial disposability. In contrast, under natural disposability, this chapter could not find such a difference between them. The fact, which BIT outperformed SUB in terms of their unified and scale efficiencies, provides the policy implication that these power plants face a managerial challenge to shift their coal combustions from SUB to BIT in the United States.

Besides the empirical finding, this chapter could not confirm the other hypothesis on whether coal-fired power plants with small operation (less than 50% of plant capacity) outperformed ones with large operation (more than 50% of plant capacity), and vice versa, in terms of their unified and scale efficiency measures under natural and managerial disposability, with the exception of *UEM(v)*. The rationale was because their plant operations were frequently monitored by US regulatory agencies regardless of their capacity factor or size.

This chapter discussed a methodological issue on DEA applied to energy and environmental assessment. This chapter used $\varepsilon = 0.5$ for the proposed non-radial approach. The use of ε is very important in DEA because it guarantees that all multipliers (dual variables) are positive, along with data ranges, so that information on production factors is fully utilized in the proposed radial and non-radial approaches[6].

This chapter used the classification of coal-fired power plants based upon their plant capacity factors. In the near future, it will be necessary for us to examine

[6] Many previous DEA studies have not paid serious attention to the requirement that DEA performance assessment should be evaluated on a basis of positive multipliers (i.e., dual variables). A problem is that the selection of ε is often subjective in the estimation of unified and scale efficiency measures under natural and managerial disposability concepts. The selection is mathematically required in the proposed environmental assessment that all unified and scale efficiencies should be between zero and one, as specified by the property of "efficiency requirement." See Chapter 6 on the efficiency requirement. The best selection of ε is still unknown. However, it is possible for us to select $\varepsilon = 0$. In the case, we need to depend upon a use of strong complementary slackness conditions (SCSCs). See Chapter 7 for a detailed description on SCSCs.

other classification methods on the performance of coal-fired power plants. Such extended investigations are useful in not only the United States but also other industrial nations. In summary, we look forward to seeing a progress in technology innovation, such as clean coal technology[7].

[7] The clean coal technology has a high potential in future energy use. See https://energy.gov/fe/science-innovation/clean-coal-research. Unfortunately, the commercialization of clean coal technology is very limited at the current moment. Thus, the coal-fired power plants examined in this chapter have not yet utilized the new technology. We do not investigate the technology innovation on coal in detail, rather focusing upon an empirical study on current coal-fired power plants by DEA environmental assessment. The proposed assessment fully utilizes the concept of managerial disposability that puts more weight on environmental performance than operational one. Thus, this chapter investigates the environmental technology of coal-fired power plants within the empirical framework that we can investigate on our best research capability.

19

MEASUREMENT IN TIME HORIZON

19.1 INTRODUCTION

This chapter[1] describes how to assess the performance of decision making units (DMUs) in a time horizon. To explore the research issue on time series data, this chapter formulates radial models under natural or managerial disposability and then extends them to time series analysis. The proposed approach uses the radial measurement discussed in Chapter 17 for time series analysis because the radial framework has an inefficiency measure (ξ) that is unrestricted in its sign. In contrast, the non-radial models discussed in Chapter 16 cannot be directly applied to the proposed framework for time series because it does not contain an inefficiency measure. Non-radial measurement needs to use another computational framework (e.g., window analysis) for a time series. See Chapter 26 that discusses non-radial measurement by window analysis. In the proposed time series analysis, a negative inefficiency measure indicates a frontier progress and a positive inefficiency exhibits the opposite case (so, no frontier progress). Thus, this chapter uses radial measurement to examine whether the frontier shift occurs among different periods.

[1] This chapter is partly based upon the article: Sueyoshi, T. and Goto, M. (2013c) DEA environmental assessment in a time horizon: Malmquist index on fuel mix, electricity and CO_2 of industrial nations. *Energy Economics*, **40**, 370–382.

Environmental Assessment on Energy and Sustainability by Data Envelopment Analysis, First Edition. Toshiyuki Sueyoshi and Mika Goto.

To document the practicality of the proposed approach, this chapter applies radial measurement to performance assessment of the electric power industry of ten Organization for Economic Co-operation and Development (OECD) nations from 1999 to 2009. The application is useful for developing green growth in economy and pollution reduction in order to develop a sustainable society.

The remainder of this chapter is organized as follows: Section 19.2 reviews Malmquist index measurement that serves as a conceptual basis for DEA time series analysis. Section 19.3 discusses how to unify desirable and undesirable outputs under natural or managerial disposability in a time horizon. Section 19.4 describes formulations under natural disposability. Section 19.5 shifts our interest to formulation under managerial disposability. Section 19.6 applies the proposed approach to the electric power industry of OECD nations. Section 19.7 summarizes this chapter. The appendix of this chapter briefly summarizes the other types of indexes in addition to the Malmquist index.

19.2 MALMQUIST INDEX

Two groups of research were closely related to DEA in a time horizon. One of the two groups was due to Malmquist (1953)[2]. His study documented how to measure production indexes between different regions. Other production economists, including Bjurek (1996)[3], Bjurek and Hjamarsson (1995)[4] and Grifell-Tatje and Lovell (1995)[5], applied the Malmquist index to DEA dynamics and efficiency analysis. The Malmquist index measurement proposed in the previous studies has methodological strengths and drawbacks. A contribution of this group was that the index measurement was introduced and separated into its subcomponents. Consequently, it was possible for us to measure both the shift of an efficiency frontier and the influence of each subcomponent in the Malmquist index. That was a contribution, indeed. In contrast, the index measurement had a drawback. That is, the previous measurement depended upon an assumption that production activity was characterized by a functional form. Furthermore, it was expected that production technology always shifted an efficiency frontier toward better performance in observed periods. However, such underlying assumptions were not always satisfied in many real performance assessments. Thus, it is necessary for us to consider the index measurement in which an efficiency frontier shift may not occur or the frontier retreats itself in the worst case.

[2] Malmquist, S. (1953) Index number and indifferences surfaces. *Trabajos de Estatistica*, **4**, 209–2042.
[3] Bjurek, H. (1996) The Malmquist total factor productivity index. *Scandinavian Journal of Economics*, **98**, 303–3013.
[4] Bjurek, H. and Hjalmarsson, L. (1995) Productivity in multiple output public services: a quadratic frontier function and the Malmquist index approach. *Journal of Public Economics*, **56**, 447–460.
[5] Grifell-Tatje, E. and Lovel, C.A.K. (1995) A note on the Malmquist productivity index. *Economic Letters*, **47**, 167–175.

In addition to the economic approach, management scientists (e.g., Bowlin 1987)[6] were interested in "DEA window analysis." The methodology examined how much an efficiency score was changed by shifting a combination of adjacent periods, often referred to as a "window." As an extension, Thore *et al.* (1994)[7] and Goto and Tsutsui (1998)[8] combined the window analysis and the Malmquist index measurement. The combined measure was referred to as the "Malmquist productivity index." An important feature of the window analysis was that it could avoid the assumption that an efficiency frontier did not retreat. Even though a crossover occurred among efficiency frontiers, the window analysis pooled observations for a few consecutive periods into a window where a new efficiency frontier was identified. Consequently, a group of efficiency scores was smoothed over time and these efficiency scores were determined by comparing these performance measures with a newly established efficiency frontier within the window. That was a contribution. A drawback of the window analysis was that it lacked an analytical scheme to decompose an efficiency change into subcomponents, as found in the Malmquist index measurement. A combination between the Malmquist index measurement and the window analysis could be found in Sueyoshi and Aoki (2001)[9] and Sueyoshi and Goto (2001)[10].

Finally, this chapter uses the term "efficiency" when a performance measure exists between 0% (full inefficiency) and 100% (full efficiency). Such an analytical feature indicates the property of efficiency requirement as discussed in Chapter 6. Meanwhile, another term "index" is used hereafter to express a measure that may be more than unity (100%). When the index is more than unity, it indicates a possible occurrence of a frontier shift due to a technology development and/or a managerial effort. In contrast, the opposite case can be found if it is equal to or less than unity.

19.3 FRONTIER SHIFT IN TIME HORIZON

19.3.1 No Occurrence of Frontier Crossover

To extend the concept of natural or managerial disposability (as discussed in Chapter 17) into a time horizon, this chapter prepares four figures to intuitively describe the measurement of the Malmquist index. The four figures visually serve

[6] Bowlin, W.F. (1987) Evaluating the efficiency of US air force real-property maintenance activities. *Journal of the Operational Research Society*, **38**, 127–135.

[7] Thore, S., Kozmetsky, G. and Phillips, F. (1994) DEA of financial statements data: the US computer industry. *Journal of Productivity Analysis*, **5**, 229–248.

[8] Goto, M. and Tsutsui, M. (1998) Comparison of productive and cost efficiencies among Japanese and US electric utilities. *OMEGA*, **26**, 177–194.

[9] Sueyoshi, T. and Aoki, S. (2001) A use of a nonparametric statistic for DEA frontier shift: the Kruskal and Wallis rank test. *OMEGA*, **29**, 1–18.

[10] Sueyoshi, T. and Goto, M. (2001) Slack-adjusted DEA for time series analysis: performance measurement of Japanese electric power generation industry in 1984–1993. *European Journal of Operational Research*, **133**, 232–259.

as an analytical basis for computing 12 subcomponents originated from the Malmquist index under two disposability concepts.

Figure 19.1 depicts a frontier shift from the t–1 th period to the t th period under natural disposability in which DMUs attempt to enhance their operational performance as the first priority and their environmental performance as the second priority. For our visual convenience, Figure 19.1 considers a simple case of a single input (x) and two desirable outputs (g_1 and g_2). An undesirable output (b) is assumed to be same on all DMUs, so dropping the influence on the frontier shift. The visual description is easily extendable to multiple inputs and desirable outputs, as found later in the proposed DEA formulations.

In Figure 19.1, a_{t-1} stands for the observed performance of a DMU at the t–1 th period. Meanwhile, c_t indicates the observed performance of the DMU at the t th period for $t = 2,\ldots,T$. The symbol (0) stands for an origin in the figure. A frontier shift occurs toward the north-east direction under the assumption that a

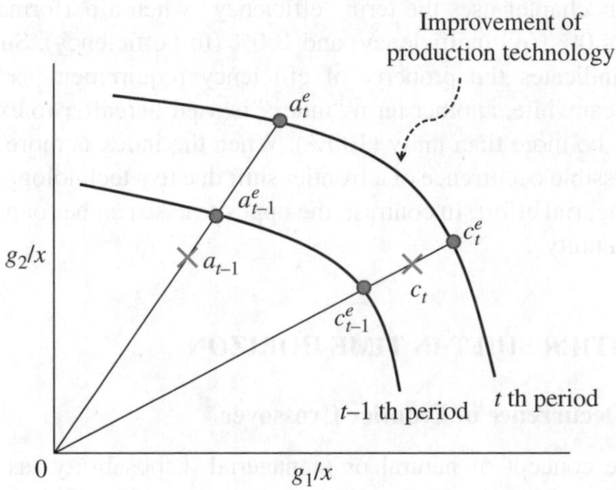

FIGURE 19.1 Frontier shift under natural disposability (a) Source: Sueyoshi and Goto (2013c). (b) The efficiency frontier shifts toward an increase in desirable outputs without any frontier crossover between the two periods. The natural disposability implies that the first priority is operational performance and the second priority is environmental performance. The figure assumes that an undesirable output (b) is same for all DMUs, so reducing the influence on a frontier shift. (c) The figure assumes that DEA does not suffer from an occurrence of multiple projections and multiple reference sets. (d) A unique projection onto the two efficiency frontiers is for our visual convenience. This chapter fully understands that a real projection in DEA is more complicated than the one depicted in the figure. (e) As discussed in Chapter 3, this chapter clearly understands that the projection should be measured by the L1-metric distance, not the L2-metric distance, as depicted in the figure. Such a metric change is for our visual convenience.

technology progress on desirable outputs occur between the two periods. A shift toward the north-east direction is due to performance betterment under natural disposability. Figure 19.1 assumes that the amount of undesirable outputs is the same between the two periods.

The performance of a_{t-1} is projected to a_{t-1}^e on the efficiency frontier in the t–1 th period and a_t^e on the efficiency frontier in the t th period, respectively, where the superscript e implies an efficiency frontier on which observed performance is projected. In the two cases, this chapter assumes a unique projection and a unique reference set in the two periods, as visually described in Figure 19.1. Under the same assumption, the performance of c_t is projected to c_{t-1}^e on the efficiency frontier for the t–1 th period and c_t^e on the efficiency frontier for the t th period, respectively.

A geometric mean, which is widely used to measure the average of the two line segments ($a_t^e - a_{t-1}^e$ and $c_t^e - c_{t-1}^e$), indicates the degree of a frontier shift between the t–1 th and t th periods. The geometric mean, indicating a "Malmquist index," is expressed by the following subcomponents:

$$IN_{t-1}^{t} = \sqrt{\frac{0a_t^e}{0a_{t-1}^e}\frac{0c_t^e}{0c_{t-1}^e}} = \sqrt{\frac{\dfrac{0a_t^e}{0a_{t-1}}\dfrac{0c_t^e}{0c_t}}{\dfrac{0a_{t-1}^e}{0a_{t-1}}\dfrac{0c_{t-1}^e}{0c_t}}}, \tag{19.1}$$

where IN_{t-1}^{t} stands for the Malmquist index between the t–1 th and t th periods under natural disposability. Equation (19.1) indicates "Malmquist" index, but the above abbreviation does not use M to express IN_{t-1}^{t} because this chapter uses M to express managerial disposability.

Shifting our description from natural disposability to managerial disposability, Figure 19.2 depicts a frontier shift from the t–1 th period to the t th period in which each DMU attempts to enhance environmental performance as the first priority and the operational performance as the second priority. For our visual description, Figure 19.2 considers a single input (x) and two undesirable outputs (b_1 and b_2). It is assumed that all the DMUs have produced a same amount of a desirable output (g), so that it is not depicted in the figure. In Figure 19.2, d_{t-1} stands for the observed performance of a DMU in the t–1 th period. Meanwhile, q_t indicates the observed performance of the DMU in the t th period. A frontier shift occurs toward the south-west direction under the assumption that an eco-technology progress (e.g. clean coal technology for electricity generation) on undesirable outputs occur between the two periods. A shift toward the south-west direction is due to managerial disposability.

The performance of d_{t-1} can be projected to d_{t-1}^e on the efficiency frontier for the t–1 th period and d_t^e on the efficiency frontier for the t th period, respectively, in Figure 19.2. Meanwhile, the performance of q_t can be projected to q_{t-1}^e on the efficiency frontier of the t–1 th period and q_t^e on the efficiency frontier of the t th period, respectively. The Malmquist index, which measures the geometric mean

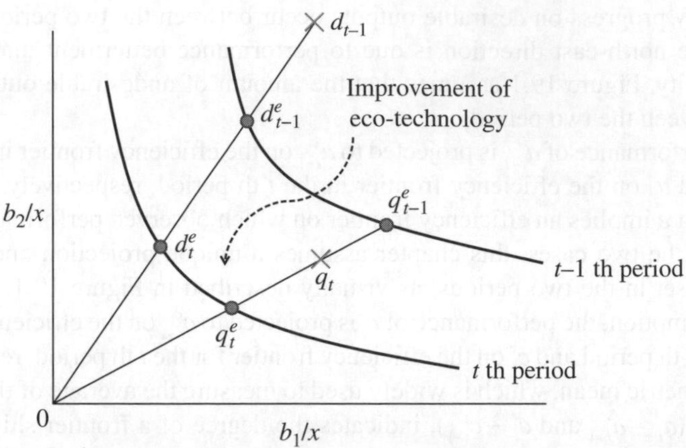

FIGURE 19.2 Frontier shift under managerial disposability (a) Source: Sueyoshi and Goto (2013c). (b) The efficiency frontier shifts toward a decrease in undesirable outputs without any frontier crossover between the two periods. The managerial disposability implies that the first priority is environmental performance and the second priority is operational performance. The figure assumes that desirable outputs are same on all DMUs, so dropping an influence on the frontier shift. (c) See the notes for Figure 19.1.

of the two line segments ($d^e_{t-1} - d^e_t$ and $q^e_{t-1} - q^e_t$), is expressed by the following subcomponents:

$$IM^t_{t-1} = \sqrt{\frac{0d^e_{t-1}}{0d^e_t}\frac{0q^e_{t-1}}{0q^e_t}} = \sqrt{\frac{\dfrac{0d^e_{t-1}}{0d_{t-1}}\dfrac{0q^e_{t-1}}{0q_t}}{\dfrac{0d^e_t}{0d_{t-1}}\dfrac{0q^e_t}{0q_t}}}, \tag{19.2}$$

where IM^t_{t-1} stands for the Malmquist index between the t–1 th and t th periods under managerial disposability.

19.3.2 Occurrence of Frontier Crossover

Figure 19.3, corresponding to Figure 19.1, depicts a possible occurrence of the frontier crossover between the t–1 th and t th periods. The performance of DMUs in Figure 19.3 is measured by natural disposability. In Figure 19.3, an efficiency frontier retreats between the two periods. As a result, it is necessary to combine the two frontiers to shape a new efficiency frontier, or the dotted line in Figure 19.3. To attain the status of efficiency, a_{t-1} and c_t need to shift their locations to $a^e_{t-1\&t}$ and $c^e_{t-1\&t}$ on the newly shaped efficiency frontier.

The geometric mean measures the average of the two line segments ($a^e_{t-1\&t} - a^e_{t-1}$ and $c^e_{t-1\&t} - c^e_t$), indicating the degree of a frontier shift between the t–1 th and t th periods. The geometric mean in this case is expressed by the following notation:

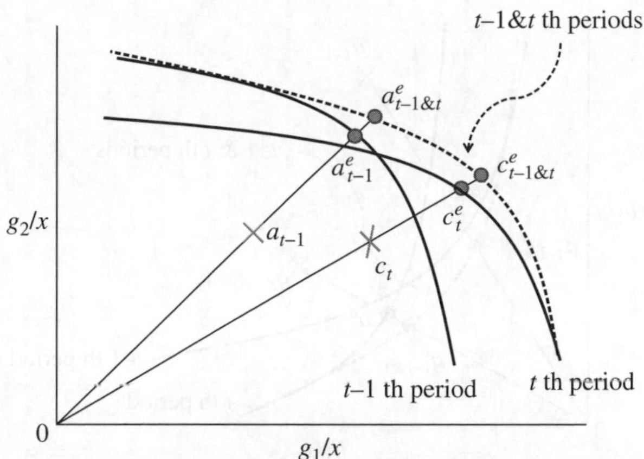

FIGURE 19.3 Frontier crossover between two periods under natural disposability (a) Source: Sueyoshi and Goto (2013c). (b) See the notes for Figure 19.1. (c) The dotted line of the upper curve indicates the frontier for the t–1 th and t th periods.

$$INC_{t-1}^{t} = \sqrt{\frac{0a^e_{t-1\&t}}{0a^e_{t-1}}\frac{0c^e_{t-1\&t}}{0c^e_t}} = \sqrt{\frac{\dfrac{0a^e_{t-1\&t}}{0a_{t-1}}\dfrac{0c^e_{t-1\&t}}{0c_t}}{\dfrac{0a^e_{t-1}}{0a_{t-1}}\dfrac{0c^e_t}{0c_t}}}, \tag{19.3}$$

where INC_{t-1}^{t} stands for the Malmquist index between the t–1 th and t th periods under natural disposability, considering a possible occurrence of the frontier crossover (C) between the two periods.

In similar manner, Figure 19.4, corresponding to Figure 19.2, visually describes a frontier crossover between the t–1 th and t th periods. In this figure, d_{t-1} and q_t need to shift their locations to the projected points, $d^e_{t-1\&t}$ and $q^e_{t-1\&t}$, on the newly shaped efficiency frontier, or the dotted line for the enhancement of their environmental performance.

The geometric mean in this case measures the average of the two line segments ($d^e_{t-1} - d^e_{t-1\&t}$ and $q^e_t - q^e_{t-1\&t}$), indicating the degree of a frontier shift between the t–1 th and t th periods. The geometric mean is expressed by the following subcomponents:

$$IMC_{t-1}^{t} = \sqrt{\frac{0d^e_{t-1}}{0d^e_{t-1\&t}}\frac{0q^e_t}{0q^e_{t-1\&t}}} = \sqrt{\frac{\dfrac{0d^e_{t-1}}{0d_{t-1}}\dfrac{0q^e_t}{0q_t}}{\dfrac{0d^e_{t-1\&t}}{0d_{t-1}}\dfrac{0q^e_{t-1\&t}}{0q_t}}}, \tag{19.4}$$

where IMC_{t-1}^{t} stands for the Malmquist index between the t–1 th and the t th periods under managerial disposability, considering a possible occurrence of the frontier crossover between the two periods.

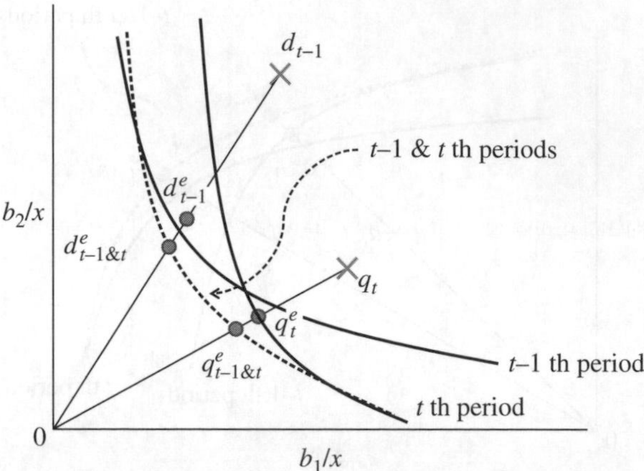

FIGURE 19.4 Frontier crossover between two periods under managerial disposability (a) Source: Sueyoshi and Goto (2013c). (b) See the notes for Figure 19.2. (c) The dotted line of the bottom curve indicates the frontier for the t–1 th and t th periods.

It is important to note two concerns. One concern is that the performance of all DMUs at the t th period depends upon only that of the t–1 th and t th periods, not the other periods, as depicted in Figures 19.3 and 19.4. Such a relationship is used for our visual convenience. However, it is not difficult for us to increase the number of periods in the proposed formulations. We can deal with an occurrence of the frontier crossover not only between the adjacent (t–1 th and t th) periods but also among all other period combinations. The other concern is that Equations (19.3) and (19.4) have an analytical benefit, compared with Equations (19.1) and (19.2). That is, we can measure the four components related to a frontier shift by Equations (19.3) and (19.4) even if a frontier crossover occurs between two periods. Meanwhile, Equations (19.1) and (19.2) do not incorporate such an occurrence of the frontier crossover.

19.4 FORMULATIONS FOR NATURAL DISPOSABILITY

The six components to measure IN^{t}_{t-1} and INC^{t}_{t-1} include

(a) UEN^{R}_{t-1} (unified efficiency in the t–1 th period),
(b) UEN^{R}_{t} (unified efficiency in the t th period),
(c) $IUIN^{R}_{t-1\rightarrow t}$ (inter-temporal unified index from the t–1 th to the t th period),
(d) $IUIN^{R}_{t\rightarrow t-1}$ (inter-temporal unified index from the t th period to the t–1 th period),
(e) $IUEN^{R}_{t-1\rightarrow t-1\&t}$ (inter-temporal unified efficiency from the t–1 th to the t–1 th & t th periods) and
(f) $IUEN^{R}_{t\rightarrow t-1\&t}$ (inter-temporal unified efficiency from the t th period to the t–1 th & t th periods.

All efficiency and index measures are determined by radial (R) measurement under natural (N) disposability and constant RTS.

19.4.1 No Occurrence of Frontier Crossover

To measure the Malmquist index under natural disposability in a time horizon (t = 2,…,T), this chapter considers n DMUs (j = 1,…,n) whose operational and environmental achievements are relatively compared with others in their performance assessment. The performance comparison among multiple periods needs to add subscripts (t) to J so that J_t stands for all DMUs in the t th period.

Returning to Figure 19.1, which depicts a frontier shift with a technology progress, the Malmquist index between the two periods can be restructured as follows:

$$IN_{t-1}^{t} = \sqrt{\frac{0a_t^e}{0a_{t-1}^e}\frac{0c_t^e}{0c_{t-1}^e}} = \sqrt{\frac{\frac{0a_t^e}{0a_{t-1}}\frac{0c_t^e}{0c_t}}{\frac{0a_{t-1}^e}{0a_{t-1}}\frac{0c_{t-1}^e}{0c_t}}} = \sqrt{\frac{\frac{0a_{t-1}}{0a_{t-1}^e}\frac{0c_t}{0c_{t-1}^e}}{\frac{0a_{t-1}}{0a_t^e}\frac{0c_t}{0c_t^e}}} \tag{19.5}$$

$$\Rightarrow IN_{t-1}^{t} = \sqrt{\frac{UEN_{t-1}^{R}}{IUIN_{t-1\to t}^{R}}\frac{IUIN_{t\to t-1}^{R}}{UEN_{t}^{R}}}$$

where UEN_{t-1}^{R} and UEN_{t}^{R} are two efficiency measures that belong to a range between zero (full inefficiency) and one (full efficiency). Meanwhile, $IUIN_{t\to t-1}^{R}$ and $IUIN_{t-1\to t}^{R}$ may become larger or smaller than unity, depending upon an occurrence of a frontier crossover between the two periods.

These measures can be formulated by the four DEA models, each of which is specified as follows: First, the degree of UEN_{t}^{R} on the k-th DMU in the t th period (t = 2,…,T) is measured by the following model under natural disposability and constant RTS and DTS:

$$\begin{aligned}
&\text{Maximize } \xi + \varepsilon_s\left(\sum_{i=1}^{m}R_i^x d_i^{x-} + \sum_{r=1}^{s}R_r^g d_r^g + \sum_{f=1}^{h}R_f^b d_f^b\right)\\
&\text{s.t.} \quad \sum_{j\in J_t} x_{ijt}\lambda_{jt} + d_i^{x-} = x_{ikt}\ \left(k\in J_t \text{ and } i=1,\dots,m\right),\\
&\qquad \sum_{j\in J_t} g_{rjt}\lambda_{jt} - d_r^g - \xi g_{rkt} = g_{rkt}\ \left(k\in J_t \text{ and } r=1,\dots,s\right),\\
&\qquad \sum_{j\in J_t} b_{fjt}\lambda_{jt} + d_f^b + \xi b_{fkt} = b_{fkt}\ \left(k\in J_t \text{ and } f=1,\dots,h\right),\\
&\qquad \lambda_{jt}\ge 0\ \left(j=1,\dots,n \text{ and } t=2,\dots,T\right), \xi: \text{URS}, d_i^{x-}\ge 0\ \left(i=1,\dots,m\right),\\
&\qquad d_r^g\ge 0\ \left(r=1,\dots,s\right) \text{ and } d_f^b\ge 0\ \left(f=1,\dots,h\right),
\end{aligned} \tag{19.6}$$

where d_i^x ($i = 1,\ldots,m$), d_r^g ($r = 1,\ldots,s$) and d_f^b ($f = 1,\ldots,h$) are all slack variables related to inputs, desirable outputs and undesirable outputs, respectively. $\lambda_{jt} = (\lambda_{1t},\ldots,\lambda_{nt})^{Tr}$ is a column vector of unknown variables, often referred to as "structural" or "intensity" variables, that are used for connecting production factors. An unknown scalar value (ξ), which is unrestricted (URS), stands for an inefficiency measure. The inefficiency measure corresponds to a distance between an efficiency frontier and an observed vector of desirable and undesirable outputs, as depicted in Figures 19.1 and 19.2. A scalar value (ε_s) is also incorporated into Model (19.6) and it stands for a very small number to indicate the relative importance between the inefficiency score and the total amount of slacks. This chapter sets $\varepsilon_s = 0.0001$ for our computation. The selection is subjective. It is possible for us to set $\varepsilon_s = 0$. In the case, it is necessary to depend upon a method for multiplier restriction (e.g., cone ratio) or the incorporation of strong complementary slackness conditions (SCSCs) because dual variables may become zero in these magnitudes. That is problematic in DEA because corresponding production factors are not fully utilized in the efficiency assessment. See Chapter 7 for a description on how to handle the difficulty by SCSCs.

The data range adjustments in Model (19.6) are determined by the upper and lower bounds of production factors. These upper and lower bounds are specified by

$$R_i^x = (m+s+h)^{-1}\left(\max\left\{x_{ij}\,\middle|\, j \in J_{t-1} \cup J_t\right\} - \min\left\{x_{ij}\,\middle|\, j \in J_{t-1} \cup J_t\right\}\right)^{-1},$$
$$R_r^g = (m+s+h)^{-1}\left(\max\left\{g_{rj}\,\middle|\, j \in J_{t-1} \cup J_t\right\} - \min\left\{g_{rj}\,\middle|\, j \in J_{t-1} \cup J_t\right\}\right)^{-1} \text{ and}$$
$$R_f^b = (m+s+h)^{-1}\left(\max\left\{b_{fj}\,\middle|\, j \in J_{t-1} \cup J_t\right\} - \min\left\{b_{fj}\,\middle|\, j \in J_{t-1} \cup J_t\right\}\right)^{-1}.$$

Here, $\cup$ stands for a union set. Thus, $J_{t-1} \cup J_t$ stands for all DMUs in the t–1 th and t th periods.

The degree of UEN_t^R of the k-th DMU in the t th period is determined by

$$UEN_t^R = 1 - \left[\xi^* + \varepsilon_s\left(\sum_{i=1}^{m} R_i^x d_i^{x-*} + \sum_{r=1}^{s} R_r^g d_r^{g*} + \sum_{f=1}^{h} R_f^b d_f^{b*}\right)\right], \qquad (19.7)$$

where the inefficiency score and all slack variables are determined on the optimality of Model (19.6). The equation within the parenthesis, obtained from the optimality of Model (19.6), indicates the level of unified inefficiency under natural disposability. The unified efficiency is obtained by subtracting the level of unified inefficiency from unity. The degree of UEN_t^R on the k-th DMU is always less than unity, indicating an existence of some level of inefficiency, or equal unity for the status of full efficiency.

Second, the degree of UEN_{t-1}^{R} regarding the k-th DMU in the t–1 th period is measured by replacing t by t–1 in Model (19.6). Then, the degree is measured by Equation (19.7) as follows:

$$UEN_{t-1}^{R} = 1 - \left[\xi^* + \varepsilon_s \left(\sum_{i=1}^{m} R_i^x d_i^{x-*} + \sum_{r=1}^{s} R_r^g d_r^{g*} + \sum_{f=1}^{h} R_f^b d_f^{b*} \right) \right],$$

where the inefficient score and slacks on optimality are measured by Model (19.6) after replacing t by t–1 in the formulation.

Third, the degree of $IUIN_{t-1\to t}^{R}$ regarding the k-th DMU from the t–1 th period to the t th period is determined by the following model:

$$\begin{aligned}
&\text{Maximize } \xi + \varepsilon_s \left(\sum_{i=1}^{m} R_i^x d_i^{x-} + \sum_{r=1}^{s} R_r^g d_r^g + \sum_{f=1}^{h} R_f^b d_f^b \right) \\
&s.t. \quad \sum_{j \in J_t} x_{ijt} \lambda_{jt} + d_i^{x-} = x_{ikt-1} \left(k \in J_{t-1} \text{ and } i = 1,\ldots,m \right), \\
&\sum_{j \in J_t} g_{rjt} \lambda_{jt} - d_r^g - \xi g_{rkt-1} = g_{rkt-1} \left(k \in J_{t-1} \text{ and } r=1,\ldots,s \right), \qquad (19.8) \\
&\sum_{j \in J_t} b_{fjt} \lambda_{jt} + d_f^b + \xi b_{fkt-1} = b_{fkt-1} \left(k \in J_{t-1} \text{ and } f = 1,\ldots,h \right), \\
&\lambda_{jt} \geq 0 \ (j = 1,\ldots,n \text{ and } t=2,\ldots,T), \ \xi : \text{URS}, \ d_i^{x-} \geq 0 \ (i = 1,\ldots,m), \\
&d_r^g \geq 0 \ (r = 1,\ldots,s) \text{ and } d_f^b \geq 0 \ (f = 1,\ldots,h),
\end{aligned}$$

where an efficiency frontier consists of DMUs in the t th period and a DMU to be examined is selected from a group of DUMs in the t–1 th period. The degree of the index is measured by Equation (19.7) where the inefficiency measure and all slack variables are determined on optimality of Model (19.8).

The degree of $IUIN_{t\to t-1}^{R}$ regarding the k-th DMU that shifts from the t th period to the t–1 th period is determined by the following model:

$$\begin{aligned}
&\text{Maximize } \xi + \varepsilon_s \left(\sum_{i=1}^{m} R_i^x d_i^{x-} + \sum_{r=1}^{s} R_r^g d_r^g + \sum_{f=1}^{h} R_f^b d_f^b \right) \\
&s.t. \quad \sum_{j \in J_{t-1}} x_{ijt-1} \lambda_{jt-1} + d_i^{x-} = x_{ikt} \left(k \in J_t \text{ and } i = 1,\ldots,m \right), \\
&\sum_{j \in J_{t-1}} g_{rjt-1} \lambda_{jt-1} - d_r^g - \xi g_{rkt} = \text{g}_{rkt} \left(k \in J_t \text{ and } r = 1,\ldots,s \right), \qquad (19.9) \\
&\sum_{j \in J_{t-1}} b_{fjt-1} \lambda_{jt-1} + d_f^b + \xi b_{fkt} = \text{b}_{fkt} \left(k \in J_t \text{ and } f = 1,\ldots,h \right), \\
&\lambda_{jt-1} \geq 0 \ (j = 1,\ldots,n \text{ and } t = 2,\ldots,T), \ \xi : \text{URS}, \ d_i^{x-} \geq 0 \ (i = 1,\ldots,m), \\
&d_r^g \geq 0 \ (r = 1,\ldots,s) \text{ and } d_f^b \geq 0 \ (f = 1,\ldots,h),
\end{aligned}$$

where an efficiency frontier consists of DMUs in the t–1 th period and a DMU to be examined is selected from a group of DUMs in the t th period. The degree of the index is measured by Equation (19.7) where the inefficiency measure and all slack variables are determined on optimality of Model (19.9).

19.4.2 Occurrence of Frontier Crossover

Returning to Figure 19.3, the conceptual framework for Equation (19.3) under natural disposability is expressed by the following measures in the case of a frontier crossover:

$$INC_{t-1}^{t} = \sqrt{\frac{0a_{t-1\&t}^{e}}{0a_{t-1}^{e}}\frac{0c_{t-1\&t}^{e}}{0c_{t}^{e}}} = \sqrt{\frac{\dfrac{0a_{t-1\&t}^{e}}{0a_{t-1}}\dfrac{0c_{t-1\&t}^{e}}{0c_{t}}}{\dfrac{0a_{t-1}^{e}}{0a_{t-1}}\dfrac{0c_{t}^{e}}{0c_{t}}}} = \sqrt{\frac{\dfrac{0a_{t-1}}{0a_{t-1}^{e}}\dfrac{0c_{t}}{0c_{t}^{e}}}{\dfrac{0a_{t-1}}{0a_{t-1\&t}^{e}}\dfrac{0c_{t}}{0c_{t-1\&t}^{e}}}} \tag{19.10}$$

$$\Rightarrow INC_{t-1}^{t} = \sqrt{\frac{UEN_{t-1}^{R}}{IUEN_{t-1\to t-1\&t}^{R}}\frac{UEN_{t}^{R}}{IUEN_{t\to t-1\&t}^{R}}},$$

where INC_{t-1}^{t} is separated into four subcomponents, all of which are used to measure a frontier shift between the two periods (i.e., UEN_{t-1}^{R}, UEN_{t}^{R}, $IUEN_{t-1\to t-1\&t}^{R}$ and $IUEN_{t\to t-1\&t}^{R}$). Among the four subcomponents, this chapter has already discussed the measurement of UEN_{t-1}^{R} and UEN_{t}^{R}, so focusing upon the remaining measurement of $IUEN_{t-1\to t-1\&t}^{R}$ and $IUEN_{t\to t-1\&t}^{R}$, hereafter.

The degree of $IUEN_{t-1\to t-1\&t}^{R}$ regarding the k-th DMU between the two periods is determined by the following model:

$$\begin{aligned}
&\text{Maximize } \xi + \varepsilon_s\left(\sum_{i=1}^{m}R_i^x d_i^{x-} + \sum_{r=1}^{s}R_r^g d_r^g + \sum_{f=1}^{h}R_f^b d_f^b\right)\\
&s.t. \quad \sum_{j\in J_{t-1\&t}} x_{ijt-1\&t}\lambda_{jt-1\&t} + d_i^{x-} = x_{ikt-1}\left(k\in J_{t-1} \text{ and } i=1,\ldots,m\right),\\
&\qquad \sum_{j\in J_{t-1\&t}} g_{rjt-1\&t}\lambda_{jt-1\&t} - d_r^g - \xi g_{rkt-1} = g_{rkt-1}\left(k\in J_{t-1} \text{ and } r=1,\ldots,s\right),\\
&\qquad \sum_{j\in J_{t-1\&t}} b_{fjt-1\&t}\lambda_{jt-1\&t} + d_f^b + \xi b_{fkt-1} = b_{fkt-1}\left(k\in J_{t-1} \text{ and } f=1,\ldots,h\right),\\
&\qquad \lambda_{jt-1\&t} \ge 0\ \left(j=1,\ldots,n \text{ and } t=2,\ldots,T\right), \xi : \text{URS}, d_i^{x-} \ge 0\left(i=1,\ldots,m\right),\\
&\qquad d_r^g \ge 0\ \left(r=1,\ldots,s\right) \text{ and } d_f^b \ge 0\ \left(f=1,\ldots,h\right),
\end{aligned} \tag{19.11}$$

where an efficiency frontier consists of DMUs in the t–1 th and t th periods and a DMU to be examined by Model (19.11) is selected from a group of DUMs in the t–1 th period. The degree of $IUEN_{t-1\to t-1\&t}^{R}$ on the k-th DMU is determined by

Equation (19.7) where the inefficiency measure and all slack variables are determined on the optimality of Model (19.11).

The degree of $IUEN^{R}_{t\to t-1\&t}$ on the k-th DMU between the two periods is determined by the following model:

$$\begin{aligned}
&\text{Maximize } \xi+\varepsilon_s\left(\sum_{i=1}^{m}R_i^x d_i^{x-}+\sum_{r=1}^{s}R_r^g d_r^g+\sum_{f=1}^{h}R_f^b d_f^b\right)\\
&s.t.\quad \sum_{j\in J_{t-1\&t}} x_{ijt-1\&t}\lambda_{jt-1\&t}+d_i^{x-} \qquad\qquad = x_{ikt}\left(k\in J_t \text{ and } i=1,\ldots,m\right),\\
&\qquad \sum_{j\in J_{t-1\&t}} g_{rjt-1\&t}\lambda_{jt-1\&t} \quad -d_r^g \quad -\xi g_{rkt}=g_{rkt}\left(k\in J_t \text{ and } r=1,\ldots,s\right),\\
&\qquad \sum_{j\in J_{t-1\&t}} b_{fjt-1\&t}\lambda_{jt-1\&t} \quad +d_f^b+\xi b_{fkt}=b_{fkt}\left(k\in J_t \text{ and } f=1,\ldots,h\right),\\
&\qquad \lambda_{jt-1\&t}\ge 0\ \left(j=1,\ldots,n \text{ and } t=2,\ldots,T\right),\ \xi:\text{URS},\ d_i^{x-}\ge 0\left(i=1,\ldots,m\right),\\
&\qquad d_r^g\ge 0\ \left(r=1,\ldots,s\right) \text{ and } d_f^b\ge 0\ \left(f=1,\ldots,h\right),
\end{aligned} \tag{19.12}$$

where an efficiency frontier consists of DMUs in the t–1 th and t th periods and a DMU to be examined is selected from DUMs in the t th period. The degree of $IUEN^{R}_{t\to t-1\&t}$ regarding the k-th DMU in the t th period is determined by Equation (19.7) where the inefficiency measure and all slack variables are determined on the optimality of Model (19.12).

At the end of this section, it is necessary to describe two differences between index and efficiency measures. One of the two differences is that as a result of incorporating a frontier crossover in Models (19.11) and (19.12), the two models always produce efficiency scores, not indexes, which belong to a range between 0 (full inefficiency) and 1 (full efficiency). Such a feature cannot be found in Models (19.8) and (19.9). The other difference is that IN^{t}_{t-1} may be more than, equal to or less than unity, but INC^{t}_{t-1} is more than or equal to unity and it cannot be less than unity. If a DMU exhibits unity in INC^{t}_{t-1}, then it indicates that there is a frontier crossover between the t–1 th and t th periods and the achievement of the DMU locates on the three efficiency frontiers for the t–1 th, t th and the combined t–1 th and t th periods. In other words, there is no technological progress between the two periods. In contrast, if a DMU exhibits more than unity in INC^{t}_{t-1}, then it indicates that the efficiency frontier of the t th period locates differently from that of the t–1 th period, as depicted in Figure 19.3. Thus, it is possible for us to identify whether the DMU has an operational improvement (e,g., due to technological progress) or an operational retreat (e.g,, due to poor business condition) between the two periods by examining INC^{t}_{t-1}. Finally, it is important to note that our description here is based upon the assumption that a data set to be examined does not contain an outlier because it drastically changes the location of an efficiency frontier. See Chapter 2 on a description on an outlier.

19.5 FORMULATIONS UNDER MANAGERIAL DISPOSABILITY

To describe the six components regarding IM_{t-1}^{t} and IMC_{t-1}^{t}, this chapter specifies their subcomponents as follows:

(a) UEM_{t-1}^{R} (unified efficiency in the t–1 th period),
(b) UEM_{t}^{R} (unified efficiency in the t th period),
(c) $IUIM_{t-1\to t}^{R}$ (inter-temporal unified index from the t–1 th to the t th period),
(d) $IUIM_{t\to t-1}^{R}$ (inter-temporal unified index from the t th to the t–1 th period),
(e) $IUEM_{t-1\to t-1\&t}^{R}$ (inter-temporal unified efficiency from the t–1 th to the t–1 th & t th periods), and
(f) $IUEM_{t\to t-1\&t}^{R}$ (inter-temporal unified efficiency from the t th to the t–1 th & t th periods).

All efficiency and index measures are determined by radial measurement under managerial (M) disposability and constant DTS.

19.5.1 No Occurrence of Frontier Crossover

The conceptual framework in Equation (19.2), or IM_{t-1}^{t} under managerial disposability, is expressed by the following subcomponents as depicted in Figure 19.2:

$$IM_{t-1}^{t}=\sqrt{\frac{0d_{t-1}^{e}}{0d_{t}^{e}}\frac{0q_{t-1}^{e}}{0q_{t}^{e}}}=\sqrt{\frac{\frac{0d_{t-1}^{e}}{0d_{t-1}}\frac{0q_{t-1}^{e}}{0q_{t}}}{\frac{0d_{t}^{e}}{0d_{t-1}}\frac{0q_{t}^{e}}{0q_{t}}}}\Rightarrow IM_{t-1}^{t}=\sqrt{\frac{UEM_{t-1}^{R}}{IUIM_{t-1\to t}^{R}}\frac{IUIM_{t\to t-1}^{R}}{UEM_{t}^{R}}} \quad (19.13)$$

As formulated in Equation (19.13), IM_{t-1}^{t} is separated into four subcomponents, all of which are measured under constant DTS.

First, the degree of UEM_{t}^{R} of the k-th DMU in the t th period is measured by the following model:

$$\begin{aligned}
&\text{Maximize } \xi+\varepsilon_s\left(\sum_{i=1}^{m}R_i^x d_i^{x+}+\sum_{r=1}^{s}R_r^g d_r^g+\sum_{f=1}^{h}R_f^b d_f^b\right)\\
&s.t.\quad \sum_{j\in J_t}x_{ijt}\lambda_{jt}-d_i^{x+} = x_{ikt}\left(k\in J_t \text{ and } i=1,\dots,m\right),\\
&\qquad \sum_{j\in J_t}g_{rjt}\lambda_{jt}-d_r^g-\xi g_{rkt}=g_{rkt}\left(k\in J_t \text{ and } r=1,\dots,s\right),\\
&\qquad \sum_{j\in J_t}b_{fjt}\lambda_{jt}+d_f^b+\xi b_{fkt}=b_{fkt}\left(k\in J_t \text{ and } f=1,\dots,h\right),\\
&\qquad \lambda_{jt}\ge 0\ \left(j=1,\dots,n \text{ and } t=2,\dots,T\right),\ \xi:\text{URS},\ d_i^{x+}\ge 0\ \left(i=1,\dots,m\right),\\
&\qquad d_r^g\ge 0\ \left(r=1,\dots,s\right) \text{ and } d_f^b\ge 0\ \left(f=1,\dots,h\right).
\end{aligned} \quad (19.14)$$

The degree of UEM_t^R regarding the k-th DMU in the t th period is determined by

$$UEM_t^R = 1 - \left[\xi^* + \varepsilon_s \left(\sum_{i=1}^{m} R_i^x d_i^{x+*} + \sum_{r=1}^{s} R_r^g d_r^{g*} + \sum_{f=1}^{h} R_f^b d_f^{b*} \right) \right], \quad (19.15)$$

where the inefficiency score and all slack variables are determined on the optimality of Model (19.14). The equation within the parenthesis, obtained from the optimality of Model (19.14), indicates the level of unified inefficiency under managerial disposability. The unified efficiency is obtained by subtracting the level of inefficiency from unity. Thus, the unified efficiency always belongs to the range between 0 (full inefficiency) and 1 (full efficiency).

Second, the degree of UEM_{t-1}^R regarding the k-th DMU at the t–1 th period is measured by replacing t by t–1 in Model (19.14). Equation (19.15) is used to measure the degree of UEM_{t-1}^R after replacing t by t–1 in Model (19.14).

Third, the degree of $IUIM_{t-1\to t}^R$ of the k-th DMU from the t–1 th period to the t th period is determined by the following model:

$$\text{Maximize } \xi + \varepsilon_s \left(\sum_{i=1}^{m} R_i^x d_i^{x+} + \sum_{r=1}^{s} R_r^g d_r^g + \sum_{f=1}^{h} R_f^b d_f^b \right)$$

$$\begin{aligned}
s.t. \quad & \sum_{j \in J_t} x_{ijt}\lambda_{jt} - d_i^{x+} = x_{ikt-1} \left(k \in J_{t-1} \text{ and } i = 1,\ldots,m \right), \\
& \sum_{j \in J_t} g_{rjt}\lambda_{jt} - d_r^g - \xi g_{rkt-1} = g_{rkt-1} \left(k \in J_{t-1} \text{ and } r = 1,\ldots,s \right), \qquad (19.16) \\
& \sum_{j \in J_t} b_{fjt}\lambda_{jt} + d_f^b + \xi b_{fkt-1} = b_{fkt-1} \left(k \in J_{t-1} \text{ and } f = 1,\ldots,h \right), \\
& \lambda_{jt} \geq 0 \ \left(j = 1,\ldots,n \text{ and } t = 2,\ldots,T \right), \ \xi : \text{URS}, \ d_i^{x+} \geq 0 \ \left(i = 1,\ldots,m \right), \\
& d_r^g \geq 0 \ \left(r = 1,\ldots,s \right) \text{ and } d_f^b \geq 0 \ \left(f = 1,\ldots,h \right),
\end{aligned}$$

where an efficiency frontier consists of DMUs in the t th periods and a DMU to be examined is selected from DUMs in the t–1 th period. The degree of $IUIM_{t-1\to t}^R$ of the k-th DMU is measured by Equation (19.15) where the inefficiency measure and all slack variables are determined on the optimality of Model (19.16).

The degree of $IUIM^{R}_{t\to t-1}$ of the k-th DMU from the t th to the t–1 th period is determined by the following model:

$$\begin{aligned}
&\text{Maximize } \xi+\varepsilon_s\left(\sum_{i=1}^{m}R_i^x d_i^{x+}+\sum_{r=1}^{s}R_r^g d_r^g+\sum_{f=1}^{h}R_f^b d_f^b\right)\\
&s.t.\quad \sum_{j\in J_{t-1}} x_{ijt-1}\lambda_{jt-1}-d_i^{x+} = x_{ikt}\left(k\in J_t \text{ and } i=1,\ldots,m\right),\\
&\qquad \sum_{j\in J_{t-1}} g_{rjt-1}\lambda_{jt-1}-d_r^g-\xi g_{rkt}=g_{rkt}\left(k\in J_t \text{ and } r=1,\ldots,s\right),\qquad (19.17)\\
&\qquad \sum_{j\in J_{t-1}} b_{fjt-1}\lambda_{jt-1}+d_f^b+\xi b_{fkt}=b_{fkt}\left(k\in J_t \text{ and } f=1,\ldots,h\right),\\
&\qquad \lambda_{jt-1}\ge 0\ (j=1,\ldots,n \text{ and } t=2,\ldots,T),\ \xi:\text{URS},\ d_i^{x+}\ge 0\ (i=1,\ldots,m),\\
&\qquad d_r^g\ge 0\ (r=1,\ldots,s) \text{ and } d_f^b\ge 0\ (f=1,\ldots,h),
\end{aligned}$$

where an efficiency frontier consists of DMUs in the t–1 th period and a DMU to be examined is selected from a group of DUMs in the t th period. The degree of $IUIM^{R}_{t\to t-1}$ on the k-th DMU between the two periods is measured by Equation (19.15) where the inefficiency measure and all slack variables are determined on optimality of Model (19.17).

19.5.2 Occurrence of Frontier Crossover

By shifting our description to the case of a frontier crossover, the conceptual framework on Equation (19.4), or IMC^{t}_{t-1}, under managerial disposability, is expressed by the following unified measures as depicted in Figure 19.4:

$$IMC^{t}_{t-1}=\sqrt{\frac{0d^{e}_{t-1}}{0d^{e}_{t-1\&t}}\frac{0q^{e}_{t}}{0q^{e}_{t-1\&t}}}=\sqrt{\frac{\dfrac{0d^{e}_{t-1}}{0d_{t-1}}\dfrac{0q^{e}_{t}}{0q_{t}}}{\dfrac{0d^{e}_{t-1\&t}}{0d_{t-1}}\dfrac{0q^{e}_{t-1\&t}}{0q_{t}}}}$$

$$\Rightarrow IMC^{t}_{t-1}=\sqrt{\frac{UEM^{R}_{t-1}}{IUEM^{R}_{t-1\to t-1\&t}}\frac{UEM^{R}_{t}}{IUEM^{R}_{t\to t-1\&t}}}\qquad (19.18)$$

The degree of UEM^{R}_{t-1} and that of UEM^{R}_{t} are measured by Model (19.14). The first measurement needs to replace the t th by the t–1 th period in the model.

The degree of $IUEM^{R}_{t-1\rightarrow t-1\&t}$ of the k-th DMU in the t–1 th period is determined by the following model:

$$\begin{aligned}
&\text{Maximize } \xi+\varepsilon_s\left(\sum_{i=1}^{m}R_i^x d_i^{x+}+\sum_{r=1}^{s}R_r^g d_r^g+\sum_{f=1}^{h}R_f^b d_f^b\right)\\
&\text{s.t.}\quad \sum_{j\in J_{t-1\&t}} x_{ijt-1\&t}\lambda_{jt-1\&t}-d_i^{x+} = x_{ikt-1}\left(k\in J_{t-1}\text{ and } i=1,\ldots,m\right),\\
&\qquad \sum_{j\in J_{t-1\&t}} g_{rjt-1\&t}\lambda_{jt-1\&t}-d_r^{g}-\xi g_{rkt-1} = g_{rkt-1}\left(k\in J_{t-1}\text{ and } r=1,\ldots,s\right),\\
&\qquad \sum_{j\in J_{t-1\&t}} b_{fjt-1\&t}\lambda_{jt-1\&t}+d_f^{b}+\xi b_{fkt-1} = b_{fkt-1}\left(k\in J_{t-1}\text{ and } f=1,\ldots,h\right),\\
&\qquad \lambda_{jt-1\&t}\geq 0\ \left(j=1,\ldots,n\text{ and } t=2,\ldots,T\right),\ \xi:\text{URS},\ d_i^{x+}\geq 0\left(i=1,\ldots,m\right),\\
&\qquad d_r^g\geq 0\ \left(r=1,\ldots,s\right)\text{ and } d_f^b\geq 0\ \left(f=1,\ldots,h\right),
\end{aligned}\tag{19.19}$$

where an efficiency frontier consists of DMUs in the t–1 th and t th periods and a DMU to be examined is selected from DUMs in the t–1 th period. The degree of $IUEM^{R}_{t-1\rightarrow t-1\&t}$ of the k-th DMU is measured by Equation (19.15) where the inefficiency score and all slack variables are determined on the optimality of Model (19.19).

The degree of $IUEM^{R}_{t\rightarrow t-1\&t}$ of the k-th DMU in the t th period is determined by the following model:

$$\begin{aligned}
&\text{Maximize } \xi+\varepsilon_s\left(\sum_{l=1}^{m}R_i^x d_i^{x+}+\sum_{r=1}^{s}R_r^g d_r^g+\sum_{f=1}^{h}R_f^b d_f^b\right)\\
&\text{s.t.}\quad \sum_{j\in J_{t-1\&t}} x_{ijt-1\&t}\lambda_{jt-1\&t}-d_i^{x+} = x_{ikt}\left(k\in J_{t}\text{ and } i=1,\ldots,m\right),\\
&\qquad \sum_{j\in J_{t-1\&t}} g_{rjt-1\&t}\lambda_{jt-1\&t}-d_r^{g}-\xi g_{rkt} = g_{rkt}\left(k\in J_{t}\text{ and } r=1,\ldots,s\right),\\
&\qquad \sum_{j\in J_{t-1\&t}} b_{fjt-1\&t}\lambda_{jt-1\&t}+d_f^{b}+\xi b_{fkt} = b_{fkt}\left(k\in J_{t}\text{ and } f=1,\ldots,h\right),\\
&\qquad \lambda_{jt-1\&t}\geq 0\ \left(j=1,\ldots,n\text{ and } t=2,\ldots,T\right),\ \xi:URS,\ d_i^{x+}\geq 0\left(i=1,\ldots,m\right),\\
&\qquad d_r^g\geq 0\ \left(r=1,\ldots,s\right)\text{ and } d_f^b\geq 0\ \left(f=1,\ldots,h\right),
\end{aligned}\tag{19.20}$$

where an efficiency frontier consists of DMUs in the t–1 th and t th periods and a DMU to be examined is selected from a group of DUMs in the t th period. The degree of $IUEM^{R}_{t\rightarrow t-1\&t}$ is measured by Equation (19.15) where the inefficiency score and all slack variables are determined on the optimality of the model in Equation (19.20). As a result of incorporating an occurrence of a frontier shift in Models (19.19) and (19.20), the two models can always produce an efficiency score on a range between 0 (full inefficiency) and 1 (full efficiency). Such a feature cannot be found in Models (19.16) and (19.17).

The proposed Malmquist index measurement under natural or managerial disposability has the four important features discussed here. First, all the models are formulated under constant RTS and DTS. It is impossible to measure the indexes under variable RTS and DTS because we must face an infeasible solution. See Chapter 6 for the mathematical property. Second, this chapter assumes that all DEA models in Sections 19.3 and 19.4 can uniquely produce a DEA solution. However, the assumption is not always true because they may suffer from an occurrence of multiple solutions (e.g., multiple projections and reference sets). If such a problem occurs, it is necessary to incorporate SCSCs to the proposed DEA models for a time horizon. Third, this chapter considers only two periods (t–1 and t) in discussing the occurrence of a frontier crossover between the two periods. It is possible for us to extend the number of periods to more than three periods ($J_p \cup \cdots \cup J_{t-1} \cup J_t$), where the subscript ($p$) indicates a starting period of the window. See Sueyoshi and Goto (2015). In the case of multiple (more than two) periods, the data ranges (R) are determined by the upper and lower bounds of production factors as follows:

$$R_i^x = (m+s+h)^{-1}\Big(\max\left\{ x_{ij} \;\middle|\; j \in J_p \cup \ldots \cup J_{t-1} \cup J_t \right\} - \min\left\{ x_{ij} \;\middle|\; j \in J_p \cup \ldots \cup J_{t-1} \cup J_t \right\}\Big)^{-1},$$

$$R_r^g = (m+s+h)^{-1}\Big(\max\left\{ g_{rj} \;\middle|\; j \in J_p \cup \ldots \cup J_{t-1} \cup J_t \right\} - \min\left\{ g_{rj} \;\middle|\; j \in J_p \cup \ldots \cup J_{t-1} \cup J_t \right\}\Big)^{-1} \text{ and}$$

$$R_f^b = (m+s+h)^{-1}\Big(\max\left\{ b_{fj} \;\middle|\; j \in J_p \cup \ldots \cup J_{t-1} \cup J_t \right\} - \min\left\{ b_{fj} \;\middle|\; j \in J_p \cup \ldots \cup J_{t-1} \cup J_t \right\}\Big)^{-1}.$$

Finally, many production economists (e.g., Zhou *et al.* 2012a) used a directional distance function to express the analytical structure for inefficiency measurement. This chapter does not follow such a conventional way of measurement because of a difficulty in mathematically describing a frontier crossover between different periods and different technology developments. Further, the use of conventional distance functions has implicitly assumed a unique solution on DEA. In contrast, this chapter formulates DEA models as linear programming, so that they may suffer from an occurrence of multiple solutions in their primal and dual models.

19.6 ENERGY MIX OF INDUSTRIAL NATIONS

This chapter sampled ten industrial nations from the OECD countries whose energy structures are characterized by three inputs (i.e., the net electrical capacity of main producer plants with respect to combustible, nuclear, hydro and renewable energies), a single desirable output (i.e, the total net production of electricity) and a single undesirable output (i.e., the amount of CO_2 emissions from fuel combustion of main electric power plants). The two units MWe and GWh stand for Megawatt electric and Gigawatt hour, respectively.

Data Accessibility: The data sources used in this chapter are OECD statistics such as (a) IEA electricity information statistics, OECD net electrical capacity, total main activity producer plants (containing inputs), (b) IEA electricity information statistics, OECD electricity/heat supply and consumption, electricity, total net production (containing a desirable output), and (c) IEA CO_2 emissions from fuel combustion statistics, detailed CO_2 estimates, main activity electricity plants, total (containing an undesirable output; CO_2 emission). Here, IEA stands for International Energy Agency.

Table 19.1 summarizes the Malmquist indexes (IN_{t-1}^{t}) of the ten nations which are measured under natural disposability under the assumption that a frontier crossover has not occurred in the observed period (2000–2009). In a similar manner, Table 19.2 summarizes the indexes (INC_{t-1}^{t}) under the assumption that a frontier crossover has occurred between the adjacent annual periods. Note that unified efficiency measures (UEN_{t}^{R}) under natural disposability are summarized in Sueyoshi and Goto (2013c).

The index measures provide two interesting findings. One of the two findings is that the electric power industry is a huge process industry whose investment in new technology requires a large amount of capital. The speed of technology innovation on generation (as a desirable output) is relatively slow so that the Malmquist index measures fluctuate around unity until a generation system can completely utilize the new technology. Consequently, Table 19.1 (under the assumption of no frontier crossover) cannot catch the influence of technology innovation. Meanwhile, Table 19.2 (under an occurrence of the crossover between two adjacent periods) can identify such a frontier shift with the indexes more than or equal to unity although the speed and degree of technology innovation are still very modest.

According to the research of Sueyoshi and Goto (2013c), France has been utilizing nuclear energy to generate electricity and to reduce the amount of CO_2 emission. The Netherlands uses hydro and renewable generations. The performance of France indicates that the nuclear power generation may be important in developing a sustainable society with low CO_2 emission. It is indeed true that we need to pay attention to renewable energy sources, as well. However, their generation outputs are significantly small, compared with that of nuclear generation.

TABLE 19.1 Malmquist index under natural disposability: no frontier crossover

	Canada	France	Germany	Japan	Korea	Mexico	Netherlands	Spain	United Kingdom	United States
2000	1.0183	1.0191	1.0706	1.0374	1.1030	1.0403	0.9915	1.0544	1.0706	1.0587
2001	0.9708	1.0576	1.0109	1.0198	0.9764	0.9986	1.0109	0.9881	1.0256	0.9479
2002	1.0191	0.9801	1.0037	1.0109	1.0244	0.9665	0.9221	1.0150	0.9992	0.9968
2003	0.9563	0.9922	1.0074	0.9952	1.0161	0.9356	0.9115	0.9648	1.0213	1.0175
2004	0.9585	1.0144	0.9954	1.0169	1.0108	1.0536	0.9812	1.0301	1.0015	1.0277
2005	1.0159	1.0193	1.0150	1.0061	1.0234	1.0378	0.9433	1.0416	0.9388	0.9828
2006	0.9826	1.0495	0.9715	0.9789	0.9164	0.9851	0.9320	1.0127	0.9947	0.9529
2007	1.0154	0.9808	1.0374	1.0306	1.0093	1.0111	1.0285	1.0111	0.9096	1.0822
2008	1.0114	0.9804	0.9774	0.9795	0.9658	1.0261	0.9451	1.0341	0.9376	0.9717
2009	0.9378	0.9018	0.9369	0.9558	0.9844	0.9835	1.0122	1.0445	0.9788	0.9872

(a) Source: Sueyoshi and Goto (2013c).

TABLE 19.2 Malmquist index under natural disposability: frontier crossover

	Canada	France	Germany	Japan	Korea	Mexico	Netherlands	Spain	United Kingdom	United States
2000	1.0000	1.0000	1.0398	1.0184	1.0242	1.0113	1.0000	1.0263	1.0321	1.0203
2001	1.0095	1.0000	1.0163	1.0171	1.0000	1.0000	1.0000	1.0191	1.0121	1.0208
2002	1.0022	1.0000	1.0014	1.0052	1.0000	1.0112	1.0000	1.0071	1.0025	1.0029
2003	1.0143	1.0000	1.0162	1.0090	1.0025	1.0309	1.0000	1.0193	1.0133	1.0169
2004	1.0000	1.0000	1.0110	1.0153	1.0000	1.0265	1.0000	1.0150	1.0160	1.0220
2005	1.0012	1.0000	1.0149	1.0116	1.0069	1.0149	1.0000	1.0214	1.0528	1.0257
2006	1.0030	1.0000	1.0154	1.0116	1.0066	1.0020	1.0000	1.0111	1.0021	1.0246
2007	1.0016	1.0038	1.0183	1.0191	1.0000	1.0000	1.0000	1.0068	1.0133	1.0410
2008	1.0000	1.0000	1.0106	1.0100	1.0025	1.0110	1.0000	1.0180	1.0228	1.0148
2009	1.0000	1.0444	1.0402	1.0250	1.0000	1.0070	1.0000	1.0229	1.0170	1.0177

(a) Source: Sueyoshi and Goto (2013c).

A shortcut to establishing a sustainable society may depend upon the use of nuclear power generation until we can access advanced technology on renewable generation with reasonable costs. This concern on electricity generation does not imply that this chapter denies the importance of renewable energy sources. As found in the Netherlands, it is possible for us to fully utilize a combination between hydro-generation and renewable energy (e.g., solar and wind) for electricity generation and reduction of CO_2 emissions. It is easily envisioned that the use of hydro and renewable energy will be able to contribute to the development of a sustainable society. However, it is important for us to consider each nation's geographic features and historical development of generation capacities for optimal fuel mix strategy.

Table 19.3 summarizes the Malmquist indexes (IM_{t-1}^{t}) of the ten nations which are measured under managerial disposability under the assumption that a crossover of efficiency frontiers has not occurred in the observed periods (2000-2009). In a similar manner, Table 19.4 summarizes the indexes (IMC_{t-1}^{t}) under the assumption that a frontier crossover has occurred between two adjacent (t–1 th and t th) periods. Note that unified efficiency measures (UEM_{t}^{R}) under managerial disposability are summarized in Sueyoshi and Goto (2013c).

The index measures in Tables 19.3 and 19.4 provide two interesting findings. One of the two findings is that the electric power industry has a time lag to establish eco-technology innovation on CO_2 reduction because the electricity generation belongs to a capital-intensive industry so that the technology innovation needs a large amount of capital investment to reduce the amount of CO_2 emission which is required by the government of each nation. Such a unique feature influences the index measures in such a manner that they fluctuate around unity, as in Table 19.3, and the technology innovation, or a frontier shift, becomes very modest, as found in Table 19.4.

19.7 SUMMARY

This chapter discussed a use of the Malmquist index to measure a frontier shift among different periods. A contribution of this chapter was that it made an analytical linkage between the index measurement and the concept of natural or managerial disposability. The Malmquist index measure, discussed in this chapter, was separated into twelve different subcomponents (6 subcomponents x 2 disposability concepts). All subcomponents provided us with information on how much the frontier shift occurred among different periods.

Contributions of the proposed Malmquist index are summarized as follows: The proposed approach incorporates both desirable and undesirable outputs in DEA formulations under natural or managerial disposability. A frontier shift is measurable under the two different disposability concepts within the conceptual

TABLE 19.3 Malmquist index under managerial disposability: no frontier crossover

	Canada	France	Germany	Japan	Korea	Mexico	Netherlands	Spain	United Kingdom	United States
2000	1.2223	1.1516	1.0881	1.0736	1.0924	1.0927	1.0203	1.1655	1.0667	1.1079
2001	1.3170	1.2622	1.2583	1.2377	1.2746	1.2589	1.1874	1.3317	1.2311	1.2565
2002	0.8953	0.9208	0.9101	0.9101	0.9101	0.9101	0.9101	0.9101	0.9101	0.9101
2003	0.9169	0.9363	0.9100	0.9100	0.9100	0.9100	0.9100	0.9100	0.9100	0.9100
2004	1.0588	1.0565	1.0776	1.0776	1.0776	1.0776	1.0776	1.0776	1.0776	1.0776
2005	0.8773	0.8509	0.8390	0.8557	0.8316	0.8584	0.8884	0.8203	0.8624	0.8715
2006	1.2951	1.2093	1.1493	1.1350	1.1564	1.1309	1.1052	1.1655	1.1271	1.1203
2007	0.9979	0.9834	0.9391	0.9342	0.9417	0.9304	0.9203	0.9435	0.9292	0.9268
2008	0.9372	0.9251	0.9964	1.0141	0.9940	1.0274	1.0632	0.9783	1.0328	1.0400
2009	0.9059	0.9054	0.9354	0.9366	0.9319	0.9375	0.9435	0.9376	0.9392	0.9400

(a) Source: Sueyoshi and Goto (2013c).

TABLE 19.4 Malmquist index under managerial disposability: frontier crossover

	Canada	France	Germany	Japan	Korea	Mexico	Netherlands	Spain	United Kingdom	United States
2000	1.1056	1.0416	1.0429	1.0363	1.0395	1.0442	1.0099	1.0769	1.0329	1.0494
2001	1.1476	1.0812	1.1201	1.1133	1.1283	1.1231	1.0878	1.1524	1.1094	1.1329
2002	1.0568	1.0258	1.0482	1.0482	1.0482	1.0482	1.0482	1.0482	1.0482	1.0482
2003	1.0443	1.0189	1.0483	1.0483	1.0483	1.0483	1.0483	1.0483	1.0483	1.0483
2004	1.0290	1.0178	1.0381	1.0381	1.0381	1.0381	1.0381	1.0381	1.0381	1.0381
2005	1.0676	1.0379	1.0930	1.0820	1.0973	1.0801	1.0596	1.1022	1.0768	1.0713
2006	1.1380	1.0627	1.0725	1.0653	1.0753	1.0642	1.0504	1.0796	1.0620	1.0586
2007	1.0010	1.0010	1.0316	1.0341	1.0307	1.0368	1.0418	1.0293	1.0374	1.0386
2008	1.0330	1.0253	1.0397	1.0364	1.0383	1.0366	1.0313	1.0453	1.0353	1.0346
2009	1.0506	1.0344	1.0322	1.0329	1.0361	1.0328	1.0287	1.0328	1.0315	1.0312

(a) Source: Sueyoshi and Goto (2013c).

framework of Stage I (A) and (B) of Figure 15-6. This type of unique feature cannot be found in the other indexes.

The frontier shift measurement has three drawbacks. First, the proposed approach belongs to only Stage I, not attaining Stages II and III, of Figure 15-6 in which disposability concepts are unified under the assumption that undesirable outputs are by-products of desirable outputs. Moreover, an occurrence of congestion under natural or managerial disposability is the most important concept in DEA environmental assessment. See Chapters 15 and 21. This chapter does not incorporate the occurrence in a time horizon. Second, the proposed approach must assume constant RTS and DTS in the formulations. If the assumption is dropped, then the approach must face an infeasible solution. See Chapter 20. Finally, the proposed approach assumes that all production factors are strictly positive, so that it cannot directly handle an occurrence of zero and/or negative values in a data set. See Chapters 26 and 27 for a description on how to deal with such a data set with zero and/or negative values. No study has been done on such a data set in a time horizon.

The proposed DEA approach identified the three empirical findings on energy, all of which were useful in the green growth development of economy to establish a sustainable society. First, the electric power industry had a time lag until eco-technology innovation really influenced the amount of generation and the reduction of CO_2 emission. That was because the industry needed a large capital investment to introduce new technology for generation and environmental protection. Therefore, it was necessary to consider the existence of a frontier crossover under the assumption that the frontier was formulated by a wider range of combinations of advanced generation technology in a time horizon. Second, nuclear generation as found in France and hydro and renewable energy as found in the Netherlands were important for the development of a sustainable society although the former was associated with a very high level of risk and the latter had a limited generation capability. Finally, the industry had been making a corporate effort to reduce the amount of CO_2 emission by utilizing nuclear power and renewable energy although the degree and speed of technology innovation was relatively modest.

Here, it is important to note that nuclear power generation is always associated with very high risk, as found in the disaster at Fukushima Daiichi in March, 2011. See Chapter 26 that discusses the fuel mix strategy of 33 OECD nations after the disaster at the Fukushima Daiichi nuclear power plant. It is a major policy issue for all nations to pay serious attention to various types of security issues in the operation of nuclear power plants. Meanwhile, the utilization of hydro power is limited in its growth because many industrial nations have already developed dams which can generate a large amount of electricity. The development opportunity for large hydro power plants is now limited in many industrial nations. As a result, many nations recently pay attention to renewable energy sources[11].

Finally, it is necessary to discuss the importance of balancing between economic prosperity and environmental protection in order to direct each nation toward a sustainable society which can attain a green growth by paying serious attention to environmental protection. To attain a sustainable society, we look forward to seeing future technology development on renewable energy. It is also expected that technology development produces an economic benefit for renewable energy in such a manner that it can compete with the other conventional generation technologies without relying on any financial support scheme such as FIT. Until the development of such a sustainable society, it may be necessary for us to utilize the nuclear power generation as an important component of a fuel mix under a rigorous safety standard in a short time horizon, not a long-term horizon, where it is hoped that renewable power generation will be more efficient than nuclear power generation.

APPENDIX

There are various types of total factor productivity (TFP) measures. They include all products and services produced and account for all of the resources used to produce these goods and services. The TFP is thus formulated by the ratio of all of desirable outputs produced over all of the inputs employed to produce them. The traditional productivity indexes, such as Laspeyres, Paasche, Fisher and Törnqvist, use price information to allow us to add up the products (desirable outputs) and resources (inputs).

Table 19.A1 briefly summarizes previous indexes. It is clearly understood that each index has strengths and drawbacks. This chapter does not provide a

[11] To promote the generation by renewable energy technologies, many European nations, the United States and Japan have introduced a FIT program in which electric power companies needed to purchase all electricity produced by renewable energy sources (e.g., wind power and solar photovoltaic generations) by a fixed price. See Sueyoshi and Goto (2014d) for their concern on FIT for solar photovoltaic generation. It is easily expected that the FIT will promote the generation by renewable energy. However, the energy density of renewable energy is low and the cost of renewable energy is still expensive. Furthermore, the energy is influenced by external factors (e.g., a weather change). The low energy density of renewable technologies implies that they need to overcome a technological difficulty to efficiently produce a large amount of electricity as a main generation source. All renewable energy technologies have not yet overcome such a difficulty at the level that they become large independent generation sources. Moreover, the nuclear power generation is always associated with a technical issue regarding how to handle nuclear waste disposal. No country promoting nuclear power generation has a clear answer on the nuclear waste disposal. An exception is Finland, in which the government has recently decided the final disposal site for nuclear waste although this chapter does not examine the performance of the nation. The issue will be a major policy problem to all nations in future which are either currently operating or planning nuclear power plants. The energy issue discussed in this chapter will be reexamined in Chapter 26 from the viewpoint of the Japanese energy mix strategy after the disaster of the Fukushima Daiichi nuclear power plant.

Table 19.A1 Popular indexes for productivity measurement

Index	Description	References
Laspeyres productivity index	Price-dependent productivity index. The index consists of an output quantity index and an input quantity index. Laspeyres quantity index between two periods (t = 0, 1) uses period 0's prices as weights.	Laspeyres (1871)[12]
Paasche productivity index	Price-dependent productivity index. The index consists of an output quantity index and an input quantity index. Paasche quantity index between two periods (t = 0 and 1) uses period 1's prices as weights.	Paasche (1874)[13]
Törnqvist productivity index	Price-dependent productivity index. The index consists of an output quantity index and an input quantity index.	Törnqvist (1936)[14], Diewert (1976)[15]
Fisher productivity index	Price-dependent productivity index. The index consists of the Fisher ideal output quantity index and the Fisher ideal input quantity index.	Diewert (1992)[16], Fisher(1922)[17]
Bennet–Bowley productivity indicator	Price-weighted arithmetic mean of the difference in output and input changes in quantity.	Bennet(1920)[18], Bowley (1919)[19]
Malmquist productivity index	Distance function-based productivity measure. Input-oriented index and output-oriented index. Explicitly links efficiency and productivity.	Caves *et al.*, (1982)[20], Färe *et al.* (1995)[21]

(Continued)

[12] Laspeyres, E. (1871) Die Berechnung einer mittleren Waarenpreissteigerung, *Jahrbücher für Nationalökonomie und Statistik*, **16**, 296–314.

[13] Paasche, H. (1874) Über die Preisentwicklung der letzten Jahre nach den Hamburger Borsennotirungen, *Jahrbücher für Nationalökonomie und Statistik*, **12**, 168–178.

[14] Törnqvist, L. (1936) The Bank of Finland's Consumption Price Index, *Bank of Finland Monthly Bulletin*, **10**, 1–8.

[15] Diewert, W.E. (1976) Exact and superlative index numbers, *Journal of Econometrics*, **4**, 115–145.

[16] Diewert, W.E. (1992) Fisher ideal output, input and productivity indexes revisited, *Journal of Productivity Analysis*, **3**(3), 211–248.

[17] Fisher, I. (1922) *The Making of Index Numbers*, Houghton-Mifflin, Boston.

[18] Bennet, T.L. (1920) The theory of measurement of change in cost of living, *Journal of the Royal Statistical Society*, **83**, 455–462.

[19] Bowley, A.L. (1919) The measurement of changes in the cost of living, *Journal of the Royal Statistical Society*, **82**, 343–372.

[20] Caves, D., Christensen, L. and Diewert, W.E. (1982) The economic theory of index numbers and the measurement of input, output, and productivity, *Econometrica*, **50**(6), 1393–1414.

[21] Färe, R., Grosskopf, S., Lindgren, B. and Roos, P. (1995) Productivity developments in Swedish hospitals: a Malmquist output index approach, in: Charnes, A., Cooper, W.W., Lewin, A.Y. and Seiford, L.M. (eds), *Data Envelopment Analysis: Theory, Methodology and Applications*. Kluwer, Boston, pp. 253–272.

Table 19.A1 (Continued)

Index	Description	References
Hicks–Moorsteen productivity index	Distance function-based productivity measure. Ratio of a Malmquist quantity index of outputs to a Malmquist quantity index of inputs. Explicitly links efficiency and productivity.	Bjurek(1994)[22], Bjurek (1996)[23]
Luenberger productivity indicator	Distance function-based productivity measure constructed from the directional distance functions of technology, which simultaneously adjusts the outputs and inputs in a direction chosen by the investigator; and has an additive structure. Expresses differences rather than ratios of distance functions. Explicitly links efficiency and productivity.	Chambers and Pope (1996)[24], Chambers (2002)[25]

(a) Outputs in the table imply desirable outputs. No previous article has separated outputs into desirable and undesirable categories. (b) This chapter does not use the econometric distance function for measuring Malmquist index. The separation between desirable and undesirable outputs is incorporated into the proposed DEA approach in a time horizon. Thus, the proposed approach follows the conceptual framework of the Malmquist index measurement, but being methodologically different from the econometric approach based upon a distance function.

conventional description on them. Rather, this chapter describes two concerns on them from the perspective of implementing the proposed DEA environmental assessment. One concern is that desirable and undesirable outputs should be incorporated in the computational framework. Then, it is necessary for us to discuss how to unify them in measuring indexes. The other concern is that the proposed formulations for the assessment need to incorporate a possible occurrence of undesirable or desirable congestion. The desirable congestion is important in measuring eco-technology development in a time horizon. The previous index theory did not incorporate such two important concerns. Thus, they are very limited in their applicability when discussing the issue of sustainability.

[22] Bjurek, H. (1994) Essays on efficiency and productivity change with application to public service production, *Economiska Studier*, **52**.

[23] Bjurek, H. (1996) The Malmquist total factor productivity index, *Scandinavian Journal of Economics*, **98**, 303–313.

[24] Chambers, R.G. and Pope, R.D. (1996) Aggregate productivity measures, *American Journal of Agricultural Economics*, **78**(5), 1360–1365.

[25] Chambers, R.G. (2002) Exact nonradial input, output, and productivity measurement, *Economic Theory*, **20**, 751–765.

20

RETURNS TO SCALE AND DAMAGES TO SCALE

20.1 INTRODUCTION

The purpose of this chapter[1] is to introduce an approach regarding how to measure returns to scale (RTS) and damages to scale (DTS) within a computational framework of DEA environmental assessment. Chapter 8 introduced concepts and methodologies regarding RTS within the conventional framework of DEA by assuming the non-existence of undesirable outputs. As discussed in Chapter 15, an important feature of DEA environmental assessment is that it incorporates undesirable outputs in the computational framework. This chapter, as an extension of Chapter 8, discusses how to measure RTS under natural disposability and DTS under managerial disposability by considering both desirable and undesirable outputs in addition to inputs. This chapter provides a new computational aspect on environmental assessment which cannot be found in Chapter 8.

To discuss the practicality of the proposed approach, this chapter compares Japanese chemical and pharmaceutical firms by considering their energy utilization and environmental protection. This chapter will be further explored in Chapter 23.

[1]This chapter is partly based upon the article: Sueyoshi, T. and Goto, M. (2014a) DEA radial measurement for environmental assessment: A comparative study between Japanese chemical and pharmaceutical firms. *Applied Energy*, **115**, 502–513.

Environmental Assessment on Energy and Sustainability by Data Envelopment Analysis,
First Edition. Toshiyuki Sueyoshi and Mika Goto.

The remainder of this chapter is organized as follows: Section 20.2 describes underlying concepts related to RTS and DTS. Section 20.3 discusses an approach on how to measure RTS and DTS by the non-radial models. Section 20.4 extends it to the radial measurement. Section 20.5 applies the proposed DEA approach to compare between Japanese chemical and pharmaceutical firms in terms of the types of RTS and DTS along with unified efficiency measures. Section 20.6 summarizes this chapter.

20.2 UNDERLYING CONCEPTS

20.2.1 Scale Elasticity

Returning to Chapter 8, this chapter describes the measurement on scale elasticity (e_g) between an input (x) and a desirable output (g) within DEA environmental assessment. Let us consider a simple case in which a supporting hyperplane is mathematically expressed by $vx - ug + wb + \sigma = 0$. For our descriptive convenience, all production factors consist of a single component. Here, the undesirable output (b) is additionally incorporated in the equation, which cannot be found in Chapter 8. Later, the supporting hyperplane is mathematically discussed in a case where each production factor has multiple components. The symbols v, u and w are parameters of a supporting hyperplane, indicating the degree of each slope regarding the supporting hyperplane. All the parameters are positive in their signs. An exception is σ that is unrestricted in the sign and indicates an intercept of the supporting hyperplane.

Using the supporting hyperplane, we obtain $\frac{dg}{dx} = \frac{v}{u}$ and $\frac{g}{x} = \frac{v}{u} + \frac{\sigma + wb}{ux}$. Consequently, the scale elasticity between g and x is measured by $e_g = \left(\frac{dg}{dx}\right) \Big/ \left(\frac{g}{x}\right) = 1 \Big/ \left(1 + \frac{\sigma + wb}{vx}\right)$. The degree of the scale elasticity (e_g) related to a desirable output is classified by σ and w by the following rule:

$$\begin{aligned} &\text{(a) } e_g > 1 \leftrightarrow \text{Increasing RTS} \leftrightarrow \sigma + wb < 0, \\ &\text{(b) } e_g = 1 \leftrightarrow \text{Constant RTS} \leftrightarrow \sigma + wb = 0 \text{ and} \\ &\text{(c) } e_g < 1 \leftrightarrow \text{Decreasing RTS} \leftrightarrow \sigma + wb > 0. \end{aligned} \tag{20.1}$$

Here, it is important to note two concerns regarding Equation (20.1). One of the two concerns is that the parameter v should be strictly positive. In the other case (e.g., zero), it is difficult for us to determine the scale elasticity. That is, if v is zero, then e_g becomes zero on a DMU to be examined. Thus, the input-related dual variable, indicating the slope of a supporting hyperplane, should be positive, not zero, in the measurement of e_g. The other concern is that e_g is determined by not only σ but also the influence (wb) of the undesirable output. Such an influence on e_g cannot be found in the conventional use of DEA, as discussed in Chapter 8, where e_g is determined by only σ. Such is the difference between the two approaches.

Next, to describe the scale elasticity (e_b) between an input and an undesirable output, let us consider that a supporting hyperplane is $-vx - ug + wb + \sigma = 0$ where the three production factors consist of a single component. Then, we have $\frac{db}{dx} = \frac{v}{w}$ and $\frac{b}{x} = \frac{v}{w} - \frac{\sigma - ug}{wx}$. Consequently, the scale elasticity related to an undesirable output is measured by $e_b = \left(\frac{db}{dx}\right) \Big/ \left(\frac{b}{x}\right) = 1 \Big/ \left(1 - \frac{\sigma - ug}{vx}\right)$. The scale elasticity ($e_b$) related to an input and an undesirable output is determined by σ and u as follows:

$$
\begin{aligned}
&\text{(a) } e_b > 1 \leftrightarrow \text{Increasing DTS} \leftrightarrow \sigma - ug > 0, \\
&\text{(b) } e_b = 1 \leftrightarrow \text{Constant DTS} \leftrightarrow \sigma - ug = 0 \text{ and} \\
&\text{(c) } e_b < 1 \leftrightarrow \text{Decreasing DTS} \leftrightarrow \sigma - ug < 0.
\end{aligned}
\tag{20.2}
$$

The above classification needs that the input-related parameter v should be positive. Otherwise, it is difficult for us to measure the scale elasticity related to the undesirable output, as discussed previously. In the case of e_b, the type is determined by not only σ but also ug, indicating an influence from a desirable output. This classification rule on e_b is different from e_g. See Equations (20.1) and (20.2).

Finally, it is important to add two concerns. One of the two concerns is that all parameters for a supporting hyperplane are expressed by dual variables and they are expected to be strictly positive in the elasticity measurement. An exception is σ because it indicates the intercept of a supporting hyperplane. The requirement on dual variables will be changed if we are interested in an occurrence of congestion. See Chapters 21 and 23. The other concern is that the concept of "scale elasticity" regarding desirable and undesirable outputs discussed in this section is related to a single desirable output and a single undesirable output. However, the concept of "scale economies" and "scale damages," which will be discussed in Sections 20.3 and 20.4, are related to multiple desirable and undesirable outputs. These concepts are measured by the proposed formulations for DEA environmental assessment.

20.2.2 Differences between RTS and DTS

Before analytically discussing RTS and DTS in the framework of radial and non-radial models for DEA environmental assessment, this chapter needs to intuitively explain mathematical similarities and differences as well as their economic implications. They have opposite implications in many economic contexts concerning scale economies and scale damages. See Chapter 8 on scale economies. For example, increasing RTS under natural disposability implies that a unit increase in inputs produces desirable outputs more than their proportional increases without worsening undesirable outputs. This result indicates that, if a DMU increases its operational size, then it becomes more productive on desirable outputs under natural disposability. In this case, a large DMU may outperform small ones in operational performance. Thus, it is recommended that the DMU should increase its operational size to enhance the unified efficiency under natural disposability.

In contrast, in the case of managerial disposability, increasing DTS implies an opposite result in such a manner that a unit increase in inputs yields undesirable outputs more than their proportional increases without worsening desirable outputs. In other words, if a DMU increases its operational size, then it produces more pollution (so, more damage) than before the size increase. Thus, it is recommended that the DMU should decrease its operational size in order to improve the unified efficiency under managerial disposability. The relationship between RTS and DTS, along with corporate strategy on the operational size, is summarized in Table 20.1.

As summarized in Table 20.1, as an alternative strategy, the type of DTS suggests whether a DMU should introduce eco-technology innovation within the operation. For example, an increasing DTS indicates that the DMU needs to introduce eco-technology innovation to enhance its environmental performance. If the DMU has a large amount of capital accumulation, the investment for

TABLE 20.1 Strategies measured by RTS and DTS

RTS	RTS-based strategy	DTS	DTS-based strategy
Increasing	It is recommended that a DMU should "increase" the current size.	Increasing	It is recommended that a DMU should "decrease" the current size. As an alternative, it is strongly recommended that the DMU should utilize eco-technology innovation for environmental protection.
Constant	It is recommended that a DMU should "maintain" the current size.	Constant	It is not recommended, but acceptable, that a DMU can "maintain" the current size. As an alternative, it is recommended that the DMU should utilize eco-technology innovation for environmental protection.
Decreasing	It is recommended that a DMU should "decrease" the current size.	Decreasing	It is not recommended, but acceptable, that a DMU can "increase" the current size.

(a) A decrease in undesirable outputs is always the best DTS-based strategy. The concept of most productive scale size (MPSS), often discussed in the DEA community, may exist in a conventional use of RTS. However, there is no such concept in DTS, corresponding to the MPSS in DEA environmental assessment. The best strategy is to produce no undesirable output. Thus, the concept of MPSS has a very limited implication in DEA environmental assessment. (b) Eco-technology innovation is essential in reducing an amount of undesirable outputs if a firm has enough capital. If the firm does not have capital accumulation, it must decrease its operation size. See Sueyoshi and Goto (2014a).

eco-technology innovation is more practical than the size reduction in terms of enhancing environmental performance. The implication on increasing DTS may be applied to the case of constant DTS, as well, because it indicates an initial stage of increasing DTS (so, more pollution). In contrast, the eco-technology innovation may be less important in the case of decreasing DTS because a size increase in inputs less proportionally produces undesirable outputs.

20.3 NON-RADIAL APPROACH

20.3.1 Scale Economies and RTS under Natural Disposability

Returning to Chapter 16, we can find the following complementary slackness conditions (CSCs) between Models (16.6) and (16.9) on optimality:

$$\begin{aligned}
&\lambda_j^*\left(\sum_{i=1}^{m} v_i^* x_{ij} - \sum_{r=1}^{s} u_r^* g_{rj} + \sum_{f=1}^{h} w_f^* b_{fj} + \sigma^*\right) = 0\ (j=1,\ldots,n),\\
&d_i^{x-*}\left(v_i^* - R_i^x\right) = 0\ (i=1,\ldots,m),\ d_r^{g*}\left(u_r^* - R_r^g\right) = 0\ (r=1,\ldots,s)\ \text{and}\\
&d_f^{b*}\left(w_f^* - R_f^b\right) = 0\ (f=1,\ldots,h).
\end{aligned} \tag{20.3}$$

For the first step to determine the type of RTS, this chapter needs to specify the location of a supporting hyperplane in the case where the three production factors have multiple components because the hyperplane is directly related to identifying the RTS measurement. Analytically, a supporting hyperplane(s) of the k-th DMU under natural disposability is expressed by $\sum_{i=1}^{m} v_i x_i - \sum_{r=1}^{s} u_r g_r + \sum_{f=1}^{h} w_f b_f + \sigma = 0$, where dual variables, v_i ($i = 1,\ldots,m$), u_r ($r = 1,\ldots,s$) and w_f ($f = 1,\ldots,h$), are parameters for indicating the direction of a supporting hyperplane(s) and σ indicates an intercept of the supporting hyperplane. The parameters are all unknown. Therefore, it is necessary to determine them by the following equation:

$$\sum_{i=1}^{m} v_i x_{ij} - \sum_{r=1}^{s} u_r g_{rj} + \sum_{f=1}^{h} w_f b_{fj} + \sigma,\ \ j \in RS_k \tag{20.4}$$

where RS_k stands for a reference set of the k-th DMU whose performance is measured by Model (16.6).

It is important to remind that Section 20.2 specifies the supporting hyperplane by $vx - ug + wb + \sigma = 0$ in the case of a single component of the three production factors. Meanwhile, Equation (20.4) specifies the hyperplane as $\sum_{i=1}^{m} v_i x_i - \sum_{r=1}^{s} u_r g_r + \sum_{f=1}^{h} w_f b_f + \sigma = 0$ in the case of their multiple components. Equation (20.4) can be easily obtained from the first group of CSCs, or Equations (20.3).

The concept of "scale economies (SEC)" corresponds to the scale electricity (e_g) if the three production factors have multiple components. If the k-th DMU is efficient in unified performance under natural disposability, then the degree of SEC is determined by

$$\begin{aligned} SEC &= \left(\sum_{i=1}^{m} v_i^* x_{ik}\right) \bigg/ \left(\sum_{r=1}^{s} u_r^* g_{rk}\right) = \left(\sum_{i=1}^{m} v_i^* x_{ik}\right) \bigg/ \left(\sum_{i=1}^{m} v_i^* x_{ik} + \sum_{f=1}^{h} w_f^* b_{fk} + \sigma^*\right) \\ &= 1 \bigg/ \left[1 + \left(\sigma^* + \sum_{f=1}^{h} w_f^* b_{fk}\right) \bigg/ \left(\sum_{i=1}^{m} v_i^* x_{ik}\right)\right] \\ &= 1 \bigg/ \left[1 + \left(\sigma^* + \zeta_k^{b*}\right) \bigg/ \left(\sum_{i=1}^{m} v_i^* x_{ik}\right)\right] \end{aligned} \tag{20.5}$$

where $\zeta_k^{b*} = \sum_{f=1}^{h} w_f^* b_{fk}$. The variable implies an influence from undesirable outputs on the degree of SEC. Equation (20.5) incorporates $\sum_{i=1}^{m} v_i^* x_{ik} - \sum_{r=1}^{s} u_r^* g_{rk} + \sum_{f=1}^{h} w_f^* b_{fk} + \sigma^* = 0$ in the reformulation, indicating the degree of unified inefficiency in Model (16.9).

In contrast, if the k-th DMU is inefficient in unified performance under natural disposability, then the inefficient DMU needs to be projected onto an efficiency frontier. In the case, the degree of SEC can be determined by

$$\begin{aligned} SEC &= \left[\sum_{i=1}^{m} v_i^* \left(x_{ik} - d_i^{x-*}\right)\right] \bigg/ \left[\sum_{r=1}^{s} u_r^* \left(g_{rk} + d_r^{g*}\right)\right] \\ &= \left[\sum_{i=1}^{m} v_i^* \left(x_{ik} - d_i^{x-*}\right)\right] \bigg/ \left[\sum_{r=1}^{s} \left(u_r^* g_{rk} + u_r^* d_r^{g*}\right)\right] \\ &= \left[\sum_{i=1}^{m} v_i^* \left(x_{ik} - d_i^{x-*}\right)\right] \bigg/ \left[\sum_{i=1}^{m} v_i^* \left(x_{ik} - d_i^{x-*}\right) + \sum_{f=1}^{h} w_f^* \left(b_{fk} - d_f^{b*}\right) + \sigma^*\right] \\ &= 1 \bigg/ \left\{1 + \left[\left(\sum_{f=1}^{h} w_f^* \left(b_{fk} - d_f^{b*}\right) + \sigma^*\right) \bigg/ \sum_{i=1}^{m} v_i^* \left(x_{ik} - d_i^{x-*}\right)\right]\right\} \\ &= 1 \bigg/ \left\{1 + \left[\left(\sigma^* + \sum_{f=1}^{h} w_f^* b_{fk}\right) \bigg/ \sum_{i=1}^{m} v_i^* \left(x_{ik} - d_i^{x-*}\right)\right]\right\} \\ &= 1 \bigg/ \left\{1 + \left[\left(\sigma^* + \zeta_k^{b*}\right) \bigg/ \sum_{i=1}^{m} v_i^* \left(x_{ik} - d_i^{x-*}\right)\right]\right\}. \end{aligned} \tag{20.6}$$

Equation (20.6) incorporates $\sum_{i=1}^{m} R_i^x d_i^{x-*} + \sum_{r=1}^{s} R_r^g d_r^{g*} + \sum_{f=1}^{h} R_f^b d_f^{b*} = \sum_{i=1}^{m} v_i^* x_{ik} - \sum_{r=1}^{s} u_r^* g_{rk} + \sum_{f=1}^{h} w_f^* b_{fk} + \sigma^*$ from the optimal objective value of Models (16.6) and (16.9). The reformulation in Equation (20.6) has three assumptions. First, the inefficient DMU has positive slacks on all production factors so that $v_i^* = R_i^x$ for all i, $u_r^* = R_r^g$ for all r and $w_f^* = R_f^b$ for all f because of CSCs, expressed by Equation (20.3). Second, the observed inputs are larger than their slacks so that their differences are positive. Finally, Equation (20.6) considers $\zeta_k^{b^*} = \sum_{f=1}^{h} w_f^* (b_{fk} - d_f^{b*}) \cong \sum_{f=1}^{h} w_f^* b_{fk}$ because this chapter is interested in the type of RTS on an efficiency frontier that is mainly determined by inputs and desirable outputs. Hence, $d_f^{b*} \cong 0$ is assumed for all f for our computational tractability, indicating that all slacks related to undesirable outputs are positive but very close to zero on the efficiency frontier. Here, the symbol $\cong$ indicates such closeness between the two parts.

Assuming both a unique projection of an inefficient DMU onto an efficiency frontier and a unique reference set for the projected DMU, this chapter determines the upper and lower bounds of the dual variable (σ) and the influence factor (ζ_k^b) from undesirable outputs by using the following model:

$$
\begin{aligned}
&\text{Maximize/Minimize} \quad \sigma + \zeta_k^b \\
&s.t. \quad \text{all constraints in Models (16.6) and (16.9),} \\
&\qquad \zeta_k^b = \sum_{f=1}^{h} w_f b_{fk}, \\
&\qquad \sum_{i=1}^{m} R_i^x d_i^x + \sum_{r=1}^{s} R_r^g d_r^g + \sum_{f=1}^{h} R_f^b d_f^b \\
&\qquad = \sum_{i=1}^{m} v_i x_{ik} - \sum_{r=1}^{s} u_r g_{rk} + \sum_{f=1}^{h} w_f b_{fk} + \sigma \text{ and} \\
&\qquad \zeta_k^b \geq 0.
\end{aligned}
\tag{20.7}
$$

The first additional constraint indicates an influence factor from undesirable outputs. The second one is because the objective function of Model (16.6) equals that of Model (16.9) on optimality.

The upper bound ($\bar{\vartheta}^*$) and the lower bound ($\underline{\vartheta}^*$) of the objective value are obtained from the maximization and minimization of Model (20.7), respectively. An optimal solution of Models (16.6) and (16.9) is feasible in Model (20.7) and vice versa.

Based upon these upper and lower bounds, the type of RTS of the k-th DMU is determined by the following classification:

$$
\begin{aligned}
&\text{(a) Decreasing RTS} \leftrightarrow \bar{\vartheta}^* \geq \underline{\vartheta}^* > 0, \\
&\text{(b) Constant RTS} \leftrightarrow \bar{\vartheta}^* \geq 0 \geq \underline{\vartheta}^* \text{ and} \\
&\text{(c) Increasing RTS} \leftrightarrow 0 > \bar{\vartheta}^* \geq \underline{\vartheta}^*.
\end{aligned}
\tag{20.8}
$$

It is important to note that the surface of an efficiency frontier is a set of piecewise linear polygonal faces under variable RTS. Therefore, the constant RTS is defined by $\bar{\vartheta}^*$ (positive) and $\underline{\vartheta}^*$ (negative), thus covering the range shaped between decreasing and increasing RTS.

20.3.2 Scale Damages and DTS under Managerial Disposability

To determine the degree and type of DTS under managerial disposability, this chapter describes the concept of scale damages (SDA). The degree of SDA depends upon the efficiency or inefficiency status of the k-th DMU. If the DMU is efficient, then the degree of SDA is determined as follows:

$$SDA = \sum_{i=1}^{m} v_i^* x_{ik} \Big/ \sum_{f=1}^{h} w_f^* b_{fk} = \sum_{i=1}^{m} v_i^* x_{ik} \Bigg/ \left(\sum_{i=1}^{m} v_i^* x_{ik} + \sum_{r=1}^{s} u_r^* g_{rk} - \sigma^* \right)$$
$$= 1 \Bigg/ \left[1 - \left(\left(\sigma^* - \zeta_k^{g*} \right) \Big/ \sum_{i=1}^{m} v_i^* x_{ik} \right) \right]. \quad (20.9)$$

The optimal condition, or $\sum_{f=1}^{h} w_f^* b_{fk} = \sum_{i=1}^{m} v_i^* x_{ik} + \sum_{r=1}^{s} u_r^* g_{rk} - \sigma^*$, of Model (16.15) is incorporated into Equation (20.9) because the optimal objective value becomes zero when the DMU is efficient. All optimal dual variables are obtained from Model (16.15). The new variable $\zeta_k^{g*} = \sum_{r=1}^{s} u_r^* g_{rk}$ is incorporated as an influence factor of desirable outputs in Equation (20.9).

If the k-th DMU is inefficient, then the DMU needs to be projected on an efficiency frontier. The degree of SDA can be determined on its projected point. That is,

$$SDA = \left[\sum_{i=1}^{m} v_i^* \left(x_{ik} + d_i^{x*} \right) \right] \Bigg/ \left[\sum_{f=1}^{h} w_f^* \left(b_{fk} - d_f^{b*} \right) \right]. \quad (20.10)$$

Following the procedure similar to SEC in the previous subsection, the degree of SDA is reformulated as follows:

$$SDA = 1 \Bigg/ \left\{ 1 - \left[\left(\sigma^* - \zeta_k^{g*} \right) \Bigg/ \sum_{i=1}^{m} v_i^* \left(x_{ik} + d_i^{x*} \right) \right] \right\}. \quad (20.11)$$

Assuming both a unique projection of an inefficient DMU onto an efficiency frontier and a unique reference set for the projected DMU, this chapter can determine the upper and lower bounds of the dual variable (σ) and the influence factor (ζ_k^g) from desirable outputs by using the following model:

$$\begin{aligned}
&\text{Maximize/Minimize } \sigma - \zeta_k^g \\
&\quad s.t. \ \text{all constraints in Models } (16.12) \text{ and } (16.15), \\
&\quad \zeta_k^g = \sum_{r=1}^{s} u_r g_{rk}, \\
&\quad \sum_{i=1}^{m} R_i^x d_i^x + \sum_{r=1}^{s} R_r^g d_r^g + \sum_{f=1}^{h} R_f^b d_f^b \qquad (20.12) \\
&\quad = -\sum_{i=1}^{m} v_i x_{ik} - \sum_{r=1}^{s} u_r g_{rk} + \sum_{f=1}^{h} w_f b_{fk} + \sigma \text{ and} \\
&\quad \zeta_k^g \geq 0.
\end{aligned}$$

The first additional constraint indicates the influence from desirable outputs. The second one is due to the condition that the objective function of Model (16.12) equals that of Model (16.15) on optimality.

The upper bound ($\bar{\upsilon}^*$) and the lower bound ($\underline{\upsilon}^*$) of the objective value are obtained from the maximization and minimization of Model (20.12), respectively. An optimal solution of Models (16.12) and (16.15) is feasible in Model (20.12) and vice versa.

Based upon the upper and lower bounds, the type of DTS of the k-th DMU is determined by the following classification:

$$\begin{aligned}&\text{(a) Increasing DTS} \leftrightarrow \bar{\upsilon}^* \geq \underline{\upsilon}^* > 0,\\&\text{(b) Constant DTS} \leftrightarrow \bar{\upsilon}^* \geq 0 \geq \underline{\upsilon}^* \text{ and}\\&\text{(c) Decreasing DTS} \leftrightarrow 0 > \bar{\upsilon}^* \geq \underline{\upsilon}^*.\end{aligned} \tag{20.13}$$

20.4 RADIAL APPROACH

20.4.1 Scale Economies and RTS under Natural Disposability

The following CSCs always satisfy between Models (17.5) and (17.7) on optimality:

$$\begin{aligned}&\lambda_j^*\left(\sum_{i=1}^{m} v_i^* x_{ij} - \sum_{r=1}^{s} u_r^* g_{rj} + \sum_{f=1}^{h} w_f^* b_{fj} + \sigma^*\right) = 0\ (j=1,\ldots,n),\\&d_i^{x*}\left(v_i^* - \varepsilon_s R_i^x\right) = 0\ (i=1,\ldots,m),\ d_r^{g*}\left(u_r^* - \varepsilon_s R_r^g\right) = 0\ (r=1,\ldots,s) \text{ and}\\&d_f^{b*}\left(w_f^* - \varepsilon_s R_f^b\right) = 0\ (f=1,\ldots,h).\end{aligned} \tag{20.14}$$

As discussed in Section 20.3, the location of a supporting hyperplane(s), in the case of multiple components of the three production factors, is directly associated with determining the type and degree of RTS. Analytically, the supporting hyperplane of the k-th DMU under natural disposability is expressed by $\sum_{i=1}^{m} v_i x_i - \sum_{r=1}^{s} u_r g_r + \sum_{f=1}^{h} w_f b_f + \sigma = 0$, where dual variables v_i ($i = 1,\ldots,m$), u_r ($r = 1,\ldots,s$) and w_f ($f = 1,\ldots,h$) are parameters for indicating the direction of a supporting hyperplane(s) and σ indicates an intercept of the supporting hyperplane. The parameters are all unknown. Therefore, they need to be determined by the following equations:

$$\begin{aligned}&\sum_{i=1}^{m} v_i x_{ij} - \sum_{r=1}^{s} u_r g_{rj} + \sum_{f=1}^{h} w_f b_{fj} + \sigma = 0,\ j \in RS_k\\&\text{and } \sum_{r=1}^{s} u_r g_{rk} + \sum_{f=1}^{h} w_f b_{fk} = 1,\end{aligned} \tag{20.15}$$

where RS_k stands for a reference set of the k-th DMU whose performance is measured by Model (17.5). A difference between the radial and non-radial measures can be found in the second term $\left(\sum_{r=1}^{s} u_r g_{rk} + \sum_{f=1}^{h} w_f b_{fk} = 1\right)$ of Equations (20.15).

It is important to note that Section 20.2.1 specifies the supporting hyperplane by $vx - ug + wb + \sigma = 0$ in the case of a single component of the three production factors. Meanwhile, Equation (20.15) specifies the supporting hyperplane as $\sum_{i=1}^{m} v_i x_i - \sum_{r=1}^{s} u_r g_r + \sum_{f=1}^{h} w_f b_f + \sigma = 0$ in the case of multiple components of the three production factors. Equation (20.15) can be easily obtained from the first group of CSCs, expressed by Equations (20.14).

To measure the degree and type of RTS under natural disposability, this chapter classifies DMUs into two groups based upon their efficiency statuses. If the k-th DMU is efficient in unified performance under natural disposability, then the degree of SEC for multiple components, corresponding to scale elasticity (e_g) related to desirable outputs, is determined by

$$
\begin{aligned}
SEC &= \left(\sum_{i=1}^{m} v_i^* x_{ik} \right) \Big/ \left(\sum_{r=1}^{s} u_r^* g_{rk} \right) = \left(\sum_{i=1}^{m} v_i^* x_{ik} \right) \Big/ \left(\sum_{i=1}^{m} v_i^* x_{ik} + \sum_{f=1}^{h} w_f^* b_{fk} + \sigma^* \right) \\
&= 1 \Big/ \left[1 + \left(\sigma^* + \sum_{f=1}^{h} w_f^* b_{fk} \right) \Big/ \left(\sum_{i=1}^{m} v_i^* x_{ik} \right) \right] \qquad (20.16) \\
&= 1 \Big/ \left[1 + \left(\sigma^* + \zeta_k^{b*} \right) \Big/ \left(\sum_{i=1}^{m} v_i^* x_{ik} \right) \right],
\end{aligned}
$$

where $\zeta_k^{b*} = \sum_{f=1}^{h} w_f^* b_{fk}$. The variable implies an influence from undesirable outputs on the degree of SEC. Equation (20.16) incorporates $\sum_{i=1}^{m} v_i^* x_{ij} - \sum_{r=1}^{s} u_r^* g_{rj} + \sum_{f=1}^{h} w_f^* b_{fj} + \sigma = 0$ in the reformulation that indicates the status of efficiency in Model (17.7).

Meanwhile, if the k-th DMU is inefficient in unified performance under natural disposability, then the inefficient DMU needs to be projected onto an efficiency frontier. In the case, the degree of SEC can be determined by

$$
\begin{aligned}
SEC &= \left[\sum_{i=1}^{m} v_i^* \left(x_{ik} - d_i^{x*} \right) \right] \Big/ \left[\sum_{r=1}^{s} u_r^* \left((1+\xi^*) g_{rk} + d_r^{g*} \right) \right] \\
&= \left[\sum_{i=1}^{m} v_i^* \left(x_{ik} - d_i^{x*} \right) \right] \Big/ \left[\sum_{r=1}^{s} \left(u_r^* g_{rk} + u_r^* d_r^{g*} \right) \right] \\
&= \left[\sum_{i=1}^{m} v_i^* \left(x_{ik} - d_i^{x*} \right) \right] \Big/ \left[\sum_{i=1}^{m} v_i^* \left(x_{ik} - d_i^{x*} \right) + \sum_{f=1}^{h} w_f^* \left(b_{fk} - d_f^{b*} \right) + \sigma^* \right] \\
&= 1 \Big/ \left(1 + \left\{ \left[\sum_{f=1}^{h} w_f^* \left(b_{fk} - d_f^{b*} \right) + \sigma^* \right] \Big/ \sum_{i=1}^{m} v_i^* \left(x_{ik} - d_i^{x*} \right) \right\} \right) \qquad (20.17) \\
&= 1 \Big/ \left\{ 1 + \left[\left(\sigma^* + \sum_{f=1}^{h} w_f^* b_{fk} \right) \Big/ \sum_{i=1}^{m} v_i^* \left(x_{ik} - d_i^{x*} \right) \right] \right\} \\
&= 1 \Big/ \left\{ 1 + \left[\left(\sigma^* + \zeta_k^{b^*} \right) \Big/ \sum_{i=1}^{m} v_i^* \left(x_{ik} - d_i^{x*} \right) \right] \right\}.
\end{aligned}
$$

In Equation (20.17), $\xi^* + \varepsilon_s\left(\sum_{i=1}^{m} R_i^x d_i^{x*} + \sum_{r=1}^{s} R_r^g d_r^{g*} + \sum_{f=1}^{h} R_f^b d_f^{b*}\right) = \sum_{i=1}^{m} v_i^* x_{ik} - \sum_{r=1}^{s} u_r^* g_{rk} + \sum_{f=1}^{h} w_f^* b_{fk} + \sigma^*$ is obtained from the optimal objective value of Models (17.5) and (17.7). Then, we incorporate $\xi^* = 0$ in the optimal condition because the k-th DMU is projected onto the efficiency frontier and it becomes zero.

The reformulation in Equation (20.17) has the following three assumptions. First, the inefficient DMU has positive slacks on all production factors so that $v_i^* = \varepsilon_s R_i^x$ for all i, $u_r^* = \varepsilon_s R_r^g$ for all r, and $w_f^* = \varepsilon_s R_f^b$ for all f. Second, the observed inputs are larger than their slacks so that their differences are positive. Finally, Equation (20.17) considers $\zeta_k^{b*} = \sum_{f=1}^{h} w_f^*(b_{fk} - d_f^{b*}) \cong \sum_{f=1}^{h} w_f^* b_{fk}$ because we are interested in the type of RTS that is mainly determined by inputs and desirable outputs. Hence, $d_f^{b*} \cong 0$ is assumed for all f for our computational tractability, indicating that all slacks related to undesirable outputs are positive but being very close to zero.

Assuming both a unique projection of an inefficient DMU onto an efficiency frontier and a unique reference set for the projected DMU, this study can determine the upper and lower bounds of the dual variable (σ) and the influence (ζ_k^b) from undesirable outputs using the following model:

$$\begin{aligned}
&\text{Maximize/Minimize} \quad \sigma + \zeta_k^b \\
&\quad s.t. \quad \text{all constraints in Models (17.5) and (17.7),} \\
&\qquad \zeta_k^b = \sum_{f=1}^{h} w_f b_{fk}, \\
&\qquad \xi + \varepsilon_s\left(\sum_{i=1}^{m} R_i^x d_i^x + \sum_{r=1}^{s} R_r^g d_r^g + \sum_{f=1}^{h} R_f^b d_f^b\right) \\
&\qquad = \sum_{i=1}^{m} v_i x_{ik} - \sum_{r=1}^{s} u_r g_{rk} + \sum_{f=1}^{h} w_f b_{fk} + \sigma \text{ and} \\
&\qquad \zeta_k^b \geq 0.
\end{aligned} \tag{20.18}$$

The first additional constraint indicates the influence from undesirable outputs. The second additional constraint is because the objective function of Model (17.5) equals that of Model (17.7) on optimality.

The upper bound ($\bar{\vartheta}^*$) and the lower bound ($\underline{\vartheta}^*$) of the objective value are obtained from the maximization and minimization of Model (20.18), respectively. An optimal solution of Models (17.5) and (17.7) is feasible in Model (20.18) and vice versa.

Based upon these upper and lower bounds, the type of RTS of the k-th DMU is determined by the following classification:

$$\begin{aligned}
&\text{(a) Decreasing RTS} \leftrightarrow \bar{\vartheta}^* \geq \underline{\vartheta}^* > 0, \\
&\text{(b) Constant RTS} \leftrightarrow \bar{\vartheta}^* \geq 0 \geq \underline{\vartheta}^* \text{ and} \\
&\text{(c) Increasing RTS} \leftrightarrow 0 > \bar{\vartheta}^* \geq \underline{\vartheta}^*.
\end{aligned} \tag{20.19}$$

20.4.2 Scale Damages and DTS under Managerial Disposability

Shifting our research interest from natural disposability to managerial disposability, where the first priority is environmental performance and the second priority is operational performance, this chapter determines SDA and DTS by the following manner: If the k-th DMU is efficient in unified efficiency under managerial disposability, then the degree of SDA can be determined by

$$SDA = \left(\sum_{i=1}^{m} v_i^* x_{ik}\right) \Big/ \left(\sum_{f=1}^{h} w_f^* b_{fk}\right) = 1 \Big/ \left[1 - \left(\sigma^* - \zeta_k^{g*}\right) \Big/ \left(\sum_{i=1}^{m} v_i^* x_{ik}\right)\right], \tag{20.20}$$

where $\zeta_k^{g*} = \sum_{r=1}^{s} u_r^* g_{rk}$. The variable implies the influence of desirable outputs on SDA.

Following the reformulation, similar to Equation (20.17), as discussed for SEC, if the k-th DMU is inefficient in unified efficiency under managerial disposability, then the degree of SDA can be determined by

$$\begin{aligned} SDA &= \left[\sum_{i=1}^{m} v_i^* \left(x_{ik} + d_i^{x*}\right)\right] \Big/ \left[\sum_{f=1}^{h} w_f^* \left(b_{fk} - d_f^{b*}\right)\right] \\ &= 1 \Big/ \left\{1 - \left[\left(\sigma^* - \zeta_k^{g*}\right) \Big/ \sum_{i=1}^{m} v_i^* \left(x_{ik} + d_i^{x*}\right)\right]\right\}. \end{aligned} \tag{20.21}$$

Assuming both a unique projection of an inefficient DMU onto an efficiency frontier and a unique reference set for the projected DMU, this chapter can determine the upper and lower bounds of the dual variable (σ) and the influence (ζ_k^g) from desirable outputs by the following model:

$$\begin{aligned} &\text{Maximize/Minimize } \sigma - \zeta_k^g \\ &\quad s.t. \ \text{all constraints in Models (17.10) and (17.12),} \\ &\qquad \zeta_k^g = \sum_{r=1}^{s} u_r g_{rk}, \\ &\qquad \xi + \varepsilon_s \left(\sum_{i=1}^{m} R_i^x d_i^x + \sum_{r=1}^{s} R_r^g d_r^g + \sum_{f=1}^{h} R_f^b d_f^b\right) \\ &\qquad = -\sum_{i=1}^{m} v_i x_{ik} - \sum_{r=1}^{s} u_r g_{rk} + \sum_{f=1}^{h} w_f b_{fk} + \sigma \text{ and} \\ &\qquad \zeta_k^g \geq 0. \end{aligned} \tag{20.22}$$

The upper bound ($\bar{\upsilon}^*$) and the lower bound ($\underline{\upsilon}^*$) of the objective value are obtained from the maximization and minimization of Model (20.22), respectively. An optimal solution of Models (17.10) and (17.12) is feasible in Model (20.22) and vice versa.

Based upon these upper and lower bounds, the type of DTS of the k-th DMU is determined by the following classification:

$$\begin{aligned}&\text{(a) Increasing DTS} \leftrightarrow \bar{\upsilon}^* \geq \underline{\upsilon}^* > 0,\\&\text{(b) Constant DTS} \leftrightarrow \bar{\upsilon}^* \geq 0 \geq \underline{\upsilon}^* \text{ and}\\&\text{(c) Decreasing DTS} \leftrightarrow 0 > \bar{\upsilon}^* \geq \underline{\upsilon}^*.\end{aligned} \tag{20.23}$$

Here, it is important to note two concerns. One of the two concerns is that the type of RTS under radial and non-radial models may be determined by dropping the influence factor (ζ_k^b) from undesirable outputs. In a similar manner, the type of DTS is determined without the influence factor (ζ_k^g) from desirable outputs. The determination may be mainly measured by only the dual variable (σ^*) that indicates the intercept of a supporting hyperplane. Such is just an approximation (i.e., quick and easy) method that may have computational practicality. The other concern is that this chapter assumes the uniqueness on a projection to an efficiency frontier and a reference set on the frontier. It is true that such an assumption may not be satisfied in the reality of DEA applications. In the case, it is necessary for us to incorporate SCSCs in the proposed approach to determine the type of RTS and DTS as discussed in Chapter 7.

20.5 JAPANESE CHEMICAL AND PHARMACEUTICAL FIRMS

This chapter applies the proposed radial approach to assess the unified performance of Japanese chemical and pharmaceutical firms, both of which are included in the chemical industry under a general classification. Due to a page limit, we do not discuss the results on the non-radial approach in this chapter. Including rubber and plastic products, the total shipment value of the chemical industry, as a whole, amounted to 40 trillion yen in 2010. This value accounted for 13.9% of the entire Japanese manufacturing industry and has increased 1.07 times since 1990. The number of employees in the industry was approximately 880,000 in 2010, which accounted for 11.5% in the entire manufacturing industry. The industry's operating profit margin remained high, compared to the other manufacturing industries. The plant investment by the industry accounted for 11.6% of all manufacturing industries. All the numbers indicate that the chemical industry plays an important role for Japanese economy. See the information source[2] that provides a detailed description on Japanese chemical industry.

The Japanese chemical industry was the third largest in global chemical shipments with US$ 338.2 billion in 2010 after China and the United States. However, the size of each company was small, compared to the other world-class chemical companies. Currently, the production activities of these chemical firms produce various environmental problems, so that the Japanese government requires them to pay serious attention to the global warming and climate change and to reduce the amount

[2] Japan Chemical Industry Association (2010) *Chemical Industry of Japan*, Japan Chemical Industry Association, Tokyo.

of chemical substances. For example, the industry is the second largest emitter of CO_2 in Japan following the steel industry, after allocating the amount of the industry's indirect emissions. Thus, environmental consideration such as improved energy efficiency and environmental protection is an important business concern for the chemical industry. The total amount of energy is treated as an input in this chapter to incorporate the influence of such energy consumption on their unified efficiency measures. The tradeoff between business prosperity and environmental protection is very important for the industry, so indicating the research importance of this study. See Chapter 14 on the history of Japanese environmental issue.

In examining the unified efficiency measures of chemical firms, this chapter compares performance between two sub-samples: pharmaceutical firms and the other chemical firms, because they have business similarities and both are classified as the chemical industry in a broad sense. The data set contains 31 Japanese chemical and pharmaceutical firms and it consists of four inputs: the amount of total assets (unit: million yen), the number of employees, the total amount of energy inputs (unit: Giga joule), and the total cost for environmental protection (unit: million yen) in order to produce a desirable output: total sales (unit: million yen), along with two undesirable outputs: the total amount of greenhouse gas emissions (unit: ton of CO_2) and the total amount of waste discharges (unit: ton). The observed annual periods were from 2007 to 2010.

Data Accessibility: Toyo Keizai Inc.[3] has published the total amount of energy inputs, the total cost for environmental protection, the total amount of greenhouse gas emissions and the total amount of waste discharges. Other production factors have been obtained from Nippon Keizai Shinbun Inc[4].

This chapter selects representative companies in the chemical industry which provide all the necessary data from 2007 to 2010. Thus, the data set used for this chapter is structured in a balanced panel data that consists of 31 firms over these four years.

To measure the unified efficiency measures and the type of RTS and DTS in these firms, this chapter first pools all data sets together and treats it as a cross-sectional data. Then, this chapter applies the proposed radial approach to the pooled data set.

The research of Sueyoshi and Goto (2014a) recorded that pharmaceutical firms outperformed chemical firms in terms of their average unified efficiency measures under natural and managerial disposability. This result implies that the former firms pay more attention to their operational and environmental performance than the latter firms.

As an extension of their work, Table 20.2 summarizes the upper bound ($\bar{\vartheta}^*$) and the lower bound ($\underline{\vartheta}^*$) of 31 firms, both of which are obtained from Model (20.18). These upper and lower bounds determine the type of RTS. As listed at the

[3] Toyo Keizai Inc. (2012) *Corporate Social Responsibility Database*. Toyo Keizai Inc., Tokyo.
[4] Nippon Keizai Shinbun Inc. (2012) *Nikkei NEEDS–FinancialQUEST*, Nippon Keizai Shinbun Inc., Tokyo.

TABLE 20.2 Returns to scale

Year		2007			2008			2009			2010		
No.	Company Name	Max	Min	Type of RTS	Max	Min	Type of RTS	Max	Min	Type of RTS	Max	Min	Type of RTS
1	Kuraray	0.2657	0.2657	D	0.3242	0.3242	D	0.3065	0.3065	D	0.0799	0.0799	D
2	Asahi-Kasei	−2.1553	−2.1553	I	−0.1114	−0.1114	I	0.9593	0.9593	D	0.9763	0.9763	D
3	Sumitomo Chemical Company	1.1027	1.1027	D	1.0685	1.0685	D	1.1782	1.1782	D	1.0915	1.0915	D
4	Tosoh	0.2046	−0.2148	C	0.0442	0.0442	D	0.1008	0.1008	D	0.0477	0.0477	D
5	Tokuyama Corporation	0.0053	0.0053	D	0.0052	0.0052	D	0.6460	0.6460	D	0.6447	0.6447	D
6	Toagosei	0.1335	0.1335	D	−0.0740	−0.0740	I	−0.0851	−0.0851	I	−0.3053	−0.3053	I
7	Taiyo Nippon Sanso	0.3645	−0.7341	C	0.1232	−0.5504	C	−0.2891	−0.2891	I	−0.5691	−0.5691	I
8	Mitsui Chemicals	0.9941	−0.0502	C	0.0322	0.0322	D	0.4643	0.4643	D	0.4202	0.4202	D
9	JSR	0.0963	0.0963	D	0.1228	0.1228	D	0.1469	0.1469	D	0.1487	0.1487	D
10	Tokyo Ohka Kogyo	−0.1179	−0.1179	I	−0.0814	−0.0814	I	−0.2965	−0.2965	I	−0.1612	−0.1612	I
11	Sekisui	0.2852	−0.0981	C	0.1036	−0.1163	C	0.0185	0.0185	D	0.0061	−0.1797	C
12	Takiron	−0.2972	−0.2972	I	−0.2900	−2.3289	I	−0.2490	−7.1154	I	−0.1732	−913.7409	I
13	Hitachi Chemical	0.3884	−0.0349	C	0.1665	0.1665	D	0.2073	0.2073	D	0.2562	−0.0501	C
14	Nippon Kayaku	0.0206	−0.2459	C	0.1027	0.1027	D	0.0983	0.0983	D	0.0950	0.0950	D
15	ADEKA	0.8771	−12.5532	C	0.0922	0.0922	D	0.0796	0.0796	D	0.0779	0.0779	D
16	Sanyo Chemical Industries	0.1210	0.1210	D	0.1240	0.1240	D	0.1418	0.1418	D	0.1327	0.1327	D
17	Takeda Pharmaceutical Company	0.9872	−0.0872	C	0.9505	−0.1137	C	0.9678	−0.1163	C	0.9492	−0.0712	C
18	Astellas	0.5544	0.5544	D	0.2480	0.2480	D	0.2946	−0.0092	C	0.2440	0.2440	D
19	Shionogi & Co.	0.4644	0.4644	D	0.6310	0.6310	D	0.5743	0.5743	D	0.5156	0.5156	D
20	Nippon Shinyaku	0.2768	0.2768	D	0.3666	−41.8633	C	0.4420	−3.8724	C	0.4026	−0.8512	C

(*Continued*)

TABLE 20.2 (Continued)

No.	Year / Company Name	2007 Max	2007 Min	2007 Type of RTS	2008 Max	2008 Min	2008 Type of RTS	2009 Max	2009 Min	2009 Type of RTS	2010 Max	2010 Min	2010 Type of RTS
21	Chugai Pharmaceutical Co.	0.7796	–0.0565	C	0.5598	0.5598	D	0.7159	–0.1570	C	–0.0294	–0.0294	I
22	Eisai	0.3155	0.3155	D	0.2116	0.2116	D	0.6646	0.1591	D	0.5347	–0.1248	C
23	Ono Pharmaceutical Co.	0.6018	–0.5896	C	0.5690	–0.5402	C	0.3505	0.3505	D	0.5114	0.2259	D
24	Tsumura & Co.	0.0265	0.0265	D	0.0256	0.0256	D	0.0254	0.0254	D	0.0203	0.0203	D
25	Kissei Pharmaceutical Co.	–1.8733	–1.8733	I	0.5243	–1.2030	C	0.3724	–5.2633	C	0.5959	–53.9556	C
26	Daiichi Sankyo Co.	0.5474	0.5474	D	0.5436	0.5436	D	0.5512	0.5512	D	0.4951	0.4951	D
27	Nippon Paint	0.6822	–4.4053	C	–0.0583	–0.3759	I	–0.1387	–0.1387	I	0.0947	–0.9554	C
28	DIC	0.1633	0.1633	D	0.1641	0.1641	D	0.9403	–2.2286	C	0.6418	–15.8280	C
29	Shiseido	0.6330	–0.3238	C	0.6029	–0.2516	C	–0.0422	–0.2301	I	0.6219	–0.2727	C
30	T. Hasegawa Co.	–0.7621	–5.2836	I	–1.0704	–1.0704	I	–0.9781	–6.6281	I	–1.0511	–1.0511	I
31	Arakawa Chemical Industries	0.1578	–1.6944	C	0.1502	–5.1439	C	–0.1640	–0.1640	I	–0.1953	–0.1953	I
Chemical firms		I	C	D									
		28.6%	23.8%	47.6%									
Pharmaceutical firms		I	C	D									
		5.0%	40.0%	55.0%									

(a) Source: Sueyoshi and Goto (2014a). (b) I, C and D stand for increasing, constant and decreasing RTS, respectively. (c) Max. and Min. indicate the upper and lower bounds of Model (20.18).

TABLE 20.3 Damages to scale

Year		2007			2008			2009			2010		
No.	Company Name	Max	Min	Type of DTS	Max	Min	Type of DTS	Max	Min	Type of DTS	Max	Min	Type of DTS
1	Kuraray	0.2171	0.2171	I	0.2366	0.2366	I	0.2430	0.2430	I	0.2367	0.2367	I
2	Asahi-Kasei	4.9969	4.1855	I	66.5688	–0.0447	C	0.0451	0.0451	I	0.0464	0.0464	I
3	Sumitomo Chemical Company	25.5812	0.1970	I	1.2689	0.5734	I	0.4082	0.4082	I	2.6965	0.0444	I
4	Tosoh	0.0175	0.0175	I	0.0364	0.0364	I	0.0365	0.0365	I	0.0349	0.0349	I
5	Tokuyama Corporation	0.0337	0.0337	I	0.0355	0.0355	I	0.0379	0.0379	I	0.0389	0.0389	I
6	Toagosei	0.1255	0.1255	I	1.5891	0.1758	I	0.5955	0.1044	I	0.1965	0.1965	I
7	Taiyo Nippon Sanso	2.8671	–0.1710	C	2.6864	–0.0788	C	0.5576	0.5576	I	3.5516	0.4821	I
8	Mitsui Chemicals	5.0876	–0.1251	C	0.2701	0.0450	I	2.2791	0.3123	I	1.6088	0.0068	I
9	JSR	0.2988	0.2988	I	0.5582	0.5582	I	0.4740	0.4740	I	0.4380	0.4380	I
10	Tokyo Ohka Kogyo	–0.2329	–0.2329	D	–0.2431	–0.2431	D	–0.2467	–0.2467	D	–0.2597	–0.2597	D
11	Sekisui	7.6427	0.2580	I	4.5479	0.3013	I	8.4312	0.5311	I	0.9649	0.9649	I
12	Takiron	–0.0490	–0.0490	D	–0.0548	–0.0548	D	–0.3452	–0.3452	D	–0.3685	–0.3685	D
13	Hitachi Chemical	1.6540	1.6540	I	1.8062	1.8062	I	1.7153	1.7153	I	1.8216	1.8216	I
14	Nippon Kayaku	–0.1328	–0.1328	D	–0.1001	–0.1001	D	–0.1247	–0.1247	D	–0.1223	–0.1223	D
15	ADEKA	80.8672	–0.9991	C	0.1503	0.1503	I	–0.0202	–0.0202	D	–0.0378	–0.0378	D
16	Sanyo Chemical Industries	0.7882	0.7882	I	0.3380	0.3380	I	0.3442	0.3442	I	0.9126	0.9126	I
17	Takeda Pharmaceutical Company	19.5288	–0.0801	C	0.4972	–0.1159	C	3.7713	–0.4342	C	28.5277	–0.4838	C
18	Astellas	0.5524	0.5524	I	0.3837	0.3837	I	0.9291	0.3824	I	0.7968	0.7968	I
19	Shionogi & Co.	1.0761	1.0761	I	1.1331	1.1331	I	1.0724	1.0724	I	1.1271	1.1271	I

(*Continued*)

TABLE 20.3 (Continued)

	Year	2007			2008			2009			2010		
No.	Company Name	Max	Min	Type of DTS	Max	Min	Type of DTS	Max	Min	Type of DTS	Max	Min	Type of DTS
20	Nippon Shinyaku	–0.5162	–0.5162	D	–0.6310	–0.6310	D	0.2010	–0.9994	C	0.3054	–0.2873	C
21	Chugai Pharmaceutical Co.	9.2419	–0.5722	C	18.3903	–0.3579	C	1.6448	–0.4788	C	0.8475	0.8475	I
22	Eisai	18.3530	4.4302	I	–0.2861	–0.2861	D	30.3043	–0.4729	C	12.3657	–0.4714	C
23	Ono Pharmaceutical Co.	2.7906	–0.5797	C	3.4447	–0.9989	C	1.3170	–0.1574	C	2.2281	–0.5970	C
24	Tsumura & Co.	0.0071	0.0071	I	–0.1634	–0.1634	D	–0.0734	–0.0734	D	–0.1711	–0.1711	D
25	Kissei Pharmaceutical Co.	–0.4758	–0.4758	D	–0.9009	–0.9009	D	–0.9033	–0.9033	D	–0.4981	–0.4981	D
26	Daiichi Sankyo Co.	19.3410	0.9919	I	1.4865	1.4865	I	3.1658	1.2318	I	4.5196	0.5157	I
27	Nippon Paint	14.2178	–0.5438	C	–0.3302	–0.3302	D	–0.3564	–0.3564	D	–0.3727	–0.3727	D
28	DIC	21.8292	0.3856	I	21.4282	0.4443	I	0.8378	0.8378	I	17.7840	0.5498	I
29	Shiseido	–0.2859	–0.2859	D	–0.4029	–0.4029	D	1.2788	1.2788	I	6.5949	–0.5371	C
30	T. Hasegawa Co.	–0.2777	–0.2777	D	–0.3264	–0.3264	D	–0.3602	–0.3602	D	–0.3212	–0.3212	D
31	Arakawa Chemical Industries	–0.0705	–0.0705	D	–0.0786	–0.0786	D	–0.0917	–0.0917	D	–0.0711	–0.0711	D
Chemical firms		I	C	D									
		59.5%	8.3%	32.1%									
Pharmaceutical firms		I	C	D									
		37.5%	37.5%	25.0%									

(a) Source: Sueyoshi and Goto (2014a). (b) I, C and D stand for increasing, constant and decreasing DTS, respectively. (c) Max. and Min. indicate the upper and lower bounds of Model (20.22).

bottom of Table 20.2, 28.6%, 23.8% and 47.6% of the chemical firms, respectively, exhibit increasing, constant and deceasing RTS. Meanwhile, 5.0%, 40.0% and 55.0% of the pharmaceutical firms exhibit increasing, constant and deceasing RTS. The two results imply that approximately 50% of firms (i.e., 47.6% of the chemical firms and 55.0% of the pharmaceutical firms) in the two groups of firms should reduce their corporate sizes to enhance their operational performance.

Table 20.3 summarizes the upper bound ($\bar{\upsilon}^*$) and lower bound ($\underline{\upsilon}^*$) of these firms, both of which are obtained from Model (20.22). These upper and lower bounds determine the type of DTS. The type of DTS, summarized in Table 20.3, is different from the type of RTS of Table 20.2. As listed at the bottom of Table 20.3, 59.5%, 8.3% and 32.1% of the chemical firms, respectively, exhibit increasing, constant and deceasing DTS. The result implies that 59.5% of chemical firms should reduce their sizes. As an alternative strategy, it is necessary for them to introduce new environmental technology for reducing their undesirable outputs. Meanwhile, 37.5%, 37.5% and 25.0% of the pharmaceutical firms exhibit increasing, constant and deceasing DTS. This result indicates that environmental strategy may depend upon each pharmaceutical firm.

20.6 SUMMARY

This chapter described how to determine the type of RTS under natural disposability and the type of DTS under managerial disposability by using both radial and non-radial models. The proposed DEA approach incorporated strategic concepts on natural or managerial disposability, discussed in Chapter 15, into the proposed computational process. The proposed measurement provided us with empirical evidence on both the operational size of each DMU and how to guide its environmental strategy by eco-technology innovation on undesirable outputs.

As an application, this chapter compared between Japanese chemical and pharmaceutical firms by their measures on RTS and DTS. This chapter identified two empirical findings. One of the two findings was that approximately 50% of firms (i.e., 47.6% of the chemical firms and 55.0% of the pharmaceutical firms) in the two groups should reduce their corporate sizes to enhance their operational performance. The other finding was that 59.5% of chemical firms should reduce their sizes to enhance their environmental performance. As an alternative strategy, it is recommended that they could introduce new environmental technology for reducing their undesirable outputs. Since Japanese chemical firms are relatively small in the global market, the resulting type of classification on RTS and DTS may become different if this chapter incorporates foreign firms (e.g., corporations in China, the United States and European nations) in the data sample used in this study. Therefore, although it is recommended in this chapter that the scale reduction for Japanese chemical and pharmaceutical firms, it does not immediately imply that they need to reduce their operation sizes in the global market. Rather, we recommend the introduction of eco-technology innovation for improving their energy and environment performance as well as to increase their operational performance.

21

DESIRABLE AND UNDESIRABLE CONGESTIONS

21.1 INTRODUCTION

This chapter[1] discusses the occurrence of congestion within the framework of DEA environmental assessment. As an extension of Chapter 9 on congestion, this chapter now separates the concept of congestion into undesirable and desirable categories in DEA environmental assessment. See Figures 9.1 to 9.4 for a visual description on an occurrence of congestion in a conventional context of DEA. After the separation, this chapter discusses how to measure an occurrence of desirable congestion (DC), or eco-technology innovation, in comparison with that of undesirable congestion (UC). The concept of DC and UC has a close linkage with, respectively, returns to scale (RTS) and damages to scale (DTS) in Chapter 20. The comparison between Chapters 20 and 21 provides us with extended implications on RTS and DTS from an occurrence of the two types of congestion. The two chapters will be further conceptually and methodologically extendable into other scale measures in Chapter 23.

A concern of this chapter is that the identification of UC is indeed important for avoiding a cost increase and a shortage of transmission capacity in the case of the electric power industry, for example. However, the identification of DC is

[1] This chapter is partly based upon Sueyoshi, T. and Goto, M. (2016) Undesirable congestion under natural disposability and desirable congestion under managerial disposability in US electric power industry measured by DEA environmental assessment. *Energy Economics*, **55**, 173–188.

Environmental Assessment on Energy and Sustainability by Data Envelopment Analysis, First Edition. Toshiyuki Sueyoshi and Mika Goto.

more important than that of UC in terms of our sustainable growth in which eco-technology innovation can effectively reduce the amount of various types of pollution. Therefore, it is necessary for this chapter to discuss the occurrence of UC under natural disposability and that of DC under managerial disposability.

As an illustrative application, this chapter applies the proposed approach to evaluate the performance of coal-fired power plants in PJM, which is a well-known regional transmission organization, in the United States. See Chapter 14 on a detailed description of PJM. This chapter finds that the occurrence of UC, due to a capacity limit of generation and transmission, is found in most of power plants. In contrast, the occurrence of DC, due to eco-technology innovation, is found in a limited number of power plants. Thus, the identification of DC assists us in determining how to facilitate eco-technology innovation for the growth of corporate sustainability. Such determination can be identified by the DEA approach proposed in this chapter.

The remainder of this chapter is organized as follows: Section 21.2 summarizes differences between UC and DC. Section 21.3 formulates radial models under natural disposability. Section 21.4 reformulates them under managerial disposability. Section 21.5 applies the proposed approach to examine the type of UC and DC of United States coal-fired power plants. Section 21.6 summarizes this chapter.

21.2 UC AND DC

Figure 21.1, originated from Figure 15.7, depicts conceptual differences between UC and DC. The occurrence of UC is usually due to a capacity limit on part or whole production facility, as depicted at the top and the bottom left-hand side of the figure, while that of DC may be due to eco-technology innovation and/or managerial challenge for pollution mitigation, as depicted at the bottom right-hand side of the figure.

The top of Figure 21.1 exhibits an efficiency frontier and the classification of UC in a desirable output (g) on the vertical axis and an input (x) on the horizontal axis. The classification has been used in a conventional framework of DEA, as discussed in Chapter 9. In a similar manner, the bottom left-hand side exhibits the classification of UC on an undesirable output (b) on the horizontal axis and a desirable output (g) on the vertical axis. The opposite case is found in the bottom right-hand side which classifies the type of DC. The two cases at the bottom are used in the computational framework of DEA environmental assessment, while an occurrence of UC in conventional DEA is depicted at the top of Figure 21.1. For our visual convenience, the three cases of Figure 21.1 consider a single component of three production factors. It is easily extendable to multiple components of these factors in the formulations proposed in this chapter. Note that each piece-wise linear contour line indicates the relationship between two production factors, depending upon a possible occurrence of UC or DC and these different combinations of production factors.

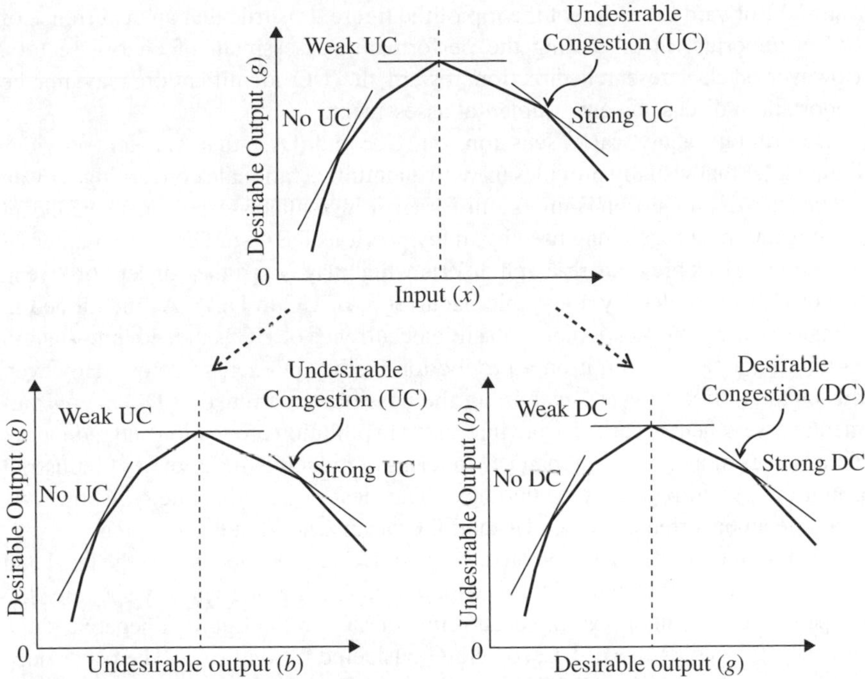

FIGURE 21.1 Undesirable congestion (UC) and desirable congestion (DC) (a) Each occurrence of congestion is classified into three categories: strong, weak and no. (b) The top of the figure indicates the possible occurrence of UC within a conventional DEA framework on x and g. This type of congestion has been long discussed in the DEA community. See Chapter 9. (c) The bottom left-hand side indicates the possible occurrence of UC in which an enhanced component(s) of an input vector increases some component(s) of the undesirable output vector, but decreases some component(s) of the desirable output vector. The equality constraints without slacks should be assigned to undesirable outputs (B) in the proposed formulations. (d) The bottom right-hand side indicates the possible occurrence of DC, or eco-technology innovation for pollution mitigation. In DEA environmental assessment, this type of occurrence indicates that an enhanced component(s) of an input vector increases some component(s) of a desirable output vector, but decreases some component(s) of an undesirable output vector. This study thinks that eco-technology innovation can solve various pollution issues, so that the occurrence of DC is important in DEA environmental assessment. For identifying the possible occurrence of DC, equality constraints without slacks are assigned to desirable outputs (G) in the proposed formulations. (e) The convex function at the bottom right-hand side depends upon the assumption that undesirable outputs (B) are "byproducts" of desirable outputs (G). It should be a concave function without such an assumption. The assumption is acceptable and realistic as long as DC occurs on B. See Chapter 15 on the assumption.

The rationale concerning why this chapter uses Figure 21.1 is that the concept of congestion, in particular DC, is very important in discussing DEA environmental assessment. However, as summarized in Chapter 28, most previous studies have paid attention to only UC as depicted in the bottom left of the figure as a straightforward extension of the top of the figure. It is true that an occurrence of UC is important in discussing the performance assessment of energy sectors. However, such a research direction toward the UC identification may not be important in discussing environmental assessment.

Before our analytical discussion on UC and DC, this chapter provides Figure 21.1 that visually provides us with an intuitive rationale concerning why an occurrence of congestion is important for environmental assessment. The concept of congestion has been long used by many previous DEA studies. See Chapter 28 that summarizes previous research efforts which have used the concept of "weak disposability" to identify a possible occurrence of UC in DEA. As mentioned in Chapter 9, the identification of a possible occurrence of UC is indeed important in examining a capacity limit on a production facility in energy sectors. However, this chapter directs toward examining the possible occurrence of DC in environmental assessment because we are interested in pollution prevention and reduction. Thus, DC is more important than UC in terms of reducing the amount of industrial pollution by utilizing eco-technology. The best research strategy is that we examine an occurrence of both UC and DC, as discussed in this chapter.

To describe the classification on the type of UC and DC, let us start the left-hand side of the bottom of Figure 21.1 in which a supporting line visually specifies the shape of a curve and a possible occurrence of UC. In the figure, a negative slope indicates the occurrence of UC, so being considered "strong UC". The occurrence implies a case in which a DMU increases an amount of an input (x). The increased input usually enlarges both desirable (g) and undesirable (b) outputs. However, such an input change increases the undesirable output (b) but decreases the desirable output vector (g), as visually specified by the negative slope of the supporting line. It is clear that such an occurrence is "undesirable" to us. In contrast, a positive slope implies the opposite case (i.e., no congestion), so being referred to as "no UC".

In addition to the two cases, it is necessary for us to consider the third situation in which the slope of the supporting line is zero; this chapter considers it as "weak UC," being between strong UC and no UC.

The description of UC is applicable to the occurrence of DC. A possible occurrence of DC, or eco-technology innovation on undesirable outputs, is depicted in the bottom right-hand side of Figure 21.1. In this chapter, the occurrence of DC is more important than that of UC, as mentioned previously, because the former case indicates green technology innovation (e.g., clean coal technology). It is believed that only the eco-technology innovation may have a potential to solve various environmental problems (e.g., air and/or water pollution), along with managerial challenges (e.g., a fuel mix strategy from coal to renewable energy). As depicted in the bottom right-hand side of Figure 21.1, the negative slope of the supporting line indicates such an occurrence of DC. More specifically speaking, the occurrence implies that an enlarged input (x) increases the desirable output (g) and decreases the undesirable output (b). A possible occurrence of DC is classified into the three categories: no, weak or strong.

Figure 21.2 visually reorganizes the bottom left-hand side of Figure 21.1 by paying attention to the relationship between a supporting line and the three types

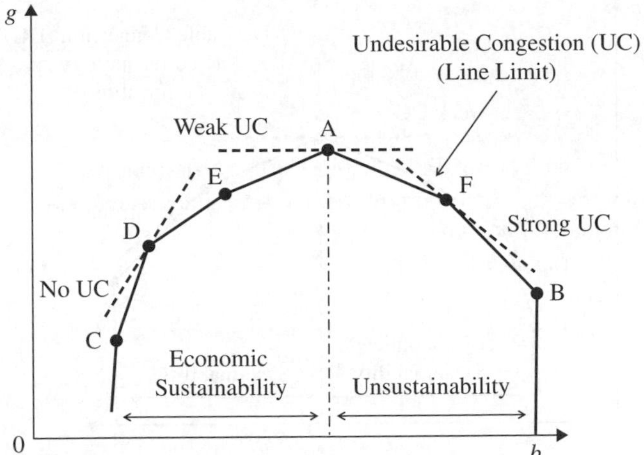

FIGURE 21.2 Undesirable congestion (UC) and unsustainability (a) Source: Sueyoshi and Goto (2016). (b) See Figure 9.5 in which x and g indicate the horizontal and vertical coordinates, respectively. Many previous studies extend Figure 9.5 to Figure 21.2 in discussing the occurrence of UC. The identification of UC is important in energy economics because it indicates a capacity limit.

of UC: no, weak and strong. The piece-wise linear contour line indicates the efficiency frontier between the two production factors (g and b). The figure depicts a possible occurrence of UC that is listed at the right-hand side of the figure, so indicating an occurrence of "unsustainability". Under such an occurrence, an increase in some component(s) of an input vector increases some components of the undesirable output vector (B) and decreases those of the desirable output vector (G) without worsening the other components. This type of congestion has been long discussed and investigated in a conventional use of DEA. For example, the strong UC indicates an occurrence of a line limit on transmission capacity between a generator and consumers. In contrast, the left-hand side of Figure 21.2 depicts "economic sustainability" where a firm can attain economic success, but lacking environmental protection. See Chapter 9, which does not consider the existence of an undesirable output. So, exactly speaking, the previous DEA studies did not consider an occurrence of UC.

Shifting our description from UC to DC, Figure 21.3 depicts the occurrence of DC between a single desirable output (g) on the horizontal axis and a single undesirable output (b) on the vertical axis. The figure depicts a possible occurrence of DC. Under such an occurrence of DC, as depicted in the right-hand side, an increase in some component(s) of an input vector increases some components of the desirable output vector (G) and decreases those of the undesirable output vector (B) without worsening the other components. This type of congestion occurs with eco-technology innovation on undesirable outputs (B). In contrast, the left-hand side of the figure indicates an occurrence of "economic sustainability," where a firm can attain economic success, but lacking an environmental protection capability.

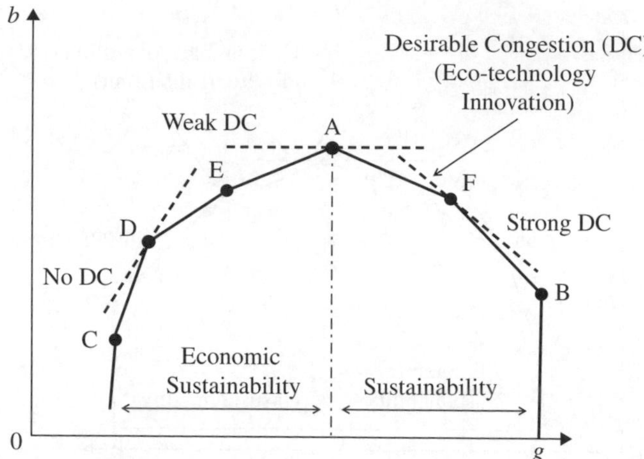

FIGURE 21.3 Desirable congestion (DC) and sustainability (a) Source: Sueyoshi and Goto (2016).

Returning to Equation (15.1) in Chapter 15, a unified (operational and environmental) production and pollution possibility set (P) to express a possible occurrence of UC under natural disposability is axiomatically expressed by

$$P(UC)_v^N(X)=\left\{(G,B): G\leq\sum_{j=1}^{n}G_j\lambda_j,\ B=\sum_{j=1}^{n}B_j\lambda_j,\ X\geq\sum_{j=1}^{n}X_j\lambda_j,\ \sum_{j=1}^{n}\lambda_j=1,\ \lambda_j\geq 0\ (j=1,\ldots,n)\right\}. \tag{21.1}$$

Here, the superscript N, stands for natural disposability with a possible occurrence of UC. Equation (21.1) incorporates $B=\sum_{j=1}^{n}B_j\lambda_j$ so that corresponding dual variables becomes unrestricted in their signs. Since each dual variable indicates a slope of a supporting hyperplane in the case of multiple components, Equation (21.1) can identify a possible occurrence of UC under natural disposability, as depicted in Figure 21.2. The subscript (v) indicates the status of variable RTS under natural disposabiity and/or variable DTS under managerial disposability. See Chapter 20.

Meanwhile, the set (P) with a possible occurrence of DC, as depicted in Figure 21.3, is formulated under managerial disposability as follows:

$$P(DC)_v^M(X)=\left\{(G,B): G=\sum_{j=1}^{n}G_j\lambda_j,\ B\geq\sum_{j=1}^{n}B_j\lambda_j,\ X\leq\sum_{j=1}^{n}X_j\lambda_j,\ \sum_{j=1}^{n}\lambda_j=1,\ \lambda_j\geq 0\ (j=1,\ldots,n)\right\}, \tag{21.2}$$

where the superscript M, stands for managerial disposability with a possible occurrence of DC. In the above case, we need to incorporate $G=\sum_{j=1}^{n}G_j\lambda_j$ into Equation (21.2) to identify a possible occurrence of DC.

At the end of this section, this chapter adds two concerns on UC and DC. The first concern is that it is impossible for us to assume constant RTS and constant DTS in the identification of UC and DC because the assumption sets a supporting hyperplane as a straight line, not a curve, passing from the origin. The slopes of the supporting hyperplane are non-negative under constant RTS and DTS. The situation is inconsistent with the identification of UC and DC. The production and pollution possibility set for UC and DC should be structured by variable RTS and DTS as specified in Equations (21.1) and (21.2). The second concern is that the equality condition (e.g., $G = \sum_{j=1}^{n} G_j \lambda_j$) in the two equations implies no slack in the formulations proposed in this chapter. Thus, the existence of slacks in equality constraints in the proposed (augmented) formulations implies that constraints are functionally inequality for DEA environmental assessment.

21.3 UNIFIED EFFICIENCY AND UC UNDER NATURAL DISPOSABILITY

As depicted in Figure 21.2, this chapter can identify a possible occurrence of UC under natural disposability. To examine such an occurrence in the radial measurement, we reorganize Model (17.5) by putting equality constraints (so, no slack variable) on undesirable outputs (B). For such a purpose, this chapter uses a radial model to measure the unified efficiency on the k-th DMU that reorganizes the original formulation, or Equation (17.5). Following Sueyoshi and Goto (2016), the restructured model becomes

$$
\begin{aligned}
\text{Maximize} \quad & \xi + \varepsilon_s \left(\sum_{i=1}^{m} R_i^{x} d_i^{x-} + \sum_{r=1}^{s} R_r^{g} d_r^{g} \right) \\
\text{s.t.} \quad & \sum_{j=1}^{n} x_{ij}\lambda_j + d_i^{x-} = x_{ik} \quad (i=1,\ldots,m), \\
& \sum_{j=1}^{n} g_{rj}\lambda_j - d_r^{g} - \xi g_{rk} = g_{rk} \quad (r=1,\ldots,s), \\
& \sum_{j=1}^{n} b_{fj}\lambda_j + \xi b_{fk} = b_{fk} \quad (f=1,\ldots,h), \\
& \sum_{j=1}^{n} \lambda_j = 1, \\
& \lambda_j \ge 0 \ (j=1,\ldots,n), \ \xi: \text{URS}, \\
& d_i^{x-} \ge 0 \ (i=1,\ldots,m) \text{ and } d_r^{g} \ge 0 \ (r=1,\ldots,s).
\end{aligned}
\tag{21.3}
$$

An important feature of Model (21.3) is that it does not have slack variables related to undesirable outputs (B), being different from Model (17.5), so that they are considered as equality constraints. The other constraints regarding inputs and desirable outputs are considered as inequality because they have slack variables in Model (21.3).

Model (21.3) has the following dual formulation:

$$
\begin{aligned}
\text{Minimize} \quad & \sum_{i=1}^{m} v_i x_{ik} - \sum_{r=1}^{s} u_r g_{rk} + \sum_{f=1}^{h} w_f b_{fk} + \sigma \\
s.t. \quad & \sum_{i=1}^{m} v_i x_{ij} - \sum_{r=1}^{s} u_r g_{rj} + \sum_{f=1}^{h} w_f b_{fj} + \sigma \geq 0 \qquad (j=1,\ldots,n), \\
& \sum_{r=1}^{s} u_r g_{rk} + \sum_{f=1}^{h} w_f b_{fk} = 1, \\
& v_i \geq \varepsilon_s R_i^x \qquad (i=1,\ldots,m), \\
& u_r \geq \varepsilon_s R_r^g \qquad (r=1,\ldots,s), \\
& w_f : \text{URS} \qquad (f=1,\ldots,h), \\
& \sigma : \text{URS}.
\end{aligned}
\tag{21.4}
$$

In Model (21.4), all the dual variables related to undesirable outputs (w_f: URS for $f = 1,\ldots,h$) are unrestricted in their signs because the constraints on B are formulated by the equality in Model (21.3).

A unified efficiency, or $UEN\left(UC\right)_v^{R*}$, of the k-th DMU under natural disposability is measured with a possible occurrence of UC. The degree is determined by

$$
\begin{aligned}
UEN\left(UC\right)_v^{R*} &= 1 - \left[\xi^* + \varepsilon_s\left(\sum_{i=1}^{m} R_i^x d_i^{x-*} + \sum_{r=1}^{s} R_r^g d_r^{g*}\right)\right] \\
&= 1 - \left[\sum_{i=1}^{m} v_i^* x_{ik} - \sum_{r=1}^{s} u_r^* g_{rk} + \sum_{f=1}^{h} w_f^* b_{fk} + \sigma^*\right],
\end{aligned}
\tag{21.5}
$$

where all variables are determined on the optimality of Models (21.3) and (21.4). The superscript R and the subscript v are related to the radial measurement with variable RTS and DTS. The UC implies a possible occurrence of undesirable congestion. The equation within parentheses, obtained from the optimality of Models (21.3) and (21.4), indicates the level of unified inefficiency under natural disposability. The unified efficiency is obtained by subtracting the level of unified inefficiency from unity as specified in Equation (21.5).

To discuss a mathematical rationale on why Models (21.3) and (21.4) can identify a possible occurrence of UC, this chapter needs to discuss the following proposition:

Proposition 21.1: A supporting hyperplane of the k-th DMU, measured by Models (21.3) and (21.4), becomes $\sum_{i=1}^{m} v_i x_i - \sum_{r=1}^{s} u_r g_r + \sum_{f=1}^{h} w_f b_f + \sigma = 0$, where $v_i \geq \varepsilon_s R_i^x$ ($i = 1,\ldots,m$), $u_r \geq \varepsilon_s R_r^g$ ($r = 1,\ldots,s$) and w_f: URS ($f = 1,\ldots,h$) are parameters (i.e., slopes) for indicating the direction of a supporting hyperplane. The σ, which is URS, indicates the intercept of the supporting hyperplane. All the parameters are unknown and need to be measured by the following equations:

$$\sum_{i=1}^{m} v_i x_{ij} - \sum_{r=1}^{s} u_r g_{rj} + \sum_{f=1}^{h} w_f b_{fj} + \sigma = 0,\ j \in RS_k \text{ and } \sum_{r=1}^{s} u_r g_{rk} + \sum_{f=1}^{h} w_f b_{fk} = 1. \quad (21.6)$$

where RS_k stands for a reference set of the k-th DMU measured by Model (21.3).

[Proof] In the $m + s + h$ dimensional space for the three production factors, a supporting hyperplane on the k-th DMU can be characterized as

$$\sum_{i=1}^{m} v_i \left(x_i - x_{ik}\right) - \sum_{r=1}^{s} u_r \left(g_r - g_{rk}\right) + \sum_{f=1}^{h} w_f \left(b_f \quad b_{fk}\right) - 0. \quad (21.7)$$

Equation (21.7) becomes

$$\sum_{i=1}^{m} v_i x_i - \sum_{r=1}^{s} u_r g_r + \sum_{f=1}^{h} w_f b_f + \sigma = 0, \quad (21.8)$$

where $\sigma = -\sum_{i=1}^{m} v_i x_{ik} + \sum_{r=1}^{s} u_r g_{rk} - \sum_{f-1}^{h} w_f b_{fk}$ is an intercept of the supporting hyperplane.

If the k-th DMU is efficient, it locates on an efficiency frontier so that the supporting hyperplane passes on the efficient DMU. In contrast, if the DMU is not efficient, it needs to be projected onto the efficiency frontier. Consequently, the k-th DMU, being either efficient or projected on an efficiency frontier, has the following conditions on optimality:

$$\sum_{i=1}^{m} v_i x_{ik}^* - \sum_{r=1}^{s} u_r g_{rk}^* + \sum_{f=1}^{h} w_f b_{fk}^* + \sigma \geq 0, \quad (21.9)$$

$$\sum_{r=1}^{s} u_r g_{rk}^* + \sum_{f=1}^{h} w_f b_{fk}^* = 1 \text{ and} \quad (21.10)$$

$$\sum_{i=1}^{m} v_i x_{ik}^* - \sum_{r=1}^{s} u_r g_{rk}^* + \sum_{f=1}^{h} w_f b_{fk}^* + \sigma = 0. \quad (21.11)$$

Here, the superscript (*) is added on the three production factors in the three equations because they are on an efficiency frontier. The first two constraints are obtained from the constraints of Model (21.4) and the last one is due to the objective value if the DMU is efficient or projected onto the efficiency frontier.

The location of the k-th DMU on an efficiency frontier is specified by

$$x_{ik}^* = \sum_{j \in RS_k} x_{ij} \lambda_j^* + d_i^{x-*} = \sum_{j \in RS_k} x_{ij} \lambda_j^*, \quad (21.12)$$

$$g_{rk}^{*} = \left(\sum_{j \in RS_k} g_{rj} \lambda_j^{*} - d_r^{g*} \right) \Big/ \left(1 + \xi^{*}\right) = \sum_{j \in RS_k} g_{rj} \lambda_j^{*} \text{ and} \tag{21.13}$$

$$b_{fk}^{*} = \left(\sum_{j \in RS_k} b_{fj} \lambda_j^{*} \right) \Big/ \left(1 - \xi^{*}\right) = \sum_{j \in RS_k} b_{fj} \lambda_j^{*}. \tag{21.14}$$

All slacks and the inefficiency measure (ξ^*) are zero on an efficiency frontier. Furthermore, considering $\sum_{j \in RS_k} \lambda_j^{*} = 1$, Equation (21.11) with Equation (21.9) becomes

$$\sum_{i=1}^{m} v_i \left(\sum_{j \in RS_k} x_{ij} \lambda_j^{*} \right) - \sum_{r=1}^{s} u_r \left(\sum_{j \in RS_k} g_{rj} \lambda_j^{*} \right) + \sum_{f=1}^{h} w_f \left(\sum_{j \in RS_k} b_{fj} \lambda_j^{*} \right) + \sum_{j \in RS_k} \sigma \lambda_j^{*} = 0. \tag{21.15}$$

In other words, Equation (21.15) becomes

$$\sum_{j \in RS_k} \left(\sum_{i=1}^{m} v_i x_{ij} - \sum_{r=1}^{s} u_r g_{rj} + \sum_{f=1}^{h} w_f b_{fj} + \sigma \right) \lambda_j^{*} = 0. \tag{21.16}$$

Since $\lambda_j^{*} > 0$ holds for all $j \in RS_k$, the supporting hyperplane on the k-th DMU is determined by the following equations:

$$\sum_{i=1}^{m} v_i x_{ij} - \sum_{r=1}^{s} u_r g_{rj} + \sum_{f=1}^{h} w_f b_{fj} + \sigma = 0 \ \left(j \in RS_k \right) \text{ and } \sum_{r=1}^{s} u_r g_{rk} + \sum_{f=1}^{h} w_f b_{fk} = 1. \tag{21.17}$$

The second part of Equation (21.17) is obtained from Equation (21.10).

Q.E.D.

The proposition implies that the supporting hyperplane measured on the optimality of Model (21.4) is expressed by $v^* x - u^* g + w^* b + \sigma^* = 0$ on an efficiency frontier in the case of a single component of the three production factors. The hyperplane is also specified by $g = \left(w^* b + v^* x + \sigma^*\right)/u^*$. Since both u^* and v^* are positive, the slope of the supporting hyperplane is determined by w^* and σ^*, both are unrestricted in their signs, on the optimality of Model (21.4).

After solving Model (21.4), we can identify the occurrence of UC by the following rule under the assumption on a unique optimal solution:

(a) If $w_f^{*} = 0$ for some (at least one) f, then "weak UC" occurs on the k-th DMU,

(b) If $w_f^{*} < 0$ for some (at least one) f, then "strong UC" occurs on the k-th DMU and

(c) If $w_f^{*} > 0$ for all f, then "no UC" occurs on the k-th DMU.

It is important to note three concerns on the classification on UC. First, this chapter assumes a unique optimal solution for our computational tractability. See Chapter 7 on the occurrence of multiple solutions. If multiple solutions occur on the k-th DMU, it is necessary for us to measure the upper and lower bounds of σ^* to prepare the decision on an occurrence of UC. See Chapters 8 and 20 for detailed descriptions on the selection on upper and lower bounds for σ^* (i.e., the intercept of the supporting hyperplane). This chapter clearly understands the drawback of our assumption on the unique solution. Second, if $w_f^* < 0$ for some f and $w_{f'}^* = 0$ for the other f', then both strong and weak UCs coexist on the k-th DMU. In this case, we consider the case as an occurrence of the strong UC on the k-th DMU. Finally, as discussed in Chapter 9, in particular Section 9.5.1, it is necessary for us to project an inefficient DMU onto an efficiency frontier in identifying the type of UC. This chapter does not consider the projection, except noting Models (9.13) and (9.14), to enhance our computational tractability.

21.4 UNIFIED EFFICIENCY AND DC UNDER MANAGERIAL DISPOSABILITY

To attain a high level of managerial disposability, a DMU needs to increase some components of the input vector in order to increase some components of the desirable output vector and simultaneously decrease those of the undesirable output vector.

As depicted in Figure 21.3, this chapter can identify a possible occurrence of DC under managerial disposability. To examine the occurrence, this chapter reorganizes Model (17.10) by eliminating slack variables on desirable outputs (G) as follows:

$$
\begin{aligned}
\text{Maximize} \quad & \xi + \varepsilon_s \left(\sum_{i=1}^{m} R_i^x d_i^{x+} + \sum_{f=1}^{h} R_f^b d_f^b \right) \\
s.t. \quad & \sum_{j=1}^{n} x_{ij}\lambda_j - d_i^{x+} = x_{ik} \quad (i=1,\ldots,m), \\
& \sum_{j=1}^{n} g_{rj}\lambda_j - \xi g_{rk} = g_{rk} \quad (r=1,\ldots,s), \\
& \sum_{j=1}^{n} b_{fj}\lambda_j + d_f^b + \xi b_{fk} = b_{fk} \quad (f=1,\ldots,h), \\
& \sum_{j=1}^{n} \lambda_j = 1, \\
& \lambda_j \geq 0 \ (j=1,\ldots,n),\ \xi : \text{URS}, \\
& d_i^{x+} \geq 0 \ (i=1,\ldots,m) \text{ and } d_f^b \geq 0 \ (f=1,\ldots,h).
\end{aligned}
\qquad (21.18)
$$

Model (21.18) drops slack variables related to desirable outputs (G) so that they are considered as equality constraints. The other groups of constraints on inputs and undesirable outputs maintain slacks so that they can be considered as inequality constraints. The description on inputs is also applicable to undesirable outputs.

Model (21.18) has the following dual formulation:

$$\begin{aligned}
\text{Minimize} \quad & -\sum_{i=1}^{m} v_i x_{ik} - \sum_{r=1}^{s} u_r g_{rk} + \sum_{f=1}^{h} w_f b_{fk} + \sigma \\
s.t. \quad & -\sum_{i=1}^{m} v_i x_{ij} - \sum_{r=1}^{s} u_r g_{rj} + \sum_{f=1}^{h} w_f b_{fj} + \sigma \ge 0 \qquad (j = 1,\ldots,n), \\
& \sum_{r=1}^{s} u_r g_{rk} + \sum_{f=1}^{h} w_f b_{fk} = 1, \qquad (21.19) \\
& v_i \ge \varepsilon_s R_i^x \qquad (i = 1,\ldots,m), \\
& u_r : \text{URS} \qquad (r = 1,\ldots,s), \\
& w_f \ge \varepsilon_s R_f^d \qquad (f = 1,\ldots,h), \\
& \sigma : \text{URS}.
\end{aligned}$$

An important feature of Model (21.19) is that the dual variables (u_r: URS for $r = 1,\ldots,s$) are unrestricted in their signs because Model (21.18) eliminates slacks related to desirable outputs.

A unified efficiency score, or $UEM\left(DC\right)_v^{R*}$, of the k-th DMU under managerial disposability becomes

$$\begin{aligned}
UEM\left(DC\right)_v^{R*} &= 1 - \left[\xi^* + \varepsilon_s \left(\sum_{i=1}^{m} R_i^x d_i^{x+*} + \sum_{f=1}^{h} R_f^b d_f^{b*}\right)\right] \\
&= 1 - \left[-\sum_{i=1}^{m} v_i^* x_{ik} - \sum_{r=1}^{s} u_i^* g_{rk} + \sum_{f=1}^{h} w_f^* b_{fk} + \sigma^*\right],
\end{aligned} \qquad (21.20)$$

where all variables are determined on the optimality of Models (21.18) and (21.19). The equation within the parenthesis, obtained from the optimality of Models (21.18) and (21.19), indicates the level of unified inefficiency under managerial disposability. The unified efficiency, along with a possible occurrence of DC, is obtained by subtracting the level of inefficiency from unity.

To discuss a mathematical rationale for why Models (21.18) and (21.19) can identify a possible occurrence of DC, this chapter needs to discuss the following proposition:

Proposition 21.2: A supporting hyperplane of the k-th DMU, measured by Models (21.18) and (21.19), under managerial disposability becomes

$-\sum_{i=1}^{m} v_i x_i - \sum_{r=1}^{s} u_r g_r + \sum_{f=1}^{h} w_f b_f + \sigma = 0$, where $v_i \geq \varepsilon_s R_i^x$ ($i = 1,\ldots,m$), u_r: URS ($r = 1,\ldots,s$) and $w_f \geq \varepsilon_s R_f^b$ ($f = 1,\ldots,h$) are parameters (slopes) for indicating the direction of the supporting hyperplane and σ indicates the intercept of the supporting hyperplane. The parameters of the supporting hyperplane are all unknown and need to be measured by the following equations:

$$-\sum_{i=1}^{m} v_i x_{ij} - \sum_{r=1}^{s} u_r g_{rj} + \sum_{f=1}^{h} w_f b_{fj} + \sigma = 0,\ j \in RS_k \text{ and } \sum_{r=1}^{s} u_r g_{rk} + \sum_{f=1}^{h} w_f b_{fk} = 1. \quad (21.21)$$

where RS_k stands for a reference set of the k-th DMU which is measured by Model (21.18).

[Proof] The proposition can be easily proved by following Proposition 21.1 with minor changes. However, this chapter documents a shortcut to prove Proposition 21.2 by complementary slackness conditions (CSCs) between primal and dual models. The proposed proof may be a straightforward manner. That is, the CSCs between Models (21.18) and (21.19) contain the following equation:

$$\left(-\sum_{i=1}^{m} v_i x_{ij} - \sum_{r=1}^{s} u_r g_{rj} + \sum_{f=1}^{h} w_f b_{fj} + \sigma \right) \lambda_j = 0 \quad \text{for} \quad j = 1,\ldots,n. \quad (21.22)$$

A reference set for the k-th DMU (in RS_k) consists of DMUs that have $\lambda_j > 0$ for $j \in RS_k$. Equation (21.22) is obtained from the CSCs between Models (21.18) and (21.19). The first part of Equations (21.21) is from Equation (21.22) and the second part is obtained from the second constraint of Model (21.19).

Q.E.D.

The supporting hyperplane is expressed by $-v^* x - u^* g + w^* b + \sigma^* = 0$ on an efficiency frontier in the case when three production factors have a single component. The supporting hyperplane in Figure 21.3 is expressed by $b = (v^* x + u^* g - \sigma^*)/w^*$. Since both w^* and v^* are positive, the slope of the supporting hyperplane is determined by u^* and σ^*, both of which are unrestricted in their signs.

After solving Model (21.19), this chapter can identify a possible occurrence of DC, or eco-technology innovation, by the following rule along with the assumption on a unique optimal solution:

(a) If $u_r^* = 0$ for some (at least one) r, then "weak DC" occurs on the k-th DMU,
(b) If $u_r^* < 0$ for some (at least one) r, then "strong DC" occurs on the k-th DMU and
(c) If $u_r^* > 0$ for all r, then "no DC" occurs on the k-th DMU.

Note that if $u_r^* < 0$ for some r and $u_{r'}^* = 0$ for the other r', then the weak and strong DCs coexist on the k-th DMU. This chapter considers it as the strong DC, so indicating technology innovation on undesirable outputs. Here, it is important to add that $u_r^* < 0$ for all r is the best case because an increase in any desirable output always decreases an amount of undesirable outputs. Meanwhile, if $u_r^* < 0$ is identified for some r, then it indicates that there is a chance to reduce an amount of undesirable output(s). Therefore, we consider the second case as an occurrence of DC. The three concerns (e.g., multiple solutions and slack adjustment) on UC are also applicable to DC.

21.5 COAL-FIRED POWER PLANTS IN UNITED STATES

21.5.1 Data

Returning to Chapter 18, this chapter uses a data set on 68 PJM's coal-fired power plants in 2010. The data source is the Environmental Protection Agency (EPA) database: eGRID year 2010, http://www.epa.gov/cleanenergy/energy-resources/egrid/index.html. The data set consists of 68 coal-fired power plants operating in the PJM Interconnection. This chapter excludes plants used for combined heat and power and those under mixed combustion with biomass fuel.

Data Accessibility: The original data set used in this chapter can be found in the supplementary section of Sueyoshi and Goto (2016).

The generation of each power plant is characterized by a single desirable output, two inputs and three undesirable outputs. The desirable output is measured by the amount of annual net generation (MWh: Megawatt hours). The two inputs are measured by (a) the nameplate capacity (MW: Megawatt) and (b) the amount of annual heat input (MMBtu). The undesirable outputs are measured by three items: (a) the annual amount of NO_x emissions (tons), (b) the annual amount of SO_2 emissions (tons) and (c) the annual amount of CO_2 emissions (tons) at each power plant. The first two belong to acid rain gases, but not GHG. The last one belongs to GHG. In this chapter, each observed data is divided by the factor average to avoid the case where a large data dominates the other small data in DEA computation. Thus, the data set used in the proposed DEA formulations is structured by index numbers so that they are unit-less.

The type of primary fuel of power plants is classified into two categories. One of the two categories is the bituminous (BIT) coal that is relatively soft coal that contains a tarlike substance, called "bitumen". The other category is the sub-bituminous (SUB) coal that is a type of coal whose properties range from those of lignite to those of bituminous coal and are used primarily as fuel for steam-electric power generation. See Chapter 18 for a a description of coal categories.

21.5.2 Occurrence of Congestion

Table 21.1 summarizes *UEN* (unified efficiency under natural disposability), dual variables and the occurrence of UC on coal-fired power plants. For example, the first power plant exhibits a degree of *UEN* of 0.9625. The power plant exhibits strong UC.

The *UEN* of the power plants becomes 0.9752 for bituminous and 0.9338 for sub-bituminous on total average, respectively. The result on UC indicates that the power plant may face an occurrence of a capacity limit on generation and a line limit on transmission. The status on "No" implies that UC does not occur, for example, in the seventeenth power plant. Thus, Table 21.1 indicates that the occurrence of UC due to a capacity constraint of generation and transmission may be found in most coal-fired power plants with both bituminous and sub-bituminous coal combustions.

Meanwhile, Table 21.2 summarizes *UEM* (unified efficiency under managerial disposability), dual variables and the occurrence of DC due to eco-technology innovation. For example, the first power plant exhibits a degree of *UEM* of 1.000 and the power plant exhibits no DC, indicating that it has no such eco-technology innovation. As found in Table 21.2, the status of strong DC implies that a power plant may have a potential on eco-technology innovation. The table indicates that such an occurrence of DC is found on a limited number of coal-fired power plants. Thus, it is suggested that a fuel mix for electricity generation should shift from coal to gas combustion because the latter plants produce more electricity and lower amounts of NO_x, SO_2 and CO_2 emissions than the former plants. Of course, this chapter clearly understands that such a fuel shift is not easy and should be considered under the fuel mix strategy of each electric power company and each nation.

Finally, the comparison between Tables 21.1 and 21.2 reveals that the number of DC occurrences, due to eco-technology innovation, is much lower than that of UC occurrences, due to the capacity limit on generation and the line limit on transmission in US coal-fired power plants. This provides us with an important policy implication on a fuel mix for electricity generation, as mentioned above.

21.6 SUMMARY

This chapter discussed a new use of DEA environmental assessment to measure an occurrence of DC, or eco-technology innovation for our sustainable growth, regrading US electric power plants in a comparison with an occurrence of UC in these generation plants. The identification of UC is important to avoid cost increases, tariff increases and shortages of electricity in the electric power industry. This type of congestion is often incurred by a capacity limit on generation and transmission that constitutes a serious obstacle to the efficient use of production factors for electricity supply to consumers. Meanwhile, the identification of DC is important in finding a possible occurrence of eco-technology innovation that can be

TABLE 21.1 Unified efficiency measures, dual variables and type of undesirable congestion (UC)

		UEN			
		Bituminous			
Plant No.	Efficiency Score	NO_x	SO_2	CO_2	UC
1	0.9625	−0.0629	0.0182	1.3073	Strong
2	0.9539	0.0458	−0.0151	0.7104	Strong
3	1.0000	1.4439	−0.5288	11.6587	Strong
4	1.0000	0.0472	−0.0168	0.4172	Strong
5	1.0000	−0.0026	0.0785	0.5281	Strong
6	1.0000	−0.0144	0.3202	0.4347	Strong
7	1.0000	1.8488	0.1051	−0.1169	Strong
8	1.0000	0.0666	−0.0220	1.0335	Strong
9	1.0000	0.0329	−0.0083	0.2796	Strong
10	0.9303	−0.0215	0.0063	0.4481	Strong
11	1.0000	−2.5143	0.5506	32.5014	Strong
12	0.8467	−0.1267	0.0708	2.9385	Strong
13	0.9752	−0.0232	−0.0037	0.4755	Strong
14	0.8860	0.1535	−0.0508	2.3827	Strong
15	0.9562	−0.0065	0.0001	0.2228	Strong
16	1.0000	−0.0044	0.0001	0.1499	Strong
17	1.0000	0.0076	0.2086	0.1000	No
18	0.9385	−0.0045	0.0676	0.6044	Strong
19	0.9693	0.0280	−0.0098	0.3119	Strong
20	1.0000	0.0370	−0.0132	0.4173	Strong
21	1.0000	−0.7091	0.1101	7.9718	Strong
22	1.0000	−1.0981	0.0226	10.4581	Strong
23	1.0000	−4.3331	0.0892	41.2671	Strong
24	0.9583	3.0147	−1.1221	35.0650	Strong
25	1.0000	2.4175	−0.8998	28.1193	Strong
26	1.0000	0.0243	−0.0091	0.2830	Strong
27	1.0000	−0.3632	−0.0502	1.5046	Strong
28	0.9254	−0.0388	0.0112	1.5896	Strong
29	1.0000	−0.0123	0.0038	0.2556	Strong
30	1.0000	−0.1063	−0.0279	1.8139	Strong
31	1.0000	−0.0110	−0.0018	0.1531	Strong
32	0.9454	−0.0253	0.3412	2.7654	Strong
33	1.0000	−0.0149	0.2605	0.2951	Strong
34	1.0000	−0.0131	−0.0075	0.2666	Strong
35	1.0000	0.0244	−0.0066	0.2043	Strong
36	0.9168	−0.3172	0.0674	4.1103	Strong
37	0.9799	0.1071	−0.0355	1.6629	Strong
38	1.0000	−0.0149	−0.0024	0.3054	Strong
39	1.0000	−0.0063	−0.0034	0.3174	Strong
40	1.0000	−0.1658	−0.0411	1.3974	Strong

TABLE 21.1 (Continued)

UEN					
Bituminous					
Plant No.	Efficiency Score	NO_x	SO_2	CO_2	UC
41	0.8300	0.1169	−0.0435	1.3592	Strong
42	0.9279	0.1333	−0.0821	3.6161	Strong
43	0.9792	−0.2411	0.0419	3.2156	Strong
44	0.9169	−0.0808	0.0451	1.8723	Strong
45	1.0000	−0.0233	0.4408	0.3708	Strong
46	0.8828	−0.3819	0.0812	4.9489	Strong
47	0.9674	−0.0174	0.4143	0.3689	Strong
48	1.0000	−0.0021	0.0279	0.2259	Strong
49	0.9778	−0.0001	0.0271	0.3324	Strong
50	1.0000	−0.0296	15.6218	191.4309	Strong
51	1.0000	0.1081	0.0020	0.1401	No
52	1.0000	−0.1915	0.0366	2.0241	Strong
53	1.0000	−0.2501	0.0478	2.6438	Strong
54	1.0000	0.0011	0.0230	0.2771	No
55	1.0000	0.0015	0.0292	0.3516	No
56	1.0000	0.0109	0.4390	0.0978	No
57	0.9616	−0.0583	0.0169	1.2134	Strong
Avg.	0.9752	−0.0283	0.2917	7.2123	
Max.	1.0000	3.0147	15.6218	191.4309	
Min.	0.8300	−4.3331	−1.1221	−0.1169	
S.D.	0.0407	0.9248	2.0823	26.3537	
Sub-bituminous					
58	0.9104	0.1258	−0.0275	1.0466	Strong
59	0.9030	0.1950	−0.0426	1.6225	Strong
60	0.8928	−0.1681	0.0422	2.2976	Strong
61	1.0000	−0.1970	−0.0264	0.8359	Strong
62	0.8980	−0.0145	−0.0035	0.3055	Strong
63	0.9075	0.0789	−0.0211	0.6529	Strong
64	0.8943	−0.0205	0.0115	0.4754	Strong
65	1.0000	−0.0089	−0.0049	0.1610	Strong
66	1.0000	−0.0745	0.0162	0.9783	Strong
67	0.9531	−0.1897	0.0460	3.8360	Strong
68	0.9131	−2.6907	0.5006	24.1199	Strong
Avg.	0.9338	−0.2695	0.0446	3.3029	
Max.	1.0000	0.1950	0.5006	24.1199	
Min.	0.8928	−2.6907	−0.0426	0.1610	
S.D.	0.0455	0.8130	0.1539	6.9851	

(a) Source: Sueyoshi and Goto (2016).

TABLE 21.2 Unified efficiency, dual variables and type of desirable congestion (DC)

UEM			
Bituminous			
Plant No.	Efficiency Score	Net Generation	DC
1	1.0000	0.2805	No
2	0.9811	0.0625	No
3	0.9991	−0.3498	Strong
4	0.9782	0.0131	No
5	1.0000	0.0197	No
6	1.0000	0.0188	No
7	1.0000	0.0149	No
8	1.0000	0.0487	No
9	1.0000	0.0847	No
10	1.0000	0.0333	No
11	1.0000	−2.5259	Strong
12	1.0000	−1.3858	Strong
13	0.9779	0.0041	No
14	1.0000	−1.5174	Strong
15	1.0000	0.0566	No
16	1.0000	0.0938	No
17	1.0000	0.0664	No
18	0.9992	0.0049	No
19	0.9995	0.0021	No
20	0.9972	0.3554	No
21	1.0000	0.5236	No
22	1.0000	−0.8648	Strong
23	0.9989	−0.8125	Strong
24	0.9759	−1.1478	Strong
25	1.0000	−2.4146	Strong
26	1.0000	0.0048	No
27	1.0000	0.0061	No
28	1.0000	0.0135	No
29	1.0000	0.0021	No
30	0.9977	0.0747	No
31	1.0000	0.0019	No
32	1.0000	0.0414	No
33	1.0000	−0.0994	Strong
34	0.9999	0.0022	No
35	1.0000	0.0019	No
36	0.9979	−0.2491	Strong
37	1.0000	0.0405	No
38	0.9997	0.0044	No
39	1.0000	0.0049	No
40	0.9976	0.0165	No

TABLE 21.2 (Continued)

Plant No.	Efficiency Score	Net Generation	DC
	UEM		
	Bituminous		
41	1.0000	−0.0353	Strong
42	0.9966	−0.1231	Strong
43	0.9974	0.1252	No
44	1.0000	−0.8406	Strong
45	1.0000	−0.0063	Strong
46	0.9983	−0.2394	Strong
47	0.9961	−0.0756	Strong
48	1.0000	0.0026	No
49	0.9996	0.0046	No
50	1.0000	2.8589	No
51	1.0000	0.0037	No
52	1.0000	0.4564	No
53	1.0000	0.7201	No
54	1.0000	0.0826	No
55	1.0000	0.1142	No
56	1.0000	0.0693	No
57	1.0000	0.3846	No
Avg.	0.9980	−0.1047	
Max.	1.0000	2.8589	
Min.	0.9759	−2.5259	
S.D.	0.0056	0.7212	
	Sub–bituminous		
58	0.9632	0.0252	No
59	1.0000	0.4667	No
60	0.9627	−0.1085	Strong
61	0.9773	0.0017	No
62	0.9627	−0.0016	Strong
63	0.9681	0.0083	No
64	0.9601	0.0017	No
65	0.9784	0.1136	No
66	0.9770	0.0233	No
67	0.9768	0.1006	No
68	1.0000	0.4334	No
Avg.	0.9751	0.0968	
Max.	1.0000	0.4667	
Min.	0.9601	−0.1085	
S.D.	0.0141	0.1840	

(a) Source: Sueyoshi and Goto (2016).

effectively used to avoid air pollution so that electric power companies can satisfy governmental standards on environmental protection.

This chapter finds that the UC incurred by a capacity limit on generation and transmission, may occur in most coal-fired power plants, regardless of the type of coal: bituminous or sub-bituminous. Meanwhile, the occurrence of DC, due to eco-technology innovation for sustainable growth, may occur on a limited number of coal-fired power plants. The result on congestion is useful in determining which technology or type of coal-fired power plant should receive investment to facilitate eco-technology innovation for the sustainable management of electric power companies.

22

MARGINAL RATE OF TRANSFORMATION AND RATE OF SUBSTITUTION

22.1 INTRODUCTION

This chapter[1] discusses the use of DEA environmental assessment to measure the marginal rate of transformation (MRT) and rate of substitution (RSU) among production factors, in particular between desirable and undesirable outputs. In the methodological concern, this chapter is an extension of Chapter 9 which discusses the quantitative approach on MRT within a conventional framework of DEA. In discussing the new approach for measuring MRT and RSU, it is necessary for us to describe that both estimates depend upon the magnitude and sign of dual variables, or multipliers, obtained from the proposed formulations. A problem associated with the estimation of MRT and RSU is that the two measures often become unstable, so often being inconsistent with economic and/or business reality, because DEA does not assume any functional form among the three production factors (*X*, *G* and *B*). The non-parametric feature makes the DEA-based MRT and RSU measurement very difficult because it often produces large or small multipliers. As a result, DEA does not attain a high level of estimation accuracy and thus has limited practicality. Consequently, almost no previous studies

[1] This chapter is partly based upon the article: Sueyoshi, T. and Yuan, Y. (2016a) Marginal rate of transformation and rate of substitution measured by DEA environmental assessment: comparison among European and North American nations. *Energy Economics*, **56**, 270–287.

Environmental Assessment on Energy and Sustainability by Data Envelopment Analysis, First Edition. Toshiyuki Sueyoshi and Mika Goto.

have explored the measurement of MRT and RSU, only paying attention to efficiency measurement. Exceptions can be found in Cooper *et al.* (2000)[2] and Asmild *et al.* (2006)[3].

To overcome such a research difficulty, the proposed measurement of MRT and RSU pays attention to two research concerns. One of the two concerns is how to restrict dual variables in such a manner that computational results are consistent with our expectation and prior information. Conventional approaches are often referred to as "multiplier restriction" methods. For example, Charnes *et al* (1989)[4] and Thompson *et al.* (1990)[5] provided their descriptions on "CR: cone ratio approach" and "AR: assurance region analysis," respectively, which have long served as well-known multiplier restriction methods. See Chapter 4 on the multiplier restriction methods. This chapter extends the multiplier restriction to environmental assessment by proposing "explorative analysis." An important feature of the proposed approach is that it can avoid prior information on multiplier restriction. It can be easily imagined that we have a difficulty in accessing prior information for restricting multipliers. The proposed explorative analysis can overcome such a difficulty related to the multiplier restriction. The other important research concern is that this chapter incorporates the concept of desirable congestion (DC), discussed in Chapter 21, into the measurement of MRT and RSU. This chapter considers that an occurrence of DC is more important than undesirable congestion (UC) because the former occurrence indicates eco-technology innovation. As mentioned previously, the underlying philosophy of this book is that we are interested in policy and business concerns for reducing the global warming and climate change by eco-technology innovation (e.g., clean coal technology in the electricity industry and hybrid cars in the automobile industry) along with related managerial challenges (e.g., a fuel mix strategy from coal to natural gas and other renewable energies in the electricity industry).

This chapter acknowledges that the research of Cooper *et al.* (2000) was the first to provide DEA-based MRT and RSU in the framework of inputs and desirable outputs. Acknowledging their contribution, however, this chapter needs to mention that they have never considered an output separation into desirable and undesirable categories. Consequently, it is impossible for us to incorporate their approach on MRT and RSU into the environmental assessment proposed in this chapter.

[2] Cooper, W.W., Park, K.S. and Ciurana, J.P. (2000) Marginal rates and elasticities of substitution with additive models in DEA. *Journal of Productivity Analysis*, **13**, 105–123.

[3] Asmild, M., Paradi, J.C. and Reese, D.N. (2006) Theoretical trade-off analysis using DEA. *OMEGA*, **34**, 337–343.

[4] Charnes, A., Cooper, W.W., Wei, Q.L. and Huang, Z.M. (1989) Cone ratio data environmental analysis and multi-objective programming. *International Journal of System Science*, **20**, 1099–1118.

[5] Thompson, R.G., Singleton, F.D., Thrall, R.M. and Smith, B.A. (1986) Comparative site evaluation for locating a high-energy physics lab in Texas. *Interface*, **16**, 35–49.

The remainder of this chapter is organized as follows: Section 22.2 discusses economic and strategic concepts used in this chapter. Section 22.3 briefly discusses the possible occurrence of DC, as an extension of Chapter 21, and extends it to the measurement of MRT and RSU. Section 22.4 discusses the analytical structure of MRT and RSU. Section 22.5 describes multiplier restriction for the two measures. Section 22.6 discusses the explorative analysis, incorporated into the proposed environmental assessment. Section 22.7 applies the proposed approach to nations in Europe and North America. Section 22.8 summarizes this chapter.

22.2 CONCEPTS

22.2.1 Desirable Congestion

Returning to Figure 15.6 in Chapter 15, Figure 22.1 depicts the unification process between desirable and undesirable outputs. Since Chapter 15 provides a description on the top of Figure 22.1, this chapter starts with the middle of Figure 22.1, which describes an occurrence of DC in the g–b space, where g and b indicate desirable and undesirable outputs in horizontal and vertical coordinates, respectively. To simplify our visual description, we consider a case of a single component of the three vectors for production factors (i.e., X, G and B).

Here, in shifting our description from the top to the middle and bottom of Figure 22.1, we need to consider that "undesirable outputs are byproducts of desirable outputs." In the middle and bottom of the figure, the negative slope of the supporting hyperplane indicates the occurrence of strong DC, or eco-technology innovation for reducing the undesirable output (i.e., industrial pollution). In contrast, a positive slope implies the opposite case (i.e., no occurrence of DC), so there is no DC. An occurrence of weak DC is identified between strong and no DC, so being weak DC.

The bottom of Figure 22.1 depicts the three types of DC: no, weak and strong DC in the g–b space. Under the occurrence of no DC, damages to return (DTR) is separated into three categories: constant, increasing and decreasing. An occurrence of eco-technology innovation under managerial disposability is identified by negative DTR, or strong DC, in the case where three production factors have a single component. When the three production factors have multiple components, the occurrence of strong DC becomes a necessary condition of negative DTR, but not sufficient. See Chapter 23 on the classification of DTR.

22.2.2 MRT and RSU

Figure 22.2 depicts RSU among an input (x), a desirable output (g) and an undesirable output (b). As depicted in Figure 22.2, for example, dg/dx measures the degree of MRT where d stands for the derivative on a functional form between

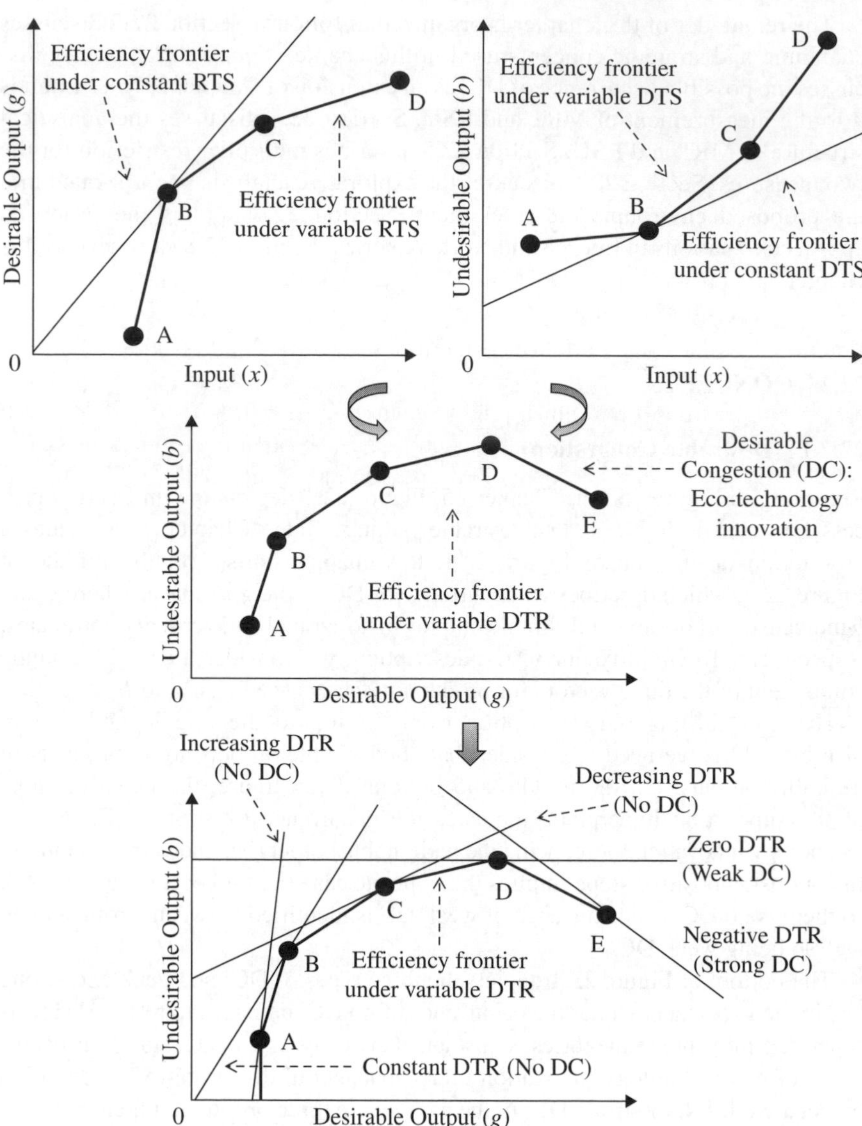

FIGURE 22.1 Damages to return under managerial disposability (a) The middle of the figure assumes that an undesirable output (*b*) is a byproduct of a desirable output (*g*). (b) The type of DTR is classified into decreasing, constant and increasing under no DC. See Chapter 23 for a description of DTR.

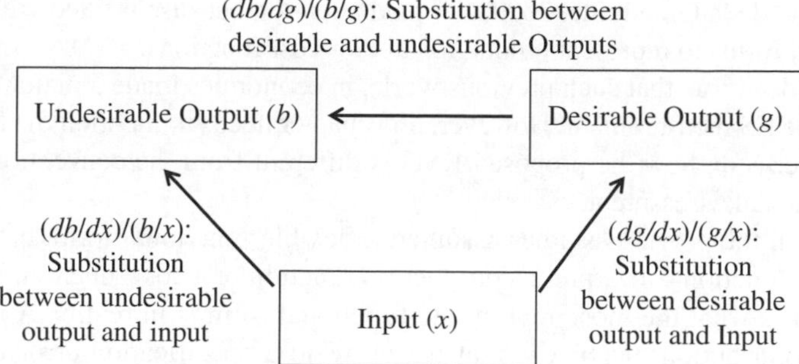

FIGURE 22.2 Marginal rate of transformation and rate of substitution

g and x. In a similar manner, the degree of MRT is applicable to other relationships among x, g and b. It can be intuitively considered that the MRT measures the rate at which a DMU willingly exchanges an amount of a production factor for another one.

The degree of $(dg/dx)/(g/x)$ measures a magnitude of RSU that is considered as "scale elasticity" when the two vectors (X and G) have a single component. The analytical relationship is applicable to the other two cases (b and x; g and b) in the manner of $(db/dx)/(b/x)$ and $(db/dg)/(b/g)$.

Here, it is important to note that RSU, explored in this chapter, is different from elasticity of substitution (ES) that serves as a fundamental concept of economics. Both measures have similarities and differences in many aspects. To describe the position of RSU more clearly, this chapter reviews ES and then compares it with RSU.

Elasticity of Substitution (ES): The concept of ES has a long history in economics, dating back to the original concept proposed by Hicks (1932)[6]. He used ES as a methodology for analyzing between capital and labor income shares in a growing economy, along with constant RTS and a natural technology change. Allen and Hicks (1934)[7] and Allen (1938)[8] extended the original concept of ES. Uzawa (1962)[9] further analyzed the concept, becoming a dominant concept of ES, conventionally referred to as "Allen–Uzawa elasticity" in modern economics. Later, Morishima (1967)[10] proposed an alternative to the Allen–Uzawa elasticity, referred to as "Morishima elasticity," to which many economists (Davis and Gauger, 1996;

[6] Hicks, J.R. (1932) *Theory of Wages*. Macmillan, London.

[7] Allen, R.G.D. and Hicks, J.R. (1934) A recommendation of the theory of Value II. *Economica* **1**, 196–219.

[8] Allen, R.G.D. (1938) *Mathematical Analysis for Economists*, Macmillan, London.

[9] Uzawa, H. (1962) Production functions with constraint elasticities of substitution. *Review of Economic Studies*, **30**, 291–299.

[10] Morishima, M. (1967) A few suggestions on the theory of elasticity (in Japanese). *Keizai Hyoron (Economic Review)*, **16**, 144–150.

Klump and de la Grandville, 2000) [11,12] paid attention because it fitted with various functional forms to represent production and/or cost functions.

It is indeed true that such previous works in economics made a major contribution on the ES measurement. However, this chapter needs to mention the following five concerns on how the proposed RSU is different from the conventional ES in environmental assessment.

First, all the previous studies assumed a flexible functional form, or so-called "distance function," to express a production function or a cost function. However, none knows what the most appropriate functional form is in reality. A methodological contribution of DEA is that it can avoid a specification error of such a distance function.

Second, another assumption associated with ES was that a functional form used for measuring the degree should be twice differentiable because both Allen–Uzawa and Morishima ES measures needed to take a derivative on a distance function that was usually a quadratic functional form. For example, Hicks (1932, pp. 241–246) proposed the following ES between two inputs:

$$ES_{12} = \left(\frac{\partial f}{\partial x_1} \frac{\partial f}{\partial x_2} \right) \Bigg/ \left(f(x_1, x_2) \frac{\partial^2 f}{\partial x_1 \partial x_2} \right), \tag{22.1}$$

where the symbol (f) stands for a functional form as well as x_1 and x_2 indicate the amount of inputs, respectively. The symbol (∂) indicates a partial derivative of a functional form. All the proceeding studies in economics followed the above definition on ES along with their extensions. Acknowledging the importance of the original definition on ES and its extension (e.g., Morishima, 1967), this chapter does not follow that definition of ES, rather measuring the derivative of a distance function by dual variables, or multipliers, of linear programming because this chapter is concerned with the measurement of RSU, so not conventional ES, within the framework of DEA environmental assessment.

Third, previous studies in economics assumed that all entities in private and public sectors worked efficiently without any loss or any inefficiency. Consequently, the previous studies could measure their ES measures on an efficiency frontier, or on the surface of a distance function to represent the efficiency frontier. That was for mathematical convenience. The reality is much different from the economic assumption. In contrast, DEA can measure the degree of efficiency or inefficiency without the assumption. Thus, the DEA-based RSU

[11] Davis, G.C. and Gauger, J. (1996) Measuring substitution in monetary-access demand systems. *Journal of Business and Economic Statistics*, **14**, 203–209.

[12] Klup, R. and de la Grandville, O. (2000) Economic growth and the elasticity of substitution: two theorems and some suggestions. *American Economic Reviewer.* **90**, 282–291.

measurement becomes conceptually more practical than the ES measurement in economics. However, nobody knows which is accurate or not in reality because all methodologies are approximations.

Fourth, previous studies in economics often imposed many assumptions such as linear homogeneity and symmetry on input prices on a functional form. In contrast, DEA can avoid such economic assumptions for specifying a functional form. Thus, DEA has more practicality than the conventional approaches proposed in economics.

Finally, Sueyoshi and Goto (2012d) discussed how to measure MRT and RSU within the computational framework of DEA environmental assessment. Their study did not measure the degree of these scale-related measures. The measurement is important because when we examine a large production process, as found in energy sectors, the environmental assessment needs to consider these scale-related measures. Furthermore, their estimates may become unstable because an efficiency frontier, measured by DEA, consists of a piece-wise linear polyhedron surface, so not assuming any functional form. Therefore, the multiplier (dual variable) restriction method is necessary for obtaining reliable MRT and RSU estimates on the frontier. The methodological issue will be discussed in this chapter, hereafter.

22.3 A POSSIBLE OCCURRENCE OF DESIRABLE CONGESTION (DC)

As depicted in Figure 22.1, this chapter incorporates a possible occurrence of DC into the proposed environmental assessment under managerial disposability because we are interested in eco-technology innovation for pollution reduction. To examine the occurrence, this chapter starts with the following model (22.2) for our descriptive convenience:

$$\begin{aligned}
&\text{Maximize} \quad \xi + \varepsilon_s\left(\sum_{i=1}^{m} R_i^x d_i^{x+} + \sum_{f=1}^{h} R_f^b d_f^b\right) \\
&s.t. \quad \sum_{j=1}^{n} x_{ij}\lambda_j - d_i^{x+} = x_{ik} \ (i=1,\dots,m), \\
&\sum_{j=1}^{n} g_{rj}\lambda_j - \xi g_{rk} = g_{rk} \ (r=1,\dots,s), \\
&\sum_{j=1}^{n} b_{fj}\lambda_j + d_f^b + \xi b_{fk} = b_{fk} \ (f=1,\dots,h), \\
&\sum_{j=1}^{n} \lambda_j = 1, \\
&\lambda_j \ge 0 \ (j=1,\dots,n), \ \xi: \text{URS}, \\
&d_i^{x+} \ge 0 \ (i=1,\dots,m) \text{ and } d_f^b \ge 0 \ (f=1,\dots,h).
\end{aligned} \tag{22.2}$$

The model has the following dual formulation:

$$
\begin{aligned}
\text{Minimize} \quad & -\sum_{i=1}^{m} v_i x_{ik} - \sum_{r=1}^{s} u_r g_{rk} + \sum_{f=1}^{h} w_f b_{fk} + \sigma \\
s.t. \quad & -\sum_{i=1}^{m} v_i x_{ij} - \sum_{r=1}^{s} u_r g_{rj} + \sum_{f=1}^{h} w_f b_{fj} + \sigma \geq 0 && (j = 1,\ldots,n), \\
& \sum_{r=1}^{s} u_r g_{rk} + \sum_{f=1}^{h} w_f b_{fk} = 1, && (22.3) \\
& v_i \geq \varepsilon_s R_i^x && (i = 1,\ldots,m), \\
& u_r: \text{URS} && (r = 1,\ldots,s), \\
& w_f \geq \varepsilon_s R_f^b && (f = 1,\ldots,h), \\
& \sigma: \text{URS}.
\end{aligned}
$$

As discussed in Chapter 21, an important feature of Model (22.3) is that the dual variables (u_r: URS for $r = 1,\ldots,s$) are unrestricted in these signs because Model (22.2) does not have slacks related to desirable outputs (G).

A unified efficiency measure, or $UEM\,(DC)_v^{R*}$, of the k-th DMU under managerial disposability becomes as follows:

$$
\begin{aligned}
UEM\,(DC)_v^{R*} &= 1 - \left[\xi^* + \varepsilon_s \left(\sum_{i=1}^{m} R_i^x d_i^{x+*} + \sum_{f=1}^{h} R_f^b d_f^{b*} \right) \right] \\
&= 1 - \left[-\sum_{i=1}^{m} v_i^* x_{ik} - \sum_{r=1}^{s} u_r^* g_{rk} + \sum_{f=1}^{h} w_f^* b_{fk} + \sigma^* \right], \qquad (22.4)
\end{aligned}
$$

where all variables are determined on the optimality of Models (22.2) and (22.3). The equation within the parenthesis, obtained from the optimality of Models (22.2) and (22.3), indicates the level of unified inefficiency under managerial disposability. The unified efficiency is obtained by subtracting the level of inefficiency from unity.

As discussed in Chapter 21, we solve Model (22.3) and then identify a possible occurrence of DC, or eco-technology innovation, by the following rule along with the assumption on a unique optimal solution:

(a) If $u_r^* = 0$ for some (at least one) r, then "weak DC" occurs on the k-th DMU,
(b) If $u_r^* < 0$ for some (at least one) r, then "strong DC" occurs on the k-th DMU and
(c) If $u_r^* > 0$ for all r, then "no DC" occurs on the k-th DMU.

If $u_r^* < 0$ for some r and $u_{r'}^* = 0$ for the other r', then the occurrence of weak and strong DCs coexist on the k-th DMU. This chapter considers it as strong DC, so indicating eco-technology innovation on undesirable outputs. It is important to note that $u_r^* < 0$ for all r is the best case because an increase in any desirable output always decreases the amount of undesirable outputs. Meanwhile, if $u_r^* < 0$ is identified for some r, then it indicates that there is a chance to reduce the amount of undesirable output(s). Therefore, this chapter also considers this case as an occurrence of DC.

22.4 MEASUREMENT OF MRT AND RSU UNDER DC

In the case of a single component of the three production factors, this chapter estimates the degrees of MRT and RSU regarding the k-th DMU under managerial disposability as follows:

$$MRT^R \text{ of } b \text{ to g} = db/dg = u^*/w^* \;\&\; RSU^R \text{ of } b \text{ to g} = (db/dg)/(b/g) = u^* g_k \big/ w^* b_k \text{ and}$$
$$MRT^R \text{ of } b \text{ to x} = db/dx = v^*/w^* \;\&\; RSU^R \text{ of } b \text{ to } x = (db/dx)/(b/x) = v^* x_k \big/ w^* b_k . \tag{22.5}$$

Since the managerial disposability has a priority order in which environmental performance comes first and operational performance comes second, this chapter considers the MRT and RSU of b to the other two production factors (g and x). The superscript R stands for radial measurement.

For efficient DMUs: This chapter can specify the degree of MRT and RSU of the k-th DMU, having multiple components of the three production factors, which are measured under managerial disposability as follows:

$$MRT_{fr}^R = \partial b_f \big/ \partial g_r = u_r^* \big/ w_f^* \;\&\; RSU_{fr}^R = \left(\partial b_f / \partial g_r\right) / \left(b_f / g_r\right) = u_r^* g_{rk} \big/ w_f^* b_{fk} \text{ and}$$
$$MRT_{fi}^R = \partial b_f \big/ \partial x_i = v_i^* \big/ w_f^* \;\&\; RSU_{fi}^R = \left(\partial b_f / \partial x_i\right) / \left(b_f / x_i\right) = v_i^* x_{ik} \big/ w_f^* b_{fk}, \tag{22.6}$$

for all $i = 1,\ldots,m$, $r = 1,\ldots,s$ and $f = 1,\ldots,h$ if the k-th DMU is efficient. The dual variables are obtained from the optimality of Model (22.3).

For inefficient DMUs: If the DMU is inefficient, the RSU measure needs to be adjusted as follows:

$$RSU_{fr}^R = \left(\partial b_f / \partial g_r\right) / \left(b_f / g_r\right) = u_r^* \bar{g}_{rk} \big/ w_f^* \bar{b}_{fk} \text{ and}$$
$$RSU_{fi}^R = \left(\partial b_f / \partial x_i\right) / \left(b_f / x_i\right) = v_i^* \bar{x}_{ik} \big/ w_f^* \bar{b}_{fk}, \tag{22.7}$$

where the upper bar indicates the amount of each production factor on an efficiency frontier after adjusting the inefficiency score and slacks. The variables are adjusted by

$$\bar{x}_{ik} = x_{ik} + d_i^{x+*} \; (i = 1,\ldots,m),\; \bar{g}_{rk} = \left(1+\xi^*\right) g_{rk} \; (r = 1,\ldots,s) \text{ and}$$
$$\bar{b}_{fk} = \left(1-\xi^*\right) b_{fk} - d_f^{b*} \; (f = 1,\ldots,h),$$

from the optimality of Model (22.2).

Here, it is necessary for us to describe three analytical concerns on MRT and RSU. First, the determination of MRT and RSU needs to project an inefficient DMU onto an efficiency frontier by slack adjustment. See Chapter 9 on slack adjustment. However, it is expected in this chapter that the influence of slack adjustment on MRT and RSU may be less than that of multiplier restriction by the proposed explorative analysis. Therefore, this chapter does not utilize the adjustment, as discussed in Chapter 9. Second, Chapter 9 describes MRT of a desirable output (g) to an input (x). So, the MRT of the k-th DMU becomes $\partial g_r / \partial x_i = v_i^* / u_r^*$. Meanwhile, the MRT of the k-th DMU on the undesirable output b to x becomes $\partial b_f / \partial x_i = v_i^* / w_f^*$ and the MRT of the k-th DMU of b to g becomes $\partial b_f / \partial g_r = u_r^* / w_f^*$. The difference can be found in the RSU measurement. See Figure 22.2. Finally, as discussed in Chapter 9, the degree of scale elasticity (for a single output) and the degree of scale economies (for multiple outputs) are defined on a projected point on an efficiency frontier. Since DEA environmental assessment has desirable and undesirable outputs, this chapter looks for the degree of scale economies. However, the degree of RSU measured by Equations (22.6) and (22.7) is related to a specific combination between two production factors. Therefore, the degree indicates "factor-specific scale economy (elasticity)." Such indicates conceptual differences between Chapter 9 for DEA and this chapter for DEA environmental assessment.

22.5 MULTIPLIER RESTRICTION

In measuring MRT and RSU, it is possible for us to incorporate prior information as multiplier restrictions so that MRT and RSU estimates may belong to an acceptable range for a user. This chapter incorporates the following restrictions on multipliers (dual variables):

$$\begin{aligned} &-1 \le u_r / w_f \le 1 \; (r = 1,\ldots,s \;\&\; f = 1,\ldots,h) \text{ and} \\ &-1 \le w_{f'} / w_f \le 1 \; (f' > f = 1,\ldots,h-1). \end{aligned} \tag{22.8}$$

The first part of Equations (22.8) indicates multiplier restriction between the r-th desirable output and the f-th undesirable output. The second part indicates the restriction between the f-th and f'-th undesirable outputs.

The proposed multiplier restriction has two analytical rationales. One rationale is that, as suggested in Chapter 2, this chapter divides each production factor by its average. Data manipulation is necessary for DEA to avoid the case where large data dominates small data in the computation. Otherwise, the computational result is often unreliable and misguiding. Thus, the data set on each production factor is unit-less. The other rationale is that the data adjustment makes it possible for the resulting multiplier to indicate the importance of each production factor in the manner that each ratio locates between –100% and 100%, as formulated in Equations (22.8).

Model (22.3), incorporating Equations (22.8), becomes

$$
\begin{aligned}
&\text{Minimize} \quad -\sum_{i=1}^{m} v_i x_{ik} - \sum_{r=1}^{s} u_r g_{rk} + \sum_{f=1}^{h} w_f b_{fk} + \sigma \\
&s.t. \quad -\sum_{i=1}^{m} v_i x_{ij} - \sum_{r=1}^{s} u_r g_{rj} + \sum_{f=1}^{h} w_f b_{fj} + \sigma \geq 0 && (j = 1,\ldots,n), \\
&\qquad \sum_{r=1}^{s} u_r g_{rk} + \sum_{f=1}^{h} w_f b_{fk} = 1, && (22.9) \\
&v_i \geq \varepsilon_s R_i^x && (i = 1,\ldots,m), \\
&\qquad u_r : \text{URS} && (r = 1,\ldots,s), \\
&\qquad\qquad w_f \geq \varepsilon_s R_f^b && (f = 1,\ldots,h), \\
&\qquad\qquad\qquad \sigma : \text{URS}, \\
&\qquad -1 \leq u_r / w_f \leq 1 && (r = 1,\ldots,s \;\&\; f = 1,\ldots,h), \\
&\qquad\qquad -1 \leq w_{f'} / w_f \leq 1 && (f' > f = 1,\ldots,h-1).
\end{aligned}
$$

In the history of DEA, the proposed multiplier restriction is often referred to as "assurance region (AR) analysis" and it has been used to reduce the number of efficient DMUs. See Chapter 4 for a description of AR analysis. That is indeed important because DEA usually produces many efficient DMUs. In contrast to the historical use, this chapter uses it for the measurement of MRT and RSU under managerial disposability. Furthermore, the proposed approach does not need any prior information on multiplier restriction, as required by AR analysis.

22.6 EXPLORATIVE ANALYSIS

Figure 22.3 visually describes Equation (22.8) as an "explorative analysis" which combines DEA-based performance assessment with multiplier restriction to obtain acceptable empirical results. It is indeed true that the proposed explorative analysis originates from the AR analysis. However, there are two differences between the approach proposed in this chapter and the AR approach.

As depicted at the top of Figure 22.3, the proposed approach adjusts a data set in such a manner that each production factor is divided by the factor average. It is possible for us to apply a DEA environmental assessment, or Model (22.9), to the adjusted data set. A problem still exists in the manner that multipliers, or dual

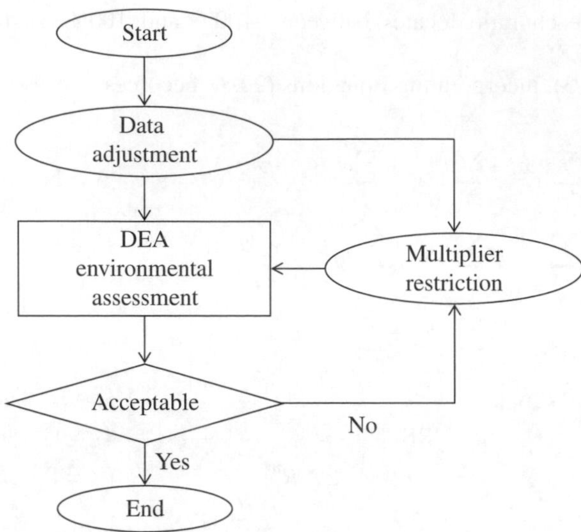

FIGURE 22.3 Explorative analysis (a) The proposed approach does not need prior information for multiplier restriction. The analytical feature is different from a conventional use of the AR analysis that needs prior information. See Chapter 4 on the AR analysis. (b) The explorative analysis implies that we investigate computational results by utilizing both a data set and additional information so that computational results can fit with our expectation and reality. (c) It is possible for us to change a combination on multiplier restriction for explorative analysis until we can satisfy MRT and RSU estimates in our expected ranges. Equations (22.8) can work most of data sets because each production factor is divided by each factor average. An exception is a data set with many zeros. In this case, we must depend on the methodologies listed in Chapters 26 and 27.

variables, become unstable in many cases. For example, they become very large and very small so that MRT and RSU estimates become very large or very small at these magnitudes. The result is often unacceptable. An approach to handle the difficulty with DEA is to restrict multipliers within an acceptable range that fits with our expectation. As discussed previously, the AR approach is one such multiplier restriction method. A problem of AR is that it needs prior information for multiplier restriction and as such is usually subjective. That is a drawback of the AR approach. In contrast, the exploration analysis does not use the prior information.

To overcome the subjectivity issue, the explorative analysis pays attention to a unique feature of the adjusted data set. That is, the data is unit-less, because each production factor is divided by the factor average, so indicating the importance of each factor. The multiplier restriction, expressed by Equations (22.8), or $-1 \le u_r / w_f \le 1$ ($r = 1,\dots,s$ and $f = 1,\dots,h$) and $-1 \le w_{f'} / w_f \le 1$ ($f' > f = 1,\dots,h-1$), specifies the range of importance among desirable and undesirable outputs by a percentile expression between –1 and 1. Thus, the proposed approach can avoid our subjectivity on multiplier restriction. It is impossible for us to utilize the proposed explorative analysis without such a data adjustment at the initial stage of DEA environment assessment.

Finally, this chapter clearly understands that the proposed approach is one of many approaches for multiplier restriction. The approach may repeat several times until the computational result on MRT and RSU satisfies our expectation, as depicted in Figure 22.3.

22.7 INTERNATIONAL COMPARISON

Using Model (22.9), this chapter compares the performance of 25 industrial nations, all of which are members of Organization for Economic Co-operation and Development (OECD). The data set used in this study is listed in the OECD library (http://www.oecd-ilibrary.org/).

Data Accessibility: the supplementary section of Sueyoshi and Yuan (2016a) provides a list of the whole data set used in this chapter.

The observed annual periods are from 2008 to 2012. All the nations are classified into the three reginal categories: Eastern European Union (EU), Western EU and North America. Eastern EU includes nations that were previously under socialism. Western EU consists of nations under capitalism[13]. North America

[13] To combat the global warming and climate change, the EU, in particular Western EU, aims to increase a share of renewable energy in its whole consumption and to achieve energy savings. Through the attainment of these targets, the EU can combat the climate change and air pollution, decrease its dependence on foreign fossil fuels and keep energy affordable. There are many energy strategies proposed in the EU. For example, the EU tries to attain more energy efficiency by accelerating investment into energy efficient buildings, eco-designed products and transport. Furthermore, the EU implements the strategic energy technology plan to accelerate the development such as on solar power, smart grids and carbon capture and storage. On the other hand, after years of economic crisis, unemployment rate is stuck near its record high of 12% and the debt mountain continues to grow. Many EU countries suffer from a sharp fall in living standards. People do not have confidence to their economic growths. Under such current serious energy and economic crises, all people in EU are more sensitive to economic and energy saving so that they pay attention to energy usage and spending. For example, people prefer small cars and electronic cars which are most popular in Europe. Many products should meet eco-design requirements for energy supply. Thus, EU countries pay more attention to environment than before even though they care economic prosperity. It can be confirmed by some EU strategies. For example, the EU countries try to reduce greenhouse gas emissions by 20% compared to 1990 levels, improve energy efficiency by 20%, produce 20% energy from renewables, by the target year of 2020. All EU countries promote new transport sector with at least 10% renewable energy. See Chapter 14 for a description on current environmental protection in EU.

contains Canada and the United States[14]. The EU nations are specified as follows: Eastern EU: Czech Republic, Estonia, Hungary, Poland, Slovak Republic, Slovenia and Western EU: Austria, Belgium, Denmark, Finland, France, Germany, Iceland, Ireland, Italy, Luxembourg, Netherlands, Norway, Portugal, Spain, Sweden, Switzerland, United Kingdom.

The data set used in this chapter consists of: (a) gross domestic product (GDP) as a single desirable output, (b) two undesirable outputs: particulate matter (PM) 2.5 and CO_2 and (c) two inputs: the number of population and the total amount of energy supply. See Sueyoshi and Yuan (2016a) for a detailed description on the data set. Descriptive statistics can be found in their article.

Table 22.1 documents the computational results obtained by the model in Equation (22.9) for 2012. For example, the degree of $UEM(DC)_v^{R*}$ regarding Austria was 0.726 in 2012. Here, $UEM(DC)_v^{R*}$ stands for unified efficiency under managerial disposability and with a possible occurrence of DC. See the first row of the table. Some countries exhibit a high level of efficiency (close to unity) in $UEM(DC)_v^{R*}$, such as France, Germany, Hungary, Iceland, Italy, Luxembourg, Portugal, Spain, Sweden, Switzerland, United Kingdom and United States. This indicates that these countries have developed balanced economy and environment efficiency, so attaining the status of social sustainability.

Table 22.2 extends Table 22.1 by summarizing average $UEM(DC)_v^{R*}$ of the three regions from 2008 to 2012. The bottom of Table 22.2 lists the p-value from

[14] The United States is the world's largest national economy. The United States has one of the world's largest and most influential financial markets. Following the financial crisis of 2007–2008, the United States economy began to recover in the second half of 2009 and as of February 2015, unemployment rate had declined from 10.0 to 5.5%. Americans have the highest average household income among OECD nations. The United States has abundant natural resources, a well-developed infrastructure and a high level of productivity. The nation is the world's second largest producer and consumer of electricity and world's third largest producer of oil and natural gas. Primary energy consumption in the United States was 18% of world total primary energy consumption in 2012. It is easy to imagine the large pollution emission produced from such a high consumption level of energy. Comparing to European people, Americans usually do not worry about their energy usages. It is easily imagined that the country pays more attention to economic recovery than environmental protection. The observation has been gradually changing in recent periods. See Chapter 14 for a description on current United States environmental protection. Canada is one of the world's wealthiest nations. Canada is unusual among developed countries in the importance of the primary sector, with the logging and oil industries being two of Canada's most important industry sectors. Canada closely resembles the United States in its market-oriented economic system and pattern of production. On the other hand, Canada is one of the few developed nations that are net exporters of energy. The vast Athabasca oil sands give Canada the world's third largest reserves of oil. Hydroelectric power is an inexpensive and relatively environmentally friendly source of abundant energy in British Columbia and Quebec. Partly because of this, Canada is also one of the world's highest per capita consumers of energy. Because of the America–Canada free trade agreement, the economies of the United States and Canada are combined together. Canada enjoys a substantial trade surplus with the United States and Canada is the American largest foreign supplier of energy including oil, gas and electricity. See Chapter 13 on energy issues. Therefore, the economies and energy consumption of two counties show the similar pattern. This proves that the two nations pay much more attention to economy, but not enough attention to the protection of environment. As discussed in Chapter 14, the situation has been gradually changing in recent years.

TABLE 22.1 Computational results for 2012: Model (21.9)

Country	$UEM(DC)_v^{R*}$	$\frac{u}{w_1}$	$\frac{u}{w_2}$	$\frac{w_1}{w_2}$	$\frac{ug}{w_1b_1}$	$\frac{ug}{w_2b_2}$
Austria	0.726	0.077	0.077	1.000	2.198	0.812
Belgium	0.779	0.116	0.116	1.000	1.552	0.661
Canada	0.674	1.000	0.000	0.000	1.258	0.000
Czech Republic	0.647	–0.057	–0.057	1.000	–0.564	–0.137
Denmark	0.720	1.000	0.014	0.014	23.924	0.245
Estonia	0.459	–1.000	0.000	0.000	–38.655	0.000
Finland	0.640	1.000	0.020	0.020	20.603	0.310
France	1.000	–0.981	–0.946	0.965	–1.393	0.909
Germany	1.000	–0.887	–0.859	0.968	–2.326	0.468
Hungary	1.000	–0.918	–0.263	0.287	–4.701	1.302
Iceland	1.000	–0.005	–0.001	0.165	–3.233	18.723
Ireland	0.707	0.020	0.020	1.000	1.399	0.393
Italy	1.000	–0.980	–0.843	0.860	–1.883	0.731
Luxembourg	1.000	0.029	0.024	0.824	6.598	64.695
Netherlands	0.915	0.116	0.116	1.000	2.961	0.299
Norway	0.852	1.000	0.087	0.087	27.240	1.560
Poland	0.693	–1.000	–0.149	0.149	–1.143	–0.099
Portugal	0.989	1.000	0.012	0.012	5.978	0.063
Slovak Republic	0.737	1.000	0.000	0.000	13.974	0.000
Slovenia	0.703	–1.000	0.000	0.000	–11.442	0.000
Spain	1.000	–0.984	–0.965	0.981	–3.077	1.081
Sweden	1.000	–0.710	–0.123	0.173	–7.982	1.542
Switzerland	1.000	–0.048	–0.046	0.944	–2.053	11.448
United Kingdom	1.000	–0.985	–0.972	0.986	–3.279	0.709
United States	1.000	–0.643	–0.613	0.953	–0.050	0.085

(a) Source: Sueyoshi and Yuan (2016a). (b) $UEM(DC)_v^{R*}$ indicates unified efficiency under managerial disposability with a possible occurrence of DC. (c) u/w_1 is MRT of PM 2.5 (w_1) to GDP (u). u/w_2 is MRT of CO_2 (w_2) to GDP (u). w_1/w_2 is MRT of CO_2 (w_2) to PM 2.5 (w_1). ug/w_1b_1 is RSU of PM 2.5 to GDP and ug/w_2b_2 is RSU of CO_2 to GDP.

the *t*-test between two regions. The table indicates that Western EU slightly outperformed North America and considerably outperformed Eastern EU during the observed period. Note that Eastern EU outperformed North America from 2008 to 2010. The mean test indicates there was a difference between the Eastern and Western EU regions in 2011 and 2012 at the significance level of 5%. Such a statistical significance could not be found from 2008 to 2010. This chapter could not find any statistically significant difference between Western EU and North America, and between Eastern EU and North America, during the observed five years.

To understand the results in Table 22.2, it is necessary for us to discuss a development on difference observed between the Western and Eastern EU

TABLE 22.2 UEM(DC) comparison among three regional blocks

		2008	2009	2010	2011	2012
Eastern EU	Ave.	0.863	0.856	0.864	0.712	0.707
	S.D.	0.155	0.132	0.153	0.175	0.174
	Min.	0.665	0.660	0.690	0.453	0.459
	Max.	1.000	1.000	1.000	1.000	1.000
Western EU	Ave.	0.893	0.900	0.901	0.894	0.902
	S.D.	0.151	0.141	0.142	0.137	0.133
	Min.	0.627	0.621	0.611	0.634	0.640
	Max.	1.000	1.000	1.000	1.000	1.000
North America	Ave.	0.847	0.849	0.838	0.826	0.837
	S.D.	0.216	0.214	0.229	0.246	0.230
	Min.	0.695	0.698	0.676	0.652	0.674
	Max.	1.000	1.000	1.000	1.000	1.000
t-test	EEU vs WEU	0.681	0.506	0.605	0.016	0.009
	EEU vs NA	0.698	0.645	0.582	0.541	0.546
	WEU vs NA	0.911	0.958	0.854	0.486	0.421

(a) Source: Sueyoshi and Yuan (2016a). (b) EEU, WEU and NA stand for Eastern European Union, Western European Union and North America, respectively.

counties. Formed in 1948 during the Cold War, the Eastern EU was a communist region outside the Soviet Union and the Western EU was a defensive ally for the United States as opposed to Soviet influence among non-communist nations. In history, the Eastern EU countries were mostly behind the Western EU countries in its economic progress. Furthermore, there is a regional difference between Eastern EU and Western EU in terms of their economic developments. That is, "imbalanced development" in economy and environment occurred between the two EU regional blocks. See the bottom of Table 22.2 which shows the difference between Eastern and Western EU nations in their averages of $UEM(DC)_v^{R*}$.

Table 22.3 summarizes MRT ($=\partial b_1/\partial g = u^*/w_1^*$) of PM 2.5 to GDP on the three regional blocks in the observed annual periods. The MRT estimates were measured under managerial disposability. The estimate indicates an increasing rate of PM 2.5 due to one unit increase of GDP. If the rate is positive, then the level of pollution, measured by PM 2.5, increases with an increase of GDP. In contrast, if the rate is negative, then the level of pollution decreases with an increase of GDP. As discussed in Chapter 21, such a case indicates the occurrence of DC. As mentioned previously, the occurrence indicates a high level of social potential that can reduce a level of pollution measured by PM 2.5.

Each number in the top row in Table 22.3 indicates the average MRT of PM 2.5 to GDP in the three regions. For example, –0.496 (so, –49.6%) of Eastern EU nations indicated that an increase by one unit in GDP could decrease their PM 2.5 by 49.6% on average in 2012. In a similar manner, such a unit increase in GDP

TABLE 22.3 Marginal rate of transformation ($\partial b_1 / \partial g = u^* / w_1^*$)

		2008	2009	2010	2011	2012
Eastern EU	Ave.	–0.784	–0.728	–0.792	–0.834	–0.496
	S.D.	0.376	0.407	0.364	0.367	0.821
	Min.	–1.000	–1.000	–1.000	–1.000	–1.000
	Max.	–0.042	–0.053	–0.073	–0.087	1.000
Western EU	Ave.	–0.183	–0.118	–0.226	–0.100	–0.072
	S.D.	0.728	0.695	0.667	0.690	0.754
	Min.	–0.992	–0.992	–0.993	–0.988	–0.985
	Max.	1.000	1.000	1.000	1.000	1.000
North America	Ave.	0.192	0.199	0.172	0.144	0.178
	S.D.	1.142	1.133	1.171	1.211	1.162
	Min.	–0.615	–0.602	–0.656	–0.712	–0.643
	Max.	1.000	1.000	1.000	1.000	1.000
t-test	EEU vs WEU	0.069	0.057	0.064	0.023	0.260
	EEU vs NA	0.517	0.568	0.461	0.661	0.675
	WEU vs NA	0.084	0.104	0.089	0.092	0.388

(a) Source: Sueyoshi and Yuan (2016a). (b) Each number indicates the average MRT of PM 2.5 to GDP.

was exchanged for a decrease of PM 2.5 by –0.072 (so, –7.2%) on average of Western EU nations. In contrast, the United States and Canada exchanged a unit increase in GDP for an increase by 0.178 (so, 17.8%) in PM 2.5. The finding was applicable to the other annual periods (2008–2011), as well, in terms of these signs and magnitudes of MRT. These results on MRT indicated that both Eastern and Western EU nations were very sensitive to the amount of PM 2.5. In contrast, North America did not have such social sensitivity because the United States and Canada paid less attention than the EU to air pollution measured by PM 2.5. As listed in the last row of Table 22.3, the *t*-test confirmed that Western EU was different from Eastern EU in terms of MRT because the null hypothesis (i.e., no difference between the two regional combinations) was rejected at the level of 10% significance in the observed years except 2012. This result is because Eastern EU is less polluted than Western EU, so that the former has more potential to reduce the amount of PM 2.5. The larger amount with a negative sign of MRT is better in terms of social potential to reduce the level of air pollution measured by PM 2.5.

Table 22.4 summarizes MRT ($= \partial b_2 / \partial g = u^* / w_2^*$) of CO_2 to GDP on the three regions during the observed annual periods. Each number indicated their averages on MRT of CO_2 to GDP. For example, –0.078 (so, –7.8%) of Eastern EU nations in 2012 indicated that each nation decreased CO_2 by 7.8% due to one unit increase in GDP on average. In a similar manner, such an increase in GDP decreased CO_2 by –0.251 (so, –25.1%) on the average of Western EU nations. The United States

TABLE 22.4 Marginal rate of transformation ($\partial b_2/\partial g = u^*/w_2^*$)

		2008	2009	2010	2011	2012
Eastern EU	Ave.	–0.248	–0.249	–0.102	–0.115	–0.078
	S.D.	0.383	0.380	0.088	0.110	0.108
	Min.	–1.000	–1.000	–0.277	–0.275	–0.263
	Max.	–0.009	–0.035	–0.034	–0.013	0.000
Western EU	Ave.	–0.245	–0.240	–0.251	–0.194	–0.251
	S.D.	0.440	0.429	0.445	0.432	0.448
	Min.	–0.988	–0.983	–0.981	–0.981	–0.972
	Max.	0.121	0.108	0.125	0.252	0.116
North America	Ave.	–0.250	–0.260	–0.262	–0.339	–0.307
	S.D.	0.354	0.367	0.371	0.479	0.434
	Min.	–0.501	–0.519	–0.524	–0.677	–0.613
	Max.	0.000	0.000	0.000	0.000	0.000
t-test	EEU vs WEU	0.990	0.968	0.432	0.669	0.366
	EEU vs NA	0.987	0.953	0.973	0.662	0.870
	WEU vs NA	0.993	0.973	0.297	0.260	0.216

(a) Source: Sueyoshi and Yuan (2016a). (b) Each number indicates the average MRT of CO_2 to GDP.

and Canada decreased CO_2 by –0.307 (so, 30.7%) due to a unit increase in GDP in 2012. The finding was applicable to the other annual periods (2008–2011). These results on MRT indicated that the three regions had the potential to decrease the amount of CO_2 along with an increase of GDP under previous eco-technology investment. The United States and Canada had been paying attention to the reduction in the amount of CO_2 emission because the United Nations had requested all nations to combat global warming and climate change. Such an international trend has influenced these two nations. Due to the prevailing international conscientiousness about global warming and climate change, as listed in the last row of Table 22.4, the *t*-test could not statistically reject the null hypothesis (i.e., no difference between any two reginal combinations) at the level of 5% significance.

Table 22.5 summarizes MRT ($= \partial b_2/\partial b_1 = w_1^*/w_2^*$) of CO_2 to PM 2.5 in the three regions during the observed annual periods. In 2012, each number indicated their average MRT of CO_2 to PM 2.5. For example, CO_2 increased 0.239 (so, 23.9%) along with a unit increase of PM 2.5 in Eastern EU. In a similar manner, such a relationship corresponded to 0.647 (so, 64.7%) in Western EU and 0.477 (so, 47.7%) in North America. The finding was applicable to the other annual periods (2008–2011). As listed in the last row of Table 22.5, the *t*-test did not reject the null hypothesis (i.e., no difference between the two reginal combinations) at the level of 5% significance.

Table 22.6 summarizes the degree of RSU [$= (db_1/dg)/(b_1/g)$] of PM 2.5 to GDP on the three regions in the observed annual periods. In the RSU measurement,

TABLE 22.5 Marginal rate of transformation ($\partial b_2/\partial b_1 = w_1^*/w_2^*$)

		2008	2009	2010	2011	2012
Eastern EU	Ave.	0.416	0.432	0.266	0.272	0.239
	S.D.	0.465	0.452	0.372	0.375	0.390
	Min.	0.009	0.035	0.034	0.013	0.000
	Max.	1.000	1.000	1.000	1.000	1.000
Western EU	Ave.	0.640	0.650	0.649	0.641	0.647
	S.D.	0.432	0.417	0.423	0.436	0.437
	Min.	0.006	0.002	0.019	0.014	0.012
	Max.	1.000	1.000	1.000	1.000	1.000
North Amcrica	Ave.	0.407	0.432	0.400	0.475	0.477
	S.D.	0.576	0.610	0.565	0.672	0.674
	Min.	0.000	0.000	0.000	0.000	0.000
	Max.	0.814	0.863	0.799	0.951	0.953
t test	EEU vs WEU	0.296	0.292	0.063	0.079	0.057
	EEU vs NA	0.490	0.507	0.452	0.631	0.622
	WEU vs NA	0.982	1.000	0.704	0.590	0.543

(a) Source: Sueyoshi and Yuan (2016a). (b) Each number indicates the average MRT of CO_2 to PM 2.5.

a ratio of PM 2.5 to GDP indicates "factor-specific scale economy (elasticity)." An increasing or decreasing rate of PM 2.5 is larger than a proportional increase or decrease in GDP if the absolute value is greater than unity. For example, in 2012, the degree of RSU on PM 2.5 to GDP in Eastern EU nations was –7.089, so that an increase in GDP might have a social potential to decrease PM 2.5 more than proportionally because of the negative sign. The result was because Eastern EU nations were not polluted in terms of PM 2.5. In contrast, the degree of RSU in Western EU and North America exhibited 3.954 and 0.604 in their average ratios, respectively. These results indicate that an increase in GDP might increase the level of air pollution, measured by PM 2.5, more than proportionally in Western EU and less than proportionally in North America. The *t*-test statistically rejected the null hypnosis (i.e., no difference between Eastern and Western EU nations) at the level of 5% significance from 2008 to 2011. There was no rejection between Eastern EU and North America as well as Western EU and North America.

Table 22.7 summarizes the degree of RSU [= $(db_2/dg)/(b_2/g)$] of CO_2 to GDP on the three regions in the observed annual periods. For example, in 2012, the degree of RSU of CO_2 to GDP was 0.178, 6.156 and 0.042, respectively, in the three regions. Thus, an increase in GDP enlarges less proportionally an amount of CO_2 in Eastern EU and North America. Thus, an increase in GDP invited a less proportional increase on the level of air pollution measured by CO_2 in the two reginal blocks. A more proportional increase can be found in Western EU. The

TABLE 22.6 Rate of substitution (ug/w_1b_1)

		2008	2009	2010	2011	2012
Eastern EU	Ave.	–13.323	–7.960	–5.902	–12.174	–7.089
	S.D.	21.417	10.008	5.008	14.463	17.565
	Min.	–56.363	–26.843	–12.715	–39.440	–38.655
	Max.	–0.389	–0.516	–0.732	–0.812	13.974
Western EU	Ave.	3.104	2.130	1.861	3.760	3.954
	S.D.	9.908	8.012	8.112	10.673	10.246
	Min.	–6.939	–7.549	–7.210	–7.803	–7.982
	Max.	25.151	22.761	23.354	28.448	27.240
North America	Ave.	0.564	0.580	0.623	0.880	0.604
	S.D.	0.864	0.884	0.953	1.322	0.925
	Min.	–0.047	–0.046	–0.051	–0.055	–0.050
	Max.	1.175	1.205	1.297	1.814	1.258
t-test	EEU vs WEU	0.019	0.021	0.041	0.009	0.074
	EEU vs NA	0.728	0.793	0.836	0.715	0.658
	WEU vs NA	0.418	0.296	0.132	0.272	0.578

(a) Source: Sueyoshi and Yuan (2016a). (b) Each number indicates the average RSU of PM 2.5 to GDP.

TABLE 22.7 Rate of substitution (ug/w_2b_2)

		2008	2009	2010	2011	2012
Eastern EU	Ave.	0.202	0.392	0.487	–0.347	0.178
	S.D.	0.921	1.523	0.930	1.181	0.554
	Min.	–0.601	–0.880	–0.764	–2.475	–0.137
	Max.	1.447	3.081	1.521	1.182	1.302
Western EU	Ave.	6.382	6.048	6.610	6.191	6.156
	S.D.	15.375	14.456	16.631	15.944	15.875
	Min.	0.030	0.010	0.122	0.066	0.063
	Max.	59.926	56.427	66.367	64.434	64.695
North America	Ave.	0.036	0.038	0.035	0.042	0.042
	S.D.	0.051	0.054	0.050	0.059	0.060
	Min.	0.000	0.000	0.000	0.000	0.000
	Max.	0.072	0.076	0.071	0.084	0.085
t-test	EEU vs WEU	0.343	0.357	0.385	0.334	0.374
	EEU vs NA	0.577	0.574	0.593	0.602	0.602
	WEU vs NA	0.818	0.766	0.539	0.674	0.755

(a) Source: Sueyoshi and Yuan (2016a). (b) Each number indicates the average RSU of CO_2 to GDP.

t-test could not reject the null hypnosis (i.e., no difference between two reginal combinations) at the level of 5% significance from 2008 to 2012.

At the end of this section, Figure 22.4 depicts a process for enhancing the level of social sustainability. First, a nation needs capital accumulation for production development that can enhance the operational efficiency. The production

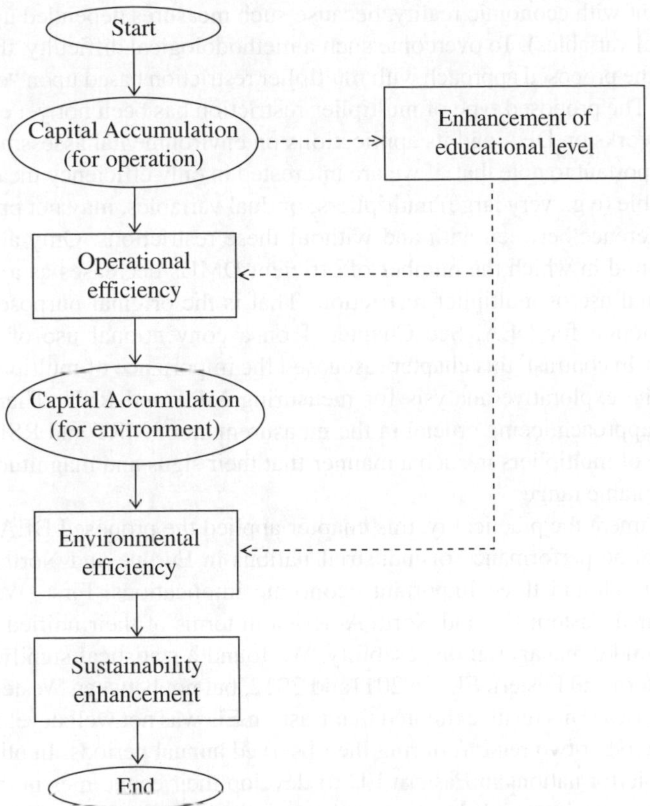

FIGURE 22.4 Sustainability enhancement

enhancement accumulates an amount of an additional capital that can be used for environmental protection. See Chapter 15 on the economic development process for sustainability. Although it is not clearly examined in this chapter, education is an essential component to support the social sustainability enhancement of each nation in such a manner that the capital accumulation is effectively used for both operational and environmental performance enhancements in a short-term horizon. Meanwhile, the investment in education is very important in a long-term horizon to accumulate human resources.

22.8 SUMMARY

This chapter discussed a new use of DEA environmental assessment to measure MRT and RSU among production factors. Considering the concept of managerial disposability, this chapter discussed how to measure the sign and magnitude of MRT and RSU. An analytical problem of the measurement on MRT and RSU was that it usually became unstable (e.g., very large) so that these results were inconsistent with economic reality, because such measures depended upon multipliers (dual variables). To overcome such a methodological difficulty, this chapter equipped the proposed approach with multiplier restriction based upon "explorative analysis." The proposed type of multiplier restriction has been not yet explored in previous works on DEA and its applications on environmental assessment.

It is important to note that, if we are interested in only efficiency measurement, unacceptable (e.g., very large) multipliers, or dual variables, may not produce any major difference between with and without these restrictions. Only a difference may be found in which the number of efficient DMUs decreases as a result of a conventional use of multiplier restriction. That is the original purpose of multiplier restriction for DEA. See Chapter 4 on a conventional use of multiplier restriction. In contrast, this chapter discussed the importance of multiplier restrictions by the explorative analysis for measuring MRT and RSU. Therefore, the proposed approach, being crucial in the measurement of MRT and RSU, reduces the degree of multipliers in such a manner that their signs and magnitudes belong to an acceptable range.

To document the practicality, this chapter applied the proposed DEA approach to evaluate the performance of industrial nations in Europe and North America. This chapter found three important economic implications. First, Western EU outperformed Eastern EU and North America in terms of their unified efficiency measures under managerial disposability. We found a statistical significance between Western and Eastern EUs in 2011and 2012, but not between Western EU and North America. This result exhibited that Eastern EU was not well developed at the level of the other two regions during the observed annual periods. In other words, it is possible for nations in Eastern EU to develop their economies in the manner that they can attain a high level of social sustainability by balancing industrial developments and pollution preventions in future. Second, Eastern EU exhibited MRT estimates which were different from those of Western EU and North America. The nations in Eastern EU had the economic potential for industrial developments because the level of their industrial pollution was less than that of the other two regions. Potential was also found in their MRT estimates, although this chapter did not find any statistical difference on MRT between the three regimes with a few exceptions. Such exceptions were found in the MRT on PM 2.5 to GDP and the estimate of CO_2 to PM 2.5. Finally, an interesting difference could be found in the RSU estimates on PM 2.5 and GDP between Eastern and Western EU nations from 2008 to 2012. They exhibited a statistically significant difference between the two regional blocks but not with North America. That is, most nations in Western EU and North America have already attained a high level of economic success, so that they have limited potential under current production and green technology innovations. In contrast, Eastern EU may be different from the other two regions in terms of attaining its industrial development and social sustainability because the region is less developed and polluted than the other two regions. It is easily imagined that Eastern EU needs capital accumulation and human resource investment for social sustainability development.

23

RETURNS TO DAMAGE AND DAMAGES TO RETURN

23.1 INTRODUCTION

Chapter 21 has discussed the two disposability concepts under a possible occurrence of "undesirable congestion (UC)," due to a capacity limit in part or all of a whole production facility, and another possible occurrence of "desirable congestion (DC)" due to eco-technology innovation. In this chapter, technology innovation contains wide implications regarding not only engineering contributions but also managerial challenges (e,g., a fuel mix shift from coal to natural gas or renewable energy for generation).

The underlying philosophy of this book is that we may overcome our current difficulty on the global warming and climate change by combining technology development with managerial challenges. It is believed that the global climate issue, influencing all over the world, can be solved by engineering capabilities and natural science that are equipped with managerial and economic wisdoms in social science. This chapter documents an analytical rationale regarding how DEA can serve as an empirical basis for connecting among such different capabilities in engineering, natural and social sciences.

As an extension of Chapters 20 and 21, the purpose of this chapter[1] is to discuss UC under natural disposability and DC under managerial disposability, along

[1] This chapter is partly based upon the article: Sueyoshi, T. and Yuan, Y. (2016b) Returns to damage under undesirable congestion and damages to return under desirable congestion measured by DEA

Environmental Assessment on Energy and Sustainability by Data Envelopment Analysis, First Edition. Toshiyuki Sueyoshi and Mika Goto.

with their linkages to returns to damage (RTD) and damages to return (DTR). This chapter also equips DEA with an analytical capability of "explorative analysis" for multiplier restriction, as discussed in Chapter 22, to improve the measurement reliability on RTD under UC and DTR under DC. After theoretically discussing our research direction, this chapter documents the practicality of the proposed approach by applying it to Chinese energy assessment and regional economic planning. The proposed application examines China's municipalities and provinces in terms of their economic and environmental developments.

The remainder of this chapter is organized as follows: Section 23.2 reviews economic concepts related to the occurrence of congestion. Section 23.3 describes these concepts with congestion under natural disposability. Section 23.4 shifts our discussion to another type of congestion under managerial disposability. Section 23.5 applies the proposed approach to discuss Chinese regional development and energy planning. Section 23.6 summarizes this chapter.

23.2 CONGESTION, RETURNS TO DAMAGE AND DAMAGES TO RETURN

23.2.1 Undesirable Congestion (UC) and Desirable Congestion (DC)

As discussed in Chapter 21, the occurrence of UC is a widely observed phenomenon in which inefficiency is identified in such a manner that an increase in some input(s) decreases some desirable output(s), without worsening other production factors, including undesirable output(s). That is an undesirable phenomenon. The straightforward expectation is that an increase in inputs leads to an increase in some components of desirable and undesirable output vectors. Such an occurrence of UC may be found in many economic activities, including business in energy sectors. See Chapter 9 that provides an illustrative example of UC in the electric power industry.

Returning to Chapter 21, we conceptually retreat our description on a possible occurrence of congestion, then connecting it to RTD and DTR. To explain the conceptual connection, this chapter presents Figure 23.1, originated from Figure 21.1. The left-hand side of the figure exhibits the desirable output (g) on the vertical axis and the undesirable output (b) on the horizontal axis. The opposite case is found in the right-hand side of the figure. For our visual convenience, Figure 23.1 considers only a single component of the three production factors. It is easily extendable to their multiple components in the mathematical formulations proposed in this chapter. The convex curve depicted in the figure indicates the relationship between two production factors.

On the left-hand side of Figure 23.1, a supporting line visually specifies the shape of a production curve and a possible occurrence of UC between g and b. For

environmental assessment with multiplier restriction: Economic and energy planning for social sustainability in China. *Energy Economics*, **56**, 288–309.

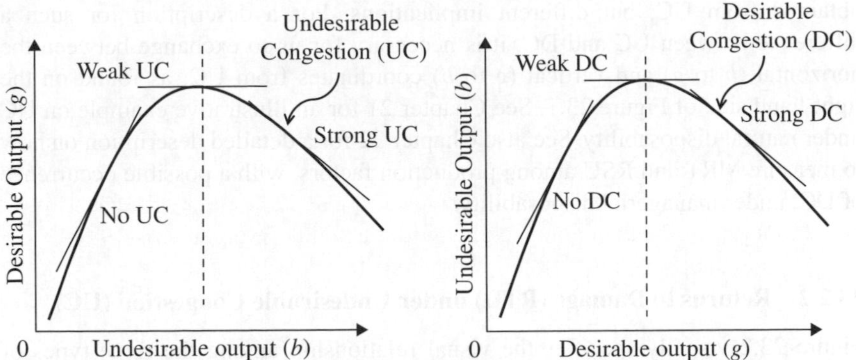

FIGURE 23.1 UC and DC (a) See Figure 21.1. Figure 23.1 changes Figure 21.1 by a smooth curve for our visual convenience and pays attention to the bottom of Figure 21.1, that is, a visual description of UC and DC. (b) For our visual convenience, part of input x is not shown in this figure (see Figure 21.1). For example, on the left-hand side of Figure 23.1, the supporting hyperplane visually specifies the shape of the production curve and the possible occurrence of UC between desirable output (g) and undesirable output (b). A negative slope indicates the occurrence of UC, so being considered as "strong UC." The occurrence of strong UC implies that enlarged input (x) increases undesirable output (b) and decreases the desirable output vector (g), as visually specified by the negative slope of the supporting line. In contrast, a positive slope implies an opposite case (i.e., no congestion), so being referred to as "no UC." When the slope of a supporting hyperplane is zero, this chapter considers it as "weak UC," being between strong UC and no UC. See Chapter 21, which explains this as the starting position of UC. The visual description on UC is extendable to a description on DC, or eco-technology innovation for pollution reduction, as found on the right-hand side of the figure. The three categories of DC include no (a positive slope), weak (a zero slope) and strong (a negative slope). (c) The occurrence of UC, or a capacity limit, becomes a serious problem in energy sectors so that it is necessary for us to identify such an occurrence in discussing energy policy. Meanwhile, the occurrence of DC, or eco-technology innovation, may be more important than UC in terms of environmental protection and assessment because it is necessary for us to reduce the level of various types of pollution. (d) For our visual description, the figure uses a smooth curve to express the relationship between two production factors. The figure also depicts a single component of the three production factors. All the DEA formulations proposed in this chapter can handle their multiple components.

example, a negative slope indicates an occurrence of UC, so being considered as "strong UC," as discussed in Chapter 21. An occurrence of strong UC implies that an enlarged input (x), not listed in Figure 23.1 for our visual description, increases an undesirable output (b) but decreases a desirable output (g). In contrast, a positive slope implies an opposite case (i.e., no congestion), so being referred to as "no UC." If the slope of a supporting line is zero, this chapter considers it as "weak UC," so being between strong UC and no UC.

The right-hand side of Figure 23.1 visually specifies the three categories (i.e., no, weak and strong) of DC whose analytical implications are directly obtained from UC, but different implications. For a description for such a difference between UC and DC, it is necessary for us to exchange between the horizontal (*b* to *g*) and vertical (*g* to *b*) coordinates from UC, as found on the right-hand side of Figure 23.1. See Chapter 21 for an illustrative example on UC under natural disposability. See also Chapter 22 for a detailed description on how to measure MRT and RSU among production factors, with a possible occurrence of DC, under managerial disposability.

23.2.2 Returns to Damage (RTD) under Undesirable Congestion (UC)

Figure 23.2 visually specifies the visual relationship among the three types of supporting hyperplane on a production and pollution possibility set. They are closely associated with the three types of UC, which are classified into "no," "weak" and "strong" UC. As mentioned previously, an occurrence of UC becomes a serious difficulty in the operation of energy sectors. See Chapter 9 for an illustrative example of such a difficulty in the electric power industry.

Figure 23.3 visually discusses the three types of RTD, including "positive," "no" and "negative" RTD. An efficiency frontier consists of the six DMUs ({C}–{D}–{E}–{A}–{F}–{B}). Positive RTD, corresponding to no UC, is further

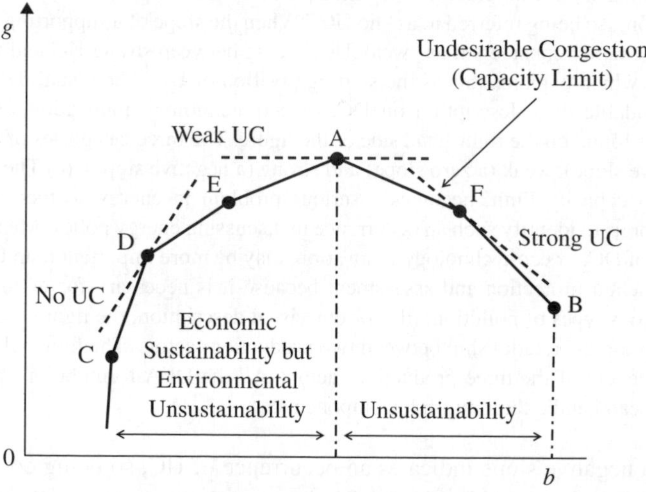

FIGURE 23.2 Undesirable Congestion (UC) (a) See Figure 21.2. (b) The figure visually specifies the relationship between the supporting hyperplane and the production and pollution possibility set. The types of UC include "no," "weak" and "strong". No UC is found at D, weak UC is found at A and strong UC is found on F. The occurrence of strong UC is identified by negative RTD if the three production factors have a single component, as depicted in the figure. (c) The figure does not incorporate the input (*x*) for our visual convenience. Hence, each dotted line can be considered as a supporting hyperplane.

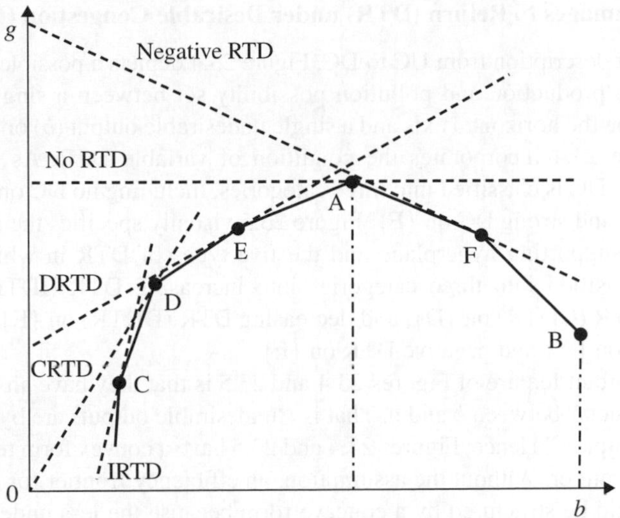

FIGURE 23.3 Returns to damage (RTD) (a) IRTD: Increasing RTD, CRTD: constant RTD and DRTD: decreasing RTD in the case of a desirable output (*g*) and undesirable output (*b*), all of which belong to "positive RTD." In addition, there are no RTD and negative RTD. The negative RTD indicates an occurrence of strong UC due to a capacity limit in part or all of a whole production system. (b) A use of scale elasticity for RTD classification indicates that *dg/db* becomes zero on {A} and negative on {F}, where *d* stands for a derivative of a functional form between two components. The derivative mathematically indicates the slope of a supporting hyperplane. The other three points ({C}, {D} and {E}) have a positive slope. Thus, Figure 23.3 indicates a mathematical linkage between strong UC and negative RTD in the case of a single component of the three production factors. The case of multiple components is discussed by the proposed formulations for DEA environmental assessment in this chapter.

separated into IRTD on {C}, CRTD on {D}, and DRTD on {E}. No RTD is found on {A}, corresponding to weak UC, and negative RTD is found on {F}, corresponding to strong UC.

Here, it is important to note that a use of point elasticity for RTD classification indicates that *dg/db* becomes zero on {A} and negative on {F}, where *d* stands for a derivative of a functional form that mathematically expresses the slope of a supporting hyperplane. The other three points {C, D and E} have a positive slope of the supporting hyperplane. Thus, Figure 23.3 indicates a mathematical linkage between strong UC and negative RTD. Such a linkage will be mathematically explored in this chapter. The identification of negative RTD is important for all industries, including energy sectors, because it indicates an occurrence of strong UC, or a capacity limit in part or all of a whole production system.

23.2.3 Damages to Return (DTR) under Desirable Congestion (DC)

Shifting our description from UC to DC, Figure 23.4 depicts a possible occurrence of DC on a production and pollution possibility set between a single desirable output (g) on the horizontal axis and a single undesirable output (b) on the vertical axis. Figure 23.4 incorporates the condition of variable DTR. As depicted in Figure 23.4, DC is classified into three categories, including no DC on {D}, weak DC on {A} and strong DC on {F}. Figure 23.5 visually specifies the relationship between a supporting hyperplane and the five types of DTR in which positive DTR is classified into three categories into increasing DTR (IDTR) on {C}, constant DTR (CDTR) on {D}, and decreasing DTR (DDTR) on {E} in addition to no DTR on {A} and negative DTR on {F}.

An important feature of Figures 23.4 and 23.5 is that they have an assumption on "byproducts" between b and g. That is, "undesirable outputs are byproducts of desirable outputs." Hence, Figures 23.4 and 23.5 have a convex form to express an efficiency frontier. Without the assumption, an efficiency frontier for undesirable output should be structured by a concave form because the less undesirable outputs are the better in environmental assessment. See Figure 15.6, which provided a visual description on the convex and concave relationships in the three production factors. In addition, it visually discussed a rationale regarding why DEA

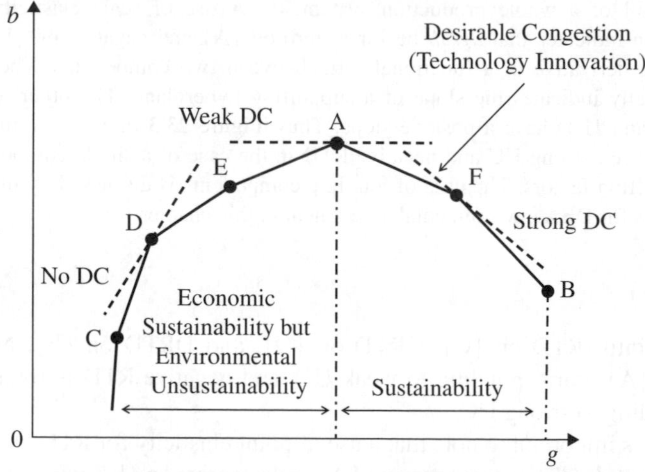

FIGURE 23.4 Desirable congestion (DC) (a) See Figure 21.3. (b) Figure 23.4 visually specifies the relationship between a supporting hyperplane and the types of DC, including "no," "weak" and "strong" DC. No DC is found on {D}, weak DC is found on {A}. Strong DC occurs on {F} and such an occurrence may be identified by negative DTR if the three production factors have a single component. (c) An important feature of Figures 23.4 and 23.5 is that they have an assumption of byproducts. That is, undesirable outputs are byproducts of the desirable outputs. Hence, the figures have a convex form to express efficiency frontiers. Without this assumption, they should be structured by a concave form to express an efficiency frontier for undesirable outputs.

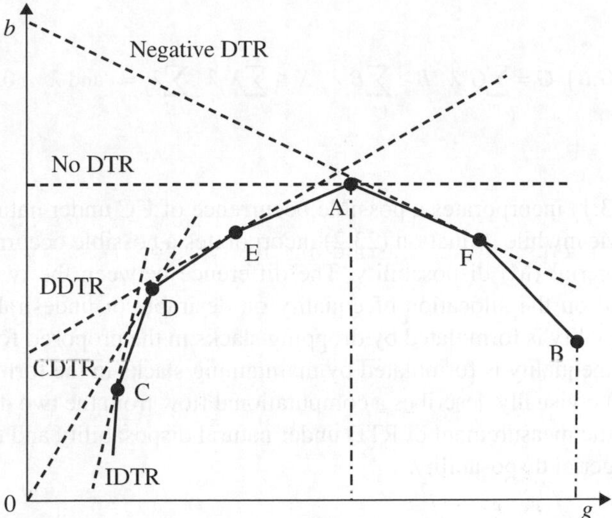

FIGURE 23.5 Damages to return (DTR) (a) IDTR: Increasing DTR, CDTR: constant DTR and DDTR: decreasing DTR in the case of an undesirable output (*b*) and desirable output (*g*). In addition, there is no DTR and negative DTR. A negative DTR indicates the occurrence of DC due to eco-technology innovation for reducing the amount of undesirable outputs.

environmental assessment is much more complicated and difficult than a conventional use of DEA. It is true that the only difference between the three production components is the existence of undesirable outputs among the production factors. However, such a difference makes the environmental assessment conceptually and analytically very different from the original use of DEA.

In Figure 23.5, the conventional use of scale elasticity for DTR classification indicates that *db/dg* becomes zero on {A} and negative on {F}. The derivative becomes positive at the other points ({C}, {D} and {E}). Thus, Figures 23.4 and 23.5 indicate a mathematical linkage between strong DC and negative DTR in the case that the three production factors have a single component. The visual description is extendable to the case of multiple components in the proposed formulations for DEA environmental assessment.

23.2.4 Possible Occurrence of Undesirable Congestion (UC) and Desirable Congestion (DC)

The possible occurrences of UC and DC are expressed by the following production and pollution set under the two disposability concepts:

$$P_{UC}^{N}(X)=\left\{(G,B): G\le\sum_{j=1}^{n}G_j\lambda_j,\ B=\sum_{j=1}^{n}B_j\lambda_j,\ X\ge\sum_{j=1}^{n}X_j\lambda_j,\ \sum_{j=1}^{n}\lambda_j=1 \text{ and } \lambda_j\ge 0, j=1,\ldots,n\right\}$$

(23.1)

and

$$P_{DC}^{M}(X)=\left\{(G,B): G=\sum_{j=1}^{n}G_j\lambda_j, B\geq\sum_{j=1}^{n}B_j\lambda_j, X\leq\sum_{j=1}^{n}X_j\lambda_j, \sum_{j=1}^{n}\lambda_j=1 \text{ and } \lambda_j\geq 0, j=1,\ldots,n\right\}. \tag{23.2}$$

Equation (23.1) incorporates a possible occurrence of UC under natural (N) disposability. Meanwhile, Equation (23.2) incorporates a possible occurrence of DC under managerial (M) disposability. The difference between the two equations can be found on the allocation of equality on desirable or undesirable outputs. Note that equality is formulated by dropping slacks in the proposed formulations. In contrast, inequality is formulated by maintaining slacks in the formulations.

Figure 23.6 visually describes a computational flow from the two disposability concepts to the measurement of RTD under natural disposability and that of DTR under managerial disposability.

23.3 CONGESTION IDENTIFICATION UNDER NATURAL DISPOSABILITY

23.3.1 Possible Occurrence of Undesirable Congestion (UC)

As depicted on the left-hand side of Figure 23.6, this chapter can identify the possible occurrence of UC under natural disposability. To examine the occurrence, this chapter uses the following model that maintains equality constraints (so, no slack variable) on undesirable outputs, which is first specified in Model (21.3):

$$\begin{aligned}
&\text{Maximize } \xi+\varepsilon_s\left(\sum_{i=1}^{m}R_i^{x-}d_i^{x-}+\sum_{r=1}^{s}R_r^{g}d_r^{g}\right)\\
&s.t. \quad \sum_{j=1}^{n}x_{ij}\lambda_j+d_i^{x-} = x_{ik}\ (i=1,\ldots,m),\\
&\sum_{j=1}^{n}g_{rj}\lambda_j - d_r^{g}-\xi g_{rk}=g_{rk}\ (r=1,\ldots,s),\\
&\sum_{j=1}^{n}b_{fj}\lambda_j + \xi b_{fk}=b_{fk}\ (\mathrm{f}=1,\ldots,h),\\
&\sum_{j=1}^{n}\lambda_j=1,\\
&\lambda_j\geq 0\ (j=1,\ldots,n),\ \xi:\text{URS},\\
&d_i^{x-}\geq 0\ (i=1,\ldots,m) \text{ and } d_r^{g}\geq 0\ (r=1,\ldots,s).
\end{aligned} \tag{23.3}$$

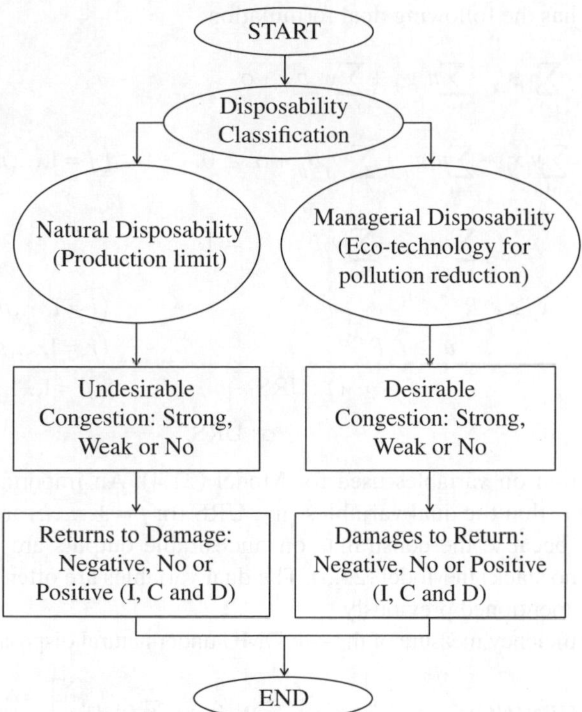

FIGURE 23.6 Computational flow from disposability to RTD and DTR (a) This figure visually describes the computational flow from the two disposability concepts to RTD under natural disposability and DTR under managerial disposability. (b) The possible occurrence of undesirable congestion is measured under natural disposability by assigning equality to the undesirable output (*B*). The occurrence indicates a technology limit on part or all of a whole production capability. (c) The possible occurrence of desirable congestion is measured under managerial disposability by assigning equality to the desirable output (*G*). The occurrence indicates an opportunity to reduce the amount of undesirable outputs by eco-technology. (d) Strong, weak or no UC correspond to negative, no or positive RTD, respectively, in the case of a single component of three production factors. The UC is a necessary condition of RTD in the case of multiple components in the three production factors. (d) Strong, weak or no DC corresponds to negative, no or positive DTR, respectively, in the case of a single component of three production factors. The DC is a necessary condition of DTR in the case of multiple components in the three production factors. (e) I, C and D indicate increasing, constant and decreasing.

See a description on variables used in Model (21.3). Model (23.3) drops slack variables related to undesirable outputs (*B*) so that they are considered as equality constraints. The other constraints regarding inputs and desirable outputs are considered as inequality because they have slack variables in Model (23.3).

Model (23.3) has the following dual formulation:

$$
\begin{aligned}
&\text{Minimize} \quad \sum_{i=1}^{m} v_i x_{ik} - \sum_{r=1}^{s} u_r g_{rk} + \sum_{f=1}^{h} w_f b_{fk} + \sigma \\
&s.t. \quad \sum_{i=1}^{m} v_i x_{ij} - \sum_{r=1}^{s} u_r g_{rj} + \sum_{f=1}^{h} w_f b_{fj} + \sigma \geq 0 \qquad (j = 1,\ldots,n), \\
&\qquad \sum_{r=1}^{s} u_r g_{rk} + \sum_{f=1}^{h} w_f b_{fk} = 1, \\
&\qquad v_i \geq \varepsilon_s R_i^x \qquad (i = 1,\ldots,m), \\
&\qquad u_r \geq \varepsilon_s R_r^g \qquad (r = 1,\ldots,s), \\
&\qquad w_f : \text{URS} \qquad (f = 1,\ldots,h), \\
&\qquad \sigma : \text{URS}.
\end{aligned}
\tag{23.4}
$$

See a description on variables used for Model (21.4). An important feature of Model (23.4) is that the dual variables (w_f: URS for $f = 1,\ldots, h$) are unrestricted in their signs because the constraints on undesirable outputs are expressed by equality (i.e., no slack) in Model (23.3). The dual variables are often referred to as multipliers as mentioned previously.

A unified efficiency measure of the k-th DMU under natural disposability become

$$
\begin{aligned}
UEN(UC)_v^{R*} &= 1 - \left[\xi^* + \varepsilon_s \left(\sum_{i=1}^{m} R_i^x d_i^{x-*} + \sum_{r=1}^{s} R_r^g d_r^{g*} \right) \right] \\
&= 1 - \left[\sum_{i=1}^{m} v_i^* x_{ik} - \sum_{r=1}^{s} u_r^* g_{rk} + \sum_{f=1}^{h} w_f^* b_{fk} + \sigma^* \right],
\end{aligned}
\tag{23.5}
$$

which incorporates the possible occurrence of UC, expressed by Equation (21.5). All variables used in Equation (23.5) are determined on the optimality of Models (23.3) and (23.4). The equation within the parenthesis, obtained from the optimal objective value of Models (23.3) and (23.4), indicates the level of unified inefficiency under natural disposability. The unified efficiency, or $UEN(UC)_v^{R*}$, is obtained by subtracting the level of inefficiency from unity as specified in Equation (23.5).

Explorative Analysis: An important advantage of Model (23.4) is that it can incorporate additional information in the form of side constraints for multiplier restrictions, as discussed in Chapter 22. For example, DEA environmental assessment usually divides the observation on each production factor by a factor average to avoid the case in which the data with a large magnitude dominates the other data with a small magnitude in the computation for assessment. Such a data adjustment is important for DEA to enhance the computational reliability. See Figure 22.3. As a result of multiplier restriction, all the observations used in this chapter are unit-less, so indicating the importance of each production factor. The data adjustment makes it possible to incorporate additional side constraints on production factors in Model (23.4) as follows:

$$\text{Inputs}: -1 \le v_{i'}/v_i \le 1\left(i' > i = 1,\ldots,m-1\right), \tag{23.6}$$

$$\text{Desirable outputs}: -1 \le u_{r'}/u_r \le 1\left(r' > r = 1,\ldots,s-1\right) \text{ and} \tag{23.7}$$

$$\text{Undesirable outputs}: -1 \le w_{f'}/w_f \le 1\left(f' > f = 1,\ldots,h-1\right). \tag{23.8}$$

See Chapter 22, which refers to the multiplier restriction as "explorative analysis." The analysis does not need any prior information so that it can avoid subjectivity in multiplier restriction, which is widely used in conventional DEA.

An important feature of the explorative analysis, used in Chapter 22, is that Equation (22.8) incorporates the following multiplier restrictions:

$$-1 \le u_r/w_f \le 1\left(r = 1,\ldots,s \text{ and } f = 1,\ldots,h\right) \text{ and}$$

$$-1 \le w_{f'}/w_f \le 1\left(f' > f = 1,\ldots,h-1\right).$$

The multiplier restriction is important because we are interested in the relationship between desirable and undesirable outputs as well as that between undesirable outputs. Meanwhile, this chapter is concerned with the measurement of *UEN*, UC and RTD under natural disposability and *UEM*, DC and DTR under managerial disposability. Thus, it is necessary for us to prepare multiplier restrictions on all the three production factors, as specified in Equations (23.6), (23.7) and (23.8).

Model (23.4), equipped with Equations (23.6) to (23.8), becomes as follows:

$$\begin{aligned}
\text{Minimize} \quad & \sum_{i=1}^{m} v_i x_{ik} - \sum_{r=1}^{s} u_r g_{rk} + \sum_{f=1}^{h} w_f b_{fk} + \sigma \\
s.t. \quad & \sum_{i=1}^{m} v_i x_{ij} - \sum_{r=1}^{s} u_r g_{rj} + \sum_{f=1}^{h} w_f b_{fj} + \sigma \ge 0 && (j = 1,\ldots,n), \\
& \sum_{r=1}^{s} u_r g_{rk} + \sum_{f=1}^{h} w_f b_{fk} = 1, \\
& v_i \ge \varepsilon_s R_i^x && (i = 1,\ldots,m), \\
& u_r \ge \varepsilon_s R_r^g && (r = 1,\ldots,s), \\
& w_f : \text{URS} && (f = 1,\ldots,h), \\
& \sigma : \text{URS}, \\
& -1 \le v_{i'}/v_i \le 1 && (i' > i = 1,\ldots,m-1), \\
& -1 \le u_{r'}/u_r \le 1 && (r' > r = 1,\ldots,s-1), \\
& -1 \le w_{f'}/w_f \le 1 && (f' > f = 1,\ldots,h-1).
\end{aligned} \tag{23.9}$$

The degree of $UEN(UC)_v^{R*}$ on the k-th DMU is determined by

$$UEN\left(UC\right)_v^{R^*} = 1 - \left(\sum_{i=1}^{m} v_i^* x_{ik} - \sum_{r=1}^{s} u_r^* g_{rk} + \sum_{f=1}^{h} w_f^* b_{fk} + \sigma^* \right), \quad (23.10)$$

where all the dual variables are identified on the optimality of Model (23.9). Equation (23.10) is different from Equation (23.5) because the side constraints in Equations (23.6) to (23.8) are additionally incorporated into Model (23.9). Therefore, Models (23.4) and (23.9) produce different $UEN(UC)_v^{R*}$ measures.

After computing Model (23.9), a possible occurrence of UC is determined by the following rule under the assumption that Model (23.9) produces a unique optimal solution as discussed in Chapter 7 (i.e., unique projection and a unique reference set):

(a) If $w_f^* < 0$ for some (at least one) f, then "strong UC" occurs on the k-th DMU,
(b) If $w_f^* > 0$ for all f, then "no UC" occurs on the k-th DMU and
(c) In the others, including $w_f^* = 0$ for some (at least one) f, then "weak UC" occurs on the k-th DMU.

It is important to note that if $w_f^* < 0$ for some f and $w_{f'}^* = 0$ for the other f', then both strong UC and weak UC may coexist on the k-th DMU. In the case, this chapter considers it as an occurrence of the strong UC on the DMU.

23.3.2 RTD Measurement under the Possible Occurrence of UC

Let the dual variables of the k-th DMU, obtained from Model (23.9), be v_i^* (i = 1, 2,…, m), u_r^* (r = 1, 2,…, s), w_f^* (f = 1, 2,…, h) and σ^* on optimality. Then, the estimated supporting hyperplane on the k-th DMU is expressed by

$$\sum_{r=1}^{s} u_r^* g_r = \sum_{i=1}^{m} v_i^* x_i + \sum_{f=1}^{h} w_f^* b_f + \sigma^*, \quad (23.11)$$

which is determined by $\sum_{i=1}^{m} v_i x_{ij} - \sum_{r=1}^{s} u_r g_{rj} + \sum_{f=1}^{h} w_f b_{fj} + \sigma$, $j \in RS_k$, where RS_k is a reference set for the k-th DMU and $\sum_{r=1}^{s} u_r g_{rk} + \sum_{f=1}^{h} w_f b_{fk} = 1$. See Proposition 21.1 in Chapter 21 that discusses mathematical implications regarding the supporting hyperplane. Note that Proposition 21.1 does not consider multiplier restriction, but Equation (23.11) has the explorative analysis. Both have the same mathematical equation, but they are different in terms of their analytical implications: congestion identification in Chapter 21 versus the determination of RTD and DTR in this chapter.

The degree (Dg) of RTD, or $DgRTD$, under a possible occurrence of UC, on the k-th DMU is measured by

$$\begin{aligned} DgRTD(UC) &= \left(\sum_{f=1}^{h} w_f^* b_f\right) \Big/ \left(\sum_{r=1}^{s} u_r^* g_r\right) \\ &= \left(\sum_{f=1}^{h} w_f^* b_f\right) \Big/ \left(\sum_{i=1}^{m} v_i^* x_i + \sum_{f=1}^{h} w_f^* b_f + \sigma^*\right) \\ &= 1 \Big/ \left[1 + \left(\sigma^* + \sum_{i=1}^{m} v_i^* x_i\right) \Big/ \left(\sum_{f=1}^{h} w_f^* b_f\right)\right]. \end{aligned} \tag{23.12}$$

As mentioned previously, this chapter assumes that the Model (23.9) has both a unique projection of an inefficient DMU onto an efficiency frontier and a unique reference set for the DMU.

Following Equation (23.12), the type of RTD on the k-th DMU is classified by the following rule on the k-th DMU:

(a) *Increasing RTD* ↔ There exists an optimal solution of Model (23.9) that satisfies all $w_f^* > 0$ ($f = 1,\ldots,h$) and $\sigma^* + \sum_{i=1}^{m} v_i^* x_i < 0$,
(b) *Constant RTD* ↔ There exists an optimal solution of Model (23.9) that satisfies all $w_f^* > 0$ ($f = 1,\ldots,h$) and $\sigma^* + \sum_{i=1}^{m} v_i^* x_i = 0$,
(c) *Decreasing RTD* ↔ There exists an optimal solution of Model (23.9) that satisfies all $w_f^* > 0$ ($f = 1,\ldots,h$) and $\sigma^* + \sum_{i=1}^{m} v_i^* x_i > 0$,
(d) *Negative RTD* ↔ There is an optimal solution of Model (23.9) that satisfies $w_f^* < 0$ for at least one $f \in \{1,\ldots,h\}$ and
(e) *No RTD* ↔ All other cases excluding (a) to (d).

Difference between UC and RTD: The type of UC is identified by the sign of dual variables (w_f^*). The type of UC is classified into the three categories. Meanwhile, these measures related to RTD are determined by not only the sign of dual variables (w_f^*) but also the sign of $\sigma^* + \sum_{i=1}^{m} v_i^* x_i$. The type of RTD is classified into the five categories. Figure 23.7 visually classifies the type of RTD under a possible occurrence of UC.

At the end of this section, it is necessary to summarize three concerns related to Model (23.9) and Equation (23.12) as well as the proposed RTD classification. First, Model (23.9) assumes a unique solution, so implying no occurrence on multiple projections and multiple reference sets. Second, Equation (23.12) is effective

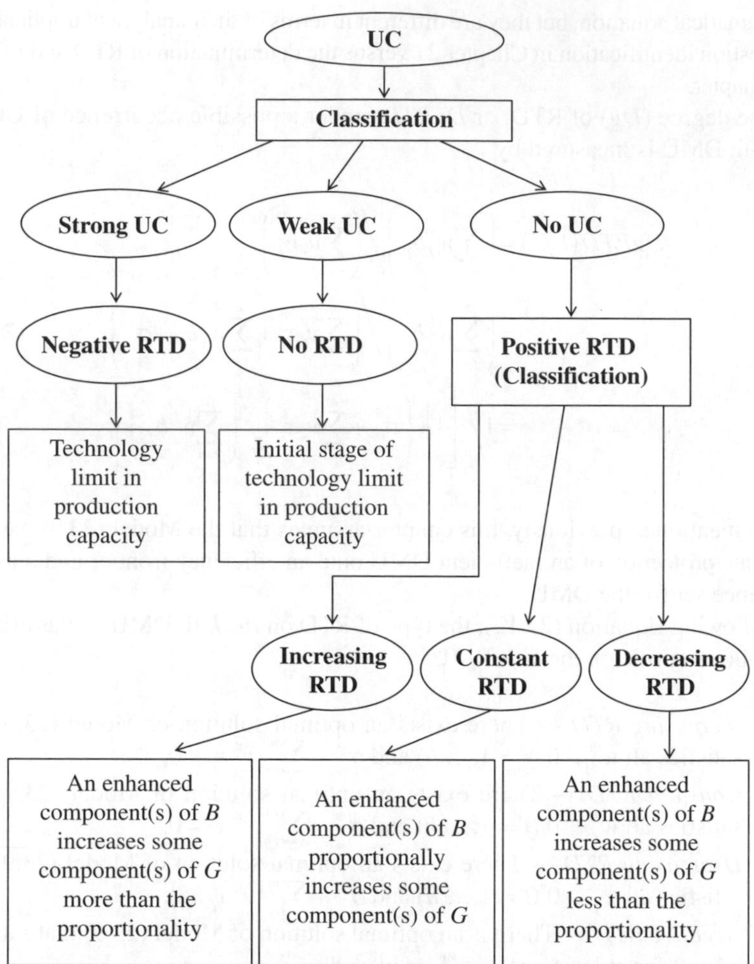

FIGURE 23.7 Returns to damage (RTD) under undesirable congestion (UC) (a) The increase in some component(s) of the input vector (X) increases the components of the other two production factors. The occurrence of UC is "undesirable" because some component(s) of the desirable output vector (G) decrease and those of the undesirable output vector (B) increase along with such a change of input vector. That is, the occurrence of UC is due to a capacity limit on part or whole in a production facility. (b) The occurrence and type of UC are identified by the sign of dual variables (w_f^*). The type of UC is classified into three categories. Meanwhile, these measures related to RTD are determined by not only the sign of dual variables (w_f^*) but also the sign of $\sigma^* + \sum_{i=1}^{m} v_i^* x_i$. The type of RTD is classified into five categories.

on only efficient DMUs, not inefficient ones. In the case of inefficiency, Equation (23.12) needs to incorporate the projection onto an efficiency frontier by both adjusting an inefficiency score and eliminating slacks from the observed production factors, as discussed in Chapter 20. The multiplier restriction by the proposed explorative analysis is more influential, so effective, on the measurement of RTD and DTR than slack adjustment. Therefore, this chapter, like Chapter 22, does not provide a detailed description on slack adjustment. Finally, the type of RTD is determined by measuring the upper and lower bound of $\sigma^* + \sum_{i=1}^{m} v_i^* x_i$. The proposed approach is just an approximation method for the RTD measurement for our descriptive convenience. Thus, the proposed RTD measurement needs to be extended further by adjusting slacks under a possible occurrence of multiple solutions (e.g., multiple references and multiple projections). See Chapter 7 on how to handle such an occurrence of multiple solutions.

23.4 CONGESTION IDENTIFICATION UNDER MANAGERIAL DISPOSABILITY

23.4.1 Possible Occurrence of Desirable Congestion (DC)

As depicted on the right-hand side of Figure 23.6, this chapter can identify an occurrence of DC under managerial disposability. To examine the occurrence, this chapter uses the following model, returning to Model (21.7), which maintains equality constraints (so, no slack variable) on desirable outputs, as specified in Equation (23.2):

$$
\begin{aligned}
&\text{Maximize } \xi + \varepsilon_s \left(\sum_{i=1}^{m} R_i^x d_i^{x+} + \sum_{f=1}^{h} R_f^b d_f^b \right) \\
&\text{s.t.} \quad \sum_{j=1}^{n} x_{ij}\lambda_j - d_i^{x+} = x_{ik} \ (i=1,\dots,m), \\
&\qquad \sum_{j=1}^{n} g_{rj}\lambda_j - \xi g_{rk} = g_{rk} \ (r=1,\dots,s), \\
&\qquad \sum_{j=1}^{n} b_{fj}\lambda_j + d_f^b + \xi b_{fk} = b_{fk} \ (f=1,\dots,h), \\
&\qquad \sum_{j=1}^{n} \lambda_j = 1, \\
&\qquad \lambda_j \ge 0 \ (j=1,\dots,n), \ \xi: \text{URS}, \\
&\qquad d_i^{x+} \ge 0 \ (i=1,\dots,m) \text{ and } d_f^b \ge 0 \ (f=1,\dots,h).
\end{aligned}
\tag{23.13}
$$

Model (23.13) drops slack variables related to desirable outputs so that they are considered as equality constraints. The other groups of constraints on inputs and undesirable outputs have slacks so that they can be considered as inequality constraints.

Model (23.13) has the following dual formulation:

$$\begin{aligned}
&\text{Minimize} && -\sum_{i=1}^{m} v_i x_{ik} - \sum_{r=1}^{s} u_r g_{rk} + \sum_{f=1}^{h} w_f b_{fk} + \sigma \\
&s.t. && -\sum_{i=1}^{m} v_i x_{ij} - \sum_{r=1}^{s} u_r g_{rj} + \sum_{f=1}^{h} w_f b_{fj} + \sigma \geq 0 \quad (j = 1,\ldots,n), \\
& && \sum_{r=1}^{s} u_r g_{rk} + \sum_{f=1}^{h} w_f b_{fk} = 1, \\
& && v_i \geq \varepsilon_s R_i^x \quad (i = 1,\ldots,m), \\
& && u_r : \text{URS} \quad (r = 1,\ldots,s), \\
& && w_f \geq \varepsilon_s R_f^b \quad (f = 1,\ldots,h), \\
& && \sigma : \text{URS}.
\end{aligned} \tag{23.14}$$

An important feature of Model (23.14) is that the dual variables (u_r: *URS* for r = 1,…, s) are unrestricted in these signs because Model (23.13) does not have slacks related to desirable outputs.

A unified efficiency score, or *UEM*(*DC*) of the k-th DMU, with a possible occurrence of DC, under managerial disposability is determined by

$$\begin{aligned}
UEM(DC)_v^{R*} &= 1 - \left[\xi^* + \varepsilon_s \left(\sum_{i=1}^{m} R_i^x d_i^{x+*} + \sum_{f=1}^{h} R_f^b d_f^{b*} \right) \right] \\
&= 1 - \left[-\sum_{i=1}^{m} v_i^* x_{ik} - \sum_{r=1}^{s} u_r^* g_{rk} + \sum_{f=1}^{h} w_f^* b_{fk} + \sigma^* \right],
\end{aligned} \tag{23.15}$$

where all variables are determined on the optimality of Models (23.13) and (23.14). The equation within the parenthesis, obtained from the optimality of Models (23.13) and (23.14), indicates the level of unified inefficiency under managerial disposability. The unified efficiency is obtained by subtracting the level of inefficiency from unity.

As discussed on Model (23.9), Model (23.14) can incorporate additional information by explorative analysis that is formulated as follows:

$$
\begin{aligned}
&\text{Minimize} \quad -\sum_{i=1}^{m} v_i x_{ik} - \sum_{r=1}^{s} u_r g_{rk} + \sum_{f=1}^{h} w_f b_{fk} + \sigma \\
&\text{s.t.} \quad -\sum_{i=1}^{m} v_i x_{ij} - \sum_{r=1}^{s} u_r g_{rj} + \sum_{f=1}^{h} w_f b_{fj} + \sigma \geq 0 \qquad (j=1,\ldots,n), \\
&\qquad \sum_{r=1}^{s} u_r g_{rk} + \sum_{f=1}^{h} w_f b_{fk} = 1, \qquad\qquad (23.16) \\
&\qquad v_i \geq \varepsilon_s R_i^x \qquad (i=1,\ldots,m), \\
&\qquad u_r : \text{URS} \qquad (r=1,\ldots,s), \\
&\qquad w_f \geq \varepsilon_s R_f^b \qquad (f=1,\ldots,h), \\
&\qquad \sigma : \text{URS}, \\
&\qquad -1 \leq v_{i'}/v_i \leq 1 \qquad (i' > i = 1,\ldots,m-1), \\
&\qquad -1 \leq u_{r'}/u_r \leq 1 \qquad (r' > r = 1,\ldots,s-1), \\
&\qquad -1 \leq w_{f'}/w_f \leq 1 \qquad (f' > f = 1,\ldots,h-1).
\end{aligned}
$$

The degree of $UEM(DC)_v^{R*}$ is determined by

$$
UEM(DC)_v^{R*} = 1 - \left(-\sum_{i=1}^{m} v_i^* x_{ik} - \sum_{r=1}^{s} u_r^* g_{rk} + \sum_{f=1}^{h} w_f^* b_{fk} + \sigma^* \right), \qquad (23.17)
$$

where all the dual variables are identified on the optimality of Model (23.16). Equation (23.17) is different from Equation (23.15) because Model (23.16) incorporates the additional side constraints in Equations (23.6) to (23.8). Thus, Equations (23.15) and (23.17) produce different $UEM(DC)_v^{R*}$ measures.

After solving Model (23.16), this chapter can identify the possible occurrence of DC, or eco-technology innovation, by the following rule under the assumption on a unique optimal solution (i. e. unique projection and a unique reference set):

(a) If $u_r^* < 0$ for some (at least one) r, then "strong DC" occurs on the k-th DMU,
(b) If $u_r^* > 0$ for all r, then "no DC" occurs on the k-th DMU and
(c) In the others, including $u_r^* = 0$ for some (at least one) r, then "weak DC" occurs on the k-th DMU.

Note that if $u_r^* < 0$ for some r and $u_{r'}^* = 0$ for the other r', then the weak and strong DCs coexist on the k-th DMU. This study considers it as the strong DC, so indicating eco-technology innovation on undesirable outputs. If $u_r^* < 0$ for all r, then it indicates the best case because an increase in any desirable output always decreases an amount of undesirable outputs. Meanwhile, if $u_r^* < 0$ is identified for some r, then it indicates that there is a chance to reduce an amount of undesirable output(s). Therefore, this chapter considers the second case as an occurrence of DC.

23.4.2 DTR Measurement under the Possible Occurrence of DC

Let the dual variables of the k-th DMU, obtained from Model (23.16), be v_i^* ($i = 1, 2,\ldots,m$), u_r^* ($r = 1, 2,\ldots,s$), w_f^* ($f = 1, 2,\ldots,h$) and σ^*. Then, an estimated supporting hyperplane on the k-th DMU is specified by

$$\sum_{f=1}^{h} w_f^* b_f = \sum_{r=1}^{s} u_r^* g_r + \sum_{i=1}^{m} v_i^* x_i - \sigma^*. \tag{23.18}$$

The equation is characterized by $-\sum_{i=1}^{m} v_i x_{ij} - \sum_{r=1}^{s} u_r g_{rj} + \sum_{f=1}^{h} w_f b_{fj} + \sigma,\ j \in RS_k$, where RS_k is a reference set of the k-th DMU along with $\sum_{r=1}^{s} u_r g_{rk} + \sum_{f=1}^{h} w_f b_{fk} = 1$.

This chapter assumes that Model (23.16) has both a unique projection of an inefficient DMU onto an efficiency frontier and a unique reference set for the projected DMU. Then, the degree (Dg) of DTR, or $DgDTR$, is measured by

$$\begin{aligned} DgDTR &= \left(\sum_{r=1}^{s} u_r^* g_r\right) \Big/ \left(\sum_{f=1}^{h} w_f^* b_f\right) \\ &= \left(\sum_{r=1}^{s} u_r^* g_r\right) \Big/ \left(\sum_{i=1}^{m} v_i^* x_i + \sum_{r=1}^{s} u_r^* g_r - \sigma^*\right) \\ &= 1 \Big/ \left[1 - \left(\sigma^* - \sum_{i=1}^{m} v_i^* x_i\right) \Big/ \left(\sum_{r=1}^{s} u_r^* g_r\right)\right]. \end{aligned} \tag{23.19}$$

Consequently, the type of DTR is classified by the following rule on the k-th DMU:

(a) *Increasing DTR* ↔ There is an optimal solution of Model (23.16) that satisfies all $u_r^* > 0$ ($r = 1,\ldots,s$) and $\sigma^* - \sum_{i=1}^{m} v_i^* x_i > 0$.
(b) *Constant DTR* ↔ There exists an optimal solution of Model (23.16) that satisfies all $u_r^* > 0$ ($r = 1,\ldots,s$) and $\sigma^* - \sum_{i=1}^{m} v_i^* x_i = 0$.
(c) *Decreasing DTR* ↔ There is an optimal solution of Model (23.16) that satisfies all $u_r^* > 0$ ($r = 1,\ldots,s$) and $\sigma^* - \sum_{i=1}^{m} v_i^* x_i < 0$.
(d) *Negative DTR* ↔ There is an optimal solution of Model (23.16) that satisfies $u_r^* < 0$ for at least one $r \in \{1,\ldots,s\}$ and
(e) *No DTR* ↔ All other cases excluding (a) to (d).

All the concerns discussed for the measurement of RTD in Section 23.3.2 are applicable to DTR. However, it is important to note that the type of DTR is determined by measuring the upper and lower bounds of $\sigma^* - \sum_{i=1}^{m} v_i^* x_i$. The proposed approach for DTR is just an approximation method for the DTR measurement.

Difference between DC and DTR: The occurrence and type of DC are identified by the sign of dual variables (u_r^*). The type of DC is classified into three categories. Meanwhile, these measures related to DTR are determined by not only the sign of dual variables (u_r^*) but also the sign of $\sigma^* - \sum_{i=1}^{m} v_i^* x_i$. The type of DTR is classified into five categories. Figure 23.8 visually describes an occurrence of DC and DTR classification.

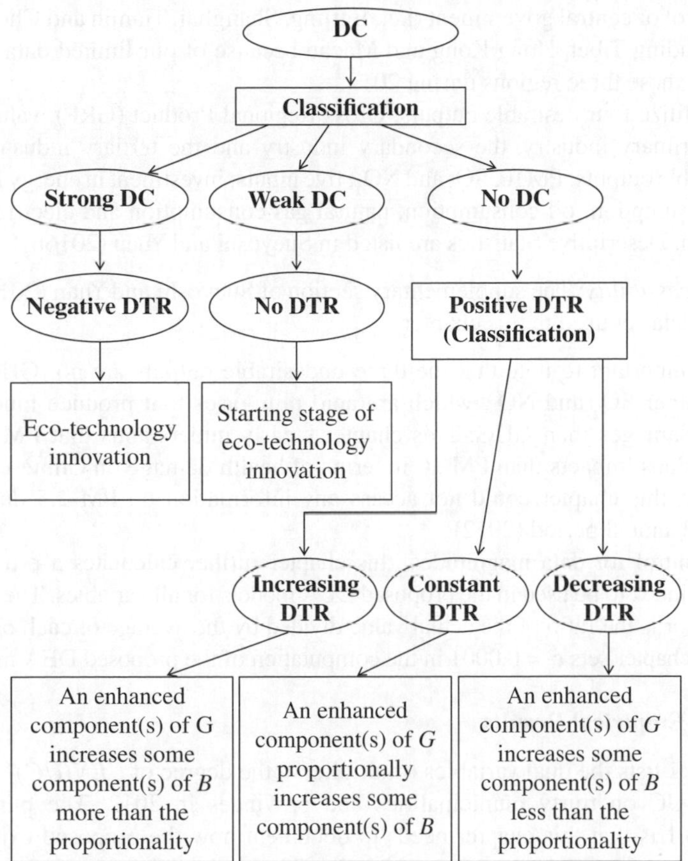

FIGURE 23.8 Damages to return (DTR) under desirable congestion (DC) (a) An increase in some component(s) of the input vector (X) increases the components of the other two production factors. The occurrence of DC is desirable because some component(s) of the desirable output vector (G) increase but those of the undesirable output vector (B) decrease along with such a change in input vector. That is an occurrence of DC. We look for such an occurrence by eco-technology innovation in a broad sense, including managerial challenges. (b) The occurrence and type of DC are identified by the sign of dual variables (u_r^*). The type of DC is classified into three categories. Meanwhile, these measures related to DTR are determined by not only the sign of dual variables (u_r^*), but also the sign of $\sigma^* - \sum_{i=1}^{m} v_i^* x_i$. The type of DTR is classified into five categories.

23.5 ENERGY AND SOCIAL SUSTAINABILITY IN CHINA

23.5.1 Data

This chapter presents a data set from the National Bureau of Statistics of the People's Republic of China (http://www.stats.gov.cn/tjsj/). The National Bureau of Statistics provides a regional data set on energy, environment and other aspects of the whole country every year. Using the data set, this chapter examines 30 provinces of China, including four well-developed municipalities directly under the control of central government (i.e., Beijing, Shanghai, Tianjin and Chongqing) but excluding Tibet, Hong Kong and Macau because of our limited data accessibility on these three regions during 2012.

We utilize four desirable outputs: Gross Regional Product (GRP), value-added of the primary industry, the secondary industry and the tertiary industry; three undesirable outputs: PM10, SO_2 and NO_2; five inputs: investment in energy industry, coal consumption, oil consumption, natural gas consumption and electricity consumption. Descriptive Statistics are listed in Sueyoshi and Yuan (2016b).

Data Accessibility: The supplementary section of Sueyoshi and Yuan (2016b) lists the full data set used in this chapter.

It is important to note that the three undesirable outputs are not GHG emissions, rather SO_2 and NO_2, which are acid rain gases that produce much more serious damages than GHG. This chapter clearly understands that PM 2.5 has more serious impacts than PM 10 in terms of health damage on Chinese people. However, this chapter could not access any information on PM 2.5 during the observed annual period (2012).

To control for data magnitudes, this chapter further calculates a provincially adjusted index to be used in the proposed DEA models for all variables. The index of each factor is the ratio of the actual value divided by the average of each one. Note that this chapter sets $\varepsilon_s = 0.0001$ in the computation of the proposed DEA models.

23.5.2 Empirical Results

Table 23.1 lists the dual variables (multipliers), the degree of $UEN(UC)_v^{R*}$ and the type of UC on thirty municipalities and provinces in 2012. The purpose of Table 23.1 is that this chapter needs to document how the proposed explorative analysis can change these measures. Model (23.4) is used for these computations. The model for $UEN(UC)_v^{R*}$ has excluded the three groups of multiplier restrictions, or Equations (23.6) to (23.8). A problem found in Table 23.1 is that the magnitude of these dual variables is relatively large so that only Inner Mongolia becomes inefficient (0.8724). The others are all efficient. The type of UC is "strong" for all provinces. These results are unacceptable in this type of empirical study because it contains many efficient provinces. Further, the empirical results are unacceptable from the perspective of energy and economic planning for Chinese sustainability development because there is no difference among them and almost all are efficient. This indicates a methodological drawback of DEA environmental assessment. This type of difficulty may often occur in DEA and its applications for environmental assessment.

Table 23.2 lists the dual variables (multipliers), the degree of $UEM(DC)_v^{R*}$ and the type of DC for the 30 municipalities and provinces in 2012. Table 23.2 also documents how the proposed multiplier restrictions can change these measures. Model (23.14) is used for these computations. The model for $UEM(DC)_v^{R*}$ does not utilize the three groups for explorative analysis, or Equations (23.6)–(23.8). The magnitude of these dual variables is relatively large so that Table 23.2 still contains many efficient municipalities and provinces in $UEM(DC)_v^{R*}$. Exceptions in these $UEM(DC)_v^{R*}$ measures are Jilin (0.8615) and Jiangxi (0.7947). All municipalities and provinces have strong DC. These results are also unacceptable from the perspective of China's sustainability development. Thus, Tables 23.1 and 23.2 clearly indicate the necessity of explorative analysis in the classification of UC and DC. The difficulty identified in Tables 23.1 and 23.2 can be overcome by incorporating the proposed explorative analysis, or Equations (23.6) to (23.8), as formulated in Models (23.9) and (23.16).

Table 23.3 summarizes $UEN(DC)_v^{R*}$ under the restriction and type of UC. Table 23.4 summarizes $UEM(DC)_v^{R*}$ under restriction and the type of DC. This chapter finds two important features in the two tables. One feature is that the number of efficient municipalities and provinces in the two tables reduces from those of Tables 23.1 and 23.2. More variations can be found in the type of UC of Table 23.3 and in the type of DC of Table 23.4. Thus, the comparison among Tables 23.3 and 23.4 indicates the importance of the proposed explorative analysis in terms of determining the type of UC and DC. Consequently, this chapter uses results from Tables 23.3 and 23.4, hereafter, for discussing Chinese energy planning and economic development.

Table 23.5 summarizes UC, RTD, DC and DTR on 30 Chinese municipalities and provinces, all of which are measured by Models (23.9) and (23.16), both of which incorporated the proposed multiplier restrictions. Table 23.5 indicates that most of provinces belonged to no or weak in UC and decreasing or no in RTD. The selection of 2012 is for our illustrative purpose. This indicates "an economic growth limit" in Chinese provinces during 2012. In other words, an increase in their energy component(s), as inputs, could increase an amount of undesirable outputs (i.e., pollution measures), but the change has either a diminishing or no effect on desirable outputs (i.e., GRP and other three economic measures) under production technology in 2012. Meanwhile, as summarized on the right-hand side of Table 23.5, the type of DC and DTR indicated that many provinces had a high level of social potential to reduce an amount of undesirable outputs because many provinces had strong DC and negative DTR. That is, the level of air pollution was very high in most of them so that the local governments could reduce the pollution level if they were very serious on the air cleanness. However, municipalities (e.g., Beijing and Shanghai) did not have such a high level of pollution mitigation capability, as specified by weak DC and no DTR. The level of air pollution in these large municipalities was beyond their eco-technology capabilities. The empirical result clearly implies that the central government should pay more serious attention to the air pollution in the municipalities.

TABLE 23.1 UEN(UC) and type of UC: Model (23.4)

Province	w_1	w_2	w_3	σ	$UEN(UC)$	UC
Beijing	–67.0832	–8.2505	–55.5696	180.9354	1.0000	Strong
Tianjin	–62.4183	–132.3779	84.9027	169.7916	1.0000	Strong
Hebei	34.4745	–194.2958	65.7054	194.4844	1.0000	Strong
Shanxi	68.9023	–160.0109	160.0633	20.4058	1.0000	Strong
Inner Mongolia	0.8787	–0.2406	0.1085	–0.4814	0.8724	Strong
Liaoning	71.9875	–210.5759	112.1153	124.0256	1.0000	Strong
Jilin	87.1350	14.4970	–171.6050	37.2246	1.0000	Strong
Heilongjiang	84.4352	–99.7433	–122.1147	163.9192	1.0000	Strong
Shanghai	237.9280	–51.9457	–201.9009	41.2487	1.0000	Strong
Jiangsu	–131.0119	7.2340	–79.3795	314.9994	1.0000	Strong
Zhejiang	116.6050	–100.2130	–178.5735	224.9997	1.0000	Strong
Anhui	–198.2258	73.5496	108.2929	97.3453	1.0000	Strong
Fujian	–13.7631	181.0110	–117.9729	41.7415	1.0000	Strong
Jiangxi	37.2990	–131.1707	10.6483	116.1614	1.0000	Strong
Shandong	–98.8847	–112.3617	24.8536	301.4538	1.0000	Strong
Henan	–37.4985	–129.0649	–28.9755	283.0348	1.0000	Strong
Hubei	20.3504	4.3239	–156.9993	200.5690	1.0000	Strong
Hunan	–72.0972	–7.6075	–106.3379	234.3867	1.0000	Strong
Guangdong	140.8545	11.6748	–208.1419	184.8141	1.0000	Strong
Guangxi	–41.9882	87.1230	–71.2541	12.4540	1.0000	Strong
Hainan	61.0643	62.2186	–94.0134	–26.7372	1.0000	Strong
Chongqing	–165.8771	–33.5683	102.1901	44.9664	1.0000	Strong
Sichuan	–185.8158	30.7800	–12.9453	282.8272	1.0000	Strong
Guizhou	–81.1661	–59.7219	135.6327	–22.3290	1.0000	Strong

Yunnan	404.0265	−138.0444	−225.1171	−4.2690	1.0000	Strong
Shaanxi	−265.8177	−11.4641	118.9462	236.2667	1.0000	Strong
Gansu	−135.2600	−31.4009	194.0356	4.8144	1.0000	Strong
Qinghai	−93.7301	−82.1600	267.1938	−19.5242	1.0000	Strong
Ningxia	83.9409	−72.0011	−17.4262	−34.0242	1.0000	Strong
Xinjiang	266.7831	−55.6519	−248.3287	−22.9935	1.0000	Strong

(a) Source: Sueyoshi and Yuan (2016b). (b) w_1 stands for the dual variable of PM10. w_2 stands for the dual variable of SO_2 and w_3 stands for the dual variable of NO_2. If $w_f^* < 0$ for some (at least one) f, then "strong UC" occurs on the k-th DMU; if $w_f^* > 0$ for all f, then "no UC" occurs on the k-th DMU. In the others, including $w_f^* = 0$ for some (at least one) f, then "weak UC" occurs on the k-th DMU. If $w_f^* < 0$ for some f and $w_{f'}^* = 0$ for other f', then both strong and weak UCs coexist on the k-th DMU. In the case, this chapter considers the occurrence as strong UC. (c) Model (23.4) does not incorporate the proposed explorative analysis. Consequently, the dual variables are large in their magnitudes. In addition, most of municipalities and provinces are rated as unity in their unified efficiencies under natural disposability and strong undesirable congestion (UC). The results are unacceptable because this chapter is concerned with policy suggestions for Chinese regional economic and energy planning.

TABLE 23.2 UEM(DC) and type of DC: Model (23.14)

Province	u_1	u_2	u_3	u_4	σ	$UEM(DC)$	DC
Beijing	–30.7337	–21.7401	–135.4404	22.5453	72.4647	1.0000	Strong
Tianjin	–15.0532	–220.4132	215.1855	–190.9422	8.2407	1.0000	Strong
Hebei	–58.3249	84.2157	–150.9658	–25.7218	91.9477	1.0000	Strong
Shanxi	–46.9406	–33.6323	–68.2970	–18.8360	69.9693	1.0000	Strong
Inner Mongolia	–58.6732	6.6154	–58.8331	–72.3189	64.7705	1.0000	Strong
Liaoning	–0.3693	–89.0342	155.5847	–225.2064	–19.9005	1.0000	Strong
Jilin	–15.0745	0.1319	12.0872	1.8615	–1.2502	0.8615	Strong
Heilongjiang	–28.2626	112.5519	–315.6184	32.5089	98.0199	1.0000	Strong
Shanghai	–30.3279	–50.6093	–59.9712	–1.6165	21.0333	1.0000	Strong
Jiangsu	–13.9364	6.0038	19.7134	–58.6522	101.8990	1.0000	Strong
Zhejiang	–26.7302	–221.2135	309.0860	–201.1764	–28.0248	1.0000	Strong
Anhui	15.7277	–14.6938	151.1975	–348.3844	–26.6403	1.0000	Strong
Fujian	–7.5001	–35.3934	93.7482	–145.4879	–26.9089	1.0000	Strong
Jiangxi	26.8480	–2.5907	–12.3890	–11.7510	–0.4537	0.7947	Strong
Shandong	–27.2371	22.9688	–40.1558	–22.7104	52.3088	1.0000	Strong
Henan	10.6726	1.5341	97.7867	–261.1320	99.8903	1.0000	Strong
Hubei	1.4970	1.4296	–2.1370	–1.2948	0.8822	1.0000	Strong
Hunan	–28.0571	4.3695	11.1387	12.3020	1.1926	1.0000	Strong
Guangdong	–12.1954	–20.8928	–23.0836	2.3836	30.3790	1.0000	Strong
Guangxi	–91.2473	107.8889	–134.0183	–127.2700	32.9213	1.0000	Strong
Hainan	–35.6506	–33.7526	–11.8598	–69.0107	–9.4207	1.0000	Strong
Chongqing	–9.5122	–213.8387	265.6208	–229.2442	11.1692	1.0000	Strong
Sichuan	–37.7014	12.6096	–44.6575	–42.2490	62.3629	1.0000	Strong
Guizhou	–19.5151	28.8248	–355.0989	45.3138	40.4201	1.0000	Strong
Yunnan	–51.1983	68.7073	–288.5694	5.0715	71.7098	1.0000	Strong

Shaanxi	–35.4637	–197.4817	389.7037	–353.5435	13.7583	1.0000	Strong
Gansu	–123.0311	21.9416	–229.7992	–138.5931	–17.9110	1.0000	Strong
Qinghai	–50.0142	–108.3042	–6.3172	–115.2217	16.3682	1.0000	Strong
Ningxia	–77.7710	–102.4935	–98.1680	–76.9879	–9.4704	1.0000	Strong
Xinjiang	–48.3400	17.8205	–168.7917	–10.8003	109.0019	1.0000	Strong

(a) Source: Sueyoshi and Yuan (2016b). (b) u_1 stands for the dual variable of GRP. u_2 stands for the dual variable of value-added of the primary industry. u_3 stands for the dual variable of value-added of the secondary industry. u_4 stands for the dual variable of value-added of the tertiary industry. If $u_r^* < 0$ for some (at least one) r, then "strong DC" occurs on the k-th DMU. If $u_r^* > 0$ for all r, then "no DC" occurs on the k-th DMU. In the others, including $u_r^* = 0$ for some (at least one) f, then "weak DC" occurs on the k-th DMU. If $u_r^* < 0$ for some r and $u_{r'}^* = 0$ for other r', then the weak and strong DCs coexist on the k-th DMU. In this case, this study considers the occurrence as strong DC. (c) Model (23.14) does not incorporate the proposed multiplier restriction. Consequently, these dual variables are large in their magnitudes. All provinces, except Heilongjiang and Jiangxi, are rated as unity in their unified efficiency measures under managerial disposability and they are all under strong DC. The results are unacceptable in terms of Chinese regional economic and energy planning.

TABLE 23.3 UEN(UC) and type of UC: Model (23.9)

Province	w_1	w_2	w_3	σ	*UEN(UC)*	UC
Beijing	0.0771	0.0771	0.0771	–0.2465	1.0000	No
Tianjin	0.1014	0.1014	0.1014	–0.3416	0.8270	No
Hebei	0.2603	0.0000	0.0000	0.1700	0.8139	Weak
Shanxi	1.0429	0.0000	0.0000	–0.3459	0.5452	Weak
Inner Mongolia	0.8797	0.0000	0.0000	–0.2771	0.5492	Weak
Liaoning	0.5293	0.0000	0.0000	–0.1894	0.6959	Weak
Jilin	0.6557	0.0000	0.0000	–0.3031	0.7249	Weak
Heilongjiang	0.5894	0.0000	0.0000	–0.2109	0.6969	Weak
Shanghai	0.1656	0.1656	0.1656	–0.2500	1.0000	No
Jiangsu	0.0574	0.0574	0.0574	0.0230	1.0000	No
Zhejiang	0.2088	0.0000	0.0000	–0.1572	1.0000	Weak
Anhui	0.1984	0.1984	0.1984	–0.2168	0.9384	No
Fujian	0.2543	0.2543	0.1650	–0.1924	1.0000	No
Jiangxi	0.4139	0.0000	0.0000	–0.5684	1.0000	Weak
Shandong	0.0533	0.0533	0.0533	0.4668	1.0000	No
Henan	0.0000	0.0000	0.0000	0.4713	0.9431	Weak
Hubei	0.4814	0.4814	–0.4814	–0.0563	1.0000	Strong
Hunan	0.1294	0.1294	0.1294	–0.0106	1.0000	No
Guangdong	0.0507	0.0507	0.0507	0.0261	1.0000	No
Guangxi	0.2411	0.2411	0.2294	–0.1928	1.0000	No
Hainan	0.7607	0.7072	0.6912	–0.5895	1.0000	No
Chongqing	0.3990	0.0141	0.0141	–0.4087	0.6208	No
Sichuan	0.1332	0.1332	0.1332	–0.1024	0.7650	No
Guizhou	0.9392	0.0000	0.0000	–0.3791	0.6111	Weak
Yunnan	0.8025	0.0000	0.0000	–0.2976	0.7743	Weak
Shaanxi	0.2505	0.2505	0.2505	–0.1964	0.4921	No
Gansu	0.2377	0.2377	0.2377	–0.2365	0.3733	No
Qinghai	0.3661	0.3661	0.3661	–22.4480	1.0000	No
Ningxia	0.9219	0.0000	0.0000	–0.3199	0.3500	Weak
Xinjiang	0.5981	0.0000	0.0000	–0.1884	0.3165	Weak

(a) Source: Sueyoshi and Yuan (2016b). (b) w_1 stands for the dual variable of PM10. w_2 stands for the dual variable of SO_2 and w_3 stands for the dual variable of NO_2. (c) If $w_f^* < 0$ for some (at least one) f, then "strong UC" occurs on the k-th DMU. If $w_f^* > 0$ for all f, then "no UC" occurs on the k-th DMU. In the others, including $w_f^* = 0$ for some (at least one) f, then "weak UC" occurs on the k-th DMU. If $w_f^* < 0$ for some f and $w_{f'}^* = 0$ for the other f', then both strong and weak UCs coexist on the k-th DMU. In the case, this study considers the occurrence as strong UC. (d) Model (23.9) incorporates the proposed multiplier restriction. Consequently, these ratios between dual variables are between –1 and +1 in their magnitudes. The results on unified efficiency measures under natural disposability and the type of undesirable congestion (UC) are acceptable in terms of Chinese regional economic and energy planning.

TABLE 23.4 UEM(DC) and type of DC: Model (23.16)

Province	u_1	u_2	u_3	u_4	σ	*UEM(DC)*	DC
Beijing	0.1462	0.0000	0.0000	0.0000	–0.2436	0.5160	Weak
Tianjin	0.0650	0.0000	0.0000	0.0000	–0.2415	0.4406	Weak
Hebei	0.0636	0.0636	0.0636	–0.0636	–0.0117	0.8386	Strong
Shanxi	0.0460	0.0460	0.0460	–0.0141	–0.0150	1.0000	Strong
Inner Mongolia	0.0559	0.0559	0.0559	–0.0559	0.2629	1.0000	Strong
Liaoning	0.0417	0.0417	0.0417	–0.0417	–0.2819	0.7306	Strong
Jilin	0.0535	0.0535	0.0535	–0.0535	–0.2781	0.6512	Strong
Heilongjiang	0.0452	0.0452	0.0000	0.0000	–0.2456	0.7103	Weak
Shanghai	0.1475	0.0000	0.0000	0.0000	–0.3827	0.6928	Weak
Jiangsu	0.1513	0.1101	0.1101	0.0149	1.0000	1.0000	No
Zhejiang	0.0846	0.0846	0.0846	–0.0846	–0.2646	0.7059	Strong
Anhui	0.0736	0.0736	0.0736	–0.0736	–0.2405	0.8794	Strong
Fujian	0.1399	0.1399	0.1399	–0.1399	–0.2629	1.0000	Strong
Jiangxi	0.0907	0.0907	0.0907	–0.0907	–0.2835	0.5939	Strong
Shandong	0.0740	0.0740	0.0679	–0.0679	0.0972	1.0000	Strong
Henan	0.1264	0.1264	0.1264	–0.1264	0.2107	0.8776	Strong
Hubei	0.0947	0.0947	0.0947	–0.0947	–0.1807	0.7368	Strong
Hunan	0.1045	0.1045	0.1045	–0.1045	–0.1994	0.8041	Strong
Guangdong	0.0551	0.0487	0.0487	0.0487	0.0951	1.0000	No
Guangxi	0.1438	0.1438	0.1438	–0.1438	–0.2744	0.8498	Strong
Hainan	0.3150	0.3150	0.0000	0.0000	–0.6491	1.0000	Weak
Chongqing	0.0541	0.0541	0.0541	–0.0541	–0.2816	0.5419	Strong
Sichuan	0.1371	0.1371	0.1371	–0.1371	0.8463	0.9455	Strong
Guizhou	0.0621	0.0621	0.0000	0.0000	–0.3388	0.7006	Weak

(*Continued*)

TABLE 23.4 (Continued)

Province	u_1	u_2	u_3	u_4	σ	*UEM(DC)*	DC
Yunnan	0.0592	0.0592	0.0592	–0.0592	–0.4002	0.8897	Strong
Shaanxi	0.0662	0.0662	0.0662	–0.0662	–0.0925	0.7197	Strong
Gansu	0.0222	0.0222	0.0000	0.0000	–0.2296	0.5000	Weak
Qinghai	0.0000	0.0000	0.0000	0.0000	–0.3188	0.4296	Weak
Ningxia	0.0000	0.0000	0.0000	0.0000	–0.2843	0.4287	Weak
Xinjiang	0.0000	0.0000	0.0000	0.0000	0.0041	0.5470	Weak

(a) Source: Sueyoshi and Yuan (2016b). (b) u_1 stands for the dual variable of GRP. u_2 stands for the dual variable of value-added of the primary industry. u_3 stands for the dual variable of value-added of the secondary industry. u_4 stands for the dual variable of value-added of the tertiary industry. If $u_r^* < 0$ for some (at least one) r, then "strong DC" occurs on the k-th DMU. If $u_r^* > 0$ for all r, then "no DC" occurs on the k-th DMU. In the others, including $u_r^* = 0$ for some (at least one) f, then "weak DC" occurs on the k-th DMU. If $u_r^* < 0$ for some r and $u_{r'}^* = 0$ for the other r', then the weak and strong DCs coexist on the k-th DMU. In the case, this study considers the occurrence as strong DC. (c) Model (23.16) incorporates the proposed multiplier restriction. Consequently, the ratios between dual variables are between –1 and +1 in their magnitudes. The results on unified efficiencies under managerial disposability and the type of desirable congestion (DC) are acceptable in terms of Chinese regional economic and energy planning.

TABLE 23.5 Classification of RTD and DTR of 30 municipalities and provinces

Province	UC	RTD	DC	DTR
Beijing	No	Decreasing	Weak	No
Tianjin	No	Decreasing	Weak	No
Hebei	Weak	No	Strong	Negative
Shanxi	Weak	No	Strong	Negative
Inner Mongolia	Weak	No	Strong	Negative
Liaoning	Weak	No	Strong	Negative
Jilin	Weak	No	Strong	Negative
Heilongjiang	Weak	No	Weak	No
Shanghai	No	Decreasing	Weak	No
Jiangsu	No	Decreasing	No	Decreasing
Zhejiang	Weak	No	Strong	Negative
Anhui	No	Decreasing	Strong	Negative
Fujian	No	Decreasing	Strong	Negative
Jiangxi	Weak	No	Strong	Negative
Shandong	No	Decreasing	Strong	Negative
Henan	Weak	No	Strong	Negative
Hubei	Strong	Negative	Strong	Negative
Hunan	No	Decreasing	Strong	Negative
Guangdong	No	Decreasing	No	Decreasing
Guangxi	No	Decreasing	Strong	Negative
Hainan	No	Increasing	Weak	No
Chongqing	No	Decreasing	Strong	Negative
Sichuan	No	Decreasing	Strong	Negative
Guizhou	Weak	No	Weak	No
Yunnan	Weak	No	Strong	Negative
Shaanxi	No	Increasing	Strong	Negative
Gansu	No	Increasing	Weak	No
Qinghai	No	Increasing	Weak	No
Ningxia	Weak	No	Weak	No
Xinjiang	Weak	No	Weak	No

(a) Source: Sueyoshi and Yuan (2016b). (b) The degree of UC, RTD, DC and DTR are determined by Models (23.9) and (23.16), both of which have the proposed multiplier restrictions, respectively. (c) Most of municipalities and provinces belong to no or weak in UC and decreasing or no in RTD. This indicates an "economic limit" in Chinese provinces during 2012. In other words, an increase in their input components, or energy sources, could enhance an amount of undesirable outputs, but the enhancement had a diminishing or no effect on desirable outputs under current production technology. (d) Meanwhile, the type of DC and DTR indicates a high potential to reduce the amount of undesirable outputs. In other words, the current level of industrial pollution is very high in most of provinces so that they could easily reduce the amount of pollution. However, large municipalities (e.g., Beijing and Shanghai) did not have such a high level of pollution mitigation potential, as specified by weak DC and no DTR. The level of air pollution in these large municipalities is beyond their eco-technology capabilities. Any reduction in pollution level is very difficult under current regulation. These large ones need strict regulation and green technology innovation.

23.6 SUMMARY

This chapter discussed the concept of natural and managerial disposability from their economic and methodological implications on social sustainability development. Then, it explored their analytical links to the concept of "congestion". The concept was further classified into UC under natural disposability and DC under managerial disposability. Considering the two groups of disposability concepts, this chapter compared RTD under UC with DTR under DC. These new scale measures (i.e., RTD and DTR) can be considered as extended concepts of returns to scale (RTS) and damages to scale (DTS).

As discussed in this chapter, UC and DC are conceptually and analytically different from RTD and DTR although they are closely related to each other. An occurrence of the former two measures is identified by a single negative multiplier (i.e., the dual variable). In contrast, the latter two measures are associated with an examination of the sign and magnitude of multiple multipliers and the intercept of the supporting hyperplane. Thus, the occurrence of UC and that of DC are a necessary condition, but not a sufficient condition on RTD and DTR, respectively, in terms of the number of negative multipliers (i.e., dual variables) on production factors.

To document the practicality of the proposed approach, this chapter applied it to Chinese economic and environmental assessments for its future economics and energy planning for social sustainability development[2]. This chapter finds three important concerns. First, the Chinese government historically paid attention to its economic prosperity, but did not pay serious attention to air pollution prevention. Second, economic and energy policy concerns focused upon well-developed municipalities (e.g., Beijing and Shanghai) and large provinces in China. Therefore, it is important for the central government to allocate economic and energy resources to other provinces so that China can reduce its regional and industrial imbalances, so enhancing the level of social

[2] See the 13th five-year plan (2016–2020). The green economy will be the most important part of the 13th five-year plan. As the world's largest carbon emitter, China's transition to a green economy will be important from an international perspective. Currently China no longer relies solely on GDP for assessing officials' performance. If environmental protection is not ensured, officials are classed as not up to standard. The Communist Party of China (CPC)'s chairman, Xi Jinping, discussed the importance of green mountains and clean waters in his major speech. This may be a good sign for environmental protection in China. The 13th five-year plan is scheduled for the end of the formation of basic ideas, reported in 2016. The energy plan will build safe, clean, efficient and sustainable modern energy systems that have proposed seven strategic tasks for increasing domestic energy productions: promoting energy conservation and efficiency, optimizing energy structure, vigorous developing of hydro, nuclear, wind, solar and geothermal, encouraging international cooperation in energy, promoting energy technologies, promoting institutional innovation and strengthening energy regulation. The goal is energy security, to find a balance between efficiency and cleanness to achieve the sustainable development of energy in China. See Zhang C. and Xu N. (2014) Green sector will boom during China's 13th five-year plan. Available at: https://www.chinadialogue.net/article/show/single/en/7280-Green-sector-will-boom-during-China-s-13th-Five-Year-Plan.

sustainability. Finally, although this chapter does not clearly discuss it, the municipalities need strict regulation on traffic control in these metropolitan areas and a fuel mix shift from coal combustion to natural gas and other renewable energies. The fuel mix strategy (e.g., from coal to natural gas, solar and nuclear generations), along with privatization[3] (i.e., from public to private energy firms), will be a major industrial policy issue for China in future.

[3] Most Chinese energy firms operate under public ownership. See, for example, the research of Sueyoshi and Goto (2012a) that documented the ownership portion of three Chinese petroleum companies. Public agencies on environmental protection usually have difficulty in monitoring and controlling public companies because their governances are connected to each other. To terminate such a political linkage, the privatization of public firms, in particular energy firms, is necessary for Chinese future. Along with such privatization, Chinese companies can change their governance structures, which are more competitive than the current ones in a world-wide global market. Consequently, China can increase the status of social sustainability.

24

DISPOSABILITY UNIFICATION

24.1 INTRODUCTION

To measure the level of "corporate sustainability" of organizations in the private sector, this chapter[1] discusses how to unify natural and managerial disposability concepts under radial and non-radial measurements. See Chapter 15 for a description of corporate sustainability. The proposed approach provides us four unified efficiency measures related to natural disposability and/or managerial disposability. These unified efficiency measures serve as a computational basis for identifying a new type of scale efficiency measures, none of which has been discussed in previous chapters, including Chapter 18.

After theoretically discussing the proposed scale efficiency measures, this chapter applies the proposed approach to examine the level of corporate sustainability among petroleum firms in the United States. The petroleum firms are generally classified into two categories. One of the two categories includes a group of "integrated companies," which have a large supply chain for spanning both upstream and downstream functions in the US petroleum industry. The other category includes "independent companies" that focus on the upstream function in their business operations.

[1]This chapter is partly based upon the article: Sueyoshi, T. and Wang, D. (2014b) Sustainability development for supply chain management in US petroleum industry by DEA environmental assessment. *Energy Economics*, **46**, 360–374.

Environmental Assessment on Energy and Sustainability by Data Envelopment Analysis, First Edition. Toshiyuki Sueyoshi and Mika Goto.

As an application of the proposed approach, this chapter examines whether the integrated companies outperform the independent companies, because a large supply chain incorporated into the former group provides them with a scale merit in their operations and gives a business opportunity to directly contact consumers[2] and identify their opinions on the operation of petroleum firms. Thus, it is easily imagined that a large supply chain can enhance the level of corporate sustainability in the US petroleum industry. Moreover, vertical integration between upstream and downstream may be a promising business trend for better performance in the petroleum industry. This chapter on US industry and Chapter 16 on world industry explore different aspects and structures regarding the petroleum business. See Chapter 13 for a description of the current trend on oil and gas businesses in the world.

The remainder of this chapter is organized as follows. Section 24.2 discusses underlying concepts for disposability unification that are incorporated in the proposed radial and non-radial approaches. Section 24.3 describes non-radial measurement under natural, managerial and natural and managerial disposability concepts. Section 24.4 describes radial measurement under the three types of disposability concepts. Section 24.5 discusses a computational flow for disposability unification. Section 24.6 applies the proposed non-radial and radial approaches to evaluate the unified (environmental and operational) performance of US petroleum companies. Section 24.7 summarizes this chapter.

24.2 UNIFICATION BETWEEN DISPOSABILITY CONCEPTS

Disposability Unification: Returning to Stage III of Figure 15.6, Figure 24.1 visually describes two functional forms in the horizontal axis (x) and in the vertical axis (g and b). One of the two efficiency frontiers, the upper line, expresses the convex relationship between the input (x) and the desirable output (g). The other efficiency frontier indicates a similar (convex) relationship between the input (x) and an undesirable output (b).

The importance of Figure 15.6 is that it visually describes the three stages (Stages I, II and III) of disposability unification. Figure 24.1 duplicates only the last stage (Stage III) for the unification of natural and managerial disposability concepts. This chapter unifies these disposability concepts based upon Stage III of Figure 15.6.

As identified from the left-hand side of Figure 24.1, the two efficiency frontiers have an increasing trend of the desirable output (g) and the undesirable output (b) along with an enhanced input (x). The rationale for the shape similarity is because the undesirable output is the “byproduct” of the desirable output. However, as shown on the right-hand side of Figure 24.1, the two frontiers have different structures.

[2] Wuyts, S., Stremersch, S., Van Den Bulte, C. and Franses, P.H. (2004) Vertical marketing systems for complex products: a triadic perspective. *Journal of Marketing Research*, **41**, 479–487.

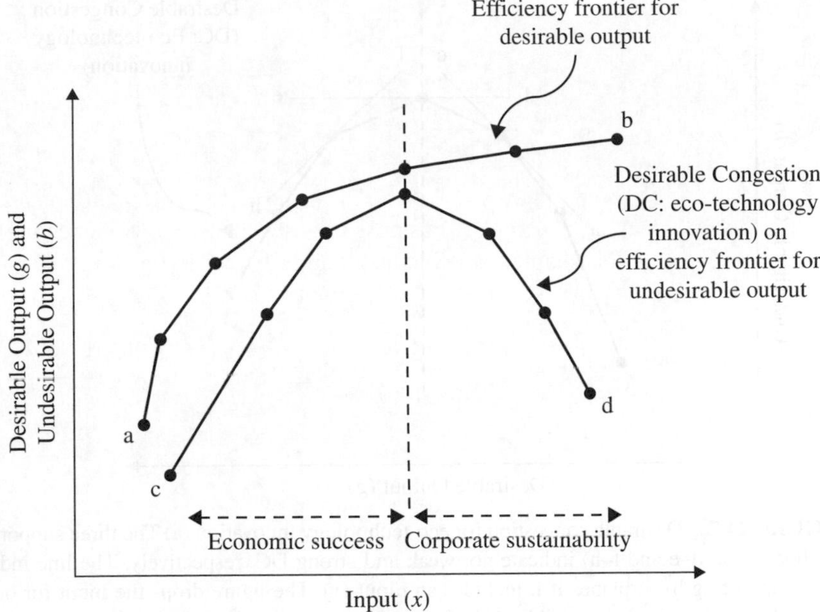

FIGURE 24.1 Disposability unification (stage III)

That is, the efficiency frontier for an undesirable output increases and then decreases with input enhancement. As discussed in Chapter 21, such a case occurs under DC, or eco-technology innovation, for pollution reduction. Thus, the final stage of disposability unification must attain the status of Stage III for the proposed unification, as depicted in both Figure 15.6 and Figure 24.1 because the unification between natural and managerial disposability concepts serves as part of the important process of DEA environmental assessment.

In discussing the proposed disposability unification, it is necessary for us to reorganize Figure 24.1 in the x and g & b space. For this purpose, Figure 24.2 visually specifies the necessary condition for the proposed unification. Figure 24.2 visually assumes that the two types of outputs (g and b) increase with increased input (x). The input change, not listed on the figure, increases the desirable output (g) but decreases the undesirable output (b) from a certain point after eco-technology is introduced into the operation of an organization. To describe the situation more clearly, Figure 24.2 shows a supporting line that we can identify on the efficiency frontier between g and b. For example, the line (a–c) indicates no occurrence of DC. The line (d–e) shows weak DC that indicates the starting situation on DC. The line (f–h) indicates strong DC, corresponding to the occurrence of eco-technology innovation on an undesirable output. See Chapter 23 for a detailed description of the three types of DC. In the case of multiple components of the three production factors, the occurrence of strong DC, or the negative slope of the supporting hyperplane, implies that an increase of the input vector leads to an increase in some

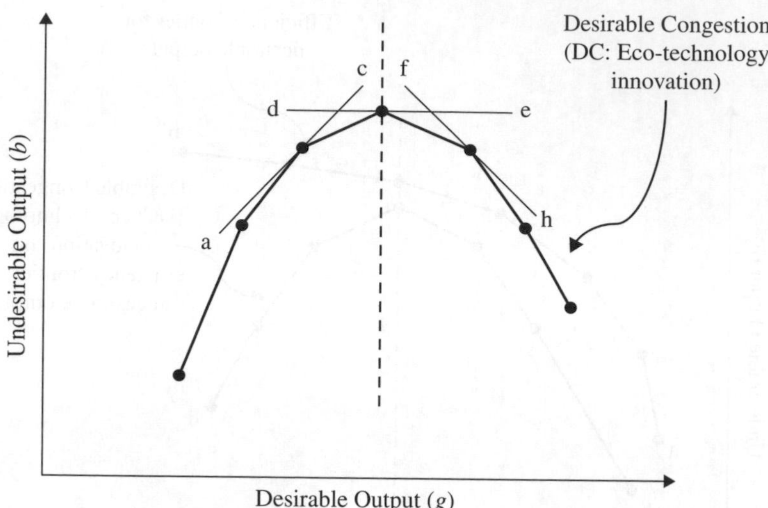

FIGURE 24.2 Desirable congestion for eco-technology innovation (a) The three supporting lines (a–c, d–e and f–h) indicate no, weak and strong DC, respectively. The line indicates supporting hyperplanes if it includes an input (x). The figure drops the input for our visual description. (b) It is possible to depict the figure by b on the horizontal axis and g on the vertical axis. In the case, the right-hand side indicates the occurrence of undesirable congestion (UC). See Chapter 21 for a detailed description on UC. UC is a conventional concept of congestion.

components of the desirable output vector, but simultaneously decreases some components of an undesirable output vector without worsening the other components. In the disposability unification, it is necessary for us, hereafter, to incorporate such a possible occurrence of strong DC in the proposed formulations.

24.3 NON-RADIAL APPROACH FOR DISPOSABILITY UNIFICATION

UENM under Variable RTS and DTS: Returning to Chapter 16 (non-radial measurement), this chapter first utilizes a non-radial model that incorporates the unification of natural and managerial disposability concepts (e.g., Goto *et al.*, 2014). The proposed model is designed to measure the level of unified efficiency under natural and managerial disposability (*UENM*) of the k-th DMU.

As mentioned in Chapter 16, the k-th DMU needs to decrease some input(s) and increase some desirable outputs to improve its operational performance under natural disposability. In contrast, the DMU may increase the inputs to improve its environmental performance under managerial disposability. To satisfy the two conflicting requirements, the proposed non-radial model separates inputs into two categories according to the two disposability concepts. Then, unification of the two disposability concepts uses the following non-radial (NR) model:

$$
\begin{aligned}
\text{Maximize} \quad & \sum_{i=1}^{m^-} R_i^x d_i^{x-} + \sum_{q=1}^{m^+} R_q^x d_q^{x+} + \sum_{r=1}^{s} R_r^g d_r^g + \sum_{f=1}^{h} R_f^b d_f^b \\
s.t. \quad & \sum_{j=1}^{n} x_{ij}^- \lambda_j + d_i^{x-} = x_{ik}^- \quad (i=1,\dots,m^-), \\
& \sum_{j=1}^{n} x_{qj}^+ \lambda_j - d_q^{x+} = x_{qk}^+ \quad (q=1,\dots,m^+), \\
& \sum_{j=1}^{n} g_{rj} \lambda_j - d_r^g = g_{rk} \quad (r=1,\dots,s), \\
& \sum_{j=1}^{n} b_{fj} \lambda_j + d_f^b = b_{fk} \quad (f=1,\dots,h), \\
& \sum_{j=1}^{n} \lambda_j = 1, \\
& \lambda_j \ge 0 \ (j=1,\dots,n),\ d_i^{x-} \ge 0 \ (i=1,\dots,m^-),\ d_q^{x+} \ge 0 \ (q=1,\dots,m^+), \\
& d_r^g \ge 0 \ (r=1,\dots,s) \text{ and } d_f^b \ge 0 \ (f=1,\dots,h).
\end{aligned} \tag{24.1}
$$

The number of original m inputs are newly separated into m^- (under natural disposability) and m^+ (under managerial disposability) in the model in Equation (24.1) in the manner that it maintains $m = m^- + m^+$. One of the two input categories uses inputs whose slacks (d_i^{x-} for $i = 1,\dots,m^-$) are formulated under natural disposability. For example, the number of employees belongs to the input category. The minus in the number of inputs in the first category produces a better performance under natural disposability. The other category contains inputs whose slacks (d_q^{x+} for $q = 1,\dots,m^+$) are formulated under managerial disposability. For example, the amount of capital investment for eco-technology innovation belongs to the input category. Another example is R&D expenditure. The amount of capital investment is used to attain technology innovation that is important for enhancing the level of production activity and that of environmental protection. The more inputs in the second category produce the better performance under managerial disposability because of green technology development.

The level of *UENM* regarding the k-th DMU under variable RTS and DTS is determined by

$$
UENM_v^{NR} = 1 - \left(\sum_{i=1}^{m^-} R_i^x d_i^{x-*} + \sum_{q=1}^{m^+} R_q^x d_q^{x+*} + \sum_{r=1}^{s} R_r^g d_r^{g*} + \sum_{f=1}^{h} R_f^b d_f^{b*} \right), \tag{24.2}
$$

where all slack variables are determined on the optimality of Model (24.1). The equation within the parenthesis, obtained Model (24.1), indicates the level of unified inefficiency under the two disposability concepts. The degree of *UENM* is obtained by subtracting the level of inefficiency from unity as specified in Equation (24.2).

Model (24.1) has the following dual formulation:

$$
\begin{aligned}
\text{Minimize} \quad & \sum_{i=1}^{m^-} v_i^- x_{ik}^- - \sum_{q=1}^{m^+} v_q^+ x_{qk}^+ - \sum_{r=1}^{s} u_r g_{rk} + \sum_{f=1}^{h} w_f b_{fk} + \sigma \\
s.t. \quad & \sum_{i=1}^{m^-} v_i^- x_{ij}^- - \sum_{q=1}^{m^+} v_q^+ x_{qj}^+ - \sum_{r=1}^{s} u_r g_{rj} + \sum_{f=1}^{h} w_f b_{fj} + \sigma \ \geq \ 0 \quad (j=1,\ldots,n), \\
& v_i^- \geq R_i^x \quad (i=1,\ldots,m^-), \\
& v_q^+ \geq R_q^x \quad (q=1,\ldots,m^+), \\
& u_r \geq R_r^g \quad (r=1,\ldots,s), \\
& w_f \geq R_f^b \quad (f=1,\ldots,h), \\
& \sigma : \text{URS}.
\end{aligned}
\tag{24.3}
$$

Here, v_i^- $(i=1,\ldots,m^-)$, v_q^+ $(q=1,\ldots,m^+)$, u_r $(r=1,\ldots,s)$ and w_f $(f=1,\ldots,h)$ are all dual variables related to the first, second, third and fourth groups of constraints in Model (24.1), respectively. The dual variable (σ), which is unrestricted (URS), is obtained from the last equation of Model (24.1).

An important feature of Model (24.1) is that it incorporates natural and managerial disposability in such a manner that inputs and two types of outputs are classified by the two disposability concepts. In the previous chapters, these disposability concepts are equally treated in the evaluation of UENM.

Problem of Model (24.3): The supporting hyperplane, obtained by Model (24.3), becomes $v^-x^- - v^+x^+ - ug + wb + \sigma = 0$ in a simple case where three production factors have a single component. The simple case is for our descriptive convenience. This chapter considers that x^+ stands for an amount for eco-technology investment, for example. The supporting hyperplane indicates $wb = -v^-x^- + v^+x^+ + ug - \sigma$ where all the dual variables, except σ, are positive. The dual variable (σ) is unrestricted in sign. It is easily thought that the supporting hyperplane does not fit with our expectation. For instance, an increase in the input (x^-) decreases the undesirable output because of $-v^-$. The sign should be opposite. The following three conditions are necessary for us to make the supporting hyperplane fit with our expectation to apply DEA to performance analysis for energy and environment. First, if x^- increases or decreases, then b increases or decreases because x^- is the input under natural disposability. Second, if x^+ increases or decreases, then b decreases or increases because x^+ indicates the input under managerial disposability (e.g., eco-technology investment). Thus, the two expectations clearly indicate that the unknown parameter (v^-) should be positive and the other (v^+) should be negative in their signs, respectively. Finally, both u and σ should be unrestricted in these signs to incorporate the possible occurrence of DC, as depicted in Figure 24.2, where the slope of the supporting hyperplane is determined by u and the intercept is determined by σ. Consequently, the supporting hyperplane should be $wb = v^-x^- - v^+x^+ + ug + \sigma$ in the simple case.

UENM(DC): To reorganize Model (24.3) in such a manner that it can fit with Figure 24.2, Sueyoshi and Goto (2014b) proposed the following non-radial model in which the k-th DMU can simultaneously attain the status of natural and managerial disposability along with the possible occurrence of desirable congestion (DC):

$$
\begin{aligned}
\text{Minimize} \quad & \sum_{i=1}^{m^-} v_i^- x_{ik}^- - \sum_{q=1}^{m^+} v_q^+ x_{qk}^+ + \sum_{r=1}^{s} u_r g_{rk} - \sum_{f=1}^{h} w_f b_{fk} + \sigma \\
s.t. \quad & \sum_{i=1}^{m^-} v_i^- x_{ij}^- - \sum_{q=1}^{m^+} v_q^+ x_{qj}^+ + \sum_{r=1}^{s} u_r g_{rj} - \sum_{f=1}^{h} w_f b_{fj} + \sigma \ \geq \ 0 && (j=1,...,n), \\
& v_i^- \geq R_i^x && (i=1,...,m^-), \\
& v_q^+ \geq R_q^x && (q=1,...,m^+), \\
& u_r : \text{URS} && (r=1,...,s), \\
& w_f \geq R_f^b && (f=1,...,h), \\
& \sigma : \text{URS},
\end{aligned}
\tag{24.4}
$$

where v_i ($i = 1,\ldots,m^-$), v_q ($q = 1,\ldots,m^+$), u_r ($r = 1,\ldots,s$) and w_f ($f = 1,\ldots,h$) are parameters to express the direction of the supporting hyperplane measured by Model (24.4). The dual variable (σ), which is unrestricted (URS), indicates the intercept of the expected supporting hyperplane.

Model (24.4), which is a dual model, has the following primal (dual's dual) formulation that unifies the two disposability concepts:

$$
\begin{aligned}
\text{Maximize} \quad & \sum_{i=1}^{m^-} R_i^x d_i^{x-} + \sum_{q=1}^{m^+} R_q^x d_q^{x+} + \sum_{f=1}^{h} R_f^b d_f^b \\
s.t. \quad & \sum_{j=1}^{n} x_{ij}^- \lambda_j + d_i^{x-} = x_{ik}^- \quad (i=1,...,m^-), \\
& \sum_{j=1}^{n} x_{qj}^+ \lambda_j - d_q^{x+} = x_{qk}^+ \quad (q=1,...,m^+), \\
& \sum_{j=1}^{n} g_{rj} \lambda_j = g_{rk} \quad (r=1,...,s), \\
& \sum_{j=1}^{n} b_{fj} \lambda_j - d_f^b = b_{fk} \quad (f=1,...,h), \\
& \sum_{j=1}^{n} \lambda_j = 1, \\
& \lambda_j \geq 0 \ (j=1,...,n),\ d_i^{x-} \geq 0 \ (i=1,...,m^-), \\
& d_q^{x+} \geq 0 \ (q=1,...,m^+) \text{ and } d_f^b \geq 0 \ (f=1,...,h).
\end{aligned}
\tag{24.5}
$$

Model (24.5) has two unique features in the structure. One of the two features is that all the inputs are classified into two categories, as mentioned previously. The first category includes a group of inputs whose slacks (d_i^{x-} for $i = 1,\ldots,m^-$) are formulated under natural disposability. The other category contains a group of inputs whose slacks (d_q^{x+} for $q = 1,\ldots,m^+$) are formulated under managerial disposability. The other unique feature is that the constraints regarding a desirable output vector $\left(\sum_{j=1}^{n} g_{rj}\lambda_j = g_{rk}\right)$ are expressed by equality because they do not have any slacks. In contrast, the other groups of constraints are expressed by inequality because they have slacks in Model (24.5). The existence of slacks indicates the status of inequality.

The unified efficiency under natural and managerial disposability, or $UENM(DC)_v^{NR}$, of the k-th DMU is measured by

$$UENM(DC)_v^{NR} = 1 - \left(\sum_{i=1}^{m^-} R_i^x d_i^{x-*} + \sum_{q=1}^{m^+} R_q^x d_q^{x+*} + \sum_{f=1}^{h} R_f^b d_f^{b*} \right), \qquad (24.6)$$

where all slack variables are determined on the optimality of Model (24.5). The equation within the parenthesis, obtained from the optimality of Model (24.5), indicates the level of unified inefficiency under natural and managerial disposability. The level of unified efficiency is obtained by subtracting the level of inefficiency from unity under the occurrence of DC, or eco-technology innovation, on undesirable outputs. It is important to note that Models (24.4) and (24.5) under non-radial measurement, corresponding to Stage III of Figure 15.6.

UENM(DC) under Constant RTS and DTS: Model (24.5) may drop $\sum_{j=1}^{n} \lambda_j = 1$ from the formulation as follows:

$$\begin{aligned}
\text{Maximize} \quad & \sum_{i=1}^{m^-} R_i^x d_i^{x-} + \sum_{q=1}^{m^+} R_q^x d_q^{x+} + \sum_{f=1}^{h} R_f^b d_f^b \\
s.t. \quad & \sum_{j=1}^{n} x_{ij}^- \lambda_j + d_i^{x-} = x_{ik}^- \ (i=1,\ldots,m^-), \\
& \sum_{j=1}^{n} x_{qj}^+ \lambda_j - d_q^{x+} = x_{qk}^+ \ (q=1,\ldots,m^+), \\
& \sum_{j=1}^{n} g_{rj} \lambda_j = g_{rk} \ (r=1,\ldots,s), \\
& \sum_{j=1}^{n} b_{fj} \lambda_j - d_f^b = b_{fk} \ (f=1,\ldots,h), \\
& \lambda_j \geq 0 \ (j=1,\ldots,n), \ d_i^{x-} \geq 0 \ (i=1,\ldots,m^-), \\
& d_q^{x+} \geq 0 \ (q=1,\ldots,m^+) \text{ and } d_f^b \geq 0 \ (f=1,\ldots,h).
\end{aligned} \qquad (24.7)$$

Model (24.7) drops $\sum_{j=1}^{n} \lambda_j = 1$ from the formulation. Then, Equation (24.6) measures the level of $UENM(DC)_c^{NR}$ under constant RTS and DTS as follows:

$$UENM(DC)_c^{NR} = \text{Equation (24.6)}, \qquad (24.8)$$

where all slacks are determined on the optimality of Model (24.7). The subscript c stands for constant RTS and DTS, so not the subscript v in Equation (24.8).

Scale Efficiency under Natural and Managerial Disposability: After obtaining the two types of *UENM(DC)* for the *k*-th DMU, this chapter uses

$$SENM(DC)^{NR} = UENM(DC)_c^{NR} / UENM(DC)_v^{NR} \quad (24.9)$$

to determine the level of scale efficiency in the case of non-radial models. The scale efficiency indicates the level at which each DMU effectively operates by considering the operational size. Thus, the degree of scale efficiency can be considered as a level of efficiency for size utilization under the possible occurrence of eco-technology innovation.

At the end of this section, it is important to note that the magnitude control factor (MCF: ε) is usually incorporated into the proposed non-radial models. See Chapter 16 which discusses the use of MCF in the non-radial approach. Since Chapter 16 has not used the MCF in its empirical study on petroleum firms, this chapter does not incorporate it in the proposed formulations, in order to maintain empirical consistency between Chapter 16 and this chapter.

24.4 RADIAL APPROACH FOR DISPOSABILITY UNIFICATION

UENM under variable RTS and DTS: Since Chapter 17 has already discussed Stage I of Figure 15.6, this chapter starts with Stage II for radial measurement. First, shifting our description from a non-radial (Section 24.3) to a radial approach, this chapter follows the previous research of Sueyoshi and Goto (e.g., Sueyoshi and Goto 2012b, 2014a) that proposed the following radial (R) model that corresponds to Model (24.1):

$$\begin{aligned}
\text{Maximize} \quad & \xi + \varepsilon_s \left(\sum_{i=1}^{m^-} R_i^x d_i^{x-} + \sum_{q=1}^{m^+} R_q^x d_q^{x+} + \sum_{r=1}^{s} R_r^g d_r^g + \sum_{f=1}^{h} R_f^b d_f^b \right) \\
s.t. \quad & \sum_{j=1}^{n} x_{ij}^- \lambda_j + d_i^{x-} = x_{ik}^- \ (i=1,\dots,m^-), \\
& \sum_{j=1}^{n} x_{qj}^+ \lambda_j - d_q^{x+} = x_{qk}^+ \ (q=1,\dots,m^+), \\
& \sum_{j=1}^{n} g_{rj} \lambda_j - d_r^g - \xi g_{rk} = g_{rk} \ (r=1,\dots,s), \qquad (24.10) \\
& \sum_{j=1}^{n} b_{fj} \lambda_j + d_f^b + \xi b_{fk} = b_{fk} \ (f=1,\dots,h), \\
& \sum_{j=1}^{n} \lambda_j = 1, \\
& \lambda_j \ge 0 \ (j=1,\dots,n), \ \xi : \text{URS}, \\
& d_i^{x-} \ge 0 (i=1,\dots,m^-), \ d_q^{x+} \ge 0 (q=1,\dots,m^+), \\
& d_r^g \ge 0 \ (r=1,\dots,s) \text{ and } d_f^b \ge 0 (f=1,\dots,h).
\end{aligned}$$

Model (24.10) combines Models (17.5) and (17.10) by a straightforward unification. The level of $UENM_v^R$ regarding the *k*-th DMU under variable RTS and DTS is determined by

$$UENM_v^R = 1 - \left[\xi^* + \varepsilon_s \left(\sum_{i=1}^{m^-} R_i^x d_i^{x-*} + \sum_{q=1}^{m^+} R_q^x d_q^{x+*} + \sum_{r=1}^{s} R_r^g d_r^{g*} + \sum_{f=1}^{h} R_f^b d_f^{b*} \right) \right], \quad (24.11)$$

where all slacks are determined on the optimality of Model (24.10). The equation within the parentheses, obtained from the optimality of Model (24.10), indicates the level of unified inefficiency. The *UENM* is obtained by subtracting the level of inefficiency from unity.

The dual formulation of Model (24.10) becomes as follows:

$$\begin{aligned}
\text{Minimize} \quad & \sum_{i=1}^{m^-} v_i^- x_{ik} - \sum_{q=1}^{m^+} v_q^+ x_{qk} - \sum_{r=1}^{s} u_r g_{rk} + \sum_{f=1}^{h} w_f b_{fk} + \sigma \\
s.t. \quad & \sum_{i=1}^{m^-} v_i^- x_{ij} - \sum_{q=1}^{m^+} v_q^+ x_{qj} - \sum_{r=1}^{s} u_r g_{rj} + \sum_{f=1}^{h} w_f b_{fj} + \sigma \geq 0 \quad (j = 1,...,n), \\
& \sum_{r=1}^{s} u_r g_{rk} + \sum_{f=1}^{h} w_f b_{fk} = 1, \qquad (24.12) \\
& v_i^- \geq \varepsilon_s R_i^x \quad (i = 1,...,m^-), \\
& v_q^+ \geq \varepsilon_s R_q^x \quad (q = 1,...,m^+), \\
& u_r \geq \varepsilon_s R_r^g \quad (r = 1,...,s), \\
& w_f \geq \varepsilon_s R_f^b \quad (f = 1,\ldots,h), \\
& \sigma : URS.
\end{aligned}$$

As discussed for *UENM* measured by the non-radial approach, Models (24.10) and (24.12) have a difficulty in measuring technology innovation, as well. A description on the rationale in Section 24.3 is applicable to Models (24.10) and (24.12).

UENM(DC): To discuss the structure of the supporting hyperplane, as depicted in Figure 24.2, this chapter reorganizes Model (24.12) whose hyperplane becomes $v^- x^- - v^+ x^+ + ug - wb + \sigma = 0$ in the case where three production factors have a single component. The corresponding formulation of Model (24.12), satisfying the requirement on the case of multiple components, becomes as follows:

$$\text{Minimize} \sum_{i=1}^{m^-} v_i^- x_{ik}^- - \sum_{q=1}^{m^+} v_q^+ x_{qk}^+ + \sum_{r=1}^{s} u_r g_{rk} - \sum_{f=1}^{h} w_f b_{fk} + \sigma$$

$$\begin{aligned}
s.t. \quad & \sum_{i=1}^{m^-} v_i^- x_{ij}^- - \sum_{q=1}^{m^+} v_q^+ x_{qj}^+ + \sum_{r=1}^{s} u_r g_{rj} - \sum_{f=1}^{h} w_f b_{fj} + \sigma \geq 0 \quad (j=1,...,n), \\
& \sum_{r=1}^{s} u_r g_{rk} = 1, \\
& v_i^- \geq \varepsilon_s R_i^x \quad (i=1,...,m^-), \\
& v_q^+ \geq \varepsilon_s R_q^x \quad (q=1,...,m^+), \\
& u_r : \text{URS} \quad (r=1,...,s), \\
& w_f \geq \varepsilon_s R_f^b \quad (f=1,...,h), \\
& \sigma : \text{URS}.
\end{aligned} \tag{24.13}$$

Along with the slight modification, Model (24.13) drops $\sum_{f=1}^{h} w_f b_{fk}$ from the second constraint of Model (24.12). The rationale is that $\sum_{r=1}^{s} u_r g_{rk} + \sum_{f=1}^{h} w_f b_{fk} = 1$ may be infeasible because u_r is unrestricted in the sign so that the dual variable cannot become negative, or the equation becomes infeasible in the worst case. To avoid such a computational difficulty, this chapter reorganizes Model (24.12) with a structure like Model (24.13).

The primal (dual's dual) formulation of Model (24.13) becomes as follows:

$$\text{Maximize} \ \xi + \varepsilon_s \left(\sum_{i=1}^{m^-} R_i^x d_i^{x-} + \sum_{q=1}^{m^+} R_q^x d_q^{x+} + \sum_{f=1}^{h} R_f^b d_f^b \right)$$

$$\begin{aligned}
s.t. \quad & \sum_{j=1}^{n} x_{ij}^- \lambda_j + d_i^{x-} = x_{ik}^- \quad (i=1,...,m^-), \\
& \sum_{j=1}^{n} x_{qj}^+ \lambda_j - d_q^{x+} = x_{qk}^+ \quad (q=1,...,m^+), \\
& \sum_{j=1}^{n} g_{rj} \lambda_j + \xi g_{rk} = g_{rk} \quad (r=1,...,s), \\
& \sum_{j=1}^{n} b_{fj} \lambda_j - d_f^b = b_{fk} \quad (f=1,...,h), \\
& \sum_{j=1}^{n} \lambda_j = 1, \\
& \lambda_j \geq 0 \ (j=1,...,n), \ \xi : \text{URS}, \ d_i^{x-} \geq 0 \ (i=1,...,m^-), \\
& d_q^{x+} \geq 0 \ (q=1,...,m^+) \ \text{and} \ d_f^b \geq 0 \quad (f=1,...,h).
\end{aligned} \tag{24.14}$$

Model (24.14) corresponds to Model (24.5) in the non-radial measurement. The level of $UENM(DC)_v^R$ regarding the k-th DMU under variable RTS and DTS is determined by

$$UENM(DC)_v^R = 1 - \left[\xi^* + \varepsilon_s \left(\sum_{i=1}^{m^-} R_i^x d_i^{x-*} + \sum_{q=1}^{m^+} R_q^x d_q^{x+*} + \sum_{f=1}^{h} R_f^b d_f^{b*} \right) \right], \quad (24.15)$$

where the inefficiency score and all slack variables are determined by the optimality of Model (24.14). The equation within the parentheses, obtained from Model (24.14), indicates the level of unified inefficiency. The degree of $UENM(DC)_v^R$ is obtained by subtracting the level of inefficiency from unity. Here, the magnitude of the inefficiency score (ξ^*) is determined by desirable outputs in Model (24.14) because these outputs are byproducts of desirable outputs. The influence of undesirable outputs exists as slacks on undesirable outputs (d_f^{b*}), as specified in Equation (24.15). It is important to note that Models (24.13) and (24.14) under radial measurement correspond to Stage III of Figure 15.6.

Meanwhile, under constant RTS and DTS, the level of $UENM(DC)_c^R$ regarding the k-th DMU is determined by

$$\begin{aligned}
&\text{Maximize} \quad \xi + \varepsilon_s \left(\sum_{i=1}^{m^-} R_i^x d_i^{x-} + \sum_{q=1}^{m^+} R_q^x d_q^{x+} + \sum_{f=1}^{h} R_f^b d_f^b \right) \\
&s.t. \quad \sum_{j=1}^{n} x_{ij}^- \lambda_j + d_i^{x-} = x_{ik}^- \quad (i = 1,\ldots,m^-), \\
&\qquad \sum_{j=1}^{n} x_{qj}^+ \lambda_j - d_q^{x+} = x_{qk}^+ \quad (q = 1,\ldots,m^+), \\
&\qquad \sum_{j=1}^{n} g_{rj} \lambda_j + \xi g_{rk} = g_{rk} \quad (r = 1,\ldots,s), \\
&\qquad \sum_{j=1}^{n} b_{fj} \lambda_j - d_f^b = b_{fk} \quad (f = 1,\ldots,h), \\
&\qquad \lambda_j \geq 0 \ (j = 1,\ldots,n), \ \xi : \text{URS}, \ d_i^{x-} \geq 0 \ (i = 1,\ldots,m^-), \\
&\qquad d_q^{x+} \geq 0 \ (q = 1,\ldots,m^+) \text{ and } d_f^b \geq 0 \ (f = 1,\ldots,h).
\end{aligned} \quad (24.16)$$

The degree of $UENM(DC)_c^R$ on the k-th DMU under constant RTS and DTS is determined by

$$UENM(DC)_c^R = \text{Equation (24.15)}. \quad (24.17)$$

All the slack variables used in Equation (24.17) are determined by the optimality of Model (24.16).

Scale Efficiency under Natural and Managerial Disposability: After obtaining the two types of UENM(DC), this chapter measures

$$SENM(DC)^R = UENM(DC)_c^R / UENM(DC)_v^R \quad (24.18)$$

as the level of scale efficiency in the case of radial measurement.

24.5 COMPUTATIONAL FLOW FOR DISPOSABILITY UNIFICATION

Figure 24.3 visually describes the computational flow for disposability unification and efficiency measurement. Although this chapter uses non-radial models for our empirical study, this description can be applied to the radial framework.

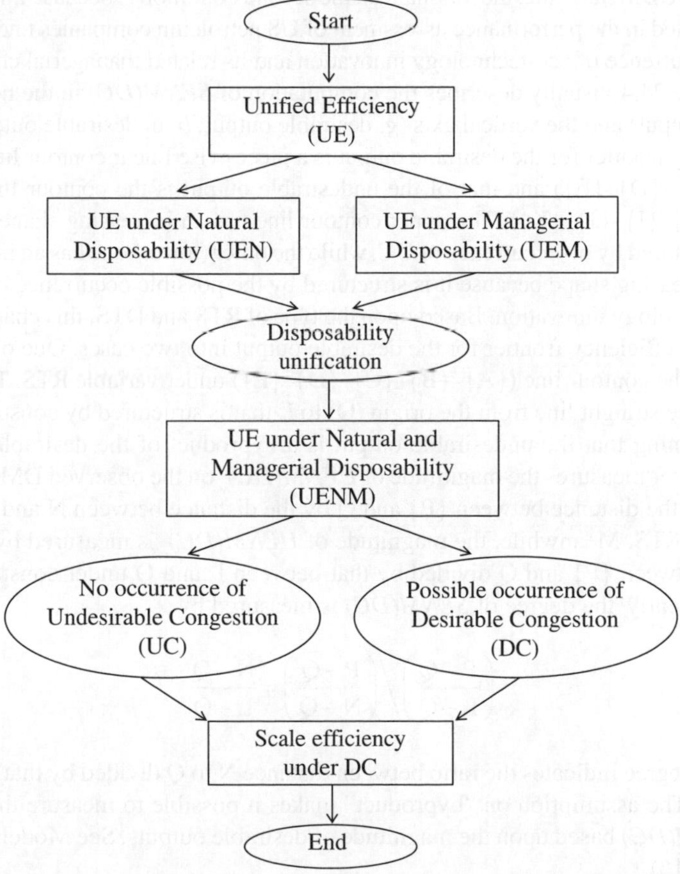

FIGURE 24.3 *UENM(DC)* and efficiency measures (a) It is possible for us to apply the proposed approach to both radial and non-radial measurements. Therefore, this chapter does not specify R (radial) and NR (non-radial) in the figure.

As depicted in Figure 24.3, the proposed approach measures the level of unified efficiency (*UE*). See Chapters 16 and 17 on *UE* measurement. As discussed in the two chapters, we measure *UE* under natural disposability (*UEN*) and *UE* under managerial disposability (*UEM*). After measuring *UE*, *UEN* and *UEM*, the proposed approach combines two disposability concepts for a unified efficiency measure (*UENM*). By unifying them, this chapter considers two conditions related to the occurrence of congestion. One of the two conditions is that it is possible for us to measure *UENM* under no occurrence of undesirable congestion (UC). The other condition is that we can consider the possible occurrence of desirable congestion (DC), or eco-technology innovation on undesirable outputs, to measure the level of *UENM(DC)*. The rationale on the first condition is that this chapter is interested in measuring the level of scale efficiency (*SENM*) on US petroleum companies. In this case, it is necessary for us to assume that no UC occurs in the transportation system from upstream to downstream. As a result, this chapter can determine the degree of *SENM(DC)*. Meanwhile, the rationale on the second condition is because this chapter is interested in the performance assessment of US petroleum companies under a possible occurrence of eco-technology innovation and its related managerial challenge.

Figure 24.4 visually describes the computation of *SENM(DC)* in the horizontal axis (*x*: input) and the vertical axis (*g*: desirable output; *b*: undesirable output). The efficiency frontier for the desirable output is a piece-wise linear contour line ({A}–{B}–{C}–{D}–{E}) and that of the undesirable output is the contour line ({F}–{G}–{H}–{I}–{J}–{K}). The upper contour line has an increasing shape because it is structured by no occurrence of UC, while the other (lower) one has an increasing and decreasing shape because it is structured by the possible occurrence of DC, or eco-technology innovation. Based upon the type of RTS and DTS, this chapter classifies the efficiency frontier for the desirable output into two cases. One of the two cases is the contour line ({A}–{B}–{C}–{D}–{E}) under variable RTS. The other case is the straight line from the origin (O) to L that is structured by constant RTS.

Assuming that the undesirable output is a byproduct of the desirable output, this chapter measures the magnitude of $UENM(DC)_v$ on the observed DMU{P} by dividing the distance between {P} and Q by the distance between N and Q under variable RTS. Meanwhile, the magnitude of $UENM(DC)_c$ is measured by the distance between {P} and Q divided by that between L and Q under constant RTS. Consequently, the degree of *SENM(DC)* is measured by

$$\left(\frac{P-Q}{L-Q}\right)\bigg/\left(\frac{P-Q}{N-Q}\right)=\frac{N-Q}{L-Q}$$

The degree indicates the ratio between distance N to Q divided by that between L to Q. The assumption on "byproduct" makes it possible to measure the degree of *SENM(DC)* based upon the magnitude of desirable outputs. See Models (24.14) and (24.16).

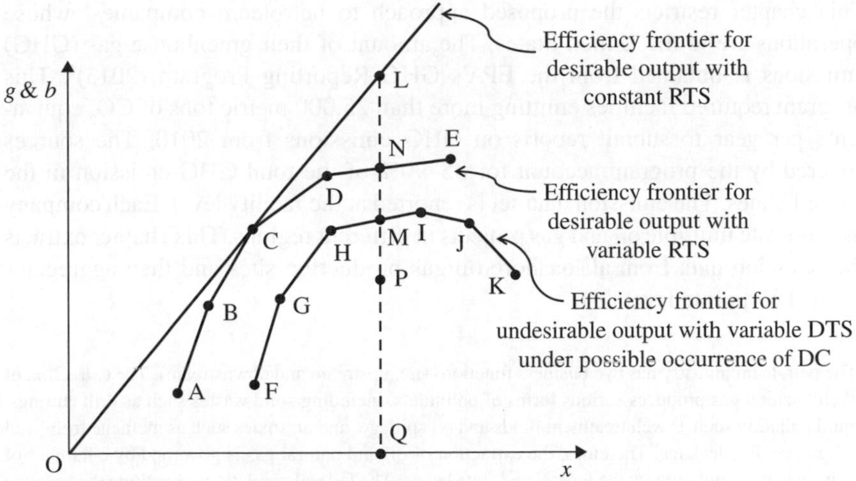

FIGURE 24.4 *SENM(DC)* measurement (a) RTS: returns to scale; DTS: damages to scale; DC: desirable congestion. (b) DMU {P} is projected onto N on the contour line segments {A}–{B}–{C}–{D}–{E} (the efficiency frontier for the desirable output under variable RTS). Meanwhile, it is projected onto L on the line from the origin O to L. The level of efficiency is determined by the distance between the observed point and the projected point. After disposability unification, the distance is measured by the desirable output. (c) The line segments, or {F}–{G}–{H}–{I}–{J}–{K}, consist of the efficiency frontier for the undesirable output. The frontier is shaped by variable DTS and the possible occurrence of desirable congestion (DC), or eco-technology innovation. (d) This figure incorporates the possible occurrence of DC, but excludes UC. As a result, the fron tier for desirable output has an increasing shape and the frontier for undesirable output has an increasing and decreasing shape. This figure incorporates the assumption that the undesirable output is a byproduct of the desirable output. Without this assumption, the efficiency frontier has a concave shape, not a convex shape as depicted in this figure. See Figure 15.6.

24.6 US PETROLEUM INDUSTRY

24.6.1 Data

This chapter utilizes a data set on US oil and natural gas production companies, all of which are listed in the New York Stock Exchange (NYSE) Energy Index (2013)[3]. The index covers 82 independent producers and 20 integrated companies.

[3] NYSE Energy Index (2010): http://www.nyse.com/about/listed/nyeid_components.shtml).

This chapter restricts the proposed approach to petroleum companies[4] whose operations are in the United States. The amount of their greenhouse gas (GHG) emissions is obtained from the EPA's GHG Reporting Program (2013)[5]. This program required facilities emitting more than 25,000 metric tons of CO_2 equivalents per year to submit reports on GHG emissions from 2010. The sources covered by the program account for 85–90% of the total GHG emission in the United States. The emission data set is reported at the facility level. Each company may operate multiple oil and gas projects in different regions. This chapter extracts the emission data from all onshore oil/gas production sites and then aggregates them at corporate level.

[4]The petroleum industry has two business functions (i.e.,upstream and downstream). The extraction of oil and natural gas produces various forms of pollutants, including solid wastes such as drill cuttings, liquid pollution such as well treatment fluids and oil spillage, and air toxics such as methane from well and process line leakage. Therefore, the extraction of oil and natural gas is governed by a number of environmental regulations at the federal and state levels. The federal regulations constitute the baseline requirements for all exploration and production activities, whereas the state governments can enact more stringent laws. At the federal level, the most prominent regulations include the National Environmental Policy Act (NEPA), Clean Air Act (CAA), Clean Water Act (CWA), Safe Drinking Water Act (SDWA), Endangered Species Act (ESA), Toxic Substance Control Act (TSCA) and Resource Conservation and Recovery Act (RCRA). All these federal statutes can impact on oil and gas production directly and indirectly. EPA is the primary federal regulatory agency responsible for environmental protection. Operations on federal land are also governed by regulations from the Bureau of Land Management. Specifically pertinent to this study are the regulations related to GHG emission in oil and gas industry. Among such regulations are the Greenhouse Gas Reporting Program (http://www.epa.gov/ghgreporting/), the National Emissions Standards for Hazardous Air Pollutants (NESHAP; http://www.epa.gov/compliance/monitoring/programs/caa/neshaps.html) and the New Source Performance Standards (NSPS; http://www.epa.gov/compliance/monitoring/programs/caa/newsource.html), all of which are designed and supervised by the EPA. The Greenhouse Gas Reporting Program, enacted as a law in 2010, requires the nation's large emitters of GHG to report the amount of GHG they emit annually. NESHAP establishes air pollutants standards for 188 different hazardous air pollutants at new and existing sources. NSPS targets new, modified and reconstructed facilities. However, there are several noticeable loopholes in the implementation of the laws. For instance, under the rule of aggregation, NESHAP considers small polluting sources in close proximity to each other as one source of emissions. However, EPA exempts oil and gas wells from aggregation and hence the stricter standards of larger emitting sources are not applicable. Prior to 2012, there was no set of well management standards at the federal level. It was the state and local regulators who primarily regulated the drilling industry. For instance, Colorado requires the usage of green completion technology to minimize gas releases during well completion, while most other states do not require green completion. In order to reduce air pollution from the drilling process, especially hydraulic fracturing for shale natural gas, EPA revised NSPS and NESHAP in 2012 to impose stricter emission rules on oil and gas production (http://www.epa.gov/airquality/oilandgas/actions.html). The focus of the new regulation is to limit the emission of volatile organic compounds (VOCs), SO_2, methane and other hazardous air pollutants. The most noticeable feature of the new regulation is a mandatory usage of green completion for most wells. A transition period until January 2015 was granted to producers for them to make cost-effective adjustments. Finally, this chapter points out that all the above regulations apply to integrated and independent petroleum companies similarly, regardless of their size and economic status.

[5]Environmental Protection Agency (2013) Greenhouse Gas Reporting Program, http://www.epa.gov/ghgreporting.

Data Accessibility: The supplementary section of Sueyoshi and Wang (2014b) contains the whole data set used in this study.

Each company's emission is closely related to its drilling activity. Therefore, this chapter extracts the number of wells drilled by a company from the annual report. In alignment with the emission data, the drilled wells used in this chapter count only those drilled onshore in the United States. We also obtain the three-year average for 2010–2012 on acquisition, finding and development (AFD) expenses from Ernst Young (2013)[6]. The AFD expenses reflect each companies' abilities to access and obtain petroleum reserves, which indirectly affect their GHG emission. A higher acquisition cost is generally paid for easy accessibility to resources which require less effort to drill, hence incurring a lower emission. A higher finding and development cost may imply more intensive exploration, hence producing more emissions from exploration. Finally, this chapter collects the companies' financial and operational data from the COMPUSTAT database. The data set used in this chapter is summarized below in detail.

(a) *GHG Emission*: This measures the amount of emissions from each firm's onshore production, including not only CO_2 but also methane (CH_4) and nitrous oxide (N_2O). The cost of adapting pollution prevention practices and the effectiveness of pollution prevention as a strategy for reducing emissions may vary with the scale of current emission.
(b) *Number of Employees*: This is regarded as a proxy for corporate size. Larger firms may have more resources to adapt GHG mitigation practices.
(c) *Capital Expenditure*: This is included to indicate the operating liquidity of a firm. Firms with higher capital expenditure may invest more in GHG mitigation.
(d) *R&D Expense*: This is a measure of a firm's technology capacity and serves as a proxy for technology innovation. It is expected that firms with higher R&D expense is more likely to acquire and implement efficient emission control technology.
(e) *Total Assets*: This includes current assets, property, plant and equipment, all of which are used as another proxy for corporate size.
(f) *Net Number of Wells Drilled*: This gives the number of wells drilled by a company for the calendar year, accounting for the fractional working interest owned by the company in each well. For instance, a 50% interest in a well is counted as 0.5 well.
(g) *Revenue*: This is income received from the sale of oil and gas and indicates the operational size of the business.

[6]Ernst Young (2013) *US Oil and Gas Reserves Study*. Accessed 23 June 2014, http://www.ey.com/Publication/vwLUAssets/US_oil_and_gas_reserves_study_2013/$FILE/US_oil_and_gas_reserves_study_2013_DW0267.pdf.

(h) *Acquisition, Finding and Development (AFD) Expense*: This is calculated as the sum of proved reserve acquisition cost and the finding and development cost. Specifically, the proved reserve acquisition cost is calculated as the reserve purchasing cost divided by the reserve purchased. The finding and development cost is calculated as the unproved reserve acquisition cost, exploration and development expenditures relative to the added reserve. Overall, this chapter computes the sum of AFD costs as the aggregate measure of each company's ability to obtain access to per barrel of oil equivalent (BOE) resources.

This chapter utilizes the amount of total revenue as a desirable output and the amount of GHG emission as an undesirable output. The five inputs under natural disposability include the number of employees, an amount of capital expenditure, a total amount of assets, the number of net wells drilled, and an amount of total AFD expense. The other input under managerial disposability is an amount of R&D expenditure.

In collecting the data set, this chapter removed companies with data missing from any of the relevant fields. Eventually, this chapter obtained a set of 50 companies ($n = 50$), including 43 independent companies and seven integrated companies. The data set consists of roughly half of all production companies in the NYSE Energy Index. Meanwhile, the total emission from these 50 companies is 82.3 MMT[7] emissions from the entire onshore production segment in the United States. Here, MMT stands for methylcyclopentadienyl manganese tricarbonyl.

The descriptive statistics of the data set used for this chapter are listed in Sueyoshi and Wang (2014b). According their study, integrated companies are larger than independent companies in all data fields, on average. The GHG emissions by integrated firms have an average of 6849.6230 thousand tons, being 8.8 times the average emission of 775.4225 thousand tons produced by independent companies. The largest emission of 24,529 thousand tons comes from Exxon Mobil, whereas EnCana Corporation has the smallest emission of 110 thousand tons. The integrated companies drill 468 wells on average, 36% more than the independent companies. However, the most active driller, Occidental Petroleum Corporation (1411.2 net wells drilled) is an independent company.

24.6.2 Unified Efficiency Measures

Table 24.1 documents the five efficiency measures, along with the p-value of the t-test at the bottom, to confirm whether integrated oil companies outperform independent ones.

[7] MMT is a gasoline octane enhancer produced by Afton Chemical Corporation (Afton), formerly known as Ethyl Corporation. MMT is allowed in US gasoline at a level equivalent to one-thirtysecond (0.03125) grams per gallon manganese (gpg Mn).

TABLE 24.1 Unified efficiency measures: non-radial approach

Company Name	*UE*	*UEN*	*UEM*	*UENM*	*UENM(DC)*
Independent companies					
Anadarko Petroleum Corporation	1.0000	0.7495	1.0000	1.0000	1.0000
Antero Resources LLC	1.0000	1.0000	0.7837	1.0000	1.0000
Apache Corporation	1.0000	0.7484	0.8878	0.9390	1.0000
Berry Petroleum Company	1.0000	1.0000	0.6758	0.9312	1.0000
BHP Billiton Group	1.0000	1.0000	1.0000	1.0000	1.0000
Bill Barrett Corporation	0.9999	0.9627	0.7163	0.9591	0.9904
Cabot Oil & Gas Corporation	1.0000	1.0000	0.6473	1.0000	1.0000
Chesapeake Energy Corporation	1.0000	0.7335	0.8467	0.8112	1.0000
Cimarex Energy Co.	0.9584	0.9683	0.7488	0.9713	0.9773
Concho Resources	1.0000	0.9156	0.8873	0.9196	0.9218
Conoco Phillips	1.0000	0.7910	0.7811	1.0000	1.0000
CONSOL Energy Inc.	0.9991	1.0000	0.8400	0.9760	0.9766
Continental Resources, Inc.	1.0000	0.9438	0.7977	0.9457	0.9495
Denbury Resources Inc.	0.9998	1.0000	0.8991	0.9758	0.9758
Devon Energy Corporation	1.0000	0.8070	0.6644	0.9479	1.0000
EnCana Corporation	1.0000	1.0000	1.0000	1.0000	1.0000
Energen Corporation	0.9999	0.9467	0.8391	0.9474	0.9503
EOG Resources, Inc.	1.0000	0.8622	0.7984	0.8836	0.8899
EP Energy LLC	1.0000	0.9725	0.7805	1.0000	1.0000
EQT Corporation	0.9967	0.9639	0.8401	0.9760	0.9781
EXCO Resources, Inc.	1.0000	0.9723	0.8253	0.9724	0.9999
Forest Oil Corporation	0.9277	0.9134	0.8418	0.9136	0.9281
Laredo Petroleum Holdings, Inc.	0.9997	1.0000	0.8886	1.0000	1.0000
Linn Energy, LLC	1.0000	1.0000	0.7266	0.9363	0.9444
National Fuel Gas Company	0.9824	0.9088	1.0000	0.9008	0.9015
Newfield Exploration Company	0.9552	0.9266	0.8115	0.9300	0.9350
Noble Energy, Inc.	1.0000	0.8978	0.7896	0.9520	1.0000
Occidental Petroleum Corporation	1.0000	0.6949	1.0000	0.8167	0.8192
PDC Energy, Inc.	1.0000	0.9878	0.7750	1.0000	1.0000
Pioneer Natural Resources Company	1.0000	0.8931	0.7334	0.9244	0.9398
Plains E & P Company	1.0000	1.0000	1.0000	1.0000	0.9142
QEP Resources, Inc.	0.9809	0.9410	0.7821	0.9410	0.9462
Quicksilver Resources, Inc.	0.9998	1.0000	0.8719	1.0000	1.0000
Range Resources Corporation	0.9999	0.9464	0.9061	1.0000	1.0000
Rosetta Resources Inc.	1.0000	1.0000	0.8031	1.0000	1.0000
SandRidge Energy, Inc.	1.0000	0.8932	0.6766	0.8917	1.0000
SM Energy Company	1.0000	0.9551	0.6941	0.9619	1.0000
Southwestern Energy Company	1.0000	1.0000	0.9359	1.0000	1.0000
Swift Energy Inc.	0.9998	0.9750	0.9152	0.9785	1.0000
Talisman Energy Inc.	0.9767	0.9146	0.8075	0.9230	0.9269

(*Continued*)

TABLE 24.1 (Continued)

Company Name	*UE*	*UEN*	*UEM*	*UENM*	*UENM(DC)*
Ultra Petroleum Corporation	1.0000	1.0000	1.0000	1.0000	1.0000
Whiting Petroleum Corporation	0.9485	0.9295	0.8398	1.0000	1.0000
WPX Energy, Inc.	0.9797	0.9311	0.7222	0.9402	0.9488
Avg.	0.9931	0.9313	0.8321	0.9574	0.9724
Max.	1.0000	1.0000	1.0000	1.0000	1.0000
Min.	0.9277	0.6949	0.6473	0.8112	0.8192
S.D.	0.0163	0.0825	0.1022	0.0473	0.0407
Integrated companies					
BP PLC	1.0000	1.0000	1.0000	1.0000	1.0000
Chevron Corporation	1.0000	1.0000	1.0000	1.0000	1.0000
Exxon Mobil Corporation	1.0000	1.0000	1.0000	1.0000	1.0000
Hess Corporation	1.0000	0.8498	0.7055	1.0000	1.0000
Marathon Oil Corporation	1.0000	1.0000	0.6733	1.0000	1.0000
Murphy Oil Corporation	1.0000	1.0000	0.8192	1.0000	1.0000
Royal Dutch Shell PLC	1.0000	1.0000	1.0000	1.0000	1.0000
Avg.	1.0000	0.9785	0.8854	1.0000	1.0000
Max.	1.0000	1.0000	1.0000	1.0000	1.0000
Min.	1.0000	0.8498	0.6733	1.0000	1.0000
S.D.	0.0000	0.0568	0.1496	0.0000	0.0000
P-value (Welch's t-test)	0.0085	0.0849	0.3937	0.0000	0.0001

(a) Source: Sueyoshi and Wang (2014b). (b) Avg., Max., Min. and S.D. stand for average, minimum, maximum, and standard deviation, respectively.

These unified efficiency scores are measured by the non-radial approach. For example, *UE*, *UEN*, *UEM*, *UENM* and *UENM(DC)* are solved by Models (16.1), (16.6), (16.12), (24.1) and (24.5), respectively. Thus, they correspond to UE_v^{NR}, UEN_v^{NR}, UEM_v^{NR}, $UENM_v^{NR}$ and $UENM(DC)_v^{NR}$, respectively. As mentioned previously, this chapter uses only the non-radial measurement, not the radial measurement, to maintain consistency with the research effort of Chapter 16 that compared national and international companies.

In Table 24.1, the worst performers among the independent companies are Forest Oil in *UE*, Occidental Petroleum in *UEN*, Cabot Oil and Gas Corporation in *UEM*, Chesapeake Energy Corporation in *UENM* and Occidental Petroleum in *UENM(DC)* on average efficiency. Among the integrated companies, the worst performers are Hess Corporation in *UEN* and Marathon Oil in *UEM*. All seven integrated companies have an efficiency of 1.0000 in *UE*, *UENM* and *UENM(DC)*. Moreover, there are seven companies with an efficiency score of 1.0000 in all five measures, including three independent companies (i.e., Billiton, EnCana and Ultra Petroleum) and four integrated companies (i.e., BP, Chevron, Exxon Mobil and Royal Dutch Shell). On average, Occidental Petroleum and Chesapeake Energy have the lowest efficiency scores. The relatively low performance of both companies can be partly attributed to their significant stake in shale productions.

The development of shale wells releases more GHG, especially methane, than conventional wells due to hydraulic fracturing[8]. The EPA has issued new rules to reduce air emissions from hydraulic fracturing[9]. In addition, the low price of natural gas in recent years has also undermined their revenues.

Overall, the efficiency score of the independent petroleum companies are 0.9931 in *UE*, 0.9313 in *UEN*, 0.8321 in *UEM*, 0.9574 in *UENM* and 0.9724 in *UENM(DC)*, on average, while the respective scores, on average, of the integrated companies are 1.0000, 0.9785, 0.8854, 1.0000 and 1.0000. Thus, on average, the integrated companies outperform the independent companies in terms of all five efficiency measures. The largest difference occurs in *UEM* and the smallest difference occurs in *UE*.

This chapter applied the *t*-test to the efficiency scores of the integrated and independent companies. The results in the bottom row of the table indicate that the integrated companies have higher *UE*, *UENM* and *UENM(DC)* than the independent companies at the 1% significance level, and higher *UEN* at the 10% significance level. Therefore, we consider that firms with a supply chain outperform those without a supply chain. An exception can be found in the *t*-test on *UEM* between the two groups because US environmental regulation is equally effective on the two types of petroleum firms.

The results in Table 24.1 also imply that the unified (operational and environmental) performance of US petroleum firms is positively affected by the size of their supply chains. In fact, the four most consistently efficient integrated companies have more branded retail outlets than the other integrated companies[10]. A larger retail network is associated with a higher visibility to consumers, hence resulting in higher market pressure on the unified performance. It can be conjectured that consumer pressure in the downstream spreads over the supply chain and has an impact on the performance of exploration and production in the upstream.

24.6.3 Scale Efficiency

As mentioned previously, the petroleum industry can be considered as a very large process industry from the exploration of oil and gas to the retail services for end users. Chakravarthy (1997)[11] and Vachon and Klassen (2006)[12] have discussed that petroleum firms may have increasing difficulties in planning and predicting their

[8] Howarth, R., Ingraffea, A. and Engelder, T. (2011) Natural gas: Should fracking stop. *Nature*, **477**, 271–275.

[9] Environmental Protection Agency (2012) http://www.epa.gov/airquality/oilandgas/actions.html.

[10] National Association of Convenience Stores (2013) *NACS Retail Fuels Report*. http://www.nacsonline.com/YourBusiness/FuelsReports/GasPrices_2012/Pages/WhoSellsGas.aspx.

[11] Chakravarthy, B. (1997) A new strategy framework for coping with turbulence. *Sloan Management Review*, **38**, 69–82.

[12] Vachon, S. and Klassen, R.D. (2006) Extending green practices across the supply chain: the impact of upstream and downstream integration. *International Journal of Operations and Production Management*, **26**, 795–821.

businesses when their corporate size grows. Thus, it is necessary for us to examine how they manage a unified performance from the perspective of their business size. In other words, it is expected that, since the integrated companies are much larger than the independent companies, their business operations are more difficult than that of the independent companies.

To investigate the magnitude of scale efficiency, this chapter measures a degree of *UENM(DC)* under the possible occurrence of DC. The degree of $UENM_v^{NR}$ under variable RTS is measured by Model (24.5). The degree of $UENM_c^{NR}$ under constant RTS is measured by Model (24.7) after dropping $\sum_{j=1}^{n}\lambda_j = 1$ into the formulation. One important feature of these two efficiency measures is that both incorporate the possible occurrence of DC, or eco-technology innovation, but exclude the occurrence of UC. The difference can be found in the type of RTS in their formulations.

After obtaining the two measures for each DMU, this chapter measures the degree of scale efficiency: *SENM(DC)* by Equation (24.9). The scale efficiency measures how each DMU effectively operates by considering the operational size. Thus, the magnitude of *SENM(DC)* indicates the level of efficiency in the effective size utilization of each DMU operation.

Table 24.2 summarizes *UENM(DC)v*, *UENM(DC)c* and *SENM(DC)* regarding all independent and integrated petroleum companies. Here, the three unified efficiencies are measured by Models (24.5) and (24.7) along with Equation (24.9), respectively. The *SENM* of independent firms is 0.9864 and that of integrated companies is 1.0000 on average. The *t*-test confirms at the 1% significance that the integrated companies outperform the independent companies in terms of their size utilization. This result implies that the integrated companies may have more managerial difficulty, along with an increase in their operational size, than the independent companies. However, by carefully controlling their large supply chains from exploration to retail services, the integrated companies can effectively increase their unified (operational and environmental) performance, so being able to enhance the level of corporate sustainability.

24.7 SUMMARY

To measure the level of corporate sustainability regarding energy companies, this chapter discussed how to unify natural and managerial disposability by radial and non-radial measurements. The unified efficiency measures included $UENM(DC)_v^{NR}$, $UENM(DC)_c^{NR}$, $UENM(DC)_v^{R}$ and $UENM(DC)_c^{R}$ as well as two scale efficiency measures: $SENM(DC)^{NR}$ and $SENM(DC)^{R}$.

As an application, this chapter used the proposed non-radial approach to examine the corporate sustainability of petroleum firms in the United States. Integrated companies have large supply chains for spanning both upstream and downstream functions. In contrast, independent companies have only the upstream function in their business operations. This empirical study found that the integrated

TABLE 24.2 Scale efficiency measures: non-radial approach

Company Name	*UENM(DC)v*	*UENM(DC)c*	Scale Efficiency (DC)
Independent companies			
Anadarko Petroleum Corporation	1.0000	1.0000	1.0000
Antero Resources LLC	1.0000	1.0000	1.0000
Apache Corporation	1.0000	1.0000	1.0000
Berry Petroleum Company	1.0000	1.0000	1.0000
BHP Billiton Group	1.0000	1.0000	1.0000
Bill Barrett Corporation	0.9904	0.9875	0.9970
Cabot Oil & Gas Corporation	1.0000	1.0000	1.0000
Chesapeake Energy Corporation	1.0000	0.8884	0.8884
Cimarex Energy Co.	0.9773	0.9693	0.9917
Concho Resources	0.9218	0.9094	0.9865
Conoco Phillips	1.0000	1.0000	1.0000
CONSOL Energy Inc.	0.9766	0.9601	0.9830
Continental Resources, Inc.	0.9495	0.9395	0.9894
Denbury Resources Inc.	0.9758	0.9576	0.9814
Devon Energy Corporation	1.0000	1.0000	1.0000
EnCana Corporation	1.0000	1.0000	1.0000
Energen Corporation	0.9503	0.9410	0.9902
EOG Resources, Inc.	0.8899	0.8817	0.9907
EP Energy LLC	1.0000	1.0000	1.0000
EQT Corporation	0.9781	0.9665	0.9881
EXCO Resources, Inc.	0.9999	0.9693	0.9694
Forest Oil Corporation	0.9281	0.9279	0.9999
Laredo Petroleum Holdings, Inc.	1.0000	0.9666	0.9666
Linn Energy, LLC	0.9444	0.9404	0.9958
National Fuel Gas Company	0.9015	0.8868	0.9836
Newfield Exploration Company	0.9350	0.9276	0.9921
Noble Energy, Inc.	1.0000	1.0000	1.0000
Occidental Petroleum Corporation	0.8192	0.8102	0.9890
PDC Energy, Inc.	1.0000	1.0000	1.0000
Pioneer Natural Resources Company	0.9398	0.9386	0.9988
Plains E & P Company	0.9142	0.8965	0.9806
QEP Resources, Inc.	0.9462	0.9378	0.9912
Quicksilver Resources, Inc.	1.0000	0.9847	0.9847
Range Resources Corporation	1.0000	1.0000	1.0000
Rosetta Resources Inc.	1.0000	1.0000	1.0000
SandRidge Energy, Inc.	1.0000	0.9168	0.9168
SM Energy Company	1.0000	0.9899	0.9899
Southwestern Energy Company	1.0000	0.9462	0.9462
Swift Energy Inc.	1.0000	0.9387	0.9387
Talisman Energy Inc.	0.9269	0.9154	0.9876

(*Continued*)

TABLE 24.2 (Continued)

Company Name	*UENM(DC)v*	*UENM(DC)c*	Scale Efficiency (DC)
Independent companies			
Ultra Petroleum Corporation	1.0000	1.0000	1.0000
Whiting Petroleum Corporation	1.0000	1.0000	1.0000
WPX Energy, Inc.	0.9488	0.9457	0.9967
Avg.	0.9724	0.9591	0.9864
Max.	1.0000	1.0000	1.0000
Min.	0.8192	0.8102	0.8884
S.D.	0.0407	0.0444	0.0233
Integrated companies			
BP PLC	1.0000	1.0000	1.0000
Chevron Corporation	1.0000	1.0000	1.0000
Exxon Mobil Corporation	1.0000	1.0000	1.0000
Hess Corporation	1.0000	1.0000	1.0000
Marathon Oil Corporation	1.0000	1.0000	1.0000
Murphy Oil Corporation	1.0000	1.0000	1.0000
Royal Dutch Shell PLC	1.0000	1.0000	1.0000
Avg.	1.0000	1.0000	1.0000
Max.	1.0000	1.0000	1.0000
Min.	1.0000	1.0000	1.0000
S.D.	0.0000	0.0000	0.0000
P-value (Welch's t-test)	0.0001	0.0000	0.0004

(a) Source: Sueyoshi and Wang (2014b).

companies outperformed the independent companies, because the large supply chain incorporated into the former group provided them with a scale merit in their operations and gave them a business opportunity to directly contact consumers. Consequently, this large supply chain can enhance the level of corporate sustainability in the US petroleum industry. Moreover, vertical integration in upstream and downstream may be a promising business trend toward better environmental performance in the petroleum industry.

25

COMMON MULTIPLIERS

25.1 INTRODUCTION

Returning to Chapter 7 (on SCSCs) and Chapter 11 (on DEA-DA), this chapter[1] discusses how to evaluate the performance of Japanese electric power companies by the combined use of DEA/SCSCs with DEA-DA. Here, DEA/SCSCs implies DEA equipped with SCSCs.

This chapter discusses their performance assessment only under natural disposability so that we can exclude the undesirable outputs from the proposed application. This type of DEA application often exists in the assessment of an electric power industry. For example, renewable energy sources (e.g., solar and wind) and nuclear power do not produce GHGs such as CO_2 when they generate electricity. Therefore, many people are interested in the full use of renewable energy. As an extension of Chapters 7 and 11, this chapter documents DEA assessment for ranking analysis whose result contains a single efficient DMU and the remaining others are inefficient.

From the perspective of DEA, the purpose of this chapter is to document how to overcome the following five difficulties, all of which are widely observed in its applications in energy sectors.

[1] This chapter is partly based upon the article: Sueyoshi, T. and Goto, M. (2012e) Efficiency efficiency-based rank assessment for electric power industry: A combined use of data envelopment analysis (DEA) and DEA-discriminant analysis (DA). *Energy Economics*, **34**, 634–644.

Environmental Assessment on Energy and Sustainability by Data Envelopment Analysis,
First Edition. Toshiyuki Sueyoshi and Mika Goto.

(a) *Reducing the number of many efficient DMUs*: A difficulty of DEA applications is that they usually produce many efficient DMUs in terms of operational efficiency (*OE*). For example, 90% of DMUs become efficient and the remaining 10% are inefficient. The result is mathematically acceptable, but managerially problematic, because of so many efficient DMUs. This type of problem often occurs when many multipliers (i.e., dual variables) become zero on optimality. The occurrence of zero in multipliers implies that DEA does not fully utilize information on production factors in a data set, even if we use them as inputs and desirable outputs. To reduce the number of efficient DMUs, previous studies proposed various types of multiplier restrictions such as assurance region (AR) analysis (Thompson *et al.*, 1986)[2] and cone ratio (CR) approach (Charnes *et al.*, 1989)[3] as discussed in Chapter 4. See also Chapters 22 and 23 on a different type of multiplier restrictions, referred to as "explorative analysis." The multiplier restrictions proposed in the previous studies often needed prior information, such as previous experience and empirical fact, to restrict multipliers in a specific range that fits with our expectation. A practical problem is that it is often difficult for us to access such prior information. As a result, DEA users have a difficulty in using conventional multiplier restrictions in a prescribed range as they expect. Furthermore, even if we can access such prior information, the multiplier restriction approaches (e.g, AR and CR) do not guarantee that DEA can always produce positive multipliers. In the worst case, they may produce an infeasible solution because of strict restriction on multipliers.

(b) *Multiple Projections and Multiple Reference Sets*: DEA often suffers from an occurrence of multiple projections and multiple reference sets. To overcome the difficulty, Sueyoshi and Sekitani (2007a, b, 2009)[4–6] theoretically discussed the use of DEA combined with SCSCs. See Chapter 7 for a detailed discussion on SCSCs. They also discussed the occurrence of multiple reference sets and multiple projections that became very problematic in the measurement of RTS. See Chapter 8 on RTS. To deal with such a problem, their studies proposed the use of DEA/SCSCs. The use of SCSCs

[2] Thompson, R.G., Singleton, F.D., Thrall, R.M. and Smith, B.A. (1986) Comparative site evaluation for locating a high-energy physics lab in Texas. *Interface*, **16**, 35–49.

[3] Charnes, A., Cooper, W.W., Wei, Q.L. and Huang, Z.M. (1989) Cone ratio data envelopment analysis and multi-objective programming. *International Journal of System Science*, **20**, 1099–1118.

[4] Sueyoshi, T. and Sekitani, K. (2007a) Measurement of returns to scale by a non-radial DEA model: A range-adjusted measure model. *European Journal of Operational Research*, **176**, 1918–1946.

[5] Sueyoshi, T. and Sekitani, K. (2007b) The Measurement of returns to scale under a simultaneous occurrence of multiple solutions in a reference set and a supporting hyperplane. *European Journal of Operational Research*, **181**, 549–570.

[6] Sueyoshi, T. and Sekitani, K. (2009) An occurrence of multiple projections in DEA-based measurement of technical efficiency: Theoretical comparison among DEA Models from desirable properties. *European Journal of Operational Research*, **196**, 764–794.

mathematically guarantees that it usually produces positive multipliers on efficient firms. However, the use of DEA/SCSCs does not guarantee that it can always reduce the number of efficient firms as required above.

(c) *Common Multipliers*: A conventional use of DEA cannot provide an industry-wide evaluation by common multipliers (weights). Rather, it evaluates each DMU by relatively comparing its performance with efficient firms belonging to a reference set. Thus, DEA provides a limited scope on performance evaluation by paying attention to only efficient DMUs in a reference set. The performance evaluation, which we deal with in this chapter, expects the DEA result which contains a single efficient DMU and the remaining other inefficient ones. This type of research is necessary for performance assessment. For example, corporate leaders and policy makers usually do not care about the magnitude of efficiency scores on their organizations (i.e., DMUs). Rather, they do care about the rank of their organizations within their industries. Thus, it is important for us to discuss how to rank all DMUs by an industry-wide assessment based upon common multipliers (weights). For such a purpose, this chapter proposes the use of DEA/SCSCs combined with DEA-DA.

(d) *Statistical Inference*: A difficulty, even in a combined use between DEA/SCSCs and DEA-DA, is that they do not have any statistical inference. That is a methodological drawback of DEA, compared with statistics and econometrics. After obtaining the efficiency scores of all DMUs, this chapter describes a use of a rank sum test to examine null hypotheses based upon their efficiency-based ranks. The combined use between DEA/SCSCs and DEA-DA, along with a rank sum test, may provide a new type of performance assessment and its related analytical capability. See Chapter 17 on the two rank sum tests (i.e., Mann–Whitney and Kruskal–Wallis rank sum tests).

(e) *Efficiency Requirement*: An efficiency score should locate between zero and one, where "zero" implies full inefficiency and "one" implies full efficiency. However, many DEA models such as radial and non-radial models, as discussed in Chapter 6, do not satisfy the efficiency requirement. Exceptions are RAM, SBM and Russell measures. See Chapter 6 for mathematical proof on the claim. This chapter discusses how the newly proposed approach can obtain such efficiency scores with the efficiency requirement.

The remainder of this chapter has the following structure: Section 25.2 visually describes the computational structure of the proposed approach. Section 25.3 describes an analytical linkage between DEA/SCSCs and DEA-DA. Section 25.4 describes a computational procedure concerning how to rank all DMUs and how to apply a rank sum test. Section 25.5 documents an illustrative example related to Japanese electric power industry. Section 25.6 provides a summary on this chapter.

25.2 COMPUTATIONAL FRAMEWORK

Figure 25.1 depicts a computational flow of the proposed approach. As mentioned previously, DEA usually produces many efficient DMUs because of many zeros in their multipliers (dual variables). Therefore, it is necessary for us to examine whether the type of difficulty occurs or not in DEA results. If the problem occurs, the proposed approach starts combining between primal and dual models in such a manner that the combination satisfies SCSCs and it may yield positive multipliers on efficient DMUs.

After the combination, the proposed approach needs to classify all DMUs into operationally efficient and inefficient groups by DEA/SCSCs. Then, we apply DEA-DA to identify common multipliers on production factors. Using the rank of all DMUs based upon their efficiency measures, we can conduct a rank sum test to examine a null hypothesis regarding whether multiple groups or periods have a same distribution or not on their efficiency measures.

25.3 COMPUTATIONAL PROCESS

First, this chapter returns to Model (7.9) in Chapter 7 that proposed the following radial model equipped with SCSCs:

$$
\begin{aligned}
&\text{Maximize} \quad \eta \\
&\quad s.t. \qquad \text{all constraints in Models } (7.1) \text{ and } (7.2), \\
&\qquad\qquad\qquad \theta = \sum_{r=1}^{s} u_r g_{rk} + \sigma, \\
&\qquad\qquad\qquad \lambda_j + \sum_{i=1}^{m} v_i x_{ij} - \sum_{r=1}^{s} u_r g_{rj} - \sigma \geq \eta \ \ (j=1,\ldots,n), \\
&\qquad\qquad\qquad\qquad\qquad\qquad v_i + d_i^x \geq \eta \ \ (i=1,\ldots,m), \\
&\qquad\qquad\qquad\qquad\qquad\qquad u_r + d_r^g \geq \eta \ \ (r=1,\ldots,s), \\
&\qquad\qquad\qquad \eta \geq 0.
\end{aligned}
\tag{25.1}
$$

Since Chapter 7 describes Model (25.1) in detail, this chapter does not describe its unique features, here. However, the following two concerns are important for discussing the analytical feature of SCSCs from environmental assessment. One of the two concerns is that Model (25.1) does not incorporate undesirable outputs, as mentioned in chapters in Section II. It is not difficult for us to extend Model (25.1) into the environmental assessment that incorporates undesirable outputs within the computational framework. The other concern is that Model (25.1) does not incorporate ε_s because the small number is not necessary under the use of SCSCs. That is a purpose of SCSCs in DEA. All chapters, except this chapter, in Section II incorporate the data range adjustments on slacks in the objective function of formulations. In the

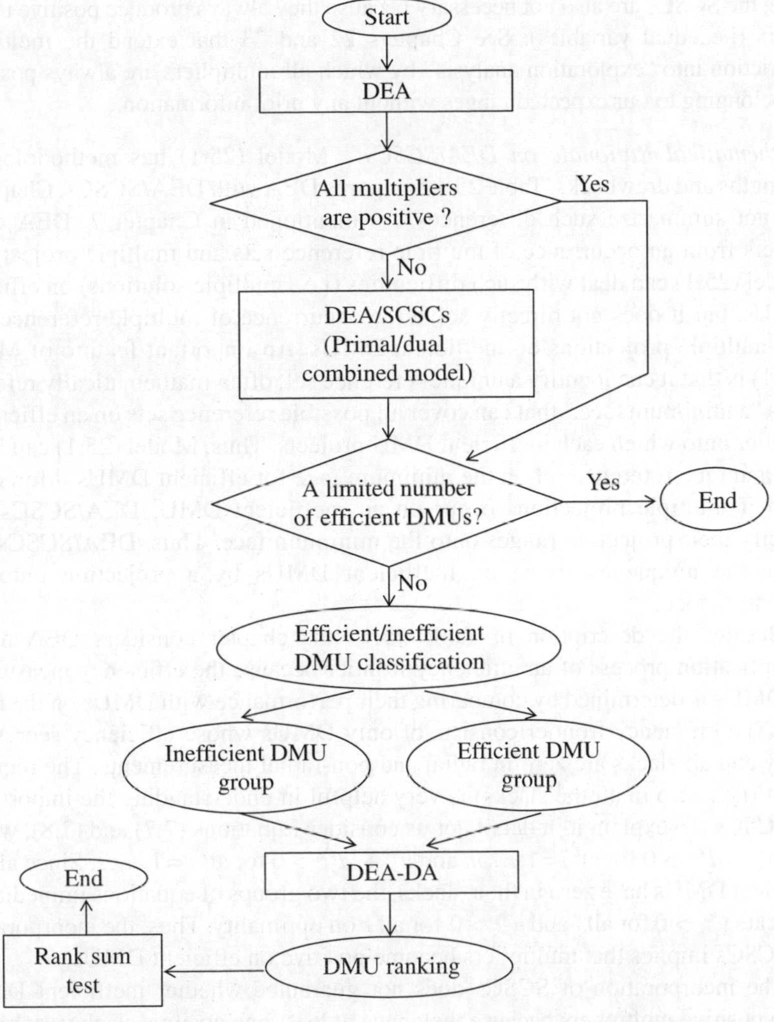

FIGURE 25.1 Computational flow of proposed approach (a) The incorporation of SCSCs into DEA is not necessary if the primal model incorporates data range adjustments on slacks, as found in radial and non-radial measurements proposed by DEA environmental assessment. See Chapters in Section II. They can always produce positive multipliers (i.e., dual variables) so that the environmental assessment discussed in this book can fully utilize all production factors. Hence, the proposed approaches can omit the computational part on DEA/SCSCs and starts from the DMU classification by DEA-DA. A problem is that the subjectivity of a user may be included in the range adjustments. The use of SCSCs can avoid such a difficulty related to a conventional use of DEA. See Chapter 7 on the use of SCSCs.

case, the SCSCs are also not necessary because they always produce positive multipliers (i.e., dual variables). See Chapters 22 and 23 that extend the multiplier restriction into "exploration analysis" by which all multipliers are always positive, so belonging to our expected ranges without any prior information.

Mathematical Rationale on DEA/SCSCs: Model (25.1) has methodological strengths and drawbacks. Table 25.1 compares DEA with DEA/SCSCs. Chapter 7 did not summarize such differences. As mentioned in Chapter 7, DEA often suffers from an occurrence of multiple reference sets and multiple projections. Model (25.1) can deal with such difficulties (i.e., multiple solutions) on efficient DMUs, but it does not directly solve the occurrence of multiple reference sets and multiple projections on inefficient DMUs. An important feature of Model (25.1) is that it can identify a unique reference set, often mathematically referred to as "a minimum face," that can cover all possible reference sets on an efficiency frontier onto which each inefficient DMU projects. Thus, Model (25.1) can identify a unique reference set on the minimum face for efficient DMUs. Moreover, even if multiple projections occur on an inefficient DMU, DEA/SCSCs can specify their projection ranges onto the minimum face. Thus, DEA/SCSCs can solve the uniqueness issue on inefficient DMUs by a projection onto the minimum face.

Besides the description in Table 25.1, this chapter considers DEA as an identification process of an efficiency frontier because the efficiency measure of all DMUs is determined by comparing their performance with DMUs on the frontier. The efficiency frontier consists of only DMUs whose efficiency scores are unity and all slacks are zero in radial and non-radial measurements. The requirement (i.e., zero in all the slacks) is very helpful in understanding the importance of SCSCs. To explain it in detail, let us consider Equations (7.7) and (7.8), which are $v_i^* + d_i^{x*} > 0$ for all $i = 1,\ldots,m$ and $u_r^* + d_r^{g*} > 0$ for all $r = 1,\ldots,s$. Since all the efficient DMUs have zero in their slacks, the two groups of equations immediately indicate $v_i^* > 0$ for all i and $u_r^* > 0$ for all r on optimality. Thus, the incorporation of SCSCs implies that multipliers become positive on efficient DMUs.

The incorporation of SCSCs does not guarantee whether inefficient DMUs have positive multipliers because they have at least one positive slack variable(s). Such a result may not be important in Model (25.1) because inefficient DMUs do not consist of an efficiency frontier. Even if we drop an inefficient DMU(s) from the computation of DEA, such elimination does not influence the whole DEA evaluation of the other DMUs. Furthermore, when DEA projects an inefficient DMU onto an efficiency frontier, the SCSCs restrict its projection in such a manner that all multipliers of the inefficient DMU become positive on the efficiency frontier. Thus, the projection of the inefficient DMU is always restricted on a reduced range, or the minimum face, shaped by the SCSCs.

After classifying all DMUs into efficient and inefficient groups by Model (25.1), the proposed approach identifies a supporting hyperplane between the two groups. Returning to Chapter 8, we can specify the supporting hyperplane by

TABLE 25.1 Differences between DEA and DEA/SCSCs

	DEA	DEA/SCSCs
Sign of multipliers (dual variables)	Multipliers may be positive or zero in their signs. The sign of many multiples is often zero. The occurrence of zero in multipliers implies that DEA evaluation does not utilize information on all production factors. Hence, the conventional use of DEA restricts multipliers by a non-Archimedean small number.	All multipliers of efficient DMUs are always positive in their signs in Model (25.1). The assessment by DEA/SCSCs can fully utilize information on all production factors.
Multiplier restriction	DEA needs multiplier restriction methods such as CR and AR. Such multiplier restriction methods proposed in the conventional DEA studies do not guarantee that all multipliers are always positive.	Model (25.1) does not need any multiplier restriction method.
Prior information	Multiplier restriction methods proposed in the previous DEA studies need prior information (e.g., experience) for the restriction. Consequently, the restriction is subjective so that different individuals use different restrictions, so producing different results on DEA performance assessment.	Model (25.1) restricts multipliers based upon the mathematical property of SCSCs. Consequently, the restriction is unique, not depending upon prior information.
Multiple reference sets and multiple projections	DEA often suffers from an occurrence of multiple reference sets and multiple projections.	Model (25.1) can uniquely identify a single reference set onto which an inefficient DMU is projected.
Reduction in the number of efficient DMUs	The multiplier restriction methods incorporated into DEA may reduce the number of efficient DMUs.	Model (25.1) cannot reduce the number of efficient DMUs.

$-\sum_{i=1}^{m} v_i x_i + \sum_{r=1}^{s} u_r g_r + \sigma$ where v_i $(i=1,\ldots,m)$ and u_r $(r=1,\ldots,s)$ are parameters for indicating the direction of a supporting hyperplane and σ indicates the intercept of the supporting hyperplane. The parameters are all unknown and need to be measured by $-\sum_{i=1}^{m} v_i x_{ij} + \sum_{r=1}^{s} u_r g_{rj} + \sigma = 0$, $j \in RS_k$, where RS_k stands for a reference set of the k-th DMU.

The above concern implies how the reference set of the k-th DMU characterizes the location of a supporting hyperplane(s). In the case, the uniqueness of a reference set is important in identifying the supporting hyperplane. Hence, this chapter fully utilizes SCSCs for DEA assessment because of the above mathematical rationale. Moreover, it is possible for us to change the location of the supporting hyperplane to separate efficient and inefficient DMUs as a discriminant hyperplane. Therefore, this chapter needs to describe the locational change of the discriminant hyperplane by using DEA-DA.

Acknowledging the importance of SCSCs in terms of identifying the supporting hyperplane, this chapter mentions that the proposed use of DEA/SCSCs has still another type of difficulty in reducing the number of efficient DMUs. To deal with the difficulty, this chapter proposes to use the following new approach that is an extended version of previous radial and non-radial approaches:

Step 1: Using results from Model (25.1), we classify all DMUs into efficient (E) and inefficient (IE) groups.

Step 2: Apply the following DEA-DA model to the two groups (E and IE):

$$
\begin{aligned}
\text{Minimize} \quad & M\sum_{j\in E}\mu_j + \sum_{j\in IE}\mu_j \\
\text{s.t.} \quad & -\sum_{i=1}^{m} v_i x_{ij} + \sum_{r=1}^{s} u_r g_{rj} + \sigma + M\mu_j \geq 0, \quad j \in E, \\
& -\sum_{i=1}^{m} v_i x_{ij} + \sum_{r=1}^{s} u_r g_{rj} + \sigma - M\mu_j \leq -\varepsilon, \quad j \in IE, \\
& \sum_{i=1}^{m} v_i + \sum_{r=1}^{s} u_r = 1, \\
& v_i \geq \varepsilon\zeta_i \; (i = 1,\ldots,m),\; u_r \geq \varepsilon\xi_r \; (r = 1,\ldots,s), \\
& \sum_{i=1}^{m} \zeta_i = m, \sum_{r=1}^{s} \xi_r = s, \\
& \sigma : \text{URS},\; v_i \geq 0 \; for\, i = 1,\ldots,m,\; u_r \geq 0 \; for\, r = 1,\ldots,s, \\
& \mu_j : \text{binary for } j = 1,\ldots,n,\; \zeta_i : \text{binary for } i = 1,\ldots m \text{ and} \\
& \xi_r : \text{binary for } r = 1,\ldots,s.
\end{aligned}
\tag{25.2}
$$

Here, M is a prescribed large number and ε is a prescribed small number. It is necessary for us to prescribe the two numbers before solving Model (25.2). The objective function minimizes the total number of incorrectly classified DMUs by counting a binary variable (μ_j). In the classification, the efficient DEA group (E) has more priority than the inefficient group (IE). Therefore, Model (25.2) adds M to the efficient group in the objective of Model (25.2). The discriminant score is expressed by $-\sigma$ $(j \in E)$ and $-\sigma - \varepsilon$ $(j \in IE)$, respectively, after changing their locations from the left hand side to the right hand side of Model (25.2). The sign of σ is not important because it is unrestricted. The small number (ε) is incorporated into the right hand side of Model (25.2) in order to avoid a case where an observation(s) exists on the estimated discriminant function. All the DMUs on desirable outputs and inputs connect to each other by a discriminant function, which we consider as the supporting hyperplane in DEA. Unknown weights v_i for $i=1,\ldots,m$ and u_r for $r=1,\ldots,s$ are determined by the minimization of incorrect classifications. The last two constraints indicate that all the variables (v_i for $i=1,\ldots,m$, and u_r for $r=1,\ldots,s$) are positive so that the discriminant function is a full model. Thus, Model (25.2) incorporates the constraints $\left(\sum_{i=1}^{m} v_i + \sum_{r=1}^{s} u_r = 1, v_i \geq \varepsilon\zeta_i \text{ for } i=1,\ldots,m, \text{and } u_r \geq \varepsilon\xi_r \text{ for } r=1,\ldots,s\right)$ to "normalize" weight estimates in the manner that all weights are expressed by a positive percentage and the sum of all weights is unity. Such a change depends upon the degree of freedom between the number of observations and that of weights.

Step 3: After applying Model (25.2) to the data set, this chapter obtains an optimal solution and compute the following score for the j-th DMU:

$$\rho_j = -\sum_{i=1}^{m} v_i^* x_{ij} + \sum_{r=1}^{s} u_r^* g_{rj} + \sigma^* \quad \text{for all} \quad j=1,\ldots,n. \tag{25.3}$$

Step 4: Using the ρ_j scores, this chapter computes their adjusted efficiency scores by the following procedure:

(a) Find the maximum and minimum values of ρ by $\max_j \rho_j$ and $\min_j \rho_j$.

(b) Find the range between them by

(b-1) Range (A) = $\max_j \rho_j - \min_j \rho_j$ if $\min_j \rho_j$ is non-negative or

(b-2) Range (B) = $\max_j \rho_j + \left|\min_j \rho_j\right|$ if $\min_j \rho_j$ is negative.

(c) The adjusted efficiency score for the j-th DMU is determined by

(c-1) Efficiency = $[\rho_j - \min_j \rho_j]/[range\ (A)]$ if $\min_j \rho_j$ is non-negative or

(c-2) Efficiency = $\left[\rho_j + \left|\min_j \rho_j\right|\right]/[range(B)]$ if $\min_j \rho_j$ is negative.

The following five concerns are important in describing the combined use between DEA-DA and DEA/SCSCs. First, Figure 25.2 visually describes the relationship between a supporting line obtained from Model (25.1) and a discriminant line determined by Model (25.2). The figure contains three efficient DMUs and the remaining inefficient DMUs. The symbol ○ stands for an efficient DMU and the symbol × stands for an inefficient DMU. For our visual convenience, the figure consists of a single input (x) on a horizontal axis and a single desirable output (g) on a vertical axis. The line s–s′ indicates a supporting line passing on DMU{B} and the line is expressed by $-vx + ug + \sigma = 0$. Meanwhile, the line d–d′ indicates a discriminant line between efficient and

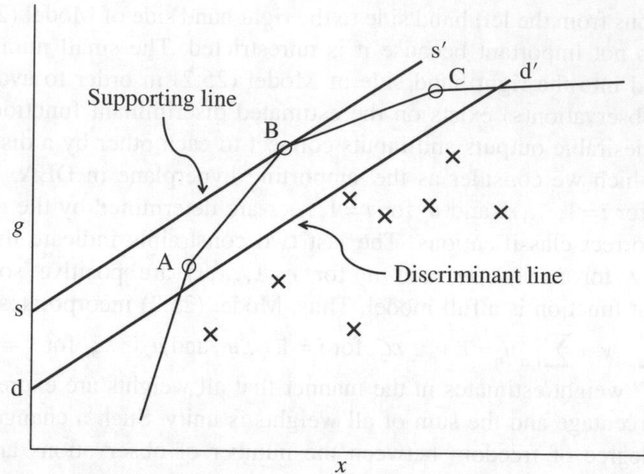

FIGURE 25.2 Visual difference between supporting line and discriminant line (a) The O indicates efficient DMUs and the X indicates inefficient DMUs. The two groups correspond to G1 and G2, respectively, in all formulations of Chapter 11. A supporting line (s–s′) can be expressed by the supporting hyperplane in the case where production factors have multiple components. In a similar manner, a supporting line (d–d′) becomes a discriminant hyperplane in the multiple components.

inefficient DMUs. Both the supporting line and the discriminant line are different in their analytical features. In the proposed approach, we consider that the supporting hyperplane determined by DEA can serve as a discriminant function for DEA-DA by shifting an intercept of the supporting line, as depicted in Figure 25.2. This is just an idea without any mathematical rationale. Second, Model (25.2) incorporates a unique feature originated from DEA. That is, DEA/SCSCs, structured by Model (25.1), provides information for distinguishing between the efficient (E) and inefficient (IE) groups. The proposed approach classifies all DMUs into two groups under the condition that Model (25.2) minimizes the number of misclassification. Third, the research of Sueyoshi and Goto (2009a, b)[7,8] has used DEA-DA as a methodology for bankruptcy and financial assessments by applying the two models. See Chapter 11. Their studies have used one of the two models for identifying an overlap between default and non-default groups. The other model classifies all observations belonging to the overlap. This chapter does not

[7] Sueyoshi, T. and Goto, M. (2009a) DEA-DA for bankruptcy-based performance assessment: misclassification analysis of the Japanese construction industry. *European Journal of Operational Research*, **199**, 576–594.

[8] Sueyoshi, T. and Goto, M. (2009b) Methodological comparison between DEA (Data Envelopment Analysis) and DEA-DA (Discriminant Analysis) from the perspective of bankruptcy assessment. *European Journal of Operational Research*, **199**, 561–575.

follow their research direction, rather making a direct linkage between DEA/SCSCs and DEA-DA. That is, this chapter uses a "standard" model, or Model (25.2), in order to classify DMUs in efficient and inefficient groups based upon their operational performance, or efficiency scores. See Chapter 11 for a description on the standard model. Thus, the proposed combination can measure the operational performance of DMUs. Fourth, since the proposed approach combines DEA/SCSCs with DEA-DA, it can provide a single criterion (i.e., an industry-wide evaluation by common multipliers) to assess the performance of all DMUs. Such a unique feature can drastically reduce the number of efficient DMUs so that we can identify a single best DMU and the remaining others with some level of inefficiency. Finally, the adjusted efficiency score of all DMUs locates between zero and unity so that it satisfies the efficiency requirement as discussed in Chapter 6.

25.4 RANK SUM TEST

To examine a null hypothesis regarding whether multiple groups or periods have a same distribution in their efficiency measures, this chapter uses the Kruskal–Wallis rank sum test[9]. Since the appendix of Chapter 17 describes the rank sum test in detail, this chapter does not discuss it further here.

25.5 JAPANESE ELECTRIC POWER INDUSTRY

25.5.1 Underlying Concepts

The Japanese industry was previously referred to as "Japan Inc.," which western nations used to admire and feared because the Japanese government (i.e., the Ministry of Economy, Trade and Industry; METI) coordinated all industrial firms like a single entity in order to pursue the national mid- and long-term goals (Johnson 1982)[10]. The current reality is different from the past. The electric power industry, as a major energy utility industry, could not escape from regulation changes by the Japanese government. It was true that the Japanese government historically had a strong influence on the management of all electric power firms so that they had similarity in their operations before the disaster of Fukushima Daiichi nuclear power plant on 11 March 2011. See Chapter 26 that discusses Japanese future strategy concerning a fuel mix after the disaster by comparing it with other industrial nations.

[9] Hollander, M. and Wolfe, D.A. (1999) *Nonparametric Statistical Methods*, John Wiley & Sons, Inc., New York.

[10] Johnson, C. (1982) *MITI and The Japanese Miracle*, Stanford University Press, Stanford.

Along with energy globalization, the Japanese government changed the direction of its energy policy toward market liberalization. According to the Federation of Electric Power Companies of Japan (2010)[11], Japan has chosen to liberalize the electric power market. In March 2000, the government partly liberalized the retail market so that power producers and suppliers could sell electricity to extra-high voltage users with a demand of ≥2000 kW. From April 2005, Japan expanded the scope of liberalization to all high-voltage users whose demand was ≥50 kW. Meanwhile, all consumers in a regulated market continued to receive electricity supplied by each regional electric power company. In April 2016, full retail competition began in Japan and all consumers are now eligible to choose their electricity suppliers. Thus, this liberalization introduced competition among electric power firms, but provided each electric power company with business freedom in which each firm can operate under less regulation than before liberalization.

Based upon the changes in regulation discussed above, this chapter raises the following two policy issues on the Japanese electric power industry. One of the two issues is whether liberalization in 2005 has influenced the operational performance of the Japanese electric power industry. The other question is whether Japanese electric power firms have exhibited a difference in their operations since liberalization in 2005.

The null hypothesis originating from the first issue is that liberalization has not yet provided a major impact on the performance of the electric power industry because Japan has suffered from economic stagnation. The null hypothesis from the second one is that business freedom after liberalization has not yet produced different business strategies among the electric power firms due to a time lag of deregulation effects after the strong governmental regulation, previously characterized by "Japan Inc."

It is important to note that electricity prices for regulated consumers are subject to public control. Government regulatory agencies commonly employ a "fair rate of return" criterion. After a firm subtracts its operating expense from gross revenue, the remaining net revenue should be sufficient to compensate the firm for its investment in plant and equipment. If the rate of return (computed as the ratio of net revenue to the value of plant and equipment; the rate base) is judged excessive, political pressure suggests the electric power firm should reduce its electricity prices. In contrast, if the rate is too low, then the firm may increase its prices. Because of the price setting mechanism, or rate of return regulation, the electric power firms tend to be motivated to over-invest on plant and equipment, often leading to an augmented rate base to produce higher revenue. This type of over-investment under regulation is often referred to as the "Averch–Johnson Effect" (Averch and Johnson, 1962)[12]. Based upon the above concern, this chapter is

[11] Federation of Electric Power Companies of Japan (2010) *Electric Review Japan 2010*, Federation of Electric Power Companies of Japan, Tokyo.

[12] Averch, H. and Johnson, L.L. (1962). Behavior of the firm under regulatory constraint. *American Economic Review*, **52**, 1052–1069.

interested in the efficient use of assets in Japanese electric power firms. Therefore, we use assets from generation to distribution as input measures in the proposed DEA evaluation.

Averch–Johnson effect: Their study discussed the economic implication of regulation, often referred to as the "Averch–Johnson effect". The concept implies that regulated companies tend to engage in an excessive amount of capital accumulation to expand the volume of their profits under no competition. As a result, there is a strong incentive to over-invest in their facilities in order to increase their profits overall. Their investments often go beyond any optimal efficiency point for capital, so producing a high level of inefficiency within their firms because of large excess facilities (i.e., assets). The concept has served as a policy basis to promote the liberalization of the electric power industry in many industrial nations.

25.5.2 Empirical Results

Sueyoshi, T. and Goto, M. (2012e) provides descriptive statistics on the data set on Japanese electric power firms used for this chapter.

Data Accessibility: The data source is Handbook of Electric Power Industry (2010).

The Japanese electric power industry consists of nine investor-owned utility firms with regional monopoly, which are listed from A to I in this chapter. One electric power company (i.e., Okinawa) is much smaller than the other firms in terms of operation size and so it is excluded from our research for this chapter. The observed period is from 2005 to 2009. This chapter measures the performance of the nine electric power firms by examining two outputs and five inputs.

The two desirable outputs are the total amount of electricity sold and the total number of customers. The five inputs are total generation asset, total transmission asset, total distribution asset, total operation cost without labor cost, and number of employees. As mentioned previously, the electric power industry is a large capital-intensive industry. Therefore, this chapter pays attention to the total assets related to generation, transmission and distribution.

Table 25.2 summarizes DEA efficiency scores and multipliers (i.e., dual variables) of the nine Japanese electric power firms from 2005 to 2009, measured by original radial formulations, or Models (7.1) and (7.2). In similar manner, Table 25.3 summarizes their efficiency scores and multipliers measured by the proposed formulation, or Model (25.1).

Comparing Table 25.2 with Table 25.3 provides us with the following two methodological implications. One of the two implications is that all efficient and inefficient firms contain zero in their multipliers in Table 25.2. However, as found in Table 25.3, the multipliers of efficient firms are all positive but those of inefficient firms contain zero. The results indicate that SCSCs change efficient DMUs to

TABLE 25.2 Efficiency and multipliers measured by Model (7.1) or (7.2)

Company	Year	Efficiency	v_1	v_2	v_3	v_4	v_5	w_1	w_2
A	2005	1.000	0.052405	0.000000	0.000000	0.102977	0.707550	0.000000	0.012166
A	2006	1.000	0.000000	0.104444	0.000000	0.177168	0.000000	0.000000	0.390412
A	2007	1.000	0.181738	0.000000	0.000000	0.092883	0.000000	0.256390	0.000000
A	2008	1.000	0.304661	0.000000	0.000000	0.009537	0.000000	0.179208	0.058398
A	2009	1.000	0.011953	0.350108	0.000000	0.000000	0.000000	0.000000	0.403690
B	2005	0.894	0.000000	0.000000	0.000000	0.005551	0.756328	0.040160	0.067629
B	2006	0.907	0.000000	0.000000	0.000000	0.001548	0.805956	0.038891	0.068280
B	2007	0.930	0.000000	0.092301	0.000000	0.000000	0.116702	0.118264	0.000259
B	2008	0.910	0.000000	0.093871	0.000000	0.000000	0.118686	0.120275	0.000263
B	2009	0.904	0.000000	0.095861	0.000000	0.000000	0.121202	0.122824	0.000269
C	2005	1.000	0.000000	0.000000	0.000000	0.023921	0.000000	0.023408	0.013826
C	2006	1.000	0.000000	0.000000	0.000000	0.023658	0.000000	0.130132	0.176102
C	2007	1.000	0.000827	0.000000	0.000000	0.000552	0.248777	0.012417	0.021457
C	2008	1.000	0.000000	0.000000	0.000000	0.000000	0.263762	0.012416	0.021675
C	2009	1.000	0.000000	0.000000	0.000000	0.000146	0.260741	0.000000	0.034596
D	2005	1.000	0.000000	0.000000	0.000000	0.059656	0.000000	0.079572	0.000000
D	2006	0.999	0.000000	0.000000	0.003864	0.014780	0.437473	0.038298	0.044606
D	2007	1.000	0.021665	0.000000	0.091193	0.000000	0.000000	0.068436	0.000000
D	2008	0.982	0.000000	0.061589	0.000000	0.000000	0.077871	0.078913	0.000173
D	2009	1.000	0.005958	0.024601	0.000000	0.035223	0.000000	0.084607	0.000000
E	2005	1.000	0.000000	0.000000	0.095255	0.216685	0.000000	0.332414	0.000000
E	2006	1.000	0.000000	0.000000	0.639116	0.000000	0.000000	0.254385	0.000000
E	2007	1.000	0.000000	0.000000	0.000000	0.000000	2.168727	0.227359	0.000000
E	2008	1.000	0.035502	0.000000	0.000000	0.000000	1.767454	0.000000	0.000000
E	2009	1.000	0.099992	0.000000	0.337524	0.000000	0.000000	0.000000	0.000000
F	2005	1.000	0.000000	0.000000	0.000000	0.051262	0.000000	0.050163	0.029630

F	2006	1.000	0.000000	0.000000	0.057198	0.019448	0.021290	0.062008	0.023033
F	2007	1.000	0.052227	0.000000	0.000000	0.017464	0.000000	0.077052	0.000000
F	2008	1.000	0.090512	0.000000	0.000000	0.000000	0.000000	0.049586	0.016209
F	2009	1.000	0.011883	0.000000	0.023242	0.032656	0.000000	0.010662	0.068193
G	2005	0.986	0.029930	0.000000	0.065278	0.067728	0.000000	0.157934	0.000000
G	2006	0.995	0.029379	0.000000	0.064077	0.066481	0.000000	0.155028	0.000000
G	2007	1.000	0.000000	0.062152	0.000000	0.069051	0.000000	0.172274	0.000000
G	2008	1.000	0.147117	0.043827	0.000000	0.000000	0.000000	0.144048	0.000000
G	2009	1.000	0.181192	0.000000	0.000000	0.014172	0.000000	0.128710	0.025773
H	2005	0.968	0.044823	0.000000	0.373325	0.008442	0.000000	0.000000	0.181610
H	2006	0.982	0.045415	0.000000	0.378256	0.008554	0.000000	0.000000	0.184009
H	2007	1.000	0.072656	0.000000	0.230278	0.061851	0.000000	0.274037	0.000000
H	2008	1.000	0.296245	0.000000	0.000000	0.039946	0.000000	0.000000	0.000000
H	2009	1.000	0.023711	0.000000	0.018924	0.221782	0.000000	0.000000	0.000000
I	2005	1.000	0.000000	0.000000	0.000000	0.096034	0.000000	0.119767	0.000000
I	2006	1.000	0.009365	0.000000	0.000000	0.029605	0.478150	0.068282	0.047391
I	2007	1.000	0.000000	0.027350	0.000000	0.043449	0.172383	0.097154	0.022450
I	2008	0.980	0.000000	0.000000	0.084605	0.010472	0.268581	0.029272	0.089285
I	2009	1.000	0.000000	0.047703	0.002675	0.048504	0.000000	0.121934	0.007735

(a) Source: Sueyoshi and Goto (2012e). (b) The results in the table are not acceptable because many multipliers (dual variables) are zero on efficient and inefficient DMUs. (c) A conventional use of DEA has limited practicality because of zero on multipliers on efficient DMUs.

TABLE 25.3 Efficiency and multipliers measured by Model (25.1)

Company	Year	Efficiency	v_1	v_2	v_3	v_4	v_5	w_1	w_2
A	2005	1.000	0.020589	0.020589	0.020589	0.205510	0.020589	0.020589	0.020589
A	2006	1.000	0.030997	0.180855	0.030997	0.076683	0.030997	0.030997	0.343905
A	2007	1.000	0.212801	0.030879	0.030879	0.030879	0.032612	0.212636	0.042047
A	2008	1.000	0.281227	0.010920	0.010920	0.010920	0.010920	0.010920	0.266833
A	2009	1.000	0.008955	0.233598	0.008955	0.008955	0.465536	0.008955	0.342835
B	2005	0.894	0.000000	0.000000	0.000000	0.005551	0.756328	0.040160	0.067629
B	2006	0.907	0.000000	0.000000	0.000000	0.001548	0.805956	0.038891	0.068280
B	2007	0.930	0.000000	0.092301	0.000000	0.000000	0.116702	0.118264	0.000259
B	2008	0.910	0.000000	0.093871	0.000000	0.000000	0.118686	0.120275	0.000263
B	2009	0.904	0.000000	0.095861	0.000000	0.000000	0.121202	0.122824	0.000269
C	2005	1.000	0.003604	0.003604	0.003604	0.015883	0.003604	0.103387	0.003604
C	2006	1.000	0.005017	0.005017	0.005017	0.013032	0.005017	0.071624	0.005017
C	2007	1.000	0.007384	0.007384	0.007384	0.007384	0.007384	0.059211	0.007384
C	2008	1.000	0.007331	0.007331	0.007331	0.007331	0.007331	0.326043	1.662427
C	2009	1.000	0.008137	0.008137	0.008137	0.008137	0.008137	0.029476	0.008137
D	2005	1.000	0.008687	0.008687	0.008687	0.033062	0.064660	0.073988	0.008687
D	2006	0.999	0.000000	0.000000	0.003864	0.014780	0.437473	0.038298	0.044606
D	2007	1.000	0.015089	0.015089	0.015089	0.015089	0.106649	0.062646	0.015089
D	2008	0.982	0.000000	0.061589	0.000000	0.000000	0.077871	0.078913	0.000173
D	2009	1.000	0.003337	0.025259	0.003337	0.032882	0.019294	0.082014	0.003337
E	2005	1.000	0.001888	0.001888	0.076872	0.218989	0.001888	0.223883	0.001888
E	2006	1.000	0.006930	0.006930	0.578665	0.006930	0.006930	0.135157	0.006930
E	2007	1.000	0.026240	0.026240	0.067217	0.026240	1.213160	0.227573	0.026240
E	2008	1.000	0.022238	0.003629	0.003629	0.003629	1.842313	0.003629	0.003629
E	2009	1.000	0.047431	0.047431	0.276749	0.047431	0.047431	0.047431	0.047431
F	2005	1.000	0.007234	0.007234	0.007234	0.034732	0.007234	0.069788	0.007234

F	2006	1.000	0.000000	0.000000	0.057198	0.019448	0.021290	0.062008	0.023033
F	2007	1.000	0.014110	0.009961	0.044080	0.009961	0.009961	0.071218	0.009961
F	2008	1.000	0.059312	0.006611	0.006611	0.006611	0.006611	0.070733	0.008646
F	2009	1.000	0.017083	0.017083	0.017083	0.017083	0.017083	0.047993	0.027522
G	2005	0.986	0.029930	0.000000	0.065278	0.067728	0.000000	0.157934	0.000000
G	2006	0.995	0.029379	0.000000	0.064077	0.066481	0.000000	0.155028	0.000000
G	2007	1.000	0.011597	0.107709	0.036126	0.011597	0.011597	0.164278	0.011597
G	2008	1.000	0.101499	0.027131	0.007173	0.007173	0.225675	0.146395	0.007173
G	2009	1.000	0.130249	0.029342	0.013232	0.016041	0.013232	0.152194	0.013232
H	2005	0.968	0.044823	0.000000	0.373325	0.008442	0.000000	0.000000	0.181610
H	2006	0.982	0.045415	0.000000	0.378256	0.008554	0.000000	0.000000	0.184009
H	2007	1.000	0.073384	0.001701	0.235524	0.057579	0.001701	0.272226	0.001701
H	2008	1.000	0.268158	0.025509	0.025509	0.025509	0.025509	0.173476	0.025509
H	2009	1.000	0.067221	0.043482	0.173092	0.071447	0.043482	0.043482	0.043482
I	2005	1.000	0.011169	0.011169	0.011169	0.068065	0.011169	0.041992	0.095030
I	2006	1.000	0.015797	0.014339	0.014339	0.041511	0.152344	0.097341	0.025754
I	2007	1.000	0.001499	0.028885	0.001499	0.041744	0.159008	0.100652	0.018739
I	2008	0.980	0.000000	0.000000	0.084605	0.010472	0.268581	0.029272	0.089285
I	2009	1.000	0.002684	0.039986	0.002684	0.049386	0.031303	0.117319	0.011446

(a) Source: Sueyoshi and Goto (2012e). (b) The results in the table are acceptable because all multipliers (dual variables) are not zero on efficient DMUs. The small number (0.000000) on efficient DMUs is not zero, rather a very small positive number near zero.

avoid their multipliers from being zero, but making them positive. Results produced by DEA/SCSCs can fully utilize information on the production factors of efficient DMUs, but not that of inefficient ones. The other implication is that even if DEA incorporates SCSCs, the combination still produces many efficient DMUs as summarized in Table 25.3. This indicates that the approach of DEA/SCSCs needs to improve the empirical effectiveness on performance assessment by combining it with DEA-DA in order to provide common multipliers.

Table 25.4 documents the efficiency scores of Japanese electric power companies, measured by a conventional use of Model (7.1), and their adjusted efficiency scores measured by the proposed approach (DEA/SCSCs combined with DEA-DA). The proposed approach produces their adjusted efficiency scores on [0, 1] so that they satisfy the efficiency requirement. In the proposed approach, the sixth electric power company {F} was the best performer (an adjusted efficiency score is 1 and rank is 1) in 2009. The second electric power company {B} was the worst performer (an adjusted efficiency score is 0 and rank is 45) in 2009. Thus, these results indicated that there was a single best performer and the remaining other firms were inefficient from the industry-wide perspective.

Table 25.4 also provides an interesting difference between the conventional use of DEA and the proposed approach. To explain such a difference, let us see the efficiency and adjusted efficiency scores of company {E}, for example. The electric power company {E}, being smaller than the other firms, has produced unity (efficiency score = 1) in the conventional efficiency measures from 2005 to 2009. Therefore, the use of DEA evaluates company {E} as an efficient electric power firm. In contrast, the proposed approach has evaluated company {E} as 0.168 (2005), 0.201 (2006), 0.194 (2007), 0.188 (2008) and 0.194 (2009) on its adjusted efficiency scores. The result indicates that company {E} is inefficient. Such an evaluation difference is because the proposed approach has an industry-wide assessment by common multipliers. In contrast, the conventional use of DEA does not have such an analytical capability for performance assessment. Thus, the proposed approach overcomes a methodological drawback of DEA, often observed in many previous studies.

Using the adjusted efficiency scores in Table 25.4, this chapter conducts the rank sum test on two null hypotheses. The first null hypothesis is whether liberalization in 2005 has not yet influenced the operational performance of Japanese electric power industry. The H statistic, measured by Equation (17.A3) in the appendix of Chapter 17, becomes 1.641 (which is less than 9.49: the χ^2 score at the 5% significance level under four degrees of freedom). Thus, this chapter cannot reject the null hypothesis, so confirming that liberalization has not yet produced any major change in the Japanese electric power industry after 2005. The result also supports the statistical validity of our computational procedure in which this chapter pools data sets from 2005 to 2009 and treats them as a single data set.

The second null hypothesis is whether Japanese electric power firms have shown a same result on their operational performance even after liberalization. The H statistic, measured by Equation (17.A3) in Chapter 17, becomes 38.282

TABLE 25.4 Efficiency (DEA) and adjusted efficiency (proposed approach)

Company	Year	Efficiency	Adjusted Efficiency	Rank
A	2005	1.000	0.342	22
A	2006	1.000	0.351	20
A	2007	1.000	0.356	19
A	2008	1.000	0.334	23
A	2009	1.000	0.330	24
B	2005	0.894	0.108	42
B	2006	0.907	0.118	40
B	2007	0.930	0.026	43
B	2008	0.910	0.005	44
B	2009	0.904	0.000	45
C	2005	1.000	0.168	33
C	2006	1.000	0.344	21
C	2007	1.000	0.578	9
C	2008	1.000	0.666	6
C	2009	1.000	0.891	4
D	2005	1.000	0.471	11
D	2006	0.999	0.556	10
D	2007	1.000	0.653	7
D	2008	0.982	0.421	16
D	2009	1.000	0.427	15
E	2005	1.000	0.168	34
E	2006	1.000	0.201	26
E	2007	1.000	0.194	27
E	2008	1.000	0.188	29
E	2009	1.000	0.194	28
F	2005	1.000	0.640	8
F	2006	1.000	0.776	5
F	2007	1.000	0.898	3
F	2008	1.000	0.934	2
F	2009	1.000	1.000	1
G	2005	0.986	0.116	41
G	2006	0.995	0.167	37
G	2007	1.000	0.168	35
G	2008	1.000	0.182	30
G	2009	1.000	0.216	25
H	2005	0.968	0.160	39
H	2006	0.982	0.167	38
H	2007	1.000	0.176	31
H	2008	1.000	0.173	32
H	2009	1.000	0.168	36
I	2005	1.000	0.373	18
I	2006	1.000	0.420	17
I	2007	1.000	0.464	12
I	2008	0.980	0.445	14
I	2009	1.000	0.450	13

(a) Source: Sueyoshi and Goto (2012e). (b) The proposed approach measures the adjusted efficiency. This chapter sets $M = 100$ and $\varepsilon = 0.001$ for the computation of DEA-DA, or Model (25.2). The resulting multipliers are $v_1 = 0.0010$, $v_2 = 0.0177$, $v_3 = 0.4468$, $v_4 = 0.0089$, $v_5 = 0.1004$, $w_1 = 0.1001$, $w2 = 0.3251$ and $\sigma = -0.083$.

(which is larger than 20.09: the χ^2 score at the 1% significance level under eight degrees of freedom). Thus, this chapter can reject the null hypothesis, so accepting the alternative hypothesis that Japanese electric power firms have taken different strategies in their operations so that they have produced different results in their efficiency measures.

25.6 SUMMARY

This chapter proposed a new type of approach to incorporate the following five analytical necessities into DEA. First, the proposed approach could reduce the number of efficient DMUs, not depending upon conventional multiplier restriction methods (e.g., CR and AR). Second, the proposed approach could provide a unique reference set and a unique projection. Third, the proposed approach provided an industry-wide evaluation by common multipliers. The proposed performance analysis provided a single efficient DMU and other inefficient ones. Fourth, the proposed approach could rank all DMUs so that it could connect to a statistical inference, or a rank sum test. Finally, the adjusted efficiency scores measured by the proposed approach located between zero and one, where "zero" implied full inefficiency and "one" implied full efficiency.

This chapter applied the proposed approach to investigate the business change of Japanese electric power industry after liberalization in 2005. The application identified two policy implications. One of the two implications was that liberalization did not influence the operational performance of the Japanese electric power industry because the Japanese economy suffered from an economic stagnation in the observed annual periods which resulted in a lower electricity demand growth than before. The other was that Japanese electric power firms exhibited different results on their operational performance after liberalization. Thus, liberalization provided them with business freedom so that the electric power firms gradually took different business strategies after 2005.

26

PROPERTY OF TRANSLATION INVARIANCE TO HANDLE ZERO AND NEGATIVE VALUES

26.1 INTRODUCTION

Acknowledging the contribution of previous DEA studies, it is necessary for this chapter[1] to discuss a difficulty on how to assess the performance of DMUs, whose production factors contain many zero and negative values in an observed data set. Such an occurrence of zero and negative values is widely observed in previous studies. For example, paying attention to energy, nuclear and renewable generations do not produce GHG emissions when they produce electricity. So, their uses contain zero in the amount of GHG emissions. Thus, in discussing a fuel mix issue for future electricity generation, we need to consider the possible usage of nuclear[2] and renewable sources with zero in their GHG emissions. For example, see Chapter 19 that used only 10 industrial nations for DEA environmental assessment

[1] This study is partly based upon the article: Sueyoshi, T. and Goto, M. (2015c) Japanese fuel mix strategy after disaster of Fukushima Daiichi's nuclear power plant: lessons from international comparison among industrial nations Measured by DEA environmental assessment in time horizon. *Energy Economics*, **52**, Part A, 87–103.

[2] Nuclear generation produces nuclear waste and other types of industrial pollution. No country knows how to solve the problem of nuclear waste. The recycle cost of nuclear energy is not officially announced but it is much larger than our expectation. For example, mixed oxide (MOX) fuel costs almost US$ 9 million per unit according to Asahi News Paper, a Japanese leading newspaper, on 28 February 2016. However, such operational costs are much less than the opportunity cost due to the major disaster which occurred at the Fukushima Daiichi nuclear power plant. That is the issue related to nuclear generation.

Environmental Assessment on Energy and Sustainability by Data Envelopment Analysis, First Edition. Toshiyuki Sueyoshi and Mika Goto.

because that chapter encountered the problem of zero in data. As a result, the data set of Chapter 19 contained only industrial nations without zero in their production factors. Other industrial nations which contained zero in their data set were excluded from Chapter 19. That is a drawback of the DEA environmental assessment discussed in previous chapters in Section II. Such a difficulty due to zero and negative values in a data set will be overcome in this chapter.

To overcome the methodological issue, this chapter first discusses an important property of non-radial models. The non-radial measures discussed in Chapter 5 have the property of "translation invariance" so that they can overcome the methodological difficulty by adding a small number into the original data set. See Chapter 6 that compares radial and non-radial models from mathematical properties required for operational efficiency. This chapter fully utilizes the property of translation invariance for the proposed environmental assessment. This chapter also reorganizes the proposed approach in such a manner that it can fit with a time horizon, often referred to as "window analysis," which is different from the Malmquist index measurement discussed in Chapter 19. Such a data adjustment due to the property of translation invariance makes it possible that this chapter can examine a shift of efficiency scores and growth measures of many industrial nations in a time horizon, not a frontier shift as found in the Malmquist index. To document the practicality of the proposed approach, this chapter examines a future fuel mix strategy of the Japanese electric power industry. This application provides us with an important implication on the practicality of translation invariance within the framework of non-radial measurement.

The remainder of this chapter has the following structure. Section 26.2 discusses the property of translation invariance in non-radial measurement. Section 26.3 extends the proposed DEA approach, equipped with the property of translation invariance, into a time horizon. Section 26.4 examines the fuel mix strategies regarding electric power industries in 33 industrial nations. Section 26.5 summarizes this chapter along with policy implications on the future fuel mix strategy after the disaster of Fukushima Daiichi nuclear power plant.

26.2 TRANSLATION INVARIANCE

To handle zero and negative values in a data set, this chapter fully utilizes the property of translation invariance. The non-radial models discussed in Chapter 16 have this desirable property because they are originated from the range-adjusted measure (RAM). See Chapter 6 on a mathematical description on the properties of RAM. No other DEA models (e.g., radial models) have this property. They must depend upon another approach that can substitute for the property of translation invariance. See, for example, Chapter 27 that discusses how to handle zero and negative values from the perspective of radial measurement for environmental assessment. The proposed approach of Chapter 27 can handle a data set, containing zero and negative values, in the framework of both radial and non-radial measurements. Note that Chapter 6 provides a description on the property of translation invariance within the conventional framework of DEA without undesirable outputs. The property works only for non-radial models originated from RAM, while the approach in Chapter 27 is straightforward and much easier than the other approach for radial and non-radial models.

For applying the property of translation invariance for DEA environmental assessment as an extension of Chapter 6, this chapter starts with a description of specifying the following data shift on all DMUs ($j=1,\ldots,n$):

$$\bar{x}_{ij} = x_{ij} + \alpha_i \ (i=1,\ldots,m),\ \bar{g}_{rj} = g_{rj} + \beta_r \ (r=1,\ldots,s) \text{ and}$$
$$\bar{b}_{fj} = b_{fj} + \delta_f \ (f=1,\ldots,h). \tag{26.1}$$

The three Greek symbols are specific positive numbers (e.g., 1 and 2) that are subjectively selected by a DEA user(s). The subjectivity does not influence DEA computational results. As a result of these data shifts, all production factors of the j-th DMU become $\bar{x}_{ij} > 0$ ($i=1,\ldots,m$), $\bar{g}_{rj} > 0$ ($r=1,\ldots,s$) and $\bar{b}_{fj} > 0$ ($f=1,\ldots,h$), where the symbol (>) implies strict positivity in all components of the three vectors for production factors.

Here, it is important to note that the symbols should be selected within the boundary of our common sense. For example, let us consider a case in which α_i = 100,000, β_r = 1 and δ_f = 1 are selected for Equations (26.1). In this case, the adjusted large data on input $\bar{x}_{ij}$ dominates the other desirable and undesirable outputs ($\bar{g}_{rj}$ and $\bar{b}_{fj}$) in the proposed computation. Consequently, DEA does not produce any reliable result on environmental assessment. This is not a mathematical problem, indeed, rather being a computational problem. In addition, the selection of $\alpha_i = \beta_r = \delta_f$ is acceptable as long as such a selection fits with the mathematical requirement that all production factors become strictly positive on their components. For instance, this chapter selects $\alpha_i = \beta_r = \delta_f = 1$ to avoid zero in the data set for Section 26.4, where the proposed approach will be applied for measuring the fuel mix strategy of industrial nations in OECD countries.

Under Natural Disposability: To examine the property of translation invariance, let us return to Model (16.6) and modify the three groups of constraints as follows:

$$\sum_{j=1}^{n} \bar{x}_{ij}\lambda_j + d_i^{x-} = \bar{x}_{ik} \leftrightarrow \sum_{j=1}^{n} (x_{ij} + \alpha_i)\lambda_j + d_i^{x-} = \sum_{j=1}^{n} x_{ij}\lambda_j + \alpha_i + d_i^{x-} = x_{ik} + \alpha_i$$
$$\leftrightarrow \sum_{j=1}^{n} x_{ij}\lambda_j + d_i^{x-} = x_{ik},$$

$$\sum_{j=1}^{n} \bar{g}_{rj}\lambda_j - d_r^{g} = \bar{g}_{rk} \leftrightarrow \sum_{j=1}^{n} (g_{rj} + \beta_r)\lambda_j - d_r^{g} = \sum_{j=1}^{n} g_{rj}\lambda_j + \beta_r - d_r^{g} = g_{rk} + \beta_r$$
$$\leftrightarrow \sum_{j=1}^{n} g_{rj}\lambda_j - d_r^{g} = g_{rk} \text{ and} \tag{26.2}$$

$$\sum_{j=1}^{n} \bar{b}_{fj}\lambda_j + d_f^{b} = \bar{b}_{fk} \leftrightarrow \sum_{j=1}^{n} (b_{fj} + \delta_f)\lambda_j + d_f^{b} = \sum_{j=1}^{n} b_{fj}\lambda_j + \delta_f + d_f^{b} = b_{fk} + \delta_f$$
$$\leftrightarrow \sum_{j=1}^{n} b_{fj}\lambda_j + d_f^{b} = b_{fk},$$

for ($i = 1,\ldots, m$), ($r = 1,\ldots, s$) and ($f = 1,\ldots, h$), respectively.

Equations (26.2) add a specific positive number to each production factor so that all observations become positive, shifting from zero or negative to positive, in their signs. A benefit of the proposed data shift is that Equations (26.2) can fit within the computational framework of Model (16.6) under natural disposability and Model (16.12) under managerial disposability, whose sign of a data set should be strictly positive in a data domain to be examined.

Equations (26.2) maintains the following three conditions: $\sum_{j=1}^{n}\alpha_i\lambda_j=\alpha_i$, $\sum_{j=1}^{n}\beta_r\lambda_j=\beta_r$ and $\sum_{j=1}^{n}\delta_f\lambda_j=\delta_f$ because of $\sum_{j=1}^{n}\lambda_j=1$. Consequently, the three constraints of Equations (26.2) become

$$\begin{aligned}&\sum_{j=1}^{n}x_{ij}\lambda_j+d_i^{x-}=x_{ik}\ (i=1,\ldots,m),\\&\sum_{j=1}^{n}g_{rj}\lambda_j-d_r^{g}=g_{rk}\ (r=1,\ldots,s)\ \text{and}\\&\sum_{j=1}^{n}b_{fj}\lambda_j+d_f^{b}=b_{fk}\ (f=1,\ldots,h).\end{aligned}\tag{26.3}$$

The above three groups of constraints are the same as Model (16.6) formulated under natural disposability. Thus, the proposed data shifts do not influence the constraints of Model (16.6).

Next, paying attention to the objective function of Model (16.6), the data shifts change the three types of slacks as follows:

$$\begin{aligned}&\sum_{i=1}^{m}R_i^x\left[(x_{ik}+\alpha_i)-\sum_{j=1}^{n}(x_{ij}+\alpha_i)\lambda_j\right]=\sum_{i=1}^{m}R_i^x\left(x_{ik}-\sum_{j=1}^{n}x_{ij}\lambda_j\right)=\sum_{i=1}^{m}R_i^x d_i^{x-},\\&\sum_{r=1}^{s}R_r^g\left[\sum_{j=1}^{n}(g_{rj}+\beta_r)\lambda_j-(g_{rk}+\beta_r)\right]=\sum_{r=1}^{s}R_r^g\left(\sum_{j=1}^{n}g_{rj}\lambda_j-g_{rk}\right)=\sum_{r=1}^{s}R_r^g d_r^g\ \text{and}\\&\sum_{f=1}^{h}R_f^b\left[(b_{fk}+\delta_f)-\sum_{j=1}^{n}(b_{fj}+\delta_f)\lambda_j\right]=\sum_{f=1}^{h}R_f^b\left(b_{fk}-\sum_{j=1}^{n}b_{fj}\lambda_j\right)=\sum_{f=1}^{h}R_f^b d_f^b.\end{aligned}\tag{26.4}$$

Equations (26.4) clearly indicates the translation invariance of the objective value of Model (16.6). Thus, the proposed data shifts from zero or negative to positive do not influence the objective value of Model (16.6) under natural disposability.

Under Managerial Disposability: In a similar manner to the discussion above, the Model (16.12) under managerial disposability has the property of translation invariance.

26.3 ASSESSMENT IN TIME HORIZON

The description of non-radial models in Chapter 16 is for a cross-sectional data set without any zero or negative values. Hereafter, this chapter needs to change them in such a manner that they can work in a time horizon even if it contains zero or negative values. To describe such a methodological change, this chapter needs to specify the data structure of all DMUs ($j=1,\ldots,n$) in the t th period ($t=1,\ldots,T$). That is, the three groups of production factors on the j-th DMU are expressed by x_{ijt} ($i=1,\ldots,m$ and $t=1,\ldots,T$), g_{rjt} ($r=1,\ldots,s$ and $t=1,\ldots,T$) and b_{fjt} ($f=1,\ldots,h$ and $t=1,\ldots,T$) in a time horizon with T periods. Since the data set to be examined in this chapter contains many zeros in the three production factors, we prepare the following data shift on the j-th DMU in the t-th period such as $\bar{x}_{ijt} = x_{ijt} + \alpha_i$ ($i=1,\ldots,m$), $\bar{g}_{rjt} = g_{rjt} + \beta_r$ ($r=1,\ldots,s$) and $\bar{b}_{fjt} = b_{fjt} + \delta_f$ ($f=1,\ldots,h$). Accordingly, the data range adjustments are restructured as follows:

$$
\begin{aligned}
R_i^x &= (m+s+h)^{-1}\Big(\max\left\{\bar{x}_{ijt} \;\middle|\; j=1,\ldots,n \text{ and } t=1,\ldots,T\right\} \\
&\quad -\min\left\{\bar{x}_{ijt} \;\middle|\; j=1,\ldots,n \text{ and } t=1,\ldots,T\right\}\Big)^{-1}, \\
R_r^g &= (m+s+h)^{-1}\Big(\max\left\{\bar{g}_{rjt} \;\middle|\; j=1,\ldots,n \text{ and } t=1,\ldots,T\right\} \\
&\quad -\min\left\{\bar{g}_{rjt} \;\middle|\; j=1,\ldots,n \text{ and } t=1,\ldots,T\right\}\Big)^{-1} \text{ and} \\
R_f^b &= (m+s+h)^{-1}\Big(\max\left\{\bar{b}_{fjt} \;\middle|\; j=1,\ldots,n \text{ and } t=1,\ldots,T\right\} \\
&\quad -\min\left\{\bar{b}_{fjt} \;\middle|\; j=1,\ldots,n \text{ and } t=1,\ldots,T\right\}\Big)^{-1}.
\end{aligned}
$$

Figure 26.1 visually describes the computational process, which incorporates DEA window analysis with the property of translation invariance, to examine a fuel mix strategy of industrial nations that incorporates the proposed data adjustment and the window analysis in a time horizon.

26.3.1 Formulations under Natural Disposability

Window Analysis: The unified efficiency measurement discussed in this chapter needs to first consider a period block, or PB, containing multiple periods in which a group of DMUs (J) are relatively compared by a distance between each observation and an efficiency frontier within PB. For example, in the case of the PB with three periods, all DMUs (J_t) in a window become $J_{PB(t)} = J_{t-2 \cup t-1 \cup t}$ for $t=3,\ldots,T$. Here, this chapter newly considers a use of window analysis that applies DEA to examine an efficiency change in a time series data within PB. Figure 26.2 visually discusses the structure of window

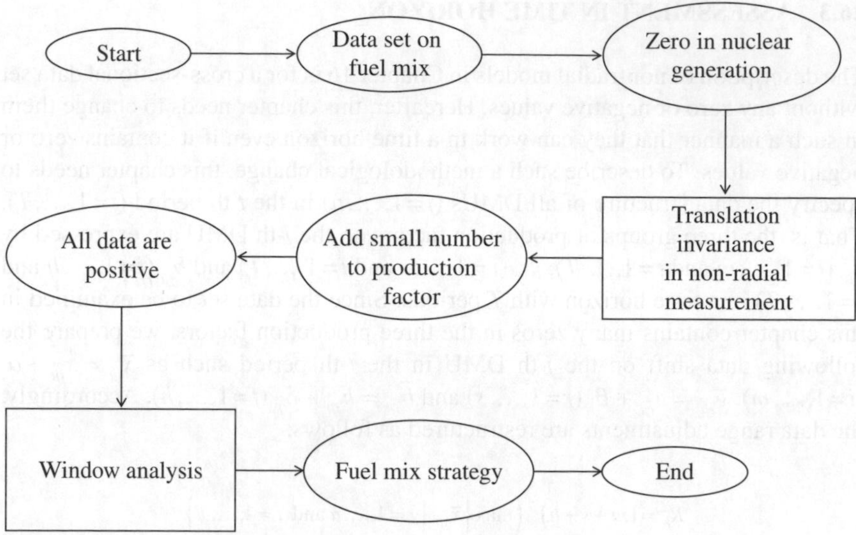

FIGURE 26.1 Computational process for fuel mix strategy (a) Many OECD nations do not have nuclear generation so that the data set used in this chapter contains many zeros. As a result, it is necessary for us to utilize the property of translation invariance. The property can be used only for the proposed non-radial model. The property indicates that an efficiency measure should be not influenced even if inputs and/or outputs are shifted toward a same direction by adding or subtracting a specific real number. The property makes it possible that we can evaluate the performance of DMUs, whose production factors contain many zeros and negative values in a data set. (b) The proposed window analysis is useful in dealing with a data set in a time horizon.

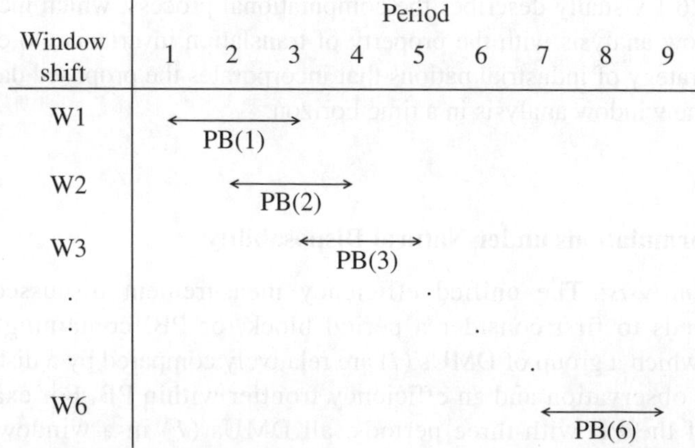

FIGURE 26.2 Shift of period block (PB) in window analysis (a) W stands for Window. W1, W2,…, W6 are related to the first, second,…, sixth window. PB stands for a period block in which we measure unified efficiency and efficiency growth. (b) The number within each set of parentheses indicates the order of window to express the corresponding PB.

analysis in the case of three periods. Note that the length of window is extendable more than the three periods.

Under the natural disposability, the unified efficiency of the k-th DMU in the t th period is measured by the following model for window analysis:

$$\begin{aligned}
&\text{Maximize} \quad \varepsilon\left(\sum_{i=1}^{m} R_i^x d_{it}^{x-} + \sum_{r=1}^{s} R_r^g d_{rt}^g + \sum_{f=1}^{h} R_f^b d_{ft}^b\right) \\
&s.t. \quad \sum_{j\in PB(t)} \bar{x}_{ijt}\lambda_{jt} + d_{it}^{x-} = \bar{x}_{ikt} \quad (k \in J_t \text{ and } i=1,\ldots,m), \\
&\qquad \sum_{j\in PB(t)} \bar{g}_{rjt}\lambda_{jt} - d_{rt}^g = \bar{g}_{rkt} \quad (k \in J_t \text{ and } r=1,\ldots,s), \\
&\qquad \sum_{j\in PB(t)} \bar{b}_{fjt}\lambda_{jt} + d_{ft}^b = \bar{b}_{fkt} \quad (k \in J_t \text{ and } f=1,\ldots,h), \\
&\qquad \sum_{j\in PB(t)} \lambda_{jt} = 1, \\
&\qquad \lambda_{jt} \geq 0 \ \left(j \in J_{PB(t)}\right), d_{it}^{x-} \geq 0 \ (i=1,\ldots,m \text{ and } t=3,\ldots,T), \\
&\qquad d_{rt}^g \geq 0 \ (r=1,\ldots,s \text{ and } t=3,\ldots,T) \text{ and} \\
&\qquad d_{ft}^b \geq 0 \ (f=1,\ldots,h \text{ and } t=3,\ldots,T),
\end{aligned} \tag{26.5}$$

where a group of DMUs, or $j \in PB(t)$, may consist of part of an efficiency frontier for the window with three periods. The k-th DMU to be evaluated is selected from the t th period. The value of ε, prescribed by a user, is used to control the magnitude of unified efficiency measured by Model (26.5). It is possible for Model (26.5) to drop the small number, but this chapter keeps it for controlling the degree of inefficiency measured by Equation (26.6).

A unified efficiency measure (UEN_{kt}^{NR}) of the k-th DMU in the t th period under natural disposability is determined by

$$UEN_{kt}^{NR} = 1 - \varepsilon\left(\sum_{i=1}^{m} R_i^x d_{it}^{x-*} + \sum_{r=1}^{s} R_r^g d_{rt}^{g*} + \sum_{f=1}^{h} R_f^b d_{ft}^{b*}\right). \tag{26.6}$$

The equation within the parenthesis, obtained from the optimality of Model (26.5), indicates the level of unified inefficiency under natural disposability. The unified efficiency is obtained by subtracting the level of inefficiency from unity. All the slacks related to the k-th DMU in the t th period are measured on the optimality of Model (26.5). This chapter adds the subscript "kt" on UEN to indicate a unified efficiency measure for the k-th DMU in the t th period, hereafter.

26.3.2 Formulations under Managerial Disposability

Window Analysis: Under managerial disposability, the unified efficiency of the k-th DMU in the t th period is measured by the following model:

$$
\begin{aligned}
&\text{Maximize } \varepsilon\left(\sum_{i=1}^{m} R_i^x d_{it}^{x+} + \sum_{r=1}^{s} R_r^g d_{rt}^g + \sum_{f=1}^{h} R_f^b d_{ft}^b\right) \\
&s.t. \quad \sum_{j \in PB(t)} \bar{x}_{ijt}\lambda_{jt} - d_{it}^{x+} = \bar{x}_{ikt}\ (k \in J_t \text{ and } i=1,\ldots,m), \\
&\qquad \sum_{j \in PB(t)} \bar{g}_{rjt}\lambda_{jt} - d_{rt}^{g} = \bar{g}_{rkt}\ (k \in J_t \text{ and } r=1,\ldots,s), \\
&\qquad \sum_{j \in PB(t)} \bar{b}_{fjt}\lambda_{jt} + d_{ft}^{b} = \bar{b}_{fkt}\ (k \in J_t \text{ and } f=1,\ldots,h), \qquad (26.7)\\
&\qquad \sum_{j \in PB(t)} \lambda_{jt} = 1, \\
&\qquad \lambda_{jt} \geq 0\left(j \in J_{PB(t)}\right), d_{it}^{x+} \geq 0\ (i=1,\ldots,m \text{ and } t=3,\ldots,T), \\
&\qquad d_{rt}^{g} \geq 0\ (r=1,\ldots,s \text{ and } t=3,\ldots,T) \text{ and} \\
&\qquad d_{ft}^{b} \geq 0\ (f=1,\ldots,h \text{ and } t=3,\ldots,T).
\end{aligned}
$$

A unified efficiency measure (UEM_{kt}^{NR}) of the k-th DMU in the t th period under managerial disposability is measured by

$$UEM_{kt}^{NR} = 1 - \varepsilon\left(\sum_{i=1}^{m} R_i^x d_{it}^{x+*} + \sum_{r=1}^{s} R_r^g d_{rt}^{g*} + \sum_{f=1}^{h} R_f^b d_{ft}^{b*}\right). \qquad (26.8)$$

The equation within the parenthesis, obtained from the optimality of Model (26.7), indicates the level of unified inefficiency under managerial disposability. The unified efficiency is obtained by subtracting the level of inefficiency from unity. All the slacks on the k-th DMU in the t th period are measured by the optimality of Model (26.7).

26.3.3 Efficiency Growth

It is necessary for us to specify the length of PB to initiate the proposed window analysis. There are many ways for specifying the length of PB. This chapter considers two specifications. One of the two specifications is to pool the data sets in all periods and treat them as cross-section data. A difficulty associated with the cross-section data may be due to the assumption that all DMUs can access future technology. For example, all countries in 1999 can access eco-technology for pollution mitigation in 2011. Such accessibility seems to be unrealistic. However, this type of assumption is often acceptable in a conventional energy sector with limited technology development. For example, the electricity, usually generated by a very large process industry, uses well-established technology so that the assumption on limited technology development may be acceptable in investigating the electric power industry.

The other specification, widely used in DEA, is to prepare a window that consists of observations in a few consecutive periods, or PB. In the analysis, this chapter shifts each window for PB to examine a change of efficiency scores. For example, the window consists of PB with three periods under the assumption that a DMU in the t th period can access recent technology in the previous two periods. The length of PB is determined by a DEA user. Shifting the location of PB (e.g., $t=3, 4,\ldots, T$), this chapter can measure a shift in unified efficiency scores of all DMUs within the t th window like a moving average method in forecasting.

Efficiency Growth (EG): The indexes of the specific k th DMU ($k=1,\ldots,n$) between the two periods are measured by the following manner under the two disposability concepts (N: natural and M: managerial), respectively:

$$EGN_{t-1}^{t} = UEN_{kt}^{NR} / UEN_{kt-1}^{NR} \text{ and} \qquad (26.9)$$
$$EGM_{t-1}^{t} = UEM_{kt}^{NR} / UEM_{kt-1}^{NR}.$$

Here, the superscript NR indicates a non radial measure. In the case of a pooled data set, this chapter changes the t th period from $t=2$ to $t=T$ where t stands for the period to be measured. It is possible for us to change the EG measurement from the pooled data to the window analysis. The window analysis with three periods, for example, follows the following computational steps:

Step 1: First, the initial window has three periods ($t=1$, 2 and 3), for example. Models (26.5) and (26.7) are applied to the data set in the window. The third period ($t=3$) becomes the first period to be measured by EG in the window analysis. All DMUs in $t=1$ and $t=2$ are selected in the window so that it can incorporate a possible occurrence of a retreat which may consist of an efficiency frontier in the window. Although $t=1$ and $t=2$ are incorporated into the window analysis, the two periods are not used in EG (the t–1 th and t th periods) in Equations (26.9), where the period starts from t–$1=3$. Thus, the fourth ($t=4$) is the initial period for EG measurement.

Step 2: Second, the window shifts to the next periods ($t=2$, 3 and 4). Models (26.5) and (26.7) are applied to the data sets in the window with a new period combination. In the second window, two periods ($t=2$ and 3) are included into the window analysis, but they are not used for EG in Equations (26.9) as the t–1 th and t th periods. The fifth ($t=5$) is the period to be measured for EG.

Step 3: Equation (26.9) for EG is used as follows: For example, the EG between $t=3$ and $t=4$ becomes $EGN_{3}^{4} = UEN_{k4}^{NR} / UEN_{k3}^{NR}$, where UEN_{k3}^{NR} is measured by the first window ($t=1$, 2 and 3) and UEN_{k4}^{NR} is measured by the second window ($t=2$, 3 and 4). The procedure on *UEN*, discussed here, is also applicable to *UEM*. Accordingly, EGM can be measured by the two windows, as well.

Step 4: The proposed window analysis continues until the PB attains the last period (T), so that the last window consists of the three periods ($t=T-2$, $T-1$ and T). Otherwise, go to Step 1 after changing the period to be measured by resetting $t=t+1$.

At the end of this section, it is important to note four concerns. First, the EG indexes indicate part of a social learning capability of the k-th DMU in the t th period. In shifting from the $t-1$ th period to the t th period, each DMU is expected to examine the performance measure by comparing it with those of the other DMUs in the $t-1$ th period. Thus, the DMU should adjust itself for better performance by utilizing its social learning capability. See Figure 15.1 about the social leaning capability in which DEA is incorporated as part of the learning capability. Second, the length of PB is subjectively selected by a DEA user. It is possible for us to select the PB with more than three periods. The best selection on PB has been not yet clearly explored until now. Third, the proposed window selection may have another alternative. For example, $EGN_3^4 = UEN_{k4}^{NR} / UEN_{k3}^{NR}$ can be measured within the second window ($t=2$, 3 and 4). In the case, however, it may be difficult for us to identify what produces a difference in the efficiency scores between the two periods. Therefore, this chapter proposes Step 3, as mentioned above, so that we can measure a change in unified efficiency between two PBs, so indicating a shift of efficiency. There is no theoretical discussion on the proposed approach. In the aspect, the Malmquist index measurement, discussed in Chapter 19, may be better than the window analysis because it can identify subcomponents of Malmquist index[3]. Thus, we can understand what produces a frontier shift of an efficiency frontier under two disposability concepts. Finally, the window analysis has more computational tractability than the Malmquist index measurement. Thus, there is a tradeoff between measurement accuracy and computational tractability on environmental assessment in a time horizon.

26.4 EFFICIENCY MEASUREMENT FOR FUEL MIX STRATEGY

This chapter examines 33 OECD nations that use the annual amount of electricity as a desirable output and the annual amount of CO_2 as an undesirable output. The inputs include the capacity of combustible generation, hydro generation, nuclear generation, pumped hydro generation and other renewable generations, including geothermal, solar, tide and wind. Descriptive statistics on a data set used in this chapter can be found in Sueyoshi and Goto (2015c).

All data are obtained from the International Energy Agency (IEA) world energy database. In particular, inputs are from "OECD – Net electricalcapacity,"

[3] See Chapter 19 for a description of "Malmquist index measurement in a time horizon." This approach is for time series analysis, but is different from the window analysis discussed in this chapter.

measured in MW (megawatt), the desirable output is from "OECD – Electricity/heat supply and consumption," measured in GWh (Gigawatt hour), and the undesirable output is from "Detailed CO_2 estimates," measured in million ton, where OECD stands for the Organization for Economic Co-operation and Development.

Data Accessibility: The supplementary section of the article: Sueyoshi and Goto (2015c) lists the data set used in this chapter.

Figure 26.3 visually summarizes the proposed computational process to identify the upper and lower bounds of a fuel mix strategy that is determined by comparing the performance of OECD nations. The international comparison may providc us with a guideline for the fuel mix strategy. An important feature of Figure 26.3 is that it identifies the fuel mix strategy from OECD nations whose unified efficiency scores and efficiency growth indexes are larger than the total averages (TA).

26.4.1 Unified Efficiency Measures

Pooled Data (1999–2011): The unified efficiency scores of 33 nations are measured under natural and managerial disposability. In this case, all observations (33 nations × 13 years = 429 observations) are pooled together as a single data set which is then solved as cross-section data by the proposed DEA approach. To focus our discussion on computational results by the window analysis, this chapter does not document a summary on the computational results, except noting that a summary on the pooled data is available in the article of Sueyoshi and Goto (2015c) along with a detailed description.

Window Analysis: Table 26.1 summarizes the unified efficiency scores of 33 nations under natural disposability which are measured by the proposed window analysis. No nation exhibits any major difference in their unified efficiency measures. All the nations exhibit a similar result and their unified efficiency measures are 0.995 on the total average at the bottom of Table 26.1.

A difference between their magnitudes can be found in Table 26.2, which summarizes the unified efficiency measures under managerial disposability. Efficient nations include Canada, Denmark, France, Germany, Italy, Japan, Norway, Spain, Sweden, Switzerland and the United States on average across the observed annual periods. The other nations have exhibited some level of inefficiency in their unified performance measures. The unified efficiency of 33 nations is 0.916 on the total average, as listed at the right-hand bottom of Table 26.2.

Tables 26.3 and 26.4, measured by window analysis, summarize the EGN indexes of 33 nations in the four annual periods (2002, 2005, 2008 and 2011) under natural disposability and these EGM indexes under managerial disposability. Table 26.3 does not exhibit any major difference among 33 nations.

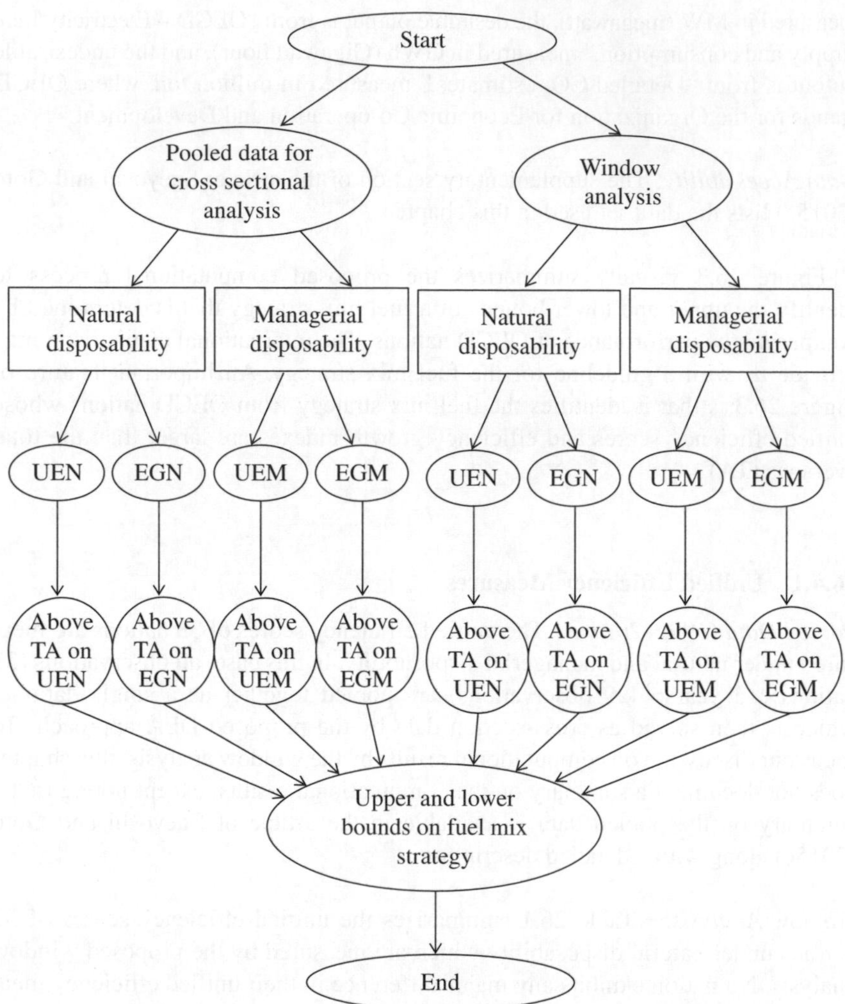

FIGURE 26.3 Upper and lower bounds of fuel mix strategy (a) TA stands for the total average that is, for example, listed at the right-hand bottom of Tables 26.1 to 26.4. Sueyoshi and Goto (2015c) lists the computational results on cross sectional analysis that uses a pooled data set. (b) This chapter does not describe on the left part (on the pooled data) of the figure because this chapter focuses upon a description on the window analysis. See Sueyoshi and Goto (2015c) that provides a detailed description on the cross section analysis.

Exceptions are Germany whose EGN index is 1.004 and the United States whose EGN index is 1.007 on average.

Table 26.4 documents their EGM indexes under managerial disposability. The nations, exhibiting EGM indexes more than unity, include Austria, Belgium,

TABLE 26.1 Unified efficiency under natural disposability: window analysis

	2002	2005	2008	2011	Average (2001–2011)
Australia	1.000	1.000	1.000	1.000	1.000
Austria	0.986	0.984	0.982	0.977	0.983
Belgium	1.000	1.000	1.000	1.000	1.000
Canada	1.000	1.000	1.000	1.000	0.998
Chile	1.000	1.000	0.999	0.998	1.000
Czech Republic	0.994	0.993	0.991	0.986	0.992
Denmark	1.000	1.000	1.000	1.000	1.000
Estonia	1.000	1.000	1.000	1.000	1.000
Finland	1.000	1.000	1.000	1.000	1.000
France	1.000	1.000	1.000	1.000	0.997
Germany	0.951	1.000	1.000	1.000	0.986
Greece	0.994	0.994	0.993	0.989	0.994
Hungary	1.000	1.000	1.000	1.000	1.000
Iceland	1.000	1.000	1.000	1.000	1.000
Ireland	1.000	0.998	0.998	0.997	0.998
Israel	1.000	1.000	1.000	1.000	1.000
Italy	1.000	1.000	1.000	1.000	1.000
Japan	0.994	1.000	0.985	0.965	0.993
Korea	1.000	1.000	1.000	1.000	0.999
Luxembourg	1.000	1.000	1.000	1.000	1.000
Mexico	1.000	1.000	1.000	1.000	1.000
Netherlands	1.000	1.000	1.000	1.000	1.000
Norway	0.994	1.000	1.000	0.999	0.999
New Zealand	1.000	1.000	1.000	1.000	1.000
Portugal	0.995	0.992	0.985	0.984	0.990
Slovak Republic	1.000	1.000	1.000	0.998	1.000
Slovenia	1.000	1.000	1.000	0.999	1.000
Spain	0.947	0.937	0.918	0.899	0.932
Sweden	1.000	1.000	1.000	1.000	0.999
Switzerland	1.000	1.000	1.000	0.996	0.999
Turkey	1.000	1.000	1.000	1.000	1.000
United Kingdom	1.000	1.000	0.997	1.000	1.000
United States	1.000	1.000	1.000	1.000	0.994
Average	0.996	0.997	0.995	0.994	0.995

(a) Source: Sueyoshi and Goto (2015c).

Finland, Hungary, Ireland, Portugal, Slovak Republic and the United Kingdom. The other nations do not exhibit any increase in their EGM indexes. The EGM index of the 33 nations is 1.000 on total average.

Here, it is important to note that the electric power industries of all OECD nations exhibit operational similarity because they have been operating under governmental

TABLE 26.2 Unified efficiency under managerial disposability: window analysis

	2002	2005	2008	2011	Average (2001–2011)
Australia	0.772	0.778	0.772	0.766	0.773
Austria	0.944	0.946	0.955	1.000	0.959
Belgium	0.852	0.864	0.882	0.883	0.868
Canada	1.000	1.000	1.000	1.000	1.000
Chile	0.889	0.880	0.823	0.803	0.855
Czech Republic	0.826	0.828	0.823	0.813	0.824
Denmark	1.000	1.000	1.000	1.000	1.000
Estonia	0.905	0.898	0.887	0.859	0.892
Finland	0.904	0.950	0.932	0.915	0.908
France	1.000	1.000	1.000	1.000	1.000
Germany	1.000	1.000	1.000	1.000	1.000
Greece	0.818	0.820	0.817	0.807	0.817
Hungary	0.877	0.902	0.890	0.897	0.890
Iceland	1.000	1.000	0.974	1.000	0.995
Ireland	0.862	0.879	0.878	0.883	0.877
Israel	0.811	0.811	0.805	0.790	0.807
Italy	1.000	1.000	1.000	1.000	0.996
Japan	0.998	1.000	1.000	1.000	1.000
Korea	0.825	0.821	0.824	0.801	0.819
Luxembourg	0.953	0.951	0.950	0.950	0.952
Mexico	0.811	0.813	0.813	0.793	0.810
Netherlands	0.833	0.852	0.841	0.803	0.835
Norway	1.000	1.000	1.000	1.000	1.000
New Zealand	0.937	0.924	0.920	0.933	0.930
Portugal	0.833	0.862	0.903	0.889	0.878
Slovak Republic	0.954	0.960	0.957	0.958	0.957
Slovenia	0.949	0.949	0.946	0.946	0.948
Spain	1.000	1.000	1.000	0.995	1.000
Sweden	0.998	1.000	1.000	1.000	1.000
Switzerland	1.000	1.000	1.000	1.000	1.000
Turkey	0.826	0.826	0.812	0.806	0.819
United Kingdom	0.815	0.815	0.816	0.821	0.816
United States	1.000	1.000	1.000	1.000	1.000
Average	0.915	0.919	0.916	0.912	0.916

(a) Source: Sueyoshi and Goto (2015c).

regulations so that they do not produce any major difference among them. However, the concern on operational similarity may be different in their environmental performance because different nations have different fuel mix policy for CO_2 emission control from electricity generations. This finding is important because it is possible for us to discuss Japanese future fuel mix strategy by comparing it with those of the other industrial nations, in particular high performance nations in terms of their unified efficiency measures and EGM indexes.

TABLE 26.3 Efficiency growth under natural disposability: window analysis

	2002	2005	2008	2011	Average (2002–2011)
Australia	1.000	1.000	1.000	1.000	1.000
Austria	1.000	1.000	0.997	0.995	0.999
Belgum	1.003	1.000	1.000	1.000	1.000
Canada	1.004	1.000	1.000	1.009	1.000
Chile	1.000	1.000	0.999	0.999	1.000
Czech Republic	0.994	0.999	0.998	1.000	0.999
Denmark	1.000	1.000	1.000	1.000	1.000
Estonia	1.000	1.000	1.000	1.000	1.000
Finland	1.000	1.000	1.000	1.000	1.000
France	1.000	1.000	1.004	1.012	1.000
Germany	0.986	1.000	1.000	1.000	1.004
Greece	1.000	0.998	0.999	0.997	0.999
Hungary	1.000	1.000	1.000	1.000	1.000
Iceland	1.000	1.000	1.000	1.000	1.000
Ireland	1.000	0.999	1.000	1.000	1.000
Israel	1.000	1.000	1.000	1.000	1.000
Italy	1.000	1.000	1.000	1.000	1.000
Japan	1.007	1.000	0.985	0.965	0.998
Korea	1.000	1.000	1.000	1.000	1.000
Luxembourg	1.000	1.001	1.000	1.000	1.000
Mexico	1.000	1.000	1.000	1.000	1.000
Netherlands	1.000	1.000	1.000	1.000	1.000
Norway	0.998	1.000	1.000	1.002	1.000
New Zealand	1.000	1.000	1.000	1.000	1.000
Portugal	0.999	0.999	0.998	0.997	0.999
Slovak Republic	1.000	1.000	1.000	0.998	1.000
Slovenia	1.000	1.000	1.000	0.999	1.000
Spain	0.994	0.996	0.994	0.955	0.994
Sweden	1.000	1.000	1.000	1.004	1.000
Switzerland	1.000	1.001	1.000	0.999	1.000
Turkey	1.001	1.000	1.000	1.000	1.000
United Kingdom	1.000	1.001	0.997	1.000	1.000
United States	1.065	1.000	1.000	1.000	1.007
Average	1.002	1.000	0.999	0.998	1.000

(a) Source: Sueyoshi and Goto (2015c).

26.4.2 Fuel Mix Strategy

Hereafter, this chapter considers the fuel mixes of 33 OECD nations, based upon which we discuss the Japanese future fuel mix strategy. This chapter clearly understands that different nations have different conditions for energy resources. For example, they are different in terms of self-sufficiency rates on their energy

TABLE 26.4 Efficiency growth under managerial disposability: window analysis

	2002	2005	2008	2011	Average (2002–2011)
Australia	0.993	0.995	1.000	0.998	0.999
Austria	1.005	1.003	1.006	1.000	1.006
Belgium	0.994	1.005	1.017	0.997	1.003
Canada	1.005	1.000	1.000	1.000	1.000
Chile	0.995	1.006	0.997	0.960	0.989
Czech Republic	1.003	1.001	1.001	0.988	0.999
Denmark	1.000	1.000	1.001	1.000	1.000
Estonia	1.000	0.996	1.004	0.972	0.995
Finland	0.988	1.066	1.037	1.009	1.001
France	1.000	1.000	1.000	1.000	1.000
Germany	1.000	1.000	1.000	1.000	1.000
Greece	1.000	1.001	1.000	0.989	0.999
Hungary	1.007	1.013	1.002	0.991	1.003
Iceland	1.000	1.000	1.001	1.000	1.000
Ireland	1.005	0.999	1.001	0.990	1.003
Israel	0.998	1.000	1.000	0.987	0.997
Italy	1.000	1.046	1.000	1.000	1.000
Japan	0.998	1.000	1.000	1.000	1.000
Korea	1.004	0.999	0.997	0.998	0.998
Luxembourg	0.994	0.999	1.000	1.000	0.999
Mexico	0.996	0.994	1.004	0.987	0.997
Netherlands	1.006	1.016	1.010	0.948	0.997
Norway	1.002	1.000	1.000	1.000	1.000
New Zealand	1.003	0.988	0.994	0.994	1.000
Portugal	0.977	0.992	1.017	0.954	1.004
Slovak Republic	1.002	1.000	0.998	0.998	1.001
Slovenia	0.998	0.999	0.999	0.998	1.000
Spain	1.000	1.000	1.000	0.995	0.999
Sweden	0.998	1.000	1.000	1.000	1.000
Switzerland	1.000	1.000	1.000	1.000	1.000
Turkey	1.006	0.997	0.998	0.989	0.998
United Kingdom	1.002	0.999	1.002	0.999	1.001
United States	1.000	1.000	1.000	1.000	1.000
Average	0.999	1.003	1.003	0.992	1.000

(a) Source: Sueyoshi and Goto (2015c).

supply capabilities. Thus, the desirable structure of fuel mixes may be different among nations. For instance, the fuel mix of the United States is different from that of Japan because the Japanese energy sufficiency rate is much lower than that of the United States.

For illustrative examples, Table 26.5 documents the fuel mixes (inputs) of nations, measured by their generation capacities, whose unified efficiency

TABLE 26.5 Fuel mix strategy under managerial disposability: window analysis

	Combustible	Hydro	Nuclear	Pumped Hydro	Renewables
Austria	26.9%	57.6%	0.0%	12.1%	3.3%
Canada	29.7%	58.2%	10.6%	0.2%	1.4%
Denmark	75.1%	0.1%	0.0%	0.0%	24.8%
France	18.1%	20.9%	53.2%	6.0%	1.9%
Germany	54.5%	7.4%	16.1%	4.2%	17.7%
Iceland	6.3%	74.5%	0.0%	0.0%	19.1%
Italy	64.7%	22.5%	0.0%	7.9%	4.9%
Japan	54.0%	17.6%	18.5%	9.8%	0.2%
Luxembourg	11.7%	44.2%	0.0%	43.0%	1.2%
Norway	1.0%	94.2%	0.0%	4.1%	0.8%
New Zealand	31.1%	60.9%	0.0%	0.0%	8.0%
Slovak Republic	30.5%	30.0%	28.7%	10.8%	0.0%
Slovenia	43.2%	32.2%	23.6%	1.0%	0.0%
Spain	46.2%	22.6%	9.3%	6.7%	15.2%
Sweden	19.5%	49.7%	28.2%	0.2%	2.5%
Switzerland	0.3%	72.8%	17.8%	9.0%	0.1%
United States	74.2%	10.6%	10.8%	2.3%	2.0%
Average	34.5%	39.8%	12.8%	6.9%	6.1%

(a) Source: Sueyoshi and Goto, (2015c) (b) See the numbers listed in the row related to [VI] in Table 26.6.

measures are more than the total average in Table 26.2 (managerial disposability and window analysis). The average fuel mix, expressed by percentages at the bottom of Table 26.5, indicates 34.5% by combustible fuel generation, 39.8% by hydro generation, 12.8% by nuclear generation, 6.9% by pumped hydro generation and 6.1% by renewable generation. These percentages consist of desirable fuel mixes determined by their unified efficiencies under managerial disposability. Their unified efficiency scores are measured by the proposed window analysis.

By combining the previous work of Sueyoshi and Goto (2015c) with the result of this chapter, we may discuss an international comparison that considers eight different approaches ($8=2\times2\times2$, where the first two are unified efficiency and EG index, the second two are natural and managerial disposability concepts and the third two are pooled data and window analysis). See Figure 26.3 that visually describes how to determine the upper and lower bounds concerning fuel mix strategy by the eight different approaches.

Table 26.6 summarizes the fuel mix strategies determined by the eight different approaches. The fuel mix strategies, expressed by percentages at the bottom of Table 26.6, indicate the range is 34.5–56.1% by combustible fuel generation, 22.4–40.5% by hydro generation, 10.4–13.7% by nuclear generation, 3.9–6.9% by pumped hydro generation and 3.7–8.4% by renewable generation.

TABLE 26.6 Fuel mix strategies measured by eight approaches

Approach		Combustible	Hydro	Nuclear	Pumped Hydro	Renewables
[I]		53.7%	26.5%	11.8%	4.2%	3.8%
[II]		34.9%	40.5%	12.9%	6.6%	5.1%
[III]		56.1%	24.8%	10.6%	4.0%	4.4%
[IV]		54.1%	22.4%	10.4%	4.7%	8.4%
[V]		51.6%	29.2%	11.3%	4.2%	3.7%
[VI]		34.5%	39.8%	12.8%	6.9%	6.1%
[VII]		55.0%	25.6%	10.9%	3.9%	4.6%
[VIII]		43.4%	32.8%	13.7%	4.4%	5.7%
Range	Max.	56.1%	40.5%	13.7%	6.9%	8.4%
	Min.	34.5%	22.4%	10.4%	3.9%	3.7%

(a) Source: Sueyoshi and Goto (2015c). (b) I: Unified efficiency measures under natural disposability (pooled data); II: Unified efficiency measures under managerial disposability (pooled data); III: Efficiency growth indexes under natural disposability (pooled data); IV: Efficiency growth indexes under managerial disposability (pooled data); V: Unified efficiency measures under natural disposability (window analysis); VI: Unified efficiency measures under managerial disposability (window analysis); VII: Efficiency growth indexes under natural disposability (window analysis); VIII: efficiency growth indexes under managerial disposability (window analysis).

26.5 SUMMARY

This chapter discussed the property of translation invariance that existed in the non-radial model for the proposed environmental assessment. This property indicates that an efficiency measure should be not influenced even if production factors are shifted toward the same direction by adding or subtracting a specific real number. The property makes it possible for us to evaluate the performance of DMUs, whose production factors contain many zeros and negative values in a data set. Such an occurrence of zero and negative values on production factors is widely observed in many DEA performance evaluations.

To document the practicality of the proposed approach, this chapter empirically investigated the policy feasibility of the Japanese future energy plan. On 1 June 2015, the Japanese government announced the new proposal of a future fuel mix that will consist of nuclear generation within a range of 20–22% and renewable generation within a range of 22–24% of total electricity generation by 2030. The difficulty in understanding the governmental claim is that it does not contain any scientific evidence concerning why Japan needs to maintain nuclear generation at the level of 20–22% and renewables at the level of 22–24%, respectively.

As a result of incorporating the property of translation invariance in the proposed DEA environmental assessment, this chapter has conducted a performance assessment on 33 OECD nations during 1999–2011 whose production factors contain many zeros in the observed annual periods. This chapter found that Japan should use 34.5–56.1% combustible fuel generation, 22.4–40.5% hydro generation, 10.4–13.7% nuclear generation, 3.9–6.9% pumped hydro generation and 3.7–8.4% renewable generation, as measured by their generation capacities. It is easily imagined that the finding is applicable to not only Japan but also the other industrial nations as a benchmark measure for future energy planning.

It may be difficult for us to directly compare between the fuel mix proposed by the Japanese government (as determined by generation amounts) and that discussed in this chapter (as measured by generation capacities). However, it is easily envisioned that the governmental future energy plan seems to be inconsistent with the range estimates identified by the proposed DEA approach. The international comparison between Japan and the other OECD nations indicates expected constraints from technology and political difficulties after the disaster of the Fukushima Daiichi nuclear power plant[4–6]. The comparison implies that the Japanese future energy plan is too ambitious to attain the policy goal. Thus, it is easily imagined that the energy plan will have considerable difficulty in its future implementation.

As a policy suggestion for enhancing the implementation capability of the Japanese future energy plan, this chapter proposes increasing the amount of combustible fuel generation by utilizing new technologies such as clean coal technologies and shifting to fuels (e.g., natural gas) with lower CO_2 emission. See Franco and Diaz (2009)[7], Kunze and Spliethoff (2010)[8] and Falcke *et al.* (2011)[9] for their descriptions of clean coal technology, a technology which may become a high potential energy source in future. See http://www.jcoal.or.jp/eng/. Unfortunately, the commercialization of clean coal technology is very limited at the current moment.

[4] Boccard, N. (2014) The cost of nuclear electricity: France after Fukushima. *Energy Policy*, **66**, 450–461.

[5] Goto, M. and Sueyoshi, T. (2015) Electric power market reform in Japan after Fukushima: Issues and future direction. *International Journal of Energy Sector Management*, **9**, 336–360.

[6] Goto, M. and Sueyoshi, T. (2016) Electricity market reform in Japan after Fukushima: Implications for climate change policy and energy market developments. *Economics of Energy and Environmental Policy*, **5**, 15–30.

[7] Franco, A. and Diaz, A.R. (2009) The future challenges for "clean coal technologies": Joining efficiency increase and pollutant emission control. *Energy*, **34**, 348–354.

[8] Kunze, C. and Spliethoff, H. (2010) Modelling of an IGCC plant with carbon capture for 2020. *Fuel Processing Technology*, **91**, 934–941.

[9] Falcke, T.J., Hoadley, A.F.A., Brennan, D.J. and Sinclair, S.E. (2011) The sustainability of clean coal technology: IGCC with/without CGS. *Process Safety and Environmental Protection*, **89**, 41–52.

27

HANDLING ZERO AND NEGATIVE VALUES IN RADIAL MEASUREMENT

27.1 INTRODUCTION

As first discussed in Chapter 2, DEA has a computational difficulty when handling a data set that contains zero and negative values. See Wang *et al.* (2014) for a description of such a difficulty in DEA environmental assessment. The purpose of this chapter[1] is to develop a new approach by which we can measure the level of corporate sustainability even if a data set contains many zero and negative values[2]. Thus, the proposed approach can improve the analytical capability of DEA environmental assessment that has been never explored by the previous chapters.

[1] This chapter is partly based upon the article: Sueyoshi, T. and Yuan, Y. (2015b) Comparison among U.S. industrial sectors by DEA environmental assessment: Equipped with analytical capability to handle zero or negative in production factors. *Energy Economics*, 52, 69–86.

[2] DEA usually treats a data set under the condition that it does not contain zero and negative values. Data sets with non-positivity (e.g. economic growth and net income) can be found in many financial and economic analyses. See, for example, Emrouznejad, A. and Amin, G.R. (2009) DEA models for a ratio data: convexity consideration. *Applied Mathematical Modelling*, 33, 486–498. The conventional use of DEA needs an additional treatment to handle non-positive values in a data set. The problem of their approach was that it could not be used for environmental assessment because it did not consider a computational scheme to deal with undesirable outputs. To overcome the methodological difficulty, Chapter 26 has applied the property of translation invariance in non-radial measurement. This chapter, following Sueyoshi and Yuan (2015b), documents a research effort to discuss how to handle zero and negative values within not only non-radial but also radial measurements.

Environmental Assessment on Energy and Sustainability by Data Envelopment Analysis,
First Edition. Toshiyuki Sueyoshi and Mika Goto.

Here, it is necessary for us to note that Chapter 26 has already discussed how to handle zero and negative values by the property of translation invariance in non-radial measurement. Unfortunately, the proposed approach does not work for radial measurement. Therefore, this chapter develops an extended approach to handle zero and negative values for both radial and non-radial measurements. The proposed approach, not depending upon such a desirable property, can handle zero and negative values in production factors. It is easily envisioned that the proposed approach will be able to enhance the quality of strategic implications derived by DEA because the analytical capability to handle zero and negative values is a necessity for environmental assessment in modern business. For example, negative values often exist in financial performance measures (e.g., return on assets: ROA; and return on equity: ROE). The DEA environmental assessment applied to private sectors needs the proposed analytical ability to discuss their corporate sustainability.

The remainder of this chapter is organized as follows. Section 27.2 discusses how to handle zero and negative values in production factors. Section 27.3 describes a unification process between two disposability concepts with non-positive vales. in data. Section 27.4 extends the discussion by considering how to incorporate eco-technology innovation to reduce the amount of undesirable outputs. Section 27.5 applies the proposed approach to evaluate the unified (environmental and operational) performance of Standard and Poors top 500 (S&P 500) companies in the United States. This chapter also summarizes empirical results obtained from the proposed application. Section 27.6 summarizes this chapter.

27.2 DISAGGREGATION

A difficulty of all the DEA models discussed in the preceding chapters, except Chapters 11, 25 and 26, is that it is necessary for us to assume strict positivity in an observed data set. Chapters 11 and 25 are related to DEA-DA and Chapter 26 uses the property of translation invariance so that they can handle zero and negative values. To avoid the assumption of strict positivity in all observations in a data set, let us consider a case, for example, in which the desirable outputs of some DMUs contain negative values. The description on negative values in desirable outputs is extendable to inputs and undesirable outputs. As mentioned previously, zero and negative values can be often observed in most of financial measures (e.g., net income in this chapter). Thus, there is a high likelihood that desirable outputs may have zero and negative values.

Disaggregation Procedure: To handle the zero and negative values in a data set, this chapter disaggregates the observed value of the r-th desirable output (g_{rj}) on the j-th DMU into positive and negative parts by the following procedure:

$$g_{rj}^{+} = \max\left(g_{rj}, \varepsilon_s\right) \text{ and } g_{rj}^{-} = \max\left(g_{rj}, \varepsilon_s\right) - g_{rj} + \varepsilon_s \text{ for all } r \text{ and } j. \quad (27.1)$$

Equations (27.1) implies the following relationships between the observations on desirable outputs:

$$\begin{aligned}
&\text{(a) If } g_{rj} > 0, \text{ then } g_{rj}^{+} = g_{rj} > 0 \text{ and } g_{rj}^{-} = \varepsilon_s > 0,\\
&\text{(b) If } g_{rj} = 0, \text{ then } g_{rj}^{+} = \varepsilon_s > 0 \text{ and } g_{rj}^{-} = 2\varepsilon_s > 0 \text{ and}\\
&\text{(c) If } g_{rj} < 0, \text{ then } g_{rj}^{+} = \varepsilon_s > 0 \text{ and } g_{rj}^{-} = 2\varepsilon_s + |g_{rj}| > 0.
\end{aligned} \tag{27.2}$$

The proposed disaggregation has five important features. First, the two parts of g_{rj} are always positive as the result of incorporating Equations (27.2) into them. Second, ε_s is a very small number (e.g., 0.001 in this chapter) so that it is acceptable for us to consider $\varepsilon_s \simeq 2\varepsilon_s$, because they do not have any large difference in terms of their magnitudes. The symbol ($\simeq$) implies that the two parts are very close in their magnitudes. Third, the second case ($g_{rj} = 0$) uses $g_{rj}^{-} = 2\varepsilon_s$, not $g_{rj}^{-} = \varepsilon_s$, so that we can avoid data independency. See Chapter 2 on the data independency. Fourth, it is important for us to use a common sense when we apply the proposed disaggregation to real assessments. For example, a single DMU has an extremely large negative value (e.g., –1 billion) in its production factor and the other DMUs have a positive data range between 10 and 20 in each factor. In this case, it is necessary for us to drop the single DMU because it is an outlier distorting the shape of the efficiency frontier. In other words, the existence of an outlier may request a special investigation on the data in production factors. Finally, the most serious issue, to be explored in the near future, is that it is necessary for us to develop the theoretical justification for the proposed data disaggregation. The purpose of Equation (27.2) is that it provides strictly positive values even if the original data set contains zero and negative values. Such a research direction is acceptable as a "quick and easy" approach, but it lacks theoretical support for the proposed aggregation, as found in the property of translation invariance.

27.3 UNIFIED EFFICIENCY MEASUREMENT

27.3.1 Conceptual Review of Disposability Unification

Returning to Chapter 15, this chapter duplicates Figure 15.6 by considering our conceptual framework for environmental assessment. Such an effort is visualized in Figures 27.1, 27.2 and 27.3, all of which are used for the proposed formulations in this chapter.

As first discussed in Chapter 15, DEA environmental assessment separates outputs into desirable and undesirable categories. The output separation cannot be found in the conventional use of DEA. Chapter 24 introduces two types of input classification under natural and managerial disposability concepts. A problem to be overcome after the output and input separations is that this chapter must consider a conceptual framework and related formulations for unifying the

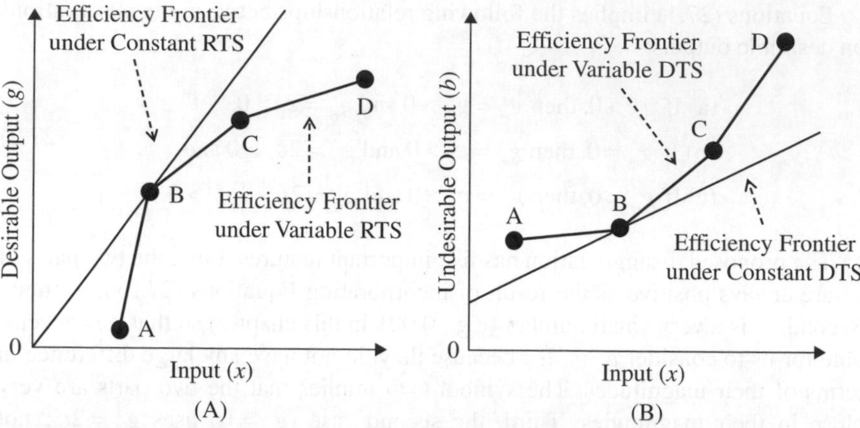

FIGURE 27.1 Efficiency frontiers for desirable and undesirable outputs (Stage I) (a) The figures do not consider the existence of zero or negative values. (b) Source: Figure 15.6. The figure corresponds to Stage I. This chapter does not list Stage II. (c) This chapter adds efficiency frontiers under constant returns to scale (RTS) and damages to scale (DTS).

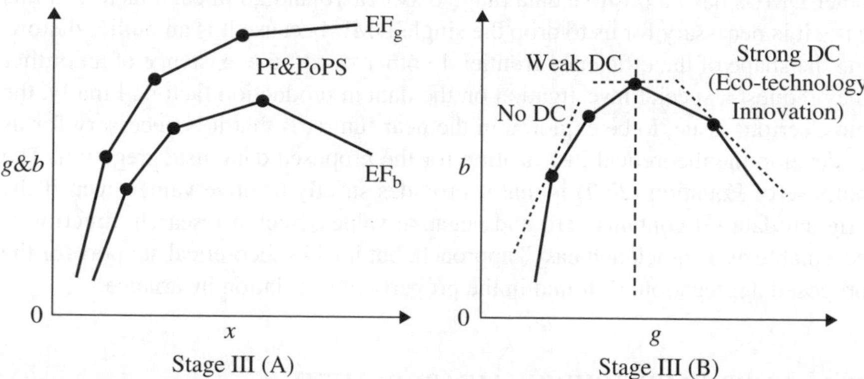

FIGURE 27.2 Efficiency frontiers under disposability unification (Stage III) (a) The right-hand side of the figure assumes a same amount of an input, so not considering the input. The figure does not consider the existence of zero or negative values in all production factors. (b) Source: Figure 15.6. (c) This chapter uses the byproduct assumption that an undesirable output is a byproduct of a desirable output. (d) Pr&PoPS stands for a production and pollution possibility set. DC indicates the occurrence of desirable congestion, or eco-technology innovation and managerial challenges to reduce the amount of undesirable outputs.

desirable and undesirable outputs by using a data set with zero and negative values. That is the purpose of this chapter.

Figures 27.1, 27.2 and 27.3, originating from Figure 15.6, visually reorganize such a unification process within the framework of DEA environmental assessment. For our visual convenience, the three figures discuss the case of a single component of three production factors.

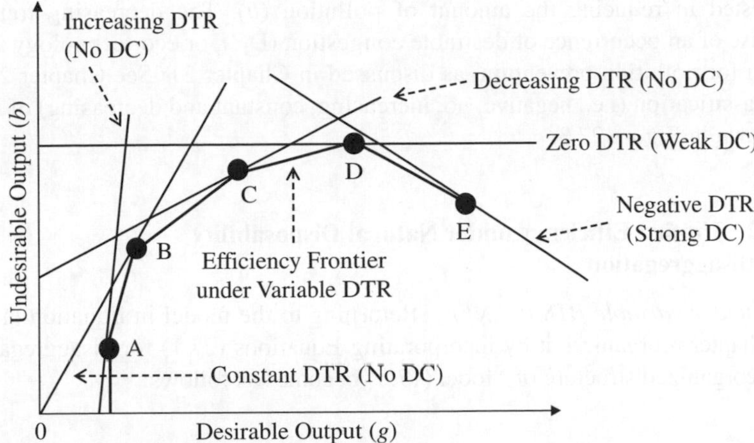

FIGURE 27.3 Efficiency frontier and damages to return (Stage III) (a) This figure does not consider the existence of zero or negative values. This chapter attempts to visually describe the type of DC. (b) This chapter uses the by-product assumption that an undesirable output is a byproduct of a desirable output. (c) DTR stands for damages to return and DC stands for desirable congestion.

Stage (I) in Figure 27.1 is separated into its two sub-sections: (A) and (B). Stage I (A) indicates a production relationship between an input (x) and a desirable output (g) under the assumption that all DMUs produce the same amount of undesirable output (b). The production possibility set locates below the efficiency frontier ({A}-{B}-{C}-{D}) in the x–g space. Stage (I) has the other sub-stage (B). A pollution possibility set locates above the efficiency frontier ({A}-{B}-{C}-{D}) in the x–b space under the assumption that they produce the same amount of desirable output (g). An important feature of Stage (I) is that the production possibility set (PrPS) in (A) is independently considered from the pollution possibility set (PoPS) in (B). The left-hand side (A) depicts two types of efficiency frontiers under variable and constant returns to scale (RTS), while the right-hand side (B) depicts two types of efficiency frontiers under variable and constant damages to scale (DTS). See Chapter 20 for a detailed description on RTS and DTS in DEA environmental assessment.

To unify the two sets related to the first stage (I), Stage III of Figure 27.2 combines them in a unified framework. The horizontal and vertical coordinates on the left-hand side of Figure 27.2 indicate x and g & b, respectively. The unification process makes it possible to identify a production and pollution possibility set (Pr & PoPS) between the two efficiency frontiers; EF_g and EF_b. The third stage (III) incorporates the assumption that undesirable outputs are byproducts of desirable outputs. Here, it is important to note that a desirable output (g) should have an increasing trend along with an increase of input (x). However, an undesirable output should have an increasing and decreasing trend because we are interested in reducing the amount of pollution (*b*). The decreasing trend is because of an occurrence of desirable congestion (DC), or eco-technology innovation for pollution prevention, as discussed in Chapter 21. See Chapter 23 on the classification (i.e., negative, no, increasing, constant and decreasing) of DTR as shown in Figure 27.3.

27.3.2 Unified Efficiency under Natural Disposability with Disaggregation

UEN under variable RTS (UEN_v^R): Returning to the model in Equation (17.5), this chapter reorganizes it by incorporating Equations (27.1) for disaggregation. The reorganized structure of Model (17.5) becomes as follows:

$$\begin{aligned}
\text{Maximize} \quad & \xi + \varepsilon_s \left[\sum_{i=1}^{m} R_i^x d_i^{x-} + \sum_{r=1}^{s} R_r^g (d_r^{g+} + d_r^{g-}) + \sum_{f=1}^{h} R_f^b d_f^b \right] \\
s.t. \quad & \sum_{j=1}^{n} x_{ij}\lambda_j + d_i^{x-} = x_{ik} \ (i=1,\ldots,m), \\
& \sum_{j=1}^{n} g_{rj}^{+}\lambda_j - d_r^{g+} - \xi g_{rk}^{+} = g_{rk}^{+} \ (r=1,\ldots,s), \\
& \sum_{j=1}^{n} g_{rj}^{-}\lambda_j + d_r^{g-} = g_{rk}^{-} \ (r=1,\ldots,s), \\
& \sum_{j=1}^{n} b_{fj}\lambda_j + d_f^{b} + \xi b_{fk} = b_{fk} \ (f=1,\ldots,h), \\
& \sum_{j=1}^{n} \lambda_j = 1, \\
& \xi : \text{URS}, \ \lambda_j \geq 0 \ (j=1,\ldots,n), \ d_i^{x-} \geq 0 \ (i=1,\ldots,m), \\
& d_r^{g+} \geq 0 \ (r=1,\ldots,s), \ d_r^{g-} \geq 0 \ (r=1,\ldots,s) \text{ and} \\
& d_f^{b} \geq 0 \ (f=1,\ldots,h),
\end{aligned} \tag{27.3}$$

where $d_r^{g+} = \sum_{j=1}^{n} g_{rj}^{+}\lambda_j - (1+\xi) g_{rk}^{+}$ $(r=1,\ldots,s)$ and $d_r^{g-} = g_{rk}^{-} - \sum_{j=1}^{n} g_{rj}^{-}\lambda_j$ $(r=1,\ldots,s)$. Model (27.3), corresponding to Stage I (A) of Figure 27.1, incorporates an analytical capability related to Equations (27.1) so that it can handle zero and negative values in desirable outputs. Each slack indicates the distance between the efficiency frontier and an observed value of the *r*-th desirable output. It is easily thought that an efficiency frontier consists of desirable outputs with positive values, not negative values. That is, the efficiency frontier indicates an upper frontier surface, consisting of efficient DMUs with positive desirable output(s), but not a lower frontier surface, consisting of DMUs with negative desirable output(s). Therefore, Model (27.3), along with the proceeding models, assigns the inefficiency score on desirable outputs with positive values.

The degree of UEN_v^R of the *k*-th DMU, where the superscript R indicates radial measurement and the subscript *v* indicates variable RTS and DTS, is measured by

$$UEN_v^R = 1 - \left[\xi^* + \varepsilon_s \left(\sum_{i=1}^{m} R_i^x d_i^{x-*} + \sum_{r=1}^{s} R_r^g (d_r^{g+*} + d_r^{g-*}) + \sum_{f=1}^{h} R_f^b d_f^{b*} \right) \right], \tag{27.4}$$

where the inefficiency score and all slack variables are identified on the optimality of Model (27.3). The equation within the outside parenthesis, obtained from Model (27.3), indicates the level of unified inefficiency under natural disposability. The degree of UEN_v^R is obtained by subtracting the level of inefficiency from unity.

UEN under constant RTS (UEN_c^R): To attain the status of constant RTS, this chapter drops the condition $\left(\sum_{j=1}^{n} \lambda_j = 1\right)$ from Model (27.3) and then measures the unified efficiency by

$$UEN_c^R = \text{Equation (27.4)}, \tag{27.5}$$

where the optimal solution is obtained from Model (27.3) without $\sum_{j=1}^{n} \lambda_j = 1$. The subscript *c* indicates constant RTS and DTS.

Scale Efficiency Measure under Natural Disposability (SEN^R): To examine how each DMU carefully manages the operational size under natural disposability, we measure the degree of SEN^R by the following equation:

$$SEN^R = UEN_c^R / UEN_v^R. \tag{27.6}$$

Since $UEN_c \le UEN_v$, the scale efficiency measure is less than or equal to unity. The higher score in *SEN* indicates the better scale management under natural disposability.

27.3.3 Unified Efficiency under Managerial Disposability with Disaggregation

UEM under variable DTS (UEM_v^R): The following radial formulation under managerial disposability, corresponding to Model (17.10), can handle zero and negative values on desirable outputs:

$$\begin{aligned}
&\text{Maximize } \xi+\varepsilon_s\left[\sum_{i=1}^{m}R_i^x d_i^{x+}+\sum_{r=1}^{s}R_r^g(d_r^{g+}+d_r^{g-})+\sum_{f=1}^{h}R_f^b d_f^b\right]\\
&s.t. \quad \sum_{j=1}^{n}x_{ij}\lambda_j - d_i^{x+} = x_{ik}\ (i=1,\ldots,m),\\
&\sum_{j=1}^{n}g_{rj}^{+}\lambda_j - d_r^{g+} - \xi g_{rk}^{+} = g_{rk}^{+}\ (r=1,\ldots,s),\\
&\sum_{j=1}^{n}g_{rj}^{-}\lambda_j + d_r^{g-} = g_{rk}^{-}\ (r=1,\ldots,s),\\
&\sum_{j=1}^{n}b_{fj}\lambda_j + d_f^{b} + \xi b_{fk} = b_{fk}\ (f=1,\ldots,h),\\
&\sum_{j=1}^{n}\lambda_j = 1,\\
&\xi:\text{URS},\ \lambda_j\ge 0\ (j=1,\ldots,n),\ d_i^{x+}\ge 0\ (i=1,\ldots,m),\\
&d_r^{g+}\ge 0\ (r=1,\ldots,s),\ d_r^{g-}\ge 0\ (r=1,\ldots,s)\text{ and}\\
&d_f^{b}\ge 0\ (f=1,\ldots,h).
\end{aligned} \tag{27.7}$$

Model (27.7) corresponds to Stage I (B) of Figure 27.1. An important feature of Model (27.7) is that it incorporates the disaggregation procedure in Equations (27.1).

The level of unified efficiency of the k-th DMU under managerial disposability is as follows:

$$UEM_v^R = 1-\left[\xi^*+\varepsilon_s\left(\sum_{i=1}^{m}R_i^x d_i^{x+*}+\sum_{r=1}^{s}R_r^g(d_r^{g+*}+d_r^{g-*})+\sum_{f=1}^{h}R_f^b d_f^{b*}\right)\right], \tag{27.8}$$

where the inefficiency score and all slack variables are identified on the optimality of Model (27.7). The equation within the parenthesis, obtained from Model (27.7), indicates a level of unified inefficiency under managerial disposability. The degree of UEM_v^R is obtained by subtracting the level of inefficiency from unity.

UEM under constant DTS (UEM_c^R): To attain the status of constant DTS, this chapter drops the condition $\left(\sum_{j=1}^{n}\lambda_j=1\right)$ from Model (27.7) and measures the level of unified efficiency by

$$UEM_c^R = \text{Equation (27.8)}, \tag{27.9}$$

where the optimal solution is obtained from Model (27.7) without $\sum_{j=1}^{n}\lambda_j=1$.

Scale Efficiency Measure under Managerial Disposability (SEM^R): To examine how each DMU carefully manages its operational size under managerial disposability, this study measures the degree of scale efficiency by

$$SEM^R = UEM_c^R / UEM_v^R. \tag{27.10}$$

Since $UEM_c^R \le UEM_v^R$, the scale efficiency is less than or equals unity. The higher score in *SEM* indicates the better scale management under managerial disposability.

27.4 POSSIBLE OCCURRENCE OF DESIRABLE CONGESTION

27.4.1 Unified Efficiency under Natural and Managerial Disposability (UENM)

As discussed in Chapter 24, the straightforward unification between Models (27.3) and (27.7) leads to the following model to measure the level of $UENM_v^R$, equipped with the disaggregation capability:

$$
\begin{aligned}
&\text{Maximize} \quad \xi + \varepsilon_s \left[\sum_{i=1}^{m^-} R_i^x d_i^{x-} + \sum_{q=1}^{m^+} R_q^x d_q^{x+} + \sum_{r=1}^{s} R_r^g (d_r^{g+} + d_r^{g-}) + \sum_{f=1}^{h} R_f^b d_f^b \right] \\
&s.t. \quad \sum_{j=1}^{n} x_{ij}^- \lambda_j + d_i^{x-} = x_{ik}^- \ (i=1,\ldots,m^-), \\
&\quad \sum_{j=1}^{n} x_{qj}^+ \lambda_j - d_q^{x+} = x_{qk}^+ \ (q=1,\ldots,m^+), \\
&\quad \sum_{j=1}^{n} g_{rj}^+ \lambda_j - d_r^{g+} - \xi g_{rk}^+ = g_{rk}^+ \ (r=1,\ldots,s), \\
&\quad \sum_{j=1}^{n} g_{rj}^- \lambda_j + d_r^{g-} = g_{rk}^- \ (r=1,\ldots,s), \\
&\quad \sum_{j=1}^{n} b_{fj} \lambda_j + d_f^b + \xi b_{fk} = b_{fk} \ (f=1,\ldots,h), \\
&\quad \sum_{j=1}^{n} \lambda_j = 1, \\
&\quad \lambda_j \ge 0 \ (j=1,\ldots,n), \ \xi : \text{URS}, \ d_i^{x-} \ge 0 \ (i=1,\ldots,m^-), \\
&\quad d_q^{x+} \ge 0 \ (q=1,\ldots,m^+), \ d_r^{g+} \ge 0 \ (r=1,\ldots,s), \\
&\quad d_r^{g-} \ge 0 \ (r=1,\ldots,s) \text{ and } d_f^b \ge 0 \ (f=1,\ldots,h).
\end{aligned}
\tag{27.11}
$$

The level of unified efficiency of the *k*-th DMU under natural and managerial disposability is as follows:

$$UENM_v^R = 1 - \left[\xi^* + \varepsilon_s \left(\sum_{i=1}^{m^-} R_i^x d_i^{x-*} + \sum_{q=1}^{m^+} R_q^x d_q^{x+*} + \sum_{r=1}^{s} R_r^g (d_r^{g+*} + d_r^{g-*}) + \sum_{f=1}^{h} R_f^b d_f^{b*} \right) \right]. \tag{27.12}$$

where the inefficiency measure and all slack variables are identified on the optimality of Model (27.11). The equation within the parenthesis, obtained from Model (27.11), indicates the level of unified inefficiency under natural and managerial disposability. The $UENM_v^R$ is obtained by subtracting the level of unified inefficiency from unity. Note that Model (27.11) corresponds to Stage II of Figure 15.6. This chapter does not depict it because we are interested in the reduction of undesirable outputs.

UENM under constant RTS and DTS ($UENM_c^R$): To attain the status of constant RTS and DTS, this chapter drops the condition $\left(\sum_{j=1}^{n} \lambda_j = 1\right)$ from Model (27.11) and measures the level of unified efficiency by

$$UENM_c^R = \text{Equation (27.12)}, \tag{27.13}$$

where the optimal solution is obtained from Model (27.11) without $\sum_{j=1}^{n} \lambda_j = 1$.

Scale Efficiency Measure under Natural and Managerial Disposability ($SENM^R$): To examine how each DMU carefully manages its operational size under natural and managerial disposability, this study measures the degree of its scale efficiency by

$$SENM^R = UENM_c^R / UENM_v^R. \tag{27.14}$$

Since $UENM_c^R \leq UENM_v^R$, the scale efficiency is less than or equals unity. The higher score in $SENM^R$ indicates the better scale management under natural and managerial disposability.

27.4.2 UENM with Desirable Congestion

A problem of Model (27.11) is that it does not incorporate a possible occurrence of DC, or eco-technology innovation for pollution mitigation. This chapter reformulates Model (27.11) so that it can measure the unified efficiency with a possible occurrence of DC. The reformulated model becomes

$$
\begin{aligned}
&\text{Maximize} \quad \xi + \varepsilon_s \left(\sum_{i=1}^{m^-} R_i^x d_i^{x-} + \sum_{q=1}^{m^+} R_q^x d_q^{x+} + \sum_{f=1}^{h} R_f^b d_f^b \right) \\
&\text{s.t.} \quad \sum_{j=1}^{n} x_{ij}^- \lambda_j + d_i^{x-} = x_{ik}^- \left(i=1,\dots,m^-\right), \\
&\sum_{j=1}^{n} x_{qj}^+ \lambda_j - d_q^{x+} = x_{qk}^+ \left(q=1,\dots,m^+\right), \\
&\sum_{j=1}^{n} g_{rj}^+ \lambda_j - \xi g_{rk}^+ = g_{rk}^+ \left(r=1,\dots,s\right), \\
&\sum_{j=1}^{n} g_{rj}^- \lambda_j = g_{rk}^- \left(r=1,\dots,s\right), \qquad (27.15) \\
&\sum_{j=1}^{n} b_{fj} \lambda_j - d_f^b = b_{fk} \left(f=1,\dots,h\right), \\
&\sum_{j=1}^{n} \lambda_j = 1, \\
&\lambda_j \ge 0 \ \left(j=1,\dots,n\right),\ \xi : \text{URS},\ d_i^{x-} \ge 0 \ \left(i=1,\dots,m^-\right), \\
&d_q^{x+} \ge 0 \ \left(q=1,\dots,m^+\right) \text{ and } d_f^b \ge 0 \ \left(f=1,\dots,h\right).
\end{aligned}
$$

Model (27.15) corresponds to Stage III of Figure 27.3 that unifies natural and managerial disposability concepts under the assumption that undesirable outputs are byproducts of desirable outputs.

The model has four unique features. First, the assumption on byproducts changes the sign of d_f^b from positive in Model (27.11) to negative $-d_f^b$ in Model (27.15). Second, Model (27.15) incorporates the possible occurrence of DC, as depicted in Figure 27.3, because the constraints related to desirable outputs (G) are expressed by equality (so, no slacks) in the formulation. Third, the formulation incorporates the disaggregation capability related to desirable outputs to deal with zero and negative values in a data set. Finally, Model (27.15) drops the inefficiency score related to undesirable outputs in the formulation because of the assumption on byproducts. Thus, it is possible for us to measure the level of unified inefficiency based upon both the inefficiency score on desirable outputs (implicitly including undesirable outputs) and slacks on inputs and undesirable outputs.

The level of unified efficiency of the k-th DMU under natural and managerial disposability is measured by

$$UENM\left(DC\right)_{v}^{R}=1-\left[\xi^{*}+\varepsilon_{s}\left(\sum_{i=1}^{m^{-}}R_{i}^{x}d_{i}^{x-*}+\sum_{q=1}^{m+}R_{q}^{x}d_{q}^{x+*}+\sum_{f=1}^{h}R_{f}^{b}d_{f}^{b*}\right)\right], \quad (27.16)$$

where the inefficiency score and all slack variables are identified on the optimality of Model (27.15). The equation within the parenthesis, obtained from Model (27.15), indicates a level of unified inefficiency with a possible occurrence of DC. The level of the unified efficiency is obtained by subtracting the level of inefficiency from unity.

The dual formulation of Model (27.15) becomes as follows:

$$\begin{aligned}
&\text{Minimize} \sum_{i=1}^{m^{-}}v_{i}^{-}x_{ik}^{-}-\sum_{q=1}^{m^{+}}v_{q}^{+}x_{qk}^{+}+\sum_{r=1}^{s}u_{r}^{+}g_{rk}^{+}+\sum_{r=1}^{s}u_{r}^{-}g_{rk}^{-}-\sum_{f=1}^{h}w_{f}b_{fk}+\sigma \\
&s.t. \quad \sum_{i=1}^{m^{-}}v_{i}^{-}x_{ij}^{-}-\sum_{q=1}^{m^{+}}v_{q}^{+}x_{qj}^{+}+\sum_{r=1}^{s}u_{r}^{+}g_{rj}^{+}+\sum_{r=1}^{s}u_{r}^{-}g_{rj}^{-}-\sum_{f=1}^{h}w_{f}b_{fj}+\sigma\geq 0 \quad \left(j=1,\ldots,n\right), \\
&\qquad \sum_{r=1}^{s}u_{r}^{+}g_{rk}^{+} = 1, \qquad (27.17) \\
&\qquad v_{i}^{-}\geq\varepsilon_{s}R_{i}^{x} \quad \left(i=1,\ldots,m^{-}\right), \\
&\qquad v_{q}^{+}\geq\varepsilon_{s}R_{q}^{x} \quad \left(q=1,\ldots,m^{+}\right), \\
&\qquad u_{r}^{+}\text{ and }u_{r}^{-}\text{: URS} \quad \left(r=1,\ldots,s\right), \\
&\qquad w_{f}\geq\varepsilon_{s}R_{f}^{b} \quad \left(f=1,\ldots,h\right), \\
&\qquad \sigma:\text{URS}.
\end{aligned}$$

UENM under constant RTS and DTS ($UENM(DC)_c^R$): To attain the status of constant RTS and DTS, this chapter drops the condition $\sum_{j=1}^{n}\lambda_j=1$ from Model (27.15) and measures the level of unified efficiency by

$$UENM(DC)_{c}^{R} = \text{Equation (27.16)}, \quad (27.18)$$

where the optimal solution is obtained from Model (27.15) without $\sum_{j=1}^{n}\lambda_j=1$.

Scale Efficiency Measure SENM(DC): To examine how each DMU carefully manages its operational size under natural and managerial disposability, this chapter measures the degree of scale efficiency by

$$SENM(DC)^{R}=UENM(DC)_{c}^{R}/UENM(DC)_{v}^{R}. \quad (27.19)$$

Since $UENM(DC)_c^R \leq UENM(DC)_v^R$, the scale efficiency is less than or equals unity. The higher score in $SENM(DC)^R$ indicates the better scale management under natural and managerial disposability. The scale efficiency considers the possible occurrence of DC, or technology innovation on industrial pollution, along with an analytical capability to deal with zero and negative values.

27.4.3 Investment Rule

After solving Model (27.17), we can identify the occurrence of DC, or eco-technology innovation for pollution mitigation, by the following rule along with the assumption of a unique optimal solution:

(a) If $u_r^{+*} = 0$ for some (at least one) r, then "weak DC" occurs on the k-th DMU,
(b) If $u_r^{+*} < 0$ for some (at least one) r, then "strong DC" occurs on the k-th DMU and
(c) If $u_r^{+*} > 0$ for all r, then "no DC" occurs on the k-th DMU.

Note that if $u_r^{+*} < 0$ for some r and $u_{r'}^{+*} = 0$ for the other r', then this chapter considers that the strong DC occurs on the k-th DMU. It is indeed true that $u_r^{+*} < 0$ for all r is the best case because an increase in any desirable output always decreases the amount of undesirable outputs. Meanwhile, if $u_r^{+*} < 0$ is identified for some r, then it indicates that there is a chance to reduce the amount of undesirable output(s). Therefore, we consider the second case as an investment opportunity because such a case may reduce the amount of industrial pollution.

Under the occurrence of strong DC (i.e., $u_r^{+*} < 0$ for at least one r), the effect of investment on undesirable outputs is determined by the following rule:

(a) If $v_q^{+*} > \varepsilon_s R_q^x$ for q in Model (27.17), then the q-th input for investment under managerial disposability can effectively decrease an amount of undesirable outputs and
(b) If $v_q^{+*} = \varepsilon_s R_q^x$ for q in Model (27.17), then the q-th input for investment has a limited effect on decreasing an amount of undesirable outputs.

The investment on inputs under managerial disposability is not recommended in the other two cases (i.e., no and weak DC). Furthermore, this chapter uses "a limited effect" in the second case. The term implies that if this study drops the data range on the q-th input in Model (27.17), then there is a high likelihood that v_q^{+*} may become zero. Moreover, $v_q^{+*} > \varepsilon R_q^x$ are required for some q, but not necessary for all q.

Finally, it is important to note that the proposed investment classification needs at least two desirable outputs because Model (27.17) has $u^+ g_k^+ = 1$ in the case of a single desirable output. Even if u is unrestricted, Model (27.17) cannot produce a negative value on the dual variable, so being unable to identify an investment opportunity. Thus, the investment rule discussed in this chapter needs multiple desirable outputs.

27.4.4 Computation Summary

Figure 27.4 depicts two approaches to handle zero or negative values in a data set. As discussed in Chapter 26, we have discussed how to handle zero and negative values by the property of translation invariance for non-radial measurement. Meanwhile, this chapter proposes data aggregation which can handle zero and negative values for both radial and non-radial measurements.

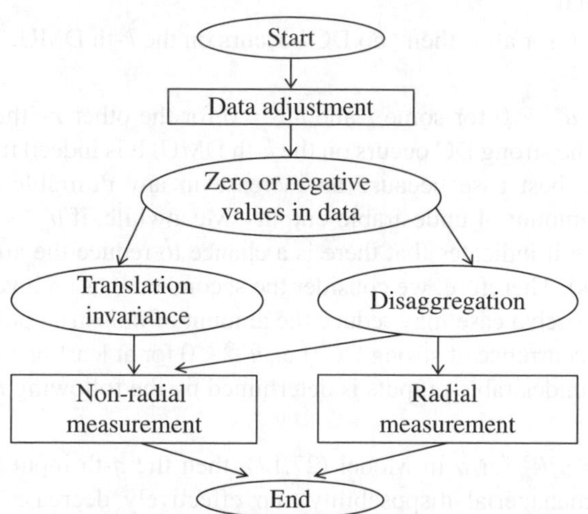

FIGURE 27.4 Two approaches for handling zero and negative values (a) Disaggregation requires more theoretical work to improve the reliability of research. For example, the selection of ε_s is subjective. None knows what the best selection is. (b) The property of translation invariance works only on non-radial measurement, as discussed in Chapter 26. (c) The disaggregation works for both radial and non-radial measurements. The proposed disaggregation needs to mathematically explore concerning how many DMUs with zero or negative values are appropriate in terms of making efficiency frontiers for desirable and undesirable outputs within the framework of DEA environmental assessment. It is necessary for us to consider which situation the proposed approach works or not. (d) It is easily imagined that the number of DMUs with positive values is larger than that of DMUs with zero or negative values. Such is just an assumption. (e) The proposed approach by disaggregation is a quick and easy method, not an exact method, to deal with zero and negative values. Mathematical justification is necessary on the property of disaggregation.

27.5 US INDUSTRIAL SECTORS

We use a data base from Carbon Disclosure Project (CDP) and COMPUSTAT. The CDP is building the world's largest database regarding corporate performance and climate change by collecting data sets via annual online questionnaire sent out to major firms across the world. This chapter utilizes the data on S&P 500 companies for 2012 and 2013, including the companies' direct and indirect GHG emission, their investment in carbon mitigation and the corresponding total estimated GHG saving. The performance of firms in the two annual periods are combined together and treated as a pooled cross-sectional data set. So, it is not necessary to specify the period (t) in this chapter. The number of observations in the previous study (e.g., Sueyoshi and Wang, 2014a) was 153 and that of this chapter is 161. Thus, eight observations contain negative values in this chapter.

Data Accessibility: The supplemental section of Sueyoshi and Yuan (2015b) lists the data set used in this chapter.

The data set consists of the following operational, environmental and financial factors: (a) Net income: This is calculated as annual revenue minus annual cost of each firm. Therefore, it may become a negative value. (b) Estimated CO_2 Saving: This indicates the annual CO_2 saving from a company's current emission level after investment in abatement technologies. The variable can be regarded as a measure of a company's technology capacity. (c) Direct CO_2 Emission: This measure indicates an amount of emissions from sources owned by a company. The cost of adapting for pollution prevention practices and the effectiveness of pollution prevention as a strategy for reducing emissions may vary with a scale of current emission. (d) Indirect CO_2 Emission: This measures an amount of emissions from generation of electricity, steam, heating and cooling purchased by a company. (e) Number of Employees: This is regarded as a proxy for a firm size. Larger firms may have more resources to adapt for CO_2 mitigation practices. (f) Working Capital: This is included to indicate the operating liquidity of a firm. Firms with higher working capital may invest more in CO_2 mitigation. (g) Total Assets: This includes current assets, property, plant and equipment, all of which are used as another proxy for a corporate size. (h) R & D Expense: This is another measure of a firm's technology capacity. It is expected that firms with higher R & D expense are more likely to acquire and implement efficient emission control technology. (i) Investment in CO_2 Abatement: This gives a total amount of investment that a company is required to achieve the estimated annual CO_2 saving. Profit maximizing firms are expected to choose technology according to their cost performance and effectiveness in mitigating the amount of CO_2 emissions.

This chapter utilizes two desirable outputs (net income, estimated annual CO_2 saving), two undesirable outputs (direct and indirect CO_2 emissions), three inputs under natural disposability (number of employees, working capital, total assets) and two inputs under managerial disposability (investment in CO_2 abatement,

R & D expense). The research of Sueyoshi and Yuan (2015b) provides descriptive statistics on the data set used for this chapter.

The desirable output (e.g., net income) contains negative values and the other data sets on production factors contain all positive values. To avoid the problem where a large data set dominates other data sets in the computation, this chapter further calculates an adjusted index to be used in the proposed DEA models for all variables. That is, each observation on a production factor is divided by the factor average. For the negative net income values, these are divided by the factor average while maintaining the negative sign. This chapter sets $\varepsilon_s = 0.001$ for the application of all DEA models.

Table 27.1 summarizes six unified efficiency scores of seven industrial sectors. On overall average, the consumer staples sector has the lowest UEN_v^R (0.3950) and the lowest UEM_v^R (0.2605). In ascending order of efficiency measures, the energy sector is the second. The consumer discretionary sector is the third with UEN_v^R (0.7460) and UEM_v^R (0.4313). The remaining four industrial sectors have higher UEN_v^R scores from 0.8195 to 0.8865. Among them, the UEM_v^R measure of the health care sector is only 0.3999. The UEM_v^R score is about 0.6132 for the industrials and information technology sector. The material sector has the highest UEM_v^R at 0.9207.

The results found in Table 27.1 indicate the following five empirical implications. First, the consumer staples sector is inefficient in unified (operational and environmental) performance measures. It is easily imagined that products of the industrial sector are less valued and the profit margin is small. Even though the sector is economically inefficient, its environmental performance is even worse. For example, tobacco can produce a wide spectrum of air pollutants as well as many substances hazardous to us. Second, UEN_v^R and UEM_v^R of the consumer discretionary sector are considerably high in their magnitudes, compared to the other sector for consumers (i.e., consumer staples). Besides automobiles, the consumer durables and retailing companies in our data set engage in house furnishings, home improvement, housewares and specialties. Third, the high-tech industries outperform the other industries, including consumer-oriented industries and energy industry. This indicates the importance of technology in the performance of all companies. Fourth, only the material industry pays more attention to environment rather than economic performance. All other industries focus on their profit improvements. Most of the materials are natural resources, which depend upon environment. Thus, it is a better strategy to pay more attention to the environment in a long-term horizon. For example, paper products are made from wood, and exhaustive exploration can definitely damage the whole industry. Finally, the level of SEN^R and SEM^R increases with the size of the industry. Most of the scale efficiency scores are above 0.900, which indicates the importance of proper scale management in these companies. The exceptions are the SEN^R measures of consumer discretionary, consumer staples and energy sectors. At the same time, scale management of environmental performance outperforms that of operational performance.

TABLE 27.1 Unified efficiency measures of US industrial sectors

Sector	Company type	UEN_v^R	UEN_c^R	SEN^R	UEM_v^R	UEM_c^R	SEM^R	# of DMUs
Consumer discretionary	Automobiles and components	0.9155	0.6913	0.7495	0.3459	0.3028	0.8743	6
	Consumer durables and apparel	0.6201	0.0785	0.1869	0.5852	0.3830	0.7073	5
	Retailing	0.5521	0.0108	0.0242	0.3026	0.0372	0.4037	2
	Overall	0.7460	0.3509	0.4215	0.4313	0.2928	0.7377	13
Consumer staples	Food, beverage and tobacco	0.3025	0.1947	0.5140	0.2511	0.2141	0.9172	9
	Household and personal products	0.6033	0.1107	0.3372	0.2818	0.2370	0.8730	4
	Overall	0.3950	0.1689	0.4596	0.2605	0.2211	0.9036	13
Energy	Energy equipment and services	0.4824	0.3720	0.7215	0.4156	0.3303	0.8742	4
	Oil and gas	0.7360	0.5200	0.7415	0.4865	0.3745	0.8425	10
	Overall	0.6636	0.4777	0.7358	0.4663	0.3619	0.8516	14
Health care	Health care equipment and services	0.9682	0.6863	0.7041	0.9992	0.8728	0.8734	2
	Biotechnology and life Sciences	0.9023	0.8611	0.9480	0.3252	0.3252	1.0000	11
	Pharmaceuticals	0.8591	0.8109	0.9219	0.3729	0.3449	0.9549	14
	Overall	0.8848	0.8221	0.9164	0.3999	0.3760	0.9672	27
Industrials	Capital goods	0.8792	0.8138	0.9261	0.6837	0.6622	0.9755	18
	Commercial and professional services	0.8696	0.8649	0.9937	0.1220	0.1220	1.0000	2
	Overall	0.8782	0.8189	0.9328	0.6276	0.6082	0.9780	20
Information technology	Semiconductors and equipment	0.8228	0.8189	0.9939	0.6103	0.6103	1.0000	20
	Software and services	0.8328	0.8205	0.9834	0.6073	0.6009	0.9895	19
	Technology hardware and equipment	0.7928	0.7910	0.9976	0.6272	0.6272	1.0000	12
	Overall	0.8195	0.8129	0.9908	0.6132	0.6108	0.9961	51
Materials	Chemicals	0.8851	0.8829	0.9962	0.9527	0.9527	1.0000	13
	Containers and packaging	0.9118	0.9105	0.9986	0.8273	0.8273	1.0000	4
	Metals and mining	0.9436	0.9436	1.0000	0.7407	0.7407	1.0000	2
	Paper and forest products	0.8372	0.8372	1.0000	1.0000	1.0000	1.0000	4
	Overall	0.8865	0.8850	0.9976	0.9207	0.9207	1.0000	23

(a) Source: Sueyoshi and Yuan (2015b). (b) *UEN*: Unified efficiency under natural disposability; *UEM*: unified efficiency under managerial disposability. (c) $SEN^R = UEN_c^R / UEN_v^R$ and $SEM^R = UEM_c^R / UEM_v^R$. The superscript **R** stands for radial measurement.

Table 27.2 summarizes the same unified efficiency scores of the seven industrial sectors. Combining their unified (operational and environmental) performance, for example, the consumer staples industry is still inefficient at $UENM_v^R$ = 0.4132 on average. The efficiency levels of the energy industry are $UENM_v^R$ = 0.6293 and $UENM_c^R$ = 0.4890. The highest efficiency measures, or $UENM_v^R$ = 0.9443 and $UENM_c^R$ = 0.9426, can be found in the materials industry. The materials, information technology and industrials sectors comprise the high-ranked group. The health care and energy sectors belong to the middle-ranked group. The consumer discretionary and consumer staples sectors form the low-ranked group. The ranking position of the industrial sectors is consistent with their scale efficiency measures. For example, the materials, information technology and industrials sectors have high $SENM^R$ measures (0.9977, 0.9904 and 0.9761, respectively).

Table 27.3 summarizes the unified efficiency scores of $UENM(DC)_v^R$, $UENM(DC)_c^R$ and $SENM(DC)^R$ for seven industrial sectors in the United States. The three efficiency measures increase drastically in the consumer discretionary, consumer staples and energy industry sectors in comparison with those of Table 27.3. Therefore, the unified performance of these industries can be improved significantly by eco-technology innovation. In particular, the unified efficiency measures of the consumer discretionary industry are increased to unity, indicating the status of full efficiency. In contrast, all efficiency scores decrease in the four industrial sectors that are ranked in the highest group in terms of efficiency measures (i.e., materials, information technology, industrials and health care) in Table 27.3. An exception may be found in the $SENM(DC)^R$ of the material industry in such a manner that $SENM^R$ is 0.9977 and $SENM(DC)^R$ is 0.9983.

Table 27.3 indicates that eco-technology innovation may not improve the performance of firms in the four industrial sectors because the companies in health care, industrials, information technology and materials sectors have already attained a high level of technology development through spending a considerable amount of capital on their engineering capabilities. Therefore, redundant investment on eco-technology innovation cannot help these companies' performance. Hence, balanced investment on eco-efficient parts of business should be emphasized and promoted. On the other hand, the degree of $UENM(DC)_v^R$ regarding two company types which are related to high technology (e.g, biotechnology and life science, and commercial and professional services) increase compared to $UENM_v^R$ in Table 27.3. This indicates that their eco-technology innovation investment can improve corporate performance due to the high-tech property of these companies.

Comparison between Tables 27.2 and 27.3 indicates two important implications. One of the two implications is that investment on eco-technology innovation in the low-tech industries such as consumer discretionary industry, consumer staples industry and energy industry is useful for improving their unified performance if desirable outputs are measured by net income and amount of CO_2 emission reduction because these industries are the largest emitters among the

TABLE 27.2 Unified efficiency measures of US industrial sectors: natural and managerial disposability

Sector	Company type	$UENM_v^R$	$UENM_c^R$	$SENM^R$	# of DMUs
Consumer	Automobiles and components	0.7463	0.5042	0.7483	6
discretionary	Consumer durables and apparel	0.5645	0.3034	0.4502	5
	Retailing	0.5532	0.0107	0.0439	2
	Overall	0.6466	0.3510	0.5253	13
Consumer	Food, beverage and tobacco	0.3259	0.2467	0.6358	9
staples	Household and personal products	0.6096	0.0929	0.2483	4
	Overall	0.4132	0.1994	0.5166	13
Energy	Energy equipment and services	0.5772	0.4855	0.7915	4
	Oil and gas	0.6501	0.4904	0.7895	10
	Overall	0.6293	0.4890	0.7900	14
Health care	Health care equipment and services	1.0000	0.9161	0.9161	2
	Biotechnology and life sciences	0.4176	0.3542	0.8545	11
	Pharmaceuticals	0.6674	0.5553	0.8250	14
	Overall	0.5903	0.5001	0.8438	27
Industrials	Capital goods	0.7970	0.7757	0.9751	18
	Commercial and professional services	0.1401	0.1360	0.9847	2
	Overall	0.7313	0.7117	0.9761	20
Information	Semiconductors and equipment	0.7127	0.7088	0.9920	20
technology	Software and services	0.7703	0.7659	0.9897	19
	Technology hardware and equipment	0.7481	0.7414	0.9889	12
	Overall	0.7425	0.7378	0.9904	51
Materials	Chemicals	0.9530	0.9525	0.9995	13
	Containers and packaging	0.9046	0.8961	0.9886	4
	Metals and mining	0.9133	0.9133	1.0000	2
	Paper and forest products	0.9714	0.9714	1.0000	4
	Overall	0.9443	0.9426	0.9977	23

(a) Source: Sueyoshi and Yuan (2015b). (b) *UENM*: Unified efficiency under natural and managerial disposability. The subscripts c and v indicate constant and variable RTS and DTS, respectively. (c) *SENM*: Scale efficiency under natural and managerial disposability. The superscript R stands for radial measurement.

TABLE 27.3 Unified efficiency measures of US industrial sectors under desirable congestion

Sector	Company type	$UENM(DC)_v^R$	$UENM(DC)_c^R$	$SENM(DC)^R$	# of DMUs
Consumer discretionary	Automobiles and components	1.0000	1.0000	1.0000	6
	Consumer durables and apparel	1.0000	1.0000	1.0000	5
	Retailing	1.0000	1.0000	1.0000	2
	Overall	1.0000	1.0000	1.0000	13
Consumer staples	Food, beverage and tobacco	0.8180	0.7384	0.8977	9
	Household and personal products	0.8173	0.7920	0.9658	4
	Overall	0.8177	0.7549	0.9187	13
Energy	Energy equipment and services	0.8251	0.7858	0.8690	4
	Oil and gas	0.7835	0.6915	0.8507	10
	Overall	0.7954	0.7184	0.8559	14
Health care	Health care equipment and services	0.8280	0.7888	0.9402	2
	Biotechnology and life sciences	0.4650	0.3288	0.6829	11
	Pharmaceuticals	0.5204	0.4138	0.7238	14
	Overall	0.5206	0.4069	0.7231	27
Industrials	Capital goods	0.5186	0.4788	0.8883	18
	Commercial and professional services	0.6374	0.5762	0.7774	2
	Overall	0.5305	0.4885	0.8772	20
Information technology	Semiconductors and equipment	0.4151	0.3924	0.9102	20
	Software and services	0.6608	0.6506	0.9573	19
	Technology hardware and equipment	0.4181	0.4146	0.9802	12
	Overall	0.5074	0.4938	0.9443	51
Materials	Chemicals	0.7345	0.7342	0.9970	13
	Containers and packaging	0.6363	0.6362	0.9998	4
	Metals and mining	0.6054	0.6054	1.0000	2
	Paper and forest products	0.9554	0.9554	1.0000	4
	Overall	0.7446	0.7444	0.9983	23

(a) Source: Sueyoshi and Yuan (2015b).

seven sectors examined in this chapter. They have historically paid more attention on operational performance than environmental performance. The other implication is that balanced investment on eco-technology innovation should be promoted. The high-tech industries, including health care, industrials, information technology and materials, do not exhibit the best performance through investment on eco-technology innovation because these companies already pay more attention to the environment than economic performance and scale management. This investment in technology innovation may absorb resources from other parts of business in these companies and it cannot lead to an immediate enhancement of their net incomes. However, technology innovation is a key factor of some sub-industries such as biotechnology and professional services where investment in eco-technology innovation is necessary.

Table 27.4, obtained by Model (27.17), lists effective and limited investment opportunities in the seven industrial sectors. On overall average, 31 observations (19.25%) are rated as effective investments and four observations (2.48%) are rated as limited investments in terms of developing corporate sustainability. The energy sector has 14.29% effective investments, rated as sixth among the seven industrial sectors. This indicates that the energy sector does not exhibit an attractive investment opportunity for developing corporate sustainability, compared with the other six industrial sectors, in terms of a short-term perspective (i.e., net income).

The result may seem inconsistent with the previous work of Wang *et al.* (2014). That research claimed that the energy industry is the best investment target in terms of a short-term concern (i.e., ROA in their case). However, the result obtained in this chapter indicates that successful energy firms can attain corporate sustainability by investing in green technology innovation, as found in their study, but unsuccessful ones cannot attain sustainability even in the short-term concern, as found in this chapter. The rationale is because unsuccessful companies (i.e.,

TABLE 27.4 Investment strategy on industrial sector

Sector	No. of effective investments	Percentage	No. of limited investments	Percentage
Consumer discretionary	2	15.38	0	0.00
Consumer staples	2	15.38	0	0.00
Energy	2	14.29	0	0.00
Health care	4	14.81	0	0.00
Industrials	2	10.00	1	5.00
Information technology	11	21.57	2	3.92
Materials	8	34.78	1	4.35
Overall	31	19.25	4	2.48

(a) Source: Sueyoshi and Yuan (2015b).

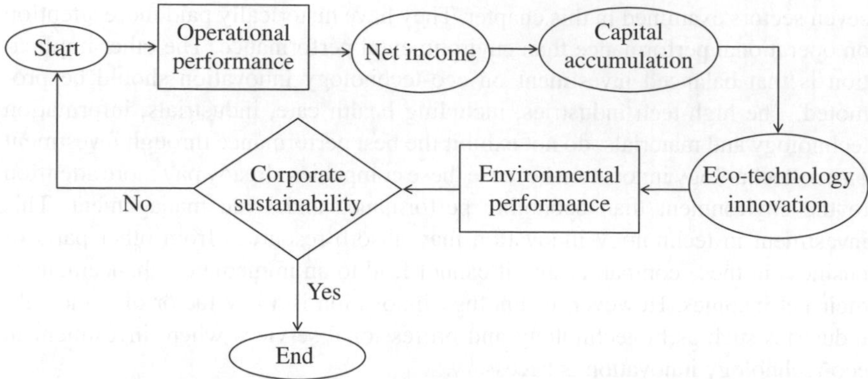

FIGURE 27.5 Corporate sustainability development (a) See Figure 15.5. The figure depicts a gradual improvement for sustainability development.

negative net incomes) do not have a large capital accumulation for green technology investment, the importance of which was first discussed by Sueyoshi and Goto (2010a). In other words, eco-technology investment for enhancing the level of corporate sustainability depends upon the net income generated by each energy firm.

Figure 27.5 visually describes such a linkage in which the enhancement in operational performance increases positive net income, so increasing the amount of capital accumulation. The eco-technology innovation, usually supported by a large amount of investment, enhances the environmental performance. The enhanced unified (operational and environmental) performance increases the level of corporate sustainability.

27.6 SUMMARY

A green image, not often reality in some cases, recently became an important component for corporate survivability in modern business. By extending previous works (e.g., Wang *et al.* 2014, Sueyoshi and Wang, 2014a,b) on environmental assessment and corporate sustainability, this chapter discussed a new use of DEA environmental assessment by incorporating an analytical capability to handle zero and negative values in a data set. The proposed approach provided us with not only a quantitative assessment on the unified performance of firms but also information regarding how to invest for eco-technology innovation for abating the amount of industrial pollution. However, we clearly acknowledge that the proposed approach is a quick and easy method, not an exact method, to deal with zero and negative values in a data set. Mathematical justification is necessary on the property of disaggregation. That will be an important future research task.

This chapter obtained the following empirical findings: First, as found in Wang *et al.* (2014), the energy industry was an attractive investment target if desirable outputs were measured by return on assets (ROA) as a short-term concern and an amount of CO_2 emission reduction because the industry was the largest emitter among the seven industrial sectors examined in their study. However, as found in Sueyoshi and Wang (2014a), if ROA was replaced by a corporate value, measured by Tobin's q ratio as a long-term concern, the energy industry was not so attractive because the industry is a very large process industry so that green investment did not immediately increase corporate value, as found in other industrial sectors.

A methodological problem of these previous studies is that they investigated only successful companies with positive ROA or positive Tobin's q ratios. The analytical feature clearly indicates the drawback of their empirical studies. In contrast, this chapter pays attention to both successful companies with positive net incomes and unsuccessful companies with negative net incomes. As a result of considering both successful and unsuccessful firms, this chapter finds that energy firms may be not attractive in terms of a short-term concern (i.e., net income) because governmental regulation on their operations and environment protections is stricter than in other industrial sectors. This chapter also finds that the energy sector may have an investment potential for eco-technology innovation in a long-term horizon until the investment becomes fully effective for pollution mitigation.

This chapter obtained the following empirical findings: First, as found in Wang et al. (2014), the energy industry was an attractive investment target if desirable outputs were measured by return on assets (ROA) as a short-term concern and an amount of CO_2 emission reduction because the industry was the largest emitter among the seven industrial sectors evaluated in their study. However, as found in Sueyoshi and Wang (2014a), if ROA was replaced by a corporate value, measured by Tobin's q ratio as a long-term concern, the energy industry was not so attractive because the industry was a very large emission industry, so that green investment did not immediately increase corporate value, as found in other industrial sectors.

A methodological problem of these previous studies is that they investigated only successful companies with positive ROA or positive Tobin's q ratios. The analytical feature clearly indicates the drawback of their empirical studies. In contrast, this chapter pays attention to both successful companies with positive net incomes and unsuccessful companies with negative net incomes. As a result of considering both successful and unsuccessful firms, this chapter finds that energy firms may be not attractive in terms of a short-term goal, or net income, because government regulation on their operations and environment protections is stricter than on other industrial sectors. This chapter also finds that the energy sector may have an investment potential for eco-technology innovation in a long-term horizon until the investment becomes fully effective for pollution mitigation.

28

LITERATURE STUDY FOR DEA ENVIRONMENTAL ASSESSMENT

28.1 INTRODUCTION

Difficulties related to the environmental protection have been gradually recognized with public awareness on the global warming and climate change. In this book, Chapter 13 discussed various energy issues and Chapter 14 discussed environmental problems which are closely related to not only our ecological systems but also economies and businesses. The gist of this book is to provide a link between the natural science and engineering efforts to reduce the amount of various pollutants and the social science efforts to support these by preparing energy and environmental policy agendas and corporate strategies which reflect our global perspective within world business. As a methodology to attain a high level of such social and corporate sustainability, this book proposes various approaches and mathematical models which originate from DEA environmental assessment for energy sectors.

In previous research efforts on DEA applied to energy and environment, Sueyoshi and Yuan (2016) summarized 407 papers[1]. As an extension of their

[1] See Sueyoshi, T. (2016) Chapter 16: DEA environmental assessment (I): concepts and methodologies, in: *Handbook of Operations Analytics Using Data Envelopment Analysis*, International Series of Operations Research and Management Science 239, (eds) Hwang *et al*., Springer Science & Business Media, pp. 413–444. See also Sueyoshi, T. and Yuan, Y. (2016) Chapter 17: DEA environmental assessment (II): a literature study, in: *Handbook of Operations Analytics Using Data Envelopment Analysis*,

Environmental Assessment on Energy and Sustainability by Data Envelopment Analysis,
First Edition. Toshiyuki Sueyoshi and Mika Goto.

literature survey, this chapter[2] summarizes previous research efforts on DEA environmental assessment during the past four decades by reorganizing their efforts into newly proposed categories of articles. To attain our research extension, this chapter returns to Chapter 13 which reviewed a general trend of world energy, in particular focusing upon various categories of primary and secondary energies.

The research classification proposed in this chapter is described by the following three steps based upon Chapters 13 and 14:

(a) All previous research efforts are separated into two groups based upon these main application areas: energy and/or environment. Then, these studies on energy are further categorized into electricity, oil, coal, gas, heat, renewable, energy efficiency or energy savings.
(b) Previous studies on environment contain articles related to environmental efficiency and sustainability.
(c) This chapter discusses other research efforts which do not clearly belong to the two categories.

The remainder of this chapter is organized as follows. Section 28.2 summarizes all articles of DEA applications on energy and environment, all of which are based upon the classification categories proposed in Chapter 13 and environment issues discussed in Chapter 14. Section 28.3 examines their publication trends and unique features, particularly regarding on energy. Section 28.4 examines articles on energy efficiency. Section 28.5 investigates articles on environment issues discussed in Chapter 14. Section 28.6 discusses other articles that are not clearly classified in any of the proposed categories. Section 28.7 summarizes this chapter.

28.2 APPLICATIONS IN ENERGY AND ENVIRONMENT

This chapter collects previous articles by using a search function of "Scopus," which is the largest abstract and citation database of peer-reviewed literature provided by Elsevier. All these research articles are classified by the application categories summarized in Table 28.1.

International Series of Operations Research and Management Science 239, (eds) Hwang *et al.*, Springer Science & Business Media, pp. 445–481. The two articles summarized 407 previous studies on DEA environmental assessment. This chapter extends their work further by adding and replacing articles, reorganizing the classification categories from previous review efforts, particularly strengthening articles regarding environment and sustainability.

[2] This chapter is partly based upon the manuscript: Sueyoshi, T., Yuan, Y. and Goto, M. (2017). A literature study for DEA applied to energy and environment. *Energy Economics* **62**, 104–124.

TABLE 28.1 Previous research efforts: application areas and periods

Application areas	Topics	1980s	1990s	2000s	2010s	All
Energy	Electricity	3	16	55	111	185
	Oil	0	2	6	15	23
	Coal	1	1	2	3	7
	Gas	0	0	4	9	13
	Heat	0	0	3	0	3
	Renewable	0	0	3	33	36
	Energy efficiency or energy saving	0	1	21	118	140
	All (Energy)	4	20	94	289	407
Environment	Environment or sustainability	1	6	41	222	270
Others	Others	0	0	3	13	16
All		5	26	138	524	693

(a) Source: Sueyoshi *et al.* (2017).

This table provides us with the following four classifications on DEA and its applications in energy and environment:

(a) The total number of articles collected in this chapter is 693, in which 407 articles are related to energy, 270 articles are associated with environment and sustainability and 16 articles are related to other issues, although the last group is also associated with energy and environment issues in a broad sense.

(b) Major research interests in previous literature were found in electricity, energy efficiency or energy saving in the area of energy. The number of the other issues or fuels in various categories (e.g., oil and coal) was much less than that of the major two areas.

(c) The total number of articles has significantly increased since the 2000s. In particular, it was only 31 before 2000, but it became 138 during the 2000s, and it became 524 only in the 2010s, or during approximately six years until May 2016.

(d) The 31 articles before 2000 (i.e., 1980s = 5, 1990s = 26) included 19 electricity-related studies (i.e., 1980s = 3, 1990s = 16) and seven environmental studies (i.e., 1980s = 1, 1990s = 6).

Meanwhile, during the early periods, there were almost no studies on other issues which were not clearly classified into the specific topics listed in the table. Thus, electricity was the most typical research area in the early stage of DEA applications for energy and environment issues. Since then, many research efforts have applied DEA for energy and environment during the 2000s. The research trend has rapidly increased during the 2010s until now.

28.3 ENERGY

28.3.1 Electricity

Electricity is a relatively old application area of performance assessment, not only by DEA but also by other popular methodologies for efficiency measurement, such as a stochastic frontier analysis (SFA). In the application of SFA, many researchers have estimated parameters regarding a production function or a cost function by assuming a specific functional form. In contrast, it is not necessary for DEA to assume any specific functional form or any error-term distribution, as discussed in Chapter 3.

The popularity of DEA, applied to an electricity industry, started from a relatively early stage. The rationale for this popularity was partly because the industry had been long regulated by local and central governments. Under such governmental regulation, it was well known among researchers that "X-inefficiency" might arise because of a lack of discipline from market competition (Leibenstein, 1966)[3]. Therefore, it was important for electric power companies to assess their operational efficiencies under regulation. It was also necessary for regulatory agencies to consider how to prepare their effective regulation schemes. Otherwise, inefficient firms could survive by receiving an economic benefit from regulation, as specified by the concept of X-inefficiency. In addition, operational data sets were well established in the electricity industry. Usually, the data sets were available in a public domain, because the electricity industry was a symbolic industry of such regulation until recently.

Hereafter, this chapter needs to mention that Leibenstein's X-inefficiency has limited implications in the context of energy and environmental assessment. The concept has four drawbacks from the perspective of modern corporate strategies. First, it is necessary for firms to consider the existence of undesirable outputs (e.g., CO_2 and other GHG emissions) in preparing their corporate strategies. The concept of X-inefficiency did not incorporate such an existence of undesirable outputs. Second, firms can increase or decrease some components of an input vector. The increase indicates an economic growth, so being under managerial disposability. The limit on input increase can be found on an efficiency frontier for undesirable outputs. In contrast, the decrease indicates an economic stability under natural disposability. The limit on input decrease can be found on

[3]According to Leibenstein, H (1966) Allocative efficiency vs. X-efficiency. *American Economic Review*, **56**, 392–415. The economic theory assumes that firms maximize their profits by minimizing inputs used to produce a given level of desirable outputs. Perfect competition enforces firms to seek a productive efficiency gain and yields less cost than efficient rivals. However, as found in utility industries such as electricity, gas and water, market structures are different from perfect competition such as monopoly and those may produce inefficiency because a lack of competition makes it possible that firms depend upon inefficient production technology but they still stay in utility businesses. In the monopolistic market, utility firms often employ more resources than are needed to produce a given level of desirable outputs. Such unused capacities result in a source of X-inefficiency.

an efficiency frontier for desirable outputs. The Leibenstein's X-inefficiency has been conventionally discussed only within the conceptual framework of the input decrease for efficiency enhancement. Third, an occurrence of innovation on production and/or green technologies was not incorporated in the framework of X-inefficiency. Finally, the cost in previous efficiency concepts implied a total amount of cost components. It is more realistic to consider marginal or average cost in discussing modern business so that a firm can increase the input vector as long as the marginal cost of a product (service) is less than the market price which is determined on an equilibrium point between supply and demand. See, for example, Sueyoshi (2010[4]) and Sueyoshi and Tadiparthi (2007[5], 2008[6]) for price setting mechanisms in the US electricity trading markets. Thus, acknowledging an academic contribution of the Leibenstein's X-inefficiency, the DEA environmental assessment proposed in this book does not follow the research direction suggested by the X-inefficiency concept. Thus, it is easily thought that this book is different from the other previous studies concerning energy, environment and sustainability in production economics.

Table 28.2 summarizes a trend of studies on DEA applied to electricity. The DEA applications are divided into three categories: (a) generation, (b) transmission and distribution network and (c) others (e.g., total operation). Such a classification is due to the business functions of the electric power industry. Recently, generation has been competitive in many countries after market liberalization began in the 1990s, while the transmission and distribution network divisions are still under regulation because it is believed that these two functions may consist of a natural monopoly[7]. The others include research on the total operation of electric power companies, electricity consumption and a literature survey on DEA applications for the industry.

Studies on the transmission and distribution network are often interested in regulation because the network sector was under regulation even after market liberalization in the electricity industry. Indeed, some countries and regions apply

[4] Sueyoshi, T. (2010) Beyond economics for guiding large public policy issue: Lessons from the Bell System divestiture and the California electricity crisis. *Decision Support Systems*, **48**, 457–469.

[5] Sueyoshi, T. and Tadiparthi, G.R. (2007) An Agent-based Approach to handle business complexity in US wholesale power trading. *IEEE Transactions on Power Systems*, **22**, 532–543.

[6] Sueyoshi, T. and Tadiparthi, G.R. (2008) An agent-based decision support system for wholesale electricity market. *Decision Support Systems*, **44**, 425–446.

[7] The concept of "natural monopoly" implies that a single company can provide operation on transmission and distribution less costly than multiple companies in the case of an electric power industry. The transmission service needs a sophistic control (e.g., dispatch planning and scheduling) via a whole grid system by fully utilizing an advanced monitoring capability. Detailed descriptions on the natural monopoly issue can be found in: (a) Charnes, A., Cooper, W.W. and Sueyoshi, T. (1988) A goal programming/constrained regression review of the Bell System breakup. *Management Science*, **34**, 1–26. (b) Sueyoshi, T. (1991) Estimation of stochastic frontier cost function using data envelopment analysis: An application to AT&T divestiture. *Journal of the Operational Research Society*, **42**, 463–477. (c) Goto, M., Inoue, T. and Sueyoshi, T. (2013) Structural reform of Japanese electric power industry: Separation between generation and transmission and distribution. *Energy Policy*, **56**, 186–200.

TABLE 28.2 Articles applied to the electricity industry

Main application areas	Topics	Sectors	Combined analysis	1980s	1990s	2000s	2010s	All	Literature group
Energy	Electricity	All		3	16	55	111	185	
		Generation	All	3	10	20	64	97	
			None	2	5	9	15	31	[A]
			Environment	1	5	11	49	66	[B]
		Network		0	4	26	31	61	[C]
		Others	All	0	2	9	16	27	
			None	0	2	8	8	18	[D]
			Environment	0	0	1	8	9	[E]

(a) Source: Sueyoshi *et al.* (2017). (b) Combined analysis means research efforts on both generation and environment, or transmission & distribution network and environment. (c) All articles on generation with combined analysis deal with various environmental issues. Among them, three articles ([63], [175], [664]) discussed both environment and energy efficiency issues. (d) Almost all articles on network are not combined analyses, but including article [432] that is combined with environment analysis and article [393] with energy efficiency. (e) Most of articles in the others examine electric power companies as a whole. In addition, they have considered electricity consumption, demand forecast, fuel mix and literature survey. (f) The number of articles increases in both categories of with and without combined analysis.

benchmarking analysis to the network regulation by DEA. In contrast, many studies on generation recently explored various environmental issues (e.g., air pollution) because of deregulation and increasing public awareness of environment issues. See Chapters 14, 17 and 18 for a detailed description on US deregulation of electricity and environmental regulation. There is an increasing trend in the number of articles. They increase in both categories (with and without combined analyses). See "None" and "Environment" in Table 28.2. The combined analysis means a joint research effort on both generation and environment or both transmission and distribution network and environment.

It is easily envisioned that more researchers tend to examine the efficiency level of utility operations, along with environmental factors such as GHG emissions and waste discharges, all of which are produced from power plants when they generate electricity by fossil fuels. All the articles in the five categories regarding electricity, as specified in Table 28.2, are classified by the following articles[8]:

(a) Category A: [14], [18], [24], [32], [49], [52], [53], [110], [120], [135], [148], [154], [155], [163], [164], [244], [251], [285], [318], [322], [364], [385], [413], [442], [466], [467], [491], [542], [563], [652] and [656].

(b) Category B: [12], [13], [37], [63], [83], [95], [146], [158], [159], [160], [162], [167], [174], [175], [194], [195], [237], [242], [257], [271], [289], [319], [324], [348], [353], [371], [374], [376], [403], [419], [435], [448], [449], [450], [459], [486], [494], [495], [499], [501], [503], [504], [507], [509], [512], [518], [520], [523], [524], [539], [553], [558], [559], [602], [612], [617], [621], [625], [630], [634], [635], [660], [661], [664], [668] and [684].

(c) Category C: [4], [5], [6], [10], [25], [26], [31], [42], [57], [88], [90], [91], [108], [119], [122], [130], [131], [132], [133], [173], [185], [210], [238], [239], [240], [270], [279], [281], [283], [299], [301], [316], [329], [356], [363], [370], [381], [387], [388], [393], [397], [402], [405], [414], [424], [431], [432], [438], [444], [447], [543], [570], [571], [579], [604], [624], [626], [627], [628], [629] and [671].

(d) Category D: [1], [2], [17], [35], [134], [183], [184], [243], [250], [258], [282], [308], [489], [502], [519], [521], [544] and [551].

(e) Category E: [256], [315], [492], [496], [497], [565], [598], [619] and [620].

28.3.2 Oil, Coal, Gas and Heat

There are 46 articles in total related to oil, coal, gas and heat. Among these efforts, 3, 7, 13 and 23 articles are related to heat, coal, gas and oil, respectively.

Table 28.3 summarizes the group classifications during four decades.

[8] Note that the numbers within square brackets [ ] in this chapter correspond to the number listed at the end of each article in the References section of this chapter.

TABLE 28.3 Articles applied to oil, coal, gas and heat

Main application areas	Topics	Sectors	Combined analysis	1980s	1990s	2000s	2010s	All	Literature group
Energy	Oil, coal, gas and heat	All		1	3	15	27	46	
		Oil	All	0	2	6	15	23	
			None	0	2	5	7	14	[F]
			Environment	0	0	1	8	9	[G]
		Coal	All	1	1	2	3	7	
			None	1	1	2	1	5	[H]
			Environment	0	0	0	2	2	[I]
		Gas	All	0	0	4	9	13	
			None	0	0	4	9	13	[J]
			Environment	0	0	0	0	0	[K]
		Heat	All	0	0	3	0	3	
			None	0	0	0	0	0	[L]
			Environment	0	0	3	0	3	[M]

(a) Source: Sueyoshi *et al.* (2017). (b) See the note for Table 28.2.

Research on the oil and gas industry examines upstream and downstream functions as well as their total business operations.

Previous articles on coal usually focused on coal mining. Research on gas mainly paid attention to gas transmission and distribution. Articles often considered gas refinery and consumption. Only three articles explored the heat industry by studying district heating systems.

All the articles in the eight categories regarding oil, coal, gas and heat (categories F–M), as specified in Table 28.3, are classified by the following articles:

(a) Category F: [29], [36], [50], [51], [143], [191], [246], [252], [253], [490], [546], [548], [622] and [659].
(b) Category G: [60], [168], [236], [352], [476], [498], [505], [517] and [526].
(c) Category H: [82], [84], [151], [273] and [546].
(d) Category I: [320] and [553].
(e) Category J: [11], [23], [27], [30], [34], [144], [145], [178], [179], [203], [340], [549] and [691].
(f) Category K: None.
(g) Category L: None.
(h) Category M: [3], [372] and [417].

As summarized in Table 28.3, DEA researchers examined the oil, coal, gas and heat industries by gradually paying attention to various environmental issues. The number of publications in the area has been increasing since the 2000s. An exception is found in the gas industry because gas is a clean fuel compared to the other fossil fuels. Consequently, research interest on gas from the perspective of environmental issues may be limited as summarized in Table 28.3.

28.3.3 Renewable Energies

It is easily imagined that we have various problems to combat the environmental and climate change issues. Those are usually caused by the use of fossil fuels. Many countries have been preparing new energy policy directions toward 2030 by supporting renewable energy so that they can replace fossil fuels. See Chapter 13 for a detailed description on renewable energy. An example of such a support scheme is the feed-in tariff (FIT) for renewable generation. See Sueyoshi and Goto (2014d) for a discussion concerning the positive and negative aspects of FIT on solar photovoltaic power generation. FIT assures fixed payment to renewable energy generation within a certain period so that it provides investors with an economic incentive. Currently, FIT is widely employed by many countries, such as Germany, the United States and Japan. Articles on the renewable energies have increased, particularly since the 2000s, along with diffusion of the financial support

TABLE 28.4 Articles applied to renewable energies

Main application areas	Topics	Sectors	1980s	1990s	2000s	2010s	All	Literature group
Energy	Renewables	All	0	0	3	33	36	
		Wind	0	0	0	7	7	[N]
		Solar	0	0	0	6	6	[O]
		Biomass	0	0	1	11	12	[P]
		Others	0	0	2	9	11	[Q]

(a) Source: Sueyoshi *et al*. (2017).

scheme and public interest on the type of renewable generation. Table 28.4 summarizes articles on renewable energies during the four decades.

All the articles in the four categories regarding renewables, as specified in Table 28.4, are classified by the following articles:

(a) Category N: [33], [217], [231], [232], [235], [325] and [404].
(b) Category O: [220], [287], [288], [297], [516] and [645].
(c) Category P: [41], [121], [186], [248], [343], [416], [430], [441], [455], [532], [560] and [631].
(d) Category Q: [70], [74], [75], [109], [192], [267], [332], [354], [485], [487] and [690].

Among renewable energies, biomass is the most popular technology in DEA applications, probably because there is a wide variety of biomass fuels, ranging from woody biomass to agricultural ones. The applications for biomass are followed by wind and solar photovoltaic, where a considerable technology progress has occurred with cost reduction in these areas. The other studies listed in Table 28.4 are related to a scenario-based analysis of renewable energy, heat pump technology and the relationship between economic growth and use of renewable energies.

28.4 ENERGY EFFICIENCY

Energy efficiency, or energy saving and conservation, is a measure for managing and restraining the growth in energy consumption. Energy efficiency can be considered as one of the easiest and most cost effective ways to combat the global warming and climate change, because a large portion of CO_2 emissions comes from energy consumption. In addition, efficient energy conservation is necessary for both improving corporate competitiveness in modern business and reducing energy costs for consumers. Such benefits of improved energy efficiency are being incorporated into corporate strategy and energy policy. For example, the 2030 climate and energy framework was adopted by European Council in October 2014, where the target of energy efficiency improvement was set at a minimum of 27%, compared to a projection of future energy consumption based on the current criteria. The Japanese government also estimated 13% conservation of final energy consumption, compared to forecasts that were calculated under the assumption of no energy saving efforts by 2030 in the proposal of the "Long Term Energy Consumption Forecast" published in July 2015.

An increasing recognition of the importance of energy efficiency has influenced a recent trend of DEA studies on energy. As summarized in Table 28.5, a large increase can be found in the 2010s. For example, 21 articles were published during the 2000s. The number increased to 118 during the 2010s. Research, which combined energy efficiency with environment, increased from 6 to 46 from the 2000s to the 2010s. Among the 140 articles in total, 52 articles examined not only energy efficiency but also environment issues. The change is because the measurement of energy efficiency is expected to improve a level of environmental protection through the effective reduction of energy usage.

All the articles in the two categories regarding energy efficiency, as specified in Table 28.5, are as follows:

(a) Category R: [7], [22], [45], [46], [47], [69], [71], [72], [77], [79], [97], [101], [116], [126], [129], [150], [166], [180], [187], [196], [197], [198], [199], [200], [205], [206], [207], [213], [215], [216], [219], [221], [224], [225], [254], [265], [277], [295], [296], [303], [304], [311], [313], [314], [327], [347], [359], [366], [367], [368], [369], [379], [380], [383], [384], [386], [389], [394], [395], [396], [399], [401], [418], [458], [460], [483], [536], [538], [555], [564], [572], [573], [576], [590], [597], [599], [600], [601], [606], [609], [623], [647], [648], [649], [670], [674], [685] and [692].

(b) Category S: [61], [73], [125], [128], [181], [189], [223], [262], [263], [264], [302], [305], [317], [349], [375], [390], [391], [392], [398], [400], [420], [421], [425], [461], [475], [479], [533], [534], [568], [578], [580], [581], [582], [583], [584], [585], [594], [596], [608], [616], [632], [633], [640], [641], [657], [662], [667], [669], [676], [678], [681] and [693].

One of the most popular research areas, in which DEA was applied to measure the level of energy efficiency, was a comparison between industries, regions and countries. Another prominent application area could be found in agriculture production because energy was a major input for production in the agriculture industry. Furthermore, many articles investigated environment issues, in which they considered an existence of GHG emissions, including CO_2, as undesirable outputs of agriculture activities.

TABLE 28.5 Articles applied to energy efficiency

Main application areas	Topics	Combined analysis	1980s	1990s	2000s	2010s	All	Literature group
Energy	Energy efficiency or energy saving	All	0	1	21	118	140	
		None	0	1	15	72	88	[R]
		Environment	0	0	6	46	52	[S]

(a) Source: Sueyoshi *et al.* (2017).

28.5 ENVIRONMENT

As discussed in Chapter 14, environment issues have become the most popular application areas for DEA researchers. Such a recent research trend is evident from the number of publications summarized in Table 28.6. The table lists previous DEA studies on environment.

During the 2010s, articles on environment occupied more than 40% of previous DEA studies in the past. These studies considered the environmental efficiency of various organizations at the level of companies, industries, regions and countries. Research efforts include exploring how to improve the degree of efficiency on operation and environment.

All the articles in the three categories regarding environment, as specified in Table 28.6, are identified by the following articles:

(a) Category T: [9], [15], [16], [19], [20], [28], [40], [48], [54], [55], [58], [62], [65], [66], [76], [78], [80], [81], [86], [87], [94], [99], [103], [105], [106], [107], [111], [112], [113], [117], [118], [124], [127], [136], [137], [142], [147], [149], [152], [153], [156], [157], [161], [165], [170], [171], [172], [176], [177], [188], [193], [201], [202], [208], [209], [211], [212], [214], [222], [226], [227], [228], [229], [234], [241], [245], [249], [255], [260], [261], [268], [269], [272], [274], [275], [276], [280], [284], [286], [294], [312], [323], [326], [331], [333], [335], [336], [337], [338], [339], [342], [344], [345], [346], [350], [351], [355], [357], [358], [360], [361], [362], [377], [378], [382], [407], [408], [409], [410], [412], [415], [423], [428], [433], [434], [436], [437], [439], [445], [446], [451], [453], [454], [456], [457], [463], [465], [468], [469], [470], [471], [472], [473], [474], [477], [478], [480], [482], [488], [493], [500], [506], [508], [510], [511], [513], [514], [515], [522], [525], [527], [528], [531], [535], [540], [541], [545], [550], [552], [557], [561], [562], [566], [567], [569], [575], [587], [588], [591], [592], [593], [605], [607], [610], [611], [613], [614], [615], [638], [639], [642], [646], [650], [651], [653], [654], [655], [658], [665], [666], [672], [673], [680], [683], [686], [687], [688] and [689].

(b) Category U: [44], [59], [64], [67], [89], [96], [98], [100], [104], [114], [138], [139], [141], [182], [190], [204], [247], [266], [300], [306], [309], [310], [328], [365], [422], [426], [427], [429], [464], [481], [556], [577], [586], [589], [595], [618], [636], [637], [675], [677] and [679].

(c) Category V: [21], [38], [39], [43], [56], [68], [85], [93], [102], [123], [140], [169], [218], [230], [233], [259], [278], [290], [298], [330], [334], [406], [411], [440], [443], [462], [529], [530], [535], [603] and [643].

The most important feature of these articles is that they incorporated undesirable outputs in their studies. Thus, their production factors included inputs and both desirable and undesirable outputs. The undesirable outputs implied the amount of CO_2 emission, chemical and pollution substances, and various types of waste discharge.

TABLE 28.6 Articles applied to environment

Main application areas	Topics	Combined analysis	1980's	1990's	2000's	2010's	All	Literature group
Environment	All		1	6	41	222	270	
	Environment	All	1	5	38	195	239	
		None	1	5	35	157	198	[T]
		Energy efficiency	0	0	3	38	41	[U]
	Sustainability		0	1	3	27	31	[V]

(a) Source: Sueyoshi *et al.* (2017).

In addition, some studies in the area focused on sustainability, including important concerns regarding sustainable development, sustainable supply chain management and eco-efficiency measurement for various entities and societies. The research trend, specified in Table 28.6, clearly indicates the importance of this research group. The number of studies regarding environment and sustainability dramatically increased during the 2010s (as exhibited in Table 28.6) from 48 (=1+6+41) articles before 2010 to a total of 222 articles after 2010.

28.6 OTHER APPLICATIONS

The category of "others" in Table 28.1 includes all articles that are not clearly classified into any specific category in the proposed classification. Table 28.7 indicates what research concerns were discussed in the articles belonging to this group.

All the articles in the four categories regarding others, as specified in Table 28.7, have the following articles:

(a) Category W: [321], [341], [663] and [682].
(b) Category X: [115], [291] and [292].
(c) Category Y: [307], [452] and [574].
(d) Category Z: [8], [92], [293], [373], [484] and [644].

The survey papers in Table 28.7 covered DEA studies applied to energy and environment as well as methodological ones. These studies did not focus upon DEA itself. Rather, they considered DEA as one of the useful methodologies for decision-making analysis on energy. Articles on research and development (R&D) examined the performance of project and technology progress. They were usually motivated by research fund allocation among projects and technologies. Articles on energy companies were interested in comparative efficiency analysis among them. These studies did not specify their industrial classifications, such as electricity, oil, coal, gas, heat and renewables. These studies were mainly

TABLE 28.7 Articles applied to other applications

Main application areas	Topics	1980s	1990s	2000s	2010s	All	Literature group
Others	All	0	0	3	13	16	
	Survey	0	0	1	3	4	[W]
	R&D	0	0	0	3	3	[X]
	Energy company	0	0	0	3	3	[Y]
	Others	0	0	2	4	6	[Z]

(a) Source: Sueyoshi *et al.* (2017).

concerned with energy efficiency and environment issues. Finally, note that articles summarized in Table 28.7 includes other research efforts regarding energy consumption analysis and the energy dependency of countries.

28.7 SUMMARY

This chapter reviewed previous research efforts that applied DEA to energy and environment issues. All previous works were classified by these applications, along with a research trend from the 1980s to the 2010s. Recently, many researchers began to pay serious attention to how to combat various difficulties related to energy and environment. The number of articles has dramatically increased, particularly after the 2000s. This chapter contains 693 articles in total, all of which were published in well-known international journals. This chapter summarizes recent research popularity on DEA environmental assessment, along with its time trend, among researchers and individuals who were interested in energy, environment and sustainability.

At the end of this chapter, it is important to note that this book is mainly concerned with methodological extensions of DEA and its applications to energy, environment and sustainability. As discussed in Chapter 15, most (approximately 90%) of the previous research used a straightforward application of conventional DEA models or those formulated under weak and strong disposability concepts. It is indeed true that many authors of DEA studies were not interested in methodological developments in their studies on energy, environment and sustainability. However, as discussed in Chapter 6, all DEA models are not perfect, and therefore have methodological strengths and drawbacks. A simple use of previous DEA models (e.g., RM and SBM) discussed in the chapters belonging to Section I have often misguided many policy agendas and business strategies on energy and environment, as discussed in Chapter 12. See also Chapter 15.

The use of weak and strong disposability concepts may be better than the straightforward use of DEA. A problem of conventional disposability concepts is that the weak disposability incorporated a possible occurrence of undesirable congestion (UC), so implying a capacity limit on part or all of a whole production facility. See Chapters 15, 21 and 23 for a detailed description on UC in DEA on energy, environment and sustainability. It is indeed important to identify such a possible occurrence of UC in guiding energy industries. However, is it important to discuss such a line limit on transmission or a capacity limit on transportation for environmental protection? The answer is clearly "no way." Therefore, this book has been discussing the concept of natural and/or managerial disposability to attain a high level of sustainability in preparing modern business strategy and energy policy. It is easily envisioned that managerial disposability, associated with a possible occurrence of DC, or eco-technology innovation, will be a major component of environmental assessment, as discussed in the chapters of Section II.

REFERENCES IN SECTION II

Abbott, M. (2005) Determining levels of productivity and efficiency in the electricity industry. *The Electricity Journal*, **18**, 62–72 [1].

Abbott, M. (2006) The productivity and efficiency of the Australian electricity supply industry. *Energy Economics*, **28**, 444–454 [2].

Agrell, P.J. and Bogetoft, P. (2005) Economic and environmental efficiency of district heating plants. *Energy Policy*, **33**, 1351–1362 [3].

Agrell, P.J., Bogetoft, P. and Tind, J. (2005) DEA and dynamic yardstick competition in scandinavian electricity distribution. *Journal of Productivity Analysis*, **23**, 173–201 [4].

Agrell, P.J. and Niknazar, P. (2014) Structural and behavioral robustness in applied best-practice regulation. *Socio-Economic Planning Sciences*, **48**, 89–103 [5].

Ajodhia, V. (2010) Integrated cost and quality benchmarking for electricity distribution using DEA. *International Journal of Energy Sector Management*, **4**, 417–433 [6].

Alp, I. and Sözen, A. (2014) Turkey's performance of energy consumption: A study making a comparison with the EU member states. *Energy Sources, Part B: Economics, Planning and Policy*, **9**, 87–100 [7].

Alp, I., Sözen, A. and Kazancioglu, S. (2013) Turkey's performance of sectoral energy consumption. *Energy Sources, Part B: Economics, Planning and Policy*, **8**, 94–105 [8].

Alsharif, K.A. and Fouad, G. (2012) Lake performance differences in response to land use and water quality: Data envelopment analysis. *Lake and Reservoir Management*, **28**, 130–141 [9].

Amado, C.A.F., Santos, S.P. and Sequeira, J.F.C. (2013) Using Data Envelopment Analysis to support the design of process improvement interventions in electricity distribution. *European Journal of Operational Research*, **228**, 226–235 [10].

Amirteimoori, A. and Kordrostami, S. (2012) A distance-based measure of super efficiency in data envelopment analysis: An application to gas companies. *Journal of Global Optimization*, **54**, 117–128 [11].

An, Q., Pang, Z., Chen, H. and Liang, L. (2015) Closest targets in environmental efficiency evaluation based on enhanced Russell measure. *Ecological Indicators*, **51**, 59–66 [12].

Arabi, B., Munisamy, S. and Emrouznejad, A. (2015) A new slacks-based measure of Malmquist–Luenberger index in the presence of undesirable outputs. *Omega*, **51**, 29–37 [13].

Arabi, B., Munisamy, S., Emrouznejad, A. and Shadman, F. (2014) Power industry restructuring and eco-efficiency changes: A new slacks-based model in Malmquist–Luenberger Index measurement. *Energy Policy*, **68**, 132–145 [14].

Arcelus, F.J. and Arocena, P. (2005) Productivity differences across OECD countries in the presence of environmental constraints. *Journal of Operational Research Society*, **56**, 1352–1362 [15].

Arjomandi, A. and Seufert, J.H. (2014) An evaluation of the world's major airlines' technical and environmental performance. *Economic Modelling*, **41**, 133–144 [16].

Arocena, P. (2008) Cost and quality gains from diversification and vertical integration in the electricity industry: A DEA approach. *Energy Economics*, **30**, 39–58 [17].

Athanassopoulos, A.D., Lambroukos, N. and Seiford, L. (1999) Data envelopment scenario analysis for setting targets to electricity generating plants. *European Journal of Operational Research*, **115**, 413–428 [18].

Avadí, Á., Vázquez-Rowe, I. and Fréon, P. (2014) Eco-efficiency assessment of the Peruvian anchoveta steel and wooden fleets using the LCA+DEA framework. *Journal of Cleaner Production*, **70**, 118–131 [19].

Azad, M.A.S. and Ancev, T. (2010) Using ecological indices to measure economic and environmental performance of irrigated agriculture. *Ecological Economics*, **69**, 1731–1739 [20].

Azad, M.A.S., Ancev, T. and Hernández-Sancho, F. (2015) Efficient water use for sustainable irrigation industry. *Water Resources Management*, **29**, 1683–1696 [21].

Azadeh, A., Amalnick, M.S., Ghaderi, S.F. and Asadzadeh, S.M. (2007) An integrated DEA PCA numerical taxonomy approach for energy efficiency assessment and consumption optimization in energy intensive manufacturing sectors. *Energy Policy*, **35**, 3792–3806 [22].

Azadeh, A., Gaeini, Z. and Moradi, B. (2014) Optimization of HSE in maintenance activities by integration of continuous improvement cycle and fuzzy multivariate approach: A gas refinery. *Journal of Loss Prevention in the Process Industries*, **32**, 415–427 [23].

Azadeh, A., Ghaderi, F. Anvari, M., Izadbakhsh, H. and Dehghan, S. (2007) Performance assessment and optimization of thermal power plants by DEA BCC and multivariate analysis. *Journal of Scientific and Industrial Research*, **66**, 860–872 [24].

Azadeh, A., Ghaderi, S.F., and Omrani, H. (2009) A deterministic approach for performance assessment and optimization of power distribution units in Iran. *Energy Policy*, **37**, 274–280 [25].

Azadeh, A., Ghaderi, S.F., Omrani, H. and Eivazy, H. (2009) An integrated DEA-COLS-SFA algorithm for optimization and policy making of electricity distribution units. *Energy Policy*, **37**, 2605–2618 [26].

Azadeh, A., Madine, M., Motevali Haghighi, S. and Mirzaei Rad, E. (2014) Continuous performance assessment and improvement of integrated HSE and maintenance systems by multivariate analysis in gas transmission units. *Journal of Loss Prevention in the Process Industries*, **27**, 32–41 [27].

Azadeh, A., Mianaei, H.S., Asadzadeh, S.M., Saberi, M. and Sheikhalishahi, M. (2015) A flexible ANN-GA-multivariate algorithm for assessment and optimization of machinery productivity in complex production units. *Journal of Manufacturing Systems*, **35**, 46–75 [28].

Azadeh, A., Mokhtari, Z., Sharahi, Z.J. and Zarrin, M. (2015) An integrated experiment for identification of best decision styles and teamworks with respect to HSE and ergonomics program: The case of a large oil refinery. *Accident Analysis and Prevention*, **85**, 30–44 [29].

Azadeh, A., Motevali Haghighi, S., Asadzadeh, S.M. and Saedi, H. (2013) A new approach for layout optimization in maintenance workshops with safety factors: The case of a gas transmission unit. *Journal of Loss Prevention in the Process Industries*, **26**, 1457–1465 [30].

Azadeh, A., Motevali Haghighi, S., Zarrin, M. and Khaefi, S. (2015) Performance evaluation of Iranian electricity distribution units by using stochastic data envelopment analysis. *International Journal of Electrical Power and Energy Systems*, **73**, 919–931 [31].

Azadeh, A. and Mousavi Ahranjani, P. (2014) The impact of job security, satisfaction and stress on performance assessment and optimization of generation companies. *Journal of Loss Prevention in the Process Industries*, **32**, 343–348 [32].

Azadeh, A., Rahimi-Golkhandan, A. and Moghaddam, M. (2014) Location optimization of wind power generation–transmission systems under uncertainty using hierarchical fuzzy DEA: A case study. *Renewable and Sustainable Energy Reviews*, **30**, 877–885 [33].

Azadeh, A., Saberi, M., Asadzadeh, S. M., Hussain, O.K. and Saberi, Z. (2013) A neuro-fuzzy-multivariate algorithm for accurate gas consumption estimation in South America with noisy inputs. *International Journal of Electrical Power and Energy Systems*, **46**, 315–325 [34].

Azadeh, A., Saberi, M., Asadzadeh, S.M. and Anvarian, N. (2013) An adaptive-network-based fuzzy inference system-data envelopment analysis algorithm for optimization of long-term electricity consumption, forecasting and policy analysis: The case of seven industrialized countries. *Energy Sources, Part B: Economics, Planning and Policy*, **8**, 56–66 [35].

Azadeh, A., Seraj, O., Asadzadeh, S.M. and Saberi, M. (2012) An integrated fuzzy regression-data envelopment analysis algorithm for optimum oil consumption estimation with ambiguous data. *Applied Soft Computing Journal*, **12**, 2614–2630 [36].

Azadeh, A. and Sheikhalishahi, M. (2015) An efficient Taguchi approach for the performance optimization of health safety, environment and ergonomics in generation companies. *Safety and Health at Work*, **6**, 77–84 [37].

Azadi, M., Jafarian, M., Farzipoor Saen, R. and Mirhedayatian, S.M. (2015) A new fuzzy DEA model for evaluation of efficiency and effectiveness of suppliers in sustainable supply chain management context. *Computers and Operations Research*, **54**, 274–285 [38].

Azadi, M., Shabani, A., Khodakarami, M. and Farzipoor Saen, R. (2014) Planning in feasible region by two-stage target-setting DEA methods: An application in green supply chain management of public transportation service providers. *Transportation Research Part E: Logistics and Transportation Review*, **70**, 324–338 [39].

Azadi, M., Shabani, A., Khodakarami, M. and Farzipoor Saen, R. (2015) Reprint of "Planning in feasible region by two-stage target-setting DEA methods: An application in green supply chain management of public transportation service providers." *Transportation Research Part E: Logistics and Transportation Review* **74**, 22–36 [40].

Babazadeh, R., Razmi, J., Pishvaee, M.S. and Rabbani, M. (2015) A non-radial DEA model for location optimization of Jatropha curcas L. cultivation. *Industrial Crops and Products*, **69**, 197–203 [41].

Bagdadioglu, N., Waddams Price, C.M. and Weyman-Jones, T.G. (1996) Efficiency and ownership in electricity distribution: A non-parametric model of the Turkish experience. *Energy Economics*, **18**, 1–23 [42].

Bai, C. and Sarkis, J. (2014) Determining and applying sustainable supplier key performance indicators. *Supply Chain Management*, **19**, 275–291 [43].

Balezentis, T., Li, T., Streimikiene, D. and Balezentis, A. (2016) Is the Lithuanian economy approaching the goals of sustainable energy and climate change mitigation? Evidence from DEA-based environmental performance index. *Journal of Cleaner Production*, **116**, 23–31 [44].

Bampatsou, C., Papadopoulos, S. and Zervas, E. (2013) Technical efficiency of economic systems of EU-15 countries based on energy consumption. *Energy Policy*, **55**, 426–434 [45].

Banaeian, N., Omid, M. and Ahmadi, H. (2012) Greenhouse strawberry production in Iran, efficient or inefficient in energy. *Energy Efficiency*, **5**, 201–209 [46].

Banaeian, N. and Zangeneh, M. (2011) Study on energy efficiency in corn production of Iran. *Energy*, **36**, 5394–5402 [47].

Barla, P. and Perelman, S. (2005) Sulphur emissions and productivity growth in industrialized countries. *Annals of Public and Cooperative Economics*, **76**, 275–300 [48].

Barros, C.P. (2008) Efficiency analysis of hydroelectric generating plants: A case study for Portugal. *Energy Economics*, **30**, 59–75 [49].

Barros, C.P. and Assaf, A. (2009) Bootstrapped efficiency measures of oil blocks in Angola. *Energy Policy*, **37**, 4098–4103 [50].

Barros, C.P. and Managi, S. (2009) Productivity assessment of Angola's oil blocks. *Energy*, **34**, 2009–2015 [51].

Barros, C.P. and Peypoch, N. (2008) Technical efficiency of thermoelectric power plants. *Energy Economics*, **30**, 3118–3127 [52].

Behera, S.K., Farooquie, J.A. and Dash, A.P. (2011) Productivity change of coal-fired thermal power plants in India: A Malmquist index approach. *IMA Journal of Management Mathematics*, **22**, 387–400 [53].

Beltrán-Esteve, M., Gómez-Limón, J.A., Picazo-Tadeo, A.J. and Reig-Martínez, E. (2014) A metafrontier directional distance function approach to assessing eco-efficiency. *Journal of Productivity Analysis*, **41**, 69–83 [54].

Beltrán-Esteve, M. and Picazo-Tadeo, A.J. (2015) Assessing environmental performance trends in the transport industry: Eco-innovation or catching-up? *Energy Economics*, **51**, 570–580 [55].

Belu, C. (2009) Ranking corporations based on sustainable and socially responsible practices. A data envelopment analysis (DEA) approach, *Sustainable Development*, **17**, 257–268 [56].

Berg, S., Lin, C. and Tsaplin, V. (2005) Regulation of state-owned and privatized utilities: Ukraine electricity distribution company performance. *Journal of Regulatory Economics*, **28**, 259–287 [57].

Berre, D., Boussemart, J.-P., Leleu, H. and Tillard, E. (2013) Economic value of greenhouse gases and nitrogen surpluses: Society vs farmers' valuation. *European Journal of Operational Research*, **226**, 325–331 [58].

Berre, D., Vayssières, J., Boussemart, J.-P., Leleu, H., Tillard, E. and Lecomte, P. (2015) A methodology to explore the determinants of eco-efficiency by combining an agronomic whole-farm simulation model and efficient frontier. *Environmental Modelling and Software*, **71**, 46–59 [59].

Bevilacqua, M. and Braglia, M. (2002) Environmental efficiency analysis for ENI oil refineries. *Journal of Cleaner Production*, **10**, 85–92 [60].

Bi, G., Wang, P., Yang, F. and Liang, L. (2014) Energy and environmental efficiency of china's transportation sector: A multidirectional analysis approach. *Mathematical Problems in Engineering*, **2014**(539596), 1–12 [61].

Bi, G.-B., Song, W. and Wu, J. (2014) A clustering method for evaluating the environmental performance based on slacks-based measure. *Computers and Industrial Engineering*, **72**, 169–177 [62].

Bi, G.-B., Song, W., Zhou, P. and Liang, L. (2014) Does environmental regulation affect energy efficiency in China's thermal power generation? Empirical evidence from a slacks-based DEA model. *Energy Policy*, **66**, 537–546 [63].

Bian, Y., He, P. and Xu, H. (2013) Estimation of potential energy saving and carbon dioxide emission reduction in China based on an extended non-radial DEA approach. *Energy Policy*, **63**, 962–971 [64].

Bian, Y., Liang, N., and Xu, H. (2015) Efficiency evaluation of Chinese regional industrial systems with undesirable factors using a two-stage slacks-based measure approach. *Journal of Cleaner Production*, **87**, 348–356 [65].

Bian, Y., Yan, S. and Xu, H. (2014) Efficiency evaluation for regional urban water use and wastewater decontamination systems in China: A DEA approach. *Resources, Conservation and Recycling*, **83**, 15–23 [66].

Bian, Y. and Yang, F. (2010) Resource and environment efficiency analysis of provinces in China: A DEA approach based on Shannon's entropy. *Energy Policy*, **38**, 1909–1917 [67].

Blancard, S. and Hoarau, J.-F. (2013) A new sustainable human development indicator for small island developing states: A reappraisal from data envelopment analysis. *Economic Modelling*, **30**, 623–635 [68].

Blancard, S. and Martin, E. (2014) Energy efficiency measurement in agriculture with imprecise energy content information. *Energy Policy*, **66**, 198–208 [69].

Blokhuis, E., Advokaat, B. and Schaefer, W. (2012) Assessing the performance of Dutch local energy companies. *Energy Policy*, **45**, 680–690 [70].

Blomberg, J., Henriksson, E. and Lundmark, R. (2012) Energy efficiency and policy in Swedish pulp and paper mills: A data envelopment analysis approach. *Energy Policy*, **42**, 569–579 [71].

Blum, H. (2015) The economic efficiency of energy-consuming equipment: a DEA approach. *Energy Efficiency*, **8**, 281–298 [72].

Bolandnazar, E., Keyhani, A. and Omid, M. (2014) Determination of efficient and inefficient greenhouse cucumber producers using data envelopment analysis approach, a case study: Jiroft city in Iran. *Journal of Cleaner Production*, **79**, 108–115 [73].

Boubaker, K. (2012a) Renewable energy in upper North Africa: Present versus 2025-horizon perspectives optimization using a data envelopment analysis (DEA) framework. *Renewable Energy*, **43**, 364–369 [74].

Boubaker, K. (2012b) A review on renewable energy conceptual perspectives in North Africa using a polynomial optimization scheme. *Renewable and Sustainable Energy Reviews*, **16**, 4298–4302 [75].

Boyd, G.A. and McClelland, J.D. (1999) The Impact of environmental constraints on productivity improvement in integrated paper plants. *Journal of Environmental Economics and Management*, **38**, 121–142 [76].

Boyd, G.A. and Pang, J.X. (2000) Estimating the linkage between energy efficiency and productivity. *Energy Policy*, **28**, 289–296 [77].

Boyd, G.A., Tolley, G. and Pang, J. (2002) Plant level productivity, efficiency, and environmental performance of the container glass industry. *Environmental and Resource Economics*, **23**, 29–43 [78].

Bozoğlu, M. and Ceyhan, V. (2009) Energy conversion efficiency of trout and sea bass production in the Black Sea, Turkey. *Energy*, **34**, 199–204 [79].

Brännlund, R., Chung, Y., Färe, R. and Grosskopf, S. (1998) Emissions trading and profitability: The Swedish pulp and paper industry. *Environmental and Resource Economics*, **12**, 345–356 [80].

Bretholt, A. and Pan, J.-N. (2013) Evolving the latent variable model as an environmental DEA technology. *Omega*, **41**, 315–325 [81].

Budeba, M.D., Joubert, J.W. and Webber-Youngman, R.C.W. (2015) A proposed approach for modelling competitiveness of new surface coal mines. *The Journal of the Southern African Institute of Mining and Metallurgy*, **115**, 1057–1064 [82].

Burnett, R.D. and Hansen, D.R. (2008) Ecoefficiency: Defining a role for environmental cost management. *Accounting, Organizations and Society*, **33**, 551–581 [83].

Byrnes, P., Färe, R., Grosskopf, S. and Lovell, C.A.K. (1988) The effect of unions on productivity: U.S. surface mining of Coal. *Management Science*, **34**, 1037–1053 [84].

Callens, I. and Tyteca, D. (1999) Towards indicators of sustainable development for firms: A productive efficiency perspective. *Ecological Economics*, **28**, 41–53 [85].

Camarero, M., Castillo, J., Picazo-Tadeo, A.J. and Tamarit, C. (2013) Eco-Efficiency and Convergence in OECD Countries. *Environmental and Resource Economics*, **55**, 87–106 [86].

Camarero, M., Castillo-Giménez, J., Picazo-Tadeo, A.J. and Tamarit, C. (2014) Is eco-efficiency in greenhouse gas emissions converging among European Union countries? *Empirical Economics*, **47**, 143–168 [87].

Cambini, C., Croce, A. and Fumagalli, E. (2014) Output-based incentive regulation in electricity distribution: Evidence from Italy. *Energy Economics*, **45**, 205–216 [88].

Camioto, F.D.C., Mariano, E.B. and Rebelatto, D.A.D.N. (2014) Efficiency in Brazil's industrial sectors in terms of energy and sustainable development. *Environmental Science and Policy*, **37**, 50–60 [89].

Çelen, A. (2013) Efficiency and productivity (TFP) of the Turkish electricity distribution companies: An application of two-stage (DEA and Tobit) analysis. *Energy Policy*, **63**, 300–310 [90].

Çelen, A. and Yalçın, N. (2012) Performance assessment of Turkish electricity distribution utilities: An application of combined FAHP/TOPSIS/DEA methodology to incorporate quality of service. *Utilities Policy*, **23**, 59–71 [91].

Chan, W. (2012) Energy benchmarking in support of low carbon hotels: Developments, challenges, and approaches in China. *International Journal of Hospitality Management*, **31**, 1130–1142 [92].

Chang, D.-S., Kuo, L.-C.R. and Chen, Y.-T. (2013) Industrial changes in corporate sustainability performance: An empirical overview using data envelopment analysis. *Journal of Cleaner Production*, **56**, 147–155 [93].

Chang, D.-S., Liu, W. and Yeh, L.-T. (2013) Incorporating the learning effect into data envelopment analysis to measure MSW recycling performance. *European Journal of Operational Research*, **229**, 496–504 [94].

Chang, D.-S. and Yang, F.-C. (2011) Assessing the power generation, pollution control, and overall efficiencies of municipal solid waste incinerators in Taiwan. *Energy Policy*, **39**, 651–663 [95].

Chang, D.-S., Yeh, L.-T. and Liu, W. (2015) Incorporating the carbon footprint to measure industry context and energy consumption effect on environmental performance of business operations. *Clean Technologies and Environmental Policy*, **17**, 359–371 [96].

Chang, M.-C. (2014) Energy intensity, target level of energy intensity, and room for improvement in energy intensity: An application to the study of regions in the EU. *Energy Policy*, **67**, 648–655 [97].

Chang, M.-C. (2015) Room for improvement in low carbon economies of G7 and BRICS countries based on the analysis of energy efficiency and environmental Kuznets curves. *Journal of Cleaner Production*, **99**, 140–151 [98].

Chang, Y.-T., Park, H.-S., Jeong, J.-B. and Lee, J.-W. (2014) Evaluating economic and environmental efficiency of global airlines: A SBM-DEA approach. *Transportation Research Part D: Transport and Environment*, **27**, 46–50 [99].

Chang, Y.-T., Zhang, N., Danao, D. and Zhang, N. (2013) Environmental efficiency analysis of transportation system in China: A non-radial DEA approach. *Energy Policy*, **58**, 277–283 [100].

Chauhan, N.S., Mohapatra, P.K.J. and Pandey, K.P. (2006) Improving energy productivity in paddy production through benchmarking? An application of data envelopment analysis. *Energy Conversion and Management*, **47**, 1063–1085 [101].

Chen, C., Zhu, J., Yu, J.-Y. and Noori, H. (2012) A new methodology for evaluating sustainable product design performance with two-stage network data envelopment analysis. *European Journal of Operational Research*, **221**, 348–359 [102].

Chen, C.C. (2015) Assessing the pollutant abatement cost of greenhouse gas emission regulation: a case study of Taiwan's freeway bus service industry. *Environmental and Resource Economics*, **61**, 477–495 [103].

Chen, J., Song, M. and Xu, L. (2015) Evaluation of environmental efficiency in China using data envelopment analysis. *Ecological Indicators*, **52**, 577–583 [104].

Chen, L., Wang, Y.-M. and Wang, L. (2016) Congestion measurement under different policy objectives: an analysis of Chinese industry. *Journal of Cleaner Production*, **112**, 2943–2952 [105].

Chen, P.-C., Chang, C.-C., Yu, M.-M. and Hsu, S.-H. (2012) Performance measurement for incineration plants using multi-activity network data envelopment analysis: The case of Taiwan. *Journal of Environmental Management*, **93**, 95–103 [106].

Chen, P.-C., Yu, M.-M., Chang, C.-C., Hsu, S.-H. and Managi, S. (2015) The enhanced Russell-based directional distance measure with undesirable outputs: Numerical example considering CO2 emissions. *Omega*, **53**, 30–40 [107].

Chien, C.-F., Lo, F.-Y. and Lin, J.T. (2003) Using DEA to measure the relative efficiency of the service center and improve operation efficiency through reorganization. *IEEE Transactions on Power Systems*, **18**, 366–373 [108].

Chien, T. and Hu, J.-L. (2007) Renewable energy and macroeconomic efficiency of OECD and non-OECD economies. *Energy Policy*, **35**, 3606–3615 [109].

Chitkara, P. (1999) A data envelopment analysis approach to evaluation of operational inefficiencies in power generating units: a case study of Indian power plants. *IEEE Transactions on Power Systems*, **14**, 419–425 [110].

Chiu, Y.-H., Huang, C.-W. and Ma, C.-M. (2011) Assessment of China transit and economic efficiencies in a modified value-chains DEA model. *European Journal of Operational Research*, **209**, 95–103 [111].

Chiu, Y.-H., Lee, J.-H., Lu, C.-C., Shyu, M.-K. and Luo, Z. (2012) The technology gap and efficiency measure in WEC countries: Application of the hybrid meta frontier model. *Energy Policy*, **51**, 349–357 [112].

Chiu, Y.-H., Lin, J.-C., Su, W.-N. and Liu, J.-K. (2015) An efficiency evaluation of the EU's allocation of carbon emission allowances. *Energy Sources, Part B: Economics, Planning and Policy*, **10**, 192–200 [113].

Choi, Y., Zhang, N. and Zhou, P. (2012) Efficiency and abatement costs of energy-related CO2 emissions in China: A slacks-based efficiency measure. *Applied Energy*, **98**, 198–208 [114].

Chun, D., Hong, S., Chung, Y., Woo, C. and Seo, H. (2016) Influencing factors on hydrogen energy R&D projects: An ex-post performance evaluation. *Renewable and Sustainable Energy Reviews*, **53**, 1252–1258 [115].

Chung, W. (2011) Review of building energy-use performance benchmarking methodologies. *Applied Energy*, **88**, 1470–1479 [116].

Chung, Y.H., Färe, R. and Grosskopf, S. (1997) Productivity and undesirable outputs: a directional distance function approach. *Journal of Environmental Management*, **51**, 229–240 [117].

Çipil, F. (2014) Performance analysis of Turkey's transport sector greenhouse gas emissions. *Energy and Environment*, **25**, 357–368 [118].

Claggett, E.T. and Ferrier, G.D. (1998) The efficiency of TVA power distributors. *Managerial and Decision Economics*, **19**, 365–376 [119].

Cook, W.D. and Green, R.H. (2005) Evaluating power plant efficiency: A hierarchical model. *Computers and Operations Research*, **32**, 813–823 [120].

Costa, A.O., Oliveira, L.B., Lins, M.P.E., Silva, A.C.M., Araujo, M.S.M., Pereira Jr., A.O. and Rosa, L.P. (2013) Sustainability analysis of biodiesel production: A review on different resources in Brazil. *Renewable and Sustainable Energy Reviews*, **27**, 407–412 [121].

Costa, M.A., Lopes, A.L.M. and de Pinho Matos, G.B.B. (2015) Statistical evaluation of Data Envelopment Analysis versus COLS Cobb–Douglas benchmarking models for the 2011 Brazilian tariff revision. *Socio-Economic Planning Sciences*, **49**, 47–60 [122].

Cracolici, M.F., Cuffaro, M. and Nijkamp, P. (2008) Sustainable tourist development in Italian holiday destinations. *International Journal of Services, Technology and Management*, **10**, 39–47 [123].

Criswell, D.R. and Thompson, R.G. (1996) Data Envelopment Analysis of space and terrestrially-based large scale commercial power systems for earth: A prototype analysis of their relative economic advantages. *Solar Energy*, **56**, 119–131 [124].

Cui, Q., Kuang, H.-B., Wu, C.-Y. and Li, Y. (2014) The changing trend and influencing factors of energy efficiency: The case of nine countries. *Energy*, **64**, 1026–1034 [125].

Cui, Q. and Li, Y. (2014) The evaluation of transportation energy efficiency: An application of three-stage virtual frontier DEA. *Transportation Research Part D: Transport and Environment*, **29**, 1–11 [126].

Cui, Q. and Li, Y. (2015a) An empirical study on the influencing factors of transportation carbon efficiency: Evidences from fifteen countries. *Applied Energy*, **141**, 209–217 [127].

Cui, Q. and Li, Y. (2015b) Evaluating energy efficiency for airlines: An application of VFB-DEA. *Journal of Air Transport Management*, **44/45**, 34–41 [128].

Cui, Y., Huang, G. and Yin, Z. (2015) Estimating regional coal resource efficiency in China using three-stage DEA and bootstrap DEA models. *International Journal of Mining Science and Technology*, **25**, 861–864 [129].

Cullmann, A. and Von Hirschhausen, C. (2008a) Efficiency analysis of East European electricity distribution in transition: Legacy of the past? *Journal of Productivity Analysis*, **29**, 155–167 [130].

Cullmann, A. and Von Hirschhausen, C. (2008b) From transition to competition: Dynamic efficiency analysis of Polish electricity distribution companies. *Economics of Transition*, **16**, 335–357 [131].

Dai, X. and Kuosmanen, T. (2014) Best-practice benchmarking using clustering methods: Application to energy regulation. *Omega*, **42**, 179–188 [132].

De Andrade, G.N., Alves, L.A., Da Silva, C.E.R.F. and De Mello, J.C.C.B.S. (2014) Evaluating electricity distributors efficiency using self-organizing map and data envelopment analysis. *IEEE Latin America Transactions*, **12**, 1464–1472 [133].

Delmas, M. and Tokat, Y. (2005) Deregulation, governance structures, and efficiency: The U.S. electric utility sector. *Strategic Management Journal*, **26**, 441–460 [134].

Dogan, N.O. and Tugcu, C.T. (2015) Energy efficiency in electricity production: A data envelopment analysis (DEA) approach for the G-20 countries. *International Journal of Energy Economics and Policy*, **5**, 246–252 [135].

Du, M., Wang, B. and Wu, Y. (2014) Sources of China's economic growth: An empirical analysis based on the BML index with green growth accounting. *Sustainability*, **6**, 5983–6004 [136].

Dyckhoff, H. and Allen, K. (2001) Measuring ecological efficiency with data envelopment analysis (DEA). *European Journal of Operational Research*, **132**, 312–325 [137].

Ebrahimi, R. and Salehi, M. (2015) Investigation of CO2 emission reduction and improving energy use efficiency of button mushroom production using data envelopment analysis. *Journal of Cleaner Production*, **103**, 112–119 [138].

Egilmez, G., Kucukvar, M. and Tatari, O. (2013) Sustainability assessment of U.S. manufacturing sectors: An economic input output-based frontier approach. *Journal of Cleaner Production*, **53**, 91–102 [139].

Egilmez, G., Kucukvar, M., Tatari, O. and Bhutta, M.K.S. (2014) Supply chain sustainability assessment of the U.S. food manufacturing sectors: A life cycle-based frontier approach. *Resources, Conservation and Recycling*, **82**, 8–20 [140].

Egilmez, G. and Park, Y.S. (2014) Transportation related carbon, energy and water footprint analysis of U.S. manufacturing: An eco-efficiency assessment. *Transportation Research Part D: Transport and Environment*, **32**, 143–159 [141].

Eguchi, S., Kagawa, S. and Okamoto, S. (2015) Environmental and economic performance of a biodiesel plant using waste cooking oil. *Journal of Cleaner Production*, **101**, 245–250 [142].

Eller, S.L., Hartley, P.R. and Medlock III, K.B. (2011) Empirical evidence on the operational efficiency of national oil companies. *Empirical Economics*, **40**, 623–643 [143].

Erbetta, F. and Rappuoli, L. (2008) Optimal scale in the Italian gas distribution industry using data envelopment analysis. *Omega*, **36**, 325–336 [144].

Ertürk, M. and Türüt-Aşık, S. (2011) Efficiency analysis of Turkish natural gas distribution companies by using data envelopment analysis method. *Energy Policy*, **39**, 1426–1438 [145].

Ewertowska, A., Galan-Martin, A., Guillen-Gosalbez, G. and Jimenez, L. (2016) Assessment of the environmental efficiency of the electricity mix of the top European economies via data envelopment analysis. *Journal of Cleaner Production*, **116**, 13–22 [146].

Falavigna, G., Manello, A. and Pavone, S. (2013) Environmental efficiency, productivity and public funds: The case of the Italian agricultural industry. *Agricultural Systems*, **121**, 73–80 [147].

Fallahi, A., Ebrahimi, R. and Ghaderi, S.F. (2011) Measuring efficiency and productivity change in power electric generation management companies by using data envelopment analysis: A case study. *Energy*, **36**, 6398–6405 [148].

Fan, J.-L., Zhang, X., Zhang, J. and Peng, S. (2014) Efficiency evaluation of CO2 utilization technologies in China: A super-efficiency DEA analysis based on expert survey. *Journal of CO2 Utilization*, **11**, 54–62 [149].

Fang, C.-Y., Hu, J.-L. and Lou, T.-K. (2013) Environment-adjusted total-factor energy efficiency of Taiwan's service sectors. *Energy Policy*, **63**, 1160–1168 [150].

Fang, H., Wu, J. and Zeng, C. (2009) Comparative study on efficiency performance of listed coal mining companies in China and the US. *Energy Policy*, **37**, 5140–5148 [151].

Färe, R. and Grosskopf, S. (2004) Modeling undesirable factors in efficiency evaluation: Comment. *European Journal of Operational Research*, **157**, 242–352 [152].

Färe, R., Grosskopf, S. and Hernandez-Sancho, F. (2004) Environmental performance: an index number approach. *Resource and Energy Economics*, **26**, 343–352 [153].

Färe, R., Grosskopf, S. and Logan, J. (1983) The relative efficiency of Illinois electric utilities. *Resources and Energy*, **5**, 349–367 [154].

Färe, R., Grosskopf, S. and Logan, J. (1985) The relative performance of publicly-owned and privately-owned electric utilities. *Journal of Public Economics*, **26**, 89–106 [155].

Färe, R., Grosskopf, S., Lovell, C.A.K. and Pasurka, C. (1989) Multilateral productivity comparisons when some outputs are undesirable: A nonparametric approach. *The Review of Economics and Statistics*, **71**, 90–98 [156].

Färe, R., Grosskopf, S., Margaritis, D. and Weber, W.L. (2012) Technological change and timing reductions in greenhouse gas emissions. *Journal of Productivity Analysis*, **37**, 205–216 [157].

Färe, R., Grosskopf, S., Noh, D.-W. and Weber, W. (2005) Characteristics of a polluting technology: Theory and practice. *Journal of Econometrics*, **126**, 469–492 [158].

Färe, R., Grosskopf, S. and Pasurka Jr, C.A. (2014) Potential gains from trading bad outputs: The case of U.S. electric power plants. *Resource and Energy Economics*, **36**, 99–112 [159].

Färe, R., Grosskopf, S. and Pasurka, C. (1986) Effects on relative efficiency in electric power generation due to environmental controls. *Resources and Energy*, **8**, 167–184 [160].

Färe, R., Grosskopf, S. and Pasurka, J.C.A. (2001) Accounting for air pollution emission in measures of state manufacturing productivity growth. *Journal of Regional Science*, **41**, 381–409 [161].

Färe, R., Grosskopf, S. and Tyteca, D. (1996) An activity analysis model of the environmental performance of firms-application to fossil-fuel-fired electric utilities. *Ecological Economics*, **18**, 161–175 [162].

Färe, R., Grosskopf, S., Yaisawarng, S., Li, S.K. and Wang, Z. (1990) Productivity growth in Illinois electric utilities. *Resources and Energy*, **12**, 383–398 [163].

Farzipoor Saen, R. (2010) Performance measurement of power plants in the existence of weight restrictions via slacks-based model. *Benchmarking: An International Journal*, **17**, 677–691 [164].

Feng, C., Chu, F., Ding, J., Bi, G. and Liang, L. (2015) Carbon Emissions Abatement (CEA) allocation and compensation schemes based on DEA. *Omega*, **53**, 78–89 [165].

Ferrier, G.D. and Hirschberg, J.G. (1992) Climate control efficiency. *The Energy Journal*, **13**, 37–54 [166].

Fleishman, R., Alexander, R., Bretschneider, S. and Popp, D. (2009) Does regulation stimulate productivity? The effect of air quality policies on the efficiency of US power plants. *Energy Policy*, **37**, 4574–4582 [167].

Francisco, C.A.C., de Almeida, M.R. and da Silva, D.R. (2012) Efficiency in Brazilian refineries under different DEA technologies. *International Journal of Engineering Business Management*, **4**, 1–11 [168].

Frota Neto, J.Q., Bloemhof-Ruwaard, J.M., van Nunen, J.A.E.E. and van Heck, E. (2008) Designing and evaluating sustainable logistics networks, *International Journal of Production Economics*, **111**, 195–208 [169].

Fu, H.-P. and Ou, J.-R. (2013) Combining PCA with DEA to improve the evaluation of project performance data: A Taiwanese bureau of energy case study. *Project Management Journal*, **44**, 94–106 [170].

Fujii, H. and Managi, S. (2015) Optimal production resource reallocation for CO2 emissions reduction in manufacturing sectors. *Global Environmental Change*, **35**, 505–513 [171].

Gerdessen, J.C. and Pascucci, S. (2013) Data envelopment analysis of sustainability indicators of European agricultural systems at regional level. *Agricultural Systems*, **118**, 78–90 [172].

Giannakis, D., Jamasb, T. and Pollitt, M. (2005) Benchmarking and incentive regulation of quality of service: an application to the UK electricity distribution networks. *Energy Policy*, **33**, 2256–2271 [173].

Golany, B., Roll, Y. and Rybak, D. (1994) Measuring efficiency of power plants in Israel by data envelopment analysis. *IEEE Transactions on Engineering Management*, **41**, 29–301 [174].

Gómez-Calvet, R., Conesa, D., Gómez-Calvet, A.R. and Tortosa-Ausina, E. (2014) Energy efficiency in the European Union: What can be learned from the joint application of directional distance functions and slacks-based measures? *Applied Energy*, **132**, 137–154 [175].

Gómez-Calvet, R., Conesa, D., Gómez-Calvet, A.R. and Tortosa-Ausina, E. (2016) On the dynamics of eco-efficiency performance in the European Union. *Computers and Operations Research*, **66**, 336–350 [176].

Gómez-Limón, J.A., Picazo-Tadeo, A.J. and Reig-Martínez, E. (2012) Eco-efficiency assessment of olive farms in Andalusia. *Land Use Policy*, **29**, 395–406 [177].

Goncharuk, A.G. (2008) Performance benchmarking in gas distribution industry, *Benchmarking: An International Journal*, **15**, 548–559 [178].

Goncharuk, A.G. (2013) What causes increase in gas prices: The case of Ukraine. *International Journal of Energy Sector Management*, **7**, 448–458 [179].

Gong, Z., Zhao, Y. and Ge, X. (2014) Efficiency assessment of the energy consumption and economic indicators in Beijing under the influence of short-term climatic factors: Based on data envelopment analysis methodology. *Natural Hazards*, **71**, 1145–1157 [180].

González-García, S., Villanueva-Rey, P., Belo, S., Vázquez-Rowe, I., Moreira, M.T., Feijoo, G. and Arroja, L. (2015) Cross-vessel eco-efficiency analysis. A case study for purse seining fishing from North Portugal targeting European pilchard. *International Journal of Life Cycle Assessment*, **20**, 1019–1032 [181].

Goto, M., Otsuka, A. and Sueyoshi, T. (2014) DEA (data envelopment analysis) assessment of operational and environmental efficiencies on Japanese regional industries. *Energy*, **66**, 535–549 [182].

Goto, M. and Sueyoshi, T. (2008) Financial ratio analysis: An application to US energy industry. In: Lee, J.-D. and Heshmati, A. (eds.) *Productivity, Efficiency, and Economic Growth in the Asia-Pacific Region*, Physica-Verlag HD, Berlin, pp. 59–79 [183].

Goto, M. and Tsutsui, M. (1998) Comparison of productive and cost efficiencies among Japanese and US electric utilities. *Omega*, **26**, 177–194 [184].

Gouveia, M.C., Dias, L.C., Antunes, C.H., Boucinha, J. and Inácio, C.F. (2015) Benchmarking of maintenance and outage repair in an electricity distribution company using the value-based DEA method. *Omega*, **53**, 104–114 [185].

Grigoroudis, E., Petridis, K. and Arabatzis, G. (2014) RDEA: A recursive DEA based algorithm for the optimal design of biomass supply chain networks. *Renewable Energy*, **71**, 113–122 [186].

Grösche, P. (2009) Measuring residential energy efficiency improvements with DEA. *Journal of Productivity Analysis*, **31**, 87–94 [187].

Guo, D. and Wu, J. (2013) A complete ranking of DMUs with undesirable outputs using restrictions in DEA models. *Mathematical and Computer Modelling*, **58**, 1102–1109 [188].

Guo, L.-L., Wu, C.-Y., Qu, Y. and Yu, J.-T. (2015) Evaluation of the energy-saving and emission-reduction potential for Chinese provinces based on regional difference coefficients. *Journal of Renewable and Sustainable Energy*, **7**(043149), 1–20 [189].

Guo, X.-D., Zhu, L., Fan, Y. and Xie, B.-C. (2011) Evaluation of potential reductions in carbon emissions in Chinese provinces based on environmental DEA. *Energy Policy*, **39**, 2352–2360 [190].

Halkos, G.E. and Tzeremes, N.G. (2011) Oil consumption and economic efficiency: A comparative analysis of advanced, developing and emerging economies. *Ecological Economics*, **70**, 1354–1362 [191].

Halkos, G.E. and Tzeremes, N.G. (2013) Renewable energy consumption and economic efficiency: Evidence from European countries. *Journal of Renewable and Sustainable Energy*, **5**(041803), 1–13 [192].

Halkos, G.E. and Tzeremes, N.G. (2014) Measuring the effect of Kyoto protocol agreement on countries' environmental efficiency in CO2 emissions: an application of conditional full frontiers. *Journal of Productivity Analysis*, **41**, 367–382 [193].

Hampf, B. (2014) Separating environmental efficiency into production and abatement efficiency: A nonparametric model with application to US power plants. *Journal of Productivity Analysis*, **41**, 457–473 [194].

Hampf, B. and Rødseth, K.L. (2015) Carbon dioxide emission standards for U.S. power plants: An efficiency analysis perspective. *Energy Economics*, **50**, 140–153 [195].

Han, X., Xue, X., Ge, J., Wu, H. and Su, C. (2014) Measuring the productivity of energy consumption of major industries in China: A DEA-based method. *Mathematical Problems in Engineering*, **2014**(121804), 1–12 [196].

Han, Y. and Geng, Z. (2014) Energy efficiency hierarchy evaluation based on data envelopment analysis and its application in a petrochemical process. *Chemical Engineering and Technology*, **37**, 2085–2095 [197].

Han, Y., Geng, Z., Gu, X. and Zhu, Q. (2015) Energy efficiency analysis based on DEA integrated ISM: A case study for Chinese ethylene industries. *Engineering Applications of Artificial Intelligence*, **45**, 80–89 [198].

Han, Y., Geng, Z. and Liu, Q. (2014) Energy efficiency evaluation based on data envelopment analysis integrated analytic hierarchy process in ethylene production. *Chinese Journal of Chemical Engineering*, **22**, 1279–1284 [199].

Han, Y., Geng, Z., Zhu, Q. and Qu, Y. (2015) Energy efficiency analysis method based on fuzzy DEA cross-model for ethylene production systems in chemical industry. *Energy*, **83**, 685–695 [200].

Haralambides, H. and Gujar, G. (2012) On balancing supply chain efficiency and environmental impacts: An eco-DEA model applied to the dry port sector of India. *Maritime Economics and Logistics*, **14**, 122–137 [201].

Hathroubi, S., Peypoch, N. and Robinot, E. (2014) Technical efficiency and environmental management: The Tunisian case. *Journal of Hospitality and Tourism Management*, **21**, 27–33 [202].

Hawdon, D. (2003) Efficiency, performance and regulation of the international gas industry – a bootstrap DEA approach. *Energy Policy*, **31**, 1167–1178 [203].

He, F., Zhang, Q., Lei, J., Fu, W. and Xu, X. (2013) Energy efficiency and productivity change of China's iron and steel industry: Accounting for undesirable outputs. *Energy Policy*, **54**, 204–213 [204].

Heidari, M.D., Omid, M. and Akram, A. (2011) Optimization of energy consumption of broiler production farms using data envelopment analysis approach. *Modern Applied Science*, **5**, 69–78 [205].

Heidari, M.D., Omid, M. and Mohammadi, A. (2012) Measuring productive efficiency of horticultural greenhouses in Iran: A data envelopment analysis approach. *Expert Systems with Applications*, **39**, 1040–1045 [206].

Hernández-Sancho, F., Molinos-Senante, M. and Sala-Garrido, R. (2011) Energy efficiency in Spanish wastewater treatment plants: A non-radial DEA approach. *Science of the Total Environment*, **409**, 2693–2699 [207].

Hernández-Sancho, F., Picazo-Tadeo, A. and Reig-Martinez, E. (2000) Efficiency and environmental regulation. *Environmental and Resource Economics*, **15**, 365–378 [208].

Hernández-Sancho, F. and Sala-Garrido, R. (2009) Technical efficiency and cost analysis in wastewater treatment processes: A DEA approach. *Desalination*, **249**, 230–234 [209].

Hess, B. and Cullmann, A. (2007) Efficiency analysis of East and West German electricity distribution companies – Do the "Ossis" really beat the "Wessis"? *Utilities Policy*, **15**, 206–214 [210].

Hoang, V.-N. and Alauddin, M. (2012) Input-orientated data envelopment analysis framework for measuring and decomposing economic, environmental and ecological efficiency: An application to OECD agriculture. *Environmental and Resource Economics*, **51**, 431–452 [211].

Hoang, V.-N. and Nguyen, T.T. (2013) Analysis of environmental efficiency variations: A nutrient balance approach. *Ecological Economics*, **86**, 37–46 [212].

Honma, S. and Hu, J.-L. (2008) Total-factor energy efficiency of regions in Japan. *Energy Policy*, **36**, 821–833 [213].

Honma, S. and Hu, J.-L. (2009a) Efficient waste and pollution abatements for regions in Japan, *International Journal of Sustainable Development and World Ecology*, **16**, 270–285 [214].

Honma, S. and Hu, J.-L. (2009b) Total-factor energy productivity growth of regions in Japan. *Energy Policy*, **37**, 394–3950 [215].

Honma, S. and Hu, J.-L. (2014) Industry-level total-factor energy efficiency in developed countries: A Japan-centered analysis. *Applied Energy*, **119**, 67–78 [216].

Hossein, S. and Meysam, A. (2015) A simulation-based data envelopment analysis (DEA) model to evaluate wind plants locations. *Decision Science Letters*, **4**, 165–180 [217].

Hou, L., Hoag, D., Keske, C.M.H. and Lu, C. (2014) Sustainable value of degraded soils in China's loess plateau: An updated approach. *Ecological Economics*, **97**, 20–27 [218].

Houshyar, E., Azadi, H., Almassi, M., Sheikh Davoodi, M.J. and Witlox, F. (2012) Sustainable and efficient energy consumption of corn production in Southwest Iran: Combination of multi-fuzzy and DEA modeling. *Energy*, **44**, 672–681 [219].

Hsiao, J.-M. (2012) Measuring the operating efficiency of solar cell companies in Taiwan with data envelopment analysis. *American Journal of Applied Sciences*, **9**, 1899–1905 [220].

Hu, J.-L. and Kao, C.-H. (2007) Efficient energy-saving targets for APEC economies. *Energy Policy*, **35**, 373–382 [221].

Hu, J.L. and Lee, Y.C. (2008) Efficient three industrial waste abatements for regions in China. *International Journal of Sustainable Development and World Ecology*, **15**, 132–144 [222].

Hu, J.-L., Lio, M.-C., Kao, C.-H. and Lin Y.-L. (2012) Total-factor energy efficiency for regions in Taiwan. *Energy Sources, Part B: Economics, Planning and Policy*, **7**, 292–300 [223].

Hu, J.-L., Lio, M.-C., Yeh, F.-Y. and Lin, C.-H. (2011) Environment-adjusted regional energy efficiency in Taiwan. *Applied Energy*, **88**, 2893–2899 [224].

Hu, J.-L. and Wang, S.-C. (2006) Total-factor energy efficiency of regions in China. *Energy Policy*, **34**, 3206–3217 [225].

Hua, Z., Bian, Y. and Liang, L. (2007) Eco-efficiency analysis of paper mills along the Huai River: An extended DEA approach. *Omega*, **35**, 578–587 [226].

Huang, C.-W., Chiu, Y.-H., Fang, W.-T. and Shen, N. (2014) Assessing the performance of Taiwan's environmental protection system with a non-radial network DEA approach. *Energy Policy*, **74**, 547–556 [227].

Huang, J., Yang, X., Cheng, G. and Wang, S. (2014) A comprehensive eco-efficiency model and dynamics of regional eco-efficiency in China. *Journal of Cleaner Production*, **67**, 228–238 [228].

Huang, R. and Li, Y. (2013) Undesirable input-output two-phase DEA model in an environmental performance audit. *Mathematical and Computer Modelling*, **58**, 971–979 [229].

Hwang, S.-N., Chen, C., Chen, Y., Lee, H.-S. and Shen, P.-D. (2012) Sustainable design performance evaluation with applications in the automobile industry: Focusing on inefficiency by undesirable factors. *Omega*, **41**, 553–558 [230].

Iglesias, G., Castellanos, P. and Seijas, A. (2010) Measurement of productive efficiency with frontier methods: A case study for wind farms. *Energy Economics*, **32**, 1199–1208 [231].

Iribarren, D., Martín-Gamboa, M. and Dufour, J. (2013) Environmental benchmarking of wind farms according to their operational performance. *Energy*, **61**, 589–597 [232].

Iribarren, D., Martín-Gamboa, M., O'Mahony, T. and Dufour, J. (2016) Screening of socio-economic indicators for sustainability assessment: A combined life cycle assessment and data envelopment analysis approach. *International Journal of Life Cycle Assessment*, **21**, 202–214 [233].

Iribarren, D., Marvuglia, A., Hild, P., Guiton, M., Popovici, E. and Benetto, E. (2015) Life cycle assessment and data envelopment analysis approach for the selection of building components according to their environmental impact efficiency: A case study for external walls. *Journal of Cleaner Production*, **87**, 707–716 [234].

Iribarren, D., Vázquez-Rowe, I., Rugani, B. and Benetto, E. (2014) On the feasibility of using energy analysis as a source of benchmarking criteria through data envelopment analysis: A case study for wind energy. *Energy*, **67**, 527–537 [235].

Ismail, Z., Tai, J.C., Kong, K.K., Law, K.H., Shirazi, S.M. and Karim, R. (2013) Using data envelopment analysis in comparing the environmental performance and technical efficiency of selected companies in their global petroleum operations. *Measurement: Journal of the International Measurement Confederation*, **46**, 3401–3413 [236].

Jahangoshai Rezaee, M., Moini, A. and Makui, A. (2012) Operational and non-operational performance evaluation of thermal power plants in Iran: A game theory approach. *Energy*, **38**, 96–103 [237].

Jamasb, T., Nillesen, P. and Pollitt, M. (2004) Strategic behavior under regulatory benchmarking. *Energy Economics*, **26**, 825–843 [238].

Jamasb, T. and Pollitt, M. (2000) Benchmarking and regulation: international electricity experience. *Utilities Policy*, **9**, 107–130 [239].

Jamasb, T. and Pollitt, M. (2003) International benchmarking and regulation: an application to European electricity distribution utilities. *Energy Policy*, **31**, 1609–1622 [240].

Jan, P., Dux, D., Lips, M., Alig, M. and Dumondel, M. (2012) On the link between economic and environmental performance of Swiss dairy farms of the alpine area. *International Journal of Life Cycle Assessment*, **17**, 706–719 [241].

Jaraite, J. and Di Maria, C. (2012) Efficiency, productivity and environmental policy: A case study of power generation in the EU. *Energy Economics*, **34**, 1557–1568 [242].

Jaunky, V.C. (2013) Divergence in technical efficiency of electric utilities: Evidence from the SAPP. *Energy Policy*, **62**, 419–430 [243].

Jiekang, W., Zhuangzhi, G. and Fan, W. (2014) Short-term multi-objective optimization scheduling for cascaded hydroelectric plants with dynamic generation flow limit based on EMA and DEA. *International Journal of Electrical Power and Energy Systems*, **57**, 189–197 [244].

Jin, J., Zhou, D. and Zhou, P. (2014) Measuring environmental performance with stochastic environmental DEA: The case of APEC economies. *Economic Modelling*, **38**, 80–86 [245].

Ju, K., Su, B., Zhou, D., Zhou, P. and Zhang, Y. (2015) Oil price crisis response: Capability assessment and key indicator identification. *Energy*, **93**, 1353–1360 [246].

Jun, G. and Xiaofei, C. (2013) A coordination research on urban ecosystem in Beijing with weighted grey correlation analysis based on DEA. *Journal of Applied Sciences*, **13**, 5749–5752 [247].

Kagawa, S., Takezono, K., Suh, S. and Kudoh, Y. (2013) Production possibility frontier analysis of biodiesel from waste cooking oil. *Energy Policy*, **55**, 362–368 [248].

Kanellopoulos, A., Berentsen, P.B.M., van Ittersum, M.K. and Oude Lansink, A.G.J.M. (2012) A method to select alternative agricultural activities for future-oriented land use studies. *European Journal of Agronomy*, **40**, 75–85 [249].

Kao, C. (2014) Efficiency decomposition for general multi-stage systems in data envelopment analysis. *European Journal of Operational Research*, **232**, 117–124 [250].

Kasap, Y. and Kiriş, Ş. (2013) An AHP-DEA approach for evaluating electricity generation firms of OECD countries. *Energy Sources, Part B: Economics, Planning and Policy*, **8**, 200–208 [251].

Kashani, H.A. (2005a) Regulation and efficiency: an empirical analysis of the United Kingdom continental shelf petroleum industry. *Energy Policy*, **33**, 915–925 [252].

Kashani, H.A. (2005b) State intervention causing inefficiency: an empirical analysis of the Norwegian Continental Shelf. *Energy Policy*, **33**, 1998–2009 [253].

Keirstead, J. (2013) Benchmarking urban energy efficiency in the UK. *Energy Policy*, **63**, 575–587 [254].

Khadivi, M.R. and Fatemi Ghomi, S.M.T. (2012) Solid waste facilities location using of analytical network process and data envelopment analysis approaches. *Waste Management*, **32**, 1258–1265 [255].

Khalili-Damghani, K. and Shahmir, Z. (2015) Uncertain network data envelopment analysis with undesirable outputs to evaluate the efficiency of electricity power production and distribution processes. *Computers and Industrial Engineering*, **88**, 131–150 [256].

Khalili-Damghani, K., Tavana, M. and Haji-Saami, E. (2015) A data envelopment analysis model with interval data and undesirable output for combined cycle power plant performance assessment. *Expert Systems with Applications*, **42**, 760–773 [257].

Kheirkhah, A., Azadeh, A., Saberi, M., Azaron, A. and Shakouri, H. (2013) Improved estimation of electricity demand function by using of artificial neural network, principal component analysis and data envelopment analysis. *Computers and Industrial Engineering*, **64**, 425–441 [258].

Khodakarami, M., Shabani, A., Farzipoor Saen, R. and Azadi, M. (2015) Developing distinctive two-stage data envelopment analysis models: An application in evaluating the sustainability of supply chain management. *Measurement: Journal of the International Measurement Confederation*, **70**, 62–74 [259].

Khodakarami, M., Shabani, A. and Saen, R.F. (2014) A new look at measuring sustainability of industrial parks: A two-stage data envelopment analysis approach. *Clean Technologies and Environmental Policy*, **16**, 1577–1596 [260].

Khoshnevisan, B., Bolandnazar, E., Shamshirband, S., Shariati, H.M., Anuar, N.B. and Mat Kiah, M.L. (2015) Decreasing environmental impacts of cropping systems using life cycle assessment (LCA) and multi-objective genetic algorithm. *Journal of Cleaner Production*, **86**, 67–77 [261].

Khoshnevisan, B., Rafiee, S., Omid, M. and Mousazadeh, H. (2013a) Applying data envelopment analysis approach to improve energy efficiency and reduce GHG (greenhouse gas) emission of wheat production. *Energy*, **58**, 588–593 [262].

Khoshnevisan, B., Rafiee, S., Omid, M. and Mousazadeh, H. (2013b) Reduction of CO2 emission by improving energy use efficiency of greenhouse cucumber production using DEA approach. *Energy*, **55**, 676–682 [263].

Khoshnevisan, B., Shariati, H.M. and Rafiee, S. (2014) Comparison of energy consumption and GHG emissions of open field and greenhouse strawberry production. *Renewable and Sustainable Energy Reviews*, **29**, 316–324 [264].

Khoshroo, A., Mulwa, R., Emrouznejad, A. and Arabi, B. (2013) A non-parametric Data Envelopment Analysis approach for improving energy efficiency of grape production. *Energy*, **63**, 189–194 [265].

Kim, K. and Kim, Y. (2012) International comparison of industrial CO2 emission trends and the energy efficiency paradox utilizing production-based decomposition. *Energy Economics*, **34**, 1724–1741 [266].

Kim, K.-T., Lee, D.J., Park, S.-J., Zhang, Y. and Sultanov, A. (2015) Measuring the efficiency of the investment for renewable energy in Korea using data envelopment analysis. *Renewable and Sustainable Energy Reviews*, **47**, 694–702 [267].

Köne, A.Ç. and Büke, T. (2012) A comparison for Turkish provinces' performance of urban air pollution. *Renewable and Sustainable Energy Reviews*, **16**, 1300–1310 [268].

Kordrostami, S., Amirteimoori, A. and Noveiri, M.J.S. (2015) Restricted variation in data envelopment analysis with undesirable factors in nature. *International Journal of Biomathematics*, **8**(1550034), 1–10 [269].

Korhonen, P. and Syrjänen, M. (2003) Evaluation of cost efficiency in finnish electricity distribution. *Annals of Operations Research*, **121**, 105–122 [270].

Korhonen, P.J. and Luptacik, M. (2004) Eco-efficiency analysis of power plants: An extension of data envelopment analysis. *European Journal of Operational Research*, **154**, 437–446 [271].

Kortelainen, M. (2008) Dynamic environmental performance analysis: A Malmquist index approach. *Ecological Economics*, **64**, 701–715 [272].

Kulshreshtha, M. and Parikh, J.K. (2002) Study of efficiency and productivity growth in opencast and underground coal mining in India: A DEA analysis. *Energy Economics*, **24**, 439–453 [273].

Kumar Mandal, S. and Madheswaran, S. (2010) Environmental efficiency of the Indian cement industry: An interstate analysis. *Energy Policy*, **38**, 1108–1118 [274].

Kumar, A., Jain, V. and Kumar, S. (2014) A comprehensive environment friendly approach for supplier selection. *Omega*, **42**, 109–123 [275].

Kumar, S. (2006) Environmentally sensitive productivity growth: A global analysis using Malmquist–Luenberger index. *Ecological Economics*, **56**, 280–293 [276].

Kumar, S., Raizada, A., Biswas, H. and Mishra, P.K. (2015) Assessing the impact of watershed development on energy efficiency in groundnut production using DEA approach in the semi-arid tropics of southern India. *Current Science*, **109**, 1831–1837 [277].

Kuosmanen, N., Kuosmanen, T. and Sipiläinen, T. (2013) Consistent aggregation of generalized sustainable values from the firm level to sectoral, regional or industry levels. *Sustainability*, **5**, 1568–1576 [278].

Kuosmanen, T. (2012) Stochastic semi-nonparametric frontier estimation of electricity distribution networks: Application of the StoNED method in the Finnish regulatory model. *Energy Economics*, **34**, 2189–2199 [279].

Kuosmanen, T., Bijsterbosch, N. and Dellink, R. (2009) Environmental cost–benefit analysis of alternative timing strategies in greenhouse gas abatement: A data envelopment analysis approach. *Ecological Economics*, **68**, 1633–1642 [280].

Kuosmanen, T., Saastamoinen, A. and Sipiläinen, T. (2013) What is the best practice for benchmark regulation of electricity distribution? Comparison of DEA, SFA and StoNED methods. *Energy Policy*, **61**, 740–750 [281].

Kwoka, J. and Pollitt, M. (2010) Do mergers improve efficiency? Evidence from restructuring the US electric power sector. *International Journal of Industrial Organization*, **28**, 645–656 [282].

Kwoka, J., Pollitt, M. and Sergici, S. (2010) Divestiture policy and operating efficiency in U.S. electric power distribution. *Journal of Regulatory Economics*, **38**, 86–109 [283].

Lábaj, M., Luptáčik, M. and Nežinský, E. (2014) Data envelopment analysis for measuring economic growth in terms of welfare beyond GDP. *Empirica: Journal of European Economics*, **41**, 407–424 [284].

Lam, P.-L. and Shiu, A. (2001) A data envelopment analysis of the efficiency of China's thermal power generation. *Utilities Policy*, **10**, 75–83 [285].

Leal Jr., I.C., de Almada Garcia, P.A. and de Almeida D'Agosto, M. (2012) A data envelopment analysis approach to choose transport modes based on eco-efficiency. *Environment, Development and Sustainability*, **14**, 767–781 [286].

Lee, A.H.I., Kang, H.-Y., Lin, C.-Y. and Shen, K.-C. (2015) An integrated decision-making model for the location of a PV solar plant. *Sustainability*, **7**, 13522–13541 [287].

Lee, A.H.I., Lin, C.Y., Kang, H.-Y. and Lee, W.H. (2012) An integrated performance evaluation model for the photovoltaics industry. *Energies*, **5**, 1271–1291 [288].

Lee, C.-Y. (2015) Distinguishing operational performance in power production: A new measure of effectiveness by DEA. *IEEE Transactions on Power Systems*, **30**, 3160–3167 [289].

Lee, K.-H. and Farzipoor Saen, R. (2012) Measuring corporate sustainability management: A data envelopment analysis approach. *International Journal of Production Economics*, **140**, 219–226 [290].

Lee, S.K., Mogi, G. and Hui, K.S. (2013) A fuzzy analytic hierarchy process (AHP)/data envelopment analysis (DEA) hybrid model for efficiently allocating energy R&D resources: In the case of energy technologies against high oil prices. *Renewable and Sustainable Energy Reviews*, **21**, 347–355 [291].

Lee, S.K., Mogi, G., Lee, S.K., Hui, K.S. and Kim, J.W. (2010) Econometric analysis of the RandD performance in the national hydrogen energy technology development for measuring relative efficiency: The fuzzy AHP/DEA integrated model approach. *International Journal of Hydrogen Energy*, **35**, 2236–2246 [292].

Lee, S.K., Mogi, G., Li, Z., Hui, K.S., Lee, S.K., Hui, K.N., Park, S.Y., Ha, Y.J. and Kim, J.W. (2011) Measuring the relative efficiency of hydrogen energy technologies for implementing the hydrogen economy: An integrated fuzzy AHP/DEA approach. *International Journal of Hydrogen Energy*, **36**, 12655–12663 [293].

Lee, T., Yeo, G.-T. and Thai, V.V. (2014) Environmental efficiency analysis of port cities: Slacks-based measure data envelopment analysis approach. *Transport Policy*, **33**, 82–88 [294].

Lee, W.-S. and Kung, C.-K. (2011) Using climate classification to evaluate building energy performance. *Energy*, **36**, 1797–1801 [295].

Lee, Y.-C., Hu, J.-L. and Kao, C.-H. (2011) Efficient saving targets of electricity and energy for regions in China. *International Journal of Electrical Power and Energy Systems*, **33**, 1211–1219 [296].

Lee, Y.-S. and Tong, L.-I. (2012) Predicting high or low transfer efficiency of photovoltaic systems using a novel hybrid methodology combining rough set theory, data envelopment analysis and genetic programming. *Energies*, **5**, 545–560 [297].

Lei, M., Zhao, X., Deng, H. and Tan, K.-C. (2013) DEA analysis of FDI attractiveness for sustainable development: Evidence from Chinese provinces. *Decision Support Systems*, **56**, 406–418 [298].

Leme, R.C., Paiva, A.P., Steele Santos, P.E., Balestrassi, P.P. and Galvão, L.d.L. (2014) Design of experiments applied to environmental variables analysis in electricity utilities efficiency: The Brazilian case. *Energy Economics*, **45**, 111–119 [299].

Li, H., Yang, W., Zhou, Z. and Huang, C. (2013) Resource allocation models' construction for the reduction of undesirable outputs based on DEA methods. *Mathematical and Computer Modelling*, **58**, 913–926 [300].

Li, J., Li, J. and Zheng, F. (2014) Unified efficiency measurement of electric power supply companies in China. *Sustainability*, **6**, 779–793 [301].

Li, K. and Lin, B. (2015a) Metafroniter energy efficiency with CO2 emissions and its convergence analysis for China. *Energy Economics*, **48**, 230–241 [302].

Li, K. and Lin, B. (2015b) The efficiency improvement potential for coal, oil and electricity in China's manufacturing sectors. *Energy*, **86**, 403–413 [303].

Li, K. and Lin, B. (2015c) The improvement gap in energy intensity: Analysis of China's thirty provincial regions using the improved DEA (data envelopment analysis) model. *Energy*, **84**, 589–599 [304].

Li, K. and Lin, B. (2016) Impact of energy conservation policies on the green productivity in China's manufacturing sector: Evidence from a three-stage DEA model. *Applied Energy*, **168**, 351–363 [305].

Li, L., Lei, Y., Pan, D. and Si, C. (2016) Research on sustainable development of resource-based cities based on the DEA approach: a case study of Jiaozuo, China. *Mathematical Problems in Engineering*, **2016**(5024837), 1–10 [306].

Li, L., Li, M. and Wu, C. (2013) Production efficiency evaluation of energy companies based on the improved super-efficiency data envelopment analysis considering undesirable outputs. *Mathematical and Computer Modelling*, **58**, 1057–1066 [307].

Li, L., Miller, D. and Schmidt, C.P. (2016) Optimizing inventory's contribution to profitability in a regulated utility: The Averch–Johnson effect. *International Journal of Production Economics*, **175**, 132–141 [308].

Li, L.-B. and Hu, J.-L. (2012) Ecological total-factor energy efficiency of regions in China. *Energy Policy*, **46**, 216–224 [309].

Li, X.-G., Yang, J. and Liu, X.-J. (2013) Analysis of Beijing's environmental efficiency and related factors using a DEA model that considers undesirable outputs. *Mathematical and Computer Modelling*, **58**, 956–960 [310].

Li, Y., Sun, L., Feng, T. and Zhu, C. (2013) How to reduce energy intensity in China: A regional comparison perspective. *Energy Policy*, **61**, 513–522 [311].

Lin, B. and Fei, R. (2015) Regional differences of CO2 emissions performance in China's agricultural sector: A Malmquist index approach. *European Journal of Agronomy*, **70**, 33–40 [312].

Lin, B. and Liu, X. (2012) Dilemma between economic development and energy conservation: Energy rebound effect in China. *Energy*, **45**, 867–873 [313].

Lin, W., Chen, B., Xie, L. and Pan, H. (2015) Estimating energy consumption of transport modes in China using DEA. *Sustainability*, **7**, 4225–4239 [314].

Lins, M.E., Oliveira, L.B., da Silva, A.C.M., Rosa, L.P. and Pereira Jr, A.O. (2012) Performance assessment of alternative energy resources in Brazilian power sector using Data Envelopment Analysis. *Renewable and Sustainable Energy Reviews*, **16**, 898–903 [315].

Lins, M.P.E., Sollero, M.K.V., Calôba, G.M. and da Silva, A.C.M. (2007) Integrating the regulatory and utility firm perspectives, when measuring the efficiency of electricity distribution. *European Journal of Operational Research*, **181**, 1413–1424 [316].

Liou, J.L., Chiu, C.R., Huang, F.M. and Liu, W.Y. (2015) Analyzing the relationship between CO2 emission and economic efficiency by a relaxed two-stage DEA model. *Aerosol and Air Quality Research*, **15**, 694–701 [317].

Liu, C.H., Lin, S.J. and Lewis, C. (2010) Evaluation of thermal power plant operational performance in Taiwan by data envelopment analysis. *Energy Policy*, **38**, 1049–1058 [318].

Liu, C.H., Lin, S.J. and Lewis, C. (2013) Evaluation of NOX, SOX and CO2 emissions of Taiwan's thermal power plants by data envelopment analysis. *Aerosol and Air Quality Research*, **13**, 1815–1823 [319].

Liu, J., Liu, H., Yao, X.-L. and Liu, Y. (2016) Evaluating the sustainability impact of consolidation policy in China's coal mining industry: A data envelopment analysis. *Journal of Cleaner Production*, **112**, 2969–2976 [320].

Liu, J.S., Lu, L.Y.Y., Lu, W.-M. and Lin, B.J.Y. (2013) A survey of DEA applications. *Omega*, **41**, 893–902 [321].

Liu, W. (2015) Measuring efficiency and scope economies of cogeneration enterprises in Northeast China under the background of energy saving and emission reduction. *Ecological Indicators*, **51**, 173–179 [322].

Liu, W., Tian, J., Chen, L., Lu, W. and Gao, Y. (2015) Environmental performance analysis of eco-industrial parks in China: a data envelopment analysis approach. *Journal of Industrial Ecology*, **19**, 1070–1081 [323].

Liu, X. and Wen, Z. (2012) Best available techniques and pollution control: A case study on China's thermal power industry. *Journal of Cleaner Production*, **23**, 113–121 [324].

Liu, Y., Ren, L., Li, Y. and Zhao, X.-G. (2015) The industrial performance of wind power industry in China. *Renewable and Sustainable Energy Reviews*, **43**, 644–655 [325].

Liu, Y., Sun, C. and Xu, S. (2013) Eco-efficiency assessment of water systems in China. *Water Resources Management*, **27**, 4927–4939 [326].

Liu, Y. and Wang, K. (2015) Energy efficiency of China's industry sector: An adjusted network DEA (data envelopment analysis)-based decomposition analysis. *Energy*, **93**, 1328–1337 [327].

Liu, Y., Wang, W., Li, X. and Zhang, G. (2010) Eco-efficiency of urban material metabolism: A case study in Xiamen, China. *International Journal of Sustainable Development and World Ecology*, **17**, 142–148 [328].

Lo, F.-Y., Chien, C.-F. and Lin, J.T. (2001) A DEA study to evaluate the relative efficiency and investigate the district reorganization of the Taiwan power company. *IEEE Transactions on Power Systems*, **16**, 170–178 [329].

Lo, S.-F. (2010a) Performance evaluation for sustainable business: A profitability and marketability framework. *Corporate Social Responsibility and Environmental Management*, **17**, 311–319 [330].

Lo, S.-F. (2010b) The differing capabilities to respond to the challenge of climate change across Annex Parties under the Kyoto Protocol. *Environmental Science and Policy*, **13**, 42–54 [331].

Longo, L., Colantoni, A., Castellucci, S., Carlini, M., Vecchione, L., Savuto, E., Pallozzi, V., Di Carlo, A., Bocci, E., Moneti, M., Cocchi, S. and Boubaker, K. (2015) DEA (data envelopment analysis)-assisted supporting measures for ground coupled heat pumps implementing in Italy: A case study. *Energy*, **90**, 1967–1972 [332].

Lorenzo-Toja, Y., Vázquez-Rowe, I., Chenel, S., Marín-Navarro, D., Moreira, M.T. and Feijoo, G. (2015) Eco-efficiency analysis of Spanish WWTPs using the LCA + DEA method. *Water Research*, **68**, 651–666 [333].

lo Storto, C. (2016) Ecological efficiency based ranking of cities: A combined DEA cross-efficiency and Shannon's entropy method. *Sustainability*, **124**, 1–29 [334].

Lozano, S. and Gutiérrez, E. (2008) Non-parametric frontier approach to modelling the relationships among population, GDP, energy consumption and CO2 emissions. *Ecological Economics*, **66**, 687–699 [335].

Lozano, S., Villa, G. and Brännlund, R. (2009) Centralised reallocation of emission permits using DEA. *European Journal of Operational Research*, **193**, 752–760 [336].

Lu, C.-C., Chiu, Y.-H., Shyu, M.-K. and Lee, J.-H. (2013) Measuring CO2 emission efficiency in OECD countries: Application of the hybrid efficiency model. *Economic Modelling*, **32**, 130–135 [337].

Lu, W.-M. and Lo, S.-F. (2007) A closer look at the economic–environmental disparities for regional development in China. *European Journal of Operational Research*, **183**, 882–894 [338].

Lu, W.-M. and Lo, S.-F. (2012) Constructing stratifications for regions in China with sustainable development concerns. *Quality and Quantity*, **46**, 1807–1823 [339].

Luo, J., Yang, Y. and Chen, Y. (2012) Optimizing the drilled well patterns for CBM recovery via numerical simulations and data envelopment analysis. *International Journal of Mining Science and Technology*, **22**, 503–507 [340].

Lv, W., Zhou, Z. and Huang, H. (2013) The measurement of undesirable output based on DEA in E&E: Models development and empirical analysis. *Mathematical and Computer Modelling*, **58**, 907–912 [341].

Macpherson, A.J., Principe, P.P. and Shao, Y. (2013) Controlling for exogenous environmental variables when using data envelopment analysis for regional environmental assessments. *Journal of Environmental Management*, **119**, 220–229 [342].

Madlener, R., Antunes, C.H. and Dias, L.C. (2009) Assessing the performance of biogas plants with multi-criteria and data envelopment analysis. *European Journal of Operational Research*, **197**, 1084–1094 [343].

Mahapatra, S., Pal, R., Hult, T. and Talluri, S. (2015) Assessment of proactive environmental initiatives: Evaluation of efficiency based on interval-scale data. *IEEE Transactions on Engineering Management*, **62**, 280–293 [344].

Mahdiloo, M., Saen, R.F. and Lee, K.-H. (2015) Technical, environmental and eco-efficiency measurement for supplier selection: An extension and application of data envelopment analysis. *International Journal of Production Economics*, **168**, 279–289 [345].

Mahdiloo, M., Tavana, M., Farzipoor Saen, R. and Noorizadeh, A. (2014) A game theoretic approach to modeling undesirable outputs and efficiency decomposition in data envelopment analysis. *Applied Mathematics and Computation*, **244**, 479–492 [346].

Makridou, G., Andriosopoulos, K. and Doumpos, M. (2016) Measuring the efficiency of energy-intensive industries across European countries. *Energy Policy*, **88**, 573–583 [347].

Mallikarjun, S. and Lewis, H.F. (2014) Energy technology allocation for distributed energy resources: A strategic technology-policy framework. *Energy*, **72**, 783–799 [348].

Mandal, S.K. (2010) Do undesirable output and environmental regulation matter in energy efficiency analysis? Evidence from Indian cement industries. *Energy Policy*, **38**, 6076–6083 [349].

Mandal, S.K. and Madheswaran, S. (2010) Environmental efficiency of the Indian cement industry: An interstate analysis. *Energy Policy*, **38**, 1108–1118 [350].

Mei, G., Gan, J. and Zhang, N. (2015) Metafrontier environmental efficiency for China's regions: A slack-based efficiency measure. *Sustainability*, **7**, 4004–4021 [351].

Mekaroonreung, M. and Johnson, A.L. (2010) Estimating the efficiency of American petroleum refineries under varying assumptions of the disposability of bad outputs. *International Journal of Energy Sector Management*, **4**, 356–398 [352].

Mekaroonreung, M. and Johnson, A.L. (2012) Estimating the shadow prices of SO2 and NOx for U.S. coal power plants: A convex nonparametric least squares approach. *Energy Economics*, **34**, 723–732 [353].

Menegaki, A.N. (2013) Growth and renewable energy in Europe: Benchmarking with data envelopment analysis. *Renewable Energy*, **60**, 363–369 [354].

Meng, F.Y., Fan, L.W., Zhou, P. and Zhou, D.Q. (2013) Measuring environmental performance in China's industrial sectors with non-radial DEA. *Mathematical and Computer Modelling*, **58**, 1047–1056 [355].

Miguéis, V.L., Camanho, A.S., Bjørndal, E. and Bjørndal, M. (2012) Productivity change and innovation in Norwegian electricity distribution companies. *Journal of the Operational Research Society*, **63**, 982–990 [356].

Mihci, H. and Mollavelioğlu, S. (2011) An assessment of sustainable agriculture in the OECD countries with special reference to Turkey, *New Mediterranean*, **10**, 4–17 [357].

Mirhedayatian, S.M., Azadi, M. and Farzipoor Saen, R. (2014) A novel network data envelopment analysis model for evaluating green supply chain management. *International Journal of Production Economics*, **147**(Part B), 544–554 [358].

Mobtaker, H.G., Akram, A., Keyhani, A. and Mohammadi, A. (2012) Optimization of energy required for alfalfa production using data envelopment analysis approach. *Energy for Sustainable Development*, **16**, 242–248 [359].

Mohammadi, A., Rafiee, S., Jafari, A., Dalgaard, T., Knudsen, M.T., Keyhani, A., Mousavi-Avval, S.H. and Hermansen, J.E. (2013) Potential greenhouse gas emission reductions in soybean farming: A combined use of life cycle assessment and data envelopment analysis. *Journal of Cleaner Production*, **54**, 89–100 [360].

Molinos-Senante, M., Hernández-Sancho, F., Mocholí-Arce, M. and Sala-Garrido, R. (2014) Economic and environmental performance of wastewater treatment plants: Potential reductions in greenhouse gases emissions. *Resource and Energy Economics*, **38**, 125–140 [361].

Molinos-Senante, M., Sala-Garrido, R. and Hernandez-Sancho, F. (2016) Development and application of the Hicks–Moorsteen productivity index for the total factor productivity assessment of wastewater treatment plants. *Journal of Cleaner Production*, **112**, 3116–3123 [362].

Moreno, P., Andrade, G.N., Angulo Meza, L. and De Mello, J.C.S. (2015) Evaluation of Brazilian electricity distributors using a network DEA model with shared inputs. *IEEE Latin America Transactions*, **13**, 2209–2216 [363].

Mou, D. (2014) Understanding China's electricity market reform from the perspective of the coal-fired power disparity. *Energy Policy*, **74**, 224–234 [364].

Mousavi-Avval, S.H., Mohammadi, A., Rafiee, S. and Tabatabaeefar, A. (2012) Assessing the technical efficiency of energy use in different barberry production systems. *Journal of Cleaner Production*, **27**, 126–132 [365].

Mousavi-Avval, S.H., Rafiee, S., Jafari, A. and Mohammadi, A. (2011a) Improving energy use efficiency of canola production using data envelopment analysis (DEA) approach. *Energy*, **36**, 2765–2772 [366].

Mousavi-Avval, S.H., Rafiee, S., Jafari, A. and Mohammadi, A. (2011b) Optimization of energy consumption for soybean production using data envelopment analysis (DEA) approach. *Applied Energy*, **88**, 3765–3772 [367].

Mukherjee, K. (2008) Energy use efficiency in the Indian manufacturing sector: An interstate analysis. *Energy Policy*, **36**, 662–672 [368].

Mukherjee, K. (2010) Measuring energy efficiency in the context of an emerging economy: The case of Indian manufacturing. *European Journal of Operational Research*, **201**, 933–941 [369].

Mullarkey, S., Caulfield, B., McCormack, S. and Basu, B. (2015) A framework for establishing the technical efficiency of electricity distribution counties (EDCs) using data envelopment analysis. *Energy Conversion and Management*, **94**, 112–123 [370].

Munisamy, S. and Arabi, B. (2015) Eco-efficiency change in power plants: Using a slacks-based measure for the meta-frontier Malmquist–Luenberger productivity index. *Journal of Cleaner Production*, **105**, 218–232 [371].

Munksgaard, J., Pade, L.-L. and Fristrup, P. (2005) Efficiency gains in Danish district heating. Is there anything to learn from benchmarking? *Energy Policy*, **33**, 1986–1997 [372].

Murillo-Zamorano, L.R. (2005) The role of energy in productivity growth: A controversial issue? *The Energy Journal*, **26**, 69–88 [373].

Murty, S., Robert Russell, R. and Levkoff, S.B. (2012) On modeling pollution-generating technologies. *Journal of Environmental Economics and Management*, **64**, 117–135 [374].

Nabavi-Pelesaraei, A., Abdi, R., Rafiee, S. and Mobtaker, H.G. (2014) Optimization of energy required and greenhouse gas emissions analysis for orange producers using data envelopment analysis approach. *Journal of Cleaner Production*, **65**, 311–317 [375].

Nag, B. (2006) Estimation of carbon baselines for power generation in India: The supply side approach. *Energy Policy*, **34**, 1399–1410 [376].

Nakano, M. and Managi, S. (2010) Productivity analysis with CO2 emissions in Japan. *Pacific Economic Review*, **15**, 708–718 [377].

Nakano, M. and Managi, S. (2012) Waste generations and efficiency measures in Japan. *Environmental Economics and Policy Studies*, **14**, 327–339 [378].

Nassiri, S. and Singh, S. (2007) Study on energy use efficiency for paddy crop using data envelopment analysis (DEA) technique. *Applied Energy*, **86**, 1320–1325 [379].

Nouri, J., Hosseinzadeh Lotfi, F., Borgheipour, H., Atabi, F., Sadeghzadeh, S.M. and Moghaddas, Z. (2013) An analysis of the implementation of energy efficiency measures in the vegetable oil industry of Iran: A data envelopment analysis approach. *Journal of Cleaner Production*, **52**, 84–93 [380].

Nykamp, S., Andor, M. and Hurink, J.L. (2012) "Standard" incentive regulation hinders the integration of renewable energy generation. *Energy Policy*, **47**, 222–237 [381].

Oggioni, G., Riccardi, R. and Toninelli, R. (2011) Eco-efficiency of the world cements industry: A data envelopment analysis. *Energy Policy*, **39**, 2842–2854 [382].

Olanrewaju, O.A., Jimoh, A.A. and Kholopane, P.A. (2012) Integrated IDA-ANN-DEA for assessment and optimization of energy consumption in industrial sectors. *Energy*, **46**, 629–635 [383].

Olanrewaju, O.A., Jimoh, A.A. and Kholopane, P.A. (2013) Assessing the energy potential in the South African industry: A combined IDA-ANN-DEA (Index Decomposition Analysis-Artificial Neural Network-Data Envelopment Analysis) model. *Energy*, **63**, 225–232 [384].

Olatubi, W.O. and Dismukes, D.E. (2000) A data envelopment analysis of the levels and determinants of coal-fired electric power generation performance. *Utilities Policy*, **9**, 47–59 [385].

Omid, M., Ghojabeige F., Delshad M. and Ahmadi, H. (2011) Energy use pattern and benchmarking of selected greenhouses in Iran using. *Energy Conversion and Management*, **52**, 153–162 [386].

Omrani, H., Azadeh, A., Ghaderi, S.F. and Aabdollahzadeh, S. (2010) A consistent approach for performance measurement of electricity distribution companies. *International Journal of Energy Sector Management*, **4**, 399–416 [387].

Omrani, H., Gharizadeh Beiragh, R. and Shafiei Kaleibari, S. (2015) Performance assessment of Iranian electricity distribution companies by an integrated cooperative game data envelopment analysis principal component analysis approach. *International Journal of Electrical Power and Energy Systems*, **64**, 617–625 [388].

Önüt, S. and Soner, S. (2006) Energy efficiency assessment for the Antalya region hotels in Turkey. *Energy and Buildings*, **38**, 964–971 [389].

Oude Lansink, A. and Bezlepkin, I. (2003) The effect of heating technologies on CO2 and energy efficiency of Dutch greenhouse firms. *Journal of Environmental Management*, **68**, 73–82 [390].

Oude Lansink, A. and Silva, E. (2003) CO2 and energy efficiency of different heating technologies in the Dutch glasshouse industry. *Environmental and Resource Economics*, **24**, 395–407 [391].

Özkara, Y. and Atak, M. (2015) Regional total-factor energy efficiency and electricity saving potential of manufacturing industry in Turkey. *Energy*, **93**, 495–510 [392].

Pacudan, R. and de Guzman, E. (2002) Impact of energy efficiency policy to productive efficiency of electricity distribution industry in the Philippines. *Energy Economics*, **24**, 41–54 [393].

Pahlavan, R., Omid, M. and Akram, A. (2011) Energy use efficiency in greenhouse tomato production in Iran. *Energy*, **36**, 6714–6719 [394].

Pahlavan, R., Omid, M., Rafiee, S. and Mousavi-Avval, S.H. (2012) Optimization of energy consumption for rose production in Iran. *Energy for Sustainable Development*, **16**, 236–241 [395].

Pahlavan, R., Rafiee, S. and Omid, M. (2012) Assessing the technical efficiency in potato production in Iran. *International Journal of Green Energy*, **9**, 229–242 [396].

Pahwa, A., Xiaoming, F. and Lubkeman, D. (2003) Performance evaluation of electric distribution utilities based on data envelopment analysis. *IEEE Transactions on Power Systems*, **18**, 400–405 [397].

Pan, H., Zhang, H. and Zhang, X. (2013) China's provincial industrial energy efficiency and its determinants. *Mathematical and Computer Modelling*, **58**, 1032–1039 [398].

Pan, X., Liu, Q. and Peng, X. (2015) Spatial club convergence of regional energy efficiency in China. *Ecological Indicators*, **51**, 25–30 [399].

Pardo Martínez, C.I. (2013) An analysis of eco-efficiency in energy use and CO2 emissions in the Swedish service industries. *Socio-Economic Planning Sciences*, **47**, 120–130 [400].

Pardo Martínez, C.I. (2015) Estimating and analyzing energy efficiency in German and Colombian manufacturing industries using DEA and data panel analysis. Part I: energy-intensive sectors. *Energy Sources, Part B: Economics, Planning and Policy*, **10**, 322–331 [401].

Park, S.-U. and Lesourd, J.-B. (2000) The efficiency of conventional fuel power plants in South Korea: A comparison of parametric and non-parametric approaches. *International Journal of Production Economics*, **63**, 59–67 [402].

Pasurka Jr, C.A. (2006) Decomposing electric power plant emissions within a joint production framework. *Energy Economics*, **28**, 26–43 [403].

Peng, X. and Cui, H. (2016) Incentive policy research on the excess profit allocation in wind power projects based on data envelopment analysis game. *Journal of Energy Engineering*, **142**(04015019), 1–6 [404].

Pereira de Souza, M.V., Souza, R.C., Pessanha, J.F.M., da Costa Oliveira, C.H. and Diallo, M. (2014) An application of data envelopment analysis to evaluate the efficiency level of the operational cost of Brazilian electricity distribution utilities. *Socio-Economic Planning Sciences*, **48**, 169–174 [405].

Pérez, V., Guerrero, F., González, M., Pérez, F. and Caballero, R. (2013) Composite indicator for the assessment of sustainability: The case of Cuban nature-based tourism destinations. *Ecological Indicators*, **29**, 316–324 [406].

Picazo-Tadeo, A.J., Beltrán-Esteve, M. and Gómez-Limón, J.A. (2012) Assessing eco-efficiency with directional distance functions. *European Journal of Operational Research*, **220**, 798–809 [407].

Picazo-Tadeo, A.J., Castillo-Giménez, J. and Beltrán-Esteve, M. (2014) An intertemporal approach to measuring environmental performance with directional distance functions: Greenhouse gas emissions in the European Union. *Ecological Economics*, **100**, 173–182 [408].

Picazo-Tadeo, A.J. and García-Reche, A. (2007) What makes environmental performance differ between firms? Empirical evidence from the Spanish tile industry. *Environment and Planning A*, **39**, 2232–2247 [409].

Picazo-Tadeo, A.J., Reig-Martínez, E. and Hernández-Sancho, F. (2005) Directional distance functions and environmental regulation. *Resource and Energy Economics*, **27**, 131–142 [410].

Pina, W.H.A. and Martinez, C.I.P. (2016) Development and urban sustainability: An analysis of efficiency using data envelopment analysis. *Sustainability*, **148**, 1–15 [411].

Piot-Lepetit, I. (2014) Technological externalities and environmental policy: How to simulate manure management regulation within a DEA framework. *Annals of Operations Research*, **214**, 31–48 [412].

Pollitt, M.G. (1996) Ownership and efficiency in nuclear power production. *Oxford Economic Papers*, **48**, 342–360 [413].

Pombo, C. and Taborda, R. (2006) Performance and efficiency in Colombia's power distribution system: Effects of the 1994 reform. *Energy Economics*, **28**, 339–369 [414].

Puri, J. and Yadav, S.P. (2015) A fully fuzzy approach to DEA and multi-component DEA for measuring fuzzy technical efficiencies in the presence of undesirable outputs. *International Journal of Systems Assurance Engineering and Management*, **6**, 268–285 [415].

Racz, V.J. and Vestergaard, N. (2016) Productivity and efficiency measurement of the Danish centralized biogas power sector. *Renewable Energy*, **92**, 397–404 [416].

Raczka, J. (2001) Explaining the performance of heat plants in Poland. *Energy Economics*, **23**, 355–370 [417].

Ramanathan, R. (2000) A holistic approach to compare energy efficiencies of different transport modes. *Energy Policy*, **28**, 743–747 [418].

Ramanathan, R. (2001) Comparative risk assessment of energy supply technologies: A data envelopment analysis approach. *Energy*, **26**, 197–203 [419].

Ramanathan, R. (2005a) An analysis of energy consumption and carbon dioxide emissions in countries of the Middle East and North Africa. *Energy*, **30**, 2831–2842 [420].

Ramanathan, R. (2005b) Estimating energy consumption of transport modes in India using DEA and application to energy and environmental policy. *Journal of Operational Research Society*, **56**, 732–737 [421].

Ramanathan, R. (2006) A multi-factor efficiency perspective to the relationships among world GDP, energy consumption and carbon dioxide emissions. *Technological Forecasting and Social Change*, **73**, 483–494 [422].

Ramli, N.A., Munisamy, S. and Arabi, B. (2013) Scale directional distance function and its application to the measurement of eco-efficiency in the manufacturing sector. *Annals of Operations Research*, **211**, 381–398 [423].

Ramos-Real, F.J., Tovar, B., Iootty, M., de Almeida, E.F. and Pinto Jr., H.Q. (2009) The evolution and main determinants of productivity in Brazilian electricity distribution 1998–2005: An empirical analysis. *Energy Economics*, **31**, 298–305 [424].

Rao, X., Wu, J., Zhang, Z. and Liu, B. (2012) Energy efficiency and energy saving potential in China: An analysis based on slacks-based measure model. *Computers and Industrial Engineering*, **63**, 578–584 [425].

Rashidi, K. and Farzipoor Saen, R. (2015) Measuring eco-efficiency based on green indicators and potentials in energy saving and undesirable output abatement. *Energy Economics*, **50**, 18–26 [426].

Rashidi, K., Shabani, A. and Farzipoor Saen, R. (2015) Using data envelopment analysis for estimating energy saving and undesirable output abatement: A case study in the Organization for Economic Co-Operation and Development (OECD) countries. *Journal of Cleaner Production*, **105**, 241–252 [427].

Rassafi, A.A. and Vaziri, M. (2007) Assessment of modal transportation sustainability: Application of data envelopment and concordance analyses. *Iranian Journal of Science and Technology, Transaction B: Engineering*, **31**, 179–193 [428].

Reinhard, S., Lovell, C.A.K. and Thijssen, G.J. (2000) Environmental efficiency with multiple environmentally detrimental variables; estimated with SFA and DEA. *European Journal of Operational Research*, **121**, 287–303 [429].

Ren, J., Tan, S., Dong, L., Mazzi, A., Scipioni, A. and Sovacool, B. K. (2014) Determining the life cycle energy efficiency of six biofuel systems in China: A data envelopment analysis. *Bioresource Technology*, **162**, 1–7 [430].

Resende, M. (2002) Relative efficiency measurement and prospects for yardstick competition in Brazilian electricity distribution. *Energy Policy*, **30**, 637–647 [431].

Reza Salehizadeh, M., Rahimi-Kian, A. and Oloomi-Buygi, M. (2015) Security-based multi-objective congestion management for emission reduction in power system. *International Journal of Electrical Power and Energy Systems*, **65**, 124–135 [432].

Riccardi, R., Oggioni, G. and Toninelli, R. (2012) Efficiency analysis of world cement industry in presence of undesirable output: Application of data envelopment analysis and directional distance function. *Energy Policy*, **44**, 140–152 [433].

Robaina-Alves, M., Moutinho, V. and MacEdo, P. (2015) A new frontier approach to model the eco-efficiency in European countries. *Journal of Cleaner Production*, **103**, 562–573 [434].

Rødseth, K.L. and Romstad, E. (2014) Environmental regulations, producer responses, and secondary benefits: Carbon dioxide reductions under the acid rain program. *Environmental and Resource Economics*, **59**, 111–135 [435].

Rogge, N. and De Jaeger, S. (2012) Evaluating the efficiency of municipalities in collecting and processing municipal solid waste: A shared input DEA-model. *Waste Management*, **32**, 1968–1978 [436].

Rosano-Peña, C., Guarnieri, P., Sobreiro, V.A., Serrano, A.L.M. and Kimura, H. (2014) A measure of sustainability of Brazilian agribusiness using directional distance functions and data envelopment analysis. *International Journal of Sustainable Development and World Ecology*, **21**, 210–222 [437].

Sadjadi, S.J. and Omrani, H. (2008) Data envelopment analysis with uncertain data: An application for Iranian electricity distribution companies. *Energy Policy*, **36**, 4247–4254 [438].

Sala-Garrido, R., Hernández-Sancho, F. and Molinos-Senante, M. (2012) Assessing the efficiency of wastewater treatment plants in an uncertain context: A DEA with tolerances approach. *Environmental Science and Policy*, **18**, 34–44 [439].

Sala-Garrido, R., Molinos-Senante, M. and Hernández-Sancho, F. (2012) How does seasonality affect water reuse possibilities? An efficiency and cost analysis. *Resources, Conservation and Recycling*, **58**, 125–131 [440].

Salazar-Ordóñez, M., Pérez-Hernández, P.P. and Martín-Lozano, J.M. (2013) Sugar beet for bioethanol production: An approach based on environmental agricultural outputs. *Energy Policy*, **55**, 662–668 [441].

Salehizadeh, M.R., Rahimi-Kian, A. and Oloomi-Buygi, M. (2015) A multi-attribute congestion-driven approach for evaluation of power generation plans. *International Transactions on Electrical Energy Systems*, **25**, 482–497 [442].

Sánchez, M.A. (2015) Integrating sustainability issues into project management. *Journal of Cleaner Production*, **96**, 319–330 [443].

Sanhueza, R., Rudnick, H. and Lagunas, H. (2004) DEA efficiency for the determination of the electric power distribution added value. *IEEE Transactions on Power Systems*, **19**, 919–925 [444].

Sanjuan, N., Ribal, J., Clemente, G. and Fenollosa, M.L. (2011) Measuring and improving eco-efficiency using data envelopment analysis: A case study of Mahón–Menorca cheese. *Journal of Industrial Ecology*, **15**, 614–628 [445].

Santana, N.B., Aparecida Do Nascimento Rebelatto, D., Périco, A.E. and Mariano, E.B. (2014) Sustainable development in the BRICS countries: An efficiency analysis by data envelopment. *International Journal of Sustainable Development and World Ecology*, **21**, 259–272 [446].

Santos, S.P., Amado, C.A.F. and Rosado, J.R. (2011) Formative evaluation of electricity distribution utilities using data envelopment analysis. *Journal of the Operational Research Society*, **62**, 1298–1319 [447].

Sarica, K. and Or, I. (2007) Efficiency assessment of Turkish power plants using DEA. *Energy*, **32**, 1484–1499 [448].

Sarkis, J. and Cordeiro, J.J. (2009) Investigating technical and ecological efficiencies in the electricity generation industry: Are there win–win opportunities. *Journal of the Operational Research Society*, **60**, 1160–1172 [449].

Sarkis, J. and Cordeiro, J.J. (2012) Ecological modernization in the electrical utility industry: An application of a bads–goods DEA model of ecological and technical efficiency. *European Journal of Operational Research*, **219**, 386–395 [450].

Sarkis, J. and Weinrach, J. (2001) Using data envelopment analysis to evaluate environmentally conscious waste treatment technology. *Journal of Cleaner Production*, **9**, 417–427 [451].

Saxena, P., Saxena, R.R. and Sehgal, D. (2016) Efficiency evaluation of the energy companies in CNX 500 Index of the NSE, India using data envelopment analysis. *Benchmarking*, **23**, 113–126 [452].

Scheel, H. (2001) Undesirable outputs in efficiency valuations. *European Journal of Operational Research*, **132**, 400–410 [453].

Serra, T., Chambers, R.G. and Oude Lansink, A. (2014) Measuring technical and environmental efficiency in a state-contingent technology. *European Journal of Operational Research*, **236**, 706–717 [454].

Sesmero, J.P., Perrin, R.K. and Fulginiti, L.E. (2012) Environmental efficiency among corn ethanol plants. *Biomass and Bioenergy*, **46**, 634–644 [455].

Shabani, A., Farzipoor Saen, R. and Torabipour, S.M.R. (2014) A new data envelopment analysis (DEA) model to select eco-efficient technologies in the presence of undesirable outputs. *Clean Technologies and Environmental Policy*, **16**, 513–525 [456].

Shabani, A., Torabipour, S.M.R., Farzipoor Saen, R. and Khodakarami, M. (2015) Distinctive data envelopment analysis model for evaluating global environment performance. *Applied Mathematical Modelling*, **39**, 4385–4404 [457].

Shabani, Z., Rafiee, S., Mobli, H. and Khanalipur, E. (2012) Optimization in energy consumption of carnation production using data envelopment analysis (DEA). *Energy Systems*, **3**, 325–339 [458].

Shakouri G, H., Nabaee, M. and Aliakbarisani, S. (2014) A quantitative discussion on the assessment of power supply technologies: DEA (data envelopment analysis) and SAW (simple additive weighting) as complementary methods for the "Grammar." *Energy*, **64**, 640–647 [459].

Shen, N., Zhou, J. and Zou, W. (2015) Energy efficiency measures and convergence in China, taking into account the effects of environmental and random factors. *Polish Journal of Environmental Studies*, **24**, 257–267 [460].

Shi, G.-M., Bi, J. and Wang, J.-N. (2010) Chinese regional industrial energy efficiency evaluation based on a DEA model of fixing non-energy inputs. *Energy Policy*, **38**, 6172–6179 [461].

Shi, P., Yan, B., Shi, S. and Ke, C. (2015) A decision support system to select suppliers for a sustainable supply chain based on a systematic DEA approach. *Information Technology and Management*, **16**, 39–49 [462].

Shi, T., Zhang, X., Du, H. and Shi, H. (2015) Urban water resource utilization efficiency in China. *Chinese Geographical Science*, **25**, 684–697 [463].

Shieh, H.-S. (2012) The greener, the more cost efficient? An empirical study of international tourist hotels in Taiwan. *International Journal of Sustainable Development and World Ecology*, **19**, 536–545 [464].

Shortall, O.K. and Barnes, A.P. (2013) Greenhouse gas emissions and the technical efficiency of dairy farmers. *Ecological Indicators*, **29**, 478–488 [465].

Singh, S.K. and Bajpai, V.K. (2013) Estimation of operational efficiency and its determinants using DEA: The case of Indian coal-fired power plants. *International Journal of Energy Sector Management*, **7**, 409–429[466].

Singh, S.K., Bajpai, V.K. and Garg, T.K. (2013) Measuring productivity change in Indian coal-fired electricity generation: 2003–2010. *International Journal of Energy Sector Management*, **7**, 46–64 [467].

Sipiläinen, T. and Huhtala, A. (2013) Opportunity costs of providing crop diversity in organic and conventional farming: Would targeted environmental policies make economic sense? *European Review of Agricultural Economics*, **40**, 441–462 [468].

Skevas, T., Lansink, A.O. and Stefanou, S.E. (2012) Measuring technical efficiency in the presence of pesticide spillovers and production uncertainty: The case of Dutch arable farms. *European Journal of Operational Research*, **223**, 550–559 [469].

Skevas, T., Stefanou, S.E. and Oude Lansink, A. (2014) Pesticide use, environmental spillovers and efficiency: A DEA risk-adjusted efficiency approach applied to Dutch arable farming. *European Journal of Operational Research*, **237**, 658–664 [470].

Song, M. and Guan, Y. (2014) The environmental efficiency of Wanjiang demonstration area: A Bayesian estimation approach. *Ecological Indicators*, **36**, 59–67 [471].

Song, M. and Wang, S. (2015) Environmental efficiency evaluation of China based on a kind of congestion and undesirable output coefficient. *Panoeconomicus*, **62**, 453–468 [472].

Song, M., Wang, S. and Liu, Q. (2013) Environmental efficiency evaluation considering the maximization of desirable outputs and its application. *Mathematical and Computer Modelling*, **58**, 1110–1116 [473].

Song, M., Wang, S. and Liu, W. (2014) A two-stage DEA approach for environmental efficiency measurement. *Environmental Monitoring and Assessment*, **186**, 3041–3051 [474].

Song, M., Yang, L., Wu, J. and Lv, W. (2013) Energy saving in China: Analysis on the energy efficiency via bootstrap-DEA approach. *Energy Policy*, **57**, 1–6 [475].

Song, M., Zhang, J. and Wang, S. (2015) Review of the network environmental efficiencies of listed petroleum enterprises in China. *Renewable and Sustainable Energy Reviews*, **43**, 65–71 [476].

Song, M.-L., Guan, Y. and Song, F. (2013) Environmental efficiency, advances in environmental technology and total factor of environmental productivity of China. *Kybernetes*, **42**, 943–954 [477].

Song, M.L. and Wang, S.H. (2014) DEA decomposition of China's environmental efficiency based on search algorithm. *Applied Mathematics and Computation*, **247**, 562–572 [478].

Song, M.-L., Zhang, L.-L., Liu, W. and Fisher, R. (2013) Bootstrap-DEA analysis of BRICS' energy efficiency based on small sample data. *Applied Energy*, **112**, 1049–1055 [479].

Song, T., Yang, Z. and Chahine, T. (2016) Efficiency evaluation of material and energy flows, a case study of Chinese cities. *Journal of Cleaner Production*, **112**, 3667–3675 [480].

Song, X., Hao, Y. and Zhu, X. (2015) Analysis of the environmental efficiency of the Chinese transportation sector using an undesirable output slacks-based measure data envelopment analysis model. *Sustainability*, **7**, 9187–9206 [481].

Sözen, A. and Alp, I. (2009) Comparison of Turkey's performance of greenhouse gas emissions and local/regional pollutants with EU countries. *Energy Policy*, **37**, 5007–5018 [482].

Sözen, A. and Alp, I. (2013) Malmquist total factor productivity index approach to modelling Turkey's performance of energy consumption. *Energy Sources, Part B: Economics, Planning and Policy*, **8**, 398–411 [483].

Sözen, A., Alp, I. and Iskender, Ü. (2014) An evaluation of Turkey's energy dependency. *Energy Sources, Part B: Economics, Planning and Policy*, **9**, 398–412 [484].

Sözen, A., Alp, I. and Kilinc, C. (2012) Efficiency assessment of the hydro-power plants in Turkey by using data envelopment analysis. *Renewable Energy*, **46**, 192–202 [485].

Sözen, A., Alp, I. and Özdemir, A. (2010) Assessment of operational and environmental performance of the thermal power plants in Turkey by using data envelopment analysis. *Energy Policy*, **38**, 6194–6203 [486].

Stallard, T., Rothschild, R. and Aggidis, G.A. (2008) A comparative approach to the economic modelling of a large-scale wave power scheme. *European Journal of Operational Research*, **185**, 884–898 [487].

Stiakakis, E. and Fouliras, P. (2009) The impact of environmental practices on firms' efficiency: The case of ICT-producing sectors, *Operational Research*, **9**, 311–328 [488].

Sueyoshi, T. (1999) Tariff structure of Japanese electric power companies: An empirical analysis using DEA. *European Journal of Operational Research*, **118**, 350–374 [489].

Sueyoshi, T. (2000) Stochastic DEA for restructure strategy: An application to a Japanese petroleum company. *Omega*, **28**, 385–398 [490].

Sueyoshi, T. and Goto, M. (2001) Slack-adjusted DEA for time series analysis: Performance measurement of Japanese electric power generation industry in 1984–1993. *European Journal of Operational Research*, **133**, 232–259 [491].

Sueyoshi, T. and Goto, M. (2009) Can environmental investment and expenditure enhance financial performance of US electric utility firms under the clean air act amendment of 1990? *Energy Policy*, **37**, 4819–4826 [492].

Sueyoshi, T. and Goto, M. (2010a) Measurement of a linkage among environmental, operational, and financial performance in Japanese manufacturing firms: A use of data envelopment analysis with strong complementary slackness condition. *European Journal of Operational Research*, **207**, 1742–1753 [493].

Sueyoshi, T. and Goto, M. (2010b) Should the US clean air act include CO2 emission control?: Examination by data envelopment analysis. *Energy Policy*, **38**, 5902–5911 [494].

Sueyoshi, T. and Goto, M. (2011a) DEA approach for unified efficiency measurement: Assessment of Japanese fossil fuel power generation. *Energy Economics*, **33**, 292–303 [495].

Sueyoshi, T. and Goto, M. (2011b) Measurement of returns to scale and damages to scale for DEA-based operational and environmental assessment: How to manage desirable (good) and undesirable (bad) outputs? *European Journal of Operational Research*, **211**, 76–89 [496].

Sueyoshi, T. and Goto, M. (2011c) Methodological comparison between two unified (operational and environmental) efficiency measurements for environmental assessment. *European Journal of Operational Research*, **210**, 684–693 [497].

Sueyoshi, T. and Goto, M. (2012a) Data envelopment analysis for environmental assessment: Comparison between public and private ownership in petroleum industry. *European Journal of Operational Research*, **216**, 668–678 [498].

Sueyoshi, T. and Goto, M. (2012b) DEA environmental assessment of coal fired power plants: Methodological comparison between radial and non-radial models. *Energy Economics*, **34**, 1854–1863 [499].

Sueyoshi, T. and Goto, M. (2012c) DEA radial and non-radial models for unified efficiency under natural and managerial disposability: Theoretical extension by strong complementary slackness conditions. *Energy Economics*, **34**, 700–713 [500].

Sueyoshi, T. and Goto, M. (2012d) DEA radial measurement for environmental assessment and planning: Desirable procedures to evaluate fossil fuel power plants. *Energy Policy*, **41**, 422–432 [501].

Sueyoshi, T. and Goto, M. (2012e) Efficiency-based rank assessment for electric power industry: A combined use of data envelopment analysis (DEA) and DEA-discriminant analysis (DA). *Energy Economics*, **34**, 634–644 [502].

Sueyoshi, T. and Goto, M. (2012f) Environmental assessment by DEA radial measurement: U.S. coal-fired power plants in ISO (independent system operator) and RTO (regional transmission organization). *Energy Economics*, **34**, 663–676 [503].

Sueyoshi, T. and Goto, M. (2012g) Returns to scale and damages to scale on U.S. fossil fuel power plants: Radial and non-radial approaches for DEA environmental assessment. *Energy Economics*, **34**, 2240–2259 [504].

Sueyoshi, T. and Goto, M. (2012h) Returns to scale and damages to scale under natural and managerial disposability: Strategy, efficiency and competitiveness of petroleum firms. *Energy Economics*, **34**, 645–662 [505].

Sueyoshi, T. and Goto, M. (2012i) Returns to scale and damages to scale with strong complementary slackness conditions in DEA assessment: Japanese corporate effort on environment protection. *Energy Economics*, **34**, 1422–1434 [506].

Sueyoshi, T. and Goto, M. (2012j) Returns to scale, damages to scale, marginal rate of transformation and rate of substitution in DEA environmental assessment. *Energy Economics*, **34**, 905–917 [507].

Sueyoshi, T. and Goto, M. (2012k) Weak and strong disposability vs. natural and managerial disposability in DEA environmental assessment: Comparison between Japanese electric power industry and manufacturing industries. *Energy Economics*, **34**, 686–699 [508].

Sueyoshi, T. and Goto, M. (2013a) A comparative study among fossil fuel power plants in PJM and California ISO by DEA environmental assessment. *Energy Economics*, **40**, 130–145 [509].

Sueyoshi, T. and Goto, M. (2013b) Data envelopment analysis as a data mining methodology for environmental assessment. *Journal of Data and Information Processing*, **1**, 19–27 [510].

Sueyoshi, T. and Goto, M. (2013c) DEA environmental assessment in a time horizon: Malmquist index on fuel mix, electricity and CO_2 of industrial nations. *Energy Economics*, **40**, 370–382 [511].

Sueyoshi, T. and Goto, M. (2013d) Returns to scale vs. damages to scale in data envelopment analysis: An impact of U.S. clean air act on coal-fired power plants. *Omega*, **41**, 164–175 [512].

Sueyoshi, T. and Goto, M. (2014a) DEA radial measurement for environmental assessment: A comparative study between Japanese chemical and pharmaceutical firms. *Applied Energy*, **115**, 502–513 [513].

Sueyoshi, T. and Goto, M. (2014b) Environmental assessment for corporate sustainability by resource utilization and technology innovation: DEA radial measurement on Japanese industrial sectors. *Energy Economics*, **46**, 295–307 [514].

Sueyoshi, T. and Goto, M. (2014c) Investment strategy for sustainable society by development of regional economies and prevention of industrial pollutions in Japanese manufacturing sectors. *Energy Economics*, **42**, 299–312 [515].

Sueyoshi, T. and Goto, M. (2014d) Photovoltaic power stations in Germany and the United States: A comparative study by data envelopment analysis. *Energy Economics*, **42**, 271–288 [516].

Sueyoshi, T. and Goto, M. (2015a) DEA environmental assessment in time horizon: Radial approach for Malmquist index measurement on petroleum companies. *Energy Economics*, **51**, 329–345 [517].

Sueyoshi, T. and Goto, M. (2015b) Environmental assessment on coal-fired power plants in U.S. north-east region by DEA non-radial measurement. *Energy Economics*, **50**, 125–139 [518].

Sueyoshi, T. and Goto, M. (2015c) Japanese fuel mix strategy after disaster of Fukushima Daiichi nuclear power plant: Lessons from international comparison among industrial nations measured by DEA environmental assessment in time horizon. *Energy Economics*, **52**, 87–103 [519].

Sueyoshi, T. and Goto, M. (2016) Undesirable congestion under natural disposability and desirable congestion under managerial disposability in U.S. electric power industry measured by DEA environmental assessment. *Energy Economics*, **55**, 173–188 [520].

Sueyoshi, T., Goto, M. and Shang, J. (2009) Core business concentration vs. corporate diversification in the US electric utility industry: Synergy and deregulation effects. *Energy Policy*, **37**, 4583–4594 [521].

Sueyoshi, T., Goto, M. and Snell, M.A. (2013) DEA environmental assessment: Measurement of damages to scale with unified efficiency under managerial disposability or environmental efficiency. *Applied Mathematical Modelling*, **37**, 7300–7314 [522].

Sueyoshi, T., Goto, M. and Sugiyama, M. (2013) DEA window analysis for environmental assessment in a dynamic time shift: Performance assessment of U.S. coal-fired power plants. *Energy Economics*, **40**, 845–857 [523].

Sueyoshi, T., Goto, M. and Ueno, T. (2010) Performance analysis of US coal-fired power plants by measuring three DEA efficiencies. *Energy Policy*, **38**, 1675–1688 [524].

Sueyoshi, T. and Wang, D. (2014a) Radial and non-radial approaches for environmental assessment by Data Envelopment Analysis: Corporate sustainability and effective investment for technology innovation. *Energy Economics*, **45**, 537–551 [525].

Sueyoshi, T. and Wang, D. (2014b) Sustainability development for supply chain management in U.S. petroleum industry by DEA environmental assessment. *Energy Economics*, **46**, 360–374 [526].

Sueyoshi, T. and Yuan, Y. (2015a) China's regional sustainability and diversified resource allocation: DEA environmental assessment on economic development and air pollution. *Energy Economics*, **49**, 239–256 [527].

Sueyoshi, T. and Yuan, Y. (2015b) Comparison among U.S. industrial sectors by DEA environmental assessment: Equipped with analytical capability to handle zero or negative in production factors. *Energy Economics*, **52**, 69–86 [528].

Sueyoshi, T. and Yuan, Y. (2016a) Marginal rate of transformation and rate of substitution measured by DEA environmental assessment: Comparison among European and North American nations. *Energy Economics*, **56**, 270–287 [529].

Sueyoshi, T. and Yuan, Y. (2016b) Returns to damage under undesirable congestion and damages to return under desirable congestion measured by DEA environmental assessment with multiplier restriction: Economic and energy planning for social sustainability in China. *Energy Economics*, **56**, 288–309 [530].

Suh, Y., Seol, H., Bae, H. and Park, Y. (2014) Eco-efficiency based on social performance and its relationship with financial performance: A cross-industry analysis of South Korea. *Journal of Industrial Ecology*, **18**, 909–919 [531].

Sun, J., Sun, D. and Guo, S. (2014) Evaluation on the efficiency of biomass power generation industry in China. *Scientific World Journal*, **2014**(831372), 1–8 [532].

Suzuki, S. and Nijkamp, P. (2016) An evaluation of energy-environment-economic efficiency for EU, APEC and ASEAN countries: design of a target-oriented DFM model with fixed factors in data envelopment analysis. *Energy Policy*, **88**, 100–112 [533].

Suzuki, S., Nijkamp, P. and Rietveld, P. (2015) A target-oriented data envelopment analysis for energy-environment efficiency improvement in Japan. *Energy Efficiency*, **8**, 433–446 [534].

Syp, A., Faber, A., Borzęcka-Walker, M. and Osuch, D. (2015) Assessment of greenhouse gas emissions in winter wheat farms using data envelopment analysis approach. *Polish Journal of Environmental Studies*, **24**, 2197–2203 [535].

Taghavifar, H., Mardani, A. and Karim-Maslak, H. (2014) Multi-criteria optimization model to investigate the energy waste of off-road vehicles utilizing soil bin facility. *Energy*, **73**, 762–770 [536].

Tajbakhsh, A. and Hassini, E. (2015) A data envelopment analysis approach to evaluate sustainability in supply chain networks. *Journal of Cleaner Production*, **105**, 74–85 [537].

Taki, M., Ajabshirchi, Y. and Mahmoudi, A. (2012) Application of parametric and non-parametric method to analyzing of energy consumption for cucumber production in Iran. *Modern Applied Science*, **6**, 75–87 [538].

Tao, Y. and Zhang, S. (2013) Environmental efficiency of electric power industry in the Yangtze river delta. *Mathematical and Computer Modelling*, **58**, 927–935 [539].

Taskin, F. and Zaim, O. (2001) The role of international trade on environmental efficiency: a DEA approach. *Economic Modelling*, **18**, 1–17 [540].

Tatari, O. and Kucukvar, M. (2012) Eco-efficiency of construction materials: Data envelopment analysis. *Journal of Construction Engineering and Management*, **138**, 733–741 [541].

Tavana, M., hakbaz, M.H. and Jafari-Songhori, M. (2009) Information technology's impact on productivity in conventional power plants. *International Journal of Business Performance Management*, **11**, 187–202 [542].

Tavassoli, M., Faramarzi, G.R. and Farzipoor Saen, R. (2015) Ranking electricity distribution units using slacks-based measure, strong complementary slackness condition, and discriminant analysis. *International Journal of Electrical Power and Energy Systems*, **64**, 1214–1220 [543].

Thakur, T., Deshmukh, S.G. and Kaushik, S.C. (2006) Efficiency evaluation of the state owned electric utilities in India. *Energy Policy*, **34**, 2788–2804 [544].

Thanh Nguyen, T., Hoang, V.-N. and Seo, B. (2012) Cost and environmental efficiency of rice farms in South Korea. *Agricultural Economics*, **43**, 369–378 [545].

Thompson, R.G., Dharmapala, P.S., Rothenberg, L.J. and Thrall, R.M. (1996) DEA/AR efficiency and profitability of 14 major oil companies in U.S. exploration and production. *Computers and Operations Research*, **23**, 357–373 [546].

Thompson, R.G., Dharmapala, P.S. and Thrall, R.M. (1995) Linked-cone DEA profit ratios and technical efficiency with application to Illinois coal mines. *International Journal of Production Economics*, **39**, 99–115. [547]

Thompson, R.G., Lee, E. and Thrall, R.M. (1992) DEA/AR-efficiency of U.S. independent oil/gas producers over time. *Computers and Operations Research*, **19**, 377–391 [548].

Toloo, M. and Babaee, S. (2015) On variable reductions in data envelopment analysis with an illustrative application to a gas company. *Applied Mathematics and Computation*, **270**, 527–533 [549].

Toma, L., March, M., Stott, A.W. and Roberts, D.J. (2013) Environmental efficiency of alternative dairy systems: A productive efficiency approach. *Journal of Dairy Science*, **96**, 7014–7031 [550].

Tone, K. and Tsutsui, M. (2007) Decomposition of cost efficiency and its application to Japanese-US electric utility comparisons. *Socio-Economic Planning Sciences*, **41**, 91–106 [551].

Triantis, K. and Otis, P. (2004) Dominance-based measurement of productive and environmental performance for manufacturing. *European Journal of Operational Research*, **154**, 447–464 [552].

Tsolas, I.E. (2010) Assessing power stations performance using a DEA-bootstrap approach. *International Journal of Energy Sector Management*, **4**, 337–355 [553].

Tsolas, I.E. (2011) Performance assessment of mining operations using nonparametric production analysis: A bootstrapping approach in DEA. *Resources Policy*, **36**, 159–167 [554].

Tu, K.-J. (2015) Establishing the dea energy management system for individual departments within universities, *Facilities*, **33**, 716–735 [555].

Tu, Z. and Shen, R. (2014) Can China's industrial SO2 emissions trading pilot scheme reduce pollution abatement costs? *Sustainability*, **6**, 7621–7645 [556].

Tyteca, D. (1996) On the measurement of the environmental performance of firms: A literature review and a productive efficiency perspective. *Journal of Environmental Management*, **46**, 281–308 [557].

Tyteca, D. (1997) Linear programming models for the measurement of environmental performance of firms: Concepts and empirical results. *Journal of Productivity Analysis*, **8**, 183–197 [558].

Tyteca, D. (1998) Sustainability indicators at the firm level. *Journal of Industrial Ecology*, **2**, 61–77 [559].

Ueasin, N., Wongchai, A. and Nonthapot, S. (2015) Performance assessment and optimization of biomass steam turbine power plants by data envelopment analysis. *International Journal of Energy Economics and Policy*, **5**, 668–672 [560].

Ullah, A. and Perret, S.R. (2014) Technical- and environmental-efficiency analysis of irrigated cotton-cropping systems in Punjab, Pakistan using data envelopment analysis. *Environmental Management*, **54**, 288–300 [561].

Van Meensel, J., Lauwers, L., Van Huylenbroeck, G. and Van Passel, S. (2010) Comparing frontier methods for economic–environmental trade-off analysis. *European Journal of Operational Research*, **207**, 1027–1040 [562].

Vaninsky, A. (2006) Efficiency of electric power generation in the United States: Analysis and forecast based on data envelopment analysis. *Energy Economics*, **28**, 326–338 [563].

Vazhayil, J.P. and Balasubramanian, R. (2012) Hierarchical multi-objective optimization of India's energy strategy portfolios for sustainable development. *International Journal of Energy Sector Management*, **6**, 301–320 [564].

Vazhayil, J.P. and Balasubramanian, R. (2013) Optimization of India's power sector strategies using weight-restricted stochastic data envelopment analysis. *Energy Policy*, **56**, 456–465 [565].

Vázquez-Rowe, I., Iribarren, D., Hospido, A., Moreira, M.T. and Feijoo, G. (2011) Computation of operational and environmental benchmarks within selected Galician fishing fleets. *Journal of Industrial Ecology*, **15**, 776–795 [566].

Vázquez-Rowe, I., Villanueva-Rey, P., Iribarren, D., Teresa Moreira, M. and Feijoo, G. (2012) Joint life cycle assessment and data envelopment analysis of grape production for vinification in the Rías Baixas appellation (NW Spain). *Journal of Cleaner Production*, **27**, 92–102 [567].

Vlontzos, G., Niavis, S. and Manos, B. (2014) A DEA approach for estimating the agricultural energy and environmental efficiency of EU countries. *Renewable and Sustainable Energy Reviews*, **40**, 91–96 [568].

Voltes-Dorta, A., Perdiguero, J. and Jiménez, J.L. (2013) Are car manufacturers on the way to reduce CO2 emissions?: A DEA approach. *Energy Economics*, **38**, 77–86 [569].

von Geymueller, P. (2009) Static versus dynamic DEA in electricity regulation: The case of US transmission system operators. *Central European Journal of Operations Research*, **17**, 397–413 [570].

von Hirschhausen, C., Cullmann, A. and Kappeler, A. (2006) Efficiency analysis of German electricity distribution utilities – Non-parametric and parametric tests. *Applied Economics*, **38**, 2553–2566 [571].

Wang, C. (2011) Sources of energy productivity growth and its distribution dynamics in China. *Resource and Energy Economics*, **33**, 279–292 [572].

Wang, C. (2013) Changing energy intensity of economies in the world and its decomposition. *Energy Economics*, **40**, 637–644 [573].

Wang, C.-N., Lin, L.-C. and Murugesan, D. (2013) Analyzing PSU's performance: A case from ministry of petroleum and natural gas of India. *Mathematical Problems in Engineering*, **2013**(802690), 1–9 [574].

Wang, D., Li, S. and Sueyoshi, T. (2014) DEA environmental assessment on U.S. Industrial sectors: Investment for improvement in operational and environmental performance to attain corporate sustainability. *Energy Economics*, **45**, 254–267 [575].

Wang, E., Shen, Z., Alp, N. and Barry, N. (2015) Benchmarking energy performance of residential buildings using two-stage multifactor data envelopment analysis with degree-day based simple-normalization approach. *Energy Conversion and Management*, **106**, 530–542 [576].

Wang, H. (2015) A generalized MCDA-DEA (multi-criterion decision analysis-data envelopment analysis) approach to construct slacks-based composite indicator. *Energy*, **80**, 114–122 [577].

Wang, H., Zhou, P. and Zhou, D.Q. (2013) Scenario-based energy efficiency and productivity in China: A non-radial directional distance function analysis. *Energy Economics*, **40**, 795–803 [578].

Wang, J.H., Ngan, H.W., Engriwan, W. and Lo, K. L. (2007) Performance based regulation of the electricity supply industry in Hong Kong: An empirical efficiency analysis approach. *Energy Policy*, **35**, 609–615 [579].

Wang, K., Lu, B. and Wei, Y.-M. (2013) China's regional energy and environmental efficiency: a range-adjusted measure based analysis. *Applied Energy*, **112**, 1403–1415 [580].

Wang, K. and Wei, Y.-M. (2014) China's regional industrial energy efficiency and carbon emissions abatement costs. *Applied Energy*, **130**, 617–631 [581].

Wang, K. and Wei, Y.-M. (2016) Sources of energy productivity change in China during 1997–2012: A decomposition analysis based on the Luenberger productivity indicator. *Energy Economics*, **54**, 50–59 [582].

Wang, K., Wei, Y.-M. and Zhang, X. (2012) A comparative analysis of China's regional energy and emission performance: Which is the better way to deal with undesirable outputs? *Energy Policy*, **46**, 574–584 [583].

Wang, K., Wei, Y.-M. and Zhang, X. (2013) Energy and emissions efficiency patterns of Chinese regions: A multi-directional efficiency analysis. *Applied Energy*, **104**, 105–116 [584].

Wang, K., Yu, S. and Zhang, W. (2013) China's regional energy and environmental efficiency: A DEA window analysis based dynamic evaluation. *Mathematical and Computer Modelling*, **58**, 1117–1127 [585].

Wang, K., Zhang, X., Wei, Y.-M. and Yu, S. (2013) Regional allocation of CO2 emissions allowance over provinces in China by 2020. *Energy Policy*, **54**, 214–229 [586].

Wang, L., Chen, Z., Ma, D. and Zhao, P. (2013) Measuring carbon emissions performance in 123 countries: Application of minimum distance to the strong efficiency frontier analysis. *Sustainability*, **5**, 5319–5332 [587].

Wang, P.-C., Lee, Y.-M. and Chen, C.-Y. (2014) Estimation of resource productivity and efficiency: An extended evaluation of sustainability related to material flow. *Sustainability*, **6**, 6070–6087 [588].

Wang, Q., Zhao, Z., Shen, N. and Liu, T. (2015) Have Chinese cities achieved the win–win between environmental protection and economic development? From the perspective of environmental efficiency. *Ecological Indicators*, **51**, 151–158 [589].

Wang, Q., Zhao, Z., Zhou, P. and Zhou, D. (2013) Energy efficiency and production technology heterogeneity in China: A meta-frontier DEA approach. *Economic Modelling*, **35**, 283–289 [590].

Wang, Q., Zhou, P. and Zhou, D. (2012) Efficiency measurement with carbon dioxide emissions: The case of China. *Applied Energy*, **90**, 161–166 [591].

Wang, W.-K., Lu, W.-M. and Wang, S.-W. (2014) The impact of environmental expenditures on performance in the U.S. chemical industry. *Journal of Cleaner Production*, **64**, 447–456 [592].

Wang, X., Huang, G., Liu, Z. and Dai, C. (2012) Hybrid inexact optimization approach with data envelopment analysis for environment management and planning in the city of Beijing, China. *Environmental Engineering Science*, **29**, 313–327 [593].

Wang, Z. and Feng, C. (2015a) A performance evaluation of the energy, environmental, and economic efficiency and productivity in China: An application of global data envelopment analysis. *Applied Energy*, **147**, 617–626 [594].

Wang, Z. and Feng, C. (2015b) Sources of production inefficiency and productivity growth in China: A global data envelopment analysis. *Energy Economics*, **49**, 380–389 [595].

Wang, Z., Feng, C. and Zhang, B. (2014) An empirical analysis of China's energy efficiency from both static and dynamic perspectives. *Energy*, **74**, 322–330 [596].

Wang, Z., Zeng, H.L., Wei, Y.M. and Zhang, Y.X. (2012) Regional total factor energy efficiency: An empirical analysis of industrial sector in China. *Applied Energy*, **97**, 115–123 [597].

Wattana, S. and Sharma, D. (2011) Electricity industry reforms in Thailand: An analysis of productivity. *International Journal of Energy Sector Management*, **5**, 494–521 [598].

Wei, C., Ni, J. and Shen, M. (2009) Empirical analysis of provincial energy efficiency in China. *China and World Economy*, **17**, 88–103 [599].

Wei, C., Ni, J. and Sheng, M. (2011) China's energy inefficiency: A cross-country comparison. *The Social Science Journal*, **48**, 478–488 [600].

Wei, Y.-M., Liao, H. and Fan, Y. (2007) An empirical analysis of energy efficiency in China's iron and steel sector. *Energy*, **32**, 2262–2270 [601].

Welch, E. and Barnum, D. (2009) Joint environmental and cost efficiency analysis of electricity generation. *Ecological Economics*, **68**, 2336–2343 [602].

West, J. (2015) Capital valuation and sustainability: a data programming approach. *Review of Quantitative Finance and Accounting*, **45**, 591–608 [603].

Weyman-Jones, T.G. (1991) Productive efficiency in a regulated industry: The area electricity boards of England and Wales. *Energy Economics*, **13**, 116–122 [604].

Whittaker, G., Barnhart, B., Färe, R. and Grosskopf, S. (2015) Application of index number theory to the construction of a water quality index: Aggregated nutrient loadings related to the areal extent of hypoxia in the northern Gulf of Mexico. *Ecological Indicators*, **49**, 162–168 [605].

Wu, A.-H., Cao, Y.-Y. and Liu, B. (2014) Energy efficiency evaluation for regions in China: An application of DEA and Malmquist indices. *Energy Efficiency*, **7**, 429–439 [606].

Wu, C., Li, Y., Liu, Q. and Wang, K. (2013) A stochastic DEA model considering undesirable outputs with weak disposability. *Mathematical and Computer Modelling*, **58**, 980–989 [607].

Wu, F., Fan, L.W., Zhou, P. and Zhou, D.Q. (2012) Industrial energy efficiency with CO2 emissions in China: A nonparametric analysis. *Energy Policy*, **49**, 164–172 [608].

Wu, F., Zhou, P. and Zhou, D.Q. (2015) Measuring energy congestion in Chinese industrial sectors: A slacks-based DEA approach. *Computational Economics*, **46**, 479–494 [609].

Wu, H., Du, S., Liang, L. and Zhou, Y. (2013) A DEA-based approach for fair reduction and reallocation of emission permits. *Mathematical and Computer Modelling*, **58**, 1095–1101 [610].

Wu, H.-Q., Shi, Y., Xia, Q. and Zhu, W.-D. (2014) Effectiveness of the policy of circular economy in China: A DEA-based analysis for the period of 11th five-year-plan. *Resources, Conservation and Recycling*, **83**, 163–175 [611].

Wu, H.T., Pineau, P.-O. and Caporossi, G. (2010) Efficiency evolution of coal-fired electricity generation in China, 1999–2007. *International Journal of Energy Sector Management*, **4**, 316–336 [612].

Wu, J., An, Q., Ali, S. and Liang, L. (2013) DEA based resource allocation considering environmental factors. *Mathematical and Computer Modelling*, **58**, 1128–1137 [613].

Wu, J., An, Q., Xiong, B. and Chen, Y. (2013) Congestion measurement for regional industries in China: A data envelopment analysis approach with undesirable outputs. *Energy Policy*, **57**, 7–13 [614].

Wu, J., An, Q., Yao, X. and Wang, B. (2014) Environmental efficiency evaluation of industry in China based on a new fixed sum undesirable output data envelopment analysis. *Journal of Cleaner Production*, **74**, 96–104 [615].

Wu, J., Lv, L., Sun, J. and Ji, X. (2015) A comprehensive analysis of China's regional energy saving and emission reduction efficiency: From production and treatment perspectives. *Energy Policy*, **84**, 166–176 [616].

Wu, J., Xiong, B., An, Q., Zhu, Q. and Liang, L. (2015) Measuring the performance of thermal power firms in China via fuzzy enhanced Russell measure model with undesirable outputs. *Journal of Cleaner Production*, **102**, 237–245 [617].

Wu, J., Zhu, Q., Ji, X., Chu, J. and Liang, L. (2016) Two-stage network processes with shared resources and resources recovered from undesirable outputs. *European Journal of Operational Research*, **251**, 182–197 [618].

Wu, Q., Jia, R. and Yu, H. (2014) Study on environmental efficiency of electric power industry based on DEA optimization model. *Journal of Information and Computational Science*, **11**, 781–795 [619].

Xie, B.-C., Fan, Y. and Qu, Q.-Q. (2012) Does generation form influence environmental efficiency performance? An analysis of China's power system. *Applied Energy*, **96**, 261–271 [620].

Xie, B.-C., Shang, L.-F., Yang, S.-B. and Yi, B.-W. (2014) Dynamic environmental efficiency evaluation of electric power industries: Evidence from OECD (Organization for Economic Cooperation and Development) and BRIC (Brazil, Russia, India and China) countries. *Energy*, **74**(c), 147–157 [621].

Xu, B. and Ouenniche, J. (2012) A data envelopment analysis-based framework for the relative performance evaluation of competing crude oil prices' volatility forecasting models. *Energy Economics*, **34**, 576–583 [622].

Xue, X., Wu, H., Zhang, X., Dai, J. and Su, C. (2015) Measuring energy consumption efficiency of the construction industry: The case of China. *Journal of Cleaner Production*, **107**, 509–515 [623].

Yadav, V.K., Chauhan, Y.K., Padhy, N.P. and Gupta, H.O. (2013) A novel power sector restructuring model based on data envelopment analysis (DEA). *International Journal of Electrical Power and Energy Systems*, **44**, 629–637 [624].

Yadav, V.K., Kumar, N., Ghosh, S. and Singh, K. (2014) Indian thermal power plant challenges and remedies via application of modified data envelopment analysis. *International Transactions in Operational Research*, **21**, 955–977 [625].

Yadav, V.K., Padhy, N.P. and Gupta, H.O. (2010) A micro level study of an Indian electric utility for efficiency enhancement. *Energy*, **35**, 4053–4063 [626].

Yadav, V.K., Padhy, N.P. and Gupta, H.O. (2011) Performance evaluation and improvement directions for an Indian electric utility. *Energy Policy*, **39**, 7112–7120 [627].

Yadav, V.K., Padhy, N.P. and Gupta, H.O. (2014) The evaluation of the efficacy of an ongoing reform initiative of an Indian electric utility. *Energy Sources, Part B: Economics, Planning and Policy*, **9**, 291–300 [628].

Yadav, V.K., Padhy, N.P. and Gupta, H.O. (2015) A holistic approach model for realistic goal-setting for efficiency enhancement with application to the Indian power sector. *Energy Sources, Part B: Economics, Planning and Policy*, **10**, 120–131 [629].

Yaisawarng, S. and Klein, J.D. (1994) The effects of sulfur dioxide controls on productivity change in the U.S. electric power industry. *The Review of Economics and Statistics*, **76**, 447–460 [630].

Yan, Q. and Tao, J. (2014) Biomass power generation industry efficiency evaluation in China. *Sustainability*, **6**, 8720–8735 [631].

Yang, F. and Yang, M. (2015) Analysis on China's eco-innovations: Regulation context, intertemporal change and regional differences. *European Journal of Operational Research*, **247**, 1003–1012 [632].

Yang, F., Yang, M. and Nie, H. (2013) Productivity trends of Chinese regions: A perspective from energy saving and environmental regulations. *Applied Energy*, **110**, 82–89 [633].

Yang, H. and Pollitt, M. (2009) Incorporating both undesirable outputs and uncontrollable variables into DEA: The performance of Chinese coal-fired power plants. *European Journal of Operational Research*, **197**, 1095–1105 [634].

Yang, H. and Pollitt, M. (2010) The necessity of distinguishing weak and strong disposability among undesirable outputs in DEA: Environmental performance of Chinese coal-fired power plants. *Energy Policy*, **38**, 4440–4444 [635].

Yang, L., Ouyang, H., Fang, K., Ye, L. and Zhang, J. (2015) Evaluation of regional environmental efficiencies in China based on super-efficiency-DEA. *Ecological Indicators*, **51**, 13–19 [636].

Yang, L. and Wang, K.-L. (2013) Regional differences of environmental efficiency of China's energy utilization and environmental regulation cost based on provincial panel data and DEA method. *Mathematical and Computer Modelling*, **58**, 1074–1083 [637].

Yang, Q., Wan, X. and Ma, H. (2015) Assessing green development efficiency of municipalities and provinces in China integrating models of super-efficiency DEA and Malmquist index. *Sustainability*, **7**, 4492–4510 [638].

Yao, L., Xu, J. and Li, Y. (2014) Evaluation of the efficiency of low carbon industrialization in cultural and natural heritage: Taking Leshan as an example. *Sustainability*, **6**, 3825–3842 [639].

Yeh, T.-L., Chen, T.-Y. and Lai, P.-Y. (2010a) A comparative study of energy utilization efficiency between Taiwan and China. *Energy Policy*, **38**, 2386–2394 [640].

Yeh, T.-L., Chen, T.-Y. and Lai, P.-Y. (2010b) Incorporating greenhouse gas effects to evaluate energy efficiency in China. *International Journal of Sustainable Development and World Ecology*, **17**, 370–376 [641].

Yin, H., He, Q., Guo, T., Zhu, J. and Yu, B. (2014) Measurement method and empirical research on the sustainable development capability of a regional industrial system based on ecological niche theory in China. *Sustainability*, **6**, 8485–8509 [642].

Yin, K., Wang, R., An, Q., Yao, L. and Liang, J. (2014) Using eco-efficiency as an indicator for sustainable urban development: A case study of Chinese provincial capital cities. *Ecological Indicators*, **36**, 665–671 [643].

Ylvinger, S. (2003) Light-duty vehicles and external impacts: Product- and policy-performance assessment. *European Journal of Operational Research*, **144**, 194–208 [644].

Yokota, S. and Kumano, T. (2013) Mega-solar optimal allocation using data envelopment analysis. *Electrical Engineering in Japan*, **183**, 24–32 [645].

You, S. and Yan, H. (2011) A new approach in modelling undesirable output in DEA model. *Journal of the Operational Research Society*, **62**, 2146–2156 [646].

Yu, F.W. and Chan, K.T. (2012a) Chiller system performance benchmark by data envelopment analysis. *International Journal of Refrigeration*, **35**, 1815–1823 [647].

Yu, F.W. and Chan, K.T. (2012b) Improved energy management of chiller systems by multivariate and data envelopment analyses. *Applied Energy*, **92**, 168–174 [648].

Yu, F.W. and Chan, K.T. (2013) Energy management of chiller systems by data envelopment analysis. *Facilities*, **31**, 106–118 [649].

Yu, Y. and Choi, Y. (2015) Measuring environmental performance under regional heterogeneity in China: A metafrontier efficiency analysis. *Computational Economics*, **46**, 375–388 [650].

Yu, Y. and Wen, Z. (2010) Evaluating China's urban environmental sustainability with data envelopment analysis. *Ecological Economics*, **69**, 1748–1755 [651].

Yunos, J.M. and Hawdon, D. (1997) The efficiency of the national electricity board in Malaysia: An intercountry comparison using DEA. *Energy Economics*, **19**, 255–269 [652].

Zaim, O. (2004) Measuring environmental performance of state manufacturing through changes in pollution intensities: a DEA framework. *Ecological Economics*, **48**, 37–47 [653].

Zaim, O. and Taskin, F. (2000a) A Kuznets curve in environmental efficiency: An application on OECD countries. *Environmental and Resource Economics*, **17**, 21–36 [654].

Zaim, O. and Taskin, F. (2000b) Environmental efficiency in carbon dioxide emissions in the OECD: A non-parametric approach. *Journal of Environmental Management*, **58**, 95–107 [655].

Zanella, A., Camanho, A.S. and Dias, T.G. (2015) Undesirable outputs and weighting schemes in composite indicators based on data envelopment analysis. *European Journal of Operational Research*, **245**, 517–530 [656].

Zha, Y., Zhao, L. and Bian, Y. (2016) Measuring regional efficiency of energy and carbon dioxide emissions in China: A chance constrained DEA approach. *Computers and Operations Research*, **66**, 351–361 [657].

Zhang, B., Bi, J., Fan, Z., Yuan, Z. and Ge, J. (2008) Eco-efficiency analysis of industrial system in China: A data envelopment analysis approach. *Ecological Economics*, **68**, 306–316 [658].

Zhang, H.-Y., Ji, Q. and Fan, Y. (2013) An evaluation framework for oil import security based on the supply chain with a case study focused on China. *Energy Economics*, **38**, 87–95 [659].

Zhang, N. and Choi, Y. (2013a) Total-factor carbon emission performance of fossil fuel power plants in China: A metafrontier non-radial Malmquist index analysis. *Energy Economics*, **40**, 549–559 [660].

Zhang, N. and Choi, Y. (2013b) A comparative study of dynamic changes in CO2 emission performance of fossil fuel power plants in China and Korea. *Energy Policy*, **62**, 324–332 [661].

Zhang, N. and Choi, Y. (2013c) Environmental energy efficiency of China's regional economies: A non-oriented slacks-based measure analysis. *The Social Science Journal*, **50**, 225–234 [662].

Zhang, N. and Choi, Y. (2014) A note on the evolution of directional distance function and its development in energy and environmental studies 1997–2013. *Renewable and Sustainable Energy Reviews*, **33**, 50–59 [663].

Zhang, N. and Kim, J.-D. (2014) Measuring sustainability by energy efficiency analysis for Korean power companies: A sequential slacks-based efficiency measure. *Sustainability*, **6**, 1414–1426 [664].

Zhang, N., Kong, F. and Choi, Y. (2014) Measuring sustainability performance for China: A sequential generalized directional distance function approach. *Economic Modelling*, **41**, 392–397 [665].

Zhang, N., Kong, F. and Kung, C.-C. (2015) On modeling environmental production characteristics: A slacks-based measure for China's Poyang Lake ecological economics zone. *Computational Economics*, **46**, 389–404 [666].

Zhang, N., Kong, F. and Yu, Y. (2015) Measuring ecological total-factor energy efficiency incorporating regional heterogeneities in China. *Ecological Indicators*, **51**, 165–172 [667].

Zhang, N., Zhou, P. and Choi, Y. (2013) Energy efficiency, CO2 emission performance and technology gaps in fossil fuel electricity generation in Korea: A meta-frontier non-radial directional distance function analysis. *Energy Policy*, **56**, 653–662 [668].

Zhang, S., Lundgren, T. and Zhou, W. (2016) Energy efficiency in Swedish industry: A firm-level data envelopment analysis. *Energy Economics*, **55**, 42–51 [669].

Zhang, X.-P., Cheng, X.-M., Yuan, J.-H. and Gao, X.-J. (2011) Total-factor energy efficiency in developing countries. *Energy Policy*, **39**, 644–650 [670].

Zhang, Y. and Bartels, R. (1998) The effect of sample size on the mean efficiency in DEA with an application to electricity distribution in Australia, Sweden and New Zealand. *Journal of Productivity Analysis*, **9**, 187–204 [671].

Zhang, Y.-J. and Da, Y.-B. (2013) Decomposing the changes of energy-related carbon emissions in China: Evidence from the PDA approach. *Natural Hazards*, **69**, 1109–1122 [672].

Zhang, Z., Chen, X. and Heck, P. (2014) Energy-based regional socio-economic metabolism analysis: An application of data envelopment analysis and decomposition analysis. *Sustainability*, **6**, 8618–8638 [673].

Zhao, X., Yang, R. and Ma, Q. (2014) China's total factor energy efficiency of provincial industrial sectors. *Energy*, **65**, 52–61 [674].

Zhou, G., Chung, W. and Zhang, X. (2013) A study of carbon dioxide emissions performance of China's transport sector. *Energy*, **50**, 302–314 [675].

Zhou, G., Chung, W. and Zhang, Y. (2014) Measuring energy efficiency performance of China's transport sector: A data envelopment analysis approach. *Expert Systems with Applications*, **41**, 709–772 [676].

Zhou, P. and Ang, B.W. (2008a) Decomposition of aggregate CO2 emissions: A production-theoretical approach. *Energy Economics*, **30**, 1054–1067 [677].

Zhou, P. and Ang, B.W. (2008b) Linear programming models for measuring economy-wide energy efficiency performance. *Energy Policy*, **36**, 2911–2916 [678].

Zhou, P., Ang, B.W. and Han, J.Y. (2010) Total factor carbon emission performance: A Malmquist index analysis. *Energy Economics*, **32**, 194–201 [679].

Zhou, P., Ang, B.W. and Poh, K.L. (2006) Slacks-based efficiency measures for modeling environmental performance. *Ecological Economics*, **60**, 111–118 [680].

Zhou, P., Ang, B.W. and Poh, K.L. (2007) A mathematical programming approach to constructing composite indicators. *Ecological Economics*, **62**, 291–297 [681].

Zhou, P., Ang, B.W. and Poh, K.L. (2008a) A survey of data envelopment analysis in energy and environmental studies. *European Journal of Operational Research*, **189**, 1–18 [682].

Zhou, P., Ang, B.W. and Poh, K.L. (2008b) Measuring environmental performance under different environmental DEA technologies. *Energy Economics*, **30**, 1–14 [683].

Zhou, P., Ang, B.W. and Wang, H. (2012) Energy and CO2 emission performance in electricity generation: A non-radial directional distance function approach. *European Journal of Operational Research*, **221**, 625–635 [684].

Zhou, P., Ang, B.W. and Zhou, D.Q. (2012) Measuring economy-wide energy efficiency performance: A parametric frontier approach. *Applied Energy*, **90**, 196–200 [685].

Zhou, P., Poh, K.L. and Ang, B.W. (2007) A non-radial DEA approach to measuring environmental performance. *European Journal of Operational Research*, **178**, 1–9 [686].

Zhou, P., Sun, Z.R. and Zhou, D.Q. (2014) Optimal path for controlling CO2 emissions in China: A perspective of efficiency analysis. *Energy Economics*, **45**, 99–110 [687].

Zhu, Z., Wang, K. and Zhang, B. (2014) Applying a network data envelopment analysis model to quantify the eco-efficiency of products: A case study of pesticides. *Journal of Cleaner Production*, **69**, 67–73 [688].

Zofio, J.L. and Prieto, A.M. (2001) Environmental efficiency and regulatory standards: The case of CO2 emissions from OECD industries. *Resource and Energy Economics*, **23**, 63–83 [689].

Zografidou, E., Petridis, K., Arabatzis, G. and Dey, P.K. (2016) Optimal design of the renewable energy map of Greece using weighted goal-programming and data envelopment analysis. *Computers and Operations Research*, **66**, 313–326 [690].

Zorić, J., Hrovatin, N. and Scarsi, G. (2009) Gas distribution benchmarking of utilities from Slovenia, the Netherlands and the UK: An application of data envelopment analysis. *South East European Journal of Economics and Business*, **4**, 113–124 [691].

Zou, G., Chen, L., Liu, W., Hong, X., Zhang, G. and Zhang, Z. (2013) Measurement and evaluation of Chinese regional energy efficiency based on provincial panel data. *Mathematical and Computer Modelling*, **58**, 1000–1009 [692].

Zou, W.-J., Cai, P.-H., Shen, N. and Lu, C.-C. (2015) The technology gap of Chinese regions' energy efficiency and spatial convergence – Based on the hybrid meta-frontier data envelopment analysis. *Journal of Renewable and Sustainable Energy*, **7**(023124), 1–14 [693].

INDEX

This index section contains the page numbers on important models, concepts and applications, all of which are used in this book. The two terms (i.e., DEA and DEA environmental assessment) are discussed in chapters of Section I and those of Section II, respectively. Therefore, this section does not list their indexes. The other important terms are listed in this section. The pages listed in this index describe their locations in this book.

Environmental Assessment on Energy and Sustainability by Data Envelopment Analysis, First Edition. Toshiyuki Sueyoshi and Mika Goto.